Karst Hydrogeology and Geomorphology

Karst Hydrogeology and Geomorphology

Derek Ford, *McMaster University, Canada*
and
Paul Williams, *University of Auckland, New Zealand*

John Wiley & Sons, Ltd

First Edition published in 1989 by Unwin Hyman Ltd.
Email (for orders and customer service enquiries): cs-books@wiley.co.uk
Visit our Home Page on www.wiley.co.uk or www.wiley.com

Reprinted with corrections September 2007

Other Wiley Editorial Offices

John Wiley & Sons Inc., 111 River Street, Hoboken, NJ 07030, USA

Jossey-Bass, 989 Market Street, San Francisco, CA 94103-1741, USA

Wiley-VCH Verlag GmbH, Boschstr. 12, D-69469 Weinheim, Germany

John Wiley & Sons Australia Ltd, 42 McDougall Street, Milton, Queensland 4064, Australia

John Wiley & Sons (Asia) Pte Ltd, 2 Clementi Loop # 02-01, Jin Xing Distripark, Singapore 129809

John Wiley & Sons Canada Ltd, 6045 Freemont Blvd, Mississauga, Ontario, L5R 4J3, Cananda

Wiley also publishes its books in a variety of electronic formats. Some content that appears in print may not be available in electronic books.

Library of Congress Cataloging-in-Publication Data

Ford, Derek (Derek C.)
    Karst hydrogeology and geomorphology / Derek Ford and Paul
    Williams. – [Rev. ed.]
      p. cm.
    ISBN: 978-0-470-84996-5 (HB : alk. paper)
    ISBN: 978-0-470-84997-2 (PB : alk. paper)
    1. Karst. 2. Hydrology, Karst. I. Williams, P. W. (Paul W.)
II. Title.
GB600.F66 2007
551.447–dc22   2006029323

Anniversary Logo Design: Richard J. Pacifico

British Library Cataloguing in Publication Data

A catalogue record for this book is available from the British Library

ISBN 978-0-470-084996-5 (HB)
ISBN 978-0-470-84997-2 (PB)

Typeset in 9/11 pt Times by Thomson Digital, India

FSC
Mixed Sources
Product group from well-managed
forests and other controlled sources

Cert no. SGS-COC-2953
www.fsc.org
© 1996 Forest Stewardship Council

We dedicate this book to our wives, Margaret Ford and Gwyneth Williams, who have sustained us throughout our student years and professional careers.

Contents

Preface

This is a substantial revision of our earlier book *Karst Geomorphology and Hydrology*, which was published in 1989. It has been recast, updated and largely rewritten, taking care to keep what we judge to be the better features of the earlier version, particularly its systems-oriented approach and its integration of hydrology and geomorphology. We have not repeated some historical material in Chapter 1 and have not cited some early authorities, because this information is still accessible to readers in the earlier book.

In Chapter 2 we review pertinent features of the medium, that is the karst rocks and their geological structure. In the following two chapters we attempt comprehensive description of the physics and chemistry of dissolution processes and global dissolution rates. In the past two to three decades the study of hydrogeology and groundwater management has become of major importance in the academic world and in practical management. This is recognized in Chapters 5 and 6 with detailed consideration of the karst hydrogeological system, emphasizing here and throughout the later text the most important fact that, in karst, the meteoric waters are routed underground by dissolutionally enlarged channels. Our understanding of the genesis of those channel patterns, i.e. of underground cave systems, has been much improved in the past two decades by broader international discussion and powerful computer modelling, which is considered in Chapter 7. Cave systems may also function as giant sediment traps, retaining evidence of all natural environmental processes occurring in the soil and vegetation above them and in their hydrological catchments. Increasingly it is recognized that caves are important natural archives that contain information on continental and marine climate change and rates of change. These are now being studied intensively because of their relevance to global warming concerns, so a broad review of recent advances in understanding is presented in Chapter 8. Chapters 9 and 10 then consider the variety of dissolutional and depositional landforms created by karst processes on the Earth's surface, first in humid temperate and tropical environments and then in the arid and cold extremes. The book concludes with broad sketches of the practical applications of karst study and the information presented in the previous chapters. Karst water resources are becoming increasingly important globally, and are the focus of Chapter 11. We consider the delineation and management of karst water supplies and the problem of groundwater pollution, which can proceed with horrific rapidity should there be spillage of the many dangerous substances being moved around the land by road, rail and pipeline. In the final chapter we review hazards arising from the juxtaposition of human activities and karst terrain. The very existence of karst forms and processes may pose hazards to construction and other economic activity and, conversely, human activity poses hazards to fragile karst ecosystems, aquifers and landform features. The chapter concludes with a consideration of environmental restoration, sustainable management and conservation.

During the past 25 years, for a variety of reasons, there has been an explosion of interest in all aspects of karst research on the part of earth scientists, other environmental scientists, civil engineers, and even in legal practice. As a consequence, there has been a huge increase in the number of relevant publications. We find it impossible to read all that is published each year in the English language alone, not to mention the large volumes of important and well-illustrated findings published in other languages. Nevertheless, in this book we have attempted a broad international perspective both in the selection of examples that we present and in the literature that we cite. By necessity, because there is limitation to the number of pages we can use if this volume is to be available at a reasonable price, we have had to work within space constraints, and so we apologize in advance to colleagues where we have not cited – or worse, overlooked – one or more of their favourite publications.

Acknowledgements

Today the international community of karst scholars is large in number, broad in its intellectual scope, vigorous and exciting. We are indebted to a great many individuals whose research, writings, discussions, company in the field and underground, have helped us. The late Joe Jennings (Australia) and Marjorie Sweeting (UK) were our mentors; their enthusiasm, scholarly example, scientific integrity and wise counsel helped set our standards and inspired our academic careers. We continue to build on the foundations they laid for us.

We deeply regret the passing, in recent years and before their time, of Jim Quinlan (USA) and Song Lin Hua (China) and recall with gratitude what they taught us. Especial appreciation to Henry Schwarcz for 40 years of collaborative study of speleothems, to Peter Crossley for 30 years technical support above and below ground, and to Steve Worthington for critical reading of the groundwater hydrology chapters and much else. Amongst the many others whose research and friendship have enriched our study and experience of karst, we wish especially to thank:

Slava Andreichouk
Tim Atkinson
Michel Bakalowicz
Pavel Bosak
Yuan Daoxian
Wolfgang Dreybrodt
Victor and Yuri Dubljansky
Ralph Ewers
Paolo Forti
Amos Frumkin
Franci Gabrovšek
Ivan Gams
John Gunn
Zupan Hajna
Russell Harmon
Carol Hill
Julia James
Sasha Klimchouk
Andrej Krancj
Wieslawa Krawczyk

Stein-Erik Lauritzen
Joyce Lundberg
Richard Maire
Alain Mangin
Andrej Mihevc
Petar Milanović
John Mylroie
Jean Nicod
Bogdan Onac
Art and Peggy Palmer
Jean-Noel Salomon
Jacques Schroeder
Yavor Shopov
Chris and Peter Smart
Tony Waltham
Bette and Will White
Zhu Xuewen and Nadja
Lu Yaoru
Laszlo Zambo

1

Introduction to Karst

Karst is the term used to describe a special style of landscape containing caves and extensive underground water systems that is developed on especially soluble rocks such as limestone, marble, and gypsum. Large areas of the ice-free continental area of the Earth are underlain by karst developed on carbonate rocks (Figure 1.1) and roughly 20–25% of the global population depends largely or entirely on groundwaters obtained from them. These resources are coming under increasing pressure and have great need of rehabilitation and sustainable management. In the chapters that follow, we show the close relationship between karst groundwater systems (hydrogeology) and karst landforms (geomorphology), both above and below the surface. And we explain the place of karst within the general realms of hydrogeology and geomorphology.

Experience shows that many hydrogeologists mistakenly assume that if karst landforms are absent or not obvious on the surface, then the groundwater system will not be karstic. This assumption can lead to serious errors in groundwater management and environmental impact assessment, because karst groundwater circulation can develop even though surface karst is not apparent. Diagnostic tests are available to clarify the situation. The prudent default situation in carbonate terrains is to assume karst exists unless proved otherwise.

We found in our first book (Ford and Williams 1989) that hydrological and chemical processes associated with karst are best understood from a systems perspective. Therefore we will continue to follow this approach here. Karst can be viewed as an open system composed of two closely integrated hydrological and geochemical subsystems operating upon the karstic rocks. Karst features above and below ground are the products of the interplay of processes in these linked subsystems.

1.1 DEFINITIONS

The word **karst** can be traced back to pre-Indoeuropean origins (Gams 1973a, 1991a, 2003; Kranjc 2001a). It stems from *karra/gara* meaning stone, and its derivatives are found in many languages of Europe and the Middle East. The district referred to as the 'Classic Karst', which is the type site where its natural characteristics first received intensive scientific investigation, is in the north-western corner of the Dinaric Karst, about two-thirds being in Slovenia and one-third in Italy. In Slovenia the word *kar(r)a* underwent linguistic evolution via *kars* to *kras*, which in addition to meaning stony, barren ground also became the regional name for the district inland of Trieste. In the Roman period the regional name appeared as Carsus and Carso but, when it became part of the Austro-Hungarian Empire, it was germanicized as the Karst. The geographical and geological schools of Vienna during that time exercised a decisive influence on the word as an international scientific term. Its technical use started in the late 18th century and by the mid-19th century it was well-established. The unusual natural features of the Kras (or Karst) region became known as 'karst phenomena' and so too, by extension, did similar features found elsewhere in the world.

We may define **karst** as comprising terrain with distinctive hydrology and landforms that arise from a combination of high rock solubility and well developed secondary (fracture) porosity. Such areas are characterized by sinking streams, caves, enclosed depressions, fluted rock outcrops, and large springs. Considerable rock solubility alone is insufficient to produce karst. Rock structure and lithology are also important: dense, massive, pure and coarsely fractured rocks develop the best karst. Soluble rocks with extremely high primary

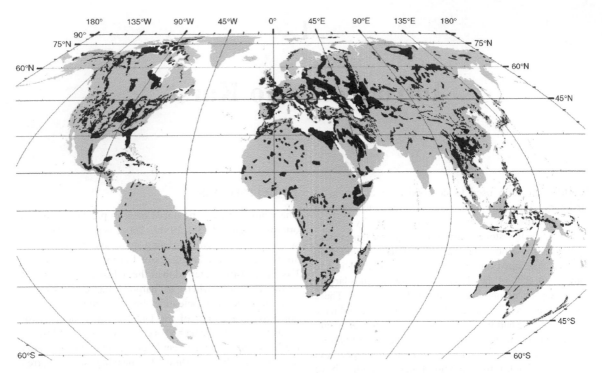

Figure 1.1   Global distribution of major outcrops of carbonate rocks. Accuracy varies according to detail of mapping. Generalization occurs in areas with interbedded lithologies and where superficial deposits mask outcrops. (Map assembled using GIS on Eckert IV equal-area projection from regional maps, many of which were subsequently published in Gunn (2000a)).

porosity (~30–50%) usually have poorly developed karst. Yet soluble rocks with negligible primary porosity (<1%) that have later evolved a large secondary porosity support excellent karst. The key to the expression of karst is the development of its unusual subsurface hydrology, the evolution of which is driven by the hydrological cycle – the 'engine' that powers karst processes. The distinctive surface and subterranean features that are a hallmark of karst result from rock dissolution by natural waters along pathways provided by the geological structure.

The main features of the karst system are illustrated in Figure 1.2. The primary division is into erosional and depositional zones. In the erosional zone there is net removal of the karst rocks, by dissolution alone and by dissolution serving as the trigger mechanism for other processes. Some redeposition of the eroded rock occurs in the zone, mostly in the form of precipitates, but this is transient. In the net deposition zone, which is chiefly offshore or on marginal (inter- and supratidal) flats, new karst rocks are created. Many of these rocks display evidence of transient episodes of dissolution within them. This book is concerned primarily with the net erosion zone, the deposition zone being the field of sedimentologists (e.g. see Alsharhan and Kendall 2003).

Within the net erosion zone, dissolution along groundwater flow paths is the diagnostic characteristic of karst. Most groundwater in the majority of karst systems is of meteoric origin, circulating at comparatively shallow depths and with short residence times underground. Deep circulating, heated waters or waters originating in igneous rocks or subsiding sedimentary basins mix with the meteoric waters in many regions, and dominate the karstic dissolution system in a small proportion of them. At the coast, mixing between seawater and fresh water can be an important agent of accelerated dissolution.

In the erosion zone most dissolution occurs at or near the bedrock surface where it is manifested as surface karst landforms. We refer to forms as being **small scale** where their characteristic dimensions (such as diameter) are commonly less than ~10 m, **intermediate scale** in the range of 10 to 1000 m, and **large scale** where dimensions are greater than 1000 m. In a general systems framework most surface karst forms can be assigned to **input**, **throughput** or **output** roles. Input landforms predominate. They discharge water into the underground and their morphology differs distinctly from landforms created by fluvial or glacial processes because of this function. Some distinctive valleys and flat-floored depressions termed **poljes** convey water across a belt of karst (and sometimes other

THE COMPREHENSIVE KARST SYSTEM

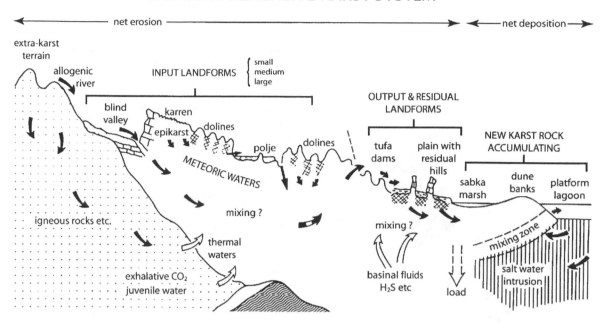

Figure 1.2 The comprehensive karst system: a composite diagram illustrating the major phenomena encountered in active karst terrains. Reproduced from Ford, D.C. and Williams, P.W. (1989) *Karst Geomorphology and Hydrology*.

rocks) at the surface and so serve in a throughput role. Varieties of erosional gorges and of precipitated or constructional landforms, such as travertine dams, may be created where karst groundwater is discharged as springs, i.e. they are output landforms. Residual karstic hills, sometimes of considerable height and abruptness may survive on the alluvial plains below receding spring lines and beside rivers.

Some karsts are buried by later consolidated rocks and are inert, i.e. they are hydrologically decoupled from the contemporary system. We refer to these as **palaeokarsts**. They have often experienced tectonic subsidence and frequently lie unconformably beneath clastic cover rocks. Occasionally they are **exhumed** and reintegrated into the active system, thus resuming a development that was interrupted for perhaps tens of millions of years. Contrasting with these are **relict karsts**, which survive within the contemporary system but are removed from the situation in which they were developed, just as river terraces – representing floodplains of the past – are now remote from the river that formed them. Relict karsts have often been subject to a major change in baselevel. A high-level corrosion surface with residual hills now located far above the modern water table is one example; drowned karst on the coast another. Drained upper level passages in multilevel cave systems are found in perhaps the majority of karsts.

Karst rocks such as gypsum, anhydrite and salt are so soluble that they have comparatively little exposure at the Earth's surface in net erosion zones, in spite of their widespread occurrence (Figure 1.3). Instead, they are protected by less soluble or insoluble cover strata such as shales. Despite this protection, circulating waters are able to attack them and selectively remove them over large areas, even where they are buried as deeply as 1000 m. The phenomenon is termed **interstratal karstification** and may be manifested by collapse or subsidence structures in the overlying rocks or at the surface. Interstratal karstification occurs in carbonate rocks also, but is of less significance. **Intrastratal karstification** refers to the preferential dissolution of a particular bed or other unit within a sequence of soluble rocks, e.g. a gypsum bed in a dolomite formation.

Reference is often made in European literature to **exokarst, endokarst** and **cryptokarst**. Exokarst refers to the suit of karst features developed at the surface. Endokarst concerns those developed underground. It is often divided into **hyperkarst**, in which the underground dissolution is by circulating meteoric waters, and **hypokarst** – dissolution by juvenile or connate waters (Figure 1.2). Some Russian authors further differentiate hypokarst into that dissolved in the soluble rocks by waters that are ascending into them from deeper formations, and that created entirely within a soluble formation by the process

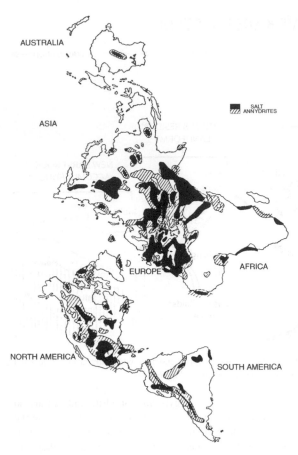

Figure 1.3 The global distribution of evaporite rocks (after Kozary, M.T. et al, Incidence of saline deposits in geologic time, Special Paper 88 © 1968 Geological Society of America). See Klimchouk and Andrejchuk (1996) for a global map of areas of gypsum and anhydrite deposition during the Precambrian and through the Palaeozoic.

of pressure solution that utilizes its contained water. Cryptokarst refers to karst forms developed beneath a blanket of permeable sediments such as soil, till, periglacial deposits and residual clays. **Karst barré** denotes an isolated karst that is impounded by impermeable rocks. **Stripe karst** is a **barré** subtype where a narrow band of limestone, etc., crops out in a dominantly clastic sequence, usually with a stratal dip that is very steep or vertical. Recently there has been an emphasis on **contact karst**, where water flowing from adjoining insoluble terrains creates exceptionally high densities or large sizes of landforms along the geological contact with the soluble strata (Kranjc 2001b).

Karst-like landforms produced by processes other than dissolution or corrosion-induced subsidence and collapse

are known as **pseudokarst**. Caves in glaciers are pseudokarst, because their development in ice involves a change in phase, not dissolution. **Thermokarst** is a related term applied to topographic depressions resulting from thawing of ground ice. **Vulcanokarst** comprises tubular caves within lava flows plus mechanical collapses of the roof into them. **Piping** is the mechanical washout of conduits in gravels, soils, loess, etc., plus associated collapse. On the other hand, dissolution forms such as **karren** (see section 9.2) on outcrops of quartzite, granite and basalt are karst features, despite their occurrence on lithologies of that are of low solubility when compared with typical karst rocks.

The extent to which karst develops on lithologies other than carbonate and evaporite rocks depends largely on the efficiency of the processes that are competing with dissolution in the particular environment. If the competitors are very weak, the small-scale (karren) solutional landforms such as flutes, pits and pans can develop on monominerallic rocks of lower solubility and even on polyminerallic rocks such as granite and basalt, although rates of development appear to be lower. Quartzites and dense siliceous sandstones can be viewed as transitional, occupying part of the continuum between karst and normal fluvial landscapes. In thermal waters their solubility approaches that of carbonate rocks and regular solutional caves may form. Given sufficient time, under normal environmental temperatures and pressures intergranular dissolution of quartz along fractures and bedding planes can permit penetration of meteoric waters underground. When there is also a sufficient hydraulic gradient, this can give rise to turbulent flow capable of flushing the detached grains and enlarging conduits by a combination of mechanical erosion and further dissolution. Thus in some quartzite terrains vadose caves develop along the flanks of escarpments or gorges where hydraulic gradients are high. The same process leads to the unclogging of embryonic passages along scarps in sandy or argillaceous limestones. Development of a phreatic zone with significant water storage and permanent water-filled caves is generally precluded. The landforms and drainage characteristics of these siliceous rocks thus can be regarded as a style of **fluvio-karst**, i.e., a landscape and subterranean hydrology that develops as a consequence of the operation of both dissolution and mechanical erosion by running water.

1.2 THE RELATIONSHIPS OF KARST WITH GENERAL GEOMORPHOLOGY AND HYDROGEOLOGY

Geomorphology is concerned with understanding the form of the ground surface, the processes that mould it, and the history of its development. Carbonate and evaporate lithologies displaying at least some karst occur over ~20% of the Earth's ice-free continental area and occupy a complete range of altitude and latitude. Thus karsts around the world are exposed to the full suite of geomorphological processes – aeolian, coastal, fluvial, glacial, physical and chemical weathering, etc. Hence to understand karst we must consider the same set of natural processes that affect all other rocks and landscape styles, including plate tectonics and climatic change. However, we must also recognize that dissolution is of paramount importance in developing karst compared with its relatively minor role in other lithologies. Chemical solution of karst rocks develops a distinctive suite of features (above and below ground) that reflect the dominance of dissolution and dissolution-induced processes, such as collapse, compared with other processes. Even so, under extreme climatic conditions other processes, such as frost-shattering, can totally mask the effects of dissolution. Thus in high mountains, glacial, periglacial, and mass-movement processes are the principal landscape-forming agents. No karst has been reported on Mount Everest (Jolmo Lungma), for example, even though it consists mainly of carbonate rocks, although karst circulation may occur in the more stable regime underground.

Karst groundwater circulation can occur only if subterranean connections are established between uplands and valley bottoms; otherwise runoff will simply flow across the surface. Where bedrock is porous, as in many sandstones, water will infiltrate and circulate underground via interconnected pores, later to discharge at the surface as springs. In such rocks, the movement of water is by laminar flow and chemical solution has no significant effect on storage capacity and transmission of groundwater. Further, long-term circulation of water has no effect on the ultimate transmissivity or storage capacity of the groundwater system. This is not the case in karst, despite the fact that karst rocks are affected by exactly the same set of forces that drive subterranean groundwater circulation in other lithologies. This is because dissolution plays a fundamental role in karst. The very act of groundwater circulation causes progressive solutional enlargement of void space and a commensurate increase in permeability. Thus although initial groundwater flow in karst is laminar, it becomes progressively more turbulent. The karst groundwater system evolves over time, distinguishing it from other groundwater systems. Consequently, the equations that can be used to describe laminar water flow in typically porous aquifers become inapplicable to karst as the flow through large subterranean conduits becomes turbulent and dominates the groundwater throughput.

The progressive evolution of karst groundwater networks and the development of turbulent flow conditions are intimately related to the evolution of karst landforms. Although karst rocks may have primary intergranular porosity and secondary fracture porosity, most water flow through them is transmitted by conduits (tertiary porosity) developed by solution. These systems receive most of their inputs from point recharge sites at the surface, such as enclosed depressions (dolines, etc.) and streamsinks, which also evolve over time as a consequence of dissolution. Thus both surface landscape and subterranean conduit system evolve together, an unusual circumstance applicable only to karst. For this reason, if one is to understand karst hydrogeology it is also necessary to understand karst geomorphology, and vice versa. This reality determines the structure and content of this book.

1.3 THE GLOBAL DISTRIBUTION OF KARST

The distribution of the principal karst rocks is shown in Figures 1.1 and 1.3. In the aggregate their surface and near-surface outcrops occupy ~20% of the planet's dry ice-free land. Regional detail depicted on these maps is of variable quality depending on the information available. The mapping is also generalized and approximate; many very small outcrops are omitted, and possibly some large ones. Thus many sites shown on Figure 1.1 in Russia, for example, are areas in which carbonates are common, but not necessarily continuous in outcrop. Carbonates are particularly abundant in the Northern Hemisphere. The old Gondwana continents expose comparatively small outcrops except around their margins, where there are some large spreads of Cretaceous or later age carbonates (post break-up of the supercontinent), such as the Nullarbor Plain in Australia. Not all carbonates display distinctive karst landforms and/or significant karst groundwater circulations because some are impure and their insoluble residues clog developing conduits; thus we estimate carbonate karst to occur over 10–15% of the continental area.

Figure 1.3 shows the maximum aggregate of gypsum, anhydrite and salt known to have accumulated over geological time. Most of it is now buried beneath later carbonate or clastic (detrital) rocks. Also, many occurrences have been partly removed by dissolution or much reduced in geographical extent by folding and thrusting, e.g. in the Andes. More than 90% of the anhydrite/gypsum and more than 99% of the salt displayed here does not crop out; nevertheless, there is gypsum and/or salt beneath ~25% of the continental surfaces. Gypsum and salt karst that is exposed at the surface is much smaller in extent than the carbonate karst, but interstratal karst is of the same order of magnitude. Hydrologically active karst within these evaporite rocks probably covers an area comparable to active carbonate karsts.

1.4 THE GROWTH OF IDEAS

The Mediterranean basin is the cradle of karstic studies. Although ancient Assyrian kings between 1100 BC and 852 BC undertook the first recorded (in carvings and bronzes) explorations of caves in the valley of the Tigris River, Greek and Roman philosophers made the first known contributions to our scientific ideas on karst, as well as contributing to a mythology that, like the River Styx, lives on in the place names given by cavers and others. Pfeiffer (1963) identified five epochs in the development of ideas about karst groundwaters, from the interval 600–400 BC until the early 20th century. Thales (624–548? BC), Aristotle (385–322 BC) and Lucretius (96–45 BC) formulated concepts on the nature of water circulation. Flavius in the first century AD described the first known attempt at karst water tracing, in the River Jordan basin (Milanović 1981). A Greek traveler and geographer of the second century AD, Pausanias, also reported experiments that were interpreted as proving the connection between a stream-sink beside Lake Stymphalia and Erasinos spring (Burdon and Papakis 1963). The conceptual understanding of hydrology established by Greek and Roman scholars remained the basis of the subject until the 17th century, when Perrault (1608–70), Mariotte (1620–84) and Halley (1656–1742) commenced its transformation into a quantitative science, showing the relationships between evaporation, infiltration and streamflow. Also in the 17th century, the understanding of karst caves was being advanced by scholars such as Xu Xiake in China (Yuan 1981; Cai *et al*. 1993) and Valvasor in Slovenia (Milanović 1981; Shaw 1992).

By the end of the 18th century, the role of carbonic acid in the dissolution of limestones was understood (Hutton 1795). Experimental work on carbonate dissolution in water followed a few decades later (Rose 1837). The concept of chemical denudation was advanced in 1854 by Bischof's calculation of the dissolved calcium carbonate load of the River Rhine and in 1875 by Goodchild's estimation of the rate of surface weathering of limestone in northern England from his observations of gravestone corrosion (Goodchild, 1890). By 1883, the first modern style study of solution denudation had been completed by Spring and Prost in the Meuse river basin in Belgium.

The mid- to late 19th century was a very significant period for the advancement of our understanding of limestone landforms. In Britain, Prestwich (1854) and Miall (1870) investigated the origin of swallow-holes, while on the European continent impressive progress was made in the study of karren by Heim (1877), Chaix (1895) and Eckert (1895), amongst others. But truly outstanding among the many excellent contributions of that time was the work of Jovan Cvijić. His 1893 exposition, *Das Karstphänomen*, laid the foundation of modern ideas in karst geomorphology, ranging over landforms of every scale from karren to poljes. His contribution to our understanding of dolines is rightly considered of 'benchmark' significance by Sweeting (1981): his thorough investigation of them provides the first instance of morphometry in geomorphology, and his conclusion that most dolines have a solutional origin has withstood the test of time.

The role of lithology became a more explicit theme in some of Cvijić's later work, best expressed in his 1925 paper in which he introduced the terms **holokarst** and **merokarst**. Holokarst is pure karst uninfluenced by other rocks, and is developed on thick limestones extending well below baselevel. It contains the full range of karst landforms and is exemplified by the Dinaric area. By contrast, merokarst (or half karst) is developed on thin sequences of limestones interbedded with other rocks, as well as on less pure carbonate formations. The landscape thus contains both fluvial and karstic elements and may be thickly covered with insoluble residues. The Mammoth Cave–Sinkhole Plain karst of Kentucky is a good example. Cvijić also identified transitional types, such as found in the French Causses, where there is extensive karstic development in thick carbonates above underlying impermeable rocks. Viewed from the perspective of our greater worldwide knowledge in the 21st century it appears that most karsts can be assigned to Cvijić's transitional type; hence his tripartite division is not particularly helpful. It is clear that as the carbonate sequence becomes thinner and perhaps overlain by and interbedded with clastic lithologies, so transitional landforms on the spectrum between karst and normal fluvial landscapes become more common. Roglić (1960) called such landscapes **fluvio-karst**. These early ideas are discussed more fully by Sweeting (1972), Roglić (1972) and Jennings (1985). Much of Cvijić's writing is now available in English translation in Stevanović and Mijatović (2005), together with his publications in French, and reviews by others.

The mid-19th century was also a time of notable advance in our understanding of groundwater flow. Although the experiments of Hagen (1839), Poiseuille (1846) and Darcy (1856) were not specifically related to karst, they nevertheless provided the theoretical foundation for later quantitative explanation of karst groundwater movement. And in 1874 the first attempt was made to analyze the hydrogeology of a large karst area. This was an investigation by Beyer, Tietze and Pilar of the 'lack of water in the karst of the military zone of Croatia'. Herak and Stringfield (1972) considered their ideas to be the forerunners of those that emerged more clearly in the early 20th century. In particular, they foreshadowed the heated and long-lasting debate that erupted on the relative importance of isolated conduit flow as opposed to integrated regional flow.

In 1903 Grund proposed that groundwater in karst terrain is regionally interconnected and ultimately controlled by sea level (Figure 1.4a). He envisaged a saturated zone within karst, the upper level of which coincides with sea level at the coast, but rises beneath the hills inland (today we call this surface the water table). Only water above sea level in the saturated zone was considered to move, and that was termed **Karstwasser**. The water body below sea level was assumed stagnant and was called **Grundwasser**. It was conceived to continue downwards until impervious rocks were ultimately encountered. Grund had a dynamic view of the karst water zone and imagined that its upper surface would rise following recharge by precipitation. Should recharge be particularly great, the saturated zone would in places rise to the surface and cause the inundation of low-lying areas. In this way he explained the flooding of poljes.

However, field evidence showing the lack of synchronous inundation in neighbouring poljes of about the same altitude was used by Grund's critics to argue against the

mechanism he proposed. Katzer (1909) observed, for example, that when springs are at different heights, it is not always the upper one that dries up first. He also noted that the responses to rainfall of springs with intermittent flow are unpredictable; some react, others appear not to. On the phenomenon of polje inundation, he stated that during their submergence phase water may sometimes still be seen flowing into adjacent stream-sinks (ponors) even when the flooding of the polje floor is becoming deeper; thus a general explanation of polje inundation through rising Karstwasser cannot hold. Katzer did not accept the division of Karstwasser and Grundwasser. Instead he interpreted karst as consisting of shallow and deep types. In the former, karstification extends down to underlying impermeable rocks, while in the latter it is contained entirely within extensive carbonate formations. Katzer was apparently influenced by results of the impressive subterranean explorations of the Czech speleologist/geographer, Schmidl (1854), and the French speleologist, Martel (1894), who exposed considerable new information on the nature of cave rivers. Deep within karst Katzer imagined water circulation to occur in essentially independent river networks (Figure 1.4b) with different water levels and with separate hydrological responses to recharge. His work therefore represents an important integration of the emerging ideas of groundwater hydrology and speleology.

One may imagine that Cvijić was stimulated by this controversy, as well as by the extra dimension added by Grund's (1914) publication on the cycle of erosion in karst. Thus we see the appearance in 1918 of another of Cvijić's now famous papers that drew together his maturing ideas on the nature of subterranean hydrology and its relation to surface karst morphology. He too rejected the division of underground water into Karstwasser and Grundwasser, although he implicitly accepted the occurrence of groundwater as we now understand it. He believed in a discontinuous water table, the level of which was controlled by lithology and structure, and he put forward the notion of three hydrographic zones in karst: a dry zone, a transitional zone, and a saturated zone with permanently circulating water. He maintained that the characteristics of these zones would change over time, the upper zones moving successively downwards and replacing those beneath as the karst develops. The idea of a dynamically evolving karst groundwater system was thus born. The very circulation of water enhances permeability and thereby progressively and continuously modifies the hydrological system. This characteristic in now recognized as an essential and unique feature of karst. Thus over a few decades around the turn of the previous century Cvijić (Figure 1.5) laid the theoretical

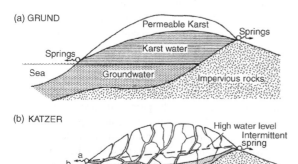

a. Highwater overflow spring b. Perennial spring c. Submarine Spring

Figure 1.4 (a) Essential features of the karst groundwater system according to Grund (1903). He envisaged a fully integrated circulation, although with stagnant water below sea level. (b) The karst water system according to Katzer (1909), who stressed the operation of essentially independent subterranean river networks.

Jovan Cvijić

Figure 1.5 Jovan Cvijić (1865–1927) was a graduate of Belgrade University and a post-graduate of Vienna where he was supervized by A. Penck. The importance of his insights into karst geomorphology and hydrology revealed in his 1893 and 1918 publications were so profound, comprehensive and far-reaching that he is widely regarded as the father of modern karst science. (Photograph provided by Karst Research Institute, Postojna.)

foundation of many of our current ideas. He synthesized the critical observations of his contemporaries with his own keen insights and thereby became the outstanding figure in the field. Without doubt he must be regarded as the father of modern karstic research.

1.5 AIMS OF THE BOOK

The main aim of this book is to demystify karst, which is often perceived to be a separate, difficult or minor (and hence conveniently dismissed) branch of hydrogeology and geomorphology, and to show its place and contribution within the wider fields of hydrology, geomorphology, and environmental science. We set about this by demonstrating that karst is explainable in terms of the

same set of natural laws and processes that apply to other landscapes and hydrogeological systems, albeit in a mix in which dissolution has an unusually important role. We place a major emphasis on understanding processes, and show how surface and subsurface karst can be linked within a systems framework. In the course of doing this we aim to reconcile linkages between 'normal' hydrogeology and karst hydrogeology, and show when karst aquifers can and should not be analysed using conventional techniques. Nevertheless, our emphasis is on the science and explanation of karst rather than on the technical aspects of its resource exploitation and management, which are expertly covered in other texts (e.g. Milanović 2004; Waltham *et al.* 2005).

We are well aware from the abundance of modern publications that we will have missed some excellent research, especially when they have been published in languages with which we are not familiar. So we apologize, especially to readers in the non-English speaking world, if we have inadvertently overlooked fundamental work from your country and have failed to discover important data and diagrams. We still have plenty to learn about karst and are well aware of the limitations of our world view.

There has been a vast increase in interest in karst over the past decade or so and a corresponding explosion in the number of publications. It is impossible for one to read all that is relevant in English, let alone the large volumes of work in other languages. Nevertheless, collected essays on karst from many regions of the world edited by Yuan and Liu (1998), and regional studies on karst in China (Yuan et al. 1991, Sweeting 1995), Siberia (Tsykin 1990), and Slovenia (Gams 2003) permit further insight into the rich international literature.

1.6 KARST TERMINOLOGY

Karst resources, especially perennial springs, caves for shelter, and shallow (readily accessible) placer mineral deposits became important in human affairs long before the advent of writing. It is no surprise that there are many different words in many different languages for a given feature such as a doline. New terms are also being introduced from time to time by writers unfamiliar with the older ones. The international karst terminology thus can be very confusing. Recent dictionaries that attempt to provide brief definitions of features, processes, etc., and list the names used for them in some of the major languages include Kósa (1995/6), Panoš (2001) and Lowe and Waltham (2002). The US Environmental Protection Agency (EPA) has also published a general lexicon, in English only (Field 1999).

2

The Karst Rocks

2.1 CARBONATE ROCKS AND MINERALS

Carbonate rocks crop out over approximately 10% of the ice-free continental areas and underlie much more. They can accumulate to thicknesses of several kilometres and volumes of thousands of cubic kilometres, accounting for about 20% of all Phanerozoic sedimentary rocks. They are forming today in tropical and temperate seas and are known from strata as old as 3.5 Ga. They host roughly 50% of the known petroleum and natural gas reserves, plus bauxite, silver, lead, zinc and many other economic deposits, even gold and diamonds. They supply agricultural lime, Portland cement, fine building stones, and are the principal source of aggregate for highways, etc. in many regions. As a consequence they are studied from many different perspectives. There are innumerable descriptive terms and many different classifications.

Figure 2.1 gives a basic classification. Carbonate rocks contain > 50% carbonate minerals by weight. There are two common, pure mineral end-members, limestone (composed of calcite or aragonite) and dolostone (dolomite). Most authors neglect 'dolostone' and describe both the mineral and the rock as **dolomite**; we shall follow this practice.

Carbonate rocks are distinctive because their accumulation is highly dependent upon organic activity and they are more prone to post-depositional alteration than other sediments. Ancient carbonate rocks preserved on the continents are mainly deposits of shallow marine platforms (< 30 m depth), including wide tracts of 'knee deep' and inter- to supratidal terrain. In areal terms 95% of modern carbonates are accumulating on deeper oceanic slopes and floors (to maximum depths as great as 4 km, below which (the **lysocline**) calcite is dissolved as fast as it settles from above), but sedimentation rates are highest on the platforms.

Carbonate deposits that are accumulating today consist of approximately equal proportions of calcite and aragonite, plus minor amounts of primary dolomite. Lithified carbonates as young as mid-Tertiary age have little aragonite remaining; it has inverted to calcite. In bulk terms, calcite:dolomite ratios are about 80:1 amongst Cretaceous strata, 3:1 in rocks of Lower Palaeozoic age and ~1:3 in the Proterozoic. However, the ratio is about 1:1 in Archaean rocks, negating any simple time trend.

2.1.1 Calcite, aragonite and dolomite mineralogy

Some important characteristics of the principal karst minerals are given in Table 2.1. In nature there is a continuous range between pure calcite or aragonite, and dolomite. These are the abundant and important carbonate minerals. Blends between dolomite and magnesite do not occur. Dolomite may become enriched in iron to form ankerite ($Ca_2FeMg(CO_3)_4$ but this is rare. Also rare are pure magnesite, iron carbonate (siderite), and more than 150 other pure carbonates that have been recognized (Railsback 1999).

In carbonate structures the CO_3^{2-} anions can be considered as three overlapping oxygen atoms with a small carbon atom tightly bound in their centre. In pure calcite, the anions are in layers that alternate with layers of calcium cations (Figure 2.2). Each Ca^{2+} ion has six CO_3^{2-} anions in octahedral co-ordination with it, building hexagonal crystals. Divalent cations smaller than Ca^{2+} may substitute randomly in the cation layers; larger cations such as Sr^{2+} can be accepted only with difficulty (Table 2.2). The basic calcite crystal forms are rhombohedrons, scalenohedrons, prisms, pinacoids or dipyramids (Figures 2.2 and 2.3). These may combine to create more than 2000 variations, associated with differing

Karst Hydrogeology and Geomorphology, Derek Ford and Paul Williams
© 2007 John Wiley & Sons, Ltd

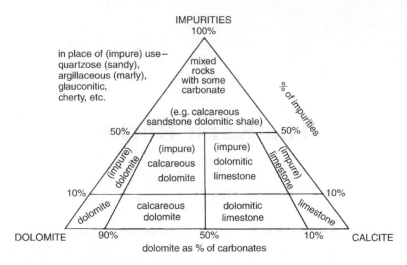

Figure 2.1 A bulk compositional classification of carbonate rocks. Reproduced with permission from Leighton, M.W. and C. Pendexter, Carbonate rock types, Mem. 1, 33–61.© 1962 American Association of Petroleum Geologists.

Table 2.1 Properties of the principal karst rock minerals

Type	Mineral	Chemical composition	Specific gravity	Hardness	Description
Carbonates	Calcite	$CaCO_3$	2.71	3	Trigonal subsystem; rhombohedral. Habit: massive, scalenohedral, rhombohedral. Twinning is rare. >2000 varieties of crystals reported. Colourless or wide range of colours. Effervesces vigorously in cold, dilute acids
	Aragonite	$CaCO_3$	2.95	3.5–4	Orthorhombic system; dipyramidal. Habit: acicular, prismatic or tabular. Frequent twinning. Metastable polymorph of calcite. Colourless, white or yellow. Effervesces in dilute acids
	Dolomite	$CaMg(CO_3)_2$	2.85	3.5–4	Hexagonal system; rhombohedral. Habit: rhombohedrons, massive, sugary (saccharoidal) or powdery. Colourless, white or brown, usually pink tinted. Effervesces slightly in dilute acid
	Magnesite	$MgCO_3$	3.0–3.2	3.5–5	Hexagonal system; rhombohedral. Habit: usually massive. White, yellowish or grey. Effervesces in warm HCl
Sulphates	Anhydrite	$CaSO_4$	2.9–3.0	3–3.5	Orthorhombic system. Habit: crystals rare, usually massive; tabular cleavage blocks. White; may be pink, brown or bluish. Slightly soluble in water
	Gypsum	$CaSO_4–2H_2O$	2.32	2	Monoclinic system. Habit: needles, or fibrous or granular or massive. Colourless, or white and silky, grey. Slightly soluble in water
	Polyhalite	$K_2Ca_2Mg(SO_4)_4.$ $2H_2O$	2.78	3–3.5	Triclinic system. Habit: granular, fibrous or foliated. White, grey, pink or red. Bitter taste
Halides	Halite	$NaCl$	2.16	2.5	Cubic euhedral crystals. Habit: usually massive or coarsely granular. Colourless or white. Very soluble in water
	Sylvite	KCl	1.99	2	Cubic system. Habit: cubic and octahedral crystals or massive. Colourless or white. Very soluble in water
	Carnallite	$KCl.MgCl_2.6H_2O$	1.6	1	Orthorhombic system. Habit: massive or granular. White. Bitter taste
Silica	Quartz	SiO_2	2.65	7	Trigonal. Habit: prismatic crystals with rhombohedral terminations most common; also massive or granular. Colourless, white, yellow, pink, brown or black; transparent
	Opal	SiO_2	2–2.25	5.5–6	Crystallites. Habit: random structure ranging from near amorphous to low cristobalite. Many different colours due to trace elements; transparent or opaque

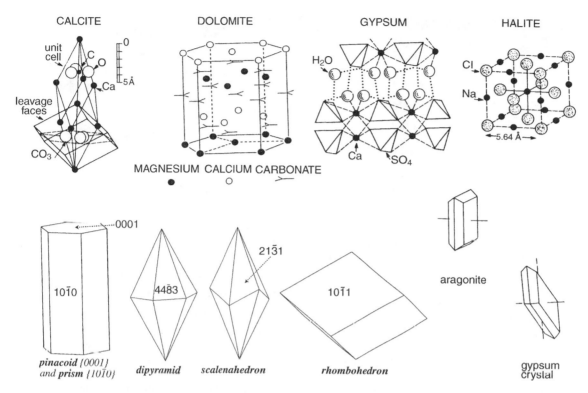

Figure 2.2 Unit cell configurations of calcite, dolomite, gypsum and halite. **Unit cell** refers to the basic building block, e.g one Ca plus one CO_3 in the case of calcite or aragonite. Calcite units can be assembled in five different configurations, **pinacoid, prism, dipyramid, scalenohedron** and **rhombohedron**. The numbers are the **Miller indices**, which measure the orientation of a crystal face to the a, b, c orientation axes. Dipyramids, scalenohedrons and rhombohedrons may be found in their pure forms or in combination with the others. Pinacoids and prisms are open-ended and so must combine with others. Combinations of these basic shapes in varying proportions create the more than 2000 different calcite crystal **habits** that are known.

concentrations of trace elements in the solid or foreign ions in the source solution (e.g. Huizing *et al.* 2003). Calcite may also be massive.

In aragonite the Ca and O atoms form unit cells in cubic co-ordination, building orthorhombic crystal

Table 2.2 Crystal cationic radii (ångströms)

Ba^{2+}	1.34	Fit into orthorhombic structures
K^+	1.33	
Pb^{2+}	1.20	
Sr^{2+}	1.12	
Ca^{2+}	0.99	
Na^+	0.97	Fit into rhombohedral structures
Mn^{2+}	0.80	
Fe^{2+}	0.74	
Zn^{2+}	0.74	
Mg^{2+}	0.66	

structures. These will not accept cations smaller than Ca. The most common substitute atom is Sr and U is also important. Aragonite is only metastable. In the presence of water it may dissolve and reprecipitate as calcite, expelling most Sr and U ions. Aragonite is 8% less in volume than calcite. Inversion to calcite therefore normally involves a reduction of porosity. Aragonite crystals display acicular (needle-like), prismatic or tabular habits; there is frequent twinning.

In ideal (or **stoichiometric**) dolomite (Figure 2.2) equimolar layers of Ca^{2+} and Mg^{2+} ions alternate regularly between the CO_3^{2-} planes. The reality is more complex. Some Ca atoms substitute into the Mg layers and trace quantities of Zn, Fe, Mn, Na and Sr atoms may be present in either Ca or Mg planes. Most dolomites are slightly Ca-rich so that the formula is properly written: $Ca_{(1+x)}Mg_{(1-x)}(CO_3)_2$. In addition, because Fe^{2+} is intermediate in size it fits readily into either Ca or Mg layers.

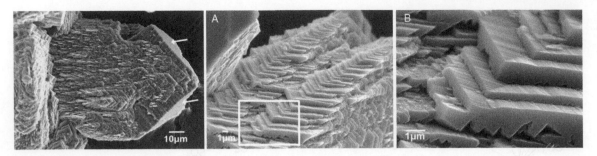

Figure 2.3   Scanning electron microscope images of a rhombohedral calcite crystal forming on a raft of calcite floating in a pool in Gorman Cave, Texas. Scale bars are in micrometres. (Images courtesy of Penny Taylor and Dr Henry Chafetz, Houston University.)

As a result, dolomite typically contains more iron than does calcite. This is why it often weathers to a pinkish or buff colour, Fe^{2+} being oxidized to Fe^{3+}. In the past such disordered dolomites have been termed 'protodolomite' or 'pseudodolomite' but these names are now out of favour. As disorder increases so does solubility.

Dolomite may be massive or powdery or sugary crystalline (**sucrose** or **saccharoidal**) in texture. The crystals are rhombic and opaque. Staining with alizarin red is the standard means of distinguishing it from calcite (Adams *et al.* 1984).

The proportions of Ca and Mg ions within a given crystal can vary between the ideal calcite and dolomite extremes, with a few per cent of Sr and Fe being substituted in as well. It follows that at the larger scale there may be variations between adjoining crystals, or larger patches of rock. This latter scale of variation tells much about initial depositional conditions in an ancient rock and so is carefully classified: calcite with zero weight per cent $MgCO_3$ is pure calcite; $> 0 < 4\%$ is 'low-Mg calcite'; $> 4 < 25\%$ is 'high-Mg calcite', which, like aragonite, is unstable and thus more likely to recrystallize during diagenesis. Some representative

bulk compositions for carbonate rocks are given in Table 2.3.

2.2 LIMESTONE COMPOSITIONS AND DEPOSITIONAL FACIES

Limestones are the most significant karst rocks. Here we consider the nature and environmental controls of their deposition. These determine much of the purity, texture, bed thickness and other properties of the final rock.

Some calcium carbonate can accumulate in almost every environment between high mountains and the deep sea. It forms transient crusts on alpine cliffs, in arctic soils and even beneath flowing glaciers. It accumulates in warmer, humid to arid, soils and can attain the status of major rock units in desert and some other lacustrine basins. At sea it precipitates locally in shallow arctic waters, in temperate waters, and can accumulate as ooze at depths as great as $-4000\,m$ in the tropical oceans. However, most that survives as consolidated older rocks was formed in shallow tropical to warm temperate marine environments, especially platforms and ramps. Figure 2.4a shows the settings and 2.4b is

Table 2.3   Some representative limestone and dolomite percentage bulk chemical compositions

Oxides	Material								
	1	2	3	4	5	6	7	8	9
CaO	56.0	55.2	40.6	42.6	37.2	54.5	30.4	29.7	34.0
MgO	—	0.2	4.5	7.9	8.6	1.7	21.9	20.3	19.0
Fe, Al oxides	—	—	2.5	0.5	1.6	—	—	0.2	0.2
SiO_2	—	0.2	14.0	5.2	8.1	0.2	—	1.5	—
CO_2	44.0	44.0	35.6	41.6	43.0	41.8	47.7	46.8	46.8

1, ideally pure limestone; 2, Holocene Coral, Bermuda; 3, average of 500 building stones; 4, average of 345 samples from Clarke 1924; 5, Hostler Limestone (Ordovician), Pennsylvania; 6, Guilin limestones (Devonian); 7, ideally pure dolomite; 8, Niagaran dolomite, Silurian; 9, Triassic dolomite, probably hydrothermal, Budapest.

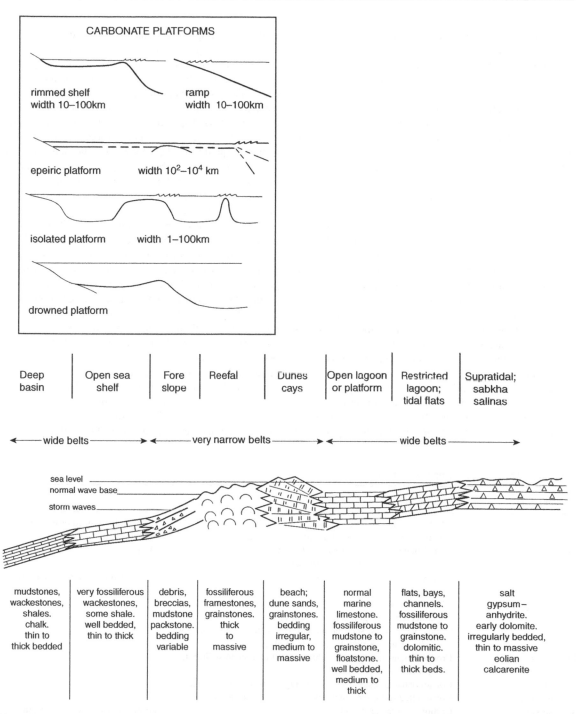

Figure 2.4 (a) Different types of carbonate platforms. Reproduced with permission from Tucker, M.E. and Wright, V.P, Carbonate Sedimentology, Blackwell Science, Oxford © 1990 Blackwell Science. (b) A composite facies model to illustrate deposition of limestone, early dolomite and evaporite rocks on a rimmed shelf and ramp. This is a generalized, simplified picture. Not all facies will be present in any given transect. Narrow belts range from a few metres to a few kilometres in width, whereas wide belts range from hundreds of metres to more than 100 km. Modified with permission from Wilson, J.L. (1974) Characteristics of carbonate platforms margins, Bulletin, 58, 810–24.© 1974 American Association of Petrologists.

an idealized, composite model that extends from supra-tidal salt marshes through lagoons, reefs and shelves out to the deep sea. Note that sulphate and halide rocks can also be formed in these model environments. It is claimed that the number of distinct limestone facies is greater than facies of all other sedimentary rocks combined: this gives rise to innumerable subtle variations in the hydrogeological and geomorphological properties of the rock.

Most of the original limestone material is aragonite or calcite precipitated by marine animals for shell and skeleton building, expelled as faeces or precipitated in the tissues of algal plants. Some aragonite is formed at the sea surface by homogeneous precipitation (**whitings**; see section 3.8) or on picoplankton nucleii. Rate of production by these various means can aggregate 500–1000 $g\,m^2\,yr^{-1}$ or more on platforms. Later, vadose encrustations and beach rock may be formed if the limestone is raised above sea level. Cementation and recrystallization during diagenesis add inorganic calcite spar that is relatively coarse. These principal components of limestones are summarized in Table 2.4; see Tucker and Wright (1990) and Carozzi *et al.* (1996) for comprehensive details.

Carbonate mud (or **micrite**) is the most important bulk constituent. It can compose entire beds or formations, or serve as matrix or infilling. Much originates as aragonite needles from algae, some is precipitated directly, the rest

is the finest fragmentary matter produced by abrasion, faunal burrowing, excretion, microbial reduction, etc.

Carbonate sand is formed mainly of faecal pellets, ooliths and fragments of skeletons and shells. It may accumulate in higher energy environments (beaches, bars, deltas) and build to sand ripples or dunes. More frequently it is dispersed within carbonate mud. Intraclasts and lithoclasts are larger erosional fragments. The former are produced when storm waves break up the local sea bed; the latter have been transported greater distances by longshore, delta or turbidity currents and may include large particles such as pebbles, cobbles and boulders, or may consist of cliff or reef talus.

Reefs make only a small volumetric contribution to the world's limestones but they can be spectacular. They range from those having a complete framework tens to many hundreds of metres in height built of successive generations of coral or algae (**framestone**) to carbonate sand, silt or mud piles containing scattered, but unlinked, corals, fragments and algal or microbial mats. Modern coral grows chiefly between 30°N and 25°S in the photic zone (upper layer of the sea where photosynthesis occurs). It grows to the sea surface and may be exposed at low tide, e.g. on coral atolls. Growth rates are commonly 1–7 $mm\,yr^{-1}$ (or 1–3 $kg\,m^{-2}$). Algal mats (stromatolites) have a greater environmental range and can flourish in the supratidal zone. Reefs may grow continuously along marine platform edges. On platforms and in

Table 2.4 The principal components of limestone

Textural type	Description	Origin
Micrite	0.5–5 µm diameter lime mud and silt particles. The largest component by volume in a majority of limestones	Clay- and silt-sized original marine grains
Peloids	Faecal pellets, micro-ooliths, 30–100 µm diameter. By volume, the most important larger particles	Ooze
Ooliths	Sand-sized spherical accretions	Sand-sized or larger original marine grains, skeletons and growths
Lumps or grapestones	Clumped peloids, ooliths	
Oncolites	Algal accretionary grains up to 8 cm diameter	
Skeletal	Corals, vertebrates, shell fauna, etc. Algal stems and other flora *in situ* or transported. Fragments of all genera	
Intraclasts	Eroded fragments of partly lithified local carbonate sediment, e.g. beach rock	
Lithoclasts	Consolidated limestone and other fragments; often allogenic	
Frameworks	Constructed reefs, etc., mounds, bioherms, biostromes	
Vadose silt	Carbonate weathering silt	Formed during vadose exposure
Pisoliths	Large ooliths or concretions, e.g. nodular caliche, cave pearls	
Stalactites	Dripstones and layered accretions, caliche crusts	
Sparite	Medium to large calcite crystals as cementing infill; drusy, blocky, fibrous, or rim cements. > 20 µm diameter	Diagenetic cements
Microspar	5–20 µm grains replacing micrite	

lagoons they may occur as fringes, as scattered, isolated mounds termed **patch reefs** when small, **mounds, bioherms** or **pinnacle reefs** when large (Riding 2003). **Biostromes** are horizontal, tabular spreads of coral or algae. A few coral species grow as isolated individuals in deep water.

When first laid down or built up these carbonate sediments contain large amounts of organic matter. This rapidly decomposes and is removed during lithification. Modern consolidated limestones contain ~1% of organic matter and ancient limestones average only 0.2%.

Most initial limestone deposits will contain some insoluble mineral impurities derived from near or distant eroding terrains. Type and proportion clearly will vary with differing environments. At the least there is volcanic ash and other dust settling into deep-sea oozes where it may supply only 0.1% by volume. At the other extreme, river clay, sand and gravel deposited in deltas and estuaries can exceed the local carbonate production rate and so create the calcareous shales, sandstones, etc. of Figure 2.1.

2.2.1 Petrological classification of limestone

Sedimentological classifications of limestone have been based upon grain size, composition and perceived facies. A scheme by Grabau (1913) with dominant grain size is still widely used, especially in Europe, and guides later schemes. It recognizes **calcilutites** (carbonate mudstones), **calcarenites** (sandstones) and **calcirudites** (conglomerates). Later, others added **calcisiltites**. Dunham (1962) refined this principle. He first divided carbonates into those preserving recognizable depositional texture and those where it has been destroyed. The latter are simply **crystalline limestone** or **crystalline dolomitic limestone**, etc. The former are subdivided into the organically bound and the loose sediment classes. An amplification of this scheme (Figure 2.5a) is now widely used. It recognizes nine original texture types, plus crystalline limestone. Some authors also designate the dominant particle kinds, e.g. **pellet wackestone, intraclast rudstone**, etc.

The other popular modern classification is that of Folk (1962), partly shown in Figure 2.5b. The two end-members are **biolithites** (framed and bound coral and algal rocks) and **micrite** or **dismicrite** (pure carbonate mud or mud disturbed by burrows). In between, other types are defined by combining different proportions of mud and later spar cement with differing kinds of original grains (allochems) having differing degrees of sorting as functions of the wave or current energies locally available. Where bottom currents are weak, occasional allochems are dispersed amongst the mud matrix, which supports them. These are **biomicrites, oomicrites**, etc. Where currents are stronger the mud is partly or entirely winnowed from the skeletal fragments; piled pellets or intraclasts, etc. produce grain-supported layers where the voids may be filled with clear calcite spar during diagenesis. These are **biosparites, pelsparites** and **intrasparites; packstone, grainstone** and **rudstone** are approximately equivalent in the Dunham classification but do not indicate the kind of grains that are most important in a rock. For excellent descriptions of carbonate fabrics in thin section, illustrated by numerous colour photomicrographs, see Adams *et al.* (1984), Adams and Mackenzie (1998) and Society of Economic Paleontologists and Mineralogists (SEPM) Photo CD-Series 1, 2, 7, 8 (Scholle and James 1995, 1996). Figure 2.6 shows a small sample of the range of fabrics.

2.2.2 Sequence stratigraphy

As carbonate debris accumulates, platforms and ramps will subside, at rates generally estimated to be between 0.05 and 2.5 m per million years on continental passive margins. If local accumulation rates are greater, the water and depositional facies become progressively shallower upwards in any given vertical section, and vice versa. At the global scale, longer term tectonic deformation of the ocean basins and shorter term withdrawal of seawater during ice ages ensure that sea levels must usually be rising or falling across sections such as shown in Figure 2.4b. Flooding produces onlap of deeper water facies, whereas withdrawal produces offlap with shallower facies. This is the basis of **sequence stratigraphy**, a modern concept widely used to interpret seismic records and reconstruct palaeoenvironments: Moore (2001) gives a comprehensive review.

As a consequence, many limestones and dolomites display strong cyclic features. Figure 2.7 portrays three sequences that are commonly found. The lithology, hydrogeological and karstic properties may vary significantly across each cyclic accumulation. Before burial in the next cycle, the two ending in supratidal exposure may well develop erosional surfaces with solutional features, or calcrete hardening and a thin clay palaeosol. After burial, such surfaces often become prominent bedding planes or slightly disconformable contacts offering preferential flow routes for all future marine and meteoric groundwater circulation; they are also favoured sites for erosional stripping, e.g. by glacier action. **Hardground** develops on the seafloor when sedimentation temporarily slows or ceases, permitting some dissolution and reprecipitation of calcite to pit and harden the top few

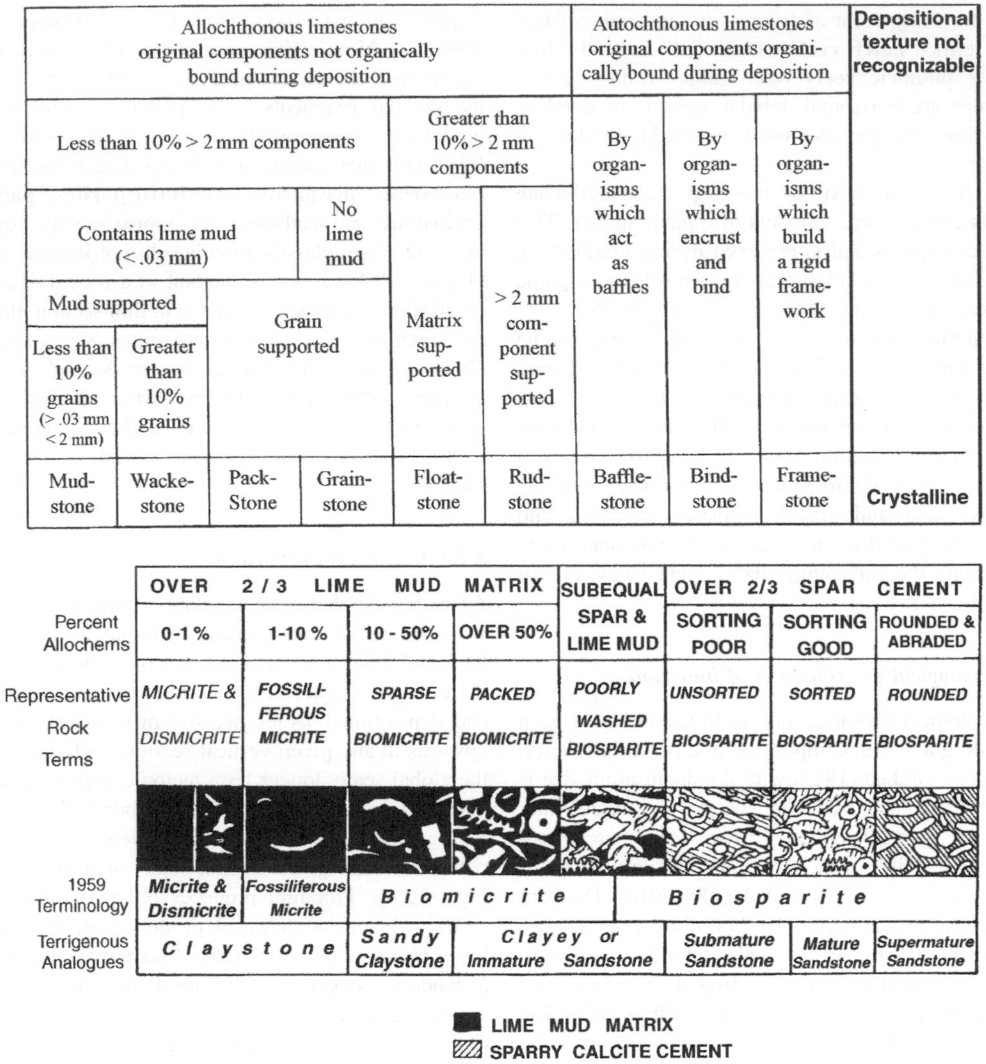

Allochthonous limestones original components not organically bound during deposition						Autochthonous limestones original components organically bound during deposition			Depositional texture not recognizable
Less than 10% > 2 mm components				Greater than 10% > 2 mm components		By organisms which act as baffles	By organisms which encrust and bind	By organisms which build a rigid framework	
Contains lime mud (<.03 mm)			No lime mud	Matrix supported	> 2 mm component supported				
Mud supported		Grain supported							
Less than 10% grains (>.03 mm <2 mm)	Greater than 10% grains								
Mud-stone	Wacke-stone	Pack-Stone	Grain-stone	Float-stone	Rud-stone	Baffle-stone	Bind-stone	Frame-stone	**Crystalline**

	OVER 2/3 LIME MUD MATRIX				SUBEQUAL SPAR & LIME MUD	OVER 2/3 SPAR CEMENT		
Percent Allochems	0-1 %	1-10 %	10 - 50%	OVER 50%		SORTING POOR	SORTING GOOD	ROUNDED & ABRADED
Representative Rock Terms	*MICRITE & DISMICRITE*	*FOSSILI-FEROUS MICRITE*	*SPARSE BIOMICRITE*	*PACKED BIOMICRITE*	*POORLY WASHED BIOSPARITE*	*UNSORTED BIOSPARITE*	*SORTED BIOSPARITE*	*ROUNDED BIOSPARITE*
1959 Terminology	*Micrite & Dismicrite*	*Fossiliferous Micrite*	*Biomicrite*			*Biosparite*		
Terrigenous Analogues	*Claystone*		*Sandy Claystone*	*Clayey or Immature Sandstone*		*Submature Sandstone*	*Mature Sandstone*	*Supermature Sandstone*

■ LIME MUD MATRIX
▨ SPARRY CALCITE CEMENT

Figure 2.5 (a) The R. J. Dunham (1962) classification of carbonate rocks, as modified by Embry and Klovan. Reproduced with permission from Embry, A.F. and Klovan, J.E., A Late Devonian reef tract in northeastern Banks Island, Northwest Territories. Canadian Petroleum Geologists Bulletin, 19, 730–81, 1971. (b) Carbonate textural spectrum. Reproduced with permission from Folk, R.L., Spectral subdivision of limestone types. Memoirs, American Association of Petroleum Geologists, 1, 62–84, 1962.

centimetres; this too may offer preferential flow routes for later karstification and stripping.

2.2.3 Terrestrial carbonates

As noted, some carbonate can be precipitated in almost every terrestrial environment. The most widespread types are tufa and travertine. There is confusion in terminology here, with some authors using 'tufa' for all surficial deposits and restricting 'travertine' to cave deposits. In this book we use **tufa** for grainy deposits accreting to algal filaments, plant stems and roots at springs, along river banks, lake edges, in cave entrances, etc. Tufa is usually a sort of framestone. It is typically dull and earthy in texture, and is highly porous once the vegetal frame rots out. In contrast, **travertine** is crystalline, dense calcite that is often well layered, quite lustrous and lacks framing plant content. That formed underground or at hot springs is largely or entirely inorganic. Other surficial travertines may be bacterially precipitated

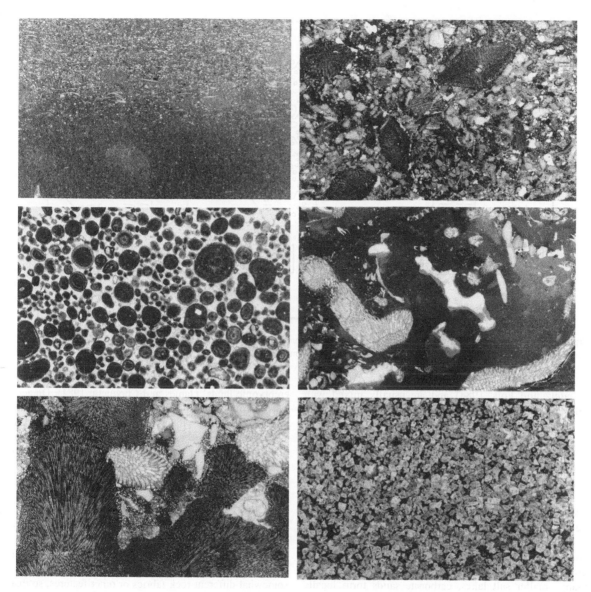

Figure 2.6 Thin sections of characteristic limestones and a dolomite. (Top left) Mudstone (Dunham 1962) or micrite (Folk 1962) coarsening upwards to microbioclastic wackestone (Dunham) or biomicrite (Folk). Ordovician, Quebec. Width (of photograph) 1 cm. (Top right) Fossiliferous packstone (Dunham) or fossiliferous micrite (Folk) composed of foraminifers (large fossils), pelmatozoans and bryozoans. Lower Cretaceous, United Arab Emirates. Width 1.5 cm. (Middle left) Oolitic grainstone (Dunham) or oosparite (Folk). Jurassic, Louisiana. Width 0.8 cm. (Middle right) Bryozoan floatstone (Embry and Klovan 1971); irregular cavities are borings filled with a second generation of geopetal mud and calcite spar. Middle Ordovician, Quebec. Width 1.5 cm. (Lower Left) Coral boundstone (Dunham), framestone (Embry and Klovan) or biolithite (Folk). Note that the large fossils are in stylolitic (pressure solution) contact. Width ~2 cm. (Lower Right) Dolomite composed of euhedral zoned crystals almost completely replacing limestone. Miocene, South Australia. Width 5.2 mm. (From photographs and descriptions kindly supplied by Professor N.P. James, Queen's University, Canada.)

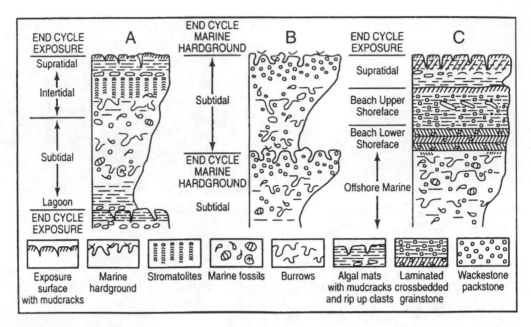

Figure 2.7 Conceptual models of common shoaling upwards sequences. Reproduced from Moore, C.H. 2001. Carbonate reservoirs: porosity, evolution and diagenesis in a sequence stratigraphic framework, Developments in Sedimentology 55. Amsterdam, p 444, © 2001 Elsevier.

(Chafetz and Folk 1984) and are usually mixed with tufas. These deposits may accumulate to tens or hundreds of metres in thickness and cover many square kilometres in area. See the Association Francaise de Karstologie (1981), Pentecost (1995), Ford and Pedley (1996) for comprehensive reviews, and Chapters 8 and 9 for further details.

Calcium carbonate is deposited in some freshwater lakes and in hypersaline water bodies such as the Dead Sea. Freshwater deposits include laminated micrites with a high silicate mud content, termed **marls**. They are common in temperate regions but rarely accumulate to great thicknesses. At Great Salt Lake, Lake Chad and many smaller salt lakes, carbonate sands form locally around the shores and out to approximately 3 m depth, where they are replaced by sulphates and halides. Algal mats are well developed on the carbonates. In the Dead Sea both gypsum and aragonite are precipitated at the water surface. The gypsum is immediately replaced by calcite which settles with the aragonite to accumulate as micrite.

2.3 LIMESTONE DIAGENESIS AND THE FORMATION OF DOLOMITE

Limestone deposits begin as unconsolidated muddy sediments with a porosity of 40–80%, that are 'bathed in their embryonic fluids' (James and Choquette 1984). Diagenesis describes their alteration to consolidated rocks with a porosity rarely more than 15% and usually less than 5%. The principal processes are compaction, dissolution, microbial micritization, calcite cementation, crystal replacement (**neomorphism**) and dolomitization. The diagenetic environment may remain the shallow to deep submarine site of deposition, or it may become vadose or phreatic with meteoric water as a consequence of offlap, uplift, etc. It can be deep subsurface because of burial by later sediments, or tectonic or hydrothermal or thermal metamorphic (Figure 2.8). In many examples several of these environments succeed one another. All introduce somewhat different rock fabrics or other features, so that there is a great deal of variation in consolidated limestone (Figure 2.9). Early alteration in shallow environments is **eogenesis**; **mesogenesis** occurs if there is deep burial, tectonic deformation, etc; **telogenesis** may then follow as cover strata are eroded away and the deeply buried carbonate becomes exposed to a new cycle of meteoric, karstic groundwater circulation, perhaps the first it has experienced. Scholle *et al.* (1983) and Purser *et al.* (1994) give comprehensive reviews.

Submarine diagenesis is slow and imperfect. There is compaction if shallow overburden is added, and some aragonite and calcite spar is precipitated into voids as seawater is expelled. Chalk is an extreme example of a

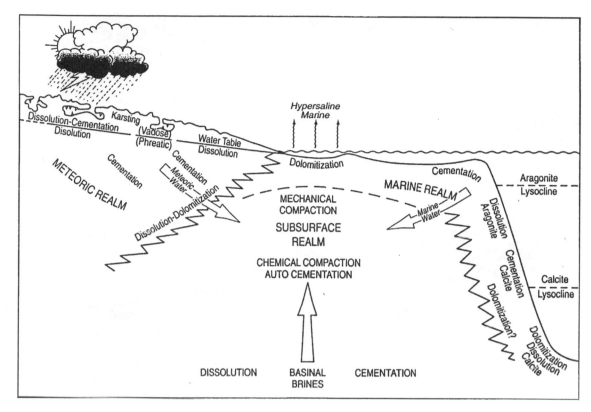

Figure 2.8 Major environments of diagenesis, calcium carbonate cementation and dolomitization. Reproduced from Moore, C.H. 2001. Carbonate reservoirs: porosity, evolution and diagenesis in a sequence stratigraphic framework, Developments in Sedimentology 55. Amsterdam, p 444, © 2001 Elsevier.

limestone from this environment, having undergone minimal diagenesis because it escaped exposure to the more acidic meteoric waters and was never deeply buried. The chalks of northwest Europe retain a porosity often > 40% and a density of only 1.5–2.0. Cementation may be weak. Freshwater marls are similar when drained.

Diagenesis in the subaerial and meteoric groundwater environment can be rapid (typically ten times faster than marine) and extensive. More than half of all ancient limestones experienced one or more such episodes. When sea level falls meteoric water invades the marine sediment and depresses the saltwater interface in the proportion, 40:1 (see section 5.8). The limestone deposit, which was stable in a seawater chemical environment, is now exposed to circulating fresh waters and all degrees of mixture, fresh:salt, as the salt solution is progressively expelled. There is now much dissolution, including the removal of most aragonite. It is replaced by calcite spar cements (Table 2.4). Slow dissolution and reprecipitation (crystal cell by cell) is fabric selective, preserving aragonite skeletal fossils, etc. as calcite

'ghosts'; this is **replacement**. Rapid dissolution obliterates the fossil and substitutes a new, coarser crystalline spar. In the vadose zone new lenticular voids may be created at bedding planes and filled with vadose silt; they appear rather like stromatolite biolithites and are termed **stromatactic**. If there is strong evaporation (e.g. a semi-arid coast) pisolites, travertine nodules and crusts (**caliche** or **calcrete**) replace the uppermost limestone: Alsharhan and Kendall (2003) provide a comprehensive review of them, and see Chapter 9. San Salvador Island, Bahamas, is a spectacular example of rapid eogenesis: carbonate sand dunes blown up during the last interglacial stage (only 120 000–130 000 year ago) are now firmly cemented **aeolianites** that host substantial karst features and caves (Mylroie and Carew 2000).

2.3.1 The formation of dolomite

Dolomite is almost as abundant as limestone. Its formation has been and remains a subject of intense study (see review by Warren 2000). The solute Mg^{2+} ion is strongly

Figure 2.9   Some representative limestones and dolomites in outcrop. (Top) Thick bedded platformal micritic limestone with regular tension joint systems. (Bottom left) Stylobedded platy limestones dissected by kluftkarren that follow tension joints. (Bottom centre) Cyclic alternation of thick and thin beds typical of shallow platform sequences. Rocks in this photograph are fully dolomitized, as suggested by the high frequency of vugs. (Bottom right) A dolomite solution breccia created by the preferential dissolution of gypsum in a sabkha sequence of dolomite and gypsum beds. It is now a firmly recemented, resistant rock.

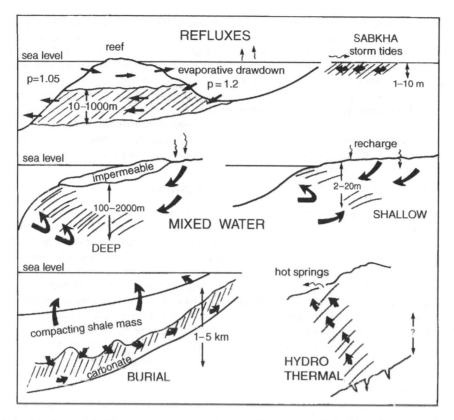

Figure 2.10   Dolomitization models, illustrating the depth to which the process may operate in different circumstances. ρ = density of salt waters.

hydrated, its electrostatic bond with H_2O molecules being 20% stronger than that of Ca^{2+}. This bond is an energy barrier that must be broken for the ion to adsorb into a crystal lattice. Where there is a high Mg:Ca ratio and abundant $(CO_3)_2$ in the solution this can occur. Standard seawater approaches these conditions because high-Mg calcite can be precipitated, but it does not attain them. That can occur in surficial waters subject to evaporation and some loss of Ca^{2+} ions to calcite or gypsum formation, permitting primary dolomite to be precipitated (e.g. in saline lakes – Last 1990). However, it is accepted that most dolomite forms by replacing earlier calcite and aragonite, i.e. it is a secondary precipitate.

The principal modern models for secondary dolomitization are shown in Figure 2.10. In **reflux** models, seawater is first concentrated to hypersaline levels (and greater density) by evaporation in lagoons and then fluxes through lagoonal, reefal or supratidal lime sediments, exchanging ions with them. The dolomite is eogenetic, formed very early in diagenesis. These models explain

the frequent association of dolomite with gypsum beds in sabkha facies.

Mixing models portray mixing of fresh and salt waters to produce conditions where calcite is soluble but dolomite is not and where Mg:Ca ratio requirements are met. Hanshaw and Back (1979) hypothesized that regional-scale dolomitization is occurring today in the mixing zone that underlies much of Florida.

It is disputed whether burial and compaction can achieve regional dolomitization in mesogenetic settings via the expulsion of basinal fluids. It may explain the preferential dolomitization of individual, more permeable, beds. Geothermal heat may play an important role, establishing upward fluxes of Mg-rich connate waters in young but deep carbonate banks such as the Bahamas (i.e. eogenetic settings; Wilson *et al.* 2001). Morrow (1998) proposed that it can also establish cellular convection systems in carbonates that are deeply buried by impermeable clastics (i.e. in mesogenetic settings), concentrating and recycling connate fluids over distances of tens to hundreds of kilometres. Some dolomite is undoubtedly of hydrothermal origin because it is seen to be localized along fractures or around solutional pipes which hot waters have ascended: a Ca:Mg molar ratio as high as 10:1 will yield dolomite at 300°C. In many instances it is apparent that a rock has been subject to two or more sequential episodes of dolomitization (e.g. see Ghazban *et al.* 1992b).

It is important to understand that as a consequence of these various modes of dolomitic replacement, the scale, extent and patterns of dolomitization within limestone masses can be highly variable. Because consolidated dolomite is normally less soluble than limestone, the effects upon local and regional groundwater circulation and karst morphogenesis can be profound.

2.3.2 Dolomite composition

Dolomicrite has grains $< 10 \, \mu m$ in diameter and is believed to be early replacement of aragonite. Much more abundant is medium crystalline or **sucrosic** dolomite, 10–20 to $100 \, \mu m$ in diameter. This is the standard dolomite. White, sparry crystalline-to-megacrystalline (centimetres in diameter) dolomite is often found in association with lead–zinc deposits and other situations where several episodes of dolomitization are possible.

Where dolomitization is complete but proceeded slowly ghosts of fossils and other allochems or framework may be well preserved. Older English-language classifications (pre-1960) define 'primary dolomite' as that preserving some original depositional texture. In 'secondary dolomite' this is entirely destroyed (e.g. Figure 2.9). The other category recognized was 'hydrothermal dolomite', broadly equivalent to the modern 'white sparry dolomite' which, however, is not always hydrothermal in origin.

As dolomitization of limestone proceeds through the range 5–75% there is, in general, progressive reduction in porosity due to infilling. Thereafter, porosity increases again because the dolomite rhombs are smaller than the calcite crystals they have now largely or entirely replaced. There is an increase in large vug porosity. The high initial and early diagenetic porosity of reef rocks (attributable to their framework) makes them preferred sites of dolomitization – which again enhances their porosity. This is why buried reefs are prime targets in petroleum exploration.

2.3.3 Dedolomitization

Dolomitization can be thrown into reverse during diagenesis if calcium ion in solution should be greatly enriched. The solution may then be supersaturated with $CaCO_3$ but undersaturated with respect to $MgCO_3$. The dolomite dissolves and crystalline calcite (pseudomorphic or new) precipitates in its place. The process creates beds or patches of etched, crumbly rock. It requires maintenance of a particular and rather delicate hydrochemical balance (see Ayora *et al.* 1998; and the discussion of **incongruent dissolution** in section 3.5). As a consequence, **dedolomite** is rare; it is most often found in association with large vugs in small reefs.

2.3.4 Breccias

Solution breccias are common and widespread throughout the geological record. The brecciation may occur entirely within limestone or dolomite or where evaporite rocks are removed in evaporite–carbonate sequences. It occurs during eogenesis when evaporites are dissolved in supratidal carbonate sequences or where caves in case-hardened carbonate sand dunes collapse. It is common during deep burial, caused by the expulsion of connate waters or by invading thermal waters, and may extend to depths of 5 km or more. It is also common in telogenetic settings, where modern meteoric waters are known to be brecciating carbonate rocks at depths as great as 2 km.

Three principal fabrics are recognized (Stanton 1966).

1. **Crackle breccia**, where beds sag apart and crack upon dissolutional removal of support but there is little displacement.
2. **Pack breccia**, where large fragments (clasts) support each other in a pile. The clasts have usually dropped and vary from partly rotated to completely disorganized (**chaotic** brecciation) in orientation. Globally, clast sizes range from small pebbles to 'cyclopean breccias' with individual blocks of $\sim 10^2 \, m^3$, but are normally more limited within any given breccia.
3. **float breccia**, where the larger fragments are separated from each other and 'float' in a matrix of fines.

A genetic association of these fabrics is often found, consisting of crackle breccias at the top and around the perimeter of a body, pack breccia within it where beds are thicker, and float breccias at the base or where beds are thin. In some Mississippi Valley Type (MVT) lead–zinc deposits, these may be underlain by a basal **trash** zone of insoluble residua (Sangster 1988; Dżulynski and Sass-Gutkiewicz 1989; and Figure 7.29a). Older breccias of all types normally display partial or complete cementation, usually with calcite or dolomite spar: in Mallorca there is an outcrop of firmly cemented chaotic breccia that is as much as 50 km in width. Solution breccias are discussed further in Chapters 7 and 9.

Mechanical breccias are also found in carbonate rocks, where they accumulated as submarine foreslope, reef-foot talus or landslide deposits, or as subaerial screes. There is usually interstitial cement. Some chaotic breccias in the Yucatán Peninsula of Mexico and neighbouring Belize may be ejecta from the celebrated Chicxulub impact crater.

2.3.5 Stylolites

Stylolites are pressure solution seams. They are a striking feature of many well-bedded limestones and dolomites. They can develop in all diagenetic environments but are most common where there is burial deeper than 500 m, with the associated high overburden pressures. Most are bedding-parallel and can give rise to pseudo-bedding, but transverse seams are also quite common. The carbonate dissolves at highest pressure points. These may be at the bottom and top, respectively, of adjoining crystals, fossils, or larger masses. The results are highly irregular, serrated dissolution seams with relief ranging from a few millimetres in homogeneous mudstones to many centimetres or more in heterogeneous rocks such as grainstones and framestones (Andrews and Railsback 1997). They stand out in exposures because they are darkened by residual insoluble minerals and organic matter concentrated along them (Figure 2.6; and see Moore 2001, p. 301). As much as 40% of the compacted mass of some carbonates may have been destroyed at the stylolite seams. The dissolved and expelled material may then be reprecipitated as cement elsewhere.

The insoluble residues reduce effective permeability at many stylolites but others may break open upon pressure release to offer preferred flow paths, increasing the effective fissure frequency. They account for much karren-scale roughness on many limestone and dolomite surfaces (Figure 2.8). In clayey limestones they contribute to nodular weathering habits. The residues are recessive in surface exposures but may protrude as thin ledges in solution cave passages.

2.3.6 Chert

Some silica will accumulate in many lime sediments, from transported quartz or from siliceous sponges, radiolaria and diatoms. If very alkaline conditions arise during diagenesis this is dissolved. It is reprecipitated as accretionary nodules or lenses of chert (**flint**). These usually accumulate along bedding planes where they may coalesce to form continuous sheets. Nodules as great as 1 m in diameter are known. Sheets are rarely thicker than ~10 cm and are normally perforated or fractured so that fluid flow across them is rarely prohibited entirely. However, it may be locally obstructed, perching or impounding the water. It also forms resistant caps in subaerial karst. Chert tends to be more abundant in the older limestones and dolomites.

2.3.7 Marble and other metacarbonates

Marble is produced when limestones or dolomites are metamorphosed by pressure and heat: pressure regimes are mostly in the range 1–10 kbar and temperatures are generally between 200 and 1000°C. The lowest grades of metamorphism are associated with deep but not hot diagenesis (< 350°C), where the smaller grains plus outer surfaces of larger ones become annealed. They recrystallize as spar upon cooling. Fossil form, lithological texture, etc. are quite well preserved. In the next grade (**greenschist** facies, ~350–500°C), partial melt or replacement (**metasomatism**) of crystals is nearly complete and the new grains may be reoriented (foliated) to accord to the ambient pressure field, destroying most fossils and sealing bedding planes. This produces hard, dense, saccharoidal crystalline rock of rather even grain size, i.e. ideal sculptors' marble. Grains are irregular in form with sinuous or zigzag surfaces. At > 400°C dolomite may be converted to periclase marble when part of the CO_2 is driven off, leaving $CaCO_3$ with **periclase** (MgO) embedded in it as insoluble octahedra. In carbonates with abundant silica (e.g. chert) and other impurities, reactions produce **wollastonite** ($CaSiO_3$, seen as big, lustrous white and insoluble crystals in many marbles), **chlorites** ($(Mg, Fe,Al)_6(OH)_6(Si,Al)_4,O_{10}$), **garnets** such as grossular ($Ca_3Al_2Si_3O_{12}$), and **talc** ($Mg_6(Si_8O_{20})(OH_4)$). Marble has very low porosity and often there is negligible permeability so that it is difficult for karst waters to penetrate. But where this is possible it tends to give the sharpest, cleanest dissolutional morphology, as in the very spectacular Ultima Esperanza karsts of Patagonia (see section 9.2; and Maire 1999).

The highest metamorphic grades, **amphibolite** (~500–700°C) and **granulite** (> 700°C) facies, are usually associated with invasion of magmatic fluids from adjoining igneous intrusions, which leach the carbonates and reconstitute them as schistose or gneissic mixtures (**skarn**) of lower solubility that are unsuitable for karst. There are always exceptions, however. In the Rudohorie Mountains of Slovakia, Mg-rich solutions from regional metamorphism produced lenticular bodies of magnesite ($MgCO_3$) within dolomites; individual masses are up to hundreds of metres in size and contain small caves (Zenis and Gaal 1986). Similar features are found in granulitic metacarbonates within a few metres of granite stocks on the west shore of Lake Baikal, and in western Tasmania. In the Carlin Trend of Nevada the leachates are reprecipitated as gold-bearing calcite veins, dikes and solution breccia fillings. It is rare for high-grade metamorphism to extend more than ~500 m or so from an igneous intrusion.

2.3.8 Carbonatites

Carbonatites are peralkaline igneous rocks composed of 60% to > 90% carbonate minerals, principally calcite. They are normally intrusive, associated with pyroxenites and amphibolites, but eruptive tuffs and lapilli are known. They are rare ($\ll$ 1% of igneous rocks in outcrop) but more than 350 examples are now described, ranging from ~1.0 to 20 km^2 in area (Bell 1989). They tend to be enriched in rare earths and base metals, suggesting that their sources are partial melts from the upper mantle.

Karst features develop well on high-calcite carbonatites. Sandvik and Erdosh (1977) describe a carbonatite intrusion containing up to 17% apatite ($Ca_5(F,Cl,OH)_3 PO_4$), where the phosphate has been concentrated as a weathering residuum in large sinkholes in the calcitic rock.

2.3.9 Conglomerates

Conglomerates are debris mounds or fans of pebble-, cobble- and/or boulder-sized clasts that have been rounded or partly rounded during transport by flowing water, waves or glaciers. The interstices may be filled with clay or other fines, and they may be cemented by calcite or silica. Conglomerates composed of limestone or dolomite clasts are common features in carbonate regions, where they usually originated as submarine delta dumps from steep mountains fronts, or detrital fans into lowland basins that later became submerged. Where the clasts are largely or entirely carbonate and the cement is calcite, they may function like other pure carbonate rocks. An example is the Dolomitic Conglomerate of the Mendip Hills, England: the clasts (up to boulder size) are from platform limestones of Lower Carboniferous age that were eroded from steep desert scarps during Permian times and swept down canyons to accumulate as fanglomerates that then became submerged. In tourist caves such as Wookey Hole solutional forms in the passage walls can be seen to pass smoothly from the undisturbed Carboniferous Limestone to the Permian conglomerate. In the Italian Fore-Alps the Montello Fan is a spectacular limestone conglomerate nearly 2000 m thick that was formed during the Miocene 'Messinian Crisis' when the Mediterranean Sea dried up (see below). Although its pebbles are small, it has quite a high clay content in the matrix and cementation is not as firm as in the diagenetically more mature case of the Mendip Hills; nevertheless, it supports a densely packed doline karst draining into well-formed caves (Ferrarese *et al.* 1997).

2.4 THE EVAPORITE ROCKS

Evaporite rocks are or have been present beneath ~25% of the continental surfaces (Figure 1.3). They are much less common in outcrop than the carbonates but extensive gypsum karsts occur in parts of China, Ukraine and the USA, and there are smaller examples in many other countries. Salt karst landscapes are limited to small patches in deserts. Interstratal solution of these rocks (with or without some surface expression) is widespread and very important (see overview by Klimchouk *et al.* 1996).

The rocks are formed by homogeneous or heterogeneous precipitation in marine, lake or ponded waters that have been concentrated by partial evaporation, or as residues left by complete evaporation. Seawater is quantitatively the most important source: restricted lagoons, plus sub-, inter- and supratidal salt flats (known collectively by the Arabic term **sabkha** (Figure 2.4b)) are the chief depositional environments. Figure 2.11 shows a good example of mingled limestone, dolomite, gypsum and anhydrite stratigraphical sequences from Canada. Playa lakes and interdune pondings in arid and semiarid environments are regionally important, and there are substantial gypsum and salt deposits in such comparatively humid regions as the Ebro Valley of Spain (e.g. Sanchez *et al.* 1998).

Figure 2.12a shows the evaporative concentrations required to induce precipitation of gypsum, anhydrite and salt. A **brine** is any water with total dissolved solids equal to or greater than those in standard seawater (~33 000 parts per million (ppm) of water). Figure 2.12b is a comprehensive summary of the variety of brines, precipitates and evaporites that may evolve, relating them to differing solute chemical concentrations in the water at the start of the processes. Calcite can precipitate from standard seawater when it is warmed (**whitings** noted above) and dolomite when there is only slight evaporative concentration. Gypsum precipitates at about three times seawater strength and salt at eleven times strength; gypsum is thus the more widespread evaporite deposit because lagoonal, sub- and intertidal waters are generally renewed before there can be much deposition

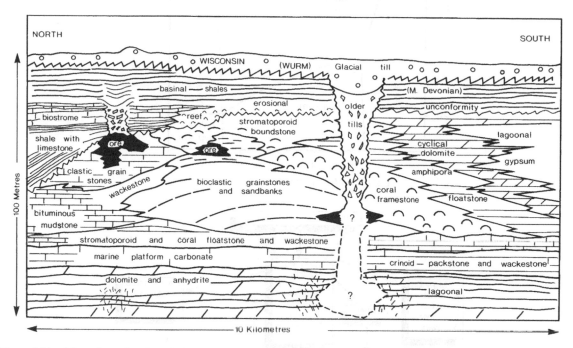

Figure 2.11 Schematic section through the Presqu'ile Reef at Pine Point, Northwest Territories, Canada, to illustrate lateral and vertical facies changes in limestone, dolomite, gypsum and anhydrite formations. The rocks are of Middle Devonian age. They display palaeokarst cavities which are believed to be younger, and which contain sulphide deposits of zinc/lead ore grade. During the Quaternary sinkholes were rejuvenated; glacial deposits from the last ice advance (Wisconsinan) are intruded through the section. Devonian section is generalized from the work of Rhodes *et al.* (1984).

of salt. The other evaporative minerals such as poly-halite, sylvite and carnallite (Table 2.1) tend to be thin or peripheral accumulations and thus are of little karst significance. In contrast to the carbonates, all of these rocks are wholly inorganic in origin and composition (Alsop *et al.* 1996; Sarg 2001).

Evaporative concentration and precipitation are powerful processes that produce rapid results. Table 2.5 gives general estimates of accumulation rates for the principal clastic, carbonate and evaporite sedimentary facies, showing gypsum and salt building up at least ten times faster than the others. Thicknesses may quickly come to exceed 1000 m. In the 'Messinian Salinity Crisis' of the late Miocene (~5.5 a) the tectonic collision between North Africa and Spain temporarily closed off the Mediterranean Sea from the Atlantic; many geologists believe that it nearly dried, depositing 1000–2000 m of shallow marine gypsum and salt at ~2000 m below modern sea level within a timespan that was probably less than 500 000 yr. The Dead Sea (surface at −400 m and floor at −730 m below sea level) is the deepest saline closed basin today; calcite and gypsum are precipitating at its surface and salt on the floor.

2.4.1 Gypsum and anhydrite

Gypsum (Table 2.1) is the mineral normally first precipitated. Primary anhydrite is rare. When buried beneath 200–300 m or more of overburden most gypsum is converted to anhydrite as the rock is dehydrated, although some is reported to survive to depths of 3000 m. If the overburden is then stripped by erosion, rehydration to gypsum normally takes place. Most gypsum that is exposed in karst has been through the cycle of dehydration and rehydration in differing temperature and pressure environments. As a consequence, there is a wide range of petrological and lithological forms; original depositional environments may be difficult to discern.

Minor gypsum occurs as isolated crystals or clusters of crystals in some carbonate rocks. It appears as rare to frequent interbeds in sequences of medium- to thin-bedded series containing dolomite, clay or shale interbeds. Where it is interbedded it is common to see dikes, diapirs and other intrusions of gypsum penetrating the other rocks.

Major gypsum deposits occur as coarsely crystalline (equant, curved, acicular, prismatic or columnar forms) to

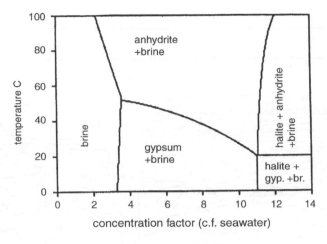

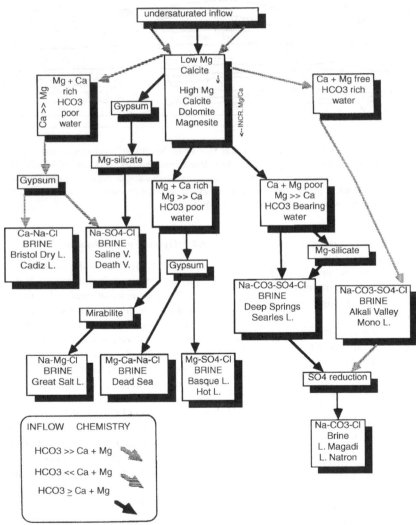

Figure 2.12 (a) The sequence of precipitation of salts when seawater is evaporated, showing the relationship to temperature. (b) The relationships between inflow chemistry, types of brines and evaporate precipitates found in lakes in continental closed basins. Reproduced from Eugster, H.P. and Hardie, L.A. 1978. Saline Lakes. in Lerman, A. (ed.) Lakes – chemistry, geology, and physics. New York, 237–93 © Springer Verlag 1978.

Table 2.5 Comparison of the various depositional rates in some sedimentary settings (Reproduced from Warren, J.K. (1989) Evaporite Sedimentology, Prentice Hall. New Jersey, 285 pp.)

Sediment type	Depositional rate ($m\,Myr^{-1}$)
Deep-sea clay	1
Deep-sea carbonate ooze	1
Shelf carbonate and mud (non-reef)	10–30
High-productivity clastic slope	40
Deep-sea turbidite	100–1000
High-productivity marine diatomite	400–1000
Shallow-water reefal limestone	1000–3000
Sabkha (supratidal) evaporate	1 m thick section progrades 1 km $1000\,yr^{-1}$
Subaqueous gypsum	10 000–40 000
Subaqueous halite	10 000–100 000
Diapir growth (Zagros Mountains, Iran)	150 000–200 000

granular or amorphous, massive beds of the pure mineral, usually translucent (**selenite**), opaque white (**alabaster**) or tinged in shades of brown, grey, yellow or pink. Individual beds 10–40 m thick are known. Sequences of them may exceed 200 m. These are rarer than the thinner, interbedded sequences but support most of the prominent surficial gypsum karsts.

Dehydration of gypsum to anhydrite occurs at pressures of 18–75 10^{-5} Pa and results in a volume reduction of approximately 38% (Warren 1989). In the field, anhydrite contains little measurable porosity, and joints and bedding planes are annealed. Flow as diapirs may occur. Overlying strata (e.g. dolomite) are often brecciated as a result of this sulphate volume reduction and compaction.

Hydration of anhydrite has been reported as deep as 2000 m (in Texas), but most of it probably occurs within 100 m of the surface and much within the topmost few tens of metres. Pechorkin (1986) stressed that it advances along reopening joints or new fractures, with the consequence that the hydration front can be highly irregular. It is common to find small patches of anhydrite surviving in gypsum cliffs where there has been rapid exposure. Quinlan (1978) argued that hydration involves dissolution with immediate reprecipitation. It generates a pressure of $\sim$20 kg cm^{-2}. Below 150–200 m this is probably dispersed in fluid flow without expansion of volume (Gorbunova 1977). Above that limit mineral volume expansion of 30% to 67% may occur, causing flow and intrusion at depth, brittle fracture towards the surface, and recreating some bulk porosity. At the surface tightly

folded (corrugated) or small-scale block-faulted topography may appear, or blisters termed **tents** or **teepees** (section 9.13). Little or no groundwater flow occurs within anhydrite.

2.4.2 Salt

The mode of occurrence of rock salt is similar to that of gypsum. It may be disseminated in carbonates, sulphates or shales, occur as thin interbeds, or as massive units up to 1000 m in thickness that contain only a few anhydrite or shale beds and often have economic deposits of potash salts (carnallite, sylvite, sylvinite ($KNaCl_2$)) at the top. Most disseminated or thin-bedded salt is dissolved during diagenesis so that its presence is signified only by voids, breccia or disconformities in surviving rocks such as dolomite. In massive salt all joints and other openings become annealed by lithostatic pressure; the rock is made quite impervious, prone to flow and intrude to form diapirs and smaller structures. The deep solution of salt can occur only where water approaches it via mechanically strong aquifer strata immediately above or below it and can sap it at the contact. Although little or no groundwater flow occurs within salt itself, where there are outcrops there can be significant vadose circulation and caves (see section 9.13).

2.5 QUARTZITES AND SILICEOUS SANDSTONES

Silica sands and pebbles cemented by silica (siliceous sandstones, conglomerates) and their metamorphosed equivalents, quartzites, can develop dissolutional karst landforms at small and intermediate scales (Mainguet 1972) because, like the carbonates and evaporites, they are monominerallic in composition or nearly so. The range of forms and hydrogeological behaviour is more limited, however. In the mineral habit of quartz, the solubility of silica is very low in meteoric waters; but quartzite strongly resists most other forms of weathering attack as well. Amorphous silica (which forms many sandstone cements) is more soluble. Solubility of all forms of silica greatly increases in water above 50°C (section 3.4).

Some measure of karst development will be found along most escarpment crests and cuestas in quartzites because groundwater hydraulic gradients are steep, permitting rapid water flow and effective flushing of residua. Massive Precambrian quartzite scarplands of the Roraima Formation in Brazil and Venezuela are perhaps the greatest example, with corridors and deep shafts drained by caves that may be several kilometres in length (see section 9.14; and e.g. Galan 1995, Correa Neto 2000). Lesser instances, plus pinnacle karsts, are found on quartzites and siliceous sandstones in many other places, e.g. Montserrat, Spain. Limited but attractive karren (chiefly pit, pan and runnel forms – see section 9.2) are even more common. There are three principal requirements for substantial development of solution features: (i) high mineral purity, so that initial surficial pits or underground dissolution channels do not become filled or blocked by grains of the insoluble aluminosilicates, etc. that are present in a majority of sandstones; (ii) thick to massive bedding with a few penetrable planes intersected by strong but widely spaced fracturing; and (iii) absence of strongly competing geomorphological processes such as frost shattering or wave attack. The absence of effective competition permits the comparatively slowly developing solution landforms to become dominant.

2.6 EFFECTS OF LITHOLOGICAL PROPERTIES UPON KARST DEVELOPMENT

The remainder of this chapter considers factors of 'rock control' on karst hydrogeology and morphogenesis. Karst specialists are concerned with a narrower range of rock types than most other hydrogeologists and geomorphologists, so it is somewhat ironic that they find it necessary to investigate local rock properties in more detail in a majority of field studies. The need is real; petrological, lithological and structural features greatly influence all aspects of karst genesis. Here we outline some main points. Necessarily, others are illustrated throughout all later chapters of the book.

Significant properties at the scale of individual crystals are summarized above. This section notes properties that are important at the scale of hand specimens. Later sections then consider local- and regional-scale factors.

2.6.1 Rock purity

Clay minerals and silica are the most common insoluble impurities in carbonate rocks. It is a widespread finding that limestones with more than 20–30% clay or silt (argillaceous limestones) form little karst. In a series of computer models Annable (2003) found that medium-grained silt was very inhibitive because it clogged protoconduits but, treated as individual particles, finer silt and clays did not; however, at these sizes there is much agglomeration of particles, producing the medium silt effect. There are no such clear-cut relationships with respect to sand content. Large and diversified carbonate karst assemblages do not commonly develop where silica

exceeds 20–30%, but shallow dolines and well-formed small caves are known in some calcareous sandstones.

The best karst rocks are > 70% pure carbonates. Studies of local limestone and dolomite specimens have been made in many countries. They have established that laboratory dissolution rates in carbonated water may vary by more than a factor of five. Fastest dissolution has been recorded where the percentage of insolubles is nil and where it was as great as 14%, although most investigations show a clear positive correlation between percentage CaO and dissolution rate. Pure dolomites are normally slowest to dissolve. But there is always much variation about any trend that cannot be explained by simple bulk purity. For example, in an exhaustive study in Pennsylvania, Rauch and White (1970) found that the greatest solubility in pure carbonates occurred where MgO = 1 to 3% and was present within silty streaks that increased the roughness (exposed area) of the dissolving surfaces. This is a textural effect. James and Choquette (1984) suggest that high-Mg calcite is normally the most soluble because of severe distortion of the calcite lattice, followed by aragonite, low-Mg calcite, pure calcite and dolomite, in descending order of solubility.

In gypsum, anhydrite and salt there is normally a simpler positive correlation between purity and solubility.

2.6.2 Grain size and texture

The finer its grain size the more soluble a rock tends to be because the area of exposed grain surfaces is increased. Many studies have found that micrites or biomicrites are most soluble and that solubility decreases substantially where sparite (coarse crystals) becomes greater than 40–50% by volume (e.g. Sweeting and Sweeting 1969; Maire 1990). In a discriminant analysis of ten different purity, grain size, texture and porosity measures applied to cavernous limestones and dolomites in Missouri, Dreiss (1982) found grain size to be the most significant, the finer grained rocks being more soluble. However, the finest grained limestones are sometimes less soluble if the grains are uniform in their size and packing because surfaces are then smooth, with exposed grain areas being reduced; such rocks are termed **porcellaneous** or **aphanitic** in texture. The Porcellaneous Band is a distinctive, very fine-grained micrite that obstructs cave genesis and perches passages in Gaping Ghyll Cave, England (Glover 1974). It is sandwiched between coarser biomicrites.

The greater the heterogeneity of grain size in a specimen, the greater is the roughness of a dissolving surface.

This increases solubility, up to a limit. Biomicrite is more soluble than pure micrite because the tiny fossil fragments protrude as roughnesses.

Karren, because they are small, are strongly affected by texture. As homogeneity increases so do the number of karren types that may be hosted and also the regularity of form displayed by any particular type. No channel karren develop well on very heterogeneous rocks such as reefs or most conglomerates. Solution scallops (round forms) and rillenkarren (linear) require fine grain size and high homogeneity. Trittkarren, a mini-cirque form, are largely confined to aphanitic rocks. The morphology of these features is discussed in section 9.2.

2.6.3 Fabric porosity

Much of the variation in erosional behaviour of carbonate rocks is due to variation in the nature, scale and distribution of voids within them. This is termed porosity. It is a subject of the greatest importance. There are now many different terms and classifications; see Moore (2001) for a modern summary.

Sedimentologists define primary porosity as that created during deposition of the rock (i.e. created first) and secondary porosity as that produced during diagenesis. For hydrogeologists, all types of bulk rock porosity are **primary**. **Fracture** (or **fissure**) and **channel** (or **conduit**) porosity are considered to be **secondary** and **tertiary** respectively. We adopt this convention throughout all later chapters – see section 5.2.

The widely adopted classification given in Figure 2.13 avoids these problems of precedence by distinguishing porosity that is related to petrofabric from that which is not. In general it is true to write that karst hydrogeology and geomorphology are concerned largely or entirely with large-scale, interconnected, non-fabric-selective porosity (penetrable bedding planes and fractures, dissolutional channels and caverns of Figure 2.13) in rocks where the fabric-selective porosity is low (<15%). This is because the hydraulic pressure gradients experienced also tend to be low, insufficient to drive significant quantities of fluids through the tiny throats separating poorly connected pores within the rock fabric itself. The converse is true in much petroleum exploration, which is concerned with high-pressure environments where many fractures are closed but fabric porosity is preserved. Large-scale karstic porosity is further discussed in section 5.1 and later.

Fabric-selective porosity can be broadly important in eogenetic conditions. In diagenetically mature rocks it helps determine such features as the form, scale and distribution of solution pits and some other karren, the distribution of many stalactites, etc. In a limestone or dolomite it tends to be positively correlated with grain size and textural heterogeneity, although there is much variation. The primary porosity of micrite is generally less than 2%, that of sparite between 5 and 10%. Dolomitization increases porosity by 5–15% in most instances. The porosity of most marble is <1%.

Anhydrite and salt anneal readily and so have negligible porosity. Where gypsum is formed by hydration, high intercrystal and breccia porosity may be created.

2.6.4 Mechanical strength

At small scale the strength of a rock is a function of its interparticle bonding. Such strength can be measured in the laboratory by compression, shear or hammer tests. At larger scale in sedimentary rocks strength is more obviously a function of the density of fissures such as joints or bedding planes. This kind of strength is not amenable to machine testing.

Compressive strength is probably the most significant of the laboratory measures (Table 2.6). It conveys some idea of how a given rock bed will respond if it must bear extra load where there is no buttressing support, as at the base of a cliff or at a cave passage junction. Weaker rocks yield by platy fracture parallel to the unsupported faces. This undermines higher parts of the cliff or cave wall, etc. and may induce more widespread block failure (Figure 7.48). A majority of carbonate rocks are quite strong and will support vertical cliffs and cave roofs for long periods unless they are thinly bedded and highly fissured. Some chalks and other poorly cemented, particle-supported limestones (e.g. aeolianites or oolites of Pleistocene age) are too weak to support big cliffs or caves of enterable dimensions.

The Schmidt hammer is a field tool designed to measure the hardness of concrete on a scale of 10–100. Its field values on natural rocks correlate quite well with compressive strength. Sample dolomitic limestones in the Caribbean area have mean Schmidt hardness, R, of 40–41 whereas limestone micrites and sparites score 34–35; there is some positive correlation with topographic ruggedness (Day, 1983). Older crystalline limestones in the USA and Canada range from 35 to 70; Tang (2002) obtained mean values of 40–60 from the thick to massive limestones supporting the great karst towers around Guilin, China.

Gypsum, anhydrite and salt are weak. In most instances they will support cliffs and cave roofs, but with excessive rates of block and slab breakdown due to mechanical failure.

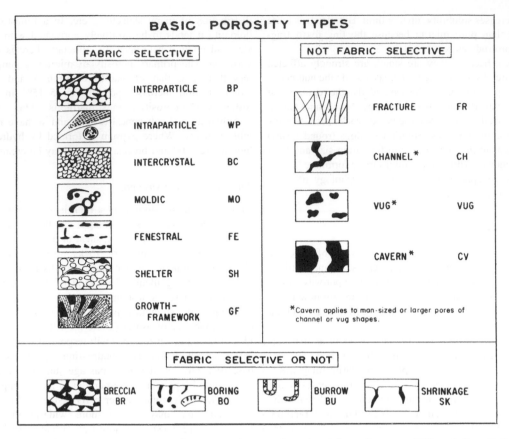

Figure 2.13 Classification of porosity in sedimentary carbonates. Reproduced from Choquette, P.W. and Pray, L.C. (1970) Geological nomenclature and classification of porosity in sedimentary carbonates. American Association of Petroleum Geologists Bulletin, 54, 207–50.

Table 2.6 Compressive strength of some common rocks (Modified with permission from Jennings, J.N. Karst Geomorpholog, ©1985 Blackwell Publishing)

Rock type	Uniaxial compressive strength (bars)*
Limestones (excluding chalk and breccias)	340–3450
Dolomites	620–3600
Marbles	460–3400
Anhydrite	220–800
Shales	300–2300
Sandstones	120–2400
Basalts	800–3600
Granites	1600–3000
Quartzites	1500–6300

*Suggested terms: very weak < 350; weak 350–700; strong 700–1750; very strong > 1750.

2.7 INTERBEDDED CLASTIC ROCKS

Here we consider the bulk solubility of rocks at the scale of geological units, members and formations. Many carbonate, sulphate and salt formations are without significant clastic interbeds for thicknesses of tens, hundreds or even several thousands of metres. These strata generally yield the best karst development. However, the geological record contains many more examples of formations with frequent beds of clay, shale, sandstone or coal between the soluble strata. These grade to shales, etc., with limestone interbeds, etc. and the geomorphological and hydrological systems grade from wholly karstic to non-karstic.

It is difficult to offer valid generalizations concerning karst development in the intermediate conditions. As frequency of shale increases in a formation, so does the likelihood that intervening limestones will be argillaceous and non-karstic but this is not always true. Groundwater penetration to initiate karst is often easier at the contact between limestone and shale than it is at bedding planes, joints, etc. within limestone. As a consequence small, independent or poorly connected, solution conduit systems may develop in adjoining, sandwiched limestones. The same is true of gypsum. Karren develop where the soluble rocks crop out. There may be small dolines. Collapse features are important where gypsum is interbedded.

2.8 BEDDING PLANES, JOINTS, FAULTS AND FRACTURE TRACES

Bedding planes, joints and faults are of the greatest importance because they host and guide almost all parts of the underground solution conduit networks that distinguish the karst system from all others. These are the planar breaks in the rock that can be significantly penetrated and modified (by dissolution or precipitation) by circulating groundwaters, past or present. When all is said and done about properties of karst rocks it is these entities where rock is absent that determine much of the variety of form and behaviour that occurs in the system. In hydrogeology it is customary to categorize all of them as **fractures**, combining to form **fracture aquifers** although, strictly speaking, this is only partly correct because many karstified bedding planes are primary depositional features without any mechanical fracturing. Structural engineers concerned with slope stability describe all as **penetrative discontinuities**.

2.8.1 Bedding planes and contacts

Bedding or parting planes in sedimentary rocks are produced by some change in sedimentation or by its temporary interruption. The change may be minor, e.g. from one size of carbonate grain to another a little larger. Major changes are represented by large differences in grain size and, more often, by the introduction of clay by a storm or flood, etc. that leaves a paper-thin or thicker parting between the successive regular carbonate layers. Interruption is usually a brief marine emergence with some erosion and the start of meteoric diagenesis or beachrock before renewed submergence, or the formation of a hardground. Subparallel pseudobedding is created by current scour where carbonate shell banks are reworked by the tides, often giving rise to platiness in weathering outcrops.

Many smaller subaerial unconformities in young rocks are diminished during diagenesis because they are preferred sites for mineral infilling or obliteration by stylolite pressure solution. More prominent interruptions persist as minor geological discontinuities and are widely used to subdivide layered sequences into **members** or lesser **units** within formations. Many authors classify them as **contacts**. The junctions between layered sediments and mounds such as reefs in carbonate deposits are another type of contact that is often preferentially penetrated by groundwaters.

In hydrogeology, bedding planes are only significant if they are sufficiently open to be penetrable by water under natural pressure gradients. Only a minority of sedimentological planes will be in most cases. In karst geomorphology, these planes are also the most important, but impenetrable planes that will rupture under mechanical stress (e.g. during cave-roof collapse) are also significant. Table 2.7 presents a standard classification of bed thickness. In karst work it defines the separation between successive bedding planes that are penetrable by water in the prevailing conditions.

The areal extent of individual penetrable bedding planes varies considerably. Where bedding is thin to very thin, they may cover only a few square metres. Where it is medium to thick the extent is normally 10^3–10^6 m^2 or much more. Truly major planes may be followed throughout a formation, sometimes for hundreds of kilometres. As a consequence major bedding planes and contacts can be considered to be **continuous** entities when solution caves are propagating through them, whereas penetrable joints and most faults are **discrete** (they terminate in comparatively short distances). This enhances the importance of bedding planes in cave genesis (Ford 1971a). Lowe (2000) termed such features **inception horizons**.

The penetrable plane itself can be considered to comprise two rock surfaces in undulatory contact, with some greater interlocked prominences and depressions due to sand ripples, hardground pitting, etc. The voids are

Table 2.7 Terminology for bed thickness and joint spacing

Bed thickness (cm)	Description	Joint spacing (cm)	Description
100–1000	Very thick bedded or massive	> 300	Very wide
30–100	Thick	100–300	Wide
10–30	Medium	30–100	Medium
3–10	Thin	5–30	Close
1–3	Very thin	< 5	Very close
< 1.0	Laminated		

shallow, sinuous, irregular in outline, and partly inter-connected via **throats** that have openings of lesser aperture. Planes most readily exploited by groundwater include those with substantial depositional disconformities, plus planes with shale laminae or thicker partings (often with disseminated pyrite) and planes with nodules or sheets of chert. Perhaps the most important are those that have served as surfaces of differential slippage during tectonic events. Even if the displacement is just a few centimetres there is some slickenside striation and brecciation that enhances openings. Most steeply tilted and all folded strata will display some measure of differential slip (Šebela 2003).

It is widely recognized that the finest karst landforms require medium to massive bedding. The solutional attack is dispersed where beds are thin: they also lack the mechanical strength to sustain steep slopes and enterable caves in many instances.

Mammoth Cave (Kentucky, USA) and Holloch (Switzerland) are amongst the most extensive limestone caves currently known. In both the great majority of conduits are guided by bedding planes. At Skocjanske Jama, a UNESCO World Heritage cavern in Slovenia, Knez (1996) recognized 62 prominent bedding planes in the thick-to-massive limestones, but showed that just three of them had guided initiation and development of the cave.

2.8.2 Joints and shear fractures; joint systems

Joints are simple pull-apart breaks in previously consolidated (or partly consolidated) rocks. In shear fractures there is some lateral or vertical displacement but it is too small to be recognized in hand specimens (Barton and Stephansson 1990). Fracturing occurs during diagenesis, later tectonism, erosional loading and unloading. It is caused by tensional or shear forces.

In regularly bedded rocks, most joints are oriented normal to bedding planes, but they may be inclined. In plan view a majority will be straight. Sinuous and curvi-linear joints predominate in reefs, however, and are quite common elsewhere. Parallel joints constitute a **joint set**. Two or more sets intersecting at regular angles compose a **joint system**. Rectangular and 60°/120° systems are the most common, caused by simple tension and shear forces respectively (e.g. Figure 2.9). Arcuate sets are frequently seen in reefs and mounds. Major joints extend through several or many beds and are often termed **master joints**. They terminate at other master joints. Cross joints are confined to one or a few beds and terminate at master joints. Master joints in thick to massive rocks may be as long as several hundred metres, exceptionally extending for many kilometres: on Anticosti Island, Quebec, a master set in thick Ordovician limestones persists for 200 km, parallel to a plate-tectonic suture in the underlying Precambrian strata.

Table 2.7 gives the scale of joint spacing. This is broadly proportional to bed thickness but the correlation is not precise. An individual bed may contain just one set or system, or several systems imposed at different times. Successive beds in a sequence often display different patterns and densities.

Joint fracture openings may be latent or tiny and impenetrable to water, or larger but filled by secondary calcite or quartz that renders them effectively impermeable. Most master joints exposed at the surface will be penetrable, however, plus many cross joints. Initially, the opposed joint surfaces display hackling and cusps around the point of breakage and fan-like (plumose) striations away from it but, in karst, this fine structure is quickly destroyed by dissolution. Under lithostatic pressure, joints are more readily closed to impenetrable dimensions than are bedding planes, reducing their significance in deeper karst situations.

It is important to understand that some new joints are created as a karst terrain is eroding, because of pressure release as rock is removed. This is not true of bedding planes or of most faulting. Large tensional joints form parallel to steep faces, especially at the rims of plateaus or along reef fronts. Small compressional joints form at the bases of cliffs and cave passage walls.

As with bedding, the best development of karst features is found where joint spacing is wide to very wide. Many caves are rectangular mazes guided rigidly by joint patterns, including Optimist's Cave (Ukraine), a gypsum cave that is second in aggregate known length in the world (Figure 7.26).

2.8.3 Faults and fracture traces

Faults are fractures with some displacement of rock up, down and/or laterally. Where this is less than about 1 cm they may be considered to grade into shear fractures or joints. At the greatest, vertical displacement extends several kilometres while lateral displacement may amount to 10^2–10^3 km.

Normal faults are produced by tension and therefore a wide opening is possible (as much as a few centimetres), although it may fill with breccia, secondary calcite, etc. **Reverse faults** and **lateral** or **transcurrent faults** are compressional features, and so may be impenetrably tight. However, formation of breccia or slickenside grooves can open them, while displacement may bring together recessed facets to create wide spaces. **Thrust** or **décollement faults** are low-angle reversed faults that are often particularly important because they are areally extensive (nearly vertical faults are not) and so emulate very penetrable bedding planes in their capacity to host interconnected solution conduits. In regions of moderate tectonic activity it is common to see a thrust fault originate in one slightly slipped bedding plane, pass through a few beds as a curvilinear surface, and terminate in a second disturbed bedding plane higher up.

Large faults are rarely represented by a single fracture surface. Usually lesser faults feather off of them at acute angles as a consequence of the wrenching of the rock. Shear fractures are often oriented parallel or close to the feathering breaks.

Fracture traces (or **linears**, **lineaments**) are narrow linear trends detectable on high-altitude and satellite images. Most karst landscapes display them (e.g. Figure 2.14). On the ground they are zones of closely spaced high-angle faults of minor displacement, plus their feathering fractures, etc.

The hydrogeological and speleogenetic role of faults and fracture traces varies with their type, size, and the diagenetic record since they were formed. At one extreme they may direct the predominant flow in a groundwater basin, like a trunk river channel on the surface, or have sinkholes aligned along them. Large normal and reversed faults often have low permeability, however, due to clayey crush fillings (**mylonite**) or precipitated calcite in them, and thus serve as barriers.

Instead, there is higher permeability in the zone of feathering fractures. Research in many regions has found that many major karst depressions are guided by these and may be centred where two traces intersect. Summarizing much hydrogeological data from tunnel works in carbonate mountains, Dublyansky and Kiknadze (1983) showed that most water intake occurred in the feathering zones, particularly in down-thrown blocks. In dolomite terrains drilling on fracture traces often yields the greatest volumes of groundwater. In many caves the fault planes themselves control local passage segments only, although the overall trend of cave development may follow the fault-trace zone. In intermediate cases cave systems extend between fault zones and beyond them, sometimes utilizing them locally and at other places being barred or deflected by them. Active thrust faults deforming passage cross-sections are reported in limestone caves in Belgium and Italy (Figure 2.15; Vandycke and Quinif 1998), and diapir normal faults in salt caves near the Dead Sea (Frumkin 1996). Šušteršic (2000) shows that cave systems and surface karst features around Postojna, Slovenia, are being torn apart (but not destroyed) by modern movement on transcurrent faults.

In the northwest Yucatan Peninsula, Mexico, a semi-circular band of **cenotes** (drowned, shaft-type sinkholes) is believed to coincide with a ring fault of the Chicxulub multiring impact crater; it might be termed 'astro-karst!' (Perry *et al.* 1996).

2.8.4 Fracture apertures

Except where they are truly gaping, as in fresh quarry walls, the physical nature of all these separations is complex and difficult to measure. In quantitative modelling it has been standard practice to treat the hypothetical fracture as a fissure with strictly parallel walls spaced a fixed distance apart. In reality this will never apply except, possibly, along joints in the shallowest karst. For hydrogeological purposes it is better to think of the water passing from more widely open areas with highly irregular shapes to others down the hydraulic gradient via constrictions (throats). Effective throat apertures appear to be log-normally distributed (e.g. Chernyshev 1983). By measuring the particulate organic matter reaching stalactites and stalagmites via the finest cracks in cave roofs worldwide, van Beynen *et al.* (2001) show that throat apertures in their tiny feedwater channels can never be less than ~0.1 μm. The minima for effective dissolutional karst genesis probably exceeds 10 μm. Most modelling assumes values of 100 μm or greater. Hanna and Rajaram

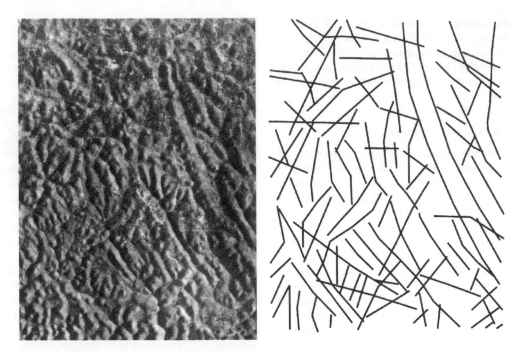

Figure 2.14 Recognition of lineaments in karst terrain. (Left) Aerial photograph of labyrinth-cone karst in Genung Sewu, Indonesia. (Right) Lineaments detected in the photograph. Reproduced from Haryono, E. and Day, M. (2004) Landform differentiation within the Gunung Kidul kegelkarst, Java, Indonesia. Journal of Cave and Karst Studies, 66(2), 62–9.

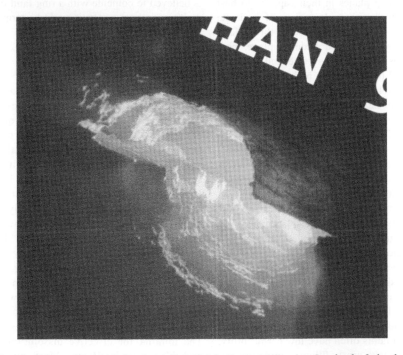

Figure 2.15 A phreatic (below water table) dissolutional conduit in Grotta del Frassino, Lombardy, Italy, that has been offset by modern slip (normal faulting) in the bedding plane in which it was initiated. (Photograph courtesy of Dr Yves Quinif.)

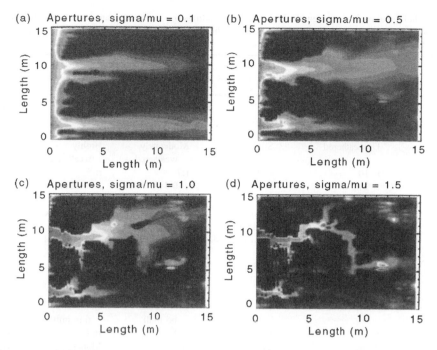

Figure 2.16 Computer modelling of the dissolutional enlargement of a fissure with an aperture that varies in a random manner. (Top left) Standard deviation of the aperture about a mean width of 400 mm is only 10%. Deviation increases to 50% (top right), 100% (bottom left) and 150% (bottom right). The upper pair of patterns are fair representations of typical dissolution in joints and faults, the lower of dissolution in bedding planes; see section 7.2 for more details. Reproduced from Hanna, R.B. and Rajaram, H. (1998) Influence of aperture variability on dissolutional growth of fissures in karst formations. *Water Resources Research*, 34(1), 2843–53.

(1998) have obtained particularly significant computer results by randomly discretizing the apertures occurring in model fissures, a standard deviation of ~50% about the mean aperture yielded realistic approximations of dissolutionally enlarged joint cavities, while deviations of 100% or greater better represented conduit development in bedding and low-angle thrust planes (Figure 2.16).

2.8.5 Geomorphological rock mass strength classification

Taking into consideration factors of rock composition, texture and compressive strength, and the frequency of penetrable bedding planes, joints and faults, the Geomorphological Rock Mass Strength Classification and Rating as proposed by Selby (1980) is a useful guide to the strength of karstifiable rocks at the scale of the principal karst landforms, i.e. cave systems, dolines, karren fields, residual hills and towers. The classification is intended to rate the strength of hillslope masses. However, it is developed from mining engineering applications (Beniawski, 1976; Brady and Brown,

1985); thus it is also pertinent to the stability of cave roofs, the likelihood of catastrophic sinkhole collapse, etc.

The Selby classification is presented in a slightly modified form in Table 2.8. It has not been applied widely in karst terrains as yet. Most that display well-developed landforms seem likely to range across the middle categories, Strong–Weak. Moon (1985), applying the classification to hillslopes of quartzite or shale in South Africa, considered that it should contain a further parameter for the roughness of fissures of all kinds (bedding planes, joints, faults). This is a particularly complex parameter in karst because the roughness (interlocking) on the fissure planes becomes progressively reduced by dissolution. In many regions, limestone and dolomite dip slopes are preferred sites for large landslides because of this factor (section 12.4).

2.9 FOLD TOPOGRAPHY

The world's karst terrains encompass every type of larger geological structure. These include plains and plateaus with horizontal or subhorizontal strata, steep and gentle

Table 2.8 Geomorphological rock mass strength classification and ratings (r = rating of parameter) (Adapted from Selby, M.J, 'A rock mass strength classification for geomorphic purposes: with tests from Antarctica and New Zealand', Zeitschrift für Geomorphologie, 24, 31–51.© 1980 © E. Schweizerbart'sche Verlagsbuchhandlung, Science Publishers

Parameter	1 Very strong	2 Strong	3 Moderate	4 Weak	5 Very weak
Intact rock strength (N-type Schmidt Hammer 'R')	100–60 r: 20	60–50 r: 18	50–40 r: 14	40–35 r: 10	35–10 r: 5
Weathering	Unweathered r: 10	Slightly weathered r: 9	Moderately weathered r:7	Highly weathered r: 5	Completely weathered r: 3
Spacing of fissures	>3 m r: 30	3–1 m r: 28	1–0.3 m r: 21	300–50 mm r: 15	<50 mm r: 8
Fissure orientations	Very favourable. Steep dips into slope, cross joints interlock r: 20	Favourable. Moderate dips into slope r: 18	Fair. Horizontal dips or nearly vertical r: 14	Unfavourable. Moderate dips out of slope r: 9	Very unfavourable. Steep dips out of slope r: 2
Width of fissures	<0.1 mm r: 7	0.1–1 mm r: 6	1–5 mm r: 5	5–20 mm r: 4	>20 mm r: 2
Continuity of fissures	None continuous r: 7	Few continuous r: 6	Continuous, no infill r: 5	Continuous, thin infill r: 4	Continuous, thick infill r: 1
Outflow of groundwater	None r: 6	Trace r: 5	Slight $<25\,L\,min^{-1}$ $10\,m^{-2}$ r: 4	Moderate $25–125\,L\,min^{-1}$ $10\,m^{-2}$ r: 3	Great $>125\,L\,min^{-1}$ $10\,m^{-2}$ r: 1
Total rating	100–91	90–71	70–51	50–26	<26

homoclines, simple and multiple fold topographies, nappe structures, diapiric domes, etc. Two examples are shown in Figure 2.17. These may create differing styles of karst at the surface and of hydrogeological organization underground.

Rock folding requires plastic deformation and so tends to occur at great depth where lithostatic pressures are high. In carbonate rocks it is thus generally associated with diagenetically mature strata of Cretaceous or greater age. It is rare to encounter significant folding in Tertiary and Quaternary limestones, although there are some spectacular examples in Papua-New Guinea. Gypsum, anhydrite and salt bodies of all ages deform, fold and flow readily, even at quite shallow depths.

The amplitude of folds ranges from a few centimetres to several kilometres. High folds may extend for hundreds of kilometres along the strike. Tensional forces tend to create strike-aligned master joint sets at the crests of anticlines and in the troughs of synclines. Differential slipping of the bedding planes is often more important on the flanks. Where cave systems extend across one or several anticlines or synclines it is common to find trunk passages centred in the troughs.

However, the converse does occur and there are instances worldwide of principal passages extending around the noses of plunging anticlines.

Where karstic beds or formations are mingled with siliciclastic strata, tilting and folding often create conditions of artesian confinement. Meteoric recharge water entering the karst rock becomes trapped beneath impermeable seals and may circulate slowly to remote springs. The longest karst groundwater flow systems that are known are created in this manner. There are well-confirmed examples in the London and Paris basins, in the Eucla and Wasa basins of Australia and in the Basin and Range country of the western USA. Recharge from the Rocky Mountains and their eastern foothills is believed to flow for more than 1000 km through carbonates beneath confining salts, sands, clays and shales in the Canadian Prairies, with underground residence times perhaps more than 30 000 yr.

2.10 PALAEOKARST UNCONFORMITIES

The lower illustration in Figure 2.17 shows a profound geological unconformity that has modern karst features

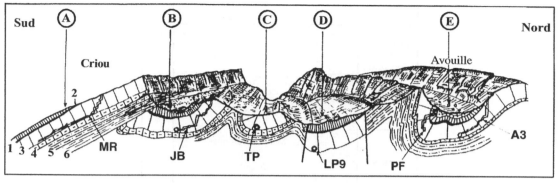

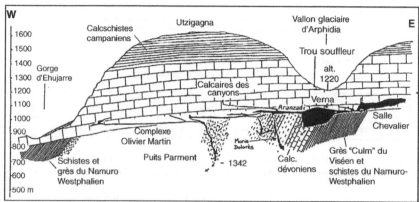

Figure 2.17 Two examples from France of karst development in complex geological structures. (Upper) The folded Alps at Samoens. Numbers identify individual formations, from lower Cretaceous (6) to mid-Cretaceous (1) in age; number 3, Urgonian, is the principal karst limestone. Letters identify individual homoclinal and synclinal structures. Réseau Mirolda (MR) and Réseau Jean Bernard (JB) are two of the world's deepest explored caves (see Figure 7.1). TP, LP9, PF and A3 are other prominent springs or shaft systems in the other structures. (Lower) Réseau Pierre St Martin is another deep system that is entered in the Alpine zone on the Franco-Spanish border in the Pyrenees. It passes through a thick, gently dipping limestone formation of late Cretaceous age, to ramify along and penetrate below a grossly unconformable contact with underlying Devonian and Carboniferous strata. Salle Verna is one of the largest known cave chambers. Reproduced with permission from Maire, R, La Haute Montagne Calcaire. Karstologia, Memoire 3, 731 p, 1990.

developing both above and below it. Many sequences of carbonate rocks are found to contain, or to be terminated by, unconformities that are karst solutional surfaces or cavities that are now inert, i.e. palaeokarst (section 1.1). Some of these are of hypogene origin, where waters ascending from lower formations were able to dissolve cavities in intra- or interstratal positions: usually, these collapsed to create breccias that are now partly or firmly cemented by calcite (see Spörli *et al.* (1992) for examples in Antarctica). Much more frequent and extensive, however, are the unconformities created when surface karst landscapes of the types described in Chapters 9 and 10 became buried by later consolidated rocks, and the karst hydrological circulation systems beneath them were also inundated and became more or less inert. The later (burial) rocks can be any type of sediment (clastic, carbonate, evaporitic, organic, e.g. coal) or even extrusive lavas and volcanic tuffs.

The buried karst landforms range in magnitude from the cyclic type displaying only shallow dissolution features developed during brief, sequence stratigraphical, episodes of marine emergence and exposure to subaerial processes (eogenetic karst – Figure 2.7), to rugged karst landscapes with local relief of tens to hundreds of metres that have taken millions of years to form (Figure 2.18). The eogenetic, disconformable or weakly unconformable, surfaces are more commonly preserved in the geological record, as would be expected; many individual geological formations contain stacked sequences of them or, where a little more rugged, they themselves may be adopted as the boundaries between individual formations (Wright *et al.* 1991). The oldest known examples are in Archaean rocks

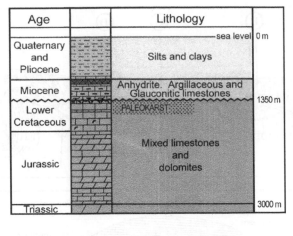

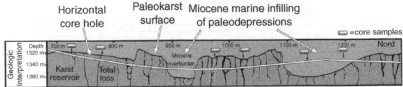

Figure 2.18   Palaeokarst interpretation of the Rospo Mare oil field, a one billion barrel field off the Adriatic coast of Italy. The erosional interval between the Lower Cretaceous and Miocene strata is approximately 70 Myr. The surface karst of subsoil karren, dolines and solutional corridors has a local relief of 10–40m and is underlain by a further 100m of vadose and shallow phreatic caves. Adapted with permission from Soudet H. J., Sorriaux P., and Rolando J. P. Relationship between Fractures and Karstification - the Oil- Bearing Paleokarst of Rospo Mare (Italy). Bulletin des Centres de Recherches Exploration-Production Elf Aquitaine 18(1), 257–297, 1994.

(> 2500 a) of the Canadian Shield, including solution caves filled with sands that are now hard sandstones: the Archaean atmosphere is believed to have had a high CO_2 content, which will have favoured vigorous karst activity. Eogenetic exposures are common in Lower Proterozoic strata (> 1000 a) in Australia, Canada and the USA, China and Russia (Bosak *et al.* 1989), and some more rugged palaeokarst surfaces are reported in Upper Proterozoic exposures in those nations. Reports of palaeokarst unconformities of all types become frequent everywhere in Palaeozoic, Mesozoic and Eocene–Oligocene limestones and dolomites because these are more widespread, better preserved and exposed than the older formations (Bosak 1989). There are now several thousand well-described examples in the literature. Surfaces younger than Oligocene in age will rarely be so firmly buried that they have become chemically inert, i.e. they are more likely to function as cryptokarst beneath loose or weakly consolidated rocks rather than as inert palaeokarst.

North America provides two good examples of palaeokarst surfaces that are subcontinental in their extent. Following a long period of sedimentary rock accumula-tion (the 'Sauk' – Sloss 1963), marine regression at ~480 a exposed large tracts of platformal and intertidal carbonates for some millions of years. Mature karst terrains with local relief of 10–60 m and subsurface dissolution to −200 m or more were able to develop before early Appalachian deformation (the Taconic Orogeny) returned them beneath the sea and net deposition recommenced. This 'Post-Sauk' palaeokarst surface can be traced in fragments from eastern Canada and the southeastern USA to west of the Mississippi River. The highly productive Knox aquifer in the southeast, important oil fields in Ohio and Oklahoma, and lead–zinc deposits in Tennessee are associated with it. Another marine regression of long duration around 325 Ma led to development of the 'Post-Kaskaskia' palaeokarst. In North America this is taken to divide strata of the Carboniferous Period into Mississippian (earlier) and Pennsylvanian (later). The karst itself is best exposed in western USA, especially in the great Jewel and Wind hydrothermal maze caves of South Dakota; some of their upper passages at least are exhumations or rejuvenations of Post-Kaskaskia meteoric water caves (Palmer and Palmer 1989).

3

Dissolution: Chemical and Kinetic Behaviour of the Karst Rocks

3.1 INTRODUCTION

When a rock dissolves in water its different minerals (or some of them) disintegrate into individual ions or molecules which diffuse into the solution. Of necessity, study of dissolution focuses upon the specific minerals rather than the aggregate rock. Hence, this chapter is concerned with mineral solubility. Because the principal karst rocks are nearly pure, monominerallic aggregates there is often little difference between discussion of, for example, calcite solubility and limestone solubility.

Dissolution is said to be **congruent** when all components of a mineral dissolve together and completely. Table 3.1 gives the dissolution reactions for a range of congruent minerals. Dissolution is **incongruent** where only a part of the components dissolve. The aluminosilicate minerals are the great example of the incongruent class, releasing Na^+, K^+, Ca^{2+}, Mg^{2+}, SiO_4^{4-}, etc., ions in reaction with water but retaining most of their atoms in reordered solids such as kaolinite, vermiculite or smectite. Incongruent dissolution of dolomite (with accompanying precipitation of calcite) may occur in some exceptional conditions mentioned later.

The sample of congruent minerals in Table 3.1 contains all the common elements of crustal rocks except Fe, and furnishes a majority of the dissolved inorganic species. It is seen that the range of solubility is enormous. Gibbsite ($Al(OH)_3$) is an example that is insoluble to all intents and purposes; even in the most favourable circumstances encountered on the surface of this planet physical processes will disaggregate it and remove it as colloids or larger grains before there is significant solution damage. Rock salt (halite) is so soluble that it is rapidly destroyed in outcrop except in the driest places; it

is principally important for its role in interstratal karstification. Sylvite and mirabilite are rarely encountered and never in great bulk. They occur as minor secondary cave minerals (section 8.4). Gypsum is quite common in outcrop. Karst features develop upon it rapidly because of its comparatively high solubility.

Limestone and dolomite are common in outcrop. Their maximum solubility varies with environmental conditions but never approaches that of gypsum. Quartzite and siliceous sandstones are equally common in outcrop. In terms of solubility and of common solute abundance in water there is a large overlap with the range exhibited by the carbonate rocks. Yet siliceous rocks are not normally considered to be karstic. This raises the question of what is the lower limit of solubility for the development of karst? The answer is that a transitional situation exists in reality, although it is rarely considered by karst specialists. Karst landforms as defined in Chapter 1 develop at all scales on siliceous sandstones and at the small scale on many rocks of yet lower specific (mineral) solubility. However, at the global scale these landforms must be considered rare and of minor importance. There are less than $30 \, mg \, L^{-1}$ SiO_2 of silica in most meteoric waters sampled on sandstones and more than $40 \, mg \, L^{-1}$ of dissolved calcite in most samples from carbonate terrains. Karst becomes abundant above the latter concentration.

Table 3.2 presents the specifications underlying some common chemical and environmental classifications of waters. On the continents solutions notably stronger than seawater are rare; most examples are in evaporating lakes or in long-resident basinal waters intercepted in deep drilling. In a majority of gypsum karsts it is unusual for concentrations to exceed $2000 \, mg \, L^{-1}$ $CaSO_4$. In carbonate terrains concentrations higher than $450 \, mg \, L^{-1}$ total

Karst Hydrogeology and Geomorphology, Derek Ford and Paul Williams
© 2007 John Wiley & Sons, Ltd

Table 3.1   Dissociation reactions and solubilities of some representative minerals that dissolve congruently in water, at 25 °C and 1 bar (105 Pa) pressure (Modified with permission from Freeze, R.A. and Cherry, J.A. Groundwater © 1979 Prentice Hall)

Mineral	Dissolution reaction	Solubility ($mg\,L^{-1}$)	Common range of abundance in waters ($mg\,L^{-1}$)
Gibbsite	$Al(OH)_3 + H_2O \rightarrow 2Al^{3+} + 6OH^-$	0.001	Trace
Quartz	$SiO_2 + H_2O \rightarrow H_4SiO_4$	12	1–12
Amorphous silica	$SiO_2 + H_2O \rightarrow H_4SiO_4$	120	1–65
Calcite	$CaCO_3 + H_2O + CO_2 \leftrightarrow Ca^{2+} + 2HCO_3^-$	60*, 400[†]	10–350
Dolomite	$CaMg(CO_3)_2 + 2H_2O + 2CO_2 \leftrightarrow Ca^{2+} + Mg^{2+} + 4HCO_3^-$	50*, 300[†]	10–300
Gypsum	$CaSO_4 \cdot 2H_2O \rightarrow Ca^{2+} + SO_4^{2-} + 2H_2O$	2400	0–1500
Sylvite	$KCl + H_2O \rightarrow K^+ + Cl^- + H^+ + OH^-$	264 000	0–10 000
Mirabilite	$NaSO_4 \cdot 10H_2O + H_2O \rightarrow Na^+ + SO_4^{2-} + H^+ + OH^-$	280 000	0–10 000
Halite	$NaCl + H_2O \rightarrow Na^+ + Cl^- + H^+ + OH^-$	360 000	0–10 000

*$P_{CO_2} = 10^{-3}$ bar.
[†]$P_{CO_2} = 10^{-1}$ bar.

dissolved solids (TDS) will almost invariably prove to be enriched by sulphates or chlorides, or nitrates if the water is polluted. The great majority of karst waters contain only a few tens or hundreds of $mg\,L^{-1}$ of dissolved solids. As a result, their chemistry is that of very dilute solutions.

3.1.1 Definition of concentration units

In the engineering and geomorphological literature mass concentrations of dissolved solids measured in water samples are commonly reported in milligrams per litre. By weight, these are equivalent to parts of solute per million parts of solution and to grams per cubic metre.

In the SI system concentrations of ions in aqueous solutions are expressed in molar units (**molarity**). A solution of 1 **mole** of calcium (atomic weight = 40.08) contains 40.08 g of calcium per litre of solution. This is a

large quantity in natural aqueous systems; thus, to avoid many zeros after the decimal point concentrations are usually reported in millimoles per litre ($mmol\,L^{-1}$ or mM) or even micromoles per litre ($\mu mol\,L^{-1}$). To convert from $mg\,L^{-1}$:

$$mol \cdot L^{-1} = \frac{mg \cdot L^{-1}}{1000 \cdot A}$$

where A denotes atomic or molecular weight. Conveniently, because the molecular weight of 1 mmol of $CaCO_3$ is equal to 100.1 mg, $1\,mmol\,L^{-1}$ of Ca^{2+} is the equivalent of $100.1\,mg\,L^{-1}$ $CaCO_3$ dissolved.

The reactions are evaluated in **equivalent** units to enable the checking of ion positive and negative charge balances: units are $eq\,L^{-1}$, $meq\,L^{-1}$ and $\mu eq\,L^{-1}$.

$$meq \cdot L^{-1} = \frac{mg \cdot L^{-1}}{1000 \cdot E}$$

where E is equivalent weight obtained by dividing atomic or molecular weight by ionic charge; in the case of ions with one charge (e.g. Na^+, K^+, Cl^-) $E = A$. For ions with two charges (e.g. Ca^{2+}, SO_4^{2-}) equivalents = moles multiplied by 2. Table 3.3 gives the factors to convert $mmol\,L^{-1}$ and $meq\,L^{-1}$ (the preferred units of chemists) to $mg\,L^{-1}$, the units of bulk weathering studies.

Readers should be careful when reading scales of calcium carbonate concentration in the literature, which may be reported as $mg\,L^{-1}$ Ca^{2+} or as $mg\,L^{-1}$ $CaCO_3$. Total hardness (dissolved bicarbonates, carbonates, sulphates, chlorides, etc. of calcium and magnesium) may be reported as $mmol\,L^{-1}$, $meq\,L^{-1}$, $mg\,L^{-1}$ $CaCO_3$ or in national units; one 'English degree' of hardness = $14.3\,mg\,L^{-1}$ $CaCO_3$,

Table 3.2   Common chemical classifications of waters

	Total dissolved solids* ($mg\,L^{-1}$)
Soft water	< 60
Hard water	> 120
Brackish water	1000–10 000
Saline water	10 000–100 000
(Seawater)	(35 000)
Brines	> 100 000
Potable water for humans	< 1000 or < 2000[†]
Potable water for livestock	< 5000

*Total dissolved solids in potable waters are presumed to be only the bicarbonates, sulphates, chlorides and their associated species as discussed in this chapter.
[†]Varies between jurisdictions: these are the two most frequent limits.

Table 3.3 Molecular and equivalent weights of the common ions and molecules encountered in karst waters

Formula	Molecular weight	Equivalent weight
Ca^{2+}	40.08	20.04
Cl^-	35.46	35.46
CO_3^{2-}	60.01	30.00
F^-	19.00	19.00
Fe^{2+}	55.85	27.93
Fe^{3+}	55.85	18.62
HCO_3^-	61.02	61.02
K^+	39.10	39.10
Mg^{2+}	24.32	12.16
Na^+	22.99	22.99
NH_4^+	18.04	18.04
NO_3^-	62.01	62.01
PO_4^{3-}	94.98	31.66
SO_4^{2-}	96.06	48.03

one 'French degree' $= 10.0\,mg\,L^{-1}$ $CaCO_3$ and one 'German degree' $= 17.8\,mg\,L^{-1}$ $CaCO_3$ (Krawczyk 1996). Example: a solution contains $250\,mg\,L^{-1}$ $CaCO_3$. The solution contains

$$100\,mg\,L^{-1}\,Ca^{2+} = 2.5\,mmol\,L^{-1}\;or\;10^{-2.60}\,mol\,L^{-1}$$
$$150\,mg\,L^{-1}\,CO_3^{2-} = 2.5\,mmol\,L^{-1}\;or\;10^{-2.60}\,mol\,L^{-1}$$

Ionic strength, I, is defined as the sum of the molar concentrations of ions in a water multiplied by the square of their charges:

$$I = \frac{1}{2}\sum m_i \cdot z_i^2$$

where m_i is the molar concentration of ion i and z_i is its charge. In most karst waters there will be only seven constituents in significant concentration:

$$I = \frac{1}{2} \cdot \Big([Na^+] + [K^+] + 4[Ca^{2+}] + 4[Mg^{2+}]$$
$$+ [HCO_3^-] + [Cl^-] + 4[SO_4^2] + [NO_3^-]\Big)$$

In limestone and dolomite areas, Na^+, Cl^-, SO_4^{2-} and NO_3^- are often present in very low concentrations and thus can be neglected as well, but it is important to establish this by measurement: it should not be assumed.

As a rule of thumb, the ionic strength of brackish waters ≈ 0.1; fresh waters > 0.01.

3.1.2 Use of negative logarithms

Because karst waters are usually very dilute solutions numbers involved in calculations may be inconveniently small. To reduce the likelihood of arithmetical errors arising from misplaced decimal points, it is conventional to do much of the calculation with negative logarithms. The symbol for a negative logarithm is lower case p. In the example given above, $100\,mg\,L^{-1}$ $Ca^{2+} = 0.0025$ $mol\,L^{-1}$. Log_{10} of this concentration is $10^{-2.6}$; thus $pCa^{2+} = 2.6$.

3.1.3 Source books

In this book we use the thermodynamic equilibrium approach and saturation indices to investigate problems of mineral dissolution. It is a comprehensive approach, giving information on the evolution of water from an initial state towards its state when sampled at a karst spring, etc. Accuracy of results is dependent on precision of pH measurements which, in the past, has been difficult to achieve in the field. Hence many karst workers have preferred bulk quantitative approaches neglecting equilibria. These yield less insight but are also less prone to error.

The 'classic' text is Garrels and Christ (1965) *Solutions, Minerals and Equilibria*. Most later works use the format and conventions adopted by these authors. A most comprehensive recent treatment is by Stumm and Morgan (1996) *Aquatic Chemistry*, 3rd edn. Other useful recent works include texts by Dreybrodt (1988), Appelo and Postma (1994), Langmuir (1996), Berner and Berner (1996), Bland and Rolls (1998) and Domenico and Schwartz (1998).

3.2 AQUEOUS SOLUTIONS AND CHEMICAL EQUILIBRIA

3.2.1 Speciation, dissociation, hydration and the Law of Mass Action

In addition to dissolution or precipitation at the solid–liquid interface, in karst studies it is necessary to consider **speciation** in the water, processes by which solute ions and molecules combine or break apart or change phase between gas and liquid. Recently there has been increasing attention paid to **redox** (reduction–oxidation) processes as well, in which species respectively gain or lose electrons: these are considered later.

Water itself is an effective conductor because it is polar. Cation–anion electric bonds are weakened in solids in contact with it. Their normal thermal agitation suffices to detach some ions, which diffuse away into the solution. For example, for halite

$$NaCl \overset{H_2O}{\Leftrightarrow} Na^+ + Cl^- \tag{3.1}$$

where $\overset{H_2O}{\Leftrightarrow}$ means 'in the presence of water'. This most simple process of solution is termed **dissociation**. It adequately describes the dissolution of rock salt and gypsum.

A more complex solution process involves the partial or complete neutralizing of either the cation or the anion charge. This unbalances the solution, requiring further dissociation (or equivalent back reaction by precipitation) to restore it.

Pure water itself dissociates to a small extent:

$$H_2O \overset{H_2O}{\Leftrightarrow} H^+ + OH^- \tag{3.2}$$

Comparatively little dissociation occurs with $CaCO_3$. But if a free proton, H^+, approaches the solid we may write the sequence of reactions:

$$CaCO_3 \Leftrightarrow Ca^{2+} + CO_3^{2-} \tag{3.3}$$

$$Ca^{2+} + CO_3^{2-} + H^+ \Leftrightarrow Ca^{2+} + HCO_3^- \tag{3.4}$$

The CO_3^{2-} ion has become **hydrated**. Unless an OH^- is within a few nanometers of the site of these reactions close to the solid–liquid interface, the solution is unbalanced there and a further CO_3^{2-} ion can dissociate to restore it. This is the process of acid dissolution. It dominates dissolution of the carbonate minerals.

Systems of such reactions proceed in a forward direction with rates proportional to the concentration of reactants. Accumulation of reaction products increases the rate of back reaction until forward and backward rates are equal. The system then has reached a dynamic equilibrium for the given set of physical conditions imposed upon it, i.e. temperature and pressure. Variation of any of these conditions induces systematic change in the concentrations of each reacting species until equilibrium is again attained. This is the 'Law of Mass Action' for reversible systems, which may be written:

$$aA + bB \Leftrightarrow cC + dD \tag{3.5}$$

where $aA = a$ moles (or mmol) of reactant species A, and $cC = c$ moles (or mmol) of product C, etc. At dynamic equilibrium this relation becomes

$$K_{eq} = \frac{[C]^c[D]^d}{[A]^a[B]^b} \tag{3.6}$$

where K_{eq} is a coefficient termed the **thermodynamic equilibrium constant** (or **solubility product** or **stability constant** or **dissociation constant** by different authors). As an example:

$$H_2O = \frac{[H^+][OH^-]}{[H_2O]} = K_w \tag{3.7}$$

By convention the value assigned to H_2O is unity, shrinking equation (3.7) to

$$K_w = [H^+] \cdot [OH^-] \tag{3.8}$$

K_w is the thermodynamic equilibrium or dissociation constant of water; it has a value of 10^{-14} at 25°C and 1 bar (10^5 Pa), and $10^{-14.9}$ at 0°C.

These are spontaneous reactions in which the energy retained in the product phases is lower than in the reactant phases. The difference is measured as the **Gibbs free energy of reaction**, $\Delta G°$, which is related to the thermodynamic equilibrium constant, K, by:

$$\Delta G° = -2.303 \cdot RT \cdot \log_{10} K \tag{3.9}$$

where R is a gas constant with the value $8.314 \, J \, mol^{-1} \, K^{-1}$ and T is the system temperature in degrees Kelvin.

Products with the lowest Gibbs free energies are the most stable. At earth surface temperatures and pressures more energetic metastable phases are frequently found, however, and may be long lasting. The concept is sketched in Figure 3.1. Reactants must surmount energy activation barriers to create products. In some instances, there is a lower barrier, E_a', for the metastable product than for the stable. In karst studies the most significant metastable example is aragonite: the activation barrier for inversion to calcite has been placed between 184 and $444 \, kJ \, mol^{-1}$ in various experiments. Aragonite may survive for millions of years at low temperatures and pressures before inverting; at 360°C it is all converted in about 16 h (see White 1997a). At the other extreme, a molecule of carbonic acid ($H_2CO_3^0$, formed from $H_2O + CO_2$, as discussed below) exists for only a fraction of a second before dissociating into H^+ and HCO_3^-. Full discussion of the Gibbs free energy concept, including tables of constants for compounds of importance in karst processes, can be found in Stumm and Morgan (1996, p. 27 *et seq.*), and other texts cited above.

3.2.2 Activity

Water with ions diffusing through it is a weak electrolyte. Some ions of opposite charge will combine to form ion

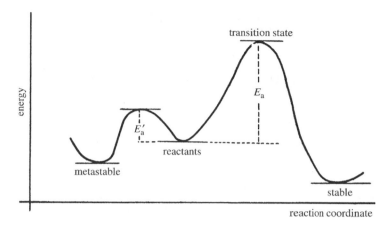

Figure 3.1 Conceptual model of the energy changes occurring in chemical reactions.

pairs with reduced charge or zero charge. Hence, the number of potentially reactive ions of a given species (e.g. Ca^{2+}) that are present in an aqueous solution is always somewhat less than the molar sum of ions of that species in the solution. The proportion of potentially reactive (or free) ions is termed the **activity** of the species. As ionic strength (I) increases from ~0 to 0.1, activity decreases. This is a reflection of the increasing opportunity for ion combination to occur. For many species it increases again between $I = 0.1$ and 1.0.

Determination of activity is fundamental to the correct computation of all equilibria for solute species. Activity itself is symbolized by 'a' in most texts; standard brackets ( ) signify that it is the activity of the contained species that is being considered; square brackets [ ] signify molarity and { } molality of a species.

The **activity coefficient** γ_i is defined as

$$\gamma_i = \frac{(a_i)}{[c_i]} \tag{3.10}$$

where $c = $ concentration

$$\gamma_i \to 1 \quad \text{as } c_i \to 0$$

Approximate values of the activity coefficients for the dissolved species of interest in most karst work are given in Figure 3.2. Normally they are not read off graphically but are computed with variants of the Debye–Hückel equations contained within larger programs computing the equilibrium state of reported dissolved species. A standard extended form of the equation is satisfactory for most purposes in normal karst waters. Consult Stumm and Morgan (1996) for better precision.

Where $I < 0.1$, the extended standard form is

$$\log \gamma_i = \frac{-Az_i^2 \cdot (I)^{\frac{1}{2}}}{1 + B \cdot r_i \cdot (I)^{\frac{1}{2}}} \tag{3.11}$$

with z the valence of the ion, A and B constants depending upon temperature and pressure ($A = 0.4883 + 8.074 \times 10^{-4} \times T$; $B = 0.3241 + 1.6 \times 10^{-4} \times T$, with T in °C) and R_i is the hydrated radius of the ith ion; the relevant radii are given in Table 3.4.

Where I is > 0.1 but < 0.5, Davies' variant (1962) is recommended by Stumm and Morgan (1996):

$$\log \gamma_i = -Az_i^2 \left(\frac{(I)^{\frac{1}{2}}}{1 + (I)^{\frac{1}{2}}} - 0.21 \right) \tag{3.12}$$

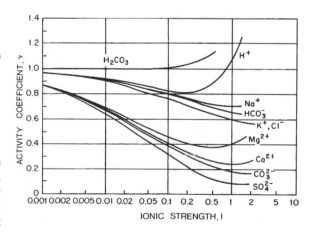

Figure 3.2 Activity coefficients and ionic strength of the common ionic constituents in karst waters. Reproduced with permission from Freeze, R.A. and J.A. Cherry, Groundwater, 604 pp © 1979 Prentice Hall, Inc.

Table 3.4 Ionic radii of ions most frequent in karst waters

r_i $(\times 10^{-8}\,\mathrm{m})$	Ion
2.5	NH_4^+
3.0	K^+, Cl^-, NO_3^-
3.5	OH^-, HS^-
4.0	SO_4^{2-}, PO_4^{3-}
4.0–4.5	Na^+, HCO_3^-
4.5	CO_3^{2-}
5.0	Sr^{2+}, Ba^{2+}, S^{2-}
6.0	Ca^{2+}, Fe^{2+}, Mn^{2+}
8.0	Mg^{2+}
9.0	H^+, Al^{3+}, Fe^{3+}

For more concentrated solutions (e.g. brines) the Pitzer equation is often applied instead (Nordstrom 2004).

3.2.3 Saturation indices

A solution containing a given mineral will be in one of three conditions.

1. Forward reaction predominates. There is net dissolution of the mineral; the solution is said to be **undersaturated** or **aggressive** with respect to the mineral.
2. There is dynamic equilibrium; the solution is **saturated** with the mineral.
3. Back reaction predominates and there may be net precipitation of the mineral. The solution is **supersaturated**.

Few sampled waters are precisely at equilibrium. Saturation indices measure the extent of their deviation, i.e. their aggressivity or supersaturated condition. The measured product of ion activity in a sample is compared with the K_{eq} value. The standard form of the **saturation index** (SI) is that of Langmuir (1971):

$$SI = \log IAP/K_{eq} \tag{3.13}$$

where K_{IAP} is the **ion activity product**. Here, a solution is at equilibrium at 0.0, aggressive waters have negative values, etc. as illustrated in Figure 3.3. An alternative index that is occasionally used is the **saturation ratio** (SR); this is simply the non-logarithmic version, where SR is 1.0 at equilibrium. Readers are urged to use the SI index in order that results can be more immediately compared.

Figure 3.3 illustrates a further point that is most important. For the mineral species of interest in karst research, the approach to dynamic equilibrium (SI = 0.0)

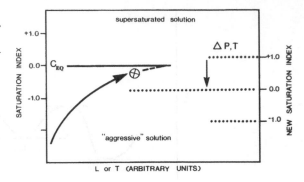

Figure 3.3 Evolutionary path of a water sample, X, approaching equilibrium with respect to a given mineral. C_{eq} = concentration at saturation = an SI value of 0.0. Change of boundary conditions will shift the SI scale, as illustrated at right.

becomes asymptotic where boundary conditions of temperature, etc. remain constant. Ideal equilibrium is difficult to impossible to attain; a comparatively long timespan or long flow path through the rock will be required to effect net addition of the last few ions. In karst, a supersaturated water almost invariably indicates that a significant change of boundary conditions has occurred. In Figure 3.3 the change is indicated by $\Delta P, T$.

3.3 THE DISSOLUTION OF ANHYDRITE, GYPSUM AND SALT

Anhydrite ($CaSO_4$) may dissociate directly in the presence of water. In field conditions it normally hydrates first, becoming gypsum which dissolves by dissociation:

$$CaSO_4 \cdot 2H_2O \Leftrightarrow Ca^{2+} + SO_4^{2-} + 2H_2O \tag{3.14}$$

The solid, gypsum, and water are both assigned values of unity. The equilibrium constant thus is

$$K_g = \frac{[Ca^{2+}][SO_4^{2-}]}{[CaSO_4]_S} \tag{3.15}$$

where K_g signifies that the constant is that of gypsum. Its value is $10^{-4.61}$ at 25°C, declining to $10^{-4.65}$ at 0°C. Similarly, the expression for halite is

$$K_h = \frac{[Na^+][Cl^-]}{[NaCl]}$$
$$K_h = [Na^+] \cdot [Cl^-] \tag{3.16}$$

$K_h = 10^{-1.52}$ at 25°C, declining to $10^{-1.58}$ at 0°C.

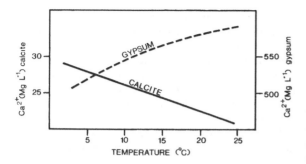

Figure 3.4 The solubility of calcite and gypsum in water and the standard atmosphere between 2° and 25°C.

The saturation index for gypsum is

$$SI_g = \log\frac{(Ca^{2+})(SO_4^{2-})}{K_g} \qquad (3.17)$$

or

$$SI_g = \log(Ca^{2+}) + \log(SO_4^{2-}) + pK_g \qquad (3.18)$$

It takes the same form for salt; however, because salt solubility is so great, even brines are strongly aggressive in most instances. The salt index is of little practical utility in karst studies.

The environmental controls of the rate and amount of gypsum or halite in solution may be summarized very simply. As with most other dissociation reactions there are positive correlations with pressure and temperature. However, the minor effects of pressure changes on solubility can be ignored even where groundwaters circulate to depths of several kilometres. The same is broadly true of temperature. The surficial environmental range from 0° to 30°C for meteoric waters has an effect that is inconsequential in the case of salt. For gypsum (Figure 3.4), it produces an increase of approximately 20% in the solubility product. It has not been shown that this is an important boost in terms of any effect on karst morphology or rate of development: see Cigna (1986) for discussion of thermal mixing effects. Solution rates and concentrations here are controlled primarily by the amount of water contacting these minerals and, to a lesser extent, by the mode of supply – as laminar or turbulent flows or as impacting raindrops or spray.

3.4 THE DISSOLUTION OF SILICA

Pure silica in its compact crystalline mode as (alpha) quartz (density = 2.65) is very resistant but does react with water, forming silicic acid by the process of **hydrolysis**:

$$SiO_2(quartz) + 2H_2O \rightarrow H_4SiO_4(aq);$$
$$K_q = 1.1 \cdot 10^{-4} \qquad (3.19)$$

Written in the alternative form, $Si(OH_4)_{aq}$, silicic acid is seen to comprise four OH groups attached to a central silicon atom. It is not very reactive. The solubility of quartz in standard temperatures is only $6{-}10\,mg\,L^{-1}$. In hotter waters it increases rapidly (e.g. $\sim60\,mg\,L^{-1}$ at 100°C).

If very alkaline conditions should chance to occur (pH > 9.0), silicic acid can pass through up to four dissociations (four 'orders') in sequence until all H^+ ions are detached, e.g.

$$H_4SiO_4 \rightarrow H_3SiO_4^- + H^+; \quad K_1 = 10^{-9.7} \qquad (3.20)$$
$$H_3SiO_4^- \rightarrow H_2SiO_4^{2-} + H^+; \quad K_2 = 10^{-13.3} \qquad (3.21)$$

etc., substantially increasing the amount of quartz that may be dissolved. However, such alkaline conditions can rarely be achieved in quartzite and siliceous sandstone terrains, where soils and waters are normally slightly acid.

If pH is lowered, for example by the addition of CO_2 to waters as discussed in the next section, the solubility of silica may be exceeded and **amorphous** (or **hydrous**) silica, $SiO_2.nH_2O$, precipitated. It is common in many dry-zone soils. The water is slowly expelled to form **opal**, a crystalline form of SiO_2 with a more open structure than quartz (opal density = 2.1). Amorphous silica is quite soluble ($K = 10^{-2.7}$), with $>100\,mg\,L^{-1}$ often measured in waters of normal karst pH, rising to $>300\,mg\,L^{-1}$ in very alkaline or hot waters.

3.5 BICARBONATE EQUILIBRIA AND THE DISSOLUTION OF CARBONATE ROCKS IN NORMAL METEORIC WATERS

3.5.1 Bicarbonate waters

The solubility of calcite and dolomite by dissociation in pure, deionized water is very low, only $14\,mg\,L^{-1}$ (as $CaCO_3$) at 25°C. This is scarcely more than the solubility of quartz.

Investigations in many countries have long established that most of the enhanced solubility of carbonate minerals that occurs is due to the hydration of atmospheric CO_2 (Roques 1962, 1964). This produces carbonic acid which, in turn, dissociates to provide H^+. Other acids may furnish additional H^+, and other complexing effects may further increase solubility. These are summarized in later sections. Here we consider the effect of CO_2 from the atmosphere and soil air. It is predominant in most carbonate karsts, which are created by meteoric waters that can circulate only to comparatively shallow depths underground and, as a consequence, have not been geothermally heated to a significant extent.

Carbon dioxide is the most soluble of the standard atmospheric gases, e.g. 64 times more soluble than N_2. Its solubility is proportional to its partial pressure (Henry's Law) and inversely proportional to temperature. **Partial pressure** is that part of the total pressure exerted by a mixture of gases that is attributable to the gas of interest.

For the dissolution of CO_2 in water, Henry's Law may be written:

$$CO_2 \text{ (aq)} = C_{ab} \cdot P_{CO_2} \cdot 1.963 \tag{3.22}$$

where CO_2 is in $g\,L^{-1}$, P_{CO_2} is the partial pressure of CO_2, 1.963 is the weight of 1 L of CO_2, in grams, at one atmosphere and 20°C, and C_{ab} is the temperature-dependent absorption coefficient, given in Table 3.5.

In the standard atmosphere, P_{CO_2} at sea level has a modern global mean value of ~0.038% or 0.00038 atmosphere (380 ppm) or a little higher. This is equivalent to ~0.6 mg CO_2 per litre of air. With increasing altitude P_{CO_2} declines slightly; e.g. Zhang (1997) measured only 120–150 ppm at 5000 m on the Tibetan Plateau, increasing to 200–300 at 4000 m. It may also be reduced a little

in forests (by assimilation) and over fresh snow. However, effects of these reductions appear to be very minor.

Of the greatest importance is the increase of P_{CO_2} that may occur in soil atmospheres as a consequence of organic compounds released in the rooting zones. In principle, CO_2 can entirely replace O_2 there, i.e. increasing P_{CO_2} to 21%. Soil CO_2 is discussed in detail in a later section.

The role of CO_2 is illustrated in Figure 3.5. Its dissolution and consequent dissociation proceed:

$$CO_2 \text{ (g)} \Leftrightarrow CO_2 \text{ (aq)} \tag{3.23}$$

$$CO_2 \text{ (aq)} + H_2O \Leftrightarrow H_2CO_3^0 \text{ (carbonic acid)} \tag{3.24}$$

Carbonic acid dissociates rapidly; nevertheless, it is conventional to combine these reactions to obtain one equilibrium expression:

$$K_{co_2} = \frac{H_2CO_3^0}{P_{CO_2}} \tag{3.25}$$

The carbonic acid dissociates:

$$H_2CO_3^0 \Leftrightarrow H^+ + HCO_3^- \tag{3.26}$$

This first-order dissociation constant is

$$K_1 = \frac{[HCO_3^-][H^+]}{[H_2CO_3^0]} \tag{3.27}$$

Table 3.5   The solubility of CO_2 (Reproduced with permission from Bögli, A, Karst Hydrology and Physical Speleology © 1980 Springer Verlag)

(a) Absorption coefficients of CO_2

Temperature of solution (°C)	0	10	20	30
Absorption coefficient C_{ab}	1.713	1.194	0.878	0.665

(b) Equilibrium solubility of CO_2 ($mg\,L^{-1}$)

P_{CO_2} (atm)	Temperature (°C)			
	0	10	20	30
0.0003	1.01	0.7	0.52	0.39
0.001	3.36	2.34	1.72	1.31
0.003	10.10	7.01	5.21	3.88
0.01	33.6	23.5	17.2	13.1
0.05	168	117	86	65.3
0.10	336	235	172	131
0.20	673	469	342	261

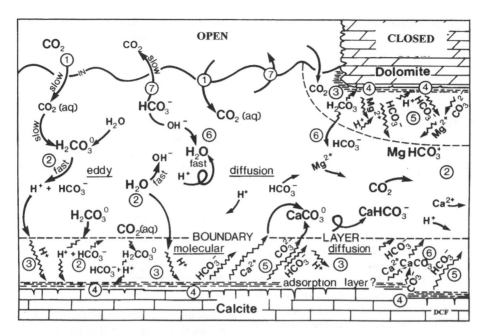

Figure 3.5 Cartoon depicting the dissolved species, reactions and comparative reaction rates that are involved in the dissolution of calcite and dolomite under open and closed conditions. Eddy diffusion dominates in the bulk liquid. There are very thin molecular diffusion boundary layers and adsorption sublayers in contact with the rock – see section 3.10 for details of their significance.

Bicarbonate ion may then dissociate to carbonate:

$$HCO_3^- \Leftrightarrow H^+ + CO_3^{2-} \tag{3.28}$$

The second-order dissociation constant is

$$K_2 = \frac{[H^+][CO_3^{2-}]}{[HCO_3^-]} \tag{3.29}$$

All fresh waters exposed to the ordinary atmosphere will contain these different species of dissolved inorganic carbon (DIC), whether or not there are carbonate rocks in the watershed.

3.5.2 Dissolution of calcite and dolomite

The pH of water in limestone and dolomite terrains usually falls between 6.5 and 8.9. In this range, HCO_3^- is the predominant species, CO_3^{2-} (aq) being negligible below pH 8.3. It is more appropriate, therefore, to approach equation (3.27) as a back reaction. This requires introducing the minerals:

$$CaCO_3(s) \Rightarrow Ca^{2+} + CO_3^{2-} \tag{3.30}$$

where $CaCO_3$ (s) is solid calcite.

$$K_{calcite\ or\ aragonite} = [Ca^{2+}][CO_3^{2-}] \tag{3.31}$$

Then,

$$CaCO_3(s) + H^+ \Leftrightarrow Ca^{2+} + HCO_3^- \tag{3.32}$$

From laboratory experiments Plummer *et al.* (1978) consider calcite dissolution to be the sum of three forward rate processes, which are reaction (3.31) plus direct reaction with carbonic acid

$$CaCO_3(s) + H_2CO_3(aq) \leftrightarrow Ca^{2+} + 2HCO_3 \tag{3.33}$$

and dissolution in water (a double dissociation)

$$CaCO_3(s) + H_2O \leftrightarrow Ca^{2+} + HCO_3^- + OH^- \tag{3.34}$$

This full sequence of reactions from equation (3.23) onwards is often summarized

$$CaCO_3 + CO_2 + H_2O \Leftrightarrow Ca^{2+} + 2HCO_3^- \tag{3.35}$$

For dolomite, the dissociation reaction is

$$CaMg(CO_3)_2 \Leftrightarrow Ca^{2+} + Mg^{2+} + 2CO_3^{2-} \qquad (3.36)$$

the dissociation constant is

$$K_d = \frac{[Ca^{2+}][Mg^{2+}][CO_3^{2-}]^2}{[CaMg(CO_3)_2]} \qquad (3.37)$$

and the summary is

$$CaMg(CO_3)_2 + 2CO_2 + 2H_2O$$
$$\Leftrightarrow Ca^{2+} + Mg^{2+} + 4HCO_3^- \qquad (3.38)$$

Equilibrium constants for these reactions at a range of temperatures are given in Table 3.6.

If the dissolution of calcite or aragonite alone is considered (and if the ion pairs, $CaHCO_3^+$ and $CaCO_3^0$, that appear in Figure 3.5 are ignored for the present), the water contains six dissolved species: Ca^{2+}, H^+, $H_2CO_3^0$, CO_3^{2-}, HCO_3^- and OH^-. These are defined by equations (3.2), (3.22), (3.27), (3.29) and (3.31). The molar concentrations of these species at equilibrium in given conditions are calculated by adding a further equation and then solving the set simultaneously. The additional equation is for charge balance:

$$m_i z_i (\text{cations}) = m_i z_i (\text{anions}) \qquad (3.39)$$

For calcium carbonate solutions this equation is

$$2m_{Ca^{2+}} + m_{H^+} = 2m_{CO_3^{2-}} + m_{HCO_3^-} + m_{OH^-} \qquad (3.40)$$

A more comprehensive charge balance equation (one that will serve for almost any natural water encountered in karst terrains) is

$$2m_{Ca^{2+}} + 2m_{Mg^{2+}} + m_{Na^+} + m_{K^+} + m_{H^+}$$
$$= 2m_{CO_3^{2-}} + 2m_{SO_4^{2-}} + m_{HCO_3^-} + m_{Cl^-} + m_{NO_3^-} + m_{OH^-}$$

Solutions are obtained by iterative approximations. Figure 3.6 shows the approximate solutions for **open** and **closed** systems (as these are defined below), with just the six species noted.

3.5.3 The saturation indices

The saturation index for calcite is

$$SI_C = \log \frac{(Ca^{2+})(CO_3^{2-})}{K_C} \qquad (3.42)$$

and has the same form for aragonite. CO_3^{2-} is present in very small amounts in pH below ~ 8.4 and so is not normally measured; instead, the standard form is

$$SI_C = \log \frac{(Ca^{2+})(HCO_3^-)K_2}{(H^+)(K_C)} \qquad (3.43)$$

or

$$SI_C = \log(Ca^{2+}) + \log(HCO_3^-) + pH - pK_2 + pK_C \qquad (3.44)$$

Table 3.6 Equilibrium constants for the carbonate dissolution system, gypsum and halite, at 1 atm pressure. (From Garrels and Christ 1965; Langmuir 1971; Plummer and Busenberg 1982)

Temperature (°C)	pK_{CO_2}	pK_1	pK_2	$pK_{calcite}$	$pK_{aragonite}$	$pK_{dolomite}$	pK_{gypsum}	pK_{halite}
0	1.12	6.58	10.63	8.38	8.22	16.56	4.65	1.52
5	1.19	6.52	10.56	8.39	8.24	16.63	—	—
10	1.27	6.46	10.49	8.41	8.26[*]	16.71	—	—
15	1.34	6.42	10.43	8.42	8.28	16.49	—	—
20	1.41	6.38	10.38	8.45	8.31	16.89	—	—
25	1.47	6.35	10.33	8.49	8.34	17.0	4.61	1.58
30	1.52	6.33	10.29	8.52[*]	8.37[*]	17.9		
50	1.72	6.29	10.17	8.66	8.54[*]	—		
70	1.85[8]	6.32[*]	10.15	8.85[*]	8.73[*]	—		
90	1.92[*]	6.38[*]	10.14	9.36	9.02	—		
100	1.97	6.42	10.14	—	—	—		

$pK_1 = 356.3094 + 0.06091964T - 21834.37/T + 126.8339 \log T + 1684915/T^2$.
$pK_2 = 107.8871 + 0.03252849T - 5151.79/T - 38.92561 \log T + 563713.9/T^2$.
$pK_C = 171.9065 + 0.077993T - 2839.319/T - 71.595 \log T$.
$\log K CaHCO_3^+ = 1.11$ at 25°C, $\log K MgHCO_3^+ = -0.95$ at 25°C, $\log K CaCO_3^0 = 3.22$ at 25°C.
*Interpolated

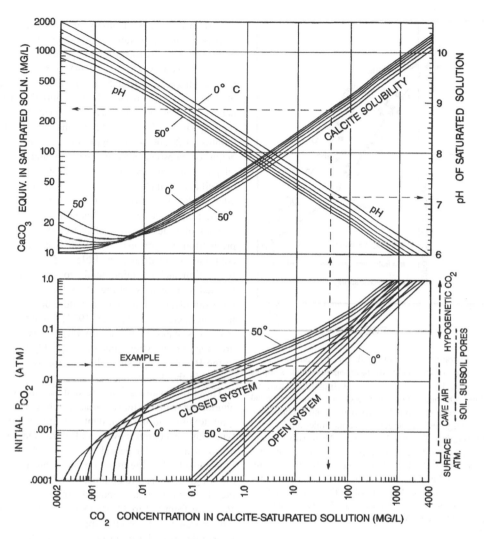

Figure 3.6 Saturation values of dissolved calcite in water at various values of CO_2 partial pressure, and for **open** and **closed** conditions as defined below. Reproduced with permission from Palmer, A.N, Geomorphic interpretation of karst features. In LaFleur, R. G. (ed.), Groundwater as a geomorphic agent. Boston, Massachusetts, 173–209 © 1984 Allen and Unwin.

For dolomite the saturation index is

$$SI_d = \log(Ca^{2+}) + \log(Mg^{2+}) + 2\log(HCO_3^-)$$
$$+ 2pH - 2pK_2 + pK_d \qquad (3.45)$$

A very significant parameter (for it reveals much of the provenance or history of a karst water) is the P_{CO_2} with which an analysed sample would be in equilibrium. This is given by

$$P_{CO_2} = \frac{(HCO_3^-)(H^+)}{K_1 \cdot K_{CO_2}} \qquad (3.46)$$

or

$$\log P_{CO_2} = \log(HCO_3^-) - pH + pK_{CO_2} + pK_1 \qquad (3.47)$$

Figure 3.7 presents a good example of the SI_c and P_{CO_2} indices put to work to explain limestone surface and groundwater geochemical behaviour changing over the course of the growing season in a cool temperate region.

3.5.4 Soil carbon dioxide

Globally, soil CO_2 is undoubtedly the most important source for enhanced solubility in carbonate rocks. The

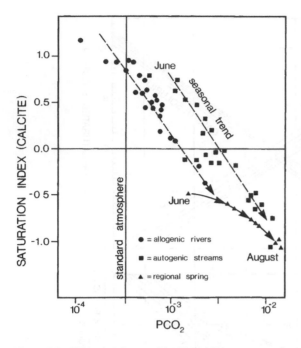

Figure 3.7 Illustrating the use of the SI_c and P_{CO_2} parameters by analysis of some simple calcium bicarbonate water samples collected between June and August in a limestone basin on Anticosti Island, Quebec. Allogenic rivers drain lime-rich glacial soils and flow for several kilometres before reaching the limestone and sinking underground. Autogenic streams drain similar local soils but sink within a few hundred metres. These two types of water plus direct subsoil percolation combine to discharge at a major regional spring. As the summer season advances a progressive increase in the effect of soil CO_2 comes to dominate the evolution of these waters. Reproduced with permission from Roberge, J., Geomorphologie du karst de la Haute-Saumons, Ile d'Anticosti, Quebec. MSc thesis, McMaster University, p. 217 1979.

porosity of normal soils is greater than 40% but part is occupied by bound water. The maximum volume available for air storage and circulation ranges from ~17% in clay soils to ~31% in very sandy soils (Drake 1984). Gases produced in soils will tend to accumulate there because rapid diffusion or drainage are retarded by the high friction of the tortuous intergranular pathways.

Into the pore spaces green plants respire approximately 40% of the CO_2 that they extract from the atmosphere above ground. Their roots are CO_2 pumps. Yet greater quantities of CO_2 are respired by soil fauna, microfauna and microflora, principally bacteria, actinomycetes and fungi. Numbers of bacteria can exceed 1–3 million per gram of soil. Their greatest densities occur in and above the rooting zone. From it the gas diffuses out to the surface (and O_2 diffuses in) and also to the soil base and below.

Carbon dioxide productivity of roots and soil bacteria increases with temperature and water availability. Optimum temperatures for different species range from 20°C to as high as 65°C (Miotke 1974). In recent work in the great subtropical karstlands of southern China, for example, Yuan (2001) measured mean monthly CO_2 abundances at −50 cm (the heart of local rooting zones) ranging from ~500 ppm during the dry, cooler winter months to 26 000–40 000 ppm ($P_{CO_2} = 2.6$–4.0%) at the height of monsoon rains in June, July and August. Cold-adapted bacteria can continue to respire down to −5°C; Cowell and Ford (1978) recorded a sharp drop in soil P_{CO_2} after the first hard frost of winter in central Canada.

The field capacity of a soil is notionally defined as the amount of water retained after free drainage. Soil CO_2 production is greatest at 50–80% capacity but may continue as soil dries to as little as 5%. Carbon dioxide production and retention tends to be greatest in fine-grained soils with swelling clays that can retain water. For example, in a loess with a mature brown earth profile in western Germany, Miotke (1974) found that maximum P_{CO_2} was stable at 0.3% atm during dry spells in the growing season. After rains it rose quickly to as much 4% in the B horizon, presumably because water sealed pore outlets in the A horizon.

These observations indicate that patterns of soil CO_2 abundance will be most variable. They vary with soil type, texture and horizon, depth, drainage and exposure, types of vegetation cover, soil flora and fauna, with seasonal and shorter period warming and wetting. There is a considerable literature on the subject because it is also of great interest to botanists, zoologists, agronomists, etc. In tropical areas reported common ranges of soil P_{CO_2} are from 0.2 to 11.0% (Smith and Atkinson 1976) with extremes > 17.5% that may be suspected of error. In temperate areas the usual range is from 0.1 to 6.0% but 10% is occasionally reported. In arctic tundra 0.2–1.0% is reported over the brief thaw season. In alpine tundras common ranges during the melt season are similar, 0.04–1.0%, rising to brief peaks greater than 3% just below the treeline.

As illustrated above, the soil CO_2 effect can be studied also by back-calculating the equilibrium P_{CO_2} of groundwaters that have drained through soil. By this means Drake and Wigley (1975) investigated limestone and dolomite spring waters in Canada and the USA that were just saturated, i.e. very close to equilibrium with respect to calcite. The sites ranged from the subarctic to Texas, a 20°C range of mean annual temperature. They obtained the linear relation:

$$\log P_{CO_2} = -2 + 0.04 \cdot T \tag{3.48}$$

where T is the mean annual temperature (°C) at a given site. This signifies a general soil enrichment effect of ×5 compared with the standard atmosphere, onto which the positive temperature effect is added.

Use of mean annual temperature will give inaccurate estimates where there are strong seasonal variations in temperature–recharge–growing-season relationships, however (Bakalowicz, 1976). Brook *et al.* (1983) investigated this question, using soil CO_2 field data from arctic to tropical locales in place of calculated P_{CO_2} of carbonate waters. Equation (3.47) was broadly confirmed. The predictive power of mean annual precipitation was found to be very poor but there was a fair correlation with the logarithm of aggregate actual evapotranspiration, which is related to the length of the growing season and thus to the period during which soil CO_2 is being produced:

$$\log P_{CO_2} = -3.47 + 2.09(1 - e^{-0.00172\,AET}) \qquad (3.49)$$

where AET is the calculated mean annual actual evapotranspiration of a site. It was concluded that ∼50% of variation of soil P_{CO_2} can be explained by temperature, 20% by precipitation, and the balance by seasonal water availability and by growth and inhibition factors.

Although it might appear that CO_2 could accumulate in a soil until all O_2 was replaced (i.e. up to $P_{CO_2} = 21\%$ by volume), this does not occur in practice. Root respiration begins to be impaired at P_{CO_2} of ∼6%. At slightly higher concentrations aerobic bacteria begin to be killed and anaerobic species are unable to replace them fast enough to maintain the CO_2 production; hence the process tends to be self-inhibiting.

3.5.5 Carbon dioxide in exposed epikarst, the vadose percolation zone and caves

Over large areas in many carbonate karst regions there is no regular soil cover. Instead, the rock is bare or mantled only with discontinuous veneers of lichen, algae and/or vegetal litter. It may lack soil because it is diagenetically young and porous, as in many tropical islands of Pleistocene age. In regions of intensive agriculture such as the Mediterranean lands, older limestones are exposed due to deforestation and overgrazing, and in glaciated areas due to scour by ice: all are indented to varying degree by dissolutional pits and troughs (karren).

In the uppermost dissolutional zone, the **epikarst**, rotting vegetation and residual soil form a rich base for roots of trees and shrubs and for bacterial activity. Although difficult to measure accurately, P_{CO_2} values up to several per cent appear to be common in the trapped detritus, which can amount to substantial volumes.

Whether the surface is bare or covered by soil, the finer particulate organic matter produced by decay filters down into fissures in the vadose percolation zone below. A part of it even penetrates into the finest cracks, those that drain to stalactites and stalagmites in underlying caves. In a worldwide analysis of stalagmites from all climates, hot to cold, wet to dry, van Beynen *et al.* (2001) found that all samples contained significant amounts of organic fragments too large to pass a 0.07 µm filter. In the vadose (aerobic) percolation zone this material produces CO_2 by bacterially mediated oxidation. In boreholes beneath bare karst in the Bahamas, for example, Whitaker and Smart (1994) calculated CO_2 concentrations of $1.6 \pm 0.8\%$ which they attributed to *in situ* oxidation.

Soil CO_2 itself may drain down into any underlying fissures or caves because it is a heavy gas. At gravitational trap sites such as pits with no outlets for air through the bottom, CO_2 from drainage and decay can accumulate to lethal levels. There have been many local studies.

In comprehensive accounts, Renault (1982) and Ek and Gewelt (1985) reviewed some thousands of measurements of CO_2 in caves worldwide. Cave air is generally enriched 2–20 times with respect to the standard atmosphere but concentrations as high as 6% have been recorded. They tend to be highest where air circulation is weakest (e.g. in the narrowest accessible fissures) or at sites closest to overlying soils. Northern Hemisphere maxima occur in July–September, when CO_2 concentrations may be two to four times as great as in winter. Bakalowicz *et al.* (1985) report precise flux rates for summer CO_2 in the large Grotte de Bedeilhac, Pyrenees. Rates approximated 4–16 kg m^{-2} day^{-1} CO_2 for the surface area of the cave. The gas was derived from gravitational drainage and the degassing of saturated infiltration waters.

Cave streams, especially floodwaters, will pick up this excess CO_2 and so boost their solvent capacity. In river caves of warm, humid regions where much vegetal debris is carried underground it may make a significant contribution to corrosional enlargement.

3.5.6 Open and closed system conditions

In the present context a system is **open** when all three phases, solid, liquid and gas, are able to react together. An ideal open system exists when such conditions are maintained until thermodynamic equilibrium is achieved. A system is **closed** when only two phases can interact at a given site.

The application to carbonate dissolution is suggested in Figures 3.5, 3.6 and 3.8. In the ideal open case, as H$^+$ and

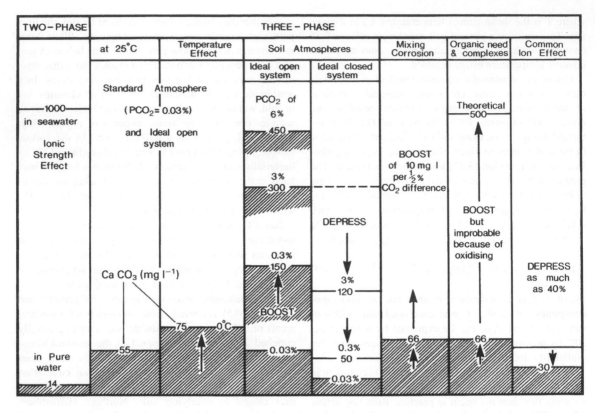

Figure 3.8 Summary of the principal complicating effects in carbonate dissolution chemistry. The numbers indicate the approximate equilibrium concentrations of $CaCO_3$ ($mg L^{-1}$) the given effects achieve. (Prepared by Ford and J.M.James.)

$H_2CO_3^0$ are converted to bicarbonate by reaction with $CaCO_3$, more CO_2 is able to dissolve from the air and so replenish the $CO_{2(aq)}$ and $H_2CO_3^0$ until equilibrium is reached. An open air pool on limestone is such a system. The value of P_{CO_2} is fixed at $\sim 0.036\%$ and is interactive until equilibrium, which at 25°C occurs when $55 mg L^{-1}$ $CaCO_3$ have been dissolved (Fig. 3.9).

In an ideal closed system, air and water alone react until the solution is saturated with dissolved CO_2 plus its derived $H_2CO_3^0$ and HCO_3^-. The water then flows away from the atmosphere and first contacts carbonate minerals where no air is present, e.g. in a water-filled capillary or fracture. The H^+ and $H_2CO_3^0$ that are withdrawn by association with $CaCO_3$ cannot be replenished. For 25°C and $P_{CO_2} = 0.036\%$, the solution becomes saturated at only $25 mg L^{-1}$ $CaCO_3$, or 40% of that achieved in the open system.

In reality, we may expect that many karst waters will evolve under hybrid conditions, i.e. where the system is part open, part closed; Figure 3.5 is an example. Drake (1984) suggests that ideal open-system conditions may

not apply in soils with low air volume or in low temperatures, because the rate of dissolution of CO_2 into recharge waters exceeds the $CO_{2(g)}$ supply rate. Very high rates of water recharge will have the same effect. However, where they have been adequately studied, it is found that karst waters at equilibrium tend toward one or another of the ideal extremes, as illustrated in Figure 3.9. The surprisingly high mean concentrations measured in some temperate regions with average annual temperatures 5–12°C can be explained by the fact that the soils are young and on glacial or thermoclastic detritus (e.g. till) that still contains many carbonate fragments dispersed in the rooting zone, yielding ideal open conditions. In the tropics, where there are deep soils on limestone they are usually dissolutional residua. These are clay-rich (favouring high P_{CO_2}) but without any surviving carbonate fragments, so that dissolution commences only at their base, where the system becomes partly or fully closed.

A fundamental point to appreciate is that the initiation and early expansion of dissolutional

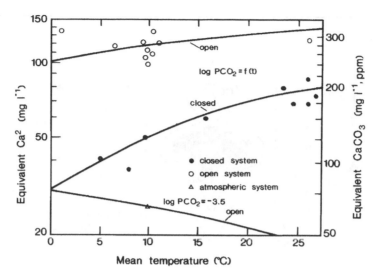

Figure 3.9 Global model for the dissolution of calcite under ideal open and ideal closed system conditions in and below soils, and for global mean open air ($P_{CO_2} \sim 350$ ppm). Open and closed functions are calculated from equation (3.47) and assume some bacterial inhibition at high P_{CO_2}. Data points are means of 20 different karst groundwater sample sets where waters are at or very close to equilibrium with respect to calcite. Geographical location of the sets ranges from arctic and alpine to tropical environments. Reproduced with permission from Drake, J.J. 1984. Theory and model for global carbonate solution by groundwater. In R. G. LaFleur (ed.). Groundwater as a geomorphic agent, pp. 210–226 © 1984 Allen and Unwin.

conduits in the rock will almost always occur under closed conditions.

3.6 THE S–O–H SYSTEM AND THE DISSOLUTION OF CARBONATE ROCKS

During the past three decades there has been increasing interest in the role of aqueous geochemical processes producing H_2S that can react to dissolve limestone and dolomite. In most localities, such processes play only a trivial quantitative role, if any, in creating surface karst landforms such as karren, solution dolines, or poljes. However, they may make important contributions to the early opening of dissolutional conduits in general. More certainly, they appear to be the predominant processes in the initiation and enlargement of one rare but remarkable class of cave, the ramiform hypogene system (Palmer 1991; see Chapter 7.8), which is represented by such large and magnificently decorated caverns as Carlsbad and Lechuguilla in the USA, Frasassi in Italy, and Novaya Afonskaya in Georgia. In addition, they are associated with the precipitation of the majority of sulphide ore bodies known in carbonate rocks, the so-called 'Mississippi Valley Type' (**MVT**) deposits. This leads many specialists to consider the S–O–H system to be of independent significance in karst studies, rather than a subsidiary contributor of complexity to the

CO_2–H_2O system such as those described in later sections.

An initial, very simple example of the processes, one that can be recognized in many ordinary meteoric water caves worldwide, is the oxidation of pyrite (FeS_2) and other iron compounds that are very common constituents of shale interbeds in limestone and dolomite. For pyrite in shale:

$$2FeS_2 + \frac{15}{2}O_2 + 4H_2O \Rightarrow Fe_2O_3 + 8H^+ + 4SO_4^{2-}$$

$$(3.50)$$

The H^+ acidifies the solution, permitting increased dissolution of calcite or dolomite (equation 3.32). Reaction (3.50) may or may not be moderated by bacteria.

As O_2 has a low solubility, these effects will be of limited importance in phreatic (anaerobic) environments, although an exception is cited below. Physical effects are readily seen in vadose (i.e. air-filled) caves. Water seeping from a limestone bedding plane that contains a little pyrite-rich shale may be conspicuously aggressive, etching a pattern of solutional microrills at its emergence. Floods that push O_2-rich water into such planes may recharge the acidity of main cave streams as they recede.

Hydrogen sulphide is liberated in many volcanic regions and may enter juvenile or meteoric groundwater as a gas.

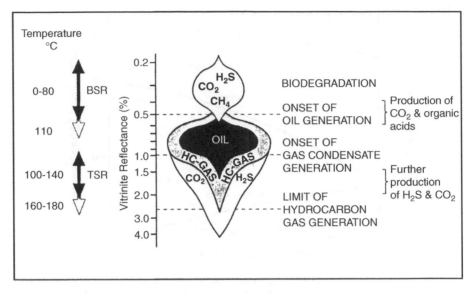

Figure 3.10   Processes of oil and gas formation, with acid release, at depth in crustal rocks. These are often associated with sulphate reduction. HC = hydrocarbons. Vitrinite reflectance is an industry standard measure of the maturation of an oil deposit. Reproduced with permission from Tucker, M.E. and Wright, V.P, Carbonate Sedimentology © 1990 Blackwell Publishing.

More important, it is also created by reduction processes operating in fluids being expelled from sedimentary basins as the sediments compact or are deformed. In particular, it is created as the abundant organic matter contained in mudstones and marls is degraded, first into **kerogen** by reduction at temperatures $< 50°C$, and then into **oil** and **natural gas** at temperatures up to 120°C. Mudstones and marls are commonly deposited with carbonate rocks (which serve as carriers for the fluids) and with sulphates and salt. Maturation is illustrated in Figure 3.10, where 'BSR' is bacterial sulphate reduction and 'TSR' is thermochemical sulphate reduction. Geothermal gradients are generally between 20 and $40°C\,km^{-1}$ in sedimentary basins, thus the processes take place principally at depths between 1.5 and 5 km.

Thermochemical reduction is noted later. Here we follow the bacterial reduction genetic model proposed by Hill (1995) for an example of comparatively shallow settings, the Delaware Basin oilfield of New Mexico–Texas and the adjoining Guadalupe Mountains that contain Carlsbad, Lechuguilla and many other remarkable caverns. The model is illustrated in Figure 3.11, where 3.11 top summarizes the complex sequence of reactions that can occur. Not all of these are likely to be present in any particular H_2S karst; it is difficult to balance all proposed components in certain of them and in the current state of knowledge should be considered as indicative modelling of the possibilities rather than proven cases.

The processes begin (Stage 1 of Hill (1995) and Figure 3.11 centre) with the flow of hydrocarbon compounds (represented here by CH_4 (**methane**, the principal component of most commercial natural gases) through reservoir rocks such as dolomites or sandstones, to the base of impermeable anhydrite cover strata:

$$Ca^{2+} + 2SO_4^{2-} + 2CH_4 + 2H^+$$
$$\Rightarrow 2H_2S(aq) + CaCO_3 + 3H_2O + CO_2 \qquad (3.51)$$

The reactions take place at substantial depth but with temperatures $< 80°C$. In the Delaware Basin they occurred during Oligocene–Miocene times. They are driven by anaerobic, sulphur-reducing bacterial species, *Desulfovibrio*. Note that calcite may be precipitated, i.e. it replaces the anhydrite here. This formed resistant masses within the Delaware Basin sulphate rocks that late Tertiary–Quaternary dissolution has exposed by removing the surrounding gypsum, creating low hills termed **castiles** (Figure 3.11 lower).

The concentrations of H_2S produced are commonly around $1\,mg\,L^{-1}$ but may be much higher. Dissolution of H_2S in water produces a weak acid that may dissociate in two steps:

$$H_2S \Leftrightarrow HS^- + H^+ \qquad (3.52)$$
$$HS^- \Leftrightarrow H^+ + S^{2-} \qquad (3.53)$$

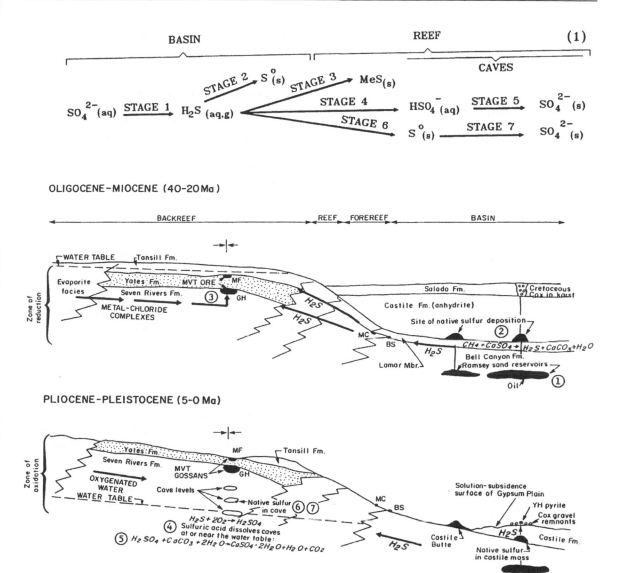

Figure 3.11 (Top) The model of Hill (1995) for the sequence of S–O–H reactions that occurred in the Delaware sedimentary basin of New Mexico–Texas. (Centre) The model during Stages 1, 2 and 3, when the basin and adjoining Guadalupe reefs were deeply buried by younger rocks. (Lower) The model during Stages 4, 5, 6 and 7, when deformation, uplift and erosion exposed the reefs as a mountain range (Guadalupe Mountains) and removed much of the evaporite rocks in the basin. Reproduced from Hill, C.A. (1995) Sulfur redox reactions: hydrocarbons, native sulfur, Mississippi Valley-type deposits sulfuric acid karst in the Delaware Basin, New Mexico and Texas. Environmental Geology, 25, 16–23.

At 25°C the dissociation constants for these reactions are 10^{-7} and 10^{-13} respectively. Calcite or dolomite may be dissolved (equation 3.32).

Oxygenated meteoric water flowing down gradient in artesian trapping structures may reach the dissolved H_2S, creating a redox interface within the phreatic zone:

$$H_2S + \frac{1}{2}O_2 \Rightarrow S^0 + H_2O \quad \text{(at pH} <6-7) \quad (3.54)$$

or

$$2HS^- + O_2 + 2H^+ \Rightarrow 2S^0 + 2H_2O \quad (\text{at } 7 < pH < 9)$$
$$(3.55)$$

S^0 is pure (or **native**) sulphur. Economic deposits of sulphur were formed during Stage 2 deep in the Delaware Basin (Figure 3.11 centre), and much smaller but striking mounds of it were formed in the aerated caves that had been created by the time of Stage 6 (Figure 3.11 lower). The processes are mediated by oxidizing bacteria (*Thiobacillus* spp), which use the elemental sulphur as a food and energy store. Native sulphur of phreatic origin appears to be rare in karst, however, due to the low solubility of O_2.

In Stage 3 the H_2S-saturated water has flowed out of the basin into adjoining carbonate reef and back-reef rocks, where it mixes with metal-rich chloride waters typical of those that are expelled from any adjoining lagoonal evaporite rocks as the latter are buried by later deposits. In the reducing zone below any redox interface there, limestone dissolution can occur with precipitates of dolomite and massive sulphides filling the voids that are created. This complex reaction may be written:

$$H_2S + CO_2 + MeCl^+ + Mg^{2+} + 2CaCO_3 + H_2O$$
$$\Rightarrow MeS + Ca^{2+} + CaMg(CO_3)_2$$
$$+ HCO_3^- + Cl^- + 3H^+ \qquad (3.56)$$

where *Me* represents heavy metals, chiefly Fe, Zn and Pb. Taking its origin in the generation of methane, this is considered to be the basic equation for deposition of MVT sulphide deposits in carbonate rocks (e.g. Anderson 1991).

Where H_2S is able to migrate to the water table or just below it, conditions permit very effective oxidation, creating sulphuric acid:

$$H_2S + 2O_2 \Rightarrow H_2SO_4 \qquad (3.57)$$

which reacts with limestone

$$H_2SO_4 + CaCO_3 \Rightarrow Ca^{2+} + SO_4^{2-} + CO_2 + H_2O$$
$$(3.58)$$

This is a strong acid, capable of dissolving very large underground chambers at the water table very quickly. If there is high P_{CO_2} in the air or water, $600 \, mg \, L^{-1}$ or more of calcite may be taken into solution at equilibrium (see Palmer, 1991, figure 22). The processes are probably moderated by *Thiobacillus* spp. It is a most potent mechanism of cave genesis; e.g. Galdenzi and Menicheti (1990) show that it played a major role in the formation of Grotta Grande del Vento (Frasassi), while Korshunov and Semikolennyh (1994) suggest that the extensive caves of Kugitangtau in Turkmenistan were largely excavated by it. Baiyun Cave, Hebei Province, China, is a small but good example developed at the edge of a coal basin instead of an oil basin.

Hill (1987) calculated that the H_2S required to generate Carlsbad Big Room (Figure 7.33) is less than 10% of one year's natural gas production from the adjoining New Mexico gas fields, so the mechanism is quantitatively feasible. In the Hill model it represents Stage 4, when the water table was being lowered through the limestones as a consequence of the uplift of the Guadalupe Mountains in late Tertiary times (Fig 3.11 lower).

From equation (3.58), a proportion of the Ca^{2+} and SO_4^{2-} ions pair together, precipitate as gypsum and settle out in the slack-water areas of any water-table ponds. There is also direct conversion of limestone to gypsum on cavern walls, both below and above the water line, the latter being attributable to vapour condensation. Once again, it is believed that sulphur-oxidizing species moderate the reaction.

In Lower Kane Cave, Wyoming, Stern *et al.* (2002) measured 1.3 ppm H_2S in the air; vapour condensed with a pH ~5.3, which was lowered to 3.0 by bacterial action, and as low as pH 1.7 in some pendular droplets. In Carlsbad Caverns the pond-floor gypsum is up to 4 m in thickness, and wall alteration crusts reach depths of 0.5–1.0 m. This is Stage 5 in Hill's model. The final stage, 7, describes a minor effect occurring at present where water droplets accumulate on the native sulphur deposits in Carlsbad and other caverns, forming encrustations of gypsum:

$$S^0 + Ca^{2+} + 2O_2 + 2H_2O \Rightarrow CaSO_4 \cdot 2H_2O \qquad (3.59)$$

Hydrogen sulphide reactions play an important quantitative role in one further setting. This is on modern coasts where caves of varying origin have been inundated by salt water as a consequence of the post-glacial rise of sea level, and a thin freshwater layer rests on the denser salt water. Open cave mouths, drowned shafts and collapses (**cenotes**) trap organic debris from the surface. Deep in the salt water sulphate-reducing organisms produce H_2S from marine sulphate and the debris, which anaerobic and aerobic oxidizers around the halocline (fresh–salt interface) convert to H_2SO_4; many divers report that the rock walls in this narrow zone are typically spongy and full of solutional holes as a consequence (see e.g. Wilson and Morris, 1994, figure 1).

3.7 CHEMICAL COMPLICATIONS IN CARBONATE DISSOLUTION

This major section could well be entitled 'boosters and depressants'. It is concerned with particular conditions and effects that can significantly increase or decrease the solubility of the carbonate minerals. Some of these effects also work on gypsum but because its solubility is so great they are of lesser importance. Most of the analysis has been concerned with calcite solubility but aragonite and dolomite are similarly affected in most cases. Figure 3.8 summarizes the principal effects.

The section begins with effects that occur within the carbonate solution systems already described, then continues with effects when foreign acids, ions or molecules are introduced.

3.7.1 Temperature and pressure effects – deep karst

From Figure 3.8 and Henry's Law, the solubility of calcite in water equilibrated to the standard atmosphere ($P_{CO_2} = 0.03\%$) at 25°C is 55 mg L^{-1}. This increases to 75 mg L^{-1} at 0°C.

Water often cools as it passes underground. This enhances its solvent potential. For a water saturated at 240 mg L^{-1} CaCO$_3$ and cooling from 20° to 10°C, Bögli (1980) cites a boost of 17.7 mg L^{-1} CaCO$_3$.

Increase of hydrostatic pressure has negligible effects on dissolved species, including gases. However, if any CO$_2$ bubbles can be introduced into water under pressure, CO$_2$ (g) solubility increases at a rate of approximately 6 mg L^{-1} per 100 m^{-1} depth of water (at 25°C) until a depth of ~400 m. At greater depths solubility increases at ~0.3 mg L^{-1} per 100 m^{-1}.

When a cave is flooding rapidly, much air may be trapped and dissolved at pressures up to several atmospheres. Bögli (1980) and others have suggested that this boost may explain the development of ceiling half-tubes in cave passages because bubbles will be dissolved against the ceiling. Much more significant, we believe, is the combination of pressure and cooling where crustal exhalative CO$_2$ in active volcanic or tectonic areas is added (initially as gas bubbles) to deeply circulating waters that emerge via karst hot springs (Yoshimura *et al.* 2004).

Added CO$_2$ may greatly boost the solvent capacity of deep, hot water, creating deep karst. As the water ascends and pressure falls, gas is released as bubbles. The solution may become supersaturated with calcite, or the cooling effect may predominate so that the gas becomes redissolved to create a second zone of boosted solvent capacity; e.g. waters tapped by boreholes near Cave of the Winds, Colorado, are highly carbonated and undersaturated by ~200 mg L^{-1} CaCO$_3$ when they emerge. The complex association of corrosional cavities with precipitated CaCO$_3$ linings in many thermal water caves is explained by changing permutations of this cooling and/or degassing relationship (see section 3.10 and Ford *et al.* (1993) for the example of Wind Cave, South Dakota).

Deep dissolution, with some reprecipitation, is important in carbonate-hosted oil and gas fields, at temperatures generally above 50°C and with commensurate pressures. Simple hydrolysis of calcite is very effective above 75°C, yielding CO$_2$ (equation 3.4). Complex reactions between clay minerals and carbonates may dissolve dolomite, precipitate calcite and yield one mole CO$_2$ per mole dolomite consumed. In the main 'oil window' between 80 and 120°C thermochemical reduction of any sulphates yields H$_2$S. Decarboxylation (destruction of fatty acids) yields CO$_2$. Catalytic degradation of other kerogen produces aliphatic acids, RCOOH (e.g. acetic acid CH$_3$COOH), which are more soluble than carbonic acid and behave in a similar manner.

Han (1998) cites a good example of the cumulative effects of these processes in his study of the Renqiu palaeokarst-hosted oil and gas deposit, part of the giant Bohai field, China. Temperature was 80°C, pressure 300 atm and P_{CO_2} 20 atm. Total dissolved solids ranged 3750–10 000 mg L^{-1} in 14 water samples, SI$_c$ being negative (−0.5 to −1.5) in 10 of them. Using rock samples from the oilfield cores, solubility was investigated across a range of temperatures and pressures, under both open and closed conditions, with 6-h experimental runs. With T fixed at 60°C there was rapid increase of solubility as P_{CO_2} was increased from 1 to 5 atm, slowing with further increase to $P_{CO_2} = 25$ atm; with P_{CO_2} fixed at 20 atm, solubility peaked at 55 mg cm^{-3} at ~50°C, with a linear decline thereafter to ~20 mg cm^{-3} at the upper limiting temperature of 120°C. Physical effects were readily seen under magnification, taking the form of collapse of earlier diagenetic porosity and introduction of new etch pits.

3.7.2 Inorganic exotic acids

Here we refer to acids generated outside the bicarbonate and sulphur systems by reaction with other minerals. This is illustrated in Figure 3.12 where the increase in solubility that is shown for HCl will be true for all acids introduced in 1 N solution.

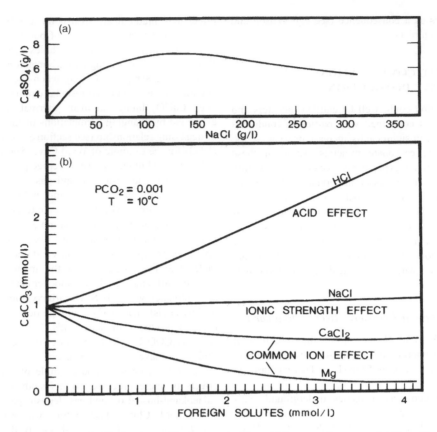

Figure 3.12   Illustrating common ion, foreign ion and ionic strength effects. (a) Increase of gypsum solubility with addition of NaCl. (b) Ionic strength and common ion effects upon calcite solubility at 10°C. Reproduced with permission from Picknett, R.G., L.G. Bray, and R.D. Stenner 1976. The chemistry of cave waters, in T. D. Ford & C.H.D. Culliingford (eds.). The Science of Speleology. pp 212–266. © 1976 Academic Press, Elsevier.

Dilute solutions of hydrochloric acid occur in nature. The reaction may be represented:

$$CaCO_3 + HCl \Rightarrow Ca^{2+} + HCO_3^- + Cl^- \qquad (3.60)$$

Most exotic production involves oxidation, e.g. the case of manganese, which is common in very low concentrations, ($< 1\,mg\,L^{-1}$) in cave waters:

$$Mn^{2+} + \frac{1}{2}O_2 + H_2O \Rightarrow MnO_2 + 2H^+ \qquad (3.61)$$

In the case of siderite ($FeCO_3$, present as thin layers or nodules in many limestones and dolomites) several reaction paths may enhance acidity, for example:

$$2\,FeCO_3 + \frac{1}{2}O_2 + 5H_2O \Rightarrow 2Fe(OH)_3 + 2HCO_3^-$$
$$+ 2H^+ \qquad (3.62)$$

or

$$FeCO_3 + H^+ \Rightarrow Fe^{2+} + HCO_3^- \qquad (3.63)$$
$$Fe^{2+} \Rightarrow Fe^{3+} + e^- \qquad (3.64)$$
$$2Fe^{3+} + 6H_2O \Rightarrow 2Fe(OH)_3 + 6H^+ \qquad (3.65)$$

The $Fe(OH)_3$ is a hydrated precipitate.

Similar reactions may occur with other metal carbonates in limestones, while sulphides will produce the H_2S reactions summarized above. Streams flowing on to limestones from siliciclastic rocks often contain these acids also. However, the quantities involved in most instances are very small.

3.7.3 'Acid rain'

The pH of normal rainwater is between 5.6 and 6.4. In industrialized regions and for hundreds of kilometres downwind of them it is now often below 4.5: indeed, it has been recorded below 3.5 in sites as remote from industry as Spitsbergen (Krawczyk *et al.* 2002). There are two principal sources of the additional acidity:

1. H_2SO_4 from atmospheric oxidation of SO_2 produced in the burning of sulphur-rich fossil fuels and the smelting of sulphide ores;
2. HNO_3, produced when atmospheric nitrogen is oxidized in internal combustion engines and then vented as exhaust gas, or from the manufacture of inorganic fertilizers.

Acid rain has been attacking limestone buildings in Europe for more than 150 year. The H_2SO_4 reaction (equation 3.58) is the most noticeable because surficial skins of stone are spalled off by the expansion involved; the process is termed 'sulphation'; (see section 12.6). Nitrates also are now raining out in amounts as great as $10 \, kg \, ha^{-1} \, a^{-1}$ in many parts of the world. Attention has been focused on the deleterious effects of acid rain on forests, rivers and lakes in regions where there are no carbonate rocks to buffer the acid, e.g. much of central Sweden or the eastern Canadian Shield. So far as we are aware, acid rain has not yet produced any notable new karst on carbonate rocks (as opposed to buildings made from them), but students of karst water equilibria and erosion rates must be alert to its potential effects in the water samples that they analyse. Theoretically, acid rain at pH 4.0 may dissolve 100 times more $CaCO_3$ than 'normal' rainwater at pH ~6.0, although this has not been established in field studies.

3.7.4 Ion pairs

Ion pairs normally found in bicarbonate waters are shown in Figure 3.5. They are weakly associated cation–anion pairs within a solution that also contains many free ions. As ionic strength increases, pairing will increase. This reduces the activity of ions of interest and so increases mineral solubility.

Chloride ions do not pair significantly, thus ion pairs in karst and saline waters will be combinations of the cations Ca^{2+}, Mg^{2+}, K^+, Na^+ and H^+, with CO_3^{2-}, HCO_3^-, OH^- and SO_4^{2-}. For example, $H_2CO_3^0$ (carbonic acid) is an ion pair. More complex pairings such as $Ca(HCO_3)_2^0$ are possible, but minor. In bicarbonate and sulphate waters the significant pairs appear to be $CaHCO_3^+$, $CaCO_3^0$, $MgHCO_3^+$, $MgCO_3^0$, $CaSO_4^0$ and $MgSO_4^0$.

As an early example, Wigley (1971) studied spring water from a gypsum and carbonate basin in British Columbia. Total dissolved solids were $1700 \, mg \, L^{-1}$. He found that 70.6% of Ca^{2+} ions were free, 26.7% paired with SO_4^{2-}, 1.7% with HCO_3^- and 1% with CO_3^{2-}. Ion pairing for Mg^{2+} was almost identical.

Although it may increase carbonate and gypsum solubility a little (generally $< 10\%$), ion pairing is truly more important for its effect on calculated saturation indices. If pairing is not allowed for, the index values are overestimated – solutions appear more saturated than they are. Standard computer programs mentioned below (PHREEQC, WATEQ4F, etc.) compute all probable pairs. It is essential that this be done where total dissolved solids exceed $100 \, mg \, L^{-1}$.

3.7.5 Common ions

The principle of the common ion effect is that if one of the ions created by dissolution of a given mineral should be introduced from some other source, the solubility of that mineral is reduced. For calcite, aragonite and dolomite, this normally implies alternative sources of Ca^{2+} or Mg^{2+} ions. Other carbonate and magnesium minerals are rare, thus the 'common ion effect' occurs chiefly where Ca^{2+} is furnished by gypsum or (to a much lesser extent) from calcic feldspars.

Addition of Ca^{2+} from gypsum decreases the activity of the ion but increases the molar product by a much greater amount. As a consequence, less calcite can be dissolved before equilibrium is attained. At 10°C addition of $100 \, mg \, L^{-1} \, Ca^{2+}$ from gypsum reduces a given calcite solubility of $100 \, mg \, L^{-1} \, CaCO_3$ to $66 \, mg \, L^{-1}$. Where total ionic strength is much less than 0.1, the solubility of calcite and dolomite is considerably reduced in waters that have already had substantial contact with gypsum; this is a principal cause of dedolomitization, the incongruent dissolution of the mineral with compensatory precipitation of calcite.

3.7.6 Mixing corrosion in meteoric waters

The important concept of **mixing corrosion** in fresh waters was introduced into karst studies by Bögli (1964; see also Bögli 1980, pp. 35–37), which is an effect produced by the mechanical mixing in the CO_2 system of two karst waters from different sources, where both are saturated with calcite and therefore, acting alone, are incapable of further dissolution. In general, in such mixing cases the resulting mixture will be somewhat aggressive if one or more of the waters contains less than $250 \, mg \, L^{-1}$ $CaCO_3$. The same effect applies in the mixing of two H_2S waters from substantially different sources.

The principle is illustrated with an example in Figure 3.13. In the dissolved $CaCO_3$ range of

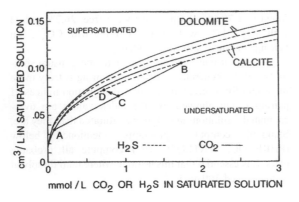

Figure 3.13 The principles of mixing corrosion in saturated solutions in the CO_2 or H_2S systems. See text for details. Reproduced with permission from Palmer, A.N., Origin and morphology of limestone caves. Geological Society of America Bulletin, 103, 1–21 © 1991 Geological Society of America.

$0-350\,mg\,L^{-1}$ that is found in most carbonate karsts the equilibrium relationships between P_{CO_2} or H_2S and calcite are non-linear. Groundwater A is saturated for open atmospheric P_{CO_2} (or a low H_2S environment) and groundwater B with a rich soil CO_2 atmosphere. They mix along the line AB and if the two waters are of equal volume (mixing ratio = 1:1, as shown at point C in the figure) the resulting mixture is undersaturated. The spare H^+ ions can react with calcite or dolomite until equilibrium is regained at point D. The angle that line C–D subtends with horizontal is determined by the mixing ratio; A:B = 1:1 = 45°; A:B = 2:1 = 60°, etc.

Bögli's proposal has been confirmed by calculations by many others. He contended that the CO_2 mixing corrosion mechanism described here is crucial for the initiation of caves in limestone and dolomite because otherwise waters are saturated before they have advanced more than a few metres along their host fissures. Subsequent determinations that higher order kinetics begin to apply when a solution becomes 80–90% saturated (section 3.10) have greatly reduced the significance of this argument. However, it or some similar effect offers a good explanation for the initiation of wall and ceiling solution pockets that are common in limestone phreatic caves where penetrable joints intersect major passages. In a computer model Gabrovšek and Dreybrodt (2000) have shown that it may relocate the locus of initial cave passage development and accelerate its rate.

3.7.7 Ionic-strength effects and seawater mixing

The ionic-strength effect is sometimes termed the **foreign-ion effect**. Addition of large quantities of foreign ions such as Na^+, K^+ and Cl^- to a bicarbonate water decreases the activity of Ca^{2+}, HCO_3^-, etc. and so increases calcite and dolomite solubility (Figure 3.12).

The ionic-strength effect is primarily associated with addition of salt. Solubility of gypsum is tripled in a seawater-strength solution (Figure 3.12a).

In the low concentrations of normal limestone fresh water (Fig. 3.12b) the effect appears modest. Approximately $250\,mg\,L^{-1}$ NaCl need to be added to boost $CaCO_3$ solubility by $10\,mg\,L^{-1}$. When thousands of $mg\,L^{-1}$ NaCl are added in salt-water mixing situations, however, the effects can be considerable. Figure 3.14 shows Plummer's (1975) analysis for seawater mixing at 25°C. With high P_{CO_2} calcite solubility can be boosted to $\sim 1000\,mg\,L^{-1}$.

On limestone coasts a mixing zone between fresh and marine groundwaters exists (see section 5.8), in which dissolution, precipitation and replacement reactions can be widespread and quantitatively very important; they are the subject of intense investigation at the present time (e.g. Martin *et al.* 2002). In an illustrative early study of the Yucatan Peninsula of Mexico (a young, permeable limestone plain), Back *et al.* (1984) showed that waters from the interior flow for as much as 100 km underground

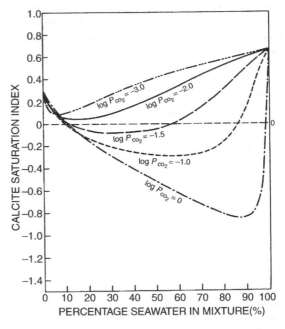

Figure 3.14 Effect of seawater–freshwater mixing upon calcite solubility, at 25°C and P_{CO_2} as specified. Reproduced with permission from Plummer, L.N., Mixing of seawater with calcium carbonate ground water: quantitative studies in the geological sciences. Geological Society of America Mem. 142, 219–236 © 1975 Geological Society of America.

and become saturated with calcite at $\sim 250\, \mathrm{mg\,L^{-1}}$. In the final 1 km of their journey to the sea, a further $120\, \mathrm{mg\,L^{-1}}$ $CaCO_3$ is added from ionic-strength effects as mixing with seawater occurs.

3.7.8 Trace-element effects

The minor element content of limestones or trace elements from other sources that are present in the water have been found to have substantial effects on calcite solubility in laboratory experiments. The principal study by Terjesen *et al.* (1961) is confirmed by some later work. The presence of tiny amounts of certain metals ($< 1\,\mathrm{\mu mol\,L^{-1}}$) reduces calcite solubility. The inhibiting effect increases with increasing trace-element content and is believed to be due to absorption of the metal ions onto dislocations in the calcite crystal surface that otherwise are the sites of dissolution, as explained in section 3.10. In decreasing order of effectiveness, important metals investigated were scandium, lead, copper, gold, zinc, manganese, nickel, barium and magnesium. As examples, $6\,\mathrm{mg\,L^{-1}}$ Cu^{2+} or $1\,\mathrm{mg\,L^{-1}}$ Pb^{2+} reduced calcite solubility by a factor of two at $P_{CO_2} = 1$ atm. No research has been conducted at more appropriate P_{CO_2} levels. No allowance for these trace-metal effects is made in Saturation Index calculations. The effects are probably insignificant in most waters.

Some magnesium is present in most calcites. Effects of differing proportions of Mg^{2+} in the solid solution are discussed in section 3.10 and Chapter 2. Phosphate ions may also be a strong inhibitor and Mn^{2+} at concentrations $> 0.5\,\mathrm{\mu m}$. Some organic compounds such as maleic and tartaric acids are also inhibitive.

3.8 BIOKARST PROCESSES

Biokarst landforms are those 'produced largely by the direct biological erosion and/or deposition of calcium carbonate' (Viles 1984) or other karst minerals. On Earth's land surfaces the most important quantitative biotic contribution to karstification is undoubtedly the production of CO_2 in soil, summarized above; once dissolved, this CO_2 is responsible for much more dissolution than the sum of all other processes mentioned below. However, its action is not direct because the carbonic acid dissolution process itself is inorganic, and usually physically distant from the CO_2-generating sources. At the base of soils plant root expansion may also break up bedrocks mechanically, increasing their exposed surficial areas and thus their dissolution rates. Humic and fulvic acids are secreted by the roots and may effect minor dissolution. They are more significant in speleothem studies, however; see section 8.7.

3.8.1 Phytokarst

Phytokarst landforms occur on rocks exposed to daylight (or artificial light in some commercial caves) that are bare except for a plant cover. They are 'produced by rock solution in which boring plant filaments are the main agent of destruction' (Folk *et al.* 1973). The principal biota are colonies of tiny **blue-green algae** (*cyanobacteria*), **red algae** (*rhodophyceae*) or **diatoms** (*bacillariophyceae*). They are created by photosynthesis; e.g.

$$6CO_2 + 6H_2O \Rightarrow C_6H_{12}O_6 + 6O_2 \qquad (3.66)$$

Secretion of water and CO_2 from the filaments that attach the cells to the rock permits general etching underneath the colony, or boring (to depths up to several millimetres) at the filament tips.

These processes are most effective on limestone and dolomite but can also play a role on gypsum. Algal colonies are found on most bare rocks in temperate and warmer climates, sometimes veneering them completely. The greatest effects are on sea coasts because the colonies are most abundant and diversified there, and also serve as food for molluscs which further grind and comminute the rock when grazing, accelerating its dissolution. For example, Tudhope and Risk (1985) estimated that the erosion rate due to these processes was $350\,\mathrm{g\,CaCO_3\,m^{-2}\,a^{-1}}$ on a lagoon floor at the Davies Reef, Australia. The principal dissolutional features are small pits, often densely packed or overlapping, and displaying an immense variety of forms in detail; see section 9.2.

3.8.2 Biokarst underground; bacterial activity

A wide variety of larger animals, ranging from mid-sized mammals down to submillimetric isopods and arthropods, can live in caves, lesser fissures and pores, including tiny **phreatophytes** below the water table. Their abundance is generally limited by lack of nutrition, however. As a consequence, their contribution to dissolution and other erosion processes appears to be quantitatively trivial except very locally. In a pioneer review, Caumartin (1963) suggested that phreatophytes consuming O_2 and respiring CO_2 might boost carbonate dissolution by as much as 10%, but this has not received support from field studies.

Modern attention is focused on the role of microbes, principally bacteria, primitive single-celled animals deriving their energy from electron transfer (redox) processes. **Heterotrophic** species consume matter carried down from the Earth's surface. **Chemolithotrophic** species reduce oxides in rocks, chiefly S, N, Fe, Mn and H oxides. **Chemoautotrophic** bacteria live on other species. All are bases of food chains for any higher animals in their environments. Anaerobic bacteria can live and multiply in temperatures up to 100°C and pressures > 1 kbar. Northup and Lavoie (2001) give a comprehensive review.

Bacterial activity takes place in three differing underground karstic environments. First and most significant is the anaerobic, i.e. below the water table to the limiting pressures and temperatures noted above. The reduction of sulphates by *Desulphovibrio* spp (discussed in section 3.6) appears to be the most important deep bacterial source of acids for carbonate dissolution, probably far exceeding the sum of all others in quantitative terms. Other chemolithotrophic species may contribute CO_2, NH_4 and H^+ for acid dissolution more locally (e.g. James 1994).

The second distinct environment is around the water table itself, including streams and flooding zones in river caves, tributary fissures, etc. Dissolved O_2 and other heterotrophic sources of energy are more abundant here, encouraging higher densities of both oxidizing and reducing species. They may mediate reactions with $FeCO_3$, etc, to boost CO_2 and H^+ in the water. One reaction for which results can be seen in most cave streams is the oxidation of Mn^{2+} to produce birnessite and other brown or black manganese coatings on pebbles (especially of silicate rocks) or on passage walls; it is not known to occur in the absence of bacteria (Ehrlich 1981).

The final environment is that of the relict cave that has been abandoned by its formative streams. Flowing waters no longer erode its walls and its clastic sediments are immobile on the floors. Acidic bacterial biofilms may then coat the walls, etching them and accelerating selective weathering to depths of several centimetres (Jones 2001; Zupan Hajna 2003), especially where they are wetted by seepage or condensation. As entrance zones become progressively illuminated by roof collapse, etc., algae compete with the bacteria, replacing them with phytokarst pittings oriented towards the light. Seepage waters from soils carry varying quantities of ammonia into caves; in fine-grained, well-drained cave sediments nitrifying bacteria convert it into saltpetre (the nitre KNO_3 component of gunpowder) in two steps

$$2NH_3 + 3O_2 \Rightarrow 2NO_2^- + 2H^+ + 2H_2O \qquad (3.67)$$
$$2NO_2^- + O_2 \Rightarrow 2NO_3^- \qquad (3.68)$$

Although most investigations have focused on bacterial activity in carbonate karst, it can also be significant in gypsum caves. Zoloushka ('Cinderella') Cave, Ukraine,

has passed through all three of the above environments during the past 50 year as a consequence of progressive drainage for quarrying purposes. In an outstanding study, Andrejchouk and Klimchouk (2001) identified seven different oxidizing and reducing species at work on rock and in sediments there. While still anaerobic, gypsum cave roofs were altered to limestone plus native sulphur (equations 3.50, 3.53) by bacterial mediation. In the dewatered cave, aerobic species quickly started to work; for example, in the air, N_2 was locally increased from the atmospheric standard concentration of ~79% to > 83% by the exotic reaction:

$$5S + 6KNO_3 + 4NaHCO_3 \Rightarrow 3K_2SO_4$$
$$+ 2Na_2SO_4 + 4CO_2 + 3N_2 + 2H_2O \qquad (3.69)$$

3.8.3 Algae, bacteria and calcite deposition

While algae and bacteria may be locally or regionally significant in initiating or accelerating dissolution, there has been much closer study of their roles in facilitating calcite and aragonite deposition surficially (tufa and travertine) and in caves. They may serve as **passive** nucleii for precipitation onto surfaces or can be **active** catalysts inducing it. Passive mineralization occurs where bacterial cell walls, sheaths, etc. are covered with anionic sites to which cations such as Ca^{2+} can bind, whether the bacterium is alive or dead. Active mineralization occurs when bacteria produce enzymes or other chemicals that induce precipitation. **Biologically controlled mineralization** is common in algae and protozoa that build calcite structures such as coccoloiths. Bacteria appear to be essential for formation of many types of moonmilk (see e.g. Northup and Lavoie 2001).

3.9 MEASUREMENTS IN THE FIELD AND LABORATORY: COMPUTER PROGRAMS

Most nations measure total hardness, calcium hardness, bicarbonate, sulphate and chloride concentrations as part of their water quality monitoring programmes and so publish handbooks of standard methods. Our purpose here is to give only the briefest summary, with some tips on practice for karst purposes, and to refer to some other useful sources.

3.9.1 Temperature and specific electrical conductivity

Measurement of the specific conductivity (SpC) of a water sample is a quick and easy process that can give a very

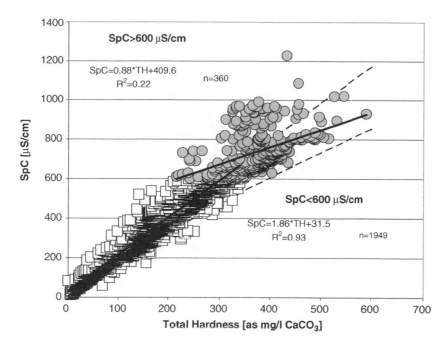

Figure 3.15　Specific electrical conductivity versus total hardness expressed as mg L^{-1} $CaCO_3$. The 2300 individual results plotted here are from 20 different, recently published, data collections from carbonate karstlands worldwide. Lower dashed line is the theoretical relationship between SpC and a pure $Ca + CO_2$ | HCO_3 solution, allowing for ion pairing, and upper dashed line is for the molar equivalent. The dolomite relationships are very similar. Solid line=best fit where SpC $> 600\,\mu S\,cm^{-1}$. Best fit for SpC $< 600\,\mu S\,cm^{-1}$ is lost in the data. The equations are for these two lines of best fit. Reproduced with permission from Krawczyk, W.E. and Ford, D.C. Correlating specific conductivity with total hardness in limestone and dolomite karst waters. Earth Surface Processes and Landforms, 31; 221–234, © 2006 John Wiley and Sons.

good first approximation of the dissolved load because, at low I, electrical conductivity is proportional to total ionic concentration. Modern solid-state digital conductivity meters are robust, small, light and cheap. They measure a wide range of conductivities on linear scales, with results corrected to a standard temperature (customarily 25°C). They include temperature probes accurate to 0.5°C or better: this is sufficient for most applications, including computation of saturation indices.

Figure 3.15 shows the theoretical relationship between SpC, and pure calcite (or aragonite) and dolomite solutions expressed as mg L^{-1} $CaCO_3$. Overlaid are the measured results from > 2300 full ionic analyses reported from many different carbonate karst regions worldwide in recent years (Krawczyk and Ford 2006). Where SpC $< 600\,\mu S\,cm^{-1}$, the best-fit line to the field relationship is

$$SpC = 1.86 \cdot TH + 31.5 \qquad R^2 = 0.93 \qquad (3.70)$$

where TH is the total hardness. Waters from some of the individual sample regions closely approximate the theoretical relationship $(R^2 > 0.995)$, signifying that there were little but Ca^{2+}, Mg^{2+} and HCO_3^- ions in solution. In contrast, from areas of substantial agricultural or industrial pollution, correlation sometimes drops below $R^2 = 0.8$. The best-fit line begins to deviate significantly from theoretical values when total hardness exceeds $\sim 250\,mg\,L^{-1}$ as $CaCO_3$. In most instances this deviation can be interpreted to indicate that sulphate rocks or salt are also present in small amounts and supplying some ions to the solution. Nevertheless, the one standard error of estimate band about the best fit is pleasingly narrow: for SpC $= 200\,\mu S\,cm^{-1}$ it gives $TH = 98 \pm 2\,mg\,L^{-1}$; for SpC $= 400\,\mu S\,cm^{-1}$, $TH = 193 \pm 2\,mg\,L^{-1}$.

Very few samples with SpC $> 600\,\mu S\,cm^{-1}$ are pure calcite or dolomite solutions, although such high conductivity can be achieved by evaporative concentration or where there is deep source CO_2. In addition to sulphates and chlorides, some of the largest positive deviations are due to nitrate contamination.

Figure 3.16 shows the same relationship for waters in gypsum karst areas, based on 140 complete ion analyses from five different sites in Europe and North America.

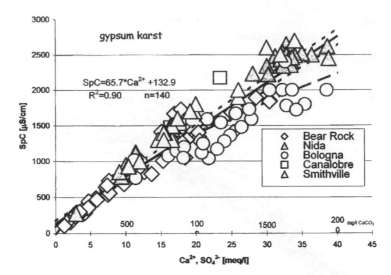

Figure 3.16   Specific electrical conductivity versus dissolved sulphates expressed as meq L^{-1} Ca^{2+}+ SO$_4^{2-}$. The lower dashed line is for the pure Ca + SO$_4$ + H$_2$O system, with ion pairing, and upper dashed line is for the molar equivalent. The equation is for the best fit, indicated by the solid line. Reproduced from Krawczyk, W.E. and Ford, D.C. Correlating specific conductivity with total hardness in limestone and dolomite karst waters. *Earth Surface Processes and Landforms*, 31; 221–234, © 2006 John Wiley and Sons.

Here total dissolved solids are expressed in meq L^{-1} Ca^{2+} + SO$_4^{2-}$. The best-fit relationship is the polynomial:

$$SpC = 65.7 \cdot Ca^{2+} + 132.9 \qquad R^2 = 0.90 \qquad (3.71)$$

It is seen that deviation from the theoretical relationship for pure gypsum becomes significant above 10 meq L^{-1}, which is equal to ~680 mg L^{-1} of Ca^{2+} + SO$_4^{2-}$ in solution. As saturation with respect to gypsum is approached (30–35 meq L^{-1} or 2100–2400 mg L^{-1}) the best-fit values are again greater than the theoretical in most instances; probably, this is due chiefly to the presence of ions from carbonates and chlorides.

Conductivity meters can be adapted for continuous recording. Where an accurate relationship between electrical conductivity and the dissolved species of interest has been established, therefore, continuous estimates of the latter can be readily obtained.

3.9.2 Field measurement of pH

The accurate measurement of the pH of a water sample has been the gravest analytical problem encountered in the carbonate equilibrium approach that we advocate. However, modern solid-state digital pH meters are small, robust and inexpensive; some are even submersible. Used with a combined (glass plus reference) electrode, and buffers as recommended by the manufacturers, field measurements are reproducible to ±0.05 pH or better. A remaining difficulty is the need to bring buffers to the ambient

temperature of the water sample before any measurement is made, which can require uncomfortable waiting in cold weather.

Where it is not practicable to take the instrument to the sampling site (e.g. deep in a cave), pH should be determined as soon as possible afterwards. Ek (1973) showed that for some karst waters a very good linear correlation exists between field pH and laboratory pH so that the former can be omitted. In any study area this fortunate circumstance must be established. It cannot be assumed.

3.9.3 Specific determination of dissolved species

Ideally, all variables should be measured at the sampling site in order to avoid disturbances, such as loss of CO$_2$, that occur during transport and storage. In practice, it is quite feasible to assemble portable apparatus for a backpack or a small field laboratory and to obtain adequately accurate results with it.

Total hardness (Ca^{2+} and Mg^{2+}) and Ca^{2+} concentrations are determined by complexometric titration with EDTA-Na. With care, results are reproducible to ±1.0 mg L^{-1}.

Carbonate alkalinity (HCO$_3^-$ plus a little CO$_3^{2-}$) can also be determined by titration using diluted hydrochloric acid (HCl) and commercial indicators or bromocresol green and methyl red indicator. It is better done by potentiometric titration, using 0.01 or 0.02 N HCl and the field pH meter, with an end-point at pH 4.5.

Chloride is best determined by ion-selective electrodes. Accurate determination of sulphate under field conditions is more difficult. A common method is turbidimetry, which measures a sulphate precipitate when a barium salt is added; a set of turbid standard solutions (e.g. 10) should be prepared (Krawczyk 1996). In many karst regions nitrates of natural origin may be present. In the modern world there can be abundant nitrates from livestock, fertilizers or acid rain. These substances will appear in SpC measurements and distort the ion balances (see below); therefore it is desirable to test for them. Standard methods use ion-selective electrodes.

3.9.4 Laboratory methods

All of the above methods can be used in a fully equipped laboratory. However, greater accuracy and rapid processing of large batches of samples can be obtained by use of AAS (atomic absorption spectrometry) and ICP–AES (inductively coupled plasma–atomic emission spectrometry) for cations or IC (ion chromatography) for both cations and anions, as in commercial water quality laboratories. Samples should be acidified for cation analysis, which may dissolve any suspended calcium carbonate. Filtration is also necessary, which can disturb the equilibria by precipitating $CaCO_3$. Dissolved silica may be determined on a spectrophotometer at a wavelength of 816 nm, by the method of reduction to a blue complex.

3.9.5 Analytical accuracy

The completeness and/or accuracy of the analysis of a water sample are checked by calculating the ion balance error (IBE):

$$IBE(\%) = 100 \frac{\sum cations - \sum anions}{\sum cations + \sum anions} \qquad (3.72)$$

where all ion concentrations are given in $meq\,L^{-1}$.

Given the problems of field science, we find any error up to 3–5% acceptable. Where error is greater, either a mistake has been made in the determinations or there are one or more major ionic species present that have not been measured.

Because of the field difficulties, saturation index values for calcite, aragonite and dolomite will normally have an error of ± 0.01–0.03.

3.9.6 Computer programs

In North America at least four major programs are currently in use – PHREEQC, WATEQ4F, MINTEQA2 and SOLMINEQ-GW. They are large, general purpose programs that contain comprehensive data on 30 or more minerals, solute species, their thermodynamic properties and equilibria in groundwaters. All have been modified through several generations and can be customized for use with Excel, etc. PHREEQC (Parkhurst and Appelo, 1999) is perhaps the most widely used.

3.10 DISSOLUTION AND PRECIPITATION KINETICS OF KARST ROCKS

Solution kinetics refers to the dynamics of dissolution. Processes are at their most vigorous when a solution is far from equilibrium. Forward processes to dissolve a mineral and back reactions to precipitate it are governed by the same rules and so may be considered together. The central problem in kinetics is to determine what controls the rate of reaction in specified conditions. Using solutions to the problem, karst specialists may then devise numerical models to estimate, for example, rates of extension of proto-caves or growth of stalagmites. Most relevant kinetic studies have been limited to calcite so it is emphasized here. Aragonite, dolomite, gypsum and salt are discussed more briefly.

Reactions are **homogeneous** when they take place in one phase, e.g. $CO_2\,(aq) + H_2O \rightarrow H_2CO_3^0$. They are **heterogeneous** when two phases are involved. All rock surface solution reactions are heterogeneous and so is all karstic precipitation.

In a static liquid, ions and molecules of dissolved species move from regions of higher concentration to lower concentrations by the process of molecular diffusion. If the liquid is flowing or is disturbed by waves or currents, dissolved species are dispersed by eddy diffusion which typically is several orders of magnitude more rapid than molecular diffusion.

In most karst situations the water is in motion and therefore eddy diffusion dominates. However, a **diffusion boundary layer (DBL)** is assumed at the liquid–solid interface where the water is static because of friction and thus molecular diffusion must operate. The boundary layer will be saturated with respect to the mineral, or nearly so. It is very thin, ranging from perhaps 1 μm to 1 mm in differing situations; mid-range values around 30 μm are often cited by experimentalists. Its thickness is determined by surface roughness, fluid viscosity and velocity of flow in the bulk liquid above it. Plummer *et al.* (1979) proposed that there may be a further 'adsorption layer' of solute ions and molecules that is loosely bound to the solid surface. This sublayer is only a few molecules deep.

Figure 3.5, the equilibria cartoon, includes these concepts. The DBL can be conceived as a weak shield between the aggressive bulk liquid and the soluble solid.

Taking the calcium bicarbonate dissolution–equilibration–precipitation sequence as our example, potential rate controls in the open system may be divided into the following four general categories.

1. CO_2 dissolves into the water, or degasses from it back into the atmosphere.
2. Speciation of solutes takes place in the bulk liquid or, to a limited extent, in the DBL.
3. Potentially aggressive species (H^+, $H_2CO_3^0$, CO_2 (aq)) diffuse in through the DBL, and the resulting reaction products diffuse out.
4. Reaction occurs at the solid–liquid interface: a Ca^{2+} or CO_3^{2-} ion dissociates from the crystal surface, or H^+ combines with a CO_3^{2-} ion to dislodge it.

The dissolution rate will be determined by whichever reaction within these categories is the slowest. Where the system is closed, the first category is eliminated.

Researchers have approached the dissolution question primarily by laboratory experiments. Classically, calcite powders or Iceland spar crystals (which have relatively few imperfections) are dissolved in acid solutions, with occasional or steady stirring. Constant pH is maintained by bubbling in CO_2 (pH static experiments) or pH is allowed to change as dissolution proceeds (pH drift). It is infeasible to measure changes of concentration in the DBL itself because it is too thin, so kinetic processes and rates are estimated by changes of concentrations in the bulk liquid. Later experiments have added spinning discs of polished crystal, limestone, etc. Under laminar flow conditions, the faster the spin, the thinner the DBL; in theory it then maintains a constant thickness after turbulent (eddy) flow commences, thus permitting diffusion rates in the DBL to be estimated. Within the past decade, application of atomic force microscopy (AFM) or optical interferometry has made it possible to observe dissolution actually occurring at the solid–liquid interface (i.e. *in aquo*) at near-molecular scales, 0.2–0.5 nm (e.g Hillner *et al.* 1992). Atomic force microscopy works in real time but provides only brief glimpses, while interferometry relies on surface mapping of changes between successive static snapshots. However, a convincing picture of the rate controls of calcite dissolution is now appearing; Morse and Arvidson (2002) give a comprehensive review.

Rates of dissolution of CO_2 and of degassing are not well known. There is rapid diffusion of the species in air and flowing water so that conditions at the interface are uniform at a scale of centimetres. Roques (1969) found that a drop of water forming at the tip of a soda straw stalactite lost 10% of its CO_2 in the first second, 30% in 90 s and 70% in 15 min. Dissolution or degassing of CO_2 do not appear to be rate-controlling in most circumstances.

The speciation reactions are homogeneous. Most are effectively instantaneous but the hydration of CO_2 takes approximately 30 s at 25°C. Great importance is now attached to this slow reaction by several authorities (see below) if it occurs within the boundary layer. Roques (1969) showed that all equilibration within the bulk liquid occurs within 5 min. This is faster than CO_2 dissolution or degassing and so is not considered to be rate-controlling.

Molecular diffusion of species in or out through a DBL is described by Fick's first law:

$$F = \frac{-D(C_{eq} - C_{bulk})}{X} \tag{3.73}$$

where F is mass flux ($M L^{-3} T^{-1}$), D is a diffusion coefficient ($L^2 T^{-1}$), C_{eq} is the equilibrium concentration of the species (which is assumed to be the concentration at the solid–liquid interface) and C_{bulk} its actual concentration in the bulk liquid; X is the thickness of the DBL. For the species of interest in karst work, values of D are $1–2 \times 10^{-5}$ cm^2 s^{-1} at 25°C, falling to about one half of these rates at 0°C. Different ionic strengths (I) have little effect. The diffusion coefficient of CO_2 in still air is ~ 1 cm^2 s^{-1} at 25°C and 0.14 cm^2 s^{-1} at 0°C. Eddy diffusivity rates in water range 10^{-1} to 10^{-3} cm^2 s^{-1}.

From these findings we see that where diffusion processes are rate-controlling they must be those of the liquid DBL and thus likely to be affected by conditions at the solid–liquid interface. This has been appreciated for a long time, leading to the adoption of conceptual models such as that shown in Figure 3.17. This suggests that diffusion of H^+ ions from the bulk liquid into the DBL will be the control where H^+ is very abundant, i.e. in very acid solutions. As concentrations of solutes increase in the bulk, conditions move to a Transition regime of more complex kinetics, succeeded by an H^+ Independent regime very close to saturation.

To understand reactions at the solid surface it is best to picture a 'step, kink and hole' model such as Figure 3.18a. Atoms and molecules in calcite, etc. are ordered in layers. Isolated atoms resting on a top layer have highest free-energy available for dissociation because it is likely that only one chemical bond attaches them to the layer. Atoms at a step have two likely bonds, at kinks three or four, and in a hole five or more. These will be the preferred sites of dissolution. A H^+ ion that has diffused to the crystal surface will move across it until encountering a CO_3^{2-} molecule at such a site. The HCO_3^- ion created then diffuses away, laying bare a Ca^{2+} atom which dissociates in its turn. Studies using AFM and interferometry have found that dissolution rates are fastest on the steepest surfaces, which are usually flights of steps one molecule in height and width. Destruction thus passes from one atomic layer to the next, much like

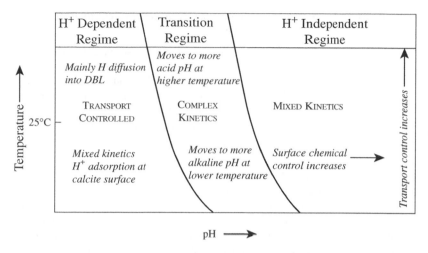

Figure 3.17   A schematic representation of rate-controlling mechanisms for calcite dissolution as a function of pH and temperature. Reproduced from Morse, J.W. and Arvidson, R.S, 'The dissolution kinetics of major sedimentary carbonate minerals'. Earth Science Reviews, 58; 51–84 © 2002 Elsevier.

unravelling successive rows of knitting. Measured migration rates have ranged $0.5–3.5\,nm\,s^{-1}$ in H^+-dependent regimes. Holes are also enlarged into etch pits that form and coalesce rapidly; e.g. 800 nm deep in 30 min. As

drawn in Figure 3.18a, nucleii possessing multiple bonds to a top layer are more important in calcite precipitation than in dissolution because they will attract ions to them from the solution.

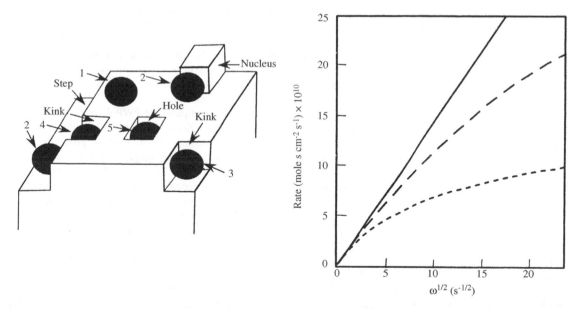

Figure 3.18   (Left) A model for dissolving or accreting calcite surfaces at the molecular scale. Steps, kinks, holes and nucleii are different types of surface sites, with numbers suggesting their likely number of chemical bonds. Adsorbed ions are shown in black. (Right) An example of experimental spinning disc calcite dissolution rates; $\omega' =$ disc rotation rate. The straight line represents the theoretical function if the rate were entirely transport-controlled. The long dashed line is for runs with Iceland spar (few imperfections), the short dash line for runs with Carrera marble (many dislocations). Effects of surface control become very apparent in this figure. Adapted from Morse, J.W. and Arvidson, R.S, 'The dissolution kinetics of major sedimentary carbonate minerals'. Earth Science Reviews, 58, 51–84. © 2002 Elsevier and Sjoberg, E.L. and Rickard, D.T. 'The influence of experimental design on the rate of calcite dissolution'. Geochimica et Cosmochimica Acta, 47; 2281–2286. © 1983 Elsevier.

Most individual crystals (and all larger aggregations of crystals) have imperfections, small faults (termed **screw dislocations**) that cut through the regularly stacked boxes of Figure 3.18a to create offsetting in all layers. These also function as steep surfaces for fastest unravelling.

We can now appreciate the dissolution-inhibiting effects of trace elements such as nickel, copper, etc. mentioned above. In eroding back, a kink or screw dislocation arrives at one of these atoms that will not dissociate or combine with H^+. Erosional exploitation of that particular kink, etc. ceases.

Using transmission electron microscopy (TEM) Schott *et al.* (1989) found that the density of surface defects such as dislocations, kinks and holes, and of inhibitors, had little effect on dissolution rates in H^+-dependent regimes, but that in Transition and H^+-Independent conditions the effects became two to three times greater than earlier estimates based on changing concentrations in the bulk liquid had suggested. This is illustrated by the spinning disc experiments shown in Figure 3.18b. In a Transition regime, as spin is increased (i.e. DBL thickness, X, is reduced) dissolution rates deviate ever further from the linear rate predicted for a pure Transport regime. Due to metamorphism, the Carrara marbles have high densities of defects and of inhibitors, and are very deviant. Iceland spars are nearly ideal calcites with few defects or inhibitors and much less affected. Findings such as these led Morse and Arvidson (2002) to write... 'The distressing result is that there can be no general equation that is applicable to all calcites that simply relates surface area and solution composition! It also implies that the influences of inhibitors will differ for different calcites.'

In many analyses DBL and surface effects are merged in a term, A/V, where A is the surface area of the solid and V is the volume of the solution. At one extreme, a speck of calcite dust settling through a pool has a very low A/V ratio: the DBL is essentially of infinite thickness, so that eddy diffusion is irrelevant and surface effects are trivial. At the other extreme is the case where water is flowing through bedding planes or joints with very small mean apertures and high surface roughness. Here, the A/V ratio is large and any effect that can thin the DBL to permit eddy diffusivity to play a role is potentially very significant: this introduces the important concept of **breakthrough**, which is discussed later. A bare limestone surface with rain falling upon it is one example of intermediate conditions that are, kinetically, very complex because A/V ratios change momentarily with the rainfall intensity and pulses of sheet flow.

With these caveats in mind, calcite solution may now be appraised.

3.10.1 Kinetics of calcite dissolution

There were comprehensive series of experiments during the 1970s, with seawater by Berner and Morse (e.g. 1974) and with carbonated distilled water by Plummer and Wigley (1976) and Plummer *et al.* (1978, 1982). These covered the complete range of expected natural temperatures and a wide range of ionic strengths.

Results were expressed as standard rate equations of the form:

$$\frac{dC}{dt} = \frac{k_c A}{V(C_{eq} - C)^n} \tag{3.74}$$

where K_c is a surface dissolution rate constant for calcite, C_{eq} is the concentration at saturation (i.e. at the solid surface) and C is the concentration in the bulk flow. Plummer *et al.* (1978) proposed that there are three significant forward rate processes – reaction with H^+, reaction with $H_2CO_3^0$, and by $CaCO_3$ dissociation alone. Their comprehensive rate equation (the 'PWP Equation') is:

$$r = k_1 \cdot a_{H^+} + k_2 \cdot a_{H_2CO_3^0} + k_3 \cdot a_{H_2O}$$
$$- k_4 \cdot a_{Ca^{2+}} \cdot a_{HCO_3^-} \tag{3.75}$$

where a represents activity. The three forward rates were fitted as functions of temperature: $\log k_1 = 0.98 - 444/T$, $\log k_2 = 2.84 - 217/T$ and $\log k_3 = -5.86 - 317/T$, where T is in °K. The final term is the expression for back reaction. The derivation of k_4 is complicated but can be approximated by standard rate equations for calcite precipitation given below.

Figure 3.19a compares the PWP theoretical rate model with some dissolutional experimental results and AFM surface measurements. Dissolution is given in moles $cm^{-2} s^{-1}$. Note that the range of pH is much larger than will be found in any regional karst waters. It is seen that there is close agreement with respect to the form of the function across this great pH range, but disagreement of about one order of magnitude in actual measured rates in the general karstic range between pH ~5 and pH 8.9. The AFM results correspond well with the PWP theory in this latter range, but there is comparatively little change of actual rates there.

An alternative perspective is given in Figure 3.19b, where data from the same or similar experiments are normalized against their initial (fastest) rates, and the evolving degree of saturation with respect to calcite (i.e. the arithmetic Saturation Ratio, where equilibrium occurs at 1.00). Curve 1 is for calcite in seawater, curve 2 for dolomite. Curves 3–7 are for calcite in pure $H_2O + CO_2$ solutions – it is seen that they are all similar, displaying

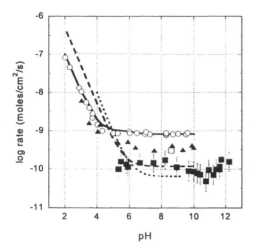

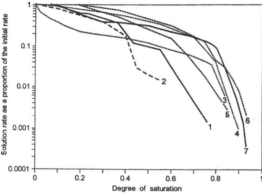

Figure 3.19 (Upper) Theoretical and experimental calcite solution rates compared. Strong dashed line=the PWP model; short dashes=model of Chou *et al.* (1989). Open circles, the square and filled triangles are from experiments with powders. Black squares with error bars are AFM results. Reproduced from Morse, J.W. and Arvidson, R.S, 'The dissolution kinetics of major sedimentary carbonate minerals'. Earth Science Reviews, 58, 51–84 © 2002 Elsevier (Lower) Dissolutional experimental approaches to calcite and dolomite saturation. Curve 1 from Berner and Morse (1974), using calcite in seawater. All other curves with distilled water and added CO_2. Curve 2, dolomite (Herman and White 1985); curves 3 and 4, Plummer and Wigley (1976) using Iceland spar; curves 5 and 6 from Svensonn and Dreybrodt (1992) using Iceland spar; curve 7 from Eisenlohr *et al.* (1997) using limestone discs. (This figure compiled by S.R.H. Worthington.) Reproduced from Morse, J.W. and Arvidson, R.S, 'The dissolution kinetics of major sedimentary carbonate minerals'. Earth Science Reviews, 58, 51–84 © 2002 Elsevier.

log-linear rates of dissolution until the solutions are 70–80% saturated, when rates drop off rapidly in power law form, i.e. the kinetics shift to a higher order (higher value of *n* in equation 3.73). No experiment could be carried to

100% saturation under fixed boundary conditions because the rates became too slow to measure. The small differences between curves 3, 4, 5, 6 and 7 can be attributed to differing experimental procedures plus the problematic surface effects.

The k_1 term of equation (3.75) describes the diffusion of H^+ through the DBL to the solid surface and is the rate control in the H^+ Dependent regime. This generally requires stronger acid solutions than are encountered in the majority of karst situations; it is most effective where acid mine drainage spills onto limestone or in very acid rains, and perhaps where organic acids drain from peat bogs. It can occur in some instances of bacteria-mediated H_2SO_4 cave genesis, e.g. Cueva de Villa Luz, Mexico (Hose *et al.* 2000).

It appears that most limestone dissolution takes place in transition regimes with complex kinetics (Figure 3.17); the processes of terms k_1, k_2 and k_3 in the PWP equation are all in play, diffusion in and out and surface effects are all variables with quantitative importance that may vary from point to point and, in some instances, with time at a given point. Rate plots against saturation, such as Figure 3.19b, give the best practical insight: SI_c increases from < 0.1 to > 0.8 across the range.

In the H^+ Independent regime solutions are close to saturation. Little free H^+ remains, pH thus approaches its upper limit for the ambient conditions. Back reaction (term k_4 in the PWP equation) may contribute to slowing the net dissolution. Density of surface defects can become the predominant control, which has led different experimentalists to derive powers of *n* ranging from 2 to 11 for their best fit. There is general support for Dreybrodt and Buhmann's (1991) contention that the comparatively slow process of hydration of the little remaining CO_2 (aq) within the DBL becomes an important control of the forward reaction rate.

In seawater SI_c is generally greater than 0.7, which is only 0.2 pH below thermodynamic equilibrium. Calcite dissolution thus takes place at the high end of the transition regime or in the H^+ Independent regime. Only phosphate functions as a major inhibitor. In phosphate-free seawater, the reaction order has been found to vary from 2 with calcite powders to 4.5 for calcite marine tests with complex void geometry (Morse and Arvidson 2002).

3.10.2 The solution kinetics of aragonite, Mg-calcites and dolomite, and quartz

There have been few investigations of the solution kinetics of aragonite. Busenberg and Plummer (1986) found that the behaviour is very similar to that of calcite.

Dolomite dissolution has also received relatively little study, here because of the slow reaction rates coupled with difficulties created by the many variations in structure and composition in the double carbonate. Uncertainties in the thermodynamic solubility of ideal dolomite remain close to ×10.

Busenberg and Plummer (1982) worked with cleavage fragments, using fixed and free-drift pH experiments, and were unable to come closer to saturation than $SI_d = -3$. Others have used reactor columns, spinning discs or scanning interferometry. Sample findings are shown in Figure 3.20, where it is seen that kinetic

behaviour is much the same as that of calcite in very acidic, H^+ Dependent conditions. There are sadly few data for pH 5.0–8.9, the normal range in karst waters. The dissolution rates there are more than one order of magnitude slower than the calcite and aragonite rates.

Busenberg and Plummer (1982) proposed that dolomite dissolution proceeds in two steps. The first is reaction of the $CaCO_3$ component with H^+, $H_2CO_3^0$ and H_2O as in equation (3.74). The second sees the same reactions occurring much more slowly with the $MgCO_3$ component: this is rate-controlling. As solute concentrations increase (but still at SI_d below -3 or 0.1%) there begins to be

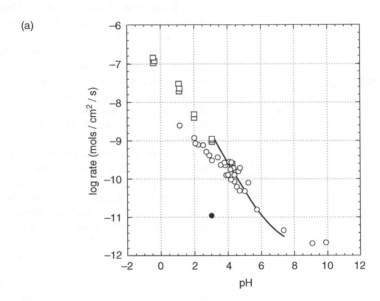

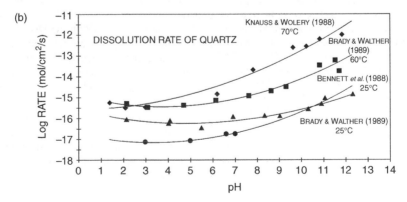

Figure 3.20 (a) Dolomite dissolution rates. The solid curve is the theoretical equation of Chou et al. (1989); open circles and squares, results of selected bulk dissolution experiments; filled circle, mean results from vertical scanning interferometry. Reproduced from Morse, J.W. and Arvidson, R.S. (2002) The dissolution kinetics of major sedimentary carbonate minerals. Earth Science Reviews, 58, 51–84. (b) Some experimental dissolution rates obtained for quartz. Reproduced with permission from Mecchia, M. and Piccini, L., Hydrogeology and SiO_2 geochemistry of the Aonda Cave System, Auyan-Tepui, Bolivar, Venezuela. Bollettino Sociedad Venezolana Espeleologia, 33; 1–18, 1999.

significant back reaction of HCO_3^-. It is being adsorbed onto positively charged sites (i.e. protruberances of Ca^{2+} and Mg^{2+}) at the surface. This explains the exponential slowing of the dissolution rate. Their comprehensive kinetic equation is of the same form as that for calcite:

$$\lambda = k_1 \cdot a_{H^+}^n + k_2 \cdot a_{H_2CO_3^0}^n + k_3 \cdot a_{H_2O}^n + k_4 \cdot a_{HCO_3^-}$$

$$(3.76)$$

$n = 0.5$ at temperatures below $45°C$, i.e. the square root of activity in a double carbonate.

We can now propose a mental picture of what happens at the solid surfaces when dolomites and Mg-calcites dissolve. Ideal (stoichiometric) dolomite comprises $CaCO_3$ and $MgCO_3$ molecules in alternate layers. A layer of $CaCO_3$ is quickly peeled away by dissolution. Magnesium carbonate is more strongly bonded. Its exposed layer resists dissolution while any residual pinnacles of $CaCO_3$ plus $MgCO_3$ protruding from it attract HCO_3^- ions in back reaction. As the density of lattice defects increases so do the opportunities to 'unravel' the crystal by following screw dislocations that breach its $MgCO_3$ layers.

Mg-calcite is a solid solution. Molecules of $MgCO_3$ are scattered from place to place in the $CaCO_3$ lattice; there is no regular alternation of layers. In 'low Mg-calcite' the greater resistance of the few $MgCO_3$ clumps creates many steps and kinks in calcite about them. Low Mg-calcite is more soluble than pure calcite as a consequence. In 'high Mg-calcite' the density of clumps may become too great; the step and kink forward effects are overwhelmed by the back reaction of HCO_3^- adsorbing onto protruberances. Most field researchers have found that high Mg-calcite is less soluble than pure calcite, although it is still more soluble than true dolomite in most instances.

Using similar experimental designs, the dissolution rates of quartz (the predominant mineral in sandstone and quartzite karst) are found to be much lower than those of dolomite, $10^{-15}\,mol\,cm^{-2}\,s^{-1}$ or less at $25°C$ (Figure 3.20b; from Mecchia and Piccini 1999). At hydrothermal spring temperatures (e.g. $70°C$ in the figure) they begin to approach the lowest dolomite rates.

3.10.3 The dissolution kinetics of salt, gypsum and anhydrite

These minerals are very soluble, dissolving by molecular dissociation alone. The dissolution rates thus are expected to be transport-controlled by diffusion out through the DBL, with possible surface effects appearing close to saturation.

Alkattan *et al.* (1997) studied salt dissolution with spinning disc experiments, obtaining a surface rate constant, k_{ss}, of $6.7 \times 10^{-1}\,mmol\,cm^{-2}\,s^{-1}$ in a standard rate equation of the form of equation (3.74). Trace amounts of Co, Cd, Cr and Pb were found to be effective inhibitors but Fe and Zn were not. Given the purity of natural salt bodies, these inhibitors can be ignored in most cases. In their absence, the dissolution rate is in linear relation to salt concentration in the bulk solution, i.e. $n = 1$ in the rate equation.

In the caves and karst of the Mount Sedom salt dome, Israel, Frumkin (2000a,b) has found that laminar flow films on shaft walls become saturated within a few minutes. When turbulent flood waters sweep through the caves, measured channel floor recession rates are as great as $0.2\,mm\,s^{-1}$, in good agreement with the rate constant given above (Frumkin and Ford, 1995).

In similar experiments with gypsum, experimenters such as Jeschke *et al.* (2001) found dissolution of natural gypsum (including selenite crystals, which have relatively few imperfections) to be in nearly linear relation to bulk solute concentration up to ∼60% saturation, following which mixed kinetic effects began to become significant (i.e. Regime 2). They obtained a surface rate constant, k_{sg}, of ∼1.1 to $1.3 \times 10^{-4}\,mmol\,cm^{-2}\,s^{-1}$. In laminar flow films approximately 1 mm in thickness dissolution rates were ∼$3 \times 10^{-6}\,mmol\,cm^{-2}\,s^{-1}$, illustrating the diffusion control. Experiments with pure synthetic gypsum found a linear rate law up to equilibrium.

At and above 94% saturation in natural gypsum, rates of dissolution dropped abruptly as surface effects came to dominate the process. A value of $n = 4.5$ was obtained, nearly identical to the result frequently found in experiments with calcite crystals.

The dissolution of anhydrite is approximately 20 times slower than that of gypsum, essentially because the latter is precipitating onto the anhydrite surface as molecules dissociate there. This can be considered to be a surface control effect that will apply until a given molecular layer has been converted, which can then be unravelled at the gypsum rate. The surface rate constant quoted for anhydrite is $k_{sa} = 6 \times 10^{-6}\,mmol\,cm^{-2}\,s^{-1}$.

Large but differing increases in the gypsum rate constants are seen where there are significant amounts of salt (e.g. $10\,g\,L^{-1}$) in solution. This is attributed to the ionic strength effect (Figure 3.12). It can be expected to be an important booster where salt beds are intercalated with anhydrite or gypsum.

3.10.4 Penetration distances and breakthrough times in limestone, dolomite and gypsum

As emphasized in Chapter 1, the essence of the karst system is that its water (and thus, energy) flows are routed underground in channels created by dissolutional widening of fissures, rather than being discharged over the Earth's surface. Fissure apertures are initially very small, offering high resistance to flow; thus two important questions in dissolution kinetics are how far from a groundwater input point a fissure may be significantly enlarged in a given time? Or, conversely, how far can underground enlargement be extended in geologically feasible timespans?

Early penetration will occur under closed-system conditions with laminar flow that prohibits eddy diffusion. From some early experiments, Weyl (1958) believed that Ca^{2+} diffusion out through the DBL became the rate control of limestone dissolution as saturation was approached, and so introduced the concept of 'L_9 penetration distance' – the distance a given solution will flow through a capillary or crack before it is 90% saturated and, in his opinion, effectively incapable of further enlargement. Given reasonable values for initial apertures and acidity he found that it was no more than 1–2 m in most geological settings. This appears to prohibit most karstification because, while necessary penetration distances are only 1–2 m for many types of karren, minimum distances are 1–100 km for the majority of regional karst drainage systems. Weyl's argument was a factor in Bogli's (1964) advocacy of mixing corrosion effects as key boosters of solvent capability in fresh waters.

In a series of experiments that successfully induced dissolutional microconduits to extend and compete with one another in fissures in artificial gypsum (Plaster of Paris), Ewers (1973, 1978, 1982) showed that key cave-pattern building events occur when the dissolutional fronts of one or more conduits break through the full length of the high-resistance, unmodified fissure to reach an unimpeded point of discharge such as a spring or earlier cave; those conduits enlarge rapidly and attract the flow in other conduits towards them (see Chapter 7.2). White (1977a) coupled this concept of hydrodynamic breakthrough with the new findings of Berner and Morse (1974) and Plummer and Wigley (1976) on rate controls of calcite dissolution, to invert Weyl's (1958) argument: **kinetic breakthrough** occurs when a fissure or conduit within it becomes sufficiently enlarged to permit Regime 2 Mixed Kinetics to operate throughout its length. Further enlargement to permit turbulent flow with eddy diffusion may then follow rapidly.

Since 1977 there has been much computer modelling of breakthrough and conduit development; Dreybrodt and Gabrovšek (2000b, 2002) and Palmer (1991, 2000) give comprehensive reviews, and the findings are further developed in Chapter 7.2. To illustrate the key importance of kinetic considerations in karst system evolution

in the carbonate and sulphate rocks here, we present only the most simple modelling case, using recent results of Dreybrodt and his associates because they have been in the forefront of such work.

The simple case is that of the 'parallel plate' fracture, i.e. the walls are straight and one fixed distance apart, as in an idealized bedrock joint. In the example given in Figure 3.21, the rock is pure calcite, the fracture is 1000 m in length and 1 m in height or breadth. Its initial aperture is 200 µm, a popular value in modelling. There is a pressure head of 50 m at the input end, giving a hydraulic gradient of 0.05, a rather high value for many karst settings. At the input end the water is in equilibrium with a P_{CO_2} of $\sim 2.10^{-2.5}$ (a common value in forest soils) and has not yet acquired any Ca^{2+}. The system is closed to further addition of CO_2 thereafter.

Figure 3.21a shows the dissolutional widening that occurs in model time. At time 1 (curve 1) 100 year after flow commences, effects are negligible. At time 2 ($\sim 13\,000$ year) the fracture is 0.5 m wide at the upstream end but tapers rapidly to < 1.0 mm at 200 m downstream. At the downstream end, 1000 m, the aperture has increased to ~ 300 µm but the fracture remains a high resistance, Regime 3 element. At time 4 (18 850 year) the downstream aperture is ~ 550 µm and kinetic breakthrough is about to occur. Within the next 150 year (to time 5) the aperture there increases $\times 10$ and the amount of water discharged from the fracture increases $\times 100$. The model terminates at 19 150 year (time 9), when the fracture has become a conduit 20 cm wide at the downstream end. Its discharge is then more than five orders of magnitude greater than at the start (Figure 3.21b) and is probably quite capable of handling the flow of great storms, i.e. the system is fully karstic. Figure 3.21c shows the evolution of saturation along the fracture (Weyl's (1958) L_9 penetration distance). As late as time 4 waters are $> 90\%$ saturated after 400 m of travel; when breakthrough is complete they are $< 3\%$ saturated at 1000 m. Evolution of the hydrostatic head behind the remaining high-resistance element in the fracture is close to a mirror image of the advancing front of undersaturation. In this particular model realization the breakthrough events all took place between 18 850 and 19 015 year after beginning, a dramatic demonstration of a kinetic trigger.

The model assumed zero dissolved Ca^{2+} at the input end. Figure 3.21d shows the effect for the same model conditions if some limestone has already been dissolved before the water enters the fracture. In cases 1, 2 and 3 the waters are, respectively, 0, 50% and 75% saturated at the input: it is seen that their evolution over time is very similar despite this large range in initial dissolved

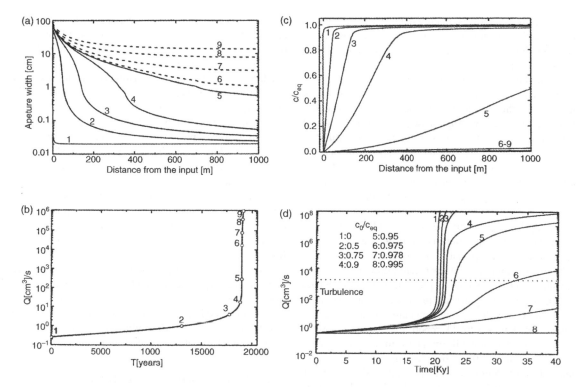

Figure 3.21 Computer modelling of the dissolutional penetration of a single, parallel plate, fracture. The fracture is initially 200 μm in aperture, 1 m in height or width, and 1000 m in length; hydraulic head is 0.05 and $P_{CO_2} = 10^{-2.5}$. (a) Curves showing dissolutional widening at nine different times between 100 year and 19 150 year after commencing the model run. Kinetic breakthrough occurs between times 4 and 5. (b) Discharge from the downstream end of the fissure over time. Numbers give temporal positions of curves 1–9 in (a) and (c). (c) Time curves for evolution of the saturation ratio. (d) Modelling the same initial conditions but with differing values of the saturation ratio at the input end. Curve 1 = 0 saturation (as in (a)–(c); curve 2 = 0.5 (50%) saturation, etc. See the text for further details. Reproduced from Dreybrodt, W., and Gabrovŝek, F, 'Basic processes and mechanisms governing the evolution of karst'. In Gabrovŝek, F. (ed). Evolution of Karst: from Prekarst to Cessation. Postojna-Lubljana, Carsologica. 115–154 © 2002 Zalozhba.

load. Where Regime 3 kinetics (> 90%) apply even at the input point, however, quite minor increases in the dissolved load can have dramatic effects. The water is 97.5% saturated in case 6 and 99.5% in case 8. After 40 000 year discharge in case 6 is nearly 10^5 times greater and karstic flow has developed; it may never develop in case 8 because surface erosion processes might remove the rock before breakthrough could be completed.

The Regime 3 rates of evolution of Figure 3.21d are most applicable in deeply circulating groundwaters, which emphasizes the potential significance over long periods of quite minor variations in their initial composition. In surficial karst, before passing underground many streams will be > 90% saturated in most weather conditions; effective dissolution occurs during storms, when SI_c generally falls below 0.01.

Breakthrough times will be significantly longer in dolomite but 10–100 times shorter in gypsum.

3.10.4 The precipitation of calcite

Calcite, in the form of travertine and tufa at the surface and speleothems underground, is the dominant karst precipitate. Its kinetic factors prove to be many and complex, and are the focus of much ongoing research in the field and laboratory.

The basic equation describing precipitation of any mineral from a solution may be written

$$R = \alpha(C - C_{eq}) \tag{3.77}$$

where $C > C_{eq}$ and α is a rate constant governed by kinetic factors. By definition a solution at thermodynamic equilibrium is in its lowest energy state. An energy barrier (the nucleation potential) therefore has to be overcome before precipitation can begin. As a consequence, in the case of calcite precipitation need not commence as soon as a solution is carried above $SI_C = 0.00$.

Homogeneous precipitation can occur in the bulk liquid, with the formation of clusters of molecules. These are enlarged by regular or irregular accretion of further molecules to form a microcrystal (or **crystallite**). Crystallites 'ripen' by regular accretion to become crystals that settle out of the liquid. The nucleation potential is high; SI_C values above $+1.5$ are required before there is any significant production. These are readily achieved in some lagoonal seawaters, etc. but more rarely in fresh waters (see Zhong and Mucci, 1993). Homogeneous precipitation is of little significance in karst studies except, possibly, in some marl lakes.

The nucleation potential for precipitation onto existing solid surfaces (heterogeneous precipitation) is much less. As a consequence significant precipitation may begin at $SI_C = +0.30$ (or lower in optimum circumstances – Contos and James (2001) report formation of calcite rafts around dust nuclei on pools in Jenolan Caves, Australia, at $SI_c \sim +0.05$, but this appears to be exceptional). Fastest deposition occurs where $SI_c > +1.0$. In heterogeneous nucleation ions, molecules and ion pairs diffuse into the adsorption layer where they are adsorbed directly at steps, kinks and other dislocations in the lattice. Slow adsorption builds the most regular crystals, layer by layer, by 2D or higher order nucleation (ion positions 2 and greater in Figure 3.18a). If the degree of supersaturation is increased by processes such as evaporation, there may be a simultaneous formation of clusters of crystallites and even small crystals within the boundary layer. These then attach to the substrate by **adhesive growth** (position 1 in Figure 3.18a) to create a solid that is less regularly ordered and of higher porosity. It lacks the lustre of crystalline surfaces, may crumble in the fingers, and is often described as 'earthy'. The fastest precipitation occurs onto earlier calcite because of its best lattice match. Micro-organisms can be important nuclei. Pentecost (1994) identified 12 different types of cyanobacteria that can live on accreting tufa surfaces. They serve as framework for the depositional microfabric, and may accelerate deposition by extracting CO_2 from the solution. Other micro-organisms can retard accumulation by respiring CO_2; at the famous travertine terraces of Huanglong, China, biofilms with diatoms reduce accumulation rates by $\sim 40\%$ where they are present (Lu *et al.* 2000).

The calcite lattice is quite robust, able to adsorb a wide variety of foreign ions and molecules without becoming totally disordered. For example, large uranyl ions, UO_2^{2+}, can be adsorbed; see Chapter 8.6. Humic and fulvic acids (long-chain molecules with atomic weights up to 30,000 daltons) can be taken up in cave calcite; they furnish much of its variety of colours. However, as in the case of dissolution, other substances inhibit precipitation by attaching to a step or kink, blocking further accretion there. They are often termed 'poisons' and include some micro-organisms, phosphates and trace metals. Most important is Mg^{2+}, which strongly inhibits calcite when present in high molar ratio. It does not adsorb onto the aragonite lattice to inhibit that polymorph, however, which is why much aragonite is precipitated from Mg-rich seawater. Some freshwater aragonite precipitation occurs where first there is deposition of gypsum to deplete Ca^{2+} and so increase the molar proportion of Mg^{2+}.

Dreybrodt and colleagues have undertaken recent theoretical and experimental studies of calcite deposition from fresh waters, with special reference to conditions in vadose (air-filled) caves; see Dreybrodt and Buhmann (1991) and Baker *et al.* (1998), for full reviews. A modified PWP equation is found to be appropriate:

$$R = k_1(H^+) + k_2(H_2CO_3^0) + k_3(H_2O)$$
$$- f.k_4(Ca^{2+})(HCO_3^-) \tag{3.78}$$

The rate control here is the back reaction, k_4, because there is net deposition; f is a factor < 1.0 to reduce it, shifting the equilibrium (minimum) Ca^{2+} concentrations necessary to higher values in response to varying kinetic controls of precipitation. Three semi-independent controls may operate:

1. surface effects, giving $f = 0.8$ on most natural crystal surfaces, falling to as low as $f = 0.5$ where the solution is flowing through fine-grained porous materials such as silt;
2. conversion of HCO_3^- and H^+ into CO_2 and H_2O, because the quantity of calcite precipitated must be balanced by CO_2 released – where V/A ratios are small (as they are in thin films on flowstones) this CO_2 production may be rate-controlling;
3. molecular diffusion of Ca^{2+}, CO_3^{2-} and HCO_3^- across the DBL.

Figure 3.22 presents the resulting growth models. In Figure 3.22a the model is for calcite being deposited from a static film of water. Rates of deposition $(mm\,a^{-1})$ are determined by Ca^{2+} concentration in the bulk solution $(mmol\,L^{-1})$, by temperature of the reactions (the cases of

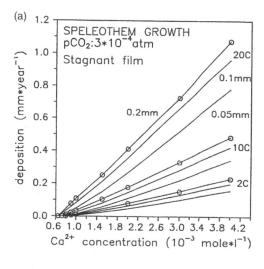

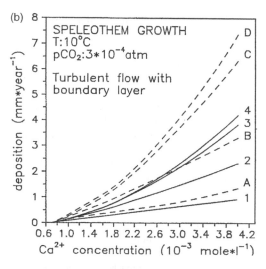

Figure 3.22 Modelling rates of deposition of calcite on speleothems. (a) From stagnant films of water; curves are for three different film thicknesses at three different temperatures. (b) From turbulent flow. Solid curves are for deposition from turbulent flow with four differing depths but a DBL fixed at 0.1 mm thickness; dashed curves for the same four depths but DBL reduced to 0.05 mm. See the text for further details. Reproduced from Dreybrodt, W. and Buhmann, D. 'A mass transfer model for dissolution and precipitation of calcite from solutions in turbulent motion'. *Chemical Geology*, 90; 107–122 © 1991, Elsevier.

2, 10 and 20°C are plotted) and by film thickness. From bottom to top, the three solid lines in each temperature cluster are for film thicknesses of 0.05, 0.1 and 0.2 mm respectively; circles plot rates for a film thickness of 0.4 mm. It is seen that, as thickness increases to ~0.2 mm, the greater number of Ca^{2+} ions made available because of the thickening are an important control at all concentrations and temperatures, but not thereafter.

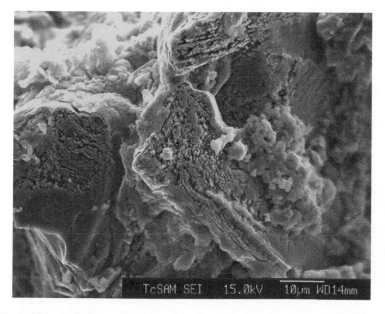

Figure 3.23 (Left) Scanning electron microscope image showing experimental dissolution of rhombohedral calcite. 10 μm scale bar. (Photograph by Henry Chafetz, with permission.) (Right) Rhombohedral calcite being precipitated to form the tip of a straw stalactite. It is fed by a capillary from a helictite erratic. The droplet is ~5 mm in diameter. (Photograph by Patrick Cabrol, with permission).

Figure 3.22b gives the model for turbulent flow, which will be more common in caves: $T = 10°C$ and pCO_2 is the same as Figure 3.22a. The DBL (where the slow molecular diffusion rates apply) is much thinner than in the static case and rates of deposition become much larger as a consequence. The solid lines are for a DBL thickness of 0.1 mm, with flow depths of 0.1 cm (1), 1.0 cm (2), 10 cm (3) and 100 cm (4). Dashed lines, A, B, C and D, are for this same sequence of depths but with a DBL only 0.05 mm in thickness. Flow depths as great as 10–100 cm are rarely achieved in streams depositing calcite in caves, so rates at 0.1–1.0 cm depths in the figure should be taken as more typical. Figure 3.23 shows an extreme instance of a rhomb growing in a water droplet beneath a calcite capillary tube in a speleothem.

Field testing of these models is in progress. For karst situations a truly static film of water is an artificial concept, but it may be approached in condensation films or in stranded ponding between flood events. For turbulent flow Baker and Smart (1995) found the model overpredicting by two to five times in two caves in southwest England, but Baker *et al.* (1998) and Genty *et al.* (2001b) report R^2 correlations of $\sim$0.7 or better from later studies for sites ranging from northern Scotland to southern France; there is minor overprediction or underprediction in different situations.

Deposition of calcite below the watertable (i.e. in phreatic settings) is generally much slower. The lowest published rate that is confirmed by independent dating is from Wind Cave, South Dakota. The cave is a backwater in a thermal water flow system and has been draining at a mean rate of $\sim$0.4 m 1000 a^{-1} during the past half million years or so. Calcite deposition commences at a depth of approximately 70 m below the water table (7 atm). The rate equation is

$$R = \acute{\alpha} \times 10^{-7}(Z - Z_0)[\text{mmol cm}^{-2}\,\text{s}^{-1}] \qquad (3.79)$$

where $\acute{\alpha}$ is determined by DBL thickness (with an estimated maximum value of 1×10^{-5} cm s^{-1}), Z_0 is the pressure-limiting depth for precipitation (-70 m in this example) and Z is any lesser depth. Actual rates at -70 m were $\sim$0.00007 mm a^{-1} or less, increasing to 0.0008 mm a^{-1} at the water table. Equilibrium concentration in the solution at -70 m is Ca^{2+} = 2 mmol L^{-1}. The value of Z_0 elsewhere will depend on temperature, Ca^{2+} and foreign ion concentrations, and hydrodynamic factors (Ford *et al.* 1993).

4

Distribution and Rate of Karst Denudation

4.1 GLOBAL VARIATIONS IN THE SOLUTIONAL DENUDATION OF CARBONATE TERRAINS

The power of rainwater to dissolve karst rocks has been appreciated for over 200 year, as is evident from James Hutton's (1795) comments on solution forms on limestones in the Alps. Estimates of natural solution rates have been made since at least 1854, when by some extraordinary computation Bischof asserted that the annual dissolved $CaCO_3$ load of the River Rhine was equivalent to '332,539 millions of oysters of the usual size'! Credit for being the pioneer of modern methods for estimating solutional denudation must go to Spring and Prost (1883) in Belgium and Ewing (1885) in the USA. After 366 days of sampling, Spring and Prost determined the annual solute load of the River Meuse at Liege to be 1 081 844 t, whereas Ewing calculated limestone denudation in a river basin in Pennsylvania to be equivalent to 1 ft in 9000 year ($34 \, mm \, ka^{-1}$). These figures are of the same order as more recent estimates in the regions concerned.

In the 1950s the French geomorphologist Jean Corbel made a great impact on conventional thinking concerning chemical denudation when he published results derived from analyses of thousands of field samples. He concluded (i) that cold high mountains provide the most favourable environment for limestone solution and (ii) that there is a factor of ten in the difference of solution rates between cold and hot regions for a given annual rainfall, hot regions having the lowest karst solution rates (Corbel 1959). He inferred from this that the principal control on solutional denudation is temperature, probably operating through its inverse effect on the solubility of carbon dioxide (section 3.4).

These conclusions really rocked the boat. They ran contrary to both morphological evidence and conventional wisdom – that weathering processes in general are most rapid in hot humid conditions. Corbel's findings were strongly disputed and so stimulated numerous other process studies on karst. The published results of some 200 later investigations were synthesized by Priesnitz (1974), who found a significant, positive relationship ($r = 0.74$) between the rate of limestone solution and the amount of runoff. This confirmed conclusions by Pulina (1971), who had identified a linear relationship between chemical denudation and precipitation, and Gams (1972), who showed solution denudation rates in Slovenia to be dependent on runoff. Bakalowicz (1992) has since reinforced this conclusion.

Smith and Atkinson (1976) used two data sets: 134 estimates of the rate of solutional denudation in different regions of the world and 231 reports on the mean hardness of spring and river waters. They confirmed the above conclusions and added more detail and critical interpretation. Runoff probably accounts for 50–77% of the variation in total solution rates, the remaining variation being mainly accounted for by solute concentration. Smith and Atkinson supported Corbel's hypothesis in that they found the greatest solutional denudation to occur in alpine and cold temperate regions, but they stressed that the climatic effect is much less marked than Corbel claimed. For a given runoff, it appears to involve an increase not of ten times between the tropics and the cool temperate alpine zones, but only 36%. The greatest rate of solutional denudation of limestone in the world occurs where it is wettest. Hence precipitation rather than temperature is the principal control. On marble in coastal Patagonia at 51°S where precipitation is about $6600 \, mm \, year^{-1}$, Maire (1999) estimated dissolutional denudation to be of the order of $160 \, mm \, ka^{-1}$ of which 25–37% contributes directly to lowering of surface outcrops. This compares with his earlier estimates from New Britain (6°S) in Papua New Guinea from five areas with rainfall ranging from 5700 to

Karst Hydrogeology and Geomorphology, Derek Ford and Paul Williams
© 2007 John Wiley & Sons, Ltd

$12000\,mm\,a^{-1}$, where he calculated solutional denudation rates of $270-760\,mm\,ka^{-1}$ (Maire 1981b).

In an investigation of the factors responsible for the observed variability in limestone denudation rates, Bakalowicz (1992) concluded that whereas climate is important because it determines water surplus and the amount of CO_2 production, it is not necessary the overriding factor. Geological and morphological conditions are also significant because they can considerably increase water transmission through karst (by encouraging surface flow to divert underground); and regional evolution and neotectonics can create conditions that can profoundly modify P_{CO_2} (e.g. a thick cover of sediments can augment groundwater CO_2 by limiting air exchange at depth, whereas significant porosity inherited from previous phases of karstification can reduce it because it enables vigorous air exchange). Under some conditions climate-independent processes related to hypogenic sources of acidity are the main factors promoting subterranean karst development, such as the S–O–H system (section 3.6).

4.1.1 Distinction between solution rates and denudation rates

The term 'solution rate' as widely used in geomorphological literature is ambiguous. The solubilities of various minerals in water are listed in Table 3.1, in which the values refer to the maximum concentration that can be achieved under a particular set of conditions given unlimited time. They represent an equilibrium state but tell us nothing about the rate at which that equilibrium was reached. This is determined by the dissolution kinetics or reaction dynamics and can be visualized as the slope of a curve plotting solute concentration or state of saturation against time (Figure 3.3). This is the meaning of **rate of solution** as understood by chemists and is to be preferred to the looser usage by geomorphologists, who also take it to mean the annual rate of chemical denudation, i.e. the **solutional denudation rate**. This latter must also be distinguished from the **karst denudation rate**, which is the sum of both chemical and mechanical erosion processes. However, since in practice it is much easier to estimate chemical than mechanical erosion, only solutional denudation rates are usually considered. This ignores an often significant part of the picture and so must be conceded as a major deficiency in karst process research.

Karst erosion occurs underground as well as on slopes and is a surface-area-dependent process, but in karst it is difficult to measure the 'true' contact surface between water and the rock (Lauritzen 1990); thus while denudation rates would be best expressed in units of volume/ volume $(m^3\,km^{-3}\,a^{-1})$, they are more commonly expressed in units of volume/area $(m^3\,km^{-2}\,a^{-1})$, which by convention and for ease of comparison are quoted as an equivalent thickness of rock removed per unit time across a horizontal surface. Millimetres per thousand years $(mm\,ka^{-1})$ are most commonly used ($1\,mm\,ka^{-1}$ is equivalent to $1\,m^3\,km^{-2}\,a^{-1}$). However, to generalize a mass transfer rate calculated for 1 year to an equivalent rate per 1000 year is only justified if there has been no significant change in environmental conditions in that period. This is an increasingly untenable assumption given the intensity of human impact on the ecosystem, although natural conditions for the past 6000 year or so can be considered metastable.

4.1.2 Separation of autogenic, allogenic and mixed denudation systems

In interpreting results of karst erosion studies, it is important to know from where the water and its load has come. An **autogenic** (or autochthonous) system is one composed entirely of karst rocks and derives its water only from that precipitated onto them (Figure 4.1a). By contrast, a purely **allogenic** (or allochthonous) system derives its water entirely from that running off a neighbouring non-karst catchment area. In practice, many if not most karst systems have a mixture of autogenic and allogenic components (Figure 4.1c). Lauritzen (1990) explained how the contribution from each may be separated in a mixed-lithology basin and the autogenic component computed.

Allogenic waters flowing into a karst area represent an import of energy capable of both chemical and mechanical work. Thus at the output boundary the autogenic and allogenic components of denudation must be separated if sense is to be made of the landform development and if valid comparisons of denudation rates are to be made with other areas. Clearly a small karst area with a massive throughput of allogenic water (such as Mulu in Borneo) will experience very much more erosion than it would if it operated only as an autogenic system. In the Riwaka basin, New Zealand, autogenic solution is about $79\,mm\,ka^{-1}$ but karst rocks cover only 46.6% of the catchment. The large allogenic input increases net solution of this karst by over 20% (Table 4.1). Meaningful comparisons of solution denudation rates from different areas thus can be made only when the relative proportions of carbonate rock are taken into account. Pitty's (1968a) and Bakalowicz's (1992) results are particularly illuminating in this respect and show that if the proportion of limestone in a basin is reduced from 100% to 50% then the specific dissolution can increase by about 60% (Figure 4.2), assuming that the

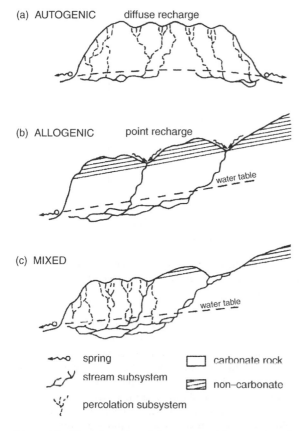

(a) AUTOGENIC   diffuse recharge

(b) ALLOGENIC   point recharge

water table

(c) MIXED

water table

⊶○ spring   ▭ carbonate rock

↱↲ stream subsystem   ▱ non–carbonate

ↆↆ percolation subsystem

Figure 4.1 Three karst denudation systems: (a) autogenic and (b) allogenic are end-members with (c) the mixed autogenic–allogenic intermediate case being the most common. Reproduced from Ford, D.C. and Williams, P.W. (1989) Karst Geomorphology and Hydrology.

non-karst rock is on the upstream side of the basin and a source of allogenic runoff.

4.1.3 Distinction between gross and net solution

Published estimates of karst denudation are not always consistent in what they represent. The reason for this is that solute load may be derived from various sources, some non-karstic (including pollution sources), and some reprecipitation of previously dissolved materials may have occurred upstream of the sampling site. The total solute discharge is the product of river discharge Q at the outflow of the basin and the corresponding solute concentration C. Each has a value that comprises the relative contributions made by its different components. Thus

$$Q = (P - E)_{\text{autogenic}} + (P - E)_{\text{allogenic}} \pm \Delta S \qquad (4.1)$$

where P is precipitation, E is evapotranspiration and ΔS is change in storage. $(P - E)_{\text{autogenic}}$ is karst runoff and $(P - E)_{\text{allogenic}}$ is non-karst runoff. Salts introduced by precipitation onto autogenic and allogenic components contribute to solute load at the outflow as do solutes derived from the corrosion of non-karst rocks by allogenic runoff. Therefore

$$
\begin{aligned}
\text{solute load} = {} & (\text{autogenic} + \text{allogenic karst corrosion} \\
& - \text{karst deposition}) + \text{allogenic non karst} \\
& \text{corrosion} + \text{solute load from rain,} \\
& \text{snowfall and atmospheric pollution}
\end{aligned}
$$

Gross karst solution comprises autogenic solution plus solution of karst by allogenic waters. Subtraction of

Table 4.1 Sources of solute load in the Riwaka basin, New Zealand (Reproduced with permission from Williams, P.W. and Dowling, R.K., Solution of marble in the karst of the Pikikiruna Range, northwest Nelson, New Zealand. Earth Surf. Proc. 4, B1010 15–36 © 1979 John Wiley and Sons)

Source of Ca + Mg load	Tonnes per year	Percentage of karst solution	Percentage of total solute load
From solution of marble by autogenic waters	1709*	79.5*	68
From solution of marble by allogenic waters	440**	20.5**	17.5
Net karst solution	2149†	100	85.5
From solution of non-karst rocks	250	—	9.9
Introduced by precipitation	116	—	4.6
Total solute load	2515	—	100

*Mainly surface lowering.
**Mainly cave conduit development.
†Equivalent to marble removal rate of $100 \pm 24 \, \text{m}^3 \, \text{km}^{-2} \, \text{a}^{-1}$.

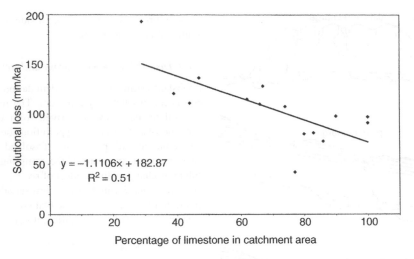

Figure 4.2   Relatonship of solutional denudation rates to the percentage of limestone in a catchment. Data from England (Pitty 1968a) and France (Bakalowicz 1992). Part of the scatter is attributable to the position of the non-karst rocks in the catchment. Solutional loss will increase where these are in the headwaters of the catchment.

karst deposition (speleothems and tufa) yields net karst solution. To gain a realistic appraisal of the rate of transformation of relief, estimates are required of gross karst solution. Where repreciptation is not important, as in most cold climates, net solution approximates the gross. But in tropical and warm temperate regimes speleothem and tufa deposition can be profuse; so net solution can significantly underestimate gross rates. Solutional denudation will be overestimated where the proportion of solutes introduced by rainfall and non-karst rocks is not subtracted. In arid regions the introduction of carbonate in dew and dust by airborne fallout may also complicate the picture (Gerson 1976).

4.1.4 Factors influencing global variations in net autogenic solution

In the presence of water, there is no dynamic threshold of calcite or dolomite solution (Ford 1980). Solutional denudation rates have also been found to depend linearly on runoff. Hence it might be supposed that because runoff values form a worldwide continuum there can be no regional discontinuities in solutional denudation. This would be so if variations in solute concentrations are minimal. This is not always the case, as illustrated for instance by the differing $CaCO_3$ values of groundwaters from various limestone lithologies in southern Britain (e.g. Paterson 1979). The underlying reasons for variations in solute concentrations need closer scrutiny.

In section 3.7 we considered factors that boost or depress the solution of carbonate minerals and stressed

the importance of open and closed system conditions. Important variables superimposed upon system conditions that influence the percentage saturation of percolating groundwaters were identified as occurrence of carbonate in the soil, rooting depth, porosity, CO_2 concentration and availability, and the residence time of the water. The model presented in Figure 3.9 helps to explain why dissolved carbonate concentrations are often much higher in some regions than others, e.g. open-system conditions in carbonate-rich glacial tills in cold northern regions explain why very high carbonate values are recorded in groundwaters there despite relatively low P_{CO_2} values often found in the soils. Porosity includes the many lithological, mineralogical and structural characteristics of the rock that will affect the surface area exposed to corrosional attack, its solubility, and the flow through time of water. The term requires further refinement and probable redefinition before the influence of different lithologies on groundwater carbonate values can be explained adequately. Theoretical chemical considerations bearing on the solubility of rocks with different mineral assemblages were discussed in sections 3.3–3.7, and geological factors in sections 2.6–2.8.

The link between chemical and environmental factors in the solutional denudation of limestones was explored by White (1984), who developed a theoretical expression for their relation:

$$D_{max} = \frac{100}{\rho(4)^{\frac{1}{3}}} \left(\frac{K_C K_1 K_{CO_2}}{K_2} \right)^{\frac{1}{3}} P_{CO_2}^{\frac{1}{3}} (P - E) \qquad (4.2)$$

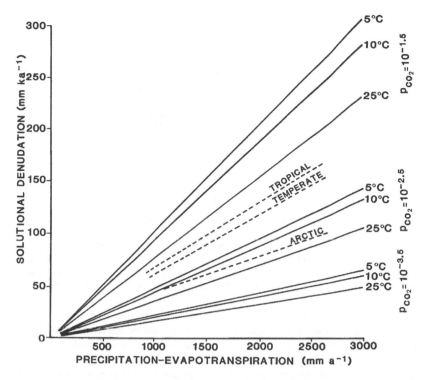

Figure 4.3   Theoretical relationships of solutional denudation of limestone to water surplus and CO_2 availability under open-system conditions. Based on White (1984) with empirical relationships derived from Smith and Atkinson (1976) superimposed as dashed lines. Reproduced from Ford, D.C. and Williams, P.W. (1989) Karst Geomorphology and Hydrology.

where D_{max} is the net autogenic solutional denudation rate (mm ka^{-1}) for the system at equilibrium, P is precipitation and E evapotranspiration (both in mm a^{-1}), ρ is rock density and K refers to the equilibrium constants cited in Table 3.5. All terms in the equation can be calculated. It combines in a single statement the rock and equilibria factors and the important climatic variables of precipitation and temperature. The equation expresses the linear variation of solutional denudation with runoff $(P - E)$, and indicates solution to vary with the cube root of P_{CO_2} the partial pressure of CO_2. Carbon dioxide volume in the atmosphere and soil varies from about 0.03% to 10% or more but, as White pointed out, because of the cube root dependence, a factor of 100 P_{CO_2} only admits a factor of five in the denudation rate. The complex effects of temperature on solutional denudation rates are incorporated in the equilibrium constants. White (1984) concluded that autogenic denudation increases by about 30% as mean temperatures decrease from 25° to 5°C, which broadly concurs with Smith and Atkinson's (1976) empirical findings. Temperature is thus the least important of the climatic variables in the equation and in practice is often more than offset by the influence of other factors.

Figure 4.3 illustrates the theoretical relationships embodied in equation (4.2), with empirical trends derived from field studies being superimposed on them. Further discussion of this point is available in White (2000).

The difference between the theoretical and observed trends of Figure 4.3 can be explained by several factors termed 'boosters and depressants' in section 3.7. First, as White pointed out, the theoretical curves assume that the water and carbonate rocks are in equilibrium, whereas in reality most karst waters are undersaturated; second, and more important, the model assumes ideal open system conditions, which often do not occur.

For a given water surplus and under open system and autogenic conditions, the effect of increasing latitude on solutional denudation in theory is to increase it, provided waters do not freeze. Increasing altitude will have a similar effect because P_{CO_2} decreases only marginally in the open atmosphere, up to elevations of 4000 m at least (Zhang 1997). However, the treeline provides an important threshold for soil CO_2 values that in turn affects calcium concentrations in groundwaters (Figure 4.4). Thus for a given runoff, solutional denudation tends to be greater below the tree-line than above it.

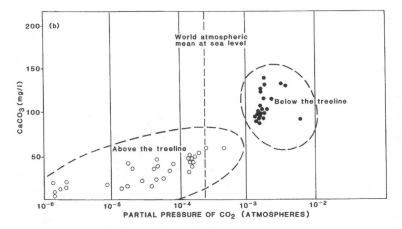

Figure 4.4   Hardness of waters from the Rocky Mountains plotted against P_{CO_2} in the gas phase in which they were equilibriated. Reproduced from Atkinson, T.C. and Smith, D.I. (1976) The erosion of limestones, in The Science of Speleology (eds T.D. Ford and C.H.D. Cullingford), Academic Press, London, pp. 151–177; after Ford, D.C. (1971a) Characteristics of limestone solution in the southern Rocky Mountains and Selkirk Mountains, Alberta and British Columbia. Canadian Journal of Earth Science, 8(6), 585–609.

In this chapter we discuss dissolution under present climatic conditions, but of course must remain cognizant of the fact that environmental conditions have changed considerably in the past, with major implications for karstification. Thus extrapolation of modern rates of dissolution into the past is fraught with difficulty. For example, over the past 60 Myear there is no doubt given fossil and isotopic records that there has been a steady decline in mean global temperatures and that atmospheric P_{CO_2} values have also changed considerably (the two probably being linked because of the greenhouse effect). Atmospheric P_{CO_2} was probably four times higher at 50 year than today, which is relevant for the rate of karst development during the Tertiary (and it was about 10 times higher in late Jurassic times, with implications for palaeokarst). During the Quaternary P_{CO_2} values have been relatively low, but have fluctuated 30% through glacial cycles, and on present Earth are in fact close to a minimum (Edmond and Huh 2003). The vigour of the hydrological cycle has also responded positively to changes in global temperature. Thus values of P_{CO_2}, P and E in equation (4.2) can be subject to considerable variation as one investigates the past.

4.2 MEASUREMENT AND CALCULATION OF SOLUTIONAL DENUDATION RATES

4.2.1 Objectives

Erosion rates are measured:

1. to obtain a generalized value for the overall rate of denudation or transformation of relief;

2. to compare denudation rates in contrasting environments and by different processes;

3. to gain more understanding of the evolution of landforms;

4. to understand the processes themselves;

5. to estimate the amount of carbon dioxide removed during limestone dissolution.

Where the principal objective is to obtain a number to compare with other karsts, perhaps in other climatic zones, then an estimate of the autogenic solutional denudation rate is required (for reasons discussed in the previous section). Where erosion by karst solution is being compared with erosion by mechanical processes, autogenic rates are also needed unless the role of allogenic water is being explicitly assessed.

The study of denudation in karst is best undertaken in a systems context, the karst drainage basin being treated as an open system with a definable boundary and identifiable inputs, throughputs and output. Secondary objectives being to define the system boundary (usually the watershed) and to measure the flux of solvent and solutes.

Hundreds of studies of solutional denudation have been completed since 1960. A major shortcoming of much of the work is that autogenic rates have often not been distinguished from mixed autogenic–allogenic rates so that there is still no unequivocal answer to the question posed long ago by climatic geomorphologists: in which climatic zone does karst evolve most rapidly? Wet regions are certainly more important than dry ones, but the field data are too 'noisy' to confirm with confidence what theory suggests (cf. Figure 4.3). Thus an important objective of solution studies over the next few years will be to compare autogenic rates in different environments with a view to clarifying this issue. At the same time it is necessary to recognize that high rates of solutional denudation do not necessarily imply rapid karstification, because that depends on where the dissolved rock comes from.

The value of denudational data is much enhanced therefore if information is available on where the erosion has taken place. The karst denudation system is three-dimensional, and if an understanding of the transformation of relief is to be acquired, the relative importance of sites contributing to the total denudation must be identified. However, to achieve a quantitative understanding of the spatial distribution of corrosion throughout the system requires a more sophisticated experimental design than that required merely to estimate total solutional denudation.

4.2.2 Experimental design and field installations

Here we assume that the objective is to measure the net autogenic solutional denudation rate in a karst basin. The most important measurements to characterize the hydrological system are rainfall input and streamflow output. Topographic complexity determines the number of rain gauges that are required to obtain an acceptable estimate of basin rainfall. Thiessen polygons or the isohyetal method can be used to calculate rainfall input (Dunne and Leopold 1978). The representativeness of the rainfall record may be assessed by comparison with long-term records from a nearby meteorological station. Under the very best conditions, the mean annual rainfall estimate is likely to have an error of at least 5%. Point measurements for individual storms can be in error up to 30%.

In many karst basins a convenient outfall point is a spring. Since the solute load (or flux) is the product of discharge and concentration both must be carefully measured. Discharge measurements are usually made in a straight channel reach a short distance downstream, but one must ensure that any discharge from underflow and overflow springs is included. The hydrological literature (e.g. Ward and Robinson 2000) provides guidance on site-selection criteria, appropriate measurement techniques and the associated errors. Ideally a solid rock channel is desirable to avoid cross-section change and to minimize leakage. Where a watertight weir or flume is used (Figure 4.5), discharge values of small streams can be accurate to 1%. Correlation of the outflow data with longer term records at nearby hydrological stations will permit assessment of the representativeness of the karst data set.

The need for secondary discharge stations depends on the objectives of the research and the nature of the

Figure 4.5 (a) A V-notch weir measuring the flow of the autogenic Cymru stream in Mangapohue Cave, New Zealand. (Photograph by J. Gunn.) (b) Rectangular section weir at Puding experimental karst basin, Guizhou Province, China.

catchment. Estimation of mean annual solutional denudation from an autogenic basin requires only a master discharge measurement site at the outfall. But in a mixed autogenic–allogenic basin, secondary sites are required to gauge the allogenic inputs. Where the spatial distribution of corrosion is being assessed, secondary sites are also desirable, but not always practicable if the system is entirely subterranean.

At the principal flow-measuring site discharge records should be continuous, e.g. with stage data accumulated by data logger (Figure 4.6). Equipment for the continuous measurement of some aspects of water quality (such as electrical conductivity) is available, but needs regular servicing, especially if flow becomes turbid and if calcite precipitation occurs. Continuous recording of electrical conductivity is valuable, because it often correlates highly with Ca^{2+} concentration (Fig. 3.15). But be warned: too

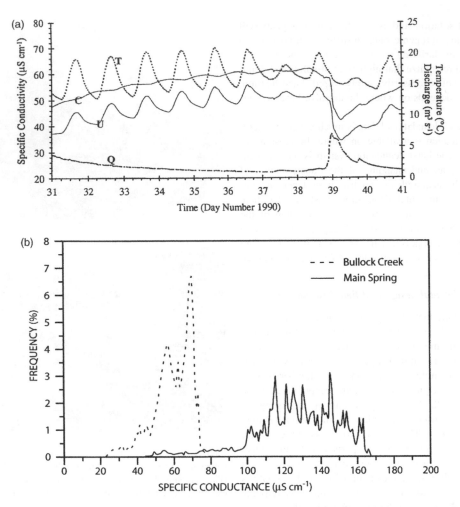

Figure 4.6 (a) Continuous records of discharge Q, specific conductivity U and temperature T of water entering Bullock Creek stream-sink (Taurus Major) in New Zealand. Note that the cyclic pattern of conductivity is suppressed after correction of C for diurnal temperature fluctuations. (b) Frequency distribution of the conductivity record at the Bullock Creek stream-sink compared with the frequency distribution at its main resurgence in Cave Creek. In the subterranean route to the spring, the allogenic flow of Bullock Creek is supplemented by about 30% extra autogenic drainage with much higher specific conductivity. Reproduced from Crawford, S.J. Hydrology and geomorphology of the Paparoa Karst, north Westland, New Zealand, 240 pp. Unpublished PhD thesis, Auckland University, New Zealand, 1994.

much reliance on automatic equipment without regular maintenance will inevitably mean data losses through power failures and other malfunctions. Conventional field sampling and laboratory analyses to check recorded values are essential. Appropriate field sampling and laboratory techniques are explained in detail by Krawczyk (1996).

A minimum requirement at the outflow site is to establish the relationship between solute concentration (Figure 3.15) and discharge. This is often achieved satisfactorily by constructing a rating curve, although the relationship of solute concentration to discharge frequently shows much scatter due, in particular, to pulses of hard water emerging at the beginning of storms (section 6.7). Errors in the rating curve also arise from the precision of the chemical techniques used to determine Ca^{2+}, etc., and from the representativeness of conditions when samples were taken, which must include a very wide range of flows in every season. The standard error of estimate on the regression assumes that the Ca^{2+} and discharge values are accurate, which may not be the case; thus a large number of samples is required to reduce the problem.

Sampling at regular intervals (e.g. weekly) seldom obtains the information needed, except over a very long period, because most flows encountered will be below mean discharge. Although continuous sampling provides the best source of water quality data, sampling based around storm-induced events is an acceptable fall-back alternative, because it focuses on sampling through the range of flows encountered prior to, during and after floods. Having constructed a statistically acceptable rating curve, one must then be alert to the probability that the general relationship may change from one year to the next (Douglas 1968), that it will not be equally valid for every season (Hellden 1973), and that there will probably be significant interannual variability (Bakalowicz 1992), i.e stationarity cannot be presumed. Thus continuous measurement over several hydrological years is essential for the derivation of reliable medium-term estimates.

4.2.3 Calculation of solutional denudation

The best known formula for calculating solutional denudation is that by Corbel (1956, 1957, 1959):

$$X = \frac{4ET}{100} \tag{4.3}$$

where X is the value of limestone solution ($m^3 \, km^{-2} \, a^{-1} \equiv mm \, ka^{-1}$), E is runoff (dm) and T is the average $CaCO_3$ content of the water ($mg \, L^{-1}$). This formula is important because it forms the basis of many of the earlier published figures of solutional denudation and its use provides insight into the methods used. The equation has been criticized for (i) assuming all carbonate rocks to have a density 2.5 (density can range from 1.5 to 2.9), (ii) ignoring $MgCO_3$ (although it would be included if total hardness was used for T) and solute accession from rainfall, (iii) ignoring the possibility that sulphate rocks contribute to Ca^{2+}, and (iv) for generalizing carbonate hardness to the mean value, thus overlooking its possible variation with flow. Nevertheless, when data are limited it is still an appropriate method for obtaining a first-order estimate of solutional denudation (e.g. see Ellaway *et al.* 1990).

Drake and Ford (1973) pointed out that a strict formulation of the solute discharge rate D could be expressed as

$$D = \frac{\int CQ dt}{\int dt} \tag{4.4}$$

where $C = C(t)$ is the solute concentration, $Q = Q(t)$ is the instantaneous runoff and t is time. This may be approximated by

$$D = \frac{\sum_{m} C_i Q_i}{m} \tag{4.5}$$

where i refers to equal time intervals in which C and Q may be considered constant, and m is the number of such periods. Since the sum of deviations of concentration and discharge from the mean is usually negative, Corbel's formula generally overestimates solutional denudation (e.g. Drake and Ford 1973, Schmidt 1979).

Very often karst rocks occupy only part of the basin being considered, in which case Corbel incorporated the fraction $1/n$ into his equation thus:

$$X = \frac{4ETn}{100} \tag{4.6}$$

However, because it is assumed that T is entirely derived from the carbonate rocks, another source of error will be introduced if some of the solute load has an allogenic origin. Using this approach, corrosion by autogenic and allogenic waters cannot be separated.

Modifications to Corbel's method by several authors resulted in only minor improvements, but significant increases in accuracy were achieved by applying mass-flux rating curves (Figure 4.7) to the flow duration curve (Williams 1970, Smith and Newson 1974, Schmidt 1979) or better still to the outflow hydrograph (Drake and Ford 1973, Julian *et al.* 1978, Gunn 1981a). Automatic equipment now permits accumulation of records of instantaneous solute concentration and discharge, and so enables annual solute flux to be determined from summation over the year. Deductions can then be made for non-karst inputs to give the net solutional denudation rate. Gunn (1981a) found that applying mass-flux rating curves to hourly discharge data and summing over a year yielded values 4% lower than those estimated from an equation using mean solute concentration (the equivalent of T in equation 4.3). This is because changes in solute concentration with discharge are not adequately taken into account in the more traditional methods which, incidentally, can also produce results differing by up to about 9% from the same data.

Errors in estimating solutional denudation can arise from many sources. The hydrological representativeness of a given sample year can be assessed by correlating runoff with rainfall data, the latter usually having by far the longer record. However, variations in seasonal distribution of rainfall from year to year can still yield differences in solute flux even if total annual rainfall

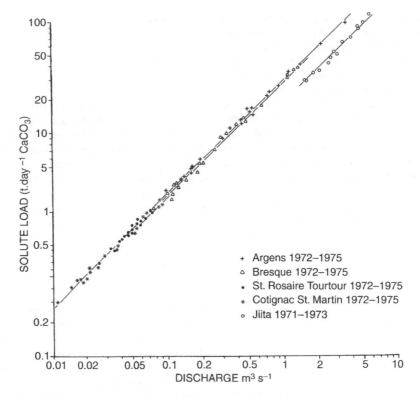

Figure 4.7 Mass-flux rating curves showing the relationships of equivalent $CaCO_3$ loads to discharge in karst springs in lower Provence, France, and Lebanon. The overlapping curves for the four French sites indicate the operation of a regional carbonate solution regime. Reproduced from Julian, M., J. Martin and J. Nicod, Les karsts Mediterraneens. Mediterranee 1 & 2, 115–31, 1978.

amount is the same. Table 4.2 illustrates potential errors in a carefully conducted study by Gunn (1981a). His best estimate of net autogenic limestone solution was $69\,m^3\,km^{-2}\,a^{-1}$, with qualitative assessment of potential errors indicating that the true value could lie between 61 and 88. The numerous sources of error that can occur in applying the Corbel technique suggest that published solutional denudation values derived by that method may be up to 100% or more in error.

In order to reduce the inherent complexities of natural systems and thus minimize potential sources of error, it is desirable to monitor entirely autogenic catchments. But geological circumstances do not always permit this, in which case autogenic denudation rates have to be deduced from mixed autogenic–allogenic systems. Should this be necessary, we recommend that a procedure similar to that described by Lauritzen (1990) should be followed. This involves a linear mixing model in which the apparent solutional denudation rates are determined from two (or more) sub-basins having different autogenic/allogenic area ratios (but otherwise similar

lithologies), and then from these the autogenic rate D_{auto} is deduced from the following equation:

$$D_{\text{auto}} = \frac{D_2 + D_1 + D_1 f_2 - D_2 f_1}{f_2 - f_1} \qquad (4.7)$$

where D_1 and D_2 are the apparent solutional denudation rates (i.e. autogenic plus allogenic solution) for each sub-basin and f_1 and f_2 are the respective fractions of the sub-basins underlain by carbonate rocks. Lauritzen illustrated this method using sub-basins from arctic Norway with annual precipitation of 2000–4000 mm, mean annual temperatures of 3 to 4°C and f values of 0.1 to 0.4. The rate of autogenic solutional denudation was determined to be $32.5 \pm 10.2\,mm\,ka^{-1}$.

4.2.4 Weight loss measurement using standard tablets

Another technique sometimes used to assess limestone solution rates is to measure the loss in weight of rock tablets of standard dimensions and lithology. Gams

Table 4.2 Data used in computing rates of solutional denudation in the autogenic Cymru basin, New Zealand (Reproduced with permission from Gunn, J, Limestone solution rates and processes in the Waitomo district, New Zealand. Earth Surf. Proc. Landforms 6, 427–445 © 1981 John Wiley and Sons)

Parameter	Measured value	Potential error (%)	Probable maximum value	Probable minimum value
Basin area (m^2)	95 350	+5, −2	100 117.5	93 443
Precipitation (mm) (water year total)	2366	±7.5	2544	2189
Discharge (L × 10^3) (water year total)	155 455	±5	163 227	147 682
Mean Ca concentration ($mg\,L^{-1}$)	48	±3	50	46
Mean Mg concentration ($mg\,L^{-1}$)	1.26	±3	1.3	1.2
Mean Ca concentration in rain ($mg\,L^{-1}$)	1.5		1.5	0.5
Mean Mg concentration in rain ($mg\,L^{-1}$)	0.32		0.56	0.32
Limestone density ($g\,cm^{-3}$)	2.66		2.66	2.5

(1981) summarized results of an international project using limestone tablets which was designed to determine:

1. the effect of different climates by means of suspending the tablets in the open air;
2. the rate of solution of bare limestone by laying the tablets on either rock or grass;
3. the corrosion rate in soil by burying the tablets at various depths;
4. the variability of corrosion within a single karst area by placing tablets in a variety of sites.

More than 1500 standardized tablets were distributed around the world. Results from nine countries were reported by Gams (1981, 1985) and a careful analysis of about 5 year data from Wisconsin was presented by Day (1984). Urushibara-Yoshino *et al.* (1999) reported results of 3 year measurements from Japan with data recorded as weight loss after different periods of exposure at the various sites, and Plan (2005) has reported data from Austria. The following international results are indicated:

1. Weight loss in tablets placed in the soil is generally greater than that from tablets in the air or on the surface and seems directly dependent on water surplus $(P - E)$ rather than on temperature.
2. Solution rates show a distinct climatic control, with generally higher rates being recorded in the humid tropics.
3. The effect of varying lithology is often greater than that of varying climate.
4. Mechanical erosion may be significant even in areas with high solution rates.
5. Solution of tablets on the surface is less at higher elevations than at lower elevations (at least in Slovenia and the French Alps).

6. In arid climates weight loss in the air is often higher than on the ground surface, although the opposite appears to hold for humid climates.
7. Weight loss at a site (in Slovenia) is considerably less than that expected from solution rates calculated from basin runoff and solute data.

This last point is reinforced by the findings of Crowther (1983), who concluded for a site in West Malaysia that rates derived from tablet weight loss are one to two orders of magnitude less than those calculated from water hardness and runoff data. Solution tablet data must therefore be interpreted cautiously. Results tend to confirm, rather than add to, our understanding of processes.

4.2.5 Short-term lowering of limestone surfaces determined by micro-erosion measurements

The micro-erosion meter (MEM) was developed by High and Hanna (1970) and improved by Coward (1975) and Trudgill *et al.* (1981). The instrument consists of a probe connected to a micrometer gauge and locks precisely into stainless steel studs set into the rock surface. Selected points in this surface can be repeatedly measured to evaluate erosional lowering. Results have been claimed accurate to the nearest 10^{-4} mm. Its use and limitations have been assessed by Spate *et al.* (1985), using a traversing version which permitted measurement of a large number of points within a triangular area of 12–200 cm^2.

Spate *et al.* (1985) undertook experiments in temperature-controlled rooms to assess three possible sources of error in the use of the traversing MEM. They found: (i) different instruments have differing temperature correction factors; (ii) differential expansion and contraction of the rock and rock–stud interface produced an apparent lowering of the surface with increasing temperature; and (iii) a considerable range in erosion of rock by the probe from point to point within a site (up to an order of magnitude). Instrument wear, particularly of the probe tip, is also a problem but no reliable data are available. Results corrected for the above factors from more than 4 year of measurements at 11 sites on Palaeozoic limestone surfaces in New South Wales, Australia (annual rainfall about 950 mm), showed a range of surface lowering rates of $0.000–0.020$ mm a^{-1}, averaging 0.007 mm a^{-1}, but with average errors of ± 0.011. Since in most cases the error term is of the same order as the solutional lowering rate, published results for other sites have to be treated with great caution. For example, the figures for bare-rock surfaces in the well-known karsts of Clare, Ireland and Yorkshire, England (Trudgill *et al.* 1981) are similar to those of New South Wales. However, Spate *et al.* (1985) concluded that the values reported for coastal sites are likely to be more reliable than those for inland sites because their rates are far greater and so the errors are not likely to be as significant. Trudgill (1976) and Spencer (1985) measured subtidal and inland rates on atolls in the range $0.1–1.8$ mm a^{-1} with a deep organic subsoil value reaching as much as 12.5 mm a^{-1}. Nevertheless, at inland sites long periods of observation should reduce uncertainties. Thus measurements over 15 year at more than 50 sites in northeastern Italy yielded average surface lowering rates of 0.02 mm a^{-1}, but with a range of $0.01–0.04$ mm a^{-1} depending on the petrographic characteristics of the carbonate rocks concerned (Cucchi and Forti 1994). Similar values were found in southeast Alaska (annual precipitation 1752–2540 mm) by Allred (2004). Bare-rock dissolution rates ranged from 0.03 mm a^{-1} under old-growth forests to 0.04 mm a^{-1} in alpine settings, but runoff from peat bogs produced dissolution rates up to 1.66 mm a^{-1}.

4.2.6 Long-term lowering of bare-rock surfaces determined by measurement of surface irregularities

Direct solutional lowering of bare-rock surfaces has been estimated from pedestals of limestone protected from corrosion by a non-carbonate caprock boulder that functions as an umbrella. These are sometimes known as **Karrentische**, and they are found on glacially scoured rock surfaces, the non-carbonate boulders being glacial erratics (Figure 4.8a). The height of the pedestal is a measure of the corrosion of the surface since the last glaciation. By this method Bogli (1961) estimated the rate of surface lowering in the Swiss Alps to be 1.51 cm $\pm 10\%$ per 1000 year and Peterson (1982) mea-

sured pedestal heights at 4300 m in the glaciated tropical mountains of West Irian (Figure 4.8b), where surface solution appears to have been twice as fast as in the Alps. In glaciated terrains currently below the treeline pedestals can be even taller. They have been measured up to 50 cm in northwest Yorkshire, England and to 51 cm in County Leitrim, Ireland (Figure 4.8c). However, recent detailed investigations using large sample populations have shown some earlier measurements to have considerably overestimated average pedestal heights, presumably because the more prominent and obvious pedestals were targeted. Thus in the Burren of County Clare, Ireland, average pedestal height has been shown to be about 60% of previous estimates (V. Williams pers. comm.) and in northern England only 10–40% of previous estimates (Goldie 2005). With the time since deglaciation now also known to have increased, this has considerably reduced estimated rates of surface lowering in those regions and now appears to confirm values measured independently by micro-erosion meter (Table 4.3).

Similar information can be obtained from the heights of emergent quartz veins and siliceous nodules (Figure 4.8d) that stand proud of a limestone surface because of differential solution, although quartz does weather and so the amount of emergence is a minimal figure unless the vein is wide (Lauritzen 1990). We must also caution that in some arctic and alpine settings severe storm blasting by grit and ice crystals can differentially abrade quartz and carbonate bedrock and thus skew results. Nevertheless, some interesting data can be obtained. From quartz veins standing proud of dolomitic limestone surfaces in Spitsbergen (latitude 78°N), Akerman (1983) estimated surface lowering to have averaged 2.5 mm ka^{-1} since isostatic uplift raised the area above sea level following the last glaciation. His data set also shows that 4000–9000 year ago the lowering rate was 3.5 mm ka^{-1} compared with 1.5 mm ka^{-1} during the past 2000–4000 year. The latter rate represents about 11% of the present total solutional denudation in the area, estimated by Hellden (1973) as $11–15$ mm ka^{-1} (although the range of chemical denudation in Spitsbergen is now considered to be wider). A similar approach using data from igneous dykes and marble pedestals led Maire *et al.* (1999) to estimate a surface lowering rate in Patagonia of about 60 mm ka^{-1} since the last glaciation around 8–10 ka; which is consistent with evidence from paint marks made in 1948 in an old quarry that were found raised 3 mm in relief by dissolution of the surrounding surface. This is perhaps the greatest rate of superficial corrosion known on bare rock and is associated with a mean annual precipitation of about 7300 mm.

Figure 4.8 (a) A glacial erratic protecting a pedestal on a limestone pavement, Inishmore, Aran Island, Ireland. Glacial retreat was about 14 ka ago. (b) Limestone pedestals formed in about 10 000 years in a glaciated valley on Mount Jaya (Carstensz) in alpine tropical Irian Jaya. (Photograph by J. Peterson.) (c) Limestone pedestal with a glacial erratic caprock protecting it from solution by rainfall, County Leitrim, Ireland. The average height of the pedestal (51 cm) suggests either an unusually rapid rate of dissolution since deglaciation about 14 ka ago or that the area was ice-free in marine oxygen isotope stage (MIS) 2 and the erratic may have been emplaced in an earlier (MIS 4?) glacial advance. (d) Chert nodules standing proud of a marble surface, Mount Owen, New Zealand. The camera case provides scale. About 10 cm lowering has occurred since deglaciation about 15 ka ago.

Table 4.3 Approximate rates of dissolutional lowering of limestone surfaces deduced from limestone pedestals and compared with micro-erosion meter measurements where available (Reproduced from Ford, D.C. and Williams, P.W. (1989) Karst Geomorphology and Hydrology)

Locality	Average height of pedestals (cm)	Time since ice retreat (year)	Rate of surface lowering (mm ka^{-1})	Extrapolated micro-erosion meter rate (mm ka^{-1})
Maren Mountains, Switzerland	15[*]	14 000	11	—
Burren, western Ireland	9[†]	14 000	6	5[††]
Leitrim, western Ireland	51[‡]	14 000?	36	—
Pennines, northern England	5–20[§]	15 000	3–13	13[††]
Mt Jaya, West Irian	30[¶]	9500	32	—
Svartisen, Norway	13[**]	9000	15	25
Patagonia	40–60[‡‡]	8000–10 000	40–75	—

[*]Bögli 1961; [†]V. Williams pers. comm. 2004; [‡]Williams 1966; [§]H. Goldie 2005; [¶]Peterson 1982; [**]Lauritzen 1990; [††]Trudgill *et al.* 1981; [‡‡]Maire *et al.* 1999, who used veins as well as pedestals.

Utilising data from pedestals, quartz veins and nodules, André (1996a) provided an interesting comparison of rates of subaerial surface lowering in polar, temperate and tropical high alpine zones. Rates ranged from < 3 mm ka^{-1} in the Arctic to > 8 mm ka^{-1} in the temperate alpine zone, and > 30 mm ka^{-1} in alpine New Guinea. She concluded that increasing rates coincide with increasing amounts of precipitation. These rates compare with < 1 mm ka^{-1} on non-carbonate quartzose rocks (André 1996b).

4.2.7 Cosmogenic chlorine-36 measurements

Cosmogenic exposure age dating has been in increasing use in geomorphology since the 1990s (Nishiikumi 1993, Cockburn and Summerfield 2004, Phillips 2004). It is most readily applied on horizontal rock surfaces at high latitude and high altitude, but with appropriate correction factors can be used at lower latitude and altitude. Exposure of calcite to cosmic radiation results in the production of ^{36}Cl from ^{40}Ca (Stone *et al.* 1998). The total production rate of ^{36}Cl is greatest at the surface, but decreases exponentially by two orders of magnitude after about 15 m. Time to saturation is about 10^5–10^6 year. Denudation by natural processes reduces the concentration of ^{36}Cl at the surface, and so there is a relationship between ^{36}Cl concentration (at any given depth) and surface erosion rate. Details are discussed by Stone *et al.* (1994).

Erosion rates of five limestone surfaces calculated from ^{36}Cl measurements of samples from Australia and Papua New Guinea were determined by Stone *et al.* (1994). Rates varied from < 5 μm a^{-1} (equivalent to < 5 mm ka^{-1}) on the arid Nullarbor Plain to 18–29 μm a^{-1} in humid southeast Australia and to 184 μm a^{-1} in the wet Strickland ranges of Papua New Guinea, although a subsequent recalibration of the ^{36}Cl production rate lowered the calculated rates by about 20%. Associated errors were estimated as about ±12%. In detailed analysis at Wombeyan ($\sim$650 m above sea level and 760 mm annual rainfall) in New South Wales, Stone *et al.* (1998) took account of surface ^{36}Cl concentration and subsurface ^{36}Cl gradient and calculated an erosion rate of 23 μm a^{-1} prior to $\sim$15 a, increasing to about 100 μm a^{-1} at present. These cosmogenic erosion rates are of the same order as those determined from previous estimates by other methods, although results are not easy to compare because the integration time of the ^{36}Cl method is 10^5–10^6 year. They yield a long-term average, which in fact is more useful in a geomorphological context than the usual short-term estimates by mass-flux methods.

4.3 SOLUTION RATES IN GYPSUM, SALT AND OTHER NON-CARBONATE ROCKS

4.3.1 Solution rates in gypsum and salt

A priori, dissolutional rates in gypsum karsts will be at least one order of magnitude more rapid than in limestone karsts because, for example, at 20°C the equilibrium solubility is 2500 mg L^{-1} compared with 60 mg L^{-1} for calcite (and 360 g L^{-1} for halite). However, where P_{CO_2} is increased in soils, etc., the dissolution of calcite is enhanced, which considerably reduces the difference in solubility between gypsum and calcite. Dissolution rates of gypsum and carbonates exposed to precipitation were compared directly in a field laboratory in Trieste, Italy. The results indicated that average rates of dissolution of gypsum samples are roughly 30 to 70 times greater than the dissolution rates of carbonates, which Klimchouk *et al.* (1996) noted accords broadly with theoretical expectations.

In purely sulphate regions it is comparatively rare to measure spring waters that are saturated; most are aggressive with respect to gypsum and anhydrite. This suggests that gypsum dissolution often does not attain the equilibrium maximum before the waters are discharged from the karst. For example, in the great interstratal gypsum karst of Ukraine, Klimchouk and Andrejchouk (1986) reported that both vadose and phreatic waters are undersaturated where the groundwaters flow with normal vigour, although where downfaulting has produced deeper artesian flow that is very sluggish, waters can be saturated. In most instances where saturated sulphate waters have been identified, it appears that saturation results from the foreign ion effect (see section 3.7 and Klimchouk 1996) due to the presence of disseminated halite or halite interbeds.

Field evidence for the rate of solutional denudation of gypsum terrains in the Ukraine, Spain and Italy has been reviewed by Klimchouk *et al.* (1996), and the results reported for the Ukraine have since been updated by Klimchouk and Aksem (2000). They reported that gypsum tablets of standard lithology and size (40–45 mm diameter, 7–8 mm thick) were installed at 53 sites, representing varying conditions of water–rock interaction in unconfined and confined aquifers, and that 644 weight-loss measurements were taken over the period 1984 to 1992. Tablets were exposed to surface precipitation ($\sim$640 mm), cave air in zones of condensation, percolation, and static or semi-static water in caves and boreholes. Aggressive groundwaters were found able to dissolve gypsum at a rate of up to 26 mm a^{-1}; the solution rate under active unconfined circulation averaged about 11 mm a^{-1}; whereas in the bulk aquifer the solution rate was about 0.1 mm a^{-1}.

In the dry Sorbas region of southern Spain (annual rainfall 250 mm) measurements were made both by standardized tablets and by micro-erosion meter (MEM), and it is interesting to note that the MEM values were about 1.5 times greater than those measured by tablets directly exposed to rainfall. In the Sorbas region, Pulido-Bosch (1986) had previously reported a surface dissolution rate of 260 mm ka^{-1}, whereas in a wet region of the French Alps with an annual rainfall exceeding 1670 mm Nicod (1976) calculated a rate of more than 1 m ka^{-1}. Other measurements made in Italy revealed different erosion rates on gypsum rocks of different lithology and a strong relationship between dissolution loss and the amount of annual precipitation.

The effect of dissolution by condensation water in cave air was measured in Spain as well as in the Ukraine and the data from the tablets were found to be essentially the same for both regions (0.004 and 0.003 mm a^{-1} respectively). However, the diurnal wetting and drying effect that some cave entrances experience in tropical coastal settings can considerably increase the rate of condensation solution. Thus Tarhule-Lips and Ford (1998a) measured 0.4–0.5 mm a^{-1} on gypsum tablets exposed in the entrance zones of caves on Cayman Brac and Isla de Mona in the Caribbean. Condensation in caves as a microclimate process involving the continuous process of vapour transfer has been examined in New Zealand by de Freitas and Schmekal (2003).

Salt karsts mainly exist in arid terrains where the low rainfall permits rock-salt outcrops to escape complete destruction by dissolution. Although salt tectonics have been the subject of considerable investigation (Jackson *et al.* 1995, Alsop *et al.* 1996), mainly because of the interest of the petroleum industry, there is relatively little information on rates of rock-salt dissolution. However, field evidence shows that the development of solution features on rock salt is very rapid because of its great solubility. For example, where allogenic streams were diverted into salt domes during mining operations in Europe, multilevel caves of enterable dimensions and profusely decorated with speleothems have developed in 200–300 year. Further, Davison *et al.* (1996) noted development of 'spectacular' karst on top of Al Salif diapir in Yemen since 1930 following the start of mining operations, despite an annual rainfall of only 80 mm. Insoluble residue often occurs as 'dissolution drapes' over the surface. In the 14 km^2 area of the salt diapir of Mount Sedom in Israel, a region with an annual rainfall of only 50 mm, Frumkin (1994) estimated the regional karst denudation rate to be about 0.50–0.75 mm a^{-1} with most of this occurring within the rock-salt mass rather than near the surface. The downcutting rate in natural

vadose cave-stream passages was estimated to be ~20 mm a^{-1} when averaged over 10 year (Frumkin and Ford 1995, Frumkin 2000a). This is in the context of a maximum diapir uplift rate of 6–7 mm a^{-1} over the past 8000 year (Frumkin 1996).

4.3.2 Solution of quartzose rocks

The dissolution of silica is explained in section 3.4 and methods for spectrometric determination of ionized silica (SiO$_2$) are explained in Krawczyk (1996). Discussions of dissolution processes affecting quartzites, quartz sandstones and silicate minerals are available in Young and Young (1992), Dove and Rimstidt (1994), Wray (1997) and Martini (2000). Simms (2004) compared denudation rates in silicate and carbonate rocks as a function of runoff (Figure 4.9).

The rate of dissolution of quartz is extremely slow under normal surface conditions, but under high temperature and high salinity the solubility and rate of quartz solution is much higher. Field observations show that waters draining from quartzose rocks have silica contents averaging about 6–7 mg L^{-1} but with a range of 1 to 30 mg L^{-1}. In the high rainfall tropical region of Venezuela, water draining from the well known karst-like relief developed in the Roraima Quartzite has silica values of 5–7 mg L^{-1}; thus quartz undersaturation is

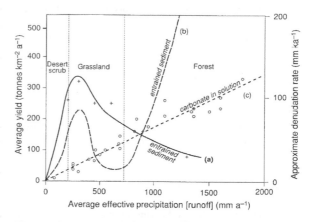

Figure 4.9 Relationship between effective precipitation, annual sediment yield for silicate outcrops and annual dissolved carbonate load from karst regions. Note that a limestone denudation rate of 1 mm ka^{-1} is equivalent to about 2.5 t km^{-2} a^{-1}. Curves (a) and (b) are estimates by Langbein and Schumm (1958) and Ohmori (1983), respectively. Carbonate data (c) are from Atkinson and Smith (1976). Reproduced with permission from Simms, M.J., Tortoises and hares: dissolution, erosion and isostasy in landscape evolution. Earth Surface Processes and Landforms 29, 477–494 © 2004 John Wiley and Sons.

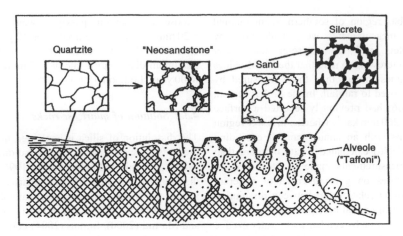

Figure 4.10   Steps in the progressive weathering of quartzite rock and the exhumation of bedrock pinnacles. Dissolution proceeds along crystal boundaries to produce a moderately weathered rock termed 'neosandstone'. Further weathering releases sand grains which are later removed mainly by mechanical erosion, thereby exposing bedrock pinnacles. In seasonally arid climates the exposed surfaces become case hardened by interstitial cementation by opaline silcrete. Small hemispherical pockets (tafoni) form in the neosandstone by granular disintegration of the softer parts of the rock. Reproduced from Martini, J.E.J, Silicate karst. In Gunn, J. (ed.) Encyclopedia of Caves and Karst Science, 649–653, © 2004 Fitzroy Dearborn, Taylor and Francis.

suggested. Low values are often associated with high organic acid content. Higher silica values tend to be measured in waters draining from silicate rocks such as granites, although sometimes higher values indicate supersaturation possibly by evaporation. Opaline silica can precipitate when groundwater becomes supersaturated. Opaline cave minerals are discussed by Hill and Forti (1997).

Microscopic observation shows that the weathering of quartzite is penetrative through the mass of the rock and consists of quartz dissolution along crystal junctions that are progressively enlarged until the rock loses its cohesion and is reduced to sand (Figure 4.10; Martini 2000, 2004). This operates preferentially down joints and along bedding planes where water access is facilitated, although the joints themselves do not necessarily enlarge. Tripathi and Rajamani (2003) found that the development of weathering rinds and karst-like topography in Proterozoic quartzites in India is explained by the chemical weathering of minor minerals such as pyrite and aluminosilicates, with the sulphate-bearing acidic solution that is released being responsible for the decomposition of the primary silicates.

Any quartz sand released can be gradually removed by dissolution, although if joints are wide enough and hydraulic gradients are high (e.g. along zones of tension near cliffs) then turbulent subsurface water flow can mechanically remove the grains. But sometimes the weathering of silicate rocks can produce such large amounts of insoluble clay residue, especially kaolinite, that interstices and joints

become clogged, and the development of subterranean drainage is inhibited. Thus while surface landforms on some sandstones may resemble karst (Figure 4.11), for example cones and ruin-form hills in some semi-arid subtropical settings, the essential hallmark of karst–subterranean drainage–is often absent. In a similar way, not all carbonate rocks develop karst if the insoluble content is too great.

The development of subsurface drainage in quartzose rocks requires them to be fairly pure and to be prepared by dissolution before enlargement of subterranean passages by physical stream erosion can occur. Thus even though mechanical processes may come to dominate underground, preconditioning by dissolution is critical. Because quartz solubility and dissolution rates are low, very long periods of time are required for the effects of quartz solution to be noticeable in the landscape regardless of climatic zone, although the presence of traces of pyrite will accelerate the process. Yanes and Briceno (1993) suggested that karst-like processes may have continued in the Roraima Quartzite for at least 70 Myear.

4.4 INTERPRETATION OF MEASUREMENTS

The large accumulated error that is likely to arise in the calculation by the mass-flux approach of a solutional denudation rate demands that caution be exercised in its interpretation and in its extrapolation over long periods of 'time'. The representativeness of individual sample years is also a significant issue. Bakalowicz (1992) showed

Figure 4.11   (a) Karst-like topography in Upper Proterozoic quartz sandstone (Bessie Creek Formation) from Arnhem Land, northern Australia. (Photograph by G. Nansen.) For a discussion of this ruiniform relief see Jennings (1983). (b) Pedestal protected by relatively insoluble ferruginous clast in quartzitic Kombologie Sandstone on the Arnhem Land escarpment, northern Australia. In this tropical monsoon climate, surface lowering is mainly achieved by physical removal of loose grains after weakening by dissolution along grain boundaries.

interannual variability in carbonate flux over five hydrological years in the Baget basin in the French Pyrenees to be about ±25%. He ascribed it as due partly to variations in annual precipitation, and partly to the seasonal distribution of flood events; floodwaters in summer being more mineralized than those in spring. Furthermore, point estimates from MEM measurements or from tablet weight loss are difficult to extrapolate over 'space' and provide less reliable areal estimates of solution than do basin studies. Consequently, the problem of the validity of karst erosion data is the problem of transforming unique values into values representative of some defined time and space. Nevertheless, acceptable convergence of estimates made by different methods can sometimes be achieved. Thus in northern Norway, Lauritzen (1990) calculated autogenic denudation by solute flux as $32.5 \pm 10.2 \, \text{mm} \, \text{ka}^{-1}$; annual surface denudation measured over 10 year by MEM was found to average $0.025 \pm 0.0027 \, \text{mm} \, \text{a}^{-1}$; and surface lowering over 9000 year since deglaciation (from maximum erratic pedestal and quartz vein heights) averaged 13.3 and $23.3 \, \text{mm} \, \text{ka}^{-1}$ respectively. These results compare quite favourably and suggest that in northern Norway 42–72% of the denudation takes place at the surface of the karst.

4.4.1 Vertical distribution of solution

The contrasts in karst landform styles in humid tropical and temperate zones appear not to be attributable to radically different solutional denudation rates (Smith and Atkinson 1976). Therefore it follows that contrasts in the three-dimensional distribution of corrosion must be

the explanatory factor – unless comparative denudation estimates are seriously in error. For this reason it is essential to gain an insight into the vertical and spatial distribution of corrosion. We will consider the vertical distribution first.

Information on the vertical distribution of corrosion can be obtained by following the evolution of the chemical characteristics of water as it runs across the surface and percolates through the soil and underlying bedrock. In this way Gams (1962) found the bulk of the corrosion of limestone in Slovenia to occur in the top 10 m of the percolation zone. This important conclusion has since been found to apply to most other places where the vertical distribution of corrosion has been studied (Table 4.4). Data on the vertical distribution of dissolution are available from Smith and Atkinson (1976), Williams and Dowling (1979), Gunn (1981a), Crowther (1989) and Zámbó and Ford (1997). Measurements show that in vegetated soil-covered karst most autogenic solution occurs near the top of the profile, i.e. in the soil and vegetation, at the soil–bedrock interface and in the uppermost bedrock. The consensus is that about 70% of autogenic solution takes place in the uppermost 10 m or so of the percolation zone, although actual figures vary from ~50 to 90% depending on lithology and other factors. The implications are that (i) most solutional denudation leads to surface lowering and (ii) solutional activity in cave conduits is of relatively minor significance in the total denudation budget, although fundamental to the development of the karst landform system. In entirely autogenic systems, much field evidence suggests that corrosion in vadose cave passages of enterable

Table 4.4 Vertical distribution of solutional denudation (Reproduced from Ford, D.C. and Williams, P.W. (1989) Karst Geomorphology and Hydrology)

Locality	Overall rate ($m^3 km^{-2} a^{-1}$)	Distribution of dissolution	Source
Fergus River, Ireland	55	60% at surface, up to 80% in the top 8 m	Williams 1963, 1968
Derbyshire, UK	83	Mostly at surface	Pitty 1968a
Northwest Yorkshire, UK	83	50% at surface	Sweeting 1966
Jura Mountains	98	33% on bare rock, 58% under soil, 37% in percolation zone, 5% in conduits	Aubert 1967, 1969
Cooleman Plain, NSW, Australia	24	75% from surface and percolation zone, 20% from conduit and river channels, 5% from covered karst	Jennings 1972a, b
Somerset Island, NWT, Canada	2	100% above permafrost layer	Smith 1972
Riwaka South Branch, New Zealand	100	80% in top 10–30 m, 18% in conduits mainly by allogenic streams	Williams and Dowling 1979
Waitomo, New Zealand	69	37% in soil profile, most of remainder in to 5–10 m of bedrock	Gunn 1981a
Caves Branch, Belize	90	60% on surface and in percolation zone, 40% in conduits (in large allogenic river passages)	Miller 1982
Svartisen, Norway	32.5	42–72% at surface	Lauritzen 1990

dimensions is negligible except during occasional floods (when corrasion may also be important). In two systems at Waitomo, New Zealand, Gunn (1981a) found water contributing to cave streams to be saturated or supersaturated in most conditions, as did Miller (1981) in caves draining a tropical cockpit karst in Belize.

Rates of surface lowering on bare rock derived from heights of pedestals and veins are compared with extrapolated MEM values in Table 4.3. The MEM results for northern England are more than double those for Clare–Galway, as also are the average pedestal heights, and the estimated surface lowering rates by both MEM and pedestal methods are of the same order. Rates of surface lowering can also be computed from rainfall volume and solute concentrations achieved in runoff over rock outcrops (Miotke 1968, Dunkerley 1983, Maire *et al.* 1999).

Even when rock appears bare, this is seldom the case, and most rock surfaces are in fact covered with a patchy thin layer of bacteria, fungi, green algae, blue-green algae and lichens. Features produced by the action of flora and fauna on karst rocks are termed **biokarst** (Viles 1984, 2003) or **phytokarst** (Folk *et al.* 1973, Bull and Laverty 1982) and processes associated with it are discussed in section 3.8. Naylor and Viles (2002) have extended the discussion to bioerosion and bioprotection in the intertidal zone, providing evidence that bioerosion, biological etching and chemical weathering are reduced once the surface is colonized by macro algae.

Blue-green algae such as *Chroococcus minutes* have been found to weaken the crystalline structure of limestone surfaces thereby facilitating physical removal of small rock fragments by raindrop impact. Most species are surface dwellers (epilithic), but in ecologically stressful environments some cyanophytes bore into rocks to depths of ~1 mm while others dwell in vacated borings or other microcavities. Borers create pits directly (Figure 4.12) whereas other species may contribute to their creation or

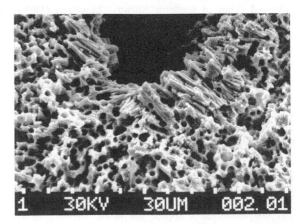

Figure 4.12 Scanning electron photomicrograph of cyanobacteria boring into a limestone surface of an inland pool, Aldabra Atoll. (Photograph by H. Viles.)

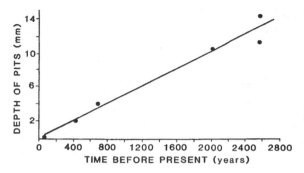

Figure 4.13 Depths of pits resulting from cyanobacteriain-duced weathering in comparable limestone walls of varying age in Jerusalem. Reproduced from Danin, A., Weathering of limestone in Jerusalem by cyanobacteria. Zeitschrift für Geomorphologic. 27(4), pp. 413–421. © 1983 E. Schweizerbart'sche Verlagsbuchhandlung, Science Publishers.

enlargement by way of the organic acids or CO_2 that they excrete. Trudgill (1985) provided a particularly good discussion of the role of organic acids as weathering agents. However, Viles and Spencer (1986) and Vilas (1987) could not relate algal activity to specific microkarst except on sea coasts, so that their role remains uncertain. Nevertheless, it is widely agreed that, once established, small pits and fissures may be preferentially deepened if fungi, lichens or mosses can establish in them and excrete CO_2. Excellent studies of the role of cyanobacteria and lichens in the weathering of limestones in have been published by Danin (1983, 1993), Danin and Garty (1983) and Darabos (2003). Cyanobacteria are shown to lead to rock weathering at a rate of $5 \, mm \, ka^{-1}$ (Figure 4.13) and cyanobacteria-induced pits have been measured to 14 mm and are associated with exfoliative weathering of limestones in Japan (Darabos 2003).

When the surface is forested, it is more obvious that biological processes will play a part in solutional denudation, but even then the role of vegetation is easily underestimated. In a study of autogenic karst solution process in peninsular Malaysia, Crowther (1989) presented a clear picture of Ca and Mg fluxes in the vegetation–soil–bedrock system (Figure 4.14). He concluded that Ca concentrations in vadose waters closely reflect P_{CO_2} and that: (i) dissolutional activity is concentrated at the limestone surface, both on rock exposures and at the soil–rock (and root–rock) interface; and (ii) humid tropical forests annually take up amounts of Ca and Mg equivalent to the net solute output in groundwaters. Vegetation was found to have an important role in solute generation through absorption of Ca and Mg by roots, canopy leaching and litter decomposition. It is probable that forests elsewhere in the world have a

similar role. In tropical forest in Belize, Miller (1982) found 60% of limestone solution to occur on the surface and in the percolation zone and, under temperate rain forest in New Zealand, Gunn (1981a) concluded that 37% of the solution occurs within the soil, at the soil–rock interface, or on occasional limestone outcrops, with most of the remainder concentrated in 5–10 m of weathered bedrock beneath. However, when significant organic material is transported underground its bacterially mediated oxidation can supply another important source of CO_2 that can sustain continued dissolution in the vadose and phreatic zone (Atkinson 1977a, Whitaker and Smart 1994).

The vertical distribution of denudation depends upon two factors: (i) the distribution of water flow and (ii) the distribution of solute concentrations. Efforts have focused mainly on the latter, with data being used in a budget approach to deduce the relative contributions to total solutional denudation (Table 4.1). Illustrations of results are provided by Jennings' (1972) research in New South Wales, Australia, Atkinson and Smith's (1976) in the Mendip Hills, England, and Williams and Dowling's (1979) in New Zealand (Figure 4.15). An estimation of mass flux through the vadose zone has been made by Crowther (1989) (Figure 4.14), but much more information is still required, especially on the spatial variation of dissolution within the epikarst, for it is there that such landforms as solution dolines are made. The most detailed study of this to date is by Zámbó (2004) and co-workers in Hungary who, since the 1970s, have measured hydrological and geochemical characteristics of the zone between a doline and an underlying cave (see further comments below).

4.4.2 Spatial distribution of solution

The solutional denudation rate is derived from the product of water flow and solute concentration. Topographic variations give rise to inequalities in runoff distribution: hills encourage runoff divergence, whereas hollows encourage runoff convergence. Soil thickness also varies spatially, tending to be thicker in depressions than on hilltops. In a karst context, spatial inequalities in the availability of solvent (water) arise from the distribution of runoff following rain, and spatial variability in soil CO_2 production arises from variations in biological activity associated with soil thickness, soil moisture and aspect. In an autogenic percolation system a wide range of flows occur on a continuum from very slow seepages through trickles to showering cascades (discussed further in section 6.3). Discharges of observed flows range over several orders of magnitude, whereas their associated solute

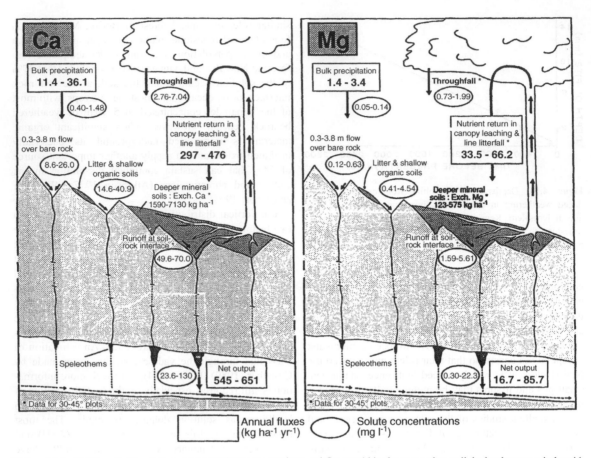

Figure 4.14   Calcium (left) and magnesium (right) concentrations and fluxes within the vegetation–soil–bedrock system in humid tropical karst in Malaysia. Reproduced from Crowther, J, Groundwater chemistry and cation budgets of tropical karst outcrops, peninsular Malaysia, 1 calcium and magnesium. Journal of Hydrology 107, 169–192. © 1989 Elsevier.

concentrations are unlikely to vary by more than one order of magnitude. Although direct relationships have been reported between flow-through time and calcium hardness (e.g. Pitty 1968b), the small volumes of highly mineralized seepages do not account for much denudation, although they make an impressive contribution to speleothem deposition. Recent observations show that in well-developed karsts the upper part of the percolation zone (the epikarst or subcutaneous zone) has a significant water storage capacity. Drainage from this zone is not uniform, but flows down preferred paths which act as foci for subcutaneous streamlines. Therefore it is probable that dissolution within the epikarst is greatest where flow paths converge above the more efficient percolation routes. Because solute concentrations do not vary widely compared with percolation discharge rates, corrosion denudation in a zone of convergent flow can be many times greater than in a zone where flow diverges. Karsts with

spatially uniform infiltration and percolation through the epikarst will experience uniform surface lowering, whereas karsts with patterns of flow divergence and convergence within the epikarst will develop accentuated relief. Inequalities in subcutaneous corrosion become increasingly manifested topographically as the surface lowers over time. Even the effect of aspect on biological activity can be important, because solute concentrations can be significantly higher in springs issuing from sunny slopes than in springs emerging from slopes that are relatively shaded (Pentecost 1992).

These points are born out by field measurements reported by Zámbó and Ford (1997) in the Aggtelek area of Hungary. They found that the capacity to dissolve limestone varies from ~3 to $30 \, \mathrm{g \, m^{-2} \, a^{-1}}$ depending on variations in soil cover and soil water, being lowest on well drained high spots and slopes where soil is thin, and highest in closed depressions where storm runoff

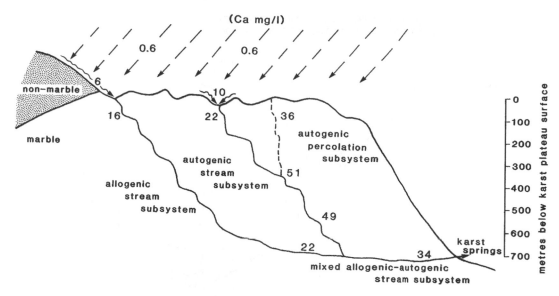

Figure 4.15 Dissolved calcium concentrations in the Pikikiruna karst, New Zealand. Reproduced with permission from Williams, P.W. and Dowling, R.K., Solution of marble in the karst of the Pikikiruna Range, northwest Nelson, New Zealand. Earth Surf. Proc. 4, B1010 15 36 © 1979 John Wiley and Sons.

converges and soil cover is thick (Figure 4.16). They calculated the mean rate of lowering of the land to be about $0.5 \, cm \, ka^{-1}$, but noted that it varies from $0.4 \, cm \, ka^{-1}$ or less on interdoline crests to $0.7–1.0 \, cm \, ka^{-1}$ in doline bottoms. Thus the differential deepening rate for the doline investigated was in the range of $0.3–0.6 \, cm \, ka^{-1}$ which, considering its depth of $20 \, m$, implies that at present rates it could have developed since the Mio-Pliocene. Given the importance of flow divergence and convergence for the development of the karst topographies modelled by Ahnert and Williams (1997), the processes measured by Zámbó and Ford (1997) are likely to be important for the development of any style of doline relief.

4.4.3 Rates of denudation and downcutting

The rates of denudation of carbonate and silicate landscapes as a function of effective precipitation are compared in Figure 4.9, although the silicate values refer mainly to mechanical erosion. Cave streams incise as a consequence of both chemical and mechanical processes. Their downcutting rate is equivalent to the rate of base-level lowering and is equal to or less than the rate of tectonic uplift. In stable alpine and cratonic areas long-term denudation and uplift will tend to balance and can be estimated by cave-passage incision rates.

Gascoyne *et al.* (1983) used the ages of speleothems in various positions above active streamways to estimate cave passage incision rates, and data on downcutting determined by such methods have been reviewed by Atkinson and Rowe (1992). However, caution must be exercised in assuming that the rate of lowering of a water table is always due to valley incision, because sometimes it is attributable to an increase in hydraulic conductivity over time, as Ford *et al.* (1993) demonstrated in the case of the Wind Cave aquifer in the Black Hills of South Dakota.

Stalagmites can accumulate only when they are not submerged or subject to erosion by flood waters, as explained in section 8.3. Hence the height of the base of a stalagmite above a bedrock stream channel divided by the stalagmite basal age yields a maximum downcutting rate for the stream. Rates of about $20–50 \, mm \, ka^{-1}$ over the past $350\,000$ year were determined in this manner for cave channels in northwest Yorkshire, England, with a mean maximum rate of valley entrenchment (aided by glaciation) in the area of 50 to $< 200 \, mm \, ka^{-1}$ (Gascoyne *et al.* 1983). This compares with: valley deepening rates of $40–70 \, mm \, ka^{-1}$ as a minimum, with maximum values to $2 \, m \, ka^{-1}$, calculated by the same method by Ford *et al.* (1981) in the Canadian Rocky Mountains; a rate of $280 \, mm \, ka^{-1}$ in the Southern Alps of New Zealand (Williams 1982b); rates of $190–510 \, mm \, ka^{-1}$ in the Qinling Mountains of China (Wang *et al.* 2004); and rates of $80–1040 \, mm \, ka^{-1}$ in the Alpi Apuane of Italy (Piccini *et al.* 2003). Sometimes incision is so slow that the use of uranium-series dating is not

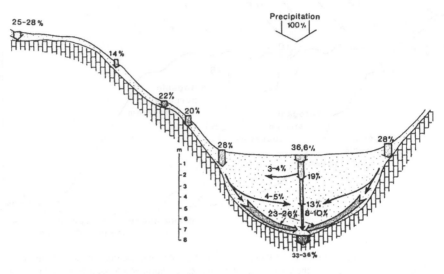

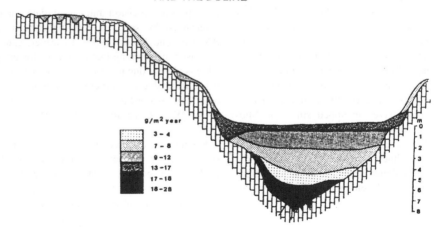

Figure 4.16   Infiltration and dissolution rates in a doline at Aggtelek, Hungary. (a) Infiltration and lateral runoff as a percentage of precipitation generalized from data for two sideslope sites and three dolines. (b) Mean annual $CaCO_3$ dissolution rates, showing an order of magnitude difference depending on location. Solution at the base of the doline is enhanced by flow convergence. Reproduced with permission from Zámbó, L. and Ford, D.C. Limestone dissolution processes in Beke doline, Aggtelek National Park, Hungary. Earth Surface Processes and Landforms, 22, 531–43 © 1997 John Wiley and Sons.

helpful, because dates exceed the upper age limit of the method. Under these circumstances the minimum ages of cave passages have been determined by magnetostratigraphy of sediment fill and ancient speleothems (Schmidt 1982, Williams *et al.* 1986, Webb *et al.* 1992, Auler *et al.* 2002). Thus in the Buchan karst of southeastern Australia, the rate of incision has averaged only 0.004 m ka^{-1} since the last reversal (Webb *et al.* 1992, Fabel *et al.* 1996). However, even when uplift and incision rates are quite rapid, magnetostratigraphy can be a valuable tool when cave passages extend over a large vertical range. This was shown at Mulu in Sarawak by Farrant *et al.* (1995),

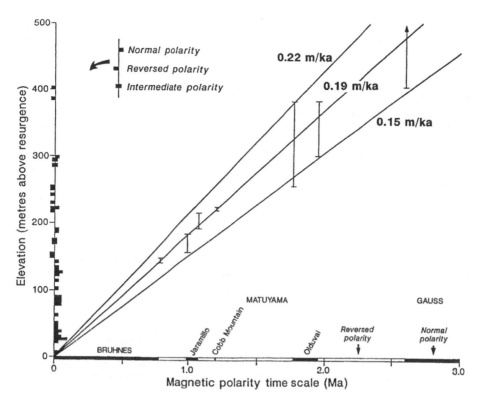

Figure 4.17   Interpretation of the Quaternary uplift rate from the magnetostratigraphy of sediments at various elevations in the Clearwater–Blackrock cave system, Mulu, Sarawak. Reproduced from Farrant, A.R., Smart, P.L., Whitaker, F.F. and Tarling, D.H, Long-term Quaternary uplift rates inferred from limestone caves in Sarawak, Malaysia. *Geology* 23(4), 357–360.© 1995 Geological Society of America.

where a stable rate of incision of about $0.19\,\mathrm{m\,ka^{-1}}$ was demonstrated over the Quaternary (Figure 4.17). Cosmogenic burial-age dating of pebbles has also been used to deduce bedrock incision rates, which at Mammoth Cave, Kentucky, averaged $<3\text{–}5\,\mathrm{m\,Ma^{-1}}$ in the late Pliocene (Granger *et al.* 2001).

From the above results we find that downcutting rates of $50\text{–}1000\,\mathrm{mm\,ka^{-1}}$ $(50\text{–}1000\,\mathrm{m\,Ma^{-1}})$ are typical for tectonically active mountainous areas, but that rates of $<5\,\mathrm{m\,Ma^{-1}}$ can be found in stable cratonic areas. However, even the highest incision rates in carbonate rocks are only a fraction of what can be achieved in salt karst, even under arid climatic conditions. Thus Frumkin (2000a) assessed long-term vadose incision rates in the Mount Sedom diapir near the Dead Sea as about $20\,\mathrm{mm\,a^{-1}}$.

4.4.4 *Vertical distribution of carbonate deposition*

Chemical processes involved in the dissolution and precipitation of calcite were discussed in Chapter 3; crystal growth of carbonate precipitates in caves is examined in

Chapter 8 and case hardening and tufa deposition in rivers and lakes in Chapter 9. This section focuses on general aspects of the vertical distribution of carbonate deposition by meteoric waters.

Most dissolution of limestone occurs in the epikarst. It may even occur entirely in the soil where this is rich in carbonate fragments, such as a calcareous glacial till. Consequently, most autogenic percolation water is close to saturation with respect to calcite as it moves downwards through the remainder of the vadose zone. Ground-air CO_2 may sustain a slight degree of undersaturation, although when percolating water leaves the soil zone it is likely that most solution will occur under essentially closed (or sequential) system conditions.

Carbonate precipitation most commonly occurs when water percolating beneath soil enters a cavity that has some connection with the outside atmosphere. Cavities may be large caverns or small vugs, but gaseous exchange with the external air is essential because it ensures that the P_{CO_2} of the cave air is broadly of the same order as that measured in the normal atmosphere (section 3.5). The percolating water is equilibrated with a higher, soil P_{CO_2} with the result that carbonate is precipitated if significant $(SI_c > +0.3)$ supersaturation is attained. Should cave ventilation also be strong, relative humidity may be less than 100%, in which case evaporation will also occur. This accelerates deposition as well as encouraging the growth of a different suite of speleothem forms. For further discussion of the kinetics of calcite precipitation see section 3.10.

In seasonally dry environments carbonate deposition can occur in the soil as a result of evaporation. In some humid environments carbonate precipitation may closely follow dissolution due to a combination of CO_2 degassing and evaporation, resulting in formation of calcrete case hardening (section 9.12). However, most carbonate deposition takes place beneath the solum and commences in the first cavity where P_{CO_2} is less than in the soil atmosphere. There are countless instances of caves profusely decorated with actively growing speleothems within a few metres of the surface. In raised atolls carbonate deposits fill interconnecting vugs in the rock matrix like butter melting down into hot toast; under such circumstances deposition reduces primary (bulk) porosity although in the standard vadose cave it reduces secondary (fissure) porosity. Case hardening is a form of eogenetic diagenesis and can reduce primary porosity by a factor of ten or more (Mylroie and Vacher 1999). In Puerto Rico, Ireland (1979) found the indurated layer to follow the topography and on average to be 2 m thick but to vary up to 10 m. Ivanovich and Ireland (1984) suggested that the formation of a case-hardened layer 1 m thick could occur within 10 000–20 000 year, assuming a constant denudation rate of between 50 and 100 mm ka^{-1}. However, work by Mylroie *et al.* (1995) on the marine oxygen isotope stage (MIS) 5e aeolianites of the Bahamas implies that much faster case-hardening rates can occur in young eogenetic rocks.

The degree of supersaturation of percolating groundwater diminishes following continued carbonate precipitation. Thus with depth through the vadose zone the amount of speleothem deposition may decrease. However, the secondary porosity of well-karstified terrains is often no more than a few per cent; thus percolating water often does not encounter aerated vadose cavities and the first opportunity to precipitate carbonate may be in the saturated zone. In a partly flooded cavern communicating directly with the outside atmosphere, heterogeneous carbonate precipitation can take place onto solid surfaces such as cave walls or other crystals. It is comparatively rare because it occurs only where currents are slow moving. Deposits are commonly destroyed if the passage is liable to turbulent floods.

These points on the vertical distribution of deposition apply to autogenic karst. Under other geological situations differing hydrological and hydrochemical conditions dictate where deposition is possible. Impervious caprocks preclude percolation and hence limit deposition to streamways, provided that mechanical action is also not too strong. Streams in allogenic systems, however, are commonly undersaturated with respect to calcite so that carbonate deposition is not usually a feature. This contrasts strongly with vadose streams in autogenic systems, which are often very favoured sites for cascades of speleothems that form rimstone pools (or gours).

The distribution of carbonate deposition in the vadose zone has received remarkably little quantitative study. However, it is generally acknowledged that the pattern of deposition discussed is much better displayed in tropical and warm temperate zones than in cold regions. This is presumed to be a consequence of the greater contrast achieved between soil and atmospheric CO_2 in the tropics compared with subarctic and alpine environments.

4.4.5 Magnitude, duration and frequency of solution

Gunn (1982) reviewed the magnitude and frequency properties of dissolved-solids transport. He compared results from 24 basins of which 10 are underlain by carbonate rocks (Table 4.5) and reached the following conclusions.

1. The greatest variation in solute transport work achieved is shown by high flows operational for only 5% of the time. In the carbonate basins these flows account for less than one-quarter of the work done (except in one case where it is 44%) compared with 24–57% in the non-carbonate catchments.
2. Flows less than mean discharge occur for 60–75% of the time and account for 20–55% of the dissolved-load transport. Flows less than median discharge usually account for less than one-quarter of the solute load.
3. High flows are less significant for dissolved-load transport in carbonate basins than in non-karst basins. Nevertheless, Gunn refuted Wolman and Miller's (1960) suggestion that a very large part of solute load is transported by flows as low as the mean or even the median discharge. In the carbonate basins flows greater than the mean account for 45–74% of the dissolved-load transport. Groves and Meiman (2005) found from a 1-year study in a partly allogenic basin in Kentucky that intense events that occurred <5% of the time were responsible for 38% of the dissolved-load transport.

Table 4.5 Magnitude and frequency parameters for dissolved solids transport (Simplified from Gunn, J., Magnitude and frequency properties of dissolved solids transport. Zeitschrift für Geomorphologie, 26(4), 505–11 © 1982 E. Schweizerbart'sche Verlagsbuchhandlung, Science Publishers)

Drainage basin		Percentage of annual solute load transported by:			
		Flows equalled or exceeded 5% of time	Flows less than the mean discharge	Flows less than the median discharge	Percentage of time required to remove 50% of solute load
Shannon	(CO$_3$)	—	32	28	—
Rickford	(CO$_3$)	5	34	23	24
Langford	(CO$_3$)	10	48	35	30
South Rockies	(CO$_3$)	13	26	19	20
	(SO$_4$)	12	35	26	23
Riwaka	(CO$_3$)	44	33	20	10
Honne	(CO$_3$)	16	—	26	26
Cymru	(CO$_3$)	18	34	21	22
Glenfield	(CO$_3$)	15	33	24	25
Cooleman Plain	(CO$_3$)	21	55	29	29
Southeast Devon (1)	(TDS)	57	20	7	5
(2)	(TDS)	29	—	17	15
(3)	(TDS)	29	46	25	18
Slapton Ley	(TDS)	28	26	12	12
Ei Creek	(TDS)	>·25	—	—	10
East Twin GP1	(TDS)	27	—	11	1.5
GP2	(TDS)	24	—	20	18
New England	(TDS)	50	—	—	5
Creedy	(TDS)	25	—	15	12

The evidence is not sufficiently comprehensive to distinguish with confidence between the magnitude and frequency behaviour of autogenic as compared with allogenic basins. However, theoretical considerations suggest that the greater the autogenic component, the less significant will be the relatively high-magnitude but low-frequency discharges in transporting solute load. This is because in autogenic basins the Ca^{2+} versus discharge relation usually has a lower slope than in allogenic basins, i.e. the dilution effect is less marked; and in autogenic basins the discharges are in any case less variable, i.e. the flow duration curve shows a narrower range. In purely autogenic karsts the solute magnitude and frequency relationships are essentially controlled by the regime of the outflowing stream.

4.4.6 Solutional denudation, karstification and inheritance

In discussing the relevance of solution rates to geomorphology, Priesnitz (1974) quite rightly observed that an average annual denudation rate does not characterize karstification. For example, there is a strong discrepancy between average lowering of the surface and the solutional

modelling of the terrain. Regions with high solutional denudation rates do not necessarily display well-developed karst. He suggested that factors important in modelling karst include the size and form of the solution front, its location, the intensity of dissolution at each point across the front, and the form and location of eventual redeposition. Priesnitz also proposed the use of a surface-lowering: surface-modelling ratio as an indicator of the morphological effectiveness of solution. Estimating the surface-modelling effect from the volume of dolines and applying this idea to an area of gypsum and limestone karst near Bad Gandersheim in Germany, he concluded that for both rock types 98–100% of the considerable solution in the Holocene has produced only surface lowering. In the decades since that investigation we have not made much more headway on this topic: the relationship between karstification and solutional denudation clearly requires much more study.

Another factor barely recognized and certainly unresolved in the interpretation of denudation rates is that of inheritance. Imagine two areas of dense limestone near baselevel that are identical in every respect except that one is already karstified and the other is not. They are uplifted an equal amount and subjected to similar climatic regimes. What would be the effect of the different geomorphological inheritance on the amount and distribution of solutional denudation and redeposition that results in these two karsts? Although this question is unresolved, different geomorphological inheritances must be significant in steering erosion processes, ventilation and degassing of groundwaters, and resultant landform development. In geomorphology as a whole, timescales are so long that there is seldom a recognizable beginning for a landscape, only an inheritance. In karst terrains more than most we can sometimes identify a beginning; perhaps a time when an impervious caprock was first breached or when a coral reef was uplifted from the sea. But the commencement of karstification in some of the world's greatest karsts, as in southern China for instance, is so far back that landforms developing in response to modern processes must depend in part on preconditioning that provides ready made avenues for solute attack.

5

Karst Hydrogeology

5.1 BASIC HYDROGEOLOGICAL CONCEPTS, TERMS AND DEFINITIONS

Karst groundwaters display many of the characteristics of underground water in other rocks. Consequently many of the concepts, principles and techniques applied to groundwater hydrology in general are applicable here. Therefore we recommend that for general detail on aspects of groundwater hydrology, texts such as those by Freeze and Cherry (1979) and Domenico and Schwartz (1998) should be consulted. However, karst groundwater systems have some features that markedly distinguish them from the rest. Thus the purpose of this chapter is to explain when universal principles of groundwater hydrology are appropriate for application in a karst context and, most importantly, when they are not.

5.1.1 Aquifers

A rock formation is regarded to be an **aquifer** when it can store, transmit and yield economically significant amounts of water. Karst aquifers like those of other rocks may be **confined**, **unconfined** and **perched** (Figure 5.1). A confined aquifer is contained like a sandwich between relatively impervious rocks that overlie and underlie it. In contrast to the formation containing the aquifer, an impermeable rock that is incapable of absorbing or transmitting significant amounts of water is known as an **aquifuge**. Other rocks such as clay and mudstone, may absorb large amounts of water, but when saturated are unable to transmit it in significant amounts. These are termed **aquicludes**. A relatively less permeable bed in an otherwise highly permeable sequence is referred to as an **aquitard**; a calcareous sandstone in a karstified limestone sequence could provide such a case.

The lower boundary of an aquifer is commonly an underlying impervious formation. But should the karst rocks be very thick, the effective lower limit of the aquifer occurs where no significant porosity has developed. This may be because the rocks have only recently been exposed to karstification or because lithostatic pressure at depth is so great that there is no penetrable fissuring and groundwater is consequently precluded. In regions of extensive or continuous permafrost, the aquifer may be restricted to a surficial 'active' layer from about 0.5 to 1 m in depth. However, in karstic rocks the permafrost aquifuge beneath the active layer is often breached by groundwater drains termed **taliks** and there may be an underlying circulation (e.g. Ford 1984).

Water within the rock occurs in pore spaces of various size, shape and origin. A distinction is made between the **porosity** n of a rock and its effective **porosity** n_e. Porosity is defined as the ratio of the aggregate volume of pores V_p to the total bulk volume V_b of the rock. Thus

$$n = V_p/V_b \qquad (5.1)$$

Effective porosity refers only to those voids that are hydrologically interconnected. For a fully saturated rock, it can be expressed as the ratio of the aggregate volume of gravitation water that will drain from the rock V_a to the total bulk volume of the rock V_b

$$n_e = V_a/V_b \qquad (5.2)$$

In unconfined aquifers, it amounts to the volume of water that will drain freely under gravity from a unit volume of the aquifer; thus bound water still remains held by molecular forces in small pores. Castany (1984b) describes measurement techniques.

Effective porosity is influenced by pore size. Thus clays with a porosity of 30–60%, but with pores of only

Karst Hydrogeology and Geomorphology, Derek Ford and Paul Williams
© 2007 John Wiley & Sons, Ltd

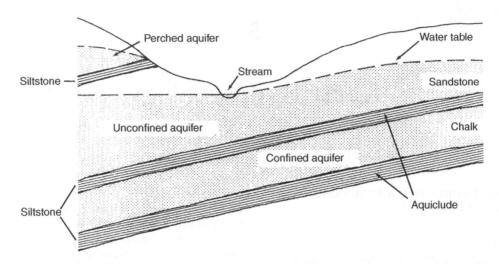

Figure 5.1   Confined, unconfined and perched aquifers. Reproduced from Dunne, T.R. and L.B. Leopold, Water in Environmental Planning, San Francisco © 1978 Freeman.

1×10^{-2} to 10^{-3} mm in width yield almost no water when able to drain freely under gravity. This is because the force of molecular adhesion involved in the adsorption of water in clay is sufficient to overcome the force of gravity. Yet massive karstified limestone with a porosity of perhaps only 3%, but with interconnected voids measuring 1–10^3 mm or more in diameter, will yield almost all the water in storage if freely drained.

We saw in Chapter 2 that when a carbonate sediment is formed it acquires a fabric-selective porosity of about 25–80% from interstitial spaces between its mix of materials. However, later chemical diagenetic processes involving dissolution and reprecipitation, dolomitization and subsequent fracturing by tectonic movements result in the modification of the original porosity. Sedimentologists define primary porosity as that created during deposition of the rock (i.e. created first) and secondary porosity as that resulting later from diagenesis. However, for the hydrogeologist, all types of bulk rock porosity are **primary** (sometimes referred to *as* **matrix porosity**), and only **fracture** (or **fissure**) porosity arising from rock folding and faulting should be considered to be **secondary.** When dissolution along penetrable fissures by circulating groundwaters develops some pathways into pipes (conduits or caves), this is referred to as **tertiary porosity**. These voids may continue to enlarge while groundwater circulation persists.

Average total porosity is a function of the reference volume of the rock considered, i.e. whether it is of regional scale (macroscopic or first-order), pumping-test scale (mesoscopic or second-order), or at hand-specimen scale (microscopic or third-order). Clearly we do not find caves

in hand specimens, and nor do we usually encounter them in boreholes. Scale of investigation is therefore an important consideration when evaluating porosity (Figure 5.2) and so too is depth, because lithostatic pressure and chemical compaction during burial and early diagenesis results in an exponential reduction of porosity with depth. Castany (1984a) cited an example from South Africa where effective porosity diminishes from 9% at 60 m below the surface to 5.5% at 75 m, 2.6% at 100 m, 2% at 125 m and 1.3% at 150 m. However, porosity–depth curves can in fact be quite variable according to the processes to which they respond (Figure 5.3).

Carbonate rocks thus acquire a range of voids of different origin that affect their capacity to store and transmit water. Karst aquifers are therefore commonly differentiated into three end-member types, according to the nature of the voids in which the water is stored and through which it is transmitted, namely **granular** (or **matrix**), **fracture** and **conduit** (Figure 5.4). In practice, most karst aquifers have components of each, but by definition must possess a significant conduit component (cave-like tubes). Table 5.1 provides examples of the distribution of porosity in four carbonate aquifers of widely differing rock type (limestone and dolostone), recharge (allogenic and autogenic) age and diagenetic maturity (Palaeozoic to Cenozoic). Matrix porosity is important in every case and is also important in providing storage capacity (Table 5.2), yet despite this and as we shall see later, in each of these cases it is the tertiary conduit (channel) porosity that dominates when it comes to providing pathways for groundwater flow.

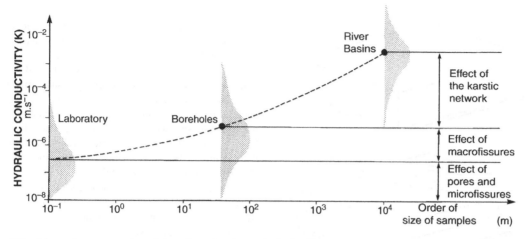

Figure 5.2 Schematic representation of the effect of scale on the hydraulic conductivity of karst. Reproduced from Kiraly, L., Rapport sur l'etat actuel des connaissances dans le domaine des caracteres physiques des roches karstiques. In A. Burger and L. Dubertret (eds.) Hydrogeology of Karstic Terrains, Series B, 3, 53–67 © 1975 International Union of Geological Sciences.

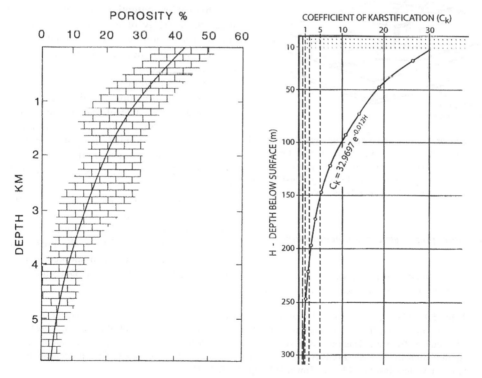

Figure 5.3 Changing porosity in limestones with depth beneath the surface. (Left) Exponential decrease of porosity with depth in shallow-marine limestones of the south Florida basin. The ornament shows the range. This illustrates the response of primary porosity to lithostatic pressure upon burial. (After Halley 1987; cited in Tucker and Wright 1990.) (Right) Exponential increase in karstification towards the surface in crystalline limestones of the Dinaric karst. This shows the response of secondary and tertiary porosity to karst dissolution. Based on data from 146 boreholes. Adapted from, Milanovic, P.T, Karst Hydrogeology Colorado, p. 434 © 1981 Water Resources Publication; and Milanovic, P. 1993. The karst environment and some consequences of reclamation projects. In Afrasiabian, A. (Ed.), Proceedings of the International Symposium on Water Resources in Karst with Special Emphasis on Arid and Semi Arid Zones, Shiraz, Iran, 409–424, 1993.

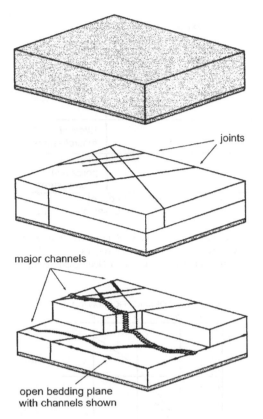

joints

major channels

open bedding plane
with channels shown

Figure 5.4   Granular, fracture and conduit aquifers. Reproduced from Worthington, S. R. H., A comprehensive strategy for understanding flow in carbonate aquifers. In A. N. Palmer, M. V. Palmer and I. D. Sasowsky (eds), Karst Modelling, Karst Waters Institute Special Publication 5, Charles Town, W. Virginia, 30–37, 1999.

5.1.2 *Water table, piezometric and potentiometric surfaces*

Water in an unconfined aquifer descends freely under gravity until it finds its own level, known as the **water**

table. This is an English language term for the surface defined by the level of free-standing water in fissures and pores that delimits the top of the saturated (or **phreatic**) zone. It is an equilibrium surface at which fluid pressure in the voids is equal to atmospheric pressure. An equivalent and completely acceptable alternative term frequently used in continental European literature is the **piezometric surface**, i.e. a surface defined by water levels in piezometers (observation wells) in an unconfined aquifer. We note that Domenico and Schwartz (1998) prefer the term water table.

The pressure at a given point below the water table is equal to the product of the depth and the unit weight of water, termed the **pressure head**, plus atmospheric pressure (Figure 5.5). When there is no flow pressures are equal in all directions and conditions are termed **hydrostatic**. The **hydraulic head** 'h' is the sum of the elevation head and the pressure head. The product of hydraulic head and acceleration due to gravity g yields the **hydraulic potential** 'Φ', an expression of the mechanical energy of water per unit mass. As gravitational acceleration is nearly constant near the Earth's surface, hydraulic head and hydraulic potential are very closely correlated.

In a confined aquifer the water in a bore tapping the water-bearing formation usually rises up the borehole under pressure to a level that is above the top of the aquifer. The theoretical surface fitted to the water levels in such bores is termed the **potentiometric surface**. When water rises up a bore, it is sometimes called an **artesian** well and the water is said to be confined under artesian conditions.

Unfortunately, confusion has arisen in the literature because the potentiometric surface, which is an imaginary surface contoured from hydraulic head data, has sometimes been referred to as the piezometric surface. The usage recommended and adopted here follows both Freeze and Cherry (1979) and Domenico and Schwartz (1998).

Table 5.1   Matrix, fracture and channel (conduit) porosity in four carbonate aquifers. The channel porosity in the chalk is equivalent to that of the 'secondary fractures' of Price *et al.* (1993) (Reproduced from Worthington, S.R.H. A comprehensive strategy for understanding flow in carbonate aquifers, in Karst Modelling (eds A.N. Palmer, M.V. Palmer and I.D. Sasowsky), Special Publication 5, Karst Waters Institute, Charles Town, WV, pp. 30–7, 1999)

Location	Porosity (%)			Age of rocks
	Matrix	Fracture	Channel	
Smithville, Ontario, Canada	6.6	0.02	0.003	Silurian
Mammoth Cave, Kentucky	2.4	0.03	0.06	Mississippian
The Chalk, England	30	0.01	0.02	Cretaceous
Nohoch Nah Chich, Yucatan, Mexico	17	0.1	0.5	Eocene

Table 5.2 Fractions of storage contributed by matrix, fracture and channel (conduit) porosity in four carbonate aquifers. (Reproduced from Worthington, S. R. H., Ford, D. C. and Beddows, P. A., Porosity and permeability enhancement in unconfined carbonate aquifers as a result of solution. In Klimchouk, A.B., Ford, D.C., Palmer, A.N. and Dreybrodt, W. (Eds.), Speleogenesis: Evolution of Karst Aquifers. National Speleological Society, Huntsville, 220–223, 2000

Location	Proportion of storage (%)		
	Matrix	Fracture	Channel
Smithville, Ontario, Canada	99.7	0.3	0.05
Mammoth Cave, Kentucky	96.4	1.2	2.4
The Chalk, England	99.9	0.03	0.07
Nohoch Nah Chich, Yucatan, Mexico	96.6	0.6	2.8

In the unsaturated (or **vadose**) zone above the water table, voids in the rock are only partially occupied by water, except after heavy rain when some fill completely. Water percolates downwards in this zone by a multiphase process, with air and water coexisting in the pores and fissures. Air bubbles may even impede percolation by blocking capillary channels. More significant impediments to downwards flow are some-times provided by localized impermeable layers, such as shale or chert bands in a limestone sequence. Ponding occurs above these layers, producing a localized saturated zone known as a **perched** aquifer, suspended above the main water table (Figure 5.1). Subdivisions of the unsaturated and saturated zones of an unconfined aquifer are listed in Table 5.3, although not all categories may be present in any given karst.

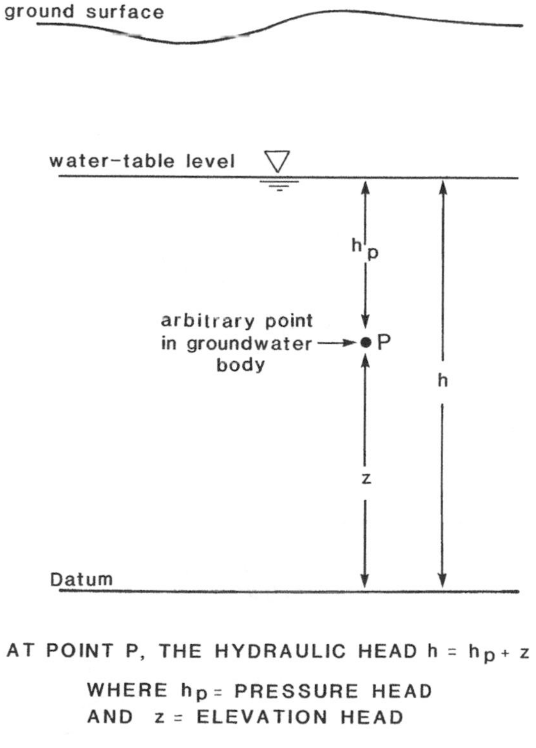

5.1.3 Flow nets

A map of the water table or of a potentiometric surface provides a two-dimensional view of the aquifer. The general movement of groundwater can be deduced as being perpendicular to the contours on these surfaces, in the direction of the steepest hydraulic gradient. However, aquifers are three-dimensional phenomena. Consequently a more complete view of groundwater movement is obtained from considering variations in **hydraulic potential** throughout the aquifer (Figure 5.6).

Points of equal fluid potential within an aquifer can be depicted by equipotential surfaces. When mapped in two dimensions on the horizontal plane these project as **equipotential lines**, and are represented in the case of

ground surface

water-table level $\triangledown$

h_p

arbitrary point in groundwater body $\rightarrow \bullet$ P

h

z

Datum

AT POINT P, THE HYDRAULIC HEAD $h = h_p + z$

WHERE h_p = PRESSURE HEAD
AND z = ELEVATION HEAD

Figure 5.5 Definition of hydraulic head, pressure head and elevation head for an unconfined aquifer. Reproduced from Ford, D.C. and Williams, P.W. (1989) Karst Geomorphology and Hydrology.

Table 5.3 Karst hydrographic zones

1 Unsaturated (vadose) zone*	1a Soil
	1b Epikarst (subcutaneous) zone
	1c Free-draining percolation (transmission) zone
2 Intermittently saturated zone*	Epiphreatic zone (zone of fluctuation of water table)
3 Saturated (phreatic) zone*	3a Shallow phreatic zone
	3b Deep phreatic (bathyphreatic) zone
	3c Stagnant phreatic zone

*Each may be traversed by caves, permanently flooded in zone 3.

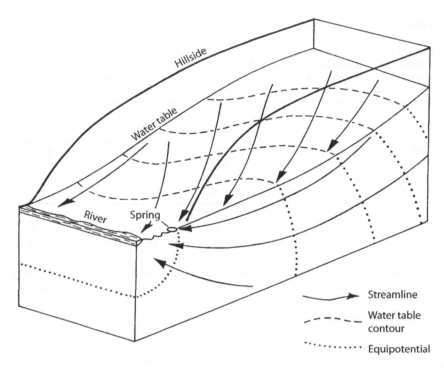

Figure 5.6   Water table contours, streamlines, equipotentials and flow net in an ideal porous, homogeneous medium. Reproduced from Ford, D.C. and Williams, P.W. (1989) Karst Geomorphology and Hydrology.

an unconfined aquifer by contours on the water table. These are lines of equal head. They can also be mapped in two dimensions in the vertical plane, when they show cross-sections of equipotential surfaces (Figure 5.6).

Water flow in an aquifer is always orthogonal to the equipotentials, and the path followed by a particle of water is known as a **streamline**. A mesh formed by a series of equipotentials and their corresponding stream-lines is known as a **flow net**. Since water flows from zones with high potential to places where it is lower, if fluid potential increases with depth then groundwater flow will be towards the surface (Hubbert 1940). Flow nets in the vertical plane parallel to the hydraulic gradient of the water table usually show flow to converge near valley floors or along the coast. If a large conduit (such as a water-filled cave) traverses the saturated zone, it may also be a zone of relatively low fluid potential and thus may cause flow to converge on it (Figure 5.7).

5.1.4 *Flow through pores and pipes*

Conventional groundwater hydrology usually considers aquifers to be porous, granular media; so we must question whether the normal laws of groundwater

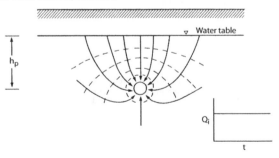

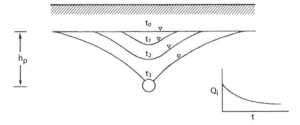

Figure 5.7   Tunnel or cave as (a) a steady-state and (b) a transient drain. Reproduced from Freeze, R.A. and J.A. Cherry, Groundwater, p. 604 © 1979 Prentice Hall.

hydrology are applicable to fractured rocks perforated by large solution pipes as found in limestones. The consideration raised by this is whether a better explanation of groundwater movement in karst will be achieved by understanding the flow of water through individual conduits or by treating the rock as an idealized continuum of saturated voids in a solid matrix. The pioneering work of Hagen (1839) and Poiseuille (1846) as compared to Darcy (1856) helps to illustrate the point.

Hagen and Poiseuille studied the flow of water through small tubes and discovered that the flow per unit cross-sectional area, or **specific discharge** u, is directly proportional to the hydraulic head loss by friction between one end of the tube and the other

$$u = \pi r^2 \left(\frac{r^2}{8} \cdot \frac{\rho g}{\mu} \cdot \frac{dh}{dl} \right) = \frac{\pi r^4}{8} \cdot \frac{\rho g}{\mu} \cdot \frac{dh}{dl} \qquad (5.3)$$

where πr^2 is the cross-sectional area of a tube of radius r, $r^2/8$ is the permeability of the tube, g is gravitational acceleration, μ is the dynamic viscosity and ρ is the fluid density of the water, and dh/dl is the change of head with distance along the tube (also termed the **hydraulic gradient**). This equation is sometimes referred to as **Poiseuille's law**. It shows specific discharge to vary directly with the pressure drop and with the fourth power of the radius, but to vary inversely with the length of the tube and the viscosity of the fluid.

By contrast, Darcy (1856) studied the flow of water through saturated sand, and the results of his experiments confirmed Hagen and Poiseuille's findings by showing that the quantity of water flowing through a granular medium is proportional to the difference in pressure, as represented by a hydraulic gradient, between the inflow and outflow points. The relationships are expressed in what is termed **Darcy's law**

$$u = -K \frac{dh}{dl} = \frac{Q}{a} \qquad (5.4)$$

where Q is the **discharge**, a is the cross-sectional **area** of granular medium through which it flows and K is the **hydraulic conductivity** (or coefficient of permeability). Because water moves down a gradient in response to changes in hydraulic potential, the negative sign on the right-hand term of the equation is a convention to indicate a loss in hydraulic potential in the direction of flow over the distance l travelled. Whereas discharge Q has the dimensions $(L^3 T^{-1})$, specific discharge u has the dimensions $(L T^{-1})$.

Hydraulic conductivity K is sometimes loosely (and incorrectly) referred to as the **permeability** k. The latter, also called the intrinsic permeability, is a measure of the

ability of a material to transmit fluids. It depends on the physical properties of the material, especially pores sizes, shapes and distribution. By contrast, the hydraulic conductivity reflects the properties of **both** the medium and the fluid. The two terms are related as follows

$$K = \frac{k \rho g}{\mu} \qquad (5.5)$$

where ρ, mass density, and μ, dynamic viscosity, are functions of the fluid alone, K has the dimensions of a velocity $(L T^{-1})$ and is commonly expressed in m s^{-1}, and k has the dimensions (L^2) and is sometimes expressed in **darcy** units, 1 darcy being about 10^{-8} cm^2 (Freeze and Cherry 1979). Figure 5.8 shows the range of values of hydraulic conductivity and permeability encountered in common earth materials. Permeability is not easy to measure in karstified aquifers, because it varies with the scale of the measurement and can be so high that standard pump tests obtain no measurable drawdown (Halihan *et al.* (1999).

Darcy's law assumes flow to be **laminar**. Under these conditions individual 'particles' of water move in parallel threads in the direction of flow, with no mixing or transverse component in their motion. This can be visualized most easily in a straight cylindrical tube of constant diameter (Figure 5.9), but it also applies to granular media. At the tube wall flow velocity v is zero, because of adhesion, and it rises to a maximum at the centre. But as the radius of the tube and flow velocity increase, so fluctuating eddies develop and transverse mixing occurs. The flow is then termed **turbulent**. Such conditions frequently arise in pipes and fissures in karst, and can be considered predominant in cave systems.

The **Reynold's number** Re is usually used to help identify the critical velocity at which laminar flow gives way to turbulent flow. It is thus valuable in helping to define the upper limit to the validity of Darcy's law. Reynold's number is expressed as

$$Re = \frac{\rho v d}{\mu} \qquad (5.6)$$

where v is the mean velocity of a fluid flowing through a pipe of diameter d. In a porous or fissured medium, the macroscopic velocity u may be substituted for v and d becomes a representative length dimension characterizing interstitial pore-space diameter or fissure width.

Under laminar flow conditions, discharge through a pipe can be evaluated by Poiseuille's law (Equation 5.3), or with the Hagen–Poiseuille equation

$$Q = \frac{\pi d^4 \rho g}{128 \mu} \cdot \frac{dh}{dl} \qquad (5.7)$$

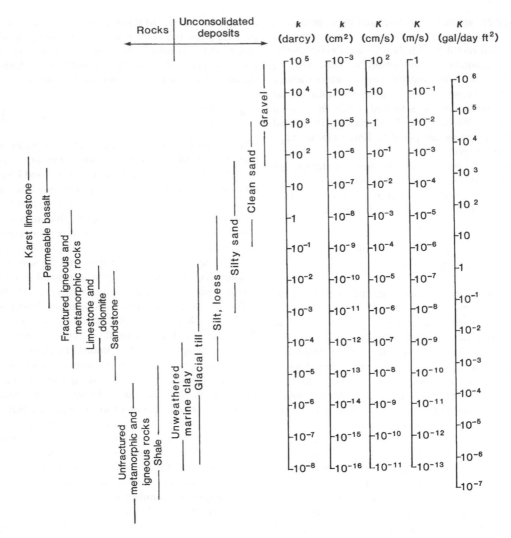

Figure 5.8 Range of values of hydraulic conductivity and permeability. Reproduced from Freeze, R.A. and J.A. Cherry, *Groundwater*, p. 604 © 1979 Prentice Hall.

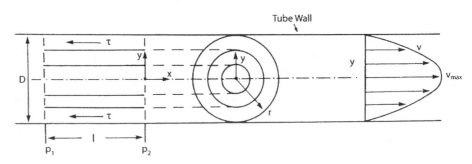

Figure 5.9 Laminar flow through a cylindrical tube. Reproduced from Hillel, D. (1982) *Introduction to Soil Physics*, Academic Press, New York.

Since the discharge is proportional to the fourth power of the diameter, large capillary tubes are very much more conductive than small ones. A tube of 2 mm diameter will conduct the water passed by 10 000 capillaries of diameter 0.2 mm. This factor dominates the processes of cave pattern construction as they are developed in Chapter 7.

Increasing velocity, sinuosity and roughness may eventually result in flow through the tube becoming turbulent. When this occurs, the specific discharge may be calculated using the Darcy–Weisbach equation (Thrailkill 1968)

$$u^2 = \frac{2dg}{f} \cdot \frac{dh}{dl} = \frac{Q^2}{a^2} \qquad (5.8)$$

and hence

$$Q = \left(\frac{2dga^2}{f}\right)^{1/2} \cdot \left(\frac{dh}{dl}\right)^{1/2} \qquad (5.9)$$

where f is a **friction factor** or coefficient. Spring and Hutter (1981a, b) explained the determination of the friction factor and discussed the relationship of the Darcy–Weisbach and Manning–Gauckler–Strickler approaches. They pointed out that in turbulent pipe flow wall friction gives rise to a quadratic dependence of shear stress τ on mean water velocity v, as expressed by the Darcy–Weisbach friction law

$$\tau = \frac{f\rho_f}{8} \cdot v^2 \qquad (5.10)$$

where ρ_f is the density of fresh water. The friction factor can thus be estimated by rearranging the equation. Relative roughness R_f on a flow boundary (e.g. walls of a pipe) can be described by the ratio of pipe radius r to a characteristic length e representing the size of the features contributing to roughness, such as projections, cavities and loose grains.

An excellent study by Lauritzen *et al.* (1985) revealed the relationship between f and Q in an active phreatic conduit in Norway (Figure 5.10a). Apparent friction shows a dramatic decrease with discharge, until attaining a constant value. A similar conclusion was reached by Crawford (1994) in New Zealand (Figure 5.10b). Friction is affected by complexity of conduit geometry, tube dimensions, roughness of walls and occurrence of breakdown. In the comparatively simple conduit studied in Norway, Lauritzen *et al.* found f to be particularly related to the hydraulic radius of the passage and the characteristic scallop size on its walls. Darcy–Weisbach f values determined in karst investigations lie in the range 0.039 to 340, with up to 208 estimated in the great drowned cave systems of Florida, whereas 0.25 was found appropriate

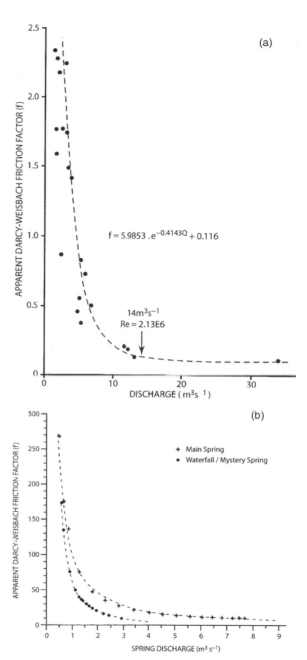

Figure 5.10   (a) Relationship between f and Q in an active phreatic conduit in Norway. Reproduced from Lauritzen, S.-E., Abbott, J., Arnesen, R. *et al.* (1985) Morphology and hydraulics of an active phreatic conduit. Cave Science, 12(4), 139–46. (b) Relationships between f and Q for conduit segments in parts of Bullock Creek cave system, New Zealand. Reproduced from Crawford, S.J., Hydrology and geomorphology of the Paparoa Karst, north Westland, New Zealand. Auckland Univ. Unpublished PhD thesis, New Zealand, 1994.

for a cave in ice (Spring and Hutter 1981b). Dreybrodt and Gabrovšek (2000a) modelled the influence of fracture roughness on the time it takes for newly developing tubes to penetrate right through the developing aquifer (termed the **break-through time** of conduits) and found that even when the flow through a smooth fracture is reduced by roughness by a factor of ten, breakthrough time is increased only by a factor of four.

5.1.5 Flow through fractured rocks

Most karst rocks are fissured because of the presence of joints, faults and bedding planes. Interconnected fissures provide routes for water flow, with the solid intervening blocks of rock constituting the matrix often (but not always) being relatively impermeable. Contemporary understanding of fluid flow through fractures in rock has been reviewed by the Committee on Fracture Characterization and Fluid Flow (1996), although they did not specifically address flow through fractures in karstified rocks. Domenico and Schwartz (1998) mentioned that at the scale of the field problem one of two approaches might be followed when trying to understand the flow of water in these rocks.

1. The continuum approach, which assumes that the fractured mass is hydraulically equivalent to a porous, granular medium (**equivalent porous medium (EPM) model**).
2. The discontinuum (or discrete) approach, which assumes that the rock cannot be characterized as a granular medium, and so considers that flow is best dealt with in individual fractures or fracture sets.

Where a continuum approach is considered inappropriate, information is required about the orientation of fracture fields, the fissure frequency (i.e. fracture density), the extent of their interconnectivity, and the size and smoothness of the fracture openings. Where laminar flow conditions exist in jointed rock, the volumetric flow rate Q through a single fracture of unit length represented by two smooth parallel plates separated by a constant distance can be modelled by what is sometimes called the 'cubic law'. Porosity increases linearly with the fracture aperture w, whereas permeability increases with the cube of the aperture

$$Q = \frac{\rho g w^3}{12\mu} \cdot \frac{dh}{dl} \tag{5.11}$$

The hydraulic conductivity of the fracture can be determined from the expression

$$K = \frac{\rho g w^2}{12\mu} \tag{5.12}$$

Brady and Brown (1985) observed that for water at 20°C, if w increases by a factor of 10 from 0.05 mm to 0.5 mm, then K increases by a factor of 1000 (from $K = 1.01 \times 10^{-7}$ to $1.01 \times 10^{-4}\,\mathrm{m\,s^{-1}}$). Values of fluid density, dynamic viscosity and kinematic viscosity for different water temperatures are given in Table 5.4.

For a set of planar fractures an equivalent hydraulic conductivity and permeability may be calculated from

$$K = \frac{\rho g N w^3}{12\mu} \tag{5.13}$$

and

$$k = \frac{N w^3}{12} \tag{5.14}$$

where N is the number of fissures per unit distance across the rock face and Nw is the planar porosity. Domenico and Schwartz (1998) pointed out that in an array of fissures with an aperture of 1 mm and with a density of one joint per metre, the equivalent hydraulic conductivity

Table 5.4 Values for fresh water of fluid density and viscosity for different water temperatures at one atmosphere pressure (1 bar = 100 kPa). Kinematic viscosity $v = m = r$. Reproduced from Handbook of Chemistry and Physics © 2003 CRC Press, Taylor and Francis

Temperature (°C)	Fluid density ρ (g cm^{-3})	Viscosity μ (μPa s)
0	0.99984	1793
10	0.99970	1307
20	0.99821	1002
30	0.99565	797.7
40	0.99222	653.2
50	0.98803	547.0

is $8.1 \times 10^{-2}\,\mathrm{cms^{-1}}$; and that just as much water will be conducted through the fissures of $1\,\mathrm{m}^2$ of such rock as through the pores of $1\,\mathrm{m}^2$ of a conventional Darcian porous medium under identical hydraulic gradients.

Natural conditions are of course more complex than the above, partly because of the uneven openness and roughness of fractures (e.g. Figure 2.15) and partly because of the three-dimensional complexity of fracture patterns. Kiraly (2002) has modelled more complex cases, where permeability and porosity are associated with (i) three orthogonal and equally developed fracture families (Figure 5.11a) and (ii) three intersecting bundles of tubes of dissolutional origin (Figure 5.11b). Plotted on the diagrams are field measurements from the karst of the Swiss Jura Mountains of permeability (about $10^{-6}\,\mathrm{m\,s^{-1}}$) and effective porosity (only 0.4–1%). He concludes that a unique fracture aperture or channel diameter cannot be found that would explain these measurement values. Thus

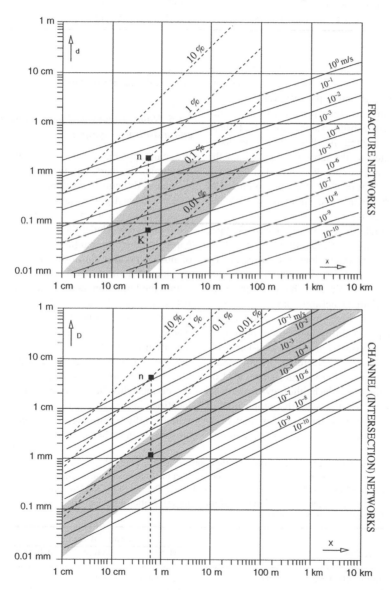

Figure 5.11 (a) Hydraulic conductivity and porosity values for various fracture networks. (b) Hydraulic conductivity and porosity values for various channel (intersection) networks. Reproduced from Kiraly, L.,. Karstification and groundwater flow. In Gabrovsek, F. (Ed.) Evolution of Karst: from Prekarst to Cessation. Postojna-Ljubljana: Institut za raziskovanje krasa, ZRC SAZU, 155–190, 2002.

Table 5.5   Matrix, fracture and channel (conduit) permeability in four carbonate aquifers (Reproduced from Worthington, S. R. H. A comprehensive strategy for understanding flow in carbonate aquifers. In A N Palmer, M V Palmer and I D Sasowsky (eds), Karst Modelling, Karst Waters Institute Special Publication 5, Charles Town, W. Virginia, 30–37, 1999)

Location	Hydraulic conductivity (m s^{-1})		
	Matrix	Fracture	Channel
Smithville, Ontario, Canada	1×10^{-10}	1×10^{-5}	3×10^{-4}
Mammoth Cave, Kentucky	2×10^{-11}	2×10^{-5}	3×10^{-3}
The Chalk, England	1×10^{-8}	4×10^{-6}	6×10^{-5}
Nohoch Nah Chich, Yucatan, Mexico	7×10^{-5}	1×10^{-3}	4×10^{-1}

the void geometry that determines the permeability is not the same as the void geometry that determines the effective porosity. The effective porosity value requires large openings in the fracture planes (up to 1 mm aperture), but the permeability value shows that these voids are not well connected. It follows that large openings are well connected to the tubes through which the groundwater flows but are poorly connected to each other. Further information on flow through fractures can be found in papers by Brush and Thomson (2003) and Konzuk and Keuper (2004).

The same set of widely differing carbonate rock types used in Table 5.1 provides a valuable illustration of how permeability and groundwater flow are affected by matrix, fracture and conduit porosity. The Tertiary and Quaternary limestones of Yucatan have large well-connected pores and channel networks and, as a consequence, have extremely high hydraulic conductivity values with channel permeability two orders of magnitude higher than that encountered in the well-karstified Mammoth Cave network in Kentucky (Table 5.5). Although at least 96% of aquifer storage is provided by the rock matrix, regardless of which lithology (from these examples) is considered (Table 5.2), the proportion of groundwater flow that passes through this matrix is minimal in all cases (Table 5.6). In every example almost

all the flow takes place through the conduit network. Such is the nature of karst.

5.1.6 Homogeneous and heterogeneous, isotropic and anisotropic aquifers

We can appreciate from the example from the Jura Mountains that karst aquifers can have a complex porosity that is difficult to describe. This contrasts with well-sorted sand and gravel aquifers that have comparatively consistent values for porosity and permeability throughout their extent. Under these latter conditions in a simple porous medium hydraulic conductivity K is independent of **position** within the formation and the aquifer is then considered **homogeneous**. But if K varies with location within the formation then it is considered **heterogeneous**. If hydraulic conductivity is the same regardless of **direction** of measurement, then the aquifer is defined as **isotropic**, but if K varies with direction then it is **anisotropic**. A characteristic of karst aquifers is that they become increasingly heterogeneous and anisotropic with time. In an unkarstified rock, heterogeneity within the permeability field may be perhaps 1 to 50, whereas karstification may increase it to perhaps 1 to 1 million. Palmer (1999) provides an interesting discussion of anisotropy in carbonate aquifers.

Table 5.6   Reproduced from Worthington, S. R. H., Ford, D. C. and Beddows, P. A., Porosity and permeability enhancement in unconfined carbonate aquifers as a result of solution. In Klimchouk, A.B., Ford, D.C., Palmer, A.N. and Dreybrodt, W. (Eds.), Speleogenesis: Evolution of Karst Aquifers. National Speleological Society, Huntsville, 220–223, 2000

Location	Proportion of flow (%)		
	Matrix	Fracture	Channel
Smithville, Ontario, Canada	0.000003	3.0	97.0
Mammoth Cave, Kentucky	0.00	0.3	99.7
The Chalk, England	0.02	6.0	94.0
Nohoch Nah Chich, Yucatan, Mexico	0.02	0.2	99.7

The ease with which water flows through a karst aquifer usually varies according to direction. This condition can be approximated by a parallel capillary tube model, where hydraulic conductivity is greatest in the direction of the tubes but very much less at right angles to them. Thus hydraulic conductivities K_x, K_y and K_z may be envisaged for different directions, although only one value for K will apply in an isotropic aquifer. Horizontal values (K_x, K_y) are of particular interest in the analysis of flow in the phreatic zone, and vertical values K_z are of importance in characterizing recharge through the vadose zone. However, the hydraulic conductivity in any particular direction is unlikely to remain constant over a large distance. Vertical hydraulic conductivity, in particular, diminishes considerably with depth below the surface in most well-karstified rock, because secondary permeability is usually greatest near the surface.

5.1.7 *Storage and transmissivity*

The proportion of water storage held in matrix, fracture and conduit porosity is illustrated for four carbonate lithologies in Table 5.2. However, not all of the water held in storage is released by gravitational flow, because that in the smallest pores is held by capillary attraction and molecular forces, and that below the level of the outflow spring is impounded (Figure 5.12). The volume of water that a unit volume of saturated rock releases following a unit decline in head is a measure of the available storage capacity of the aquifer. In an unconfined aquifer it is termed **specific yield** S_y and in a confined aquifer **specific storage** S_s (Figure 5.13). The **storativity** S of a confined aquifer is defined as the product of specific storage and aquifer thickness b

$$S = S_s b \qquad (5.15)$$

Alternative approaches to determining aquifer thickness are shown in Figure 5.12.

The ability of an aquifer to transmit water is defined by its *transmissivity T*, which is dependent on the thickness b of the aquifer and its hydraulic conductivity

$$T = Kb \qquad (5.16)$$

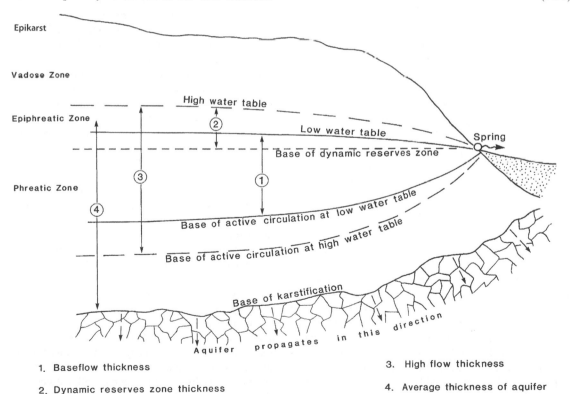

1. Baseflow thickness
2. Dynamic reserves zone thickness
3. High flow thickness
4. Average thickness of aquifer

Figure 5.12 Alternative approaches to determining the thickness of an unconfined karst aquifer. For the calculation of reserves, themaximum dynamic volume is dependent on the thickness of zone 2. Reproduced from Ford, D.C. and Williams, P.W. (1989) Karst Geomorphology and Hydrology.

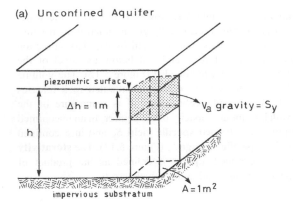

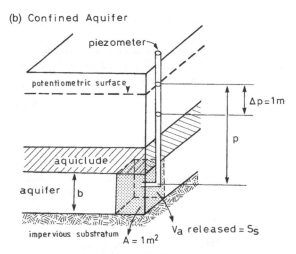

Figure 5.13 Specific storage defined for (a) unconfined and (b) confined aquifers. Reproduced from Castany, G., Hydrogeological features of carbonate rocks. In P. E. LaMoreaux, B. M. Wilson and B. A. Memon (eds.), Guide to the Hydrology of Carbonate Rocks. Studies and Reports in Hydrology 41, Paris: UNESCO, 47–67, 1984.

Clearly, T varies with direction in an anisotropic aquifer.

Specific yield and transmissivity values for some karstic limestone aquifers are presented in Table 5.7. The specific yields of unconfined aquifers are considerably larger than the storativities of confined aquifers (Freeze and Cherry 1979). Both T and S_y for karst aquifers can be estimated from spring hydrograph recession characteristics (Atkinson 1977, Sauter 1992, Baedke and Krothe 2001).

Transmissivity values will vary between the rock matrix, fissure system and conduits, and specific yield can be expected to vary according to vertical position in the karst rock, because effective porosity also varies

vertically. In most cases it will be highest in the most weathered zone of limestone near the surface (the **epikarst**, Table 5.1) and usually diminishes exponentially with depth, but there are exceptions (Figure 5.3). Eventually lithostatic pressure prevents the penetration of groundwater and precludes the development of secondary porosity.

5.2 CONTROLS ON THE DEVELOPMENT OF KARST HYDROLOGICAL SYSTEMS

5.2.1 Boundary conditions and related factors

Kiraly (2002) pointed out that groundwater flow depends on hydraulic parameters and on boundary conditions and, consequently, that other factors such as geology, geomorphology and climate exert their influence on groundwater movement solely through hydraulic parameter fields and boundary conditions. The relationship of the above factors to the physical characteristics of an aquifer, such as porosity, hydraulic conductivity and storage capacity, is illustrated schematically in Figure 5.14. There is considerable interaction amongst the principal factors as well as between them and the chemical and mechanical processes whose rates of activity they determine. Boundary conditions are determined especially by geological, topographic, climatic and biological influences, which control sites and quantities of recharge and discharge, including altitude of recharge, altitude of discharge, rainfall and infiltration rate. Thus one may identify **flow boundaries**, where flow either enters or leaves the karst, as well as **no-flow boundaries**, such as provided by an aquiclude beneath the aquifer or a downfaulted impervious block along one side of the karst.

For a given set of boundary conditions, hydraulic gradient and specific discharge can be estimated. But in the long-term the very process of karst groundwater circulation modifies effective porosity, specific storage and hydraulic conductivity, and lowering of the outlet spring modifies hydraulic potential. Consequently, the karst circulation system undergoes more feed-back giving rise to continuous self-adjustment than occurs in any other type of groundwater system.

The most abrupt changes to boundary conditions are brought about by geomorphologically rapid events (often associated with climatic change) that culminate in major alterations to hydraulic gradient because of modifications to outflow conditions. For example, valley deepening by glaciation increases the hydraulic potential of the system, whereas submergence of coastal springs by glacio-eustatic sea-level rise reduces it.

Table 5.7 Specific yield Sy and transmissivity T values for some karstic aquifers (Reproduced from Castany, G., Hydrogeological features of carbonate rocks. In P. E. LaMoreaux, B. M. Wilson and B. A. Memon (eds.), Guide to the Hydrology of Carbonate Rocks. Studies and Reports in Hydrology 41, Paris: UNESCO, 47–67, 1984)

Carbonate rock category	Age	Location	$S_y(\%)$	$T\ (\mathrm{m^2\ s^{-1}})$
Fissured limestone	Turonian–Cenomanian	Israel		0.1×10^{-2} to 1.3×10^{-1}
	Upper Cretaceous	Tunisia	0.5–1	1
	Jurassic	Lebanon	0.1–2.4	0.1×10^{-2} to 6×10^{-2}
Karstic fissured	Urgonian	Salon (France)	1–5	10^{-3}
limestone	Jurassic	Parnassos (Greece)	5	1 to 2×10^{-3}
Fissured dolomite	Jurassic	Grandes Causses (France)		10^{-3}
	Lias	Morocco		10^{-2} to 10^{-4}
	Jurassic	Parnassos (Greece)		3×10^{-5}
		Murcie (Spain)	7	
Fractured marble		Almeria (Spain)	10–12	
Marly limestone	Jurassic	Grandes Causses (France)		10^{-3}

Geological control influences karst aquifers in several ways (Table 5.8) and is discussed in more detail in Chapter 2. Boundary conditions are influenced at the regional scale through the definition of outcrop pattern, thickness and properties of karst rocks and their relationship to other lithologies. Tectonism affects the balance between rates of uplift and denudation and thus has a major influence on hydraulic potential. Regional structure is also important for its control of folding and faulting.

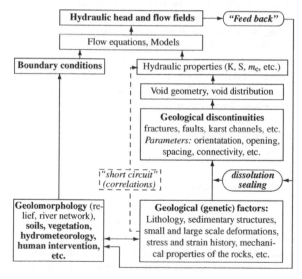

Figure 5.14 Schematic representation of the relations between groundwater flow field, hydraulic properties and geological factors in karst aquifers. Reproduced from Kiraly, L., Karstification and groundwater flow. In Gabrovsek, F. (Ed.) Evolution of Karst: from Prekarst to Cessation. Postojna-Ljubljana: Institut za raziskovanje krasa, ZRC SAZU, 155–190, 2002.

5.2.2 *Input control*

A karst aquifer can be envisaged as an open system with a boundary defined by the catchment limits and with input, throughput and output flows, mechanisms and controls. In the simplest case, only karst rocks are found within the catchment and recharge is derived solely from precipitation falling directly on them—termed **autogenic recharge** (Figure 4.1a). However, commonly more complex geological circumstances occur and runoff from neighbouring or overlying non-karst rocks drains into the karst aquifer—termed **allogenic recharge** (Figure 4.1b). Whereas autogenic recharge is often quite diffuse, down many fissures across the karst outcrop, allogenic recharge normally occurs as concentrated point-inputs of sinking streams. Both the water chemistry and the recharge volume per unit area are different in these two styles of recharge, with considerable consequences for the scale and distribution of the development of secondary permeability.

Emerged coral reefs provide natural examples of simple autogenic systems. Recharge is spatially uniform and distributed through innumerable pores and fissures across the outcrop. Where thick soils cover the bedrock recharge conditions are modified. If the soil is less permeable than the rock beneath, then it provides a recharge regulator, limiting recharge to the infiltration capacity of the soil. Permeable rock formations overlying the karst also act as percolation 'governors' in much the same way, their vertical saturated hydraulic conductivity being the principal control. Percolation input from a permeable soil is considered autogenic, whereas that from a permeable non-karstic caprock is considered diffuse allogenic in origin.

Table 5.8   Effect of hydrogeological setting on carbonate aquifers (Reproduced from White, W.B., Conceptual models for carbonate aquifers: revisited. In R.R. Dilamarter and S.C. Csallany (eds.) Hydrologic Problems in Karst Regions. Western Kentucky Univ., Bowling Green, 176–187, 1977)

Geological element	Control
Macrostructure (folds, faults)	Placement of carbonate rock units relative to other rocks
Topographic setting	Placement of recharge and discharge regions
Stratigraphical sequence	Thickness and chemical character of aquifer
Ministructure (joints, fractures)	Orientation and transmissibility of primary flow paths
Relief	Defines hydraulic gradients

Relatively concentrated recharge occurs in autogenic systems only where solution dolines are well developed (Figure 5.15), as in carbonate and sulphate polyonal karsts (sections 9.6 and 9.13). This is because solution dolines reflect underlying spatial inequalities in vertical hydraulic conductivity that result in the development of preferred percolation paths or zones. The funnelling of rainwater by enclosed depressions positively reinforces the significance of the underlying percolation path by an autocatalytic process (Williams 1985, 1993). However, the volumes of point-input recharge are small compared with those derived from allogenic basins, because of the relatively small surface areas of individual dolines.

Concentrated inflows of water from allogenic sources sink underground at **swallow holes** (also known as swallets, stream-sinks or ponors). They are of two main types: vertical point-inputs from perforated overlying beds and lateral point-inputs from adjacent impervious rocks. The flow may come from: (i) a retreating overlying caprock, (ii) the updip margin of a stratigraphically lower impermeable formation that is tilted, or (iii) an impermeable rock across a fault boundary (Figure 5.16). A perforated impermeable caprock will funnel water into the karst in much the same way as solution dolines, except that the recharge point is likely to be defined more precisely and the peak inflow larger. Inputs of this kind favour the development of large shafts beneath. Lateral-point inputs are usually much greater in volume, often being derived from large catchments, and are commonly associated with major river caves. The capacity of many ponors in the Dinaric karst exceeds $10\,\text{m}^3\,\text{s}^{-1}$ and the capacity of the largest in Biograd-Nevesinjko polje is more than $100\,\text{m}^3\,\text{s}^{-1}$ (Milanović 1993). When the capacity of the swallow hole is exceeded, back-flooding occurs and surface overflow may result.

Discharge through a conduit depends either on the amount of available recharge or on the hydraulic capacity of the passage, termed **catchment control** and **hydraulic control**, respectively, by Palmer (1984, 1991). The capacity of the input passages is the ultimate regulator of the

volume of recharge; thus if the instantaneous inflow from surface streams is too great then ponding occurs, giving rise to overflow via surface channels or to surface flooding in blind valleys and poljes. White (2002) refers to the discharge 'carrying capacity' for the conduit drainage system, from stream-sink to spring.

A more unusual form of input control is provided by head fluctuations caused by floods. This may result in discharge conduits reversing their function temporarily to become inflow passages. This occurs where outflow springs discharge at or beneath river level into a major river flowing through the karst. If a flood wave generated by heavy rain in an upstream part of its basin passes down the major river, the springs will be more deeply submerged and the hydraulic gradient in the karst will reverse, especially if the tributary karst catchment was unaffected by the storm. Inflow into the karst will then occur, as a form of bank storage. Water intruded by back-flooding will later be withdrawn from storage as the flood wave in the main river passes and the hydraulic gradient reverts to normal. Reversing springs of this sort that can temporarily function as sink points are known as **estavelles**. There are many examples along the Green River of Kentucky (Quinlan 1983). The converse is also common: some stream-sinks become temporary springs as the water table rises. These are also estavelles.

5.2.3 Output control

Most of the largest springs in the world are karst springs (Table 5.9). Only those from volcanic rocks rival their discharge output. They represent the termination of underground river systems and mark the point at which surface fluvial processes become dominant. The vertical position of the spring controls the elevation of the water table at the output of the aquifer, whereas the hydraulic conductivity and throughput discharge determine the slope of the water table upstream and its variation under different discharge conditions.

The difference in elevation between the spring and the water table upstream determines the head in the

(a) Surface doline topography

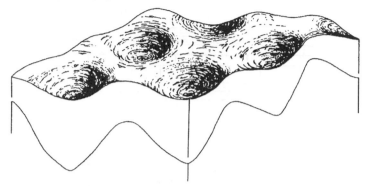

(b) Topography of the subcutaneous water-table

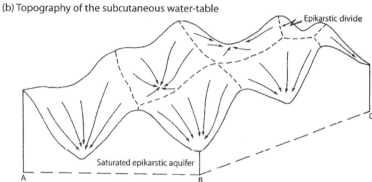

(c) Vertical hydraulic conductivity (m/day)

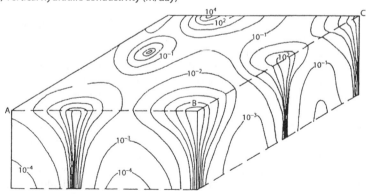

Figure 5.15 The relationship between (a) surface solution doline topography, (b) underlying relief on the subcutaneous (epikarst) water table and (c) vertical hydraulic conductivity near the base of the epikarst. Reproduced with permission from Williams, P.W., Subcutaneous hydrology and the development of doline and cockpit karst. Zeit. Geomorph. 29(4), 463–482, Zeitschrift für Geomorphologie © 1985 E. Schweizerbart'sche Verlagsbuchhandlung, Science Publishers.

system and thus the energy available to drive a deep circulation. Hence springs exercise considerable control on the operation of karst groundwater bodies. Furthermore, that control can vary markedly, because springs are most susceptible to geomorphological events such as glacio-eustatic sea-level fluctuations, valley aggradation, and valley deepening by glacial scour.

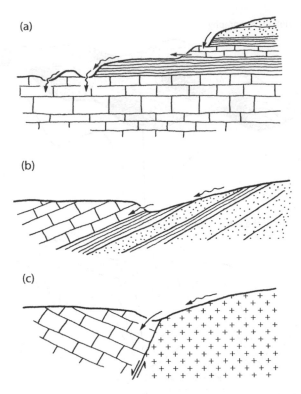

Figure 5.16   Recharge by allogenic streams flowing (a) from overlying beds, (b) from underlying beds exposed updip and (c) across a faulted contact with impervious rocks. Reproduced from Ford, D.C. and Williams, P.W. (1989) Karst Geomorphology and Hydrology.

The influence that springs exert on the aquifer which they drain depends principally upon the topographic and structural context of the spring. Springs may be classified in several ways (Sweeting 1972, Bögli 1980, Smart and Worthington 2004), but when considering their hydrological control function the following perspective is important.

1. **Free Draining Springs** (Figure 5.17a and b). In these cases, the karst rock slopes towards and lies above the adjacent valley, into which karst water drains freely under gravity. The karst system is entirely or dominantly vadose, and is sometimes termed shallow karst (Bögli 1980). Complications may arise where the underlying impermeable rock is folded or has an irregular surface, because then subterranean ponding can occur with the consequent development of isolated phreatic zones.

2. **Dammed Springs** (Figure 5.17c–e). These are the most common type of karst springs. They result from the location of a major barrier in the path of underground drainage. Impoundment may be by another lithology, either faulted or in conformable contact, or

be caused by valley aggradation, such as by glacio-fluvial deposits. The denser salt water of the sea also forms a barrier to freshwater outflow. In each case, temporary overflow springs may form in response to high water tables. The type of cave upstream of the spring will determine whether its discharge spills from a flat passage developed close to the water table or wells up from a great depth within the phreatic zone. Thus a dammed karst outflow site typically consists of a main low-water spring with one or more associated high-water relief springs; Smart (1983a, b) has termed these overflow–underflow systems. In some situations water escapes via a distributary system of several springs at about the same level, as described by Quinlan and Ewers (1981) in the case of the Sinkhole Plain–Mammoth Cave–Green River system.

3. **Confined Springs** (Figure 5.17f and g). Artesian conditions prevail where karst rocks are confined by an overlying impervious formation. Fault planes sometimes provide exit routes for the water; elsewhere it may escape where the caprock is breached by erosion. Since the emerging water is usually under hydrostatic

Table 5.9 Discharges of some of the world's largest karst springs

Spring	Discharge (m³ s⁻¹) Mean	Maximum	Minimum	Basin area (km²)	Reference
Tobio, Papua New Guinea	85–115	—	—	—	Maire 1981c
Matali, Papua New Guinea	90	>240	20	350	Maire 1981b
Trebišnjica, Herzegovina	80	—	—	1140	Milanović 2000
Bussento, Italy	>76	117	76	—	Bakalowicz 1973
Dumanli, Turkey*	50	—	25	2800	Karanjac and Gunay 1980
Galowe, Papua New Guinea	40	—	—	—	Maire 1981a
Ljubljanica, Slovenia	39	132	4.25	1100	Gospodarič and Habić 1976
Ras-el-Ain, Syria	39	—	—	—	Burdon and Safadi 1963
Disu, China	33	390	4	1050	Yuan 1981
Stella, Italy	37	—	23	—	Burdon and Safadi 1963
Chingshui, China	33	390	4	1040	Yuan 1981
Spring Creek, Florida, USA	33	—	—	>1500	Smart and Worthington 2004
Oluk Köprü, Turkey	>30	—	—	>1000	Smart and Worthington 2004
Timavo, Italy	30	138	9	>1000	Smart and Worthington 2004,
Frió, Mexico	28	515	6	>1000	Fish 1977
Ombla, Croatia	27	110	4	800–900	Bonacci 1995, Bonacci 2001
Yedi Miyarlar, Turkey	>25	—	—	>1000	Smart and Worthington 2004
Mchishta, Georgia	25	—	—	—	Smart and Worthington 2004
Coy, Mexico	24	200	13	>1000	Fish 1977
Buna, Herzegovina	24	—	—	110	Smart and Worthington 2004
Liulongdong, China	24	75	9	900	Yuan 1981
Kirkgozler, Turkey	24	—	—	—	Smart and Worthington 2004
Silver, Florida, USA	23	37	15	1900	Faulkner 1976
Rainbow, Florida, USA	22	—	—	>1500	Smart and Worthington 2004
Vaucluse, France	21	100	4	1115	Blavoux *et al.* 1992
Sinjac (Piva), Yugoslavia	21	—	—	500	Smart and Worthington 2004
Bunica, Herzegovina	20	—	—	510	Smart and Worthington 2004
Grab-Ruda, Croatia	20	—	—	390	Smart and Worthington 2004
Trollosen, Spitzbergen	20	—	—	—	Smart and Worthington 2004

*Dumanli spring is the largest of a group of Manavgat River springs that collectively yield a mean flow of 125–130 m³ s⁻¹.

pressure, an updomed turbulent 'boil' is particularly characteristic of spring pools in this class, although dammed springs that are semi-confined by a particularly thick bed may also 'boil', especially during flood. Artesian springs are also sometimes termed **vauclusian** after the type example, La Fontaine de Vaucluse (Tables 5.5 and 5.9), in the south of France (Durozoy and Paloc 1973, Blavoux *et al.* 1992).

The discharge capacity of the artesian spring determines the elevation of the potentiometric surface in the aquifer and hence the depth of the phreatic zone. Artesian springs may also have associated high-water relief springs.

Many confined springs are also thermal because of the deep circulation of the water prior to emergence. Numerous examples of geothermal karst springs are found in China, Hungary and Turkey (Günay and Şimşek 2000).

Other characteristics have also been used to classify karst springs, as noted by Bögli (1980). These include:

1. according to the outflow
   - perennial
   - periodic
   - rhythmic (ebb and flow)
   - episodic
2. according to the supposed origin of the water
   - emergence (no evidence of origin)
   - resurgence (re-emergence of a known swallet stream)
   - exsurgence (autogenic seepage water)

Periodic and rhythmic springs, sometimes referred to as ebbing and flowing wells, are particularly interesting natural phenomena. They are usually dammed springs with a siphoning reservoir system controlling their outflow. They are discussed by Trombe (1952), Mangin (1969), Gavrilovic (1970) and Bonacci and Bojanić (1991).

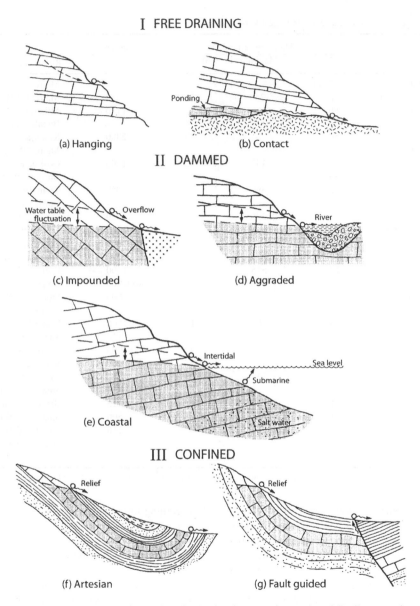

Figure 5.17   Types of springs encountered in karst. Note that in plan there may be a series of distributary springs at about the same level, especially in type II dammed aggraded springs. Reproduced from Ford, D.C. and Williams, P.W. (1989) Karst Geomorphology and Hydrology.

5.2.4 Throughput control

At the scale of the basin flow direction is determined by the direction of the hydraulic gradient, but locally flow direction is determined by pathways made available by interconnected fissures and pores. The factors which determine the density, size and distribution of voids are of fundamental importance in controlling the throughput and storage of water in a karst aquifer, because they dictate the potential flow paths. Anticlinal and synclinal structures are associated with tension and compression, respectively, and thus with joint patterns that reflect these conditions. Joints are most readily penetrated by percolating waters when under tension; thus anticlines (and domes) are potentially important sites for aquifer recharge. On the other hand, synclinal troughs favour flow convergence and the accumulation of groundwater (Lattman and Parizek 1964, Parizek 1976; Figure 2.17). Nevertheless, the relationship between the specific capacity of boreholes and their association with lineaments (joints, faults, etc.) can be complex (Tam *et al.* 2004).

Bedding planes also play a role in linking joint-dominated routes for downward percolation in the vadose zone, but are more important in the phreatic zone because of their great lateral continuity. As dip becomes steeper, so bedding-plane partings can increasingly provide recharge routes. Water confined in major bedding planes between dense, thick sheets of rock may be led to great depths in what amounts to artesian conditions, before cross-joints permit lateral movement.

Faults often operate hydrologically like major joints. Their vertical and lateral continuity can render them particularly important features in the orientation of water flow in both the vadose and phreatic zones. However, many fault planes are highly compressed or filled with secondary calcite of low porosity; they are then barriers to groundwater flow. Further, they are also sometimes important in introducing blocks of other lithologies that may act as barriers to water movement. This may arise from normal or reverse faulting of adjacent non-karst rocks or may involve fault-plane-guided intrusions of igneous material. These impose an impervious barrier across an aquifer, considerably interrupting groundwater flow and aquifer development.

The influence of tectonic structures on karst flow patterns in the fold and thrust belt of the eastern Jura Mountains of Switzerland has been investigated intensively by Herold *et al.* (2000), who conducted three extended multitracer experiments followed by three months of monitoring at around 95 springs. They found that fast long-distance transport along fold axes in crest and limb areas of anticlines is related to extension joints resulting from synorogenic folds, and that concentrated lateral drainage of water flow from anticline limbs is exclusively related to pre-orogenic normal faults that were transtensively reactivated during folding. Water flow through otherwise impermeable layers was suspected to take place through porous calcite fault gouges or breccias. By contrast, transpressively reactivated normal faults and synorogenic reverse faults were found to have no influence on groundwater circulation. The results of this outstanding study are probably applicable to many other karst areas.

Effective porosity (equation 5.2) is strongly influenced by pore size and determines specific storage and storativity (equation 5.14). Void space (diameter or width) available for water movement ranges over seven or eight orders of magnitude up to tens of metres, and since permeability is a function of void size it also varies over a wide range. As void size and continuity increase, permeability increases and resistance to flow diminishes; thus hydraulic conductivity (equation 5.5) is enhanced and, for a given aquifer thickness, transmissivity (equation 5.15) also grows.

The intrinsic properties of some rocks, such as young, highly porous corals and calcareous aeolianites, immediately permit significant aquifer storage and throughput even before a significant amount of dissolution has occurred, whereas at the other extreme marble and evaporites may be almost impermeable before karstification takes place. Rocks with an initially low void density will tend eventually to develop high conduit permeability with minimal fissure storage and diffuse phreatic flow. In contrast, rocks that are quite porous, thinly bedded and highly jointed can still develop conduit permeability if point recharge occurs, but will always have a relatively high diffuse component in the saturated zone.

The development of secondary voids is very strongly influenced by characteristics of recharge; so much so that simply because of this effect, throughput conditions in allogenic and autogenic systems are often radically different, even in the same rock type. Allogenic point-recharge favours the growth of large stream passages, whereas spatially diffuse autogenic recharge enhances pore and fissure porosity, but has little impact on the development of conduits. Thus for a given lithology, in the first case rapid turbulent throughput can occur, while in the second case flow may be mainly laminar and diffuse.

As karstification proceeds and large secondary cavities (i.e. cave systems) develop in the phreatic zone, there is a progressive decoupling of flow between that passing relatively rapidly through the karst pipes and that in the surrounding porous and fissured matrix (White 1977b). Water movement may be rapid and turbulent in one, while slow and laminar in the other. This makes the analysis of the aquifer and its response to recharge particularly difficult. Thus, for example, the speed with which a spring responds to recharge is not a simple reflection of the velocity of water flow through the aquifer; there will be a range of velocities through different subsystems within the aquifer. Constrictions within the system, for example, can produce local ponding and so exercise local control on its response to recharge that may dominate regional effects (Halihan and Wicks 1998, Halihan *et al.* 1998).

As a consequence of the effects of porosity, fissuring and differential solution, permeability may be greater in some directions than in others, as well as in certain preferred stratigraphical horizons. However, while geo-logical factors dictate where storage is greatest, local relief normally exerts a still greater influence on the direction in which groundwater flows, because hydraulic gradient is strongly influenced by it. It is local relief that determines both the highest positions where recharge can occur and the lowest points at which groundwater outflow can take place. Where the horizontal distance between the points of input and output is minimized, the hydraulic gradient is steepest. Thus, the shortest flow path from input to output boundaries determines the direction of groundwater movement in an isotropic aquifer, because its innumerable pores and fissures provide pathways for water flow in any direction; i.e. their control on flow is merely of secondary importance, a rate control. However, where strongly preferred penetrable fissure patterns cause the aquifer to be markedly anisotropic, the orientation of maximum hydraulic gradient will reflect a balance between the direction in which resistance to flow is least (i.e. where hydraulic conductivity maximized) and the direction in which the rate of energy loss is max-imized (i.e. the shortest and steepest route). In discussing the origin and morphology of limestone caves, Palmer (1991) expressed essentially the same notion in a slightly different way by pointing out that passages of vadose origin are formed by gravitational flow and so trend continuously downward along the steepest available openings, whereas phreatic passages originate along routes of greatest hydraulic efficiency (least expenditure of head per unit discharge) that conduct water to the outflow spring. Thus passages in a typical dendritic branchwork system might commence by plunging down the dip in the vadose zone, later to follow the strike in the phreatic zone.

5.3 ENERGY SUPPLY AND FLOW-NETWORK DEVELOPMENT

5.3.1 Energy supply

The development of flow paths in karst aquifers depends upon the energy supply available and its spatial distribu-tion. This derives mainly from:

1. the throughput volume of water;
2. the difference in elevation between the recharge and discharge areas;
3. the spatial distribution of recharge, i.e. on whether it is evenly distributed (as is characteristic of diffuse autogenic recharge) or is focused (as is characteristic of allogenic point recharge);
4. the aggressivity of the recharging waters.

Close to volcanic areas and in some other hot-spring regions the geothermal heat flux may also be important. Elsewhere the magnitude of the thermal flux is only sufficient to warm the groundwater flux by about $0.1°C$ a^{-1} (Bögli 1980), an effect which generally can be neglected.

The amount of dissolution that is accomplished is directly related to the amount of rainfall, as discussed in Chapter 4, but in addition to chemical energy, the principal forms of fluid energy are potential energy, kinetic energy and internal energy. Most of the potential energy is realized as kinetic energy as the water descends through the vadose zone, where much mechanical work can be done by fluvial processes. Flowing water can be regarded as a transporting machine, the stream power of which can be determined by Bagnold's (1966) equation

$$\Omega = \rho g Q \theta \qquad (5.17)$$

where Ω is the gross stream power, ρ the fluid mass density, g gravitational acceleration, Q discharge and θ is slope. From this, the energy available per unit area of channel bed – termed **specific stream power** – can be derived from

$$\omega = \frac{\Omega}{W} = \tau v \qquad (5.18)$$

where ω is specific stream power, W is stream width, τ is mean shear stress at the bed and v is mean stream velocity (see also equation 8.1). Most of this power is dissipated in overcoming fluid shear resistance to flow; so relatively little energy surplus is available for erosion and transport by the stream.

The velocity of water flow in karst varies considerably both within a given aquifer and between aquifers. Within a given aquifer and for a given hydraulic gradient, it varies over several orders of magnitude between water movement in the matrix, fissures and conduits, as indi-cated by hydraulic conductivity values in Table 5.5. Within conduits in different aquifers it also varies con-siderably, as illustrated by the wide range of groundwater conduit velocities determined by dye-tracing experiments (Figure 5.18a); a histogram of 2877 tracer tests in many different conduit systems shows the global average chan-nel flow velocity to be $0.022\,\mathrm{m\,s^{-1}}$, but to vary over more than four orders of magnitude (Figure 5.18b). Milanović (1981) reported that from 281 dye tests over distances of 10–15 km or more in the Dinaric karst, 70% of cases had a flow velocity of $< 0.05\,\mathrm{m\,s^{-1}}$, although they varied over a range from 0.002 to $0.5\,\mathrm{m\,s^{-1}}$. Within a given conduit,

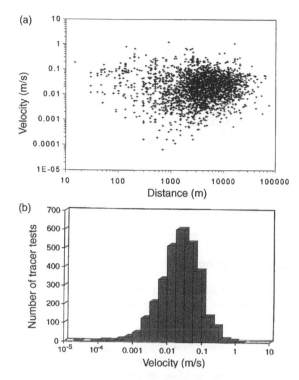

Figure 5.18 (a) Karst groundwater conduit velocities as a function of straight line distance travelled determined from dye tracing different systems. Reproduced from Worthington, S.R.H. (1994) Flow velocities in unconfined carbonate aquifers. Cave and Karst Science, 21(1), 21–2. (b) Frequency distribution of 2877 dye traces in karst conduit systems. Reproduced from Worthington, S.R.H., Ford, D.C. and Beddows, P.A. (2000) Porosity and permeability enhancement in unconfined carbonate aquifers as a result of solution, in Speleogenesis; Evolution of Karst Aquifers (eds A.V. Klimchouk, D.C. Ford, A.N. Palmer and W. Dreybrodt), National Speleological Society of America, Huntsville, AL, pp. 220–3.

flow velocity normally increases substantially with discharge, as determined by Kruse (1980) in the Maligne River system in Canada and by Stanton and Smart (1981) for swallet to resurgence systems on the Mendip Hills in the UK. Figure 5.19 provides other examples.

The rate of throughput of water through a karst system can be measured in two ways: **flow-through** time and **pulse-through** time. The previous paragraph discussed the rate of movement of tracer dyes through the system. When their density is about the same as water, they travel at the same speed as the molecules of water. They therefore measure its flow-through time. However, when a recharge pulse from heavy rain reaches a stream-sink it injects water and sediment into an aquifer. The flood passes as a **kinematic wave** down the vadose

cave passages. But once the saturated zone is reached, the recharge wave causes a rise in the water table and a **pressure pulse** is forced through phreatic conduits, giving a hydrograph peak at the spring. Kinematic waves in open channels travel about 30% faster than the water itself, in the order of tens to thousands of metres per hour, whereas pressure pulses through flooded pipes are propagated almost instantaneously (at the speed of sound). The flow-through time of the water responsible for the hydrograph rise at a spring is much longer than either and can be estimated from the travel velocity of the turbidity peak or of dyes injected into the flooded stream. The pressure pulse mechanism is also known expressively as **piston flow**.

Internal energy, best expressed by rock and water temperature, is another form of energy of significance for aquifer development. However, temperature is a secondary factor that moderates mechanical processes through its influence on the dynamic (and kinematic) viscosity of water (Table 5.4), which is more than twice as viscous at 0°C as at 30°C. The lower viscosity at higher temperatures permits a greater discharge through capillary tubes (Poiseuille's law, equation 5.3) and increases hydraulic conductivity (equation 5.5) in granular, porous media. The influence of temperature on the physics of flow therefore helps to explain some of the differences in karst encountered in the cool temperate and tropical zones, by influencing penetration distances of capillary water and the consequent work that can be done by chemical processes. Dreybrodt *et al.* (1999) noted that considering the temperature dependence of viscosity, breakthrough times for conduits in the initial stages of karstification are lower by about a factor of six in tropical karst compared with cool environment karsts, other things being equal. Worthington (2001) contends that this is an important factor in deep-cave genesis.

The temperature distribution in karst systems and the role of air and water fluxes have been reviewed by Leutscher and Jeannin (2004). Geothermal heat flux as measured in mines and boreholes shows that rock temperatures increase with depth; away from volcanic areas the standard gradient is $\sim$3°C $100\,\text{m}^{-1}$. However, numerous temperature measurements made in deep caves and at karst springs show that temperatures usually are close to the mean annual values in the outside atmosphere. Cave-air circulation within the highly permeable vadose zone results in temperature gradients that are similar to the lapse rate of humid air. Leutscher and Jeannin (2004) present a synthetic conceptual model of heat fluxes in a karst massif that is built on numerous field observations (Figure 5.20). Within 50 m or so of the surface they identify a **heterothermic** zone in which

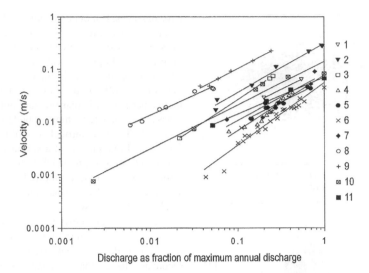

Figure 5.19   Karst groundwater conduit velocities as a fraction of maximum annual discharge in 11 different systems. Reproduced from Worthington, S.R.H. Flow velocities in unconfined carbonate aquifers. Cave and Karst Science, 21(1), 21–2. © 1994 British Cave Research Association.

seasonal variations are observed (see section 7.11 for further discussion). But deeper into the vadose zone (which can be up to 2000 m thick) conditions become **homothermic**, being characterized by high temperature stability. Rock, air and water are almost in thermal equilibrium, although water and rock temperatures are always slightly lower (∼0.15°C) than air. Observed gradients usually vary from 0.4 to 0.6°C 100 m^{-1}, although highly ventilated conduits near the top of the homothermic zone show steeper temperature gradients than the more poorly ventilated deeper parts. The temperature gradient is close to zero in a well-karstified phreatic zone down to the bottom of the main conduit network; Benderitter *et al.* (1993) have modelled the heat transfer conditions there. Below the conduits or where all of the saturated zone has poorly developed permeability, the temperature gradient may be strongly affected by the geothermal heat flux (Liedl and Sauter 1998).

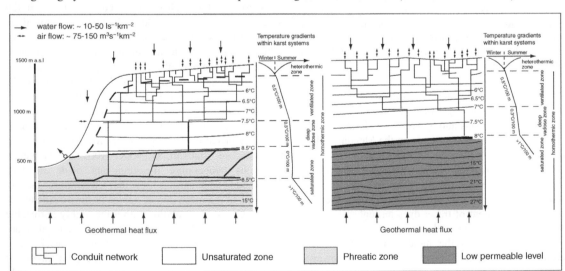

Figure 5.20   A conceptual model of the temperature distribution within a karstified massif. Reproduced from Leutscher, M. and Jeannin, P.-Y. Temperature distribution in karst systems: the role of air and water fluxes. Terra Nova, 16, 344–50 © 2004 Blackwell Publishing.

Lismonde (2002, cited by Leutscher and Jeannin 2004) calculated that the loss of potential energy of water during its vertical transit is $9.81\,J\,kg^{-1}\,m^{-1}$ and showed that if this energy is fully transformed into heat by friction, then it leads to a temperature increase of $0.234°C\,100\,m^{-1}$. By this means, the heat supplied annually by the work of gravity to the homothermic zone is proportional to the annual recharge. In a vadose zone 1000 m thick with a recharge of $10–50\,L\,s^{-1}\,km^{-2}$, this amounts to between 3×10^9 and $1.5 \times 10^{10}\,kJ\,km^{-3}$ annually. When considering the relative importance of the roles of air and water in heat transfers within the vadose zone, Leutscher and Jeannin (2004) concluded that air circulation plays the dominant role, although the relative effect of water on heat transfers is likely to increase at greater depth where air circulation is reduced. Air flow is also relatively less important in promoting heat flux in areas such as the tropics where there is little seasonal contrast in outside air temperatures. Leutscher and Jeannin's model is based on the assumption that the 'chimney effect' is at the origin of most air circulation observed within alpine caves, although they recognized that barometric fluctuations are also important in promoting air flows in larger systems. Such fluctuations will increase the influence of air flow in the flux of heat. A classic example is Castleguard Cave, Canada, a lengthy relict conduit which passes through a glacierized mountain; it displays both chimney effect and barometric air flows, but temperatures rise to $+3°C$ in the centre due to the geothermal flux, while water temperatures in the phreatic conduits underneath are only $0.5–1.0°C$ (Ford *et al.* 1976).

Temperature plays a role in influencing total chemical energy for dissolution, which can be viewed as the product of the volume and aggressivity of the solvent throughput. Kinetic factors, which are largely temperature dependent, determine the rate at which dissolution occurs; while equilibrium factors determine the ultimate solute concentrations that can be achieved given sufficient time (as explained in Chapter 3). In a conduit-dominated groundwater system, throughput is fast and there is likely to be insufficient time for most of the water to reach equilibrium solute concentrations; thus for a given discharge kinetic factors dictate both the location and the total amount of work done by chemical processes. In a preponderantly porous carbonate rock with a diffuse, largely laminar flow system, kinetic factors still determine where most chemical energy is expended (in the upper part of the vadose zone), but for a given water throughput it is equilibrium considerations that determine the total chemical load discharged from the system.

So far the discussion has centred on meteoric water and carbonic acid derived principally from soil CO_2, but it should be noted that hot hypogean waters from deep-seated sources, often associated with H_2S and CO_2, can also sometimes be a source of aggressive water capable of enhancing porosity at depth (Figure 5.3) and dissolving cave networks (Palmer 1991).

Where water is stored under confined artesian conditions, the influence of another form of energy can sometimes be observed. Water levels in artesian bores can fluctuate twice daily by several centimetres as a consequence of **earth tides** (Figure 5.21). These are caused by the same mechanism that produces marine tides, the moving tidal bulge in the solid earth causing the reservoir rocks to compress and relax and thus the fissures to close and open. The pumping action of earth tides may be especially significant in the earliest stages of karstification, when it is difficult to initiate a secondary permeability because of the short penetration distance of groundwater before it becomes fully saturated with respect to calcium carbonate (Davis 1966).

As relief is lowered by karstic denudation, regional precipitation and hence recharge may also be reduced, thus further diminishing available total energy and the depth of active circulation. Ancient peneplained karsts with only a small local relief have relatively little energy for water circulation, but they may possess very considerable storage in ancient flooded cavities.

5.3.2 *Flow network development*

We may conclude from the above discussion that, given sufficient hydraulic potential, the style of recharge has a strong influence on the occurrence, density and size of conduit permeability (although not on the **process** of conduit development). The following end-member conditions occur in the development of flow networks.

1. Diffuse recharge onto a carbonate rock with high primary porosity, e.g.rain on aeolian calcarenite or uplifted coral, when few or no karst conduits form until subaerial diagenesis (case hardening) focuses recharge;
2. Widely spaced, large volume, point recharge into a dense carbonate rock with well-developed fissures, e.g. recharge windows in a breached caprock over massive limestones, when a few very large diameter conduits form commensurate in size with their throughput discharge. Competition is limited to corridors downstream of recharge points that are separated by unkarstified rock beneath the umbrella of still intact caprock.

Doline karst falls between these two recharge extremes because, although the recharge is autogenic in origin, the flow is internally focused through a large number of point inputs of modest volume (Figure 5.15).

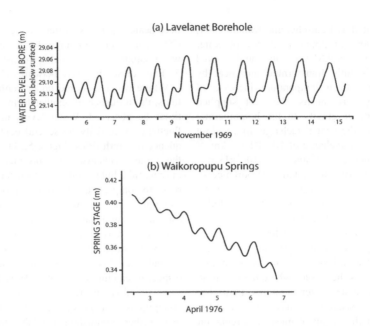

Figure 5.21 Earth tide effects as measured (a) in a borehole and (b) in an artesian karst spring. Adapted from Mangin, A. Contribution a l'etude hydrodynamique des aquiferes karstiques. Univ. Dijon These Doct. es © 1975 and Sci. Annales et Speleologie, 29(3), 283–332; 29(4), 495–601, 1974; 30(1), 21–124. © 1974 Taylor and Francis; and Williams, P.W, Hydrology of the Waikoropupu Springs: a major tidal karst resurgence in northwest Nelson (New Zealand). Journal of Hydrology, 35, 73–92. © 1977 Elsevier.

Once water has recharged the karst, the various forms of available energy mentioned above are expended in the enlargement of selected primary pores and fractures into secondary conduit networks. This process is discussed in detail in Chapter 7, so suffice it to say here that White (2002) has noted that in the early stages of this process three thresholds are crossed when fracture apertures exceed about 0.01 m: (i) a hydrodynamic threshold that permits the breakdown of laminar flow and the onset of turbulent flow; (ii) a kinetic threshold that marks a shift in dissolution rate from the fourth-order kinetics to linear kinetics (see Chapter 3); and (iii) a transport threshold that enables flow velocities to be sufficient for the physical entrainment and transport of insoluble clastic material. White pointed out that the coincidence of all three thresholds at an aperture of 0.01 m provides a natural dividing line between fracture aquifers and conduit aquifers and, further, that it separates the process of conduit development into an initiation phase as the protoconduit grows to critical threshold dimensions and an enlargement phase as it then expands to the size of typical cave passages. After breakthrough of the protoconduit at the outflow boundary (and consequently the development of a spring from what was previously just a seepage), there is a fairly sudden transition to rapid dissolution throughout the entire subterranean flow path. The entire route then enlarges

rapidly. Most caves formed in this way by meteoric waters are dendritic branchworks (Figure 5.22), but many other cave network patterns can develop, depending on the style of recharge and the kind of pre-solutional porosity that these processes act upon (Figure 5.23).

Thus conduit networks integrated from recharge points to springs extend like a plumbing system right through the karst and drain it effectively. This may seem obvious enough when rivers are observed to disappear underground into caves, but efficient, integrated drainage networks also underlie fields of dolines, for dolines and conduit networks develop simultaneously. And even in carbonate terrains where stream-sinks and dolines are rare or absent, as in much of the European chalk, caves with turbulent flow are known and borehole evidence reveals many solutionally enlarged voids (Banks *et al.* 1995, Waters and Banks 1997, Matthews *et al.* 2000). In the Chalk of northern France a fracture maze cave of over 1 km has been mapped (Rodet 1996) and numerous conduits have developed under the edge of overlying clastic deposits that shed allogenic runoff (Crampon *et al.* 1993). Only in argillaceous limestones are enlarged fissure networks uncommon or ineffective, largely because of their clogging by insoluble residues.

The continued dissolution associated with groundwater circulation ensures that the void volume of the network

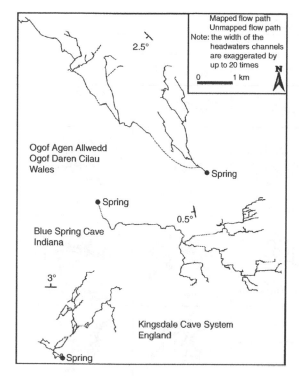

Figure 5.22 Typical dendritic flow networks in carbonate karsts formed by meteoric waters. Reproduced by permission of Worthington, S. R. H., Ford, D. C. and Beddows, P. A. 2000. 'Porosity and permeability enhancement in unconfined carbonate aquifers as a result of solution'. In Klimchouk, A.B., Ford, D.C., Palmer, A.N. and Dreybrodt, W. (Eds), Speleogenesis: Evolution of Karst Aquifers. 220–223 © 2000 National Speleological Society.

increases over time and so, as a result, and provided that recharge throughput remains constant, the level of the zone of groundwater saturation gradually falls. We can appreciate from the foregoing that there are considerable variations in the size and distribution of secondary voids in karst, and therefore in the ease with which water is transmitted through them. As a consequence the water table is seldom flat, as the term might suggest, but is more often irregular – even sometimes disjointed – and slopes towards the discharge boundary.

5.4 DEVELOPMENT OF THE WATER TABLE AND PHREATIC ZONES

5.4.1 Pre-established water tables

Nevertheless, there are some circumstances when the effective primary porosity is so high that when the limestones are first occupied by fresh waters a saturated zone with a very low-gradient water table readily estab-

lishes beneath the surface. On Niue Island for example, which is a coral atoll in the Pacific uplifted during the Pleistocene and with no rocks other than carbonates, the water table is always within a few metres of sea level, and even 6.5 km inland from the coast has an elevation of only 1.6 m (Williams 1992). The permeability is so high (typically > 25%) that in the centre of the island the water table 60 m below the surface responds diurnally by a few centimetres to marine tidal fluctuations. There is little scope for lowering the groundwater level because it is already close to sea level and continued groundwater circulation will result in minimal increase to the already high effective porosity because of high saturation levels. However, dissolution at the freshwater–saltwater interface has enhanced secondary porosity in the mixing zone and has encouraged the focusing of freshwater discharge at springs around the coast (see Figure 5.37); thus despite the high porosity the aquifer has developed conduits within the zone of palaeo-sea-level variation.

5.4.2 Karstic enhancement of porosity

Massive limestone typically has very low primary porosity and consequently very low permeability, but groundwater circulation increases it over time. If a typically anisotropic karst aquifer is conceptualized as behaving hydrologically like rock pierced by straight, parallel capillary tubes, it is possible to illustrate the way in which increasing tube diameter and percentage porosity affect hydraulic conductivity in the direction of the tubes, assuming laminar flow (Figure 5.24) (Smith *et al.* 1976). Secondary enlargement of the fissure network can lead to at least an order of magnitude increase in porosity and more than 10^5 increase in hydraulic conductivity. For example, there is about 10^7 difference in hydraulic conductivity between matrix and secondary channels in the Mammoth Cave region (Table 5.5). Thus the phreatic zone can acquire considerable storage capacity and fast throughput potential.

5.4.3 Changes in the phreatic zone

Once percolating water reaches the water table, the energy of water per unit mass, the hydraulic potential, is determined by gravitational acceleration times the local head, i.e. by the height of the water table above the outflow spring. It is this energy coupled with the vertical hydraulic conductivity characteristics of the phreatic zone that determines the depth to which groundwater can circulate, and hence the thickness of the phreatic zone: the larger the hydraulic potential the deeper the circulation. It tends to be particularly great when massively bedded, steeply dipping

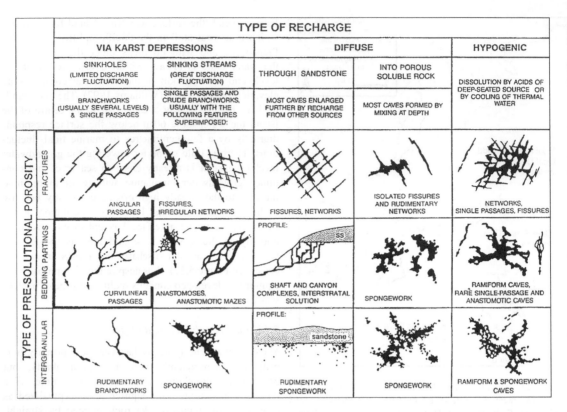

Figure 5.23   Summary of cave patterns and their relationship to types of recharge and porosity. Reproduced from Palmer, A.N., Origin and morphology of limestone caves. Geological Society of America Bulletin 103, 1–21. © 1991 Geological Society of America.

limestones slope towards the outflow boundary. Under such circumstances flow paths tend to become trapped in plunging bedding planes with only occasional cross-joints to lead water back to higher levels. The outcome is conduit systems with deep phreatic loops that can penetrate in excess of 400 m below the water table. At the other end of the spectrum are thinly bedded, fractured limestones with shallow dips. Under these circumstances and for the same throughput discharge, very high permeability develops and a lower hydraulic gradient, and so consequently a shallower circulation is supported and the phreatic zone is not very deep. Worthington (2001, 2004) provides empirical evidence that the depth of the conduit flow path is directly proportional to flow-path length, stratal dip, and fracture anisotropy, thus

$$D = 0.18(L\theta)^{0.79} \qquad (5.19)$$

where D is the mean depth of flow in metres below the water table, θ is the dimensionless dip (equal to the sine of the dip in degrees) and L is the flow-path length in metres. However, Ford (2002) contested this relationship on several grounds.

When significant secondary porosity develops in the zone where the water table fluctuates above its basal level (the **epiphreatic zone**), two things happen: (i) the increased horizontal hydraulic conductivity favours greater water movement in the shallow phreatic zone rather than at depth and (ii) the increased storage lowers the water table gradient and thereby extends the vadose zone. Circulation in the deep phreatic zone becomes less vigorous and, as karstification continues, the increasingly active shallow phreatic zone may lead to relatively stagnant conditions in the deepest phreatic voids. A stagnant phreas can also be produced as a consequence of positive baselevel changes. A rising sea level, valley aggradation or tectonic subsidence can submerge a previously active phreatic region beyond the zone of contemporary circulation. It then becomes a variety of palaeokarst and may be simply a passive store or vessel for the receipt of the finest suspended sediment. Such features are known to depths greater than 3 km in oil wells.

Estimates of phreatic zone thickness are required for the calculation of transmissivity (equation 5.10) and storativity (equation 5.11). It is not necessarily the entire depth of the phreatic zone to the base of karstification that

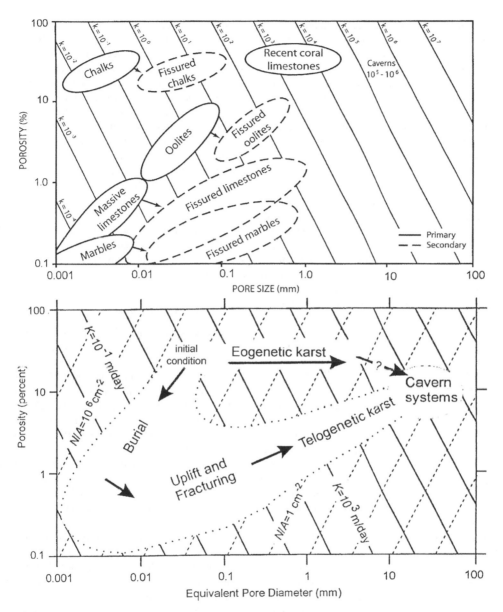

Figure 5.24 (Upper) Relationship of primary and secondary porosity and pore size to the hydraulic conductivity of karst rocks. The limestone is considered an equivalent porous medium consisting of straight tubes. Reproduced from Smith, D.I., Atkinson, T.C. and Drew, D.P. (1976). The hydrology of limestone terrains, in The Science of Speleology (eds T.D. Ford and C.H.D. Cullingford), Academic Press, London, pp. 179–212. (Lower) Model of the changes in porosity and equivalent pore diameter in limestones from their initial state after deposition, through burial, uplift and exposure to surface karstification (telogenetic karst). A short circuit can occur when young limestones are karstified during diagenesis (eogenetic karst). The labels N/A on the dashed lines refer to equivalent tube density in number of tubes per square centimetre. Reproduced from Vacher, H.L. and Mylroie, J.E. (2002) Eogenetic karst from the perspective of an equivalent porous medium. Carbonates and Evaporites, 17(2), 182–96.

should be used, because only the upper part of the phreatic zone may be involved in active groundwater circulation over the span of a few years. Atkinson's (1977b) solution to this problem in the Mendip Hills, England, was to estimate the depth of active circulation from the elevation range of the phreatic looping of a trunk conduit leading to Wookey Hole spring – one of the main resurgences in the region (Figure 7.15).

However, information of this kind is available only if springs have been dived for long distances. Alternatively it could be estimated using Worthington's (2001, 2004) approach (equation 5.19).

Transmissivity calculations for the entire phreatic zone will take into account the aquifer thickness from the water table to the base of karstification. However, since porosity and hydraulic conductivity usually diminish with depth through the aquifer, total transmissivity can be envisaged as comprising the sum of the transmissivities of different subzones within the phreas (Figure 5.12).

Active aquifer thickness varies spatially and temporally. The depth of active circulation may be less when close to the spring than when further into the aquifer and it may also be less under baseflow than under flood conditions. Where there is a stagnant phreatic zone, the karstified rock may have a high permeability and storage capacity, but zero specific discharge if there is no pressure gradient to drive the flow. Hence total aquifer thickness can be appreciated to be a sum of several parts; the different dynamic zones of the aquifer.

5.5 DEVELOPMENT OF THE VADOSE ZONE

5.5.1 The vadose zone as a resultant following phreatic drawdown

Most unkarstified crystalline carbonate rocks have a very low primary porosity (typically < 2%). So when they are first invaded by fresh waters the standing water level in the rock is close to the surface. But the porosity and permeability of karst increases over time, so progressively more void space becomes available to store and transmit the groundwater. As a result the standing water level gradually falls and so the aerated zone becomes deeper. However, while this is happening the surface also lowers by denudation. Consequently, the resultant thickness of the vadose zone is the outcome of the relative rates of movement of its upper and lower boundaries: the rate of surface lowering and the rate of water table lowering (which also responds to outflow boundary lowering caused by valley deepening). Speleological exploration in the western Caucasus Mountains shows that the vadose zone in well-karstified rocks can sometimes extend as much as 2 km below the surface.

Variability in lithology, structure and geomorphology causes contrasts in the course of evolution of the vadose zone. Thus in coral reefs interconnected primary porosity can exceed 20% and so, following reef emergence, a freshwater lens establishes near sea level with a very permeable vadose zone above it. In active alpine zones rapid rates of uplift, denudation and valley incision can give rise to rapid unloading of exposed rocks and to rearrangement of stress fields. This creates a series of unloading fractures subparallel to the surface that are superimposed upon older joint and fault systems that may also gape. As a consequence, all these fissures are relatively open near the surface where pressure is least, but close with depth. Even the early water table in such cases may be hundreds of metres below the surface, especially where climates are relatively arid. With subsequent dissolutional enlargement of the fissure systems' void space gradually increases at depth and the zone of saturation in the rock falls, thus deepening the vadose zone.

Sometimes heavy rain causes deep back-flooding and temporary pressure flow in subvertical drains passing through deep vadose zones, giving rise to processes that interrupt the evolution of regular vadose conduit morphology. This contrasts with the normal drawdown effect of a gradually falling water table by providing short-lived episodes of rising water table – a flood water invasion effect.

5.5.2 Epikarst development

The **epikarst** (also referred to as the **subcutaneous zone**) occupies the top of the vadose zone (Table 5.1). It consists usually of a particularly weathered zone of limestone that lies immediately beneath the soil, and it gradually gives way to the main body of the vadose zone that comprises largely unweathered bedrock. The epikarst is typically 3–10 m deep, but its characteristics can vary considerably. Sometimes there is little or no soil, for example in the arid zone and in glacially scoured regions, and if the rock is especially massive and has a low density of fissuring it can be 30 m or more deep. The development of the epikarst is explained by Mangin (1975), Williams (1983), Klimchouk (2000a) and by several contributors to Jones *et al.* (2004).

Chapters 3 and 4 explained that the greatest expenditure of chemical energy on the dissolution of carbonate rocks occurs near the surface, because of proximity to the main source of CO_2 production in the soil. These dissolutional processes act on fissure systems that may themselves be evolving, because of stress release following unloading. We also noted in Chapter 4 that up to about 80% of solutional denudation in the catchment is accomplished within the top 10 m or so of the limestone outcrop and that the effectiveness of corrosional attack gradually diminishes with distance from the surface (and from the CO_2 supply). The outcome of this is that the network of fissures through which percolation water passes is widened by solution near the surface, but the extent and frequency of widening diminishes gradually with depth. Solutionally widened joints taper downwards and become less numerous. This can be readily observed in quarries (Figure 5.25). A consequence of this is that

Figure 5.25 'Pinnacle and cutter' relief exposed in a limestone quarry face in Kentucky. This illustrates the tapering closure of solutionally enlarged joints with depth in the epikarst.

permeability also diminishes with depth (Figure 5.26). Porosity in the epikarst typically exceeds 20%, whereas in the relatively unweathered rock beneath it is commonly < 2%. This gives rise to a contrast in hydraulic conductivity and so, as a consequence, water tends to accumulate at the base of the epikarst, because it cannot escape as freely as it came in. However, not all fissures close and a few of them can be observed to penetrate as major openings right through the rock (Figure 5.27). As a

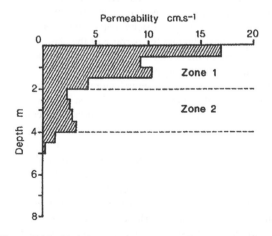

Figure 5.26 Variation of permeability with depth in the epikarst at Corconne, France. Gouisset divided the epikarst into two zones based on contrast in permeability. Reproduced from Gouisset 1981, cited in Smart, P.L. and Friederich, H. Water movement and storage in the unsaturated zone of a maturely karstified carbonate aquifer, Mendip Hills, England, Proceedings of Conference on Environmental Problems in Karst Terranes and their Solutions 59–87, © 1987 National Water Well Association.

Figure 5.27 Very high fissure frequency exposed in a quarry face near Te Kuiti, New Zealand. The vertical feature to the right of the figure is a minor fault and is the kind of discontinuity that permits doline initiation by providing a preferred drainage path for infiltrating rainwater. Reproduced from Williams, P.W., Polygonal karst and palaeokarst of the King Country, North Island, New Zealand. Zeitschrift für Geomorphologie, Suppl. Vol 136, 45–67, Zeitschrift für Geomorphologie © 2004.

result, these become the main drainage routes for water passing downwards to the vadose zone. They act as centripetal flow foci that funnel water from the fissured epikarst. These sites of flow concentration become the distal tributaries of drains through the vadose zone and are the main routes by which diffuse autogenic recharge is transmitted to conduits in the unsaturated zone and thence to the phreatic zone. Smart and Freiderich (1987) estimated that in the epikarst of the Mendip Hills in England as much as 77% of annual recharge is transmitted via the highest capacity flow routes (shaft flow to subcutaneous flow), whereas only 23% percolates via the narrow low-capacity seepage and vadose flow routes.

The epikarst is therefore underdrained. Klimchouk (1995) explained how the concentration of flow at the base of the epikarst encourages the formation of shafts and, in a particularly well-illustrated paper, Klimchouk *et al.* (1996) provided convincing field evidence of the efficacy of the process from the Sette Communi Plateau in the Italian Pre-Alps. These blind vertical shafts are known as avens by cavers. Although they develop downwards from the base of the epikarst, they can eventually be exposed by collapse as the surface lowers, and many become the main drainage route beneath dolines.

If the rate of recharge during heavy rain is greater than the maximum rate of vertical throughput, then excess recharge is stored in the void space of the epikarst, i.e. in the widened fissures and in the intergranular porosity of any fill they may contain. Water stored in that way constitutes an **epikarstic aquifer** that is perched above a leaky capillary barrier (Figure 5.28). Its piezometric surface is drawn down over the main leakage paths afforded by shafts developed down major joints (Figure 5.15), and the direction of subcutaneous flow is down the hydraulic gradient. After a long dry period the epikarst drains almost entirely, although some water remains held by capillary tension. The processes of infiltration and percolation through the epikarst are discussed more fully in section 6.3.

In many alpine regions, where the carbonate rock has been tectonically stressed and deformed during uplift and then later unloaded by rapid erosion and deep valley incision, fissures can be deep and their apertures can be relatively wide. Conditions are then favourable for the development of deep vertical shafts. As a result surface drainage is facilitated and there may be relatively little water storage capacity in the epikarst except under patches of karrenfeld (Figure 9.6), although there may be considerable seasonal storage of snow. Klimchouk (2000a) also pointed out that in alpine regions in summer condensation recharge can be significant as air is drawn into the karst at high altitudes and cools to dew point in the epikarst zone.

5.5.3 *Case-hardening and the reduction of vadose porosity*

Although the top of the vadose zone in dense crystalline carbonate rocks generally experiences the enhancement of secondary permeability as a result of intense dissolution, where the rocks have high primary porosity the opposite may occur because porosity is reduced (Figure 5.3) during meteoric zone diagenesis by a process known as **case-hardening**. This involves redeposition in primary pores further down the profile of calcite dissolved near the surface, redeposition usually occurring within a few metres. The effect is to produce a concrete-like calcrete crust and to reduce the primary porosity of the surface rock by a factor of 10 or more. It is especially common in porous rocks such as aeolian calcarenite (carbonate dune sands), coral and some chalky limestones, particularly in warm temperate to tropical zones where strong drying cycles follow rain. Carbonate deposition closely follows dissolution because of evaporation and CO_2 degassing in cavities well connected to the outside atmosphere. Frequent wetting and drying accelerates the process. The significance of case-hardening for karst is discussed further in section 9.12.

5.6 CLASSIFICATION AND CHARACTERISTICS OF KARST AQUIFERS

Significant characteristics that might be incorporated into a classification of karst aquifers include: style of recharge, flow media, flow type, conduit network topology, stores and storage capacity, and outflow response to recharge. Depending on the purpose of the classification, one or more of these may be emphasized or neglected.

An important starting point for the more useful general classifications is the conceptualization of recharge and flow type. Burdon and Papakis (1963) first drew attention to this, distinguishing **diffuse** infiltration from **concentrated** infiltration and diffuse circulation from concentrated (or localized) circulation. They also pointed out that the style of recharge may not precondition the type of flow in the saturated zone: thus concentrated circulation can occur even if recharge (infiltration) is diffuse and vice versa. White (1969) termed these flow styles 'diffuse' and 'conduit'. He also suggested how the occurrence of these contrasting forms of circulation might be deduced from readily observable geological variables. Thus his classification of carbonate aquifers (Table 5.10) is of value in identifying the range of aquifer types that occur.

Chemical contrasts in groundwaters draining from aquifers that are dominantly porous, fissured or cavernous (conduit) have encouraged the view that to consider

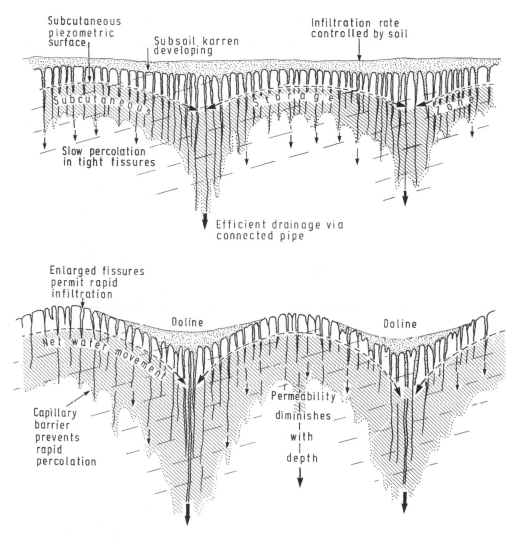

Figure 5.28 Drawdown doline initiation in the subcutaneous (epikarst) zone. Reproduced from Williams, P.W., The role of the subcutaneous zone in karst hydrology. J. Hydrol. 61, 45–67.© 1983 Elsevier.

carbonate flow media as being essentially bimodal is too simple (Bakalowicz 1977). Thus Atkinson (1985), also citing data from China (Yuan 1981, 1983), suggested that a more appropriate way of representing the older concept of granular, fracture and conduit aquifers would be its visualization by means of a ternary or three end-member spectrum (Figure 5.29a). This conceptual classification of flow media is then related to presumed phreatic flow regimes (Figure 5.29b). Hobbs and Smart (1986) elaborated this approach. They proposed a model in which the three fundamental attributes of recharge, storage and transmission are ranged between end-members; thus giving a three-dimensional field into which carbonate aquifers may be conceptually plotted (Figure 5.30).

Work in France by Mangin (1975) and Bakalowicz (1977) brought closer attention to the structure and transfer functions of drainage systems within karst (see section 6.5). Spring hydrograph response to recharge was used as a measure of aquifer karstification (Bakalowicz and Mangin 1980), four groups of karst systems being recognized that range from aquifers with extremely well-developed speleological networks to those carbonate terrains which can barely be considered karstified. However, external factors that can also be responsible for variation in spring discharge (e.g. the influence on hydrograph form of climate, recharge variability, recharge regulation by overburden, and outflow distributaries) were not fully considered.

Table 5.10 Hydrological classification of carbonate aquifers (Reproduced from White, W.B Conceptual models for carbonate aquifers. Ground Water 7(3), B97515 21 © 1969 Blackwell Publishing)

Flow type	Hydrological control	Associated cave type
Diffuse flow	Gross lithology. Shaley limestones; crystalline dolomites; high primary porosity	Caves rare, small, have irregular patterns
Free flow	Thick, massive soluble rocks	Integrated conduit cave systems
Perched	Karst system underlain by impervious rocks near or above base level	Cave streams perched, often have free air surface
Open	Soluble rocks extend upwards to surface	Sinkhole inputs; heavy sediment load; short-channel-morphology caves
Capped	Aquifer overlain by impervious rock	Vertical shaft inputs; lateral flow under capping beds; long integrated caves
Deep	Karst system extends to considerable depth below base level.	Flow is through submerged conduits
Open	Soluble rocks extend to land surface	Short, tubular abandoned caves likely to be sediment-choked
Capped	Aquifer overlain by impervious rocks	Long, integrated conduits under caprock. Active level of system inundated
Confined flow	Structural and stratigraphical controls	
Artesian	Impervious beds which force flow below regional base level	Inclined three-dimensional network of caves
Sandwich	Thin beds of soluble rock between impervious beds	Horizontal two-dimensional network caves

Figure 5.31 expresses our understanding of possible structural linkages within karst groundwater systems, although not all the characteristics shown are possessed by any one system. The dynamic relationships between the components and aquifer properties are best examined by computer modelling techniques, which we discuss in section 6.11. However, it is now common in computer modelling to represent karst aquifers being examined in terms of single, double and triple porosity types (Table 5.11), although most karst aquifers are best considered triple porosity.

5.7 APPLICABILITY OF DARCY'S LAW TO KARST

5.7.1 When and when not does Darcy's Law apply

It is of fundamental importance to establish whether or not it is justified to treat a karst aquifer as a granular, porous medium in the Darcian sense (i.e. a single-continuum porous equivalent) or even a double-continuum porous equivalent with a fractured continuum as well as a porous matrix. An implication of accepting the Darcian approach is that the rock is considered as a continuum of voids and solid matter for which certain generalized macroscopic parameters (such as K) can be defined, that represent and in some sense describe the true microscopic behaviour. In karst this means that the fractured rock penetrated by solution conduits would be replaced by a conceptual representative continuum for which it is assumed possible to determine hydrologically meaningful macroscopic parameters.

In Darcy's experiment, discharge was measured from a given cross-sectional area a of the saturated medium. Hence in equation (5.4) Q/a is an expression of the discharge per unit area. It therefore has the dimensions of a velocity and is sometimes simply denoted by u, the specific discharge (**filtration velocity** or **Darcy flux**). However, the flow does not issue from the entire cross-sectional area, but only from the voids between the solid grains. It follows then that the real microscopic velocities of flow through the interstitial spaces must be considerably larger than the averaged, macroscopic velocity denoted by u and that at some stage laminar flow will give way to turbulent flow. When this occurs is shown in Figure 5.32. Freeze and Cherry (1979) pointed out that Darcy's Law is a linear law and that if it were universally valid a plot of the specific discharge against hydraulic gradient would reveal a straight line gradient for all hydraulic gradients. This is not the case and at relatively high rates of flow Darcy's Law breaks down. Darcy's Law imposes a statistical homogeneity, but since the distribution of karst voids has a hierarchical nature, it cannot be treated as random, and the average total permeability is a function of the volume of the rock (aquifer) considered.

Since the specific discharge defines the macroscopic velocity through the medium, the average microscopic velocity $u*$ can be determined by taking into account the

(a)

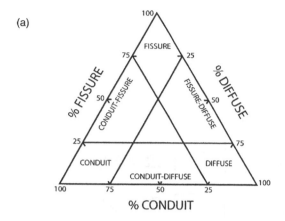

(b)

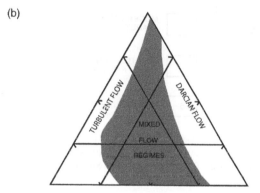

Figure 5.29 (a) Conceptual classification of karst aquifers and (b) their presumed relationship to predominant flow regimes. Modified from Atkinson, T.C., Present and future directions in karst hydrogeology. Annales de la Societe Géologique de Belgique, 108, 293–96, 1985.

actual cross-sectional area of voids through which the flow occurs. This depends upon the porosity n (equation 5.1) and hence

$$u^* = \frac{Q}{na} \qquad (5.20)$$

If pore spaces through which water flows comprise 20% of the rock ($n = 0.2$), then u^* is about five times the Darcy flux. However, since water follows relatively long tortuous flow paths through the rock, the actual velocities must be greater still.

Bear (1972) concluded from experimental evidence that Darcy's Law remains valid provided Re does not exceed 1 to 10. Since fully turbulent flow does not occur until velocities are high and Re is in the range 10^2 to 10^3, there is an interval between the turbulent and linear laminar regimes characterized by non-linear laminar flow (Figure 5.32). It should also be noted that dynamic

viscosity μ varies markedly with temperature, in the tropics being about half that encountered in the cool temperate zone (Table 5.3). Thus under some conditions, turbulent flow will occur in the tropics when it would be laminar in a cooler groundwater environment.

In a comprehensive series of experiments, Ford and Ewers (1978) showed that the law applies strictly when solutional protoconduits up to 1 mm diameter are first extending through a fissure (Figure 7.5a), but that it ceases to apply once the extension is completed and the protoconduit is connected to others. Reviewing recent evidence, White (2002) concluded that for normal hydraulic gradients the onset of non-Darcian behaviour occurs in conduits when apertures exceed about 1 cm. Nevertheless, conduit flow can remain in the laminar regime in pipes up to about 0.5 m diameter provided velocity does not exceed $1 \, mm \, s^{-1}$ (Figure 5.33). It is therefore evident as Mangin (1975) noted that in karst the range of conditions under which Darcy's Law can be considered valid is very restricted. It only applies in conditions that permit velocities to be low, and this usually involves some combination of relatively small aperture and low hydraulic gradient. Darcy's Law also assumes isotropic conditions and will not apply to an aquifer that is anisotropic and heterogeneous. Nevertheless, this variability may become relatively less important as scale increases.

We can see from Table 5.2 that even in aquifers as highly karstified as Mammoth Cave most storage is in the rock matrix. In spite of there being 550 km of cave passages in the area, the porosity attributable to the cave is less than 0.1% and that the chance of a drilled well hitting a conduit is only about 1.4% (Worthington *et al.* 2000). Thus most bores would miss the cave and well tests would be undertaken in conditions acceptable for the application of Darcy's Law. But this manifestly does not mean that the aquifer as a whole is Darcian, because 99.7% of flow passes through the conduit system (Table 5.6). In most karstified aquifers, it is probable that well testing using Darcian assumption is more-or-less acceptable in many cases, but it would be totally erroneous to conclude from that that the management of water resources (and pollution transport) in the groundwater basin as a whole can be based on Darcian principles. Even the Cretaceous Chalk of Britain and northern France, which has long been considered an archetypal porous aquifer, is now known to possess significant conduit porosity (Crampon *et al.* 1993, Banks *et al.* 1995, Rodet 1996, Waters and Banks 1997). Flow is Darcian in parts of the aquifer, but turbulent and non-Darcian elsewhere.

Wide international experience now shows that with respect to unconfined carbonate aquifers 'the prudent

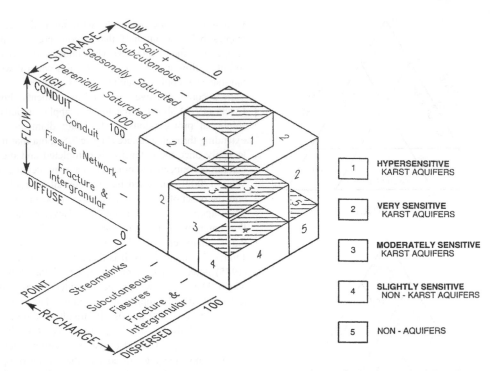

Figure 5.30   A conceptual three-dimensional field model of recharge, storage and transmission in karst aquifers. Reproduced from Hobbs, S.L. and Smart, P.L. Characterization of carbonate aquifers: a conceptual base. Proc. 9th International Speleological Congress, Barcelona 1, 43–46, 1986.

default position should be to assume karstification'. However, if diagnostic tests (see section 6.4) fail to support this, then it is justified to assume that Darcy's Law is applicable within the locality considered (although not necessarily over the wider region).

5.7.2   A question of scale of investigation?

Karst aquifers often comprise a thick, fissured mass of rock, often layered and traversed by a large branching pipe network that transmits the flow of an underground river and its tributaries to a spring or springs. Within the areally extensive body of rock the continuum approach using Darcy's Law can sometimes be justified to obtain spatially defined values of hydraulic characteristics, but this is totally inappropriate for describing flow in individual major conduits. The continuum representation of the aquifer is often an acceptable generalization when hydraulic gradients are low and storage very large. It is largely a matter of void/fissure frequency and scale. It is evident that small samples, as used in core and pump tests, yield inadequate estimates of basin-wide hydraulic characteristics (Figure 5.2). Halihan *et al.* (1999) undertook a very thorough investigation of the effect of scale of measurement on permeability (*k*) estimates in the karsti-

fied Edwards aquifer in Texas, and found permeability estimates to vary by up to nine orders of magnitude, depending on the scale and direction of measurement. Worthington *et al.* (2002) collated data from six carbonate rocks that illustrate the difficulty of obtaining representative hydraulic conductivity (*K*) estimates from conventional techniques (Figure 5.34). From core tests, two Palaeozoic carbonates have very low *K* values for the matrix (10^{-11}–10^{-10} m s^{-1}), three Mesozoic aquifers have moderate values (10^{-8} m s^{-1}) and the Cenozoic aquifer has the highest value (10^{-4} m s^{-1}). All estimates are higher when evaluated at larger scales (using pump tests and regional evaluations) and show a much smaller range (from 10^{-4} to 10^{-1} m s^{-1}).

5.7.3   Alternative approaches

An alternative to Darcian-based techniques of aquifer analysis is to accept the anisotropic and heterogeneous nature of a karst aquifer and to treat it as an interconnected system of conduits and fractures embedded in a more-or-less porous matrix. This is closer to the Hagen–Poiseuille approach in which the hydraulics of flow in individual fractures and pipes are considered. Methods used to simulate flow and transport through

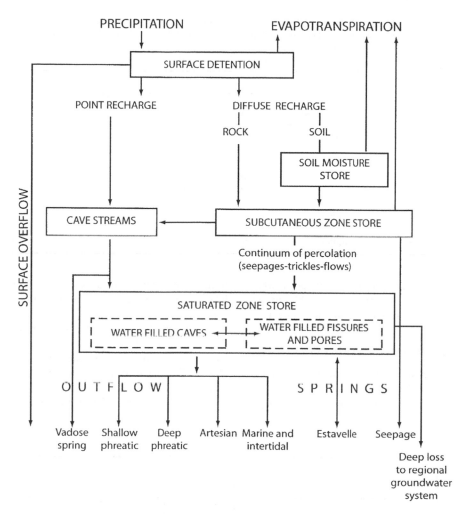

Figure 5.31 Stores and linkages in karst drainage systems. Reproduced from Ford, D.C. and Williams, P.W. (1989) Karst Geomorphology and Hydrology.

karst in this way have varying degrees of complexity. For example, black-box models (input–output models) examine the relationship between input characteristics, such recharge by rainfall, and output response, such as water quantity and quality variations, and focus particularly on spring hydrographs (Atkinson 1977b, Bakalowicz and Mangin 1980). The assumption made is that a spring hydrograph provides an integrated representation of the network of stores and passages delivering water to the aquifer outflow point, and that both the quality and the quantity of the water are of diagnostic importance for understanding the functioning of the system. However, as Teutsch and Sauter (1998) pointed out, due to their non-physical nature black-box models can describe only the flow and transport behaviour within observed ranges of input and output functions

and therefore often lack predictive power. By contrast, distributed parameter models integrate the physics of flow and transport processes in conduits and in porous media and so have more predictive power (see section 6.11 for further discussion of hydrogeological modelling).

A wide range of aquifer types are encountered in karst and, further, investigations are on difference scales and for different purposes. So we must conclude that the most appropriate tool for quantitative description and simulation of flow and transport through karst groundwater systems must be chosen with the nature of the aquifer and the objective of the investigation in mind. Different approaches to aquifer analysis have their place as will be seen in Chapter 6, but whereas analyses using Darcy's Law are usually appropriate for individual wells, they are

Table 5.11 Principal differences between single-, double- and triple-porosity aquifers. Most karst aquifers have triple-porosity characteristics Reproduced from Worthington, S. R. H., and Ford, D. C., Chemical hydrogeology of the carbonate bedrock at Smithville. Smithville Phase IV Bedrock Remediation Program. Ministry of the Environment, Ontario, 2001

	Aquifer type		
Parameter	Single porosity (porous medium)	Double porosity	Triple porosity (karst)
Flow components	Matrix	Matrix Fracture	Matrix Fracture Channel
Flow laws	Darcy	Darcy Hagen–Poiseulle	Darcy Hagen–Poiseulle Darcy–Weisbach
Flow modes	Laminar	Laminar	Laminar Turbulent
Flow lines are	Parallel	Mostly parallel	Convergent to channels

inappropriate for the groundwater basin as a whole, because most of the assumptions applicable to flow through Darcian media are violated in karst systems.

5.8 FRESHWATER–SALTWATER INTERFACE

We conclude this chapter by considering the unusual conditions encountered near the coast. The water table declines towards sea level and water quality analyses from samples taken at various depths from bores just inland show that fresh water overlies salt water, which penetrates the aquifer at depth. This interesting phenomenon was first investigated by two European scientists, Ghyben(1889) and Herzberg (1901), whose names are lent to the relationship they found (Reilly and Goodman 1985). The depth below sea level Z_s at which the freshwater–saltwater interface (the **halocline**) occurs is related to the elevation of the water table above sea level

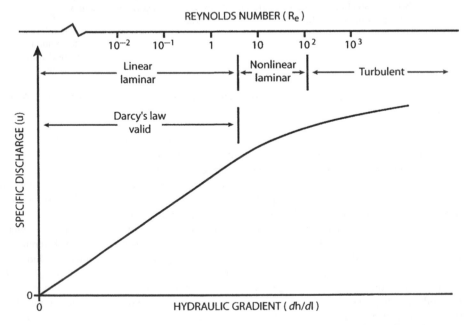

Figure 5.32 Range of validity of Darcy's Law. Reproduced from Freeze, R.A. and J.A. Cherry, Groundwater, p. 604 © 1979 Prentice Hall, Inc.

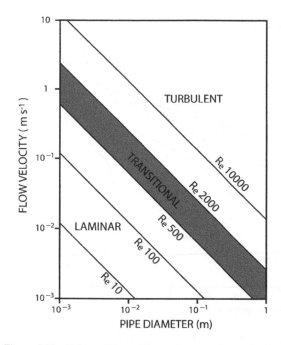

Figure 5.33 Values of Reynold's number at various velocities and pipe diameters, with fields of different flow regimes. Reproduced with permission from Smith, D.I. and T.C. Atkinson, 'Process, landforms and climate in limestone regions'. In E. Derbyshire (ed.) Geomorphology and Climate 369–409. © 1976 John Wiley and Sons.

h_f and to the density of the fresh ρ_f and salt ρ_s waters respectively. The **Ghyben–Herzberg** principle can be expressed as

$$Z_s = \frac{\rho_f}{\rho_s - \rho_f} \cdot h_f \qquad (5.21)$$

Thus if the density of the fresh water is 1.0 and that of the salt water is 1.025, then under hydrostatic equilibrium the depth to the saltwater interface is 40 times the height of the water table above sea level. A practical consequence of this is that if the pumping of a bore in a coastal aquifer causes the water table to be drawn down by 1 m, then salt water will intrude upwards beneath the well by a distance of 40 m. Excessive pumping can therefore risk contamination by saline water, as discussed by Mijatovic (1984a) in cases of seawater intrusion into aquifers of the coastal Dinaric karst and elsewhere.

The interface between fresh and salt water can be seen from isochlor values in Figure 5.35 to be a zone of transition rather than an abrupt discontinuity. It is also evident from the equipotential lines in Figure 5.35 that much of the fresh water must escape through the sea bed in the nearshore zone. The existence of submarine springs

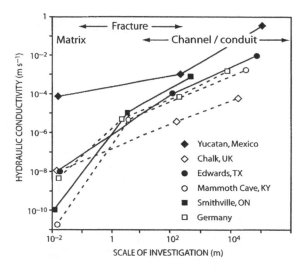

Figure 5.34 Scaling of hydraulic conductivity in six carbonate aquifers. The key lists carbonates in increasing order of age from Cenozoic (Yucatan) to Palaeozoic (Smithville). Reproduced from Worthington, S. R. H., Schindel, G. M., and Alexander, E. C, Techniques for investigating the extent of karstification in the Edwards Aquifer, Texas. In Martin, J B, Wicks, C M and Sasowsky, I D (eds), *Hydrogeology and Biology of Post-Paleozoic Carbonate Aquifers* © 2002 Karst Waters Institute.

(**vruljas**) in karst terrains is a well-known phenomenon recognized from at least the first century BC (Herak 1972). Their occurrence implies confined pipe flow at depth and their location below present sea level may be partly related to the position of springs developed during low stands of the sea in glacial episodes (section 10.3). High secondary porosity below present sea level in the zone of mixing is also a consequence of the geochemistry of brackish water (Back *et al.* 1984) – see section 3.6.

The Ghyben–Herzberg principle of gravitational equilibrium simplifies the relationship usually found in nature, because the two fluids are treated as immiscible and groundwater conditions are normally dynamic rather than static. Hubbert (1940) showed that the interface is deeper under dynamic conditions than under static. He treated the interface as a boundary surface that couples two separate flow fields, with continuity of pressure being maintained across the interface (Figure 5.36). Assumptions of essentially horizontal flow (termed Dupuit approximations) are combined with gravitational equilibrium in so-called Dupuit–Ghyben–Herzberg analysis (Bear 1972) to determine the position of the saltwater–freshwater interface.

Within simple carbonate islands such as uplifted coral reefs, the freshwater body is often described as an idealized freshwater lens. Vacher (1988) provided an

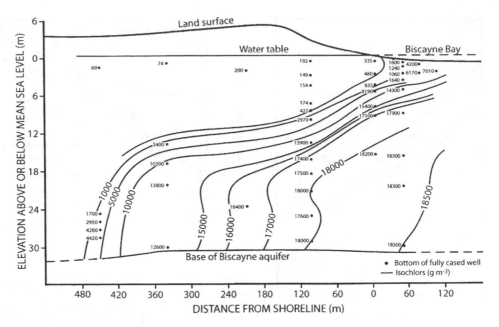

Figure 5.35 Freshwater–saltwater mixing zone in the Biscayne aquifer, Florida. Reproduced from Kohout and Klein, in Ward, R. C., Principles of Hydrology, 2nd edn, p. 403 © 1975 McGraw Hill.

authoritative explanation of the process and explained why groundwater lenses beneath islands can vary from their idealized lenticular form. The unexpected complexity of the apparently simple carbonate island case is amply illustrated by numerous contributions in *Geology and Hydrogeology of Carbonate Islands* (Vacher and Quinn 1997). For example, differences in porosity and permeability of the fore-reef, main reef and back-reef facies of the carbonate host rock can be considerable, and this heterogeneity is reflected in the topography of the piezometric surface of the lens (Figure 5.37). Furthermore, Elkhatib and Günay (1993) pointed out that in anisotropic and heterogeneous karst aquifers the form of the interface is difficult to predict and concluded that the configuration of the saltwater wedge depends mainly on the structure of zones of high karst permeability and the geology of the coastal region.

The dynamic circulation of fresh waters towards the coast is thought to induce entrainment of the underlying saline water, which results in an inflow of saline water at depth, a mixing zone of saline with fresh water and the discharge of brackish water along the coast. The Waikoropupu Springs in New Zealand provide evidence for this (Williams 1977). The springs are artesian, have a mean flow of 15 m³ s⁻¹ and are located 2.6 km inland from the head of tide. The main vent is 7.1 m above sea level and draws water from a marble aquifer that

extends well below sea level and continues beyond the coastline. The spring water contains from 0.4 to 0.6% salt water with values increasing as discharge increases, indicating that the more dynamic the outflow the greater the entrainment of saline water (note that no evaporite beds are known in the region that could provide an alternative source of salt). Other examples are provided by Drogue (1989, 1990, 1993), Ghannam *et al.* (1998) and Arfib *et al.* (2002) for cases around the Mediterranean coast.

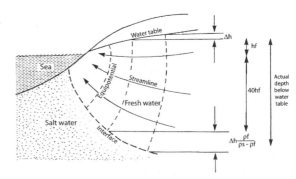

Figure 5.36 Illustration of the Ghyben–Herzberg principle under hydrodynamic conditions. Reproduced from Hubbert, M.K., The theory of groundwater motion. J. Geol., 48, 785–944 © 1940 University of Chicago Press.

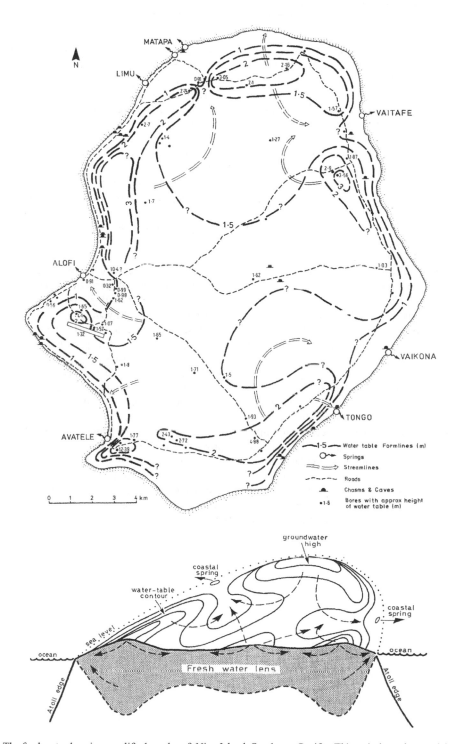

Figure 5.37 The freshwater lens in an uplifted coral reef, Niue Island, Southwest Pacific. This entirely carbonate island is built on a volcanic base 400m below sea level. Uplift has raised the former atoll rim almost 70m and the atoll floor 35m above present sea level. The configuration of the piezometric surface mainly reflects variations in phreatic hydraulic conductivity related to variations in reef facies and dolomitization. Conductivity measurement in boreholes determined the deepest transition to saline water (1000 mS) at 39m below sea level. Reproduced from Williams, P.W. 2004b Karst systems, in *Freshwaters of New Zealand*, (eds J. Harding, P. Mosley, C. Pearson and B. Sorrell), New Zealand Hydrological Society and New Zealand Limnological Society, Christchurch, pp. 31.1–31.20.

Recent interesting work in the low-lying carbonate platform of the Yucatan Peninsula between the Gulf of Mexico and the Caribbean Sea sheds further light on these ideas. The Cenozoic limestones are comparatively porous, have both high permeability and hydraulic conductivity (Table 5.5 and Figure 5.34), and contain well-developed cave systems that convey 99.7% of groundwater flow (Table 5.6). Beddows *et al.* (2002) noted that cave-diving exploration along the Caribbean coast has identified at least 74 horizontally extensive submerged cave systems with a combined length of more than 400 km and at an average depth of −16 m. They also observed that the Yucatan caves contain numerous speleothems and so must have been resubmerged after having formed during previous high sea-level stands. They reported that during diving the mixing zone can be observed as a defined interface in the middle of some of the passages, and comment that the enhanced aggressivity of water in the mixing zone (section 3.6) supports the hypothesis that mixing corrosion is responsible for much cavern enlargement at this level. They found that some coastal springs on the Yucatan Peninsula discharge water up to 75% marine salinity, but their observations indicated that not all of the saline water comes from entrainment. Dye-tracing experiments and current-meter observations showed that at least some saline–water is recharging inland via flow in caves beneath the saline–freshwater interface. They observed that ocean water circulates to more than 9 km inland and that tidal rise can still be measured at that distance. Incorporation of saline groundwater into the overlying fresh water occurs progressively seawards at a relatively slow rate, but becomes more rapid within 1 km of the coast. Recharge of saline water beneath the freshwater body was found to be driven by rising net ocean level, with discharge back to the sea occurring as net ocean level fell over several days. These processes were generated even though the amplitude between net high and low ocean levels was only a matter of decimetres. Saline groundwater flow velocities were measured up to 2 cm s^{-1}.

James (1992) also reported visually distinct freshwater–saltwater interfaces some 25 km inland from the coast of the Great Australia Bight in large flooded caves in late Eocene limestones of the Nullarbor Plain. She also provided evidence for the operation of three mixing zones:

1. close to the water table in cave lakes and canals where low ionic strength recharge from storms rests as a layer up to 2.5 m thick on underlying high ionic strength (brackish) water;
2. close to the water table in the rock matrix where variable ionic strength percolation waters encounter brackish groundwater;
3. at the halocline, which is usually about 20 m below the surface of cave lakes, where brackish water rests on water of ionic strength similar to sea water.

In conclusion, the Dupuit–Ghyben–Herzberg model provides a theoretical basis for understanding the configuration and hydrology of coastal karst aquifers, but field studies have revealed that secondary porosity distribution and changing aquifer boundary conditions (sea level) can be important controls on the fresh and saline groundwater flows.

6

Analysis of Karst Drainage Systems

6.1 THE 'GREY BOX' NATURE OF KARST

Determining the structure and properties of karst aquifers presents severe practical problems because of the anisotropic and severely heterogeneous nature of most karst, and the sparseness of information about it. Yet it is essential for water resources estimation, planning and management to be able to answer such questions as how much water can be used, where is it coming from, and what are the physical parameters characterizing the aquifer? The ability to be able to predict is important both for the management of water resources and for the development of protection strategies against pollution. But it is difficult to generalize about karst aquifers, because so many different geological contexts, storage and flow conditions exist. However, there is sometimes the opportunity to make direct observations underground, in caves, even though accessible passages penetrate only a small part of any given karst terrain.

With direct observations limited to caves, boreholes, inputs and outputs, the remaining aquifer characteristics must necessarily be deduced. Sometimes a choice is made when modelling aquifer functions in terms of rainfall–response (input–output) relationships to treat the system as a 'black box' (Knisel 1972, Dreiss 1982, 1989), but this can be fairly far removed from physical reality and tell us little about the structure of the aquifer and how it really operates. More realistic is a 'grey box' approach that uses such information as is available on the subterranean conditions to clarify the structure of the system and help explain its observed behaviour; such model representations of karst aquifers are discussed later in the chapter. Although each karst aquifer is unique in its individual characteristics, some structural components are widely found (Figure 5.31), although they vary in relative significance in different systems.

Comprehensive analysis of karst drainage systems involves determining the following:

- the areal and vertical extent of the system
- its boundary conditions
- input and output sites and volumes
- the interior structure of linkages and stores
- the capacities and physical characteristics of the stores
- the relative importance of the flow paths
- throughput rates
- the response of storage and output to recharge
- the system's response under different flow conditions.

The information required can be obtained by taking five complementary approaches: water balance estimation, borehole analysis, spring hydrograph analysis, water tracing and aquifer modelling. Water balance estimation and borehole analysis apply conventional techniques of water resources surveys to karst, whereas water tracing and spring hydrograph analysis have been mainly developed for karst. Aquifer modelling is widely applied in groundwater hydrology, but most of the conventional techniques are inappropriate for karstified aquifers because Darcian porous media assumptions do not apply. Consequently, an alternative set of models designed specifically for karst have been developed (section 6.11).

Decisions are often made about the most appropriate form of analysis that depend on certain initial assumptions about flow and aquifer characteristics (Table 6.1). Such assumptions, for example that the aquifer is isotropic and flow is laminar, should be recognized as such and treated as working hypotheses that may be modified in the light of field results. This chapter elaborates these points and discusses methods available for exploration, survey, data analysis and interpretation – the clarification of the grey box.

Karst Hydrogeology and Geomorphology, Derek Ford and Paul Williams
© 2007 John Wiley & Sons, Ltd

Table 6.1 Assumptions and decisions made about the nature of a karst aquifer and appropriate methods for its analysis (After Ford, D.C. and Williams, P.W. (1989) Karst Geomorphology and Hydrology)

Flow conditions	Boundary conditions	Aquifer characteristics	Scale and state	Form of analysis
Linear-laminar (diffuse Darcian)	Infinite areal extent	Confined/unconfined	Site specific	Borehole dilution
Mixed laminar and turbulent	Impermeable/leaky upper and lower boundaries	Constant/variable thickness	Local	Borehole recharge
Turbulent (conduit flow)	Spatially uniform/variable vertical recharge	Homogeneous/ heterogeneous	Regional	Borehole pumping
	Constant/variable potential recharge boundary	Isotropic/anisotropic	Groundwater basin	Recharge response modelling
	Constant/variable potential discharge boundary		Steady state	Water budget
	Fixed/mobile phreatic divides		Transient state	Spring hydrograph and chemograph
				Aquifer modelling

6.2 SURFACE EXPLORATION AND SURVEY TECHNIQUES

Practical methods for the evaluation of groundwater resources in general are explained in a wide variety of excellent publications, for example by Freeze and Cherry (1979), Todd (1980), Castany (1982), UNESCO/IAHS (1983), Hötzl and Werner (1992), and Domenico and Schwartz (1998). Repetition of material presented there is unnecessary, but evaluation of its applicability to karst is essential, because most groundwater texts deal inadequately with karst aquifers. However, engineering geology texts by Milanović (2000, 2004) are devoted exclusively to karst.

6.2.1 Defining the limits of the system

All aquifers have boundaries that modify flow conditions as discussed in section 5.2. For example, transmissivity is limited by the confining beds of an artesian aquifer, and by both the water table and lower limit of karstification in an unconfined aquifer. These factors determine vertical boundary conditions. Also very important are those that limit the horizontal extent of an aquifer; here, discharge and impermeable boundaries should be distinguished from recharge boundaries.

In non-karstic terrains, groundwater divides are assumed to directly underlie the surface topographic divides as determined from contour maps, aerial photographs, etc. This approach is acceptable in karst only as an initial working hypothesis, because experience in innumerable karst catchments has shown phreatic and vadose divides to deviate significantly in plan position from overlying surface watersheds. For example, the

Continental Divide between Atlantic and Pacific catchments is breached in this manner in the Rocky Mountains. Furthermore, phreatic divides may migrate laterally in response to changing water-flow conditions. A groundwater divide determined at low water-table levels may not be valid when the piezometric surface is higher.

Within a simple karst aquifer there may be several groundwater basins with minimal hydraulic connection, each draining to a different spring (or set of springs). In an unconfined aquifer, the limits of each system can be determined:

1. by mapping piezometric contours and thus establishing regions of divergence of flow (groundwater divides);
2. by water tracing, perhaps using fluorescent dyes or isotopes (section 6.10).

In a study in Kentucky, Thrailkill (1985) showed that dye tracing can reveal narrow groundwater basins that cannot be identified by interpretation of groundwater contour patterns. This and many other dye-tracing studies indicate that water tracing should be used to verify groundwater divides determined from water-table maps. Groundwater divides may also be deduced from the potentiometric surface in confined aquifers, but since flow-through times are likely to be much longer than under unconfined conditions, water tracing by dyes may not be practical. Instead, the provenance of groundwaters may have to be determined using environmental isotopes or by pulse-train analysis (section 6.10).

The impervious basin area that sustains any allogenic inputs along recharge boundaries can be determined accurately by conventional plan mapping of surface watersheds. However, the swallow holes where surface streams sink are less easily located, especially if the sites are small or under forest. Influent rivers that gradually lose their flow over several kilometres are also difficult to detect, especially if not all of the flow is lost. Field mapping and discharge measurement is the only sure way of obtaining accurate information, but aerial photograph interpretation is of considerable assistance.

6.2.2 Estimating the water budget

An important first step in the management of water resources is to estimate the size of the reserves and the gains and losses to the system. Water-balance calculation provides an order of magnitude estimate of reserves and storage changes. This is undertaken by preparing a hydrological budget, which is a quantitative evaluation of inflows, outflows and changes in storage over a specified period of time, usually a hydrological year, although it can be applied to longer or shorter periods. A hydrological year runs from dry season to dry season, i.e. it starts and ends when storage is at a minimum, and so it often does not correspond to a calendar year.

With respect to surface streams, the water balance equation in its simplest form is usually written

$$Q = P - E \pm \Delta S + e \qquad (6.1)$$

Where Q is runoff, P is precipitation, E is evapotranspiration, $\pm \Delta S$ represents withdrawal from or replenishment of storage and e represents error. Evapotranspiration includes interception losses, transpiration and evaporation from wet surfaces, and is the most difficult term to estimate. Average catchment precipitation can also be quite difficult to evaluate, because the distribution of rain gauges may not provide a representative sample of average precipitation across the catchment. Nevertheless, knowing the catchment area A, the acceptability of the estimates can be checked by comparing water surplus calculated from $P - E$ with measured Q. Dunne and Leopold (1978) expand on this approach. However, in a groundwater context the perspective may sometimes differ; thus the US Department of the Interior (1981) suggested that the hydrological budget may be summarized as

$$P - E \pm R \pm U = \Delta S \qquad (6.2)$$

where R is the difference between stream outflow ($-$) and inflow ($+$), and U is the difference between deep groundwater outflow ($-$) and inflow ($+$); with ΔS being the resultant change in total storage in groundwater, soil moisture, channels and reservoirs. The groundwater component of this is ΔS_g and can be estimated from

$$\Delta S_g = G - D \qquad (6.3)$$

in which G represents recharge to groundwater and D the discharge from it. Recharge to groundwater can be assessed from

$$G = P - (E + Q_s) \qquad (6.4)$$

in which Q_s represents surface runoff.

Blavoux *et al.* (1992) provide a particularly good example of water-budget estimation for the basin that sustains the flow of the famous Fontaine de Vaucluse in southeastern France. Its mountainous catchment rises to 1909 m, has an area of 1115 km^2, and an average altitude of 870 m. The boundaries of the basin area were determined from dye tracing and geological considerations. An altitude-belt model was used to calculate the moisture balance, using local precipitation and temperature gradients. Because precipitation increases by about 55 mm 100 m^{-1} and temperature decreases about 0.5°C 100 m^{-1}, actual evapotranspiration loses are relatively uniform with altitude. The weighted effective rainfall ($P - E$) is ~570 mm yr^{-1} for the whole basin. About 75% of the total and the whole of the dry season effective rainfall occur at elevations higher than the average altitude. Weak summer rainfalls do not influence discharge, but the spring responds to heavy rain in 1–4 days. The mean spring discharge of 21 m^3 s^{-1} (range of 3.7 to >100 m^3 s^{-1}) is consistent with the calculated effective rainfall and the estimated basin area.

Whereas the above example is from a well-researched catchment, many other karst areas have been less well studied and do not have good background data. Two cases from the 'classic' karst of Slovenia illustrate why derivation of a water budget in karst is not always straight forward. Trišič (1997) investigated the water balance of the area draining to the karstic springs of the Vipava River. The catchment area was estimated as 125.25 km^2, but could not be determined accurately because of its karstic nature. Average annual rainfall is 2024 mm but evapotranspiration withdraws about 31%. The measured average discharge of the springs is 6.78 m^3 s^{-1}, which proved to be more than could be accounted for by the estimated size of the catchment. Increasing its area to ~150 km^2 made the precipitation and discharge data compatible but, to accept a basin size adjustment of this magnitude would require independent support by water tracing. It also assumes that the other terms of the water balance were not in significant error. A similar problem was encountered by Trišič *et al.* (1997) in the water balance in the Bohinj region of the Julian Alps. In this mountainous area it is particularly difficult to estimate precipitation accurately because it falls as snow and there are few measurement stations at high altitude. Having estimated the water balance components, they encountered major discrepancies between calculated runoff $(P - E)$ and measured runoff at gauging stations. They concluded that the main problem was incorrect data on catchment area size, which had been estimated as $94 \, \text{km}^2$, because compared with calculated runoff the measured discharge in the Bohinjka catchment was 10% too small. Dye tracing of stream-sink to spring linkages was therefore undertaken to improve catchment definition.

Bonacci (2001) pointed out that in many small karst catchments only limited precipitation data are available and yet, for engineering practice and water resources management, water-balance estimates are required. Consequently it is valuable to have a reliable estimate of the regional effective infiltration coefficient, defined as the ratio of effective precipitation to total precipitation, because this knowledge permits transfer of hydrological information from catchments having adequate hydrological and meteorological measurements to those where they are insufficient. Bonacci (2001) reviewed methods for estimating monthly and annual effective infiltration coefficients, and concluded that antecedent conditions must be taken into account and that rainfall distribution during the year has a significant effect on the coefficients calculated.

The two most important problems encountered in calculating the water balance in karst areas are, firstly, determination of effective precipitation and, secondly, determination of the catchment boundaries, not least because in karst the recharge area often varies in time depending on groundwater levels. These examples show that derivation of accurate water-balance equations with acceptable and known errors can be challenging. This issue is taken up again in section 6.3, when we discuss recharge and percolation in the vadose zone.

A good example of large-scale karst water-balance estimation was given by Bocker (1977). The aim was to investigate the environmental impacts of coal and bauxite mine dewatering operations in the $15\,000 \, \text{km}^2$ of the Transdanubian Mountains, Hungary. A finite-element model using $4 \, \text{km}^2$ elements was developed. Infiltration into each element was estimated from an algorithm derived from a 15-year field experiment, and data from 480 waterworks, 93 mines, 270 observation

wells and 155 meteorological stations were included in this major computation.

6.2.3 Remote sensing using multispectral techniques

Remote sensing using conventional stereoscopic aerial photographs as well as multispectral imagery from aircraft and satellites is an important reconnaissance tool in hydrology and water resources management (Schultz and Engman 2000), including groundwater investigations (Farnsworth *et al.* 1984, Meijerink 2000). For example, Lopez Chicano and Pulido-Bosch (1993) undertook a detailed analysis of fracture patterns in the Sierra Gorda karst of Spain using aerial photographs and field mapping (Figure 6.1); Kresic (1995) provided evidence of the control of groundwater flow in the Dinaric Karst by tectonic fabric interpreted from aerial photographs and satellite images; and Tam *et al.* (2004) investigated the relationship between lineaments and borehole capacity in Vietnam.

A general review of multispectral remote sensing in karst was presented by Milanović (1981). LaMoreaux and Wilson (1984) demonstrated the particular importance of thermal infrared imagery in identifying recharge and especially discharge points, including submarine springs (Gandino and Tonelli 1983). Thermographic techniques have great (and still largely unrealized) potential for identifying springs, especially in winter in areas of deciduous forest, with modern instrumentation capable of resolution to a few metres.

6.2.4 Electrical resistivity, ground-penetrating radar and other geophysical techniques

Remote sensing using geophysical techniques is a long-established and widely accepted tool in groundwater hydrology. Milanović (1981), Arandjelovic (1984), Astier (1984) and Stierman (2004) have explained the general application of the methods to karst.

Deep exploration typically uses a combination of techniques such as three-dimensional seismic imagery supported by drilling. In this way an Oligo-Miocene palaeokarst surface buried beneath $1200 \, \text{m}$ of Mio-Pliocene sediments was mapped at $1320–1360 \, \text{m}$ depth in Cretaceous limestones beneath the Adriatic Sea (Soudet *et al.* 1994). Within the palaeokarst, a palaeo-epikarst zone of $35 \, \text{m}$ thickness was distinguished from an underlying palaeopercolation zone of $15–45 \, \text{m}$ thickness above a palaeophreatic zone with conduits of $35–79 \, \text{m}$ thickness. The karstified limestones host an oil column about $140 \, \text{m}$ high.

For shallower exploration, electrical resistivity surveys have proved to be particularly important, especially for

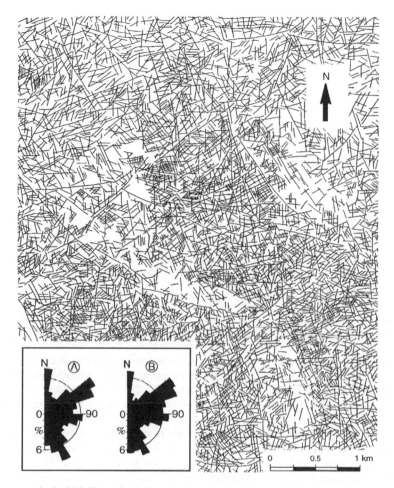

Figure 6.1　Fracture pattern in the high Sierra Gorda karst, Spain. The cumulative length of the 2904 fractures mapped is 358.6 km with fractures having an average length of 165 m. Inset: orientation direction of (A) number of fractures and (B) cumulative length. Reproduced from López-Chicano, M. and Pulido-Bosch, A., The fracturing in the Sierra Gorda karstic system (Granada), in Some Spanish Karstic Aquifers (ed. A. Pulido-Bosch), University of Granada, pp. 95–116, 1993.

establishing the vertical dimensions of a karst aquifer, because the method can distinguish between compact limestone, water-saturated karstified limestone and dry karstified limestone (Figure 6.2). Resistivity imaging introduces artificially generated electric currents into the ground and resulting potential differences are measured at the surface. Dry air-filled caves and fissures will have relatively high resistivity values, whereas values will be relatively low if they are water-filled. Resistivity surveys use a grid of survey points, and the procedure involves measuring a series of traverses with constant separation between the electrodes and then repeating the traverses with increased electrode separation several times. The increasing separation leads to information being obtained from successively greater depth. Vertical contoured sections are then produced that display the variation of resistivity both laterally and vertically over the surveyed section. Arandjelovic's (1966) work in Bosnia-Herzegovina, interpreting the base of karstification in the Trebisnjica valley (Figure 6.3), provides an especially good illustration of the value of resistivity surveys. From geophysical and other data, Milanović (1981) considered the base of karstification in the Dinaric region to be usually no deeper than 250 m.

Resistivity and microgravity methods are often used to map shallow subsurface karst features (Patterson *et al.* 1995, Rodriguez 1995, Crawford *et al.* 1999, McGrath *et al.* 2002). Gravity surveying measures variations in Earth's gravitational field produced by density differences in subsurface rocks. Caves and depressions cause

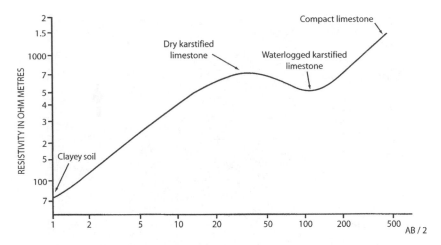

Figure 6.2   Typical electrical resistivity sounding in a calcareous zone. The depth of rock mass investigated increases with the electrode spacing *AB/2*. Reproduced from Ford, D.C. and Williams, P.W. (1989) Karst Geomorphology and Hydrology, after Astier 1984.

a reduction in the gravitational acceleration over them because of the missing mass associated with the void, whereas topographic highs on the bedrock surface will exert greater gravity pull than the surrounding less dense material. However, microgravity variations are very small, so very high-precision instruments and meticulous field techniques are essential. The survey involves establishment of a regular grid of observation points over the survey area with station spacing depending on the possible size and depth of the cavities that are to be detected. The microgravity data recorded are then presented as a residual Bouger anomaly map (e.g. McGrath *et al.* 2002).

Rodriguez (1995) described what he called the Integrated Geophysical Approach in which three geophysical methods are combined to optimize data quantity and quality acquisition, and yet to be cost effective. He recommended an initial strategic investigation of the site using a microgravity survey followed by a resistivity survey over the same grid. This is then followed by a high-resolution seismic survey and microresistivity survey in combination to map in detail the karst features found in the first stage.

McGrath *et al.* (2002) compared the results of microgravity surveying and resistivity imaging over a spring conduit that had been explored by cave diving in an area of dense Carboniferous Limestone in South Wales, UK. They used an observation-station spacing of 2.5 m on a 30 × 50 m grid to detect conduits a few metres wide at depths up to 10 m below the surface. A residual microgravity map of the resurgence zone was produced that successfully identified subsurface karst channels and

probable feeder conduits. They then followed this up with a resistivity survey in which profiles were aligned perpendicular to the already identified negative microgravity anomalies. Fifty electrodes were deployed with 1 m spacing so that a depth penetration of 7 m could be attained. The result showed a low-resistivity layer around 2–4 m depth that corresponded well with the negative amplitude microgravity anomaly. The depth scale from the electrical resistivity section could be used as an independent constraint allowing inversion of the microgravity data to give the thickness of the underground cave system. These results are reassuring, because they demonstrate that voids can be detected in areas using remote geophysical techniques where other direct methods have shown them to exist.

With recent attention being focused on the near-surface zone of karst, because of problems of building foundations and spillage of pollutants, the potential of ground-penetrating radar (GPR) is becoming more widely realized. Al-fares *et al.* (2002) provided a particularly good illustration of its capabilities. They investigated a site on the Hortus karst plateau near Montpellier in southern France, and concluded that GPR is particularly well-adapted to the analysis of the near-surface (< 30 m depth) structure of karst, especially when clay or soil that would otherwise absorb and attenuate the radar signal is thin or discontinuous. They conducted six parallel 120 m long radar traverses spaced at 15 m intervals over the course of a shallow karst conduit leading to Lamalou spring, and used a low frequency (50 MHz) signal. The radar interpretation was verified by comparing it with data available from 10 boreholes (of 32–80.5 m depth)

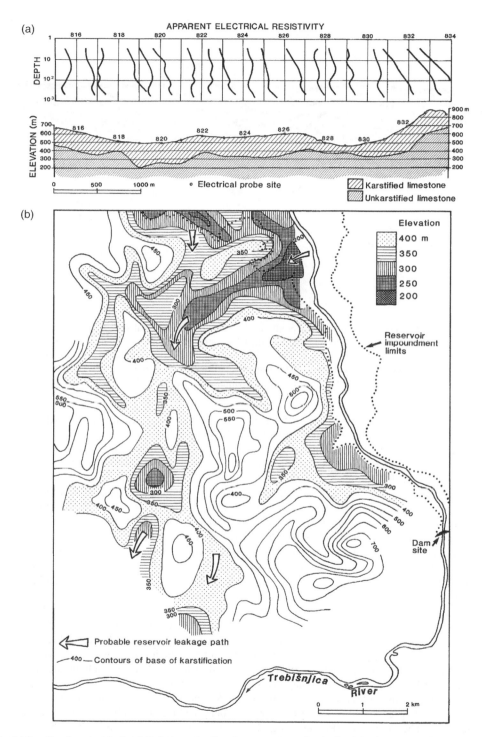

Figure 6.3   (a) Results of an electrical resistivity investigation along a cross-section in the right abutment of the Bileca reservoir and (b) contour map of the base of karstification beneath the Trebisnjica valley, Croatia/Bosnia-Herzegovina. Reproduced from Ford, D.C. and Williams, P.W. (1989) Karst Geomorphology and Hydrology, after Arandjelovic 1966.

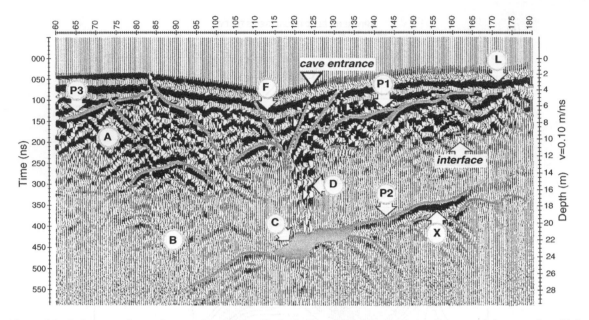

Figure 6.4 Radargram of ground-penetrating radar profile in the Lamalou experimental site, France, with interpretation aided by data from neighbouring borehole: A, fractured and karstified limestone in the epikarst; B, massive and compact limestone; C, Lamalou Cave; D, pothole inlet to cave; F, fault; L, karren; P1, 2, 3, bedding planes; X, unknown cave. The 'interface' marks the approximate base of the epikarst. Reproduced from Al-fares W., Bakalowicz M., Guerin R. and Dukhan M, Analysis of the karst aquifer structure of the Lamalou area (Herault, France) with ground penetrating radar. Journal of Applied Geophysics 51(2–4), 97–106. © 2002 Elsevier.

and from a vertical shaft giving access to the underlying cave (Figure 6.4). The conclusions were that the relative absence of electrically conductive sediments on the surface made the use of a low GPR frequency efficient, and that discontinuities in the bedrock such as bedding planes, faults and fractures could be distinguished. A strongly fissured epikarst of 8–12 m thickness could be readily separated from the underlying more massive limestone, and a cave could be clearly identified 20 m below the surface. They also considered that the results can probably be generalized to karstic aquifers of the Mediterranean type.

Although thick soils and cover deposits attenuate the GPR signal, a study by Collins *et al.* (1994) clearly revealed solution pipes in limestone bedrock covered with a clay layer in Florida. Further, Mellett and Maccarillo (1995) provided a good demonstration of how useful GPR can still be in areas with particularly thick drift cover over karstified limestone. They investigated a site adjacent to a major highway that had been experiencing cover collapse, probably exacerbated by runoff from the road. They repeated GPR surveys at 2-year intervals across the same traverse and detected very clearly the progressive upwards development of cylindrical subsidence structures through thick sediment

overburden. Radar penetration was achieved to a depth of about 6 m.

Currently the only proven geophysical technique that responds to the movement of fluids, rather than just their presence, is the natural-potential (NP) method. Lange and Barner (1995) explained that this method measures natural voltage at the Earth's surface resulting from ambient DC electric currents that occur everywhere on land and sea. Such currents can arise from a variety of sources, but especially from the flow of water through pores, fissures and conduits in the ground, the electrokinetic effect. In a survey to locate karst conduits on the island of Guam in the western Pacific, they identified ten NP anomalies as targets for drilling. Drill depths ranged from 9 to 28 m and eight intercepted cavernous zones and water-filled cavities up to 2 m high. Lange and Barner (1995) also noted that experience elsewhere showed that surface NP signals related to caverns could be identified through up to 280 m of rock.

We must caution, nevertheless, that all these remote geophysical techniques have their limitations, especially with respect to the size of cavities that can be detected and the depth below the surface that can be explored with confidence. Geophysical methods usually become more difficult to apply in areas of rugged topography, and whereas resistivity requires electrodes to have good ground contact in soil, too great a thickness of soil reduces the effectiveness of GPR by attenuating the signal. In all cases, geophysical results must be treated with caution and only be accepted for planning purposes where confirmed by independent methods such as drilling.

In spite of that, it must be said that a problem particularly amenable to geophysical study is found in coastal aquifers: that of determining the depth to the saltwater–freshwater interface (section 5.8). Since its position cannot be located very accurately using the Ghyben–Herzberg principle (equation 5.20), electrical resistivity surveys (usually combined with drilling) are often conducted to determine the distance to the interface. This technique is appropriate because in a saturated rock of given density and porosity, the resistivity is largely dependent on the salinity of the saturating fluid. Investigations of the groundwater hydrology of Niue and Nauru islands, two uplifted subcircular coral atolls in the southwest Pacific of 259 and 22 km² respectively, provide excellent illustrations of the technique. These islands have been raised 60–70 m above sea level and consist of about 400–500 m of carbonate rock on basaltic seamount pedestals. In their survey of Niue Island, Jacobson and Hill (1980) determined the depth to the saltwater–freshwater interface by electrical resistivity, its general location having been ascertained by electrical conductivity profiles down deep boreholes. Their work showed the existence of a broad transition zone rather than a sharply defined boundary, and a freshwater body that did not conform to the ideal lenticular shape, but had the form of a ring-doughnut (Figure 5.37). Further detail has been added to their work by Williams (1992) and Wheeler and Aharon (1997). A similar investigation was made on Nauru Island by Jacobson *et al.* (1997), who showed that the island is underlain by a discontinuous layer of fresh water averaging 4.7 m thick distributed in two main lenses. The water table has an average elevation of only 0.2 m above sea level, yet the underlying mixing zone of brackish water is up to 60 m thick. They attributed the unusually thick mixing zone to high permeability in the limestone with open karst fissures allowing intrusion of sea water throughout the island's substructure. Cavities over 5 m deep were encountered below sea level when drilling. Seismic refraction, resistivity and salinity profiles were also used successfully to determine the base of the freshwater lens on Pingelap Atoll by Ayers and Vacher (1986).

We conclude that for the karst hydrogeologist, geophysical techniques need to be useful to depths of a few hundred metres in most instances, but drilling is always necessary to confirm interpretation.

6.3 INVESTIGATING RECHARGE AND PERCOLATION IN THE VADOSE ZONE

6.3.1 Autogenic recharge

When discussing **input control** in section 5.2 it was noted that recharge into karst can be autogenic or allogenic (Figure 4.1). Autogenic recharge is often diffuse and slow (although it is more concentrated and rapid when focused by dolines), whereas allogenic recharge normally occurs as concentrated, very rapid point inputs of sinking streams from adjacent non-karst terrains. It follows therefore that recharge is highly variable in space, and because of the changeability of weather and climate it is also highly variable in time.

In this section we confine our attention to the autogenic component of recharge. Only a proportion of rain that falls on the surface recharges the groundwater system, because part is intercepted by vegetation, where it subsequently evaporates from leaves and branches, and some is transpired by plants (equations 6.1 and 6.2). These processes return water to the atmosphere as vapour. Sometimes rainfall is so intense that not all the water reaching the ground can be absorbed because the infiltration capacity of the surface or subsoil horizons is exceeded. This infiltration excess component therefore runs off across the surface or passes laterally through the soil as throughflow and ultimately reaches surface streams. It is the component that does not vaporize or runoff, but percolates down through the soil and epikarst that concerns us here, because this is usually the major source of groundwater recharge.

A conceptual model by Sauter (1992) for calculating components of autogenic recharge is illustrated in Figure 6.5. It takes account of rainwater passing through a canopy store, a tree-trunk store and a soil-moisture store before recharging the underlying limestone. Appropriate techniques for calculating potential evaporation and soil-water balance are discussed by Fowler (2002). The field capacity of soils over karst varies considerably from almost zero in thin skeletal soils to 150 mm or more of water in deep loams and clays, and so must be assessed on a site-by-site basis. This is important, because according to Sauter's model groundwater recharge occurs only if field capacity is exceeded. However, even when there is a soil-moisture deficit in summer, heavy rain has often been observed to stimulate a percolation response in stalactite drips in caves; thus bypasses must occur by way of macropores and desiccation cracks in the soil. This is recognized in the term RR in the model, and comes into effect when a threshold effective daily rainfall amount has been exceeded. This was set at 6 mm by Sauter (1992) for the Swabian Alb karst in southwest Germany. Other parameter values used are shown in Table 6.2.

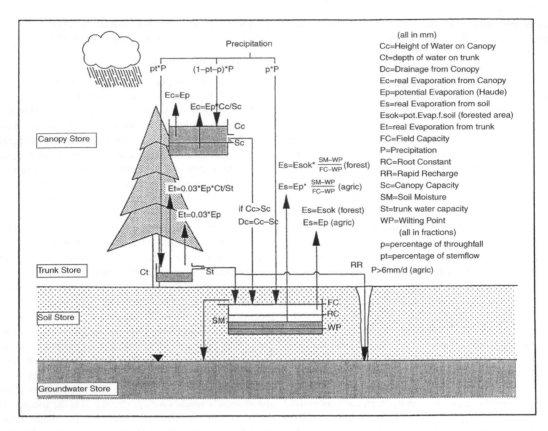

Figure 6.5   Conceptual model of procedure for calculating autogenic recharge. Reproduced from Sauter, M., Quantification and forecasting of regional groundwater flow and transport in a karst aquifer. Tübinger Geowissenschaftlichen Abhandlungen, Reihe C, 13, p. 150, 1992.

The region has soils of 20–80 cm thickness over Jurassic limestones, and about 60% of the area studied is forested (70% coniferous), the rest being used for agriculture.

Calculated results from recharge models can be verified, assuming that the basin area is known accurately, by comparing them with measured outflow (spring discharge) from the basin. This would normally be done for a hydrological year between dry seasons, when the difference in the volume of stored groundwater between years is at a minimum. Sauter (1992) found discrepancies of up to ±10%, and attributed this mainly to inaccuracies in determining the stored groundwater and/or the unknown quantity stored within the epikarst and the unsaturated zone, although discharge measurements and aerial rainfall estimates were recognized as other possible sources of error. Deriving a representative estimate of basin rainfall is a well-known problem in hydrology, because the distribution of rain gauges is usually fortuitous and seldom optimal for water-balance purposes. So all things considered, an error of ±10% is a reasonable result and ±5% would be considered exceptional.

A largely unappreciated form of recharge in karstified mountains is provided by condensation induced by air exchange through conduits in the massif. This is explained by Dublyansky and Dublyansky (2000), who calculated that annual condensation in the karsts of the western Carpathians and western Caucasus Mountains can average 54 mm (range 1–149 mm), although it does not normally exceed 9% of the annual precipitation in any given region. Since it occurs mainly in summer, it can provide a significant contribution to dry-season recharge particularly in arid zones.

In addition to water-balance techniques of estimating recharge, water quality techniques are also available and are especially useful when estimating recharge on islands. Jones and Banner (2000) review techniques in the context of Barbados and conclude that comparison of groundwater chloride concentrations and $\delta^{18}O$ values with those in rainfall permits accurate estimation of recharge. The chloride ion is conservative. It accumulates in the soil when evapotranspiration occurs, later to be flushed down to groundwater during recharge

Table 6.2 Input parameters for recharge calculation for agricultural and forested areas (Reproduced from Sauter, M., Quantification and forecasting of regional groundwater flow and transport in a karst aquifer. Tübinger Geowissenschaftlichen Abhandlungen, Reihe C, 13, p. 150, 1992)

Parameter	Description	Value
P	Precipitation	Variable
p	Percentage throughfall of precipitation	25%
Pt	Percentage stemflow of precipitation	1.6%
Sc	Canopy capacity	4.7 mm
St	Trunk water capacity	0.014 mm
FC	Field capacity	75 mm
Ep	Potential evaporation (grass)	Variable
Esok	Potential evaporation (forest)	Variable
RC	Root constant	50 mm
WP	Wilting point	10 mm
RR	Rapid recharge	Forest: none Agricultural: 6 mm

events. Thus the ratio of the concentrations of chloride in groundwater and rainfall gives a measure of the proportion of rainfall that is lost to evapotranspiration; the remainder providing recharge. In the case of $\delta^{18}O$, comparison of $\delta^{18}O$ values of groundwater with seasonal fluctuations of rainwater $\delta^{18}O$ values shows that groundwaters have relatively low $\delta^{18}O$, similar to that of rainfall in the wet season (June to December). The average groundwater $\delta^{18}O$ is equal to the weighted-average $\delta^{18}O$ values of rainfall when monthly $P > 195$ mm, which in Barbados is usually between August and November. Jones and Banner concluded that the chloride and oxygen isotope methods of estimating recharge (i) have fewer uncertainties than recharge estimates based on direct measurement of hydrological parameters, (ii) have the advantage of providing some insight into the spatial and seasonal distribution of recharge to the aquifer, and (iii) require fewer field measurements than alternative techniques.

6.3.2 Infiltration, percolation and the epikarst

Autogenic recharge infiltrates a surface that can have widely differing characteristics from bare karrenfeld to thick-soil-covered bedrock, although even in a karrenfeld there is usually some detrital debris and insoluble residue down the solutionally widened joints. We saw in section 4.4 when discussing the vertical distribution of karst

dissolution that most of the corrosional attack is near the surface close to the major source of biogenic CO_2, measurements showing that usually 50–90% occurs within 10 m of the carbonate bedrock surface. This imparts a very high secondary porosity, permeability and storage capacity to the epikarst.

The epikarst is at the top of the zone of infiltration (section 5.5). The highly variable nature of water flows (hydraulic behaviour) and karstic void distribution (structure) within it distinguishes the epikarst from the rest of the vadose zone, sometimes called the **transmission zone**, that lies beneath it (Bakalowicz 1995). Although the characteristics of the epikarst illustrated in Figures 5.28 and 6.4 convey as representative a model as any for what is normally conceptualized for the epikarst, it does in fact vary considerably from place to place. The reason for this is that every karst has its unique combination of lithology, structure, geomorphological history and climate. A few examples will demonstrate the range of conditions that can be encountered.

The karst surface of the Hortus plateau in Mediterranean France (the example in Figure 6.4) has low relief and thin patchy soil, with outcrops of karren and a few closed depressions. The epikarst is 8–12 m deep. This situation contrasts strongly, for example, with areas of intense surface dissection such the 'stone forest' regions of Mount Api in Sarawak, Lunan in Yunnan, China, the giant grikelands of the Kimberly Ranges in the Northern Territory of Australia, the 'tsingy' of Madagascar, the labyrinth karst of Nahanni, Canada, or the arête and pinnacle karst of Mount Kaijende in Papua New Guinea (Figure 6.6a). These areas have wide open joints that can be 10–100 m or more deep and several metres wide. Consequently, the epikarst is also very deep. However, sometimes (as at Lunan) the epikarst terminates at the water table with no intervening vadose transmission zone (Figure 6.6b). In other places widened joints are deep, but are largely full of weathered residue such porous dolomitic sand, as in parts of the Grands Causses of southern France (such as at Montpellier-le-Vieux). Elsewhere some karsts are thickly blanketed by combinations of weathering residue and allogenic deposits (e.g. alluvium, loess, tephra, etc.), as in the Sinkhole Plain of Kentucky. By contrast, in many karsts in high latitudes, great Pleistocene glaciers have stripped the soil and truncated the epikarst, reducing it in places to only a few metres in thickness beneath a glacio-karstic pavement surface, as in parts of Manitoba, Ontario, western Ireland and the Pyrénées (Figure 6.7). In alpine zones, faulting and deep joints opened by tension may so readily drain the vadose zone that the epikarst may not be a particularly strong feature of the groundwater system, although it will still store snow and ice.

Figure 6.6 (a) Arête-and-pinnacle landscape of Mount Kaijende, Papua New Guinea, at around 3000 m above sea level. Local relief on the pinnacle ridges is up to about 120 m. (b) In parts of the Stone Forest of Yunnan widened joints reach the zone of water-table fluctuation. Consequently there is direct communication between the epikarst and the epiphreatic zone. Horizontal banding on the rocks shows the range of water-table fluctuation at this site.

6.3.3 Transmission and storage in the epikarst

The variable characteristics described above strongly influence the capacity of the epikarst to absorb and store precipitation. Where the karst surface is largely bare, the uptake of water is determined by the characteristics of the rock (its vertical hydraulic conductivity); but where it is covered, it is controlled by the nature of the soil (its infiltration capacity). The storativity of the epikarst is determined by three factors: (i) the thickness of the epikarst, (ii) its average porosity (these first two together determining the available storage space) and (iii) the

Figure 6.7   (Upper) Glacially stripped limestone pavements in the Pyrenees Mountains between France and Spain. (Lower) Eroded limestone surface in the Taurus Mountains, Turkey, exposed in a road cut.

relative rate of inflow and outflow of water. The epikarst is like a colander: the capacity of the vessel to hold water is determined by the balance between the rate at which water comes in and the rate at which it drains. Whereas the average porosity is determined by the karst void space less the volume of granular fill, the drainage rate is controlled by the vertical hydraulic conductivity of the underlying vadose zone (Figure 5.15). This varies because of the uneven pattern of opened joints and faults and their variable permeability. Thus some epikarsts have a large storage potential but drain rapidly, others are frequently replenished by rain and are usually at least partially saturated, and some are low lying and always partly flooded.

Diffuse autogenic recharge leads to surges of percolation through the vadose zone. The hydraulic pressure of recharge produces a pressure pulse that stimulates a **transfer** of water. This process is distinct from the **transit** of individual molecules of water through the system (Bakalowicz 1995). Recalling section 5.3, these effects lead to different **pulse-through** and **flow-through** times following a recharge event, the latter being significantly longer.

Various investigations have been made of water movement through the epikarst and vadose zone by following natural and artificial water tracers and by making observations in caves. We now recognize that flow follows a range of paths from extremely slow seepages down capillary-sized openings to variable and sometimes high-volume cascades down open shafts. Some of the earliest empirical classifications of vadose waters were by Gunn (1978, 1981b), Friederich and Smart (1982) and Smart and Friederich (1987). They came to complementary conclusions by recognizing (i) a spectrum of discharges from slow low-volume seepages to variable, sometimes large, flows down open shafts (Figure 6.8a), and (ii) a range of discharge volumes and variabilities from almost unvarying low-volume seepages to extremely variable flows that responded rapidly to recharge (Figure 6.8b). These characteristics have since become better defined with improved instrumentation. Thus, for example, we now also know that high variability can occur at low discharge and that percolation from speleothems can be sensitive to air-pressure changes (Genty and Defladre 1998). Nevertheless, the interpretations made by Smart and Friederich (1987) in their outstanding study of water movement and storage in the epikarst of the Mendip Hills in England remain valid (Figure 6.9), and their recognition of flow switching when recharge exceeds certain values and non-linearity of percolation response has been confirmed by other workers (e.g. Baker *et al.* 2000, Baker and Brunsdon 2003, Sondag *et al.* 2003).

The time it takes for a molecule of water to pass through the system is its flow-through time, which can be measured by dye tracing. Friederich and Smart (1981) placed fluorescent dyes at several sites at the base of the soil above GB Cave in the Mendip Hills. It first appeared in the cave close to the injection site, but spread rapidly until most sites sampled in the cave were positive (Figure 6.10). This indicated that lateral diffusion of dye occurred. Since some positive sample sites were 80 m from the injection point and at a shallow depth, it was also evident that this diffusion took place within the

top 10 m of the epikarst. The majority of the injected tracer was discharged as a high concentration pulse via shaft flow adjacent to the injection site, but part was still detectable elsewhere 13 months later. Under conditions of slow recharge, seepage inlets had the highest concentrations, but following increased recharge after rain, a sharp high-concentration response was obtained again from shaft flow, thus indicating flushing from storage. At any particular time, concentrations varied significantly between adjacent inlets, indicating that they were not fed from a homogeneous store but from one that was imperfectly mixed.

Similar dye-tracing experiments were conducted by Bottrell and Atkinson (1992) in the Pennine karst in England. Four different fluorescent dyes were placed beneath the soil above White Scar Cave, where 24 water inlets were monitored. Ten traces were performed from seven injection sites. Weather conditions ranged from extremely wet to very dry. The dyes traversed the 45–90-m-thick vadose zone and were detected in the cave, sometimes within 24 h. It was found that water did not necessarily flow to the closest inlet below the input point, but could appear at inlets over 100 m away without also appearing at others apparently below the intervening path. These observations thus suggested that, in this karst, flow in the unsaturated zone was through discrete systems of isolated fractures. However, what happened also depended on hydrological conditions, with greatly increased spatial dispersion of dye occurring during flood events, indicating lateral flow switching. Dye concentration at inlet points showed exponential decay over time, as might be expected in a notional mixing tank. However, at some sites after a rainfall event, rather than being more diluted the dye concentration increased again, indicating a pulsed flushing effect from a dye store. This produced a 'sawtooth' pattern of gradual decay in dye concentration over several months. Bottrell and Atkinson (1992) deduced three components of flow:

1. rapid through-flow with a characteristic residence time of approximately 3 days;
2. a component with short-term storage and residence time of 30–70 days;
3. a long-residence stored component with a characteristic time of 160 days or more.

Storage components 2 and 3 were considered probably to be in water-filled voids. Those corresponding to type 2 are flushed slowly and constantly, whereas those corresponding to type 3 are flushed only for short periods during high states of flow when water (and dye) is released into type 2 storage.

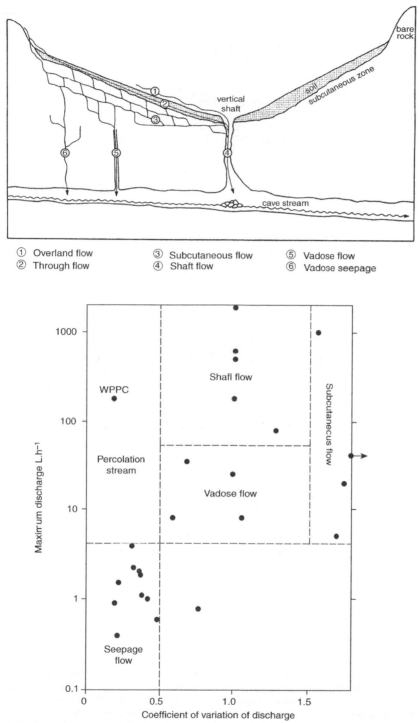

① Overland flow   ③ Subcutaneous flow   ⑤ Vadose flow
② Through flow   ④ Shaft flow   ⑥ Vadose seepage

Figure 6.8 (Upper) Six flow components in the vadose zone of karst. Reproduced from Gunn, J., Hydrological processes in karst depressions. Zeitschrift für Geomorphologie, NF, 25(3), 313–31, Zeitschrift für Geomorphologie © 1981 E. Schweizerbart'sche Verlagsbuchhandlung, Science Publishers. (Lower) Variability of flow components in the vadose zone. Reproduced from Smart, P.L. and Friederich, H. (1987) Water movement and storage in the unsaturated zone of a maturely karstified carbonate aquifer, Mendip Hills, England, Proceedings of Conference on Environmental Problems in Karst Terranes and their Solutions. National Water Well Association, Dublin, Ohio, 59–87. Since this work was done, we now know that high variability can also occur at low discharge (Genty and Deflandre 1998).

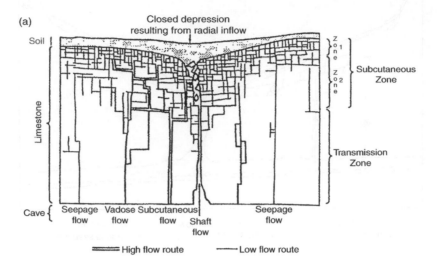

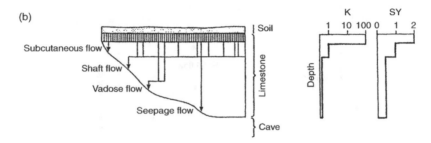

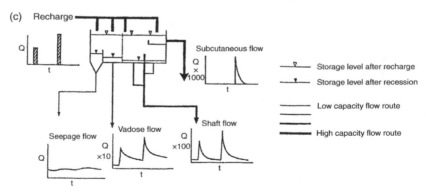

Figure 6.9 (a) Generalized nature of the unsaturated zone in the Mendip Hills. (b) Representational model of the unsaturated zone based on variable fissure frequency. Graphs K and SY represent permeability and storativity. (c) Functional model of the unsaturated zone based on an analogy with pipes and partitioned storage tanks. Reproduced from Smart, P.L. and Friederich, H., Water movement and storage in the unsaturated zone of a maturely karstified carbonate aquifer, Mendip Hills, England. Proceedings of Conference on Environmental Problems in Karst Terranes and their Solutions. National Water Well Association, Dublin, Ohio, 59–87, 1987.

Similar conclusions were reached by Kogovsek (1997) following dye tracing in the epikarst of Slovenia. Three flow components were recognized in the vadose zone: (i) rapid flow-through with velocities from 0.5 to 2 cm s^{-1};

(ii) slower velocities of the order of 10^{-2} cm s^{-1}; (iii) the slowest velocities of < 0.001 cm s^{-1}.

Sometimes naturally occurring tracers such as environmental isotopes are used to investigate processes in the

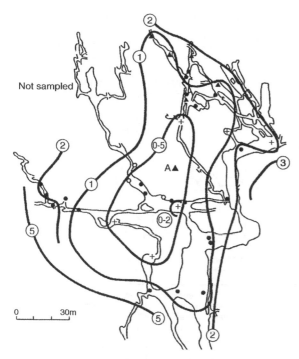

Figure 6.10 First-arrival times in days for fluorescent dyes injected at site A above GB Cave in the Mendip Hills, England. Reproduced from Smart, P.L. and Friederich, H., Water movement and storage in the unsaturated zone of a maturely karstified carbonate aquifer, Mendip Hills, England. Proceedings of Conference on Environmental Problems in Karst Terranes and their Solutions, 59–87. © 1987 National Water Well Association.

epikarst. Bakalowicz and Jusserand (1987) found from a comparison of $\delta^{18}O$ values in precipitation and percolation waters that about 18 weeks was required for the transit of water through about 300 m of limestone above Niaux Cave in southern France. Using stable isotopes and tritium, Chapman *et al.* (1992) deduced flow rates of between 7 and 15 m yr^{-1} (4.8×10^{-5} cm s^{-1}) through 250–300 m of vadose zone at Carlsbad Caverns in New Mexico. In another semi-arid region, in Israel, Even *et al.* (1986) found isotopic homogenization to occur quickly after infiltration, but then some waters percolated rapidly while others were detained for decades in the epikarst.

Tooth and Fairchild (2003) investigated the chemistry of drip waters in a cave in western Ireland, presenting a series of plumbing diagrams from soil zone to bedrock to help explain the geochemical evolution of water during percolation (Figure 6.11). Variations of the water chemistry with discharge were used to deduce the hydrogeochemical processes occurring in the unsaturated zone, and to shed light on whether increases in drip rate are a

result of direct inflow of storm water from soil macropores or due to piston flow from epikarst storage. They concluded that karst water response to recharge is dictated by the flow route taken through the soil zone (in this case comprising glacial till), the contrast between soil-matrix flow and well-connected macropore flow being particularly important, with soil-matrix flow being the dominant water source during dry periods.

This raises another related issue: whether most of the water that sustains percolation should be attributed to moisture stored in the soil or to water stored in the epikarst. It is well known that thick soils can have a large soil-water capacity but in sites with only thin or skeletal soils, the importance of epikarstic storage is unambiguous when percolation is sustained throughout a long dry season. This is the case, for example, in the semi-arid Carlsbad Cavern region in New Mexico (Williams 1983, Chapman *et al.* 1992) and in a Brazilian site studied by Sondag *et al.* (2003). We also see percolation sustained in caves beneath alpine karrenfeld essentially devoid of soil. However, in most karsts water is stored in both the soil and the epikarst, with the two stores interdigitating at the weathering front.

Another problem that confronts us when trying to understand the operation of the epikarst (in all its varieties) is whether it is best described as a well-mixed aquifer or a system of neighbouring but essentially separate compartments. The evidence is contradictory. The distinctive geochemistry of percolation waters from different drip points in the same cave described by Tooth and Fairchild (2003) indicates that even if some mixing occurs it is incomplete, and the fact that separate flow paths can exist is demonstrated by the dye tracing discussed above. Separate flow routes are even sometimes oblique rather than vertical through the vadose zone. But we have also seen that an introduced dye mass can spread during wet conditions and appear at a wider range of percolation sites in a cave than during dry conditions. This implies that there is limited horizontal dispersal and mixing in the epikarst when the level of saturation rises during a wet period, probably by a process of lateral decanting (or flow switching) into adjacent voids. Yet other evidence goes very much further and suggests almost perfect mixing. This comes from measurements of stable isotopes in percolation waters. Goede *et al.* (1982), Yonge *et al.* (1985), Even *et al.* (1986) and Williams and Fowler (2002) have shown that $\delta^{18}O$ values of cave drip waters, regardless of sample site in the cave, are very close to the average annual $\delta^{18}O$ values of the regional rainfall. This indicates homogenization of recharge waters in the stores and pathways that ultimately deliver water to underlying caves. An example from New Zealand illustrates the point (Figure 6.12):

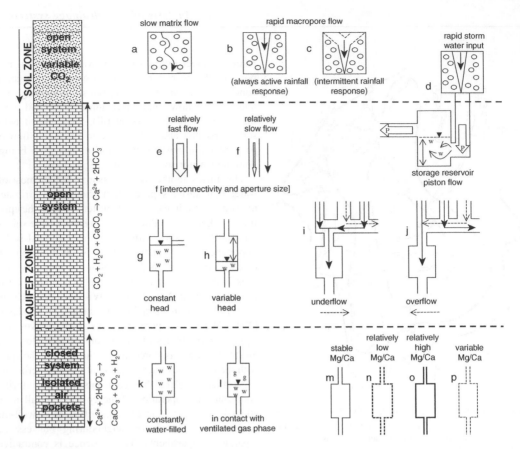

Figure 6.11 Model illustrating the potential soil and aquifer zone flow pathways and conditions which may control karst water evolution. Reproduced from Tooth, A.F. and Fairchild, I.J. Soil and karst aquifer hydrologic controls on the geochemical evolution of speleothem-forming drip waters, Crag Cave, southwest Ireland. Journal of Hydrology, 273; 51–68. © 2003 Elsevier.

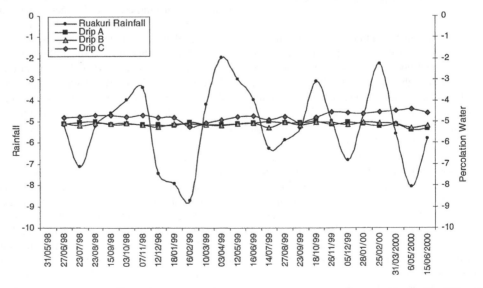

Figure 6.12 Relationship between $\delta^{18}O$ values in rainfall and cave percolation waters at Waitomo, New Zealand. Note that rainfall δ-values show considerable variation, whereas drip waters are almost invariant. Reproduced from Williams, P.W. and Fowler, A., Relationship between oxygen isotopes in rainfall, cave percolation waters and speleothem calcite at Waitomo, New Zealand. Journal of Hydrology 41(1), 53–70 © 2002 New Zealand Hydrological Society.

while the $\delta^{18}O$ of rainfall varied widely, the $\delta^{18}O$ of drips in the cave showed little variation over two years, being within analytical error of each other. In spite of that, their electrical conductivities were significantly different, indicating separate geochemical evolution within the 40–60 m percolation zone. Their drip rates and responsiveness to recharge also varied considerably.

This apparently conflicting evidence could be resolved if most of the homogenization of the stable isotope signal were to occur in the soil or in the upper, most porous part of the epikarst before the recharge is captured in percolation cells or pathways through the lower epikarst, and before most geochemical evolution of percolation water in contact with limestone occurs. Since isotopic homogenization can occur in epikarsts with very thin soils (Even *et al.* 1986), it seems that the upper part of the epikarst can be the main homogenization zone. Similar evidence from semi-arid karsts also indicates that storage time in the epikarst can be of the order of years or even decades, although in humid zones storage is usually much shorter, of the order of months but sometimes more than a year. Storage volume in the epikarst is not easy to calculate, although Smart and Friederich (1987) suggested that vadose storage in the Mendip Hills could be as much as 49% of the total karst water stored, as compared with an earlier estimate of 11% for the same region by Atkinson (1977b) based on spring-flow separation. Sauter (1992) subdivided the subcutaneous zone of the Swabian Alb of southern Germany into fast and slow subsystems and estimated storage within them. Fast subcutaneous storage (water that can be mobilized quickly within fractures and fissures) he estimated to vary between 0.3 and 2 mm with a possible maximum of 3 mm, the sub-zone having a storage coefficient of approximately 0.1%, whereas in slow subcutaneous storage the maximum stored quantities range between 20 and 30 mm and the storage coefficient is about 1%.

The evolution of the epikarst and the transfer of water to the underlying phreatic zone have been modelled by Clemens *et al.* (1999). They showed that the development of karst conduits in the saturated (phreatic) zone is partly dependent on the temporal evolution of the distribution of recharge from the epikarst. With the enlargement of paths of rapid percolation from the epikarst, the amount of undersaturated water flowing into the underlying conduit system increases, and hence the growth of phreatic conduits is accelerated. Modelling of flow networks is discussed further in section 6.11.

6.4 BOREHOLE ANALYSIS

The basic method for determining the properties of most types of aquifers is to drill one or a number of boreholes,

to be used for a variety of physical tests. Bedrock cores may or may not be taken during the drilling. Freeze and Cherry (1979) and Domenico and Schwartz (1998) provide full explanations of conventional borehole analysis in hydrogeology, and Chapellier (1992) has described the purpose and nature of well-logging techniques. In-depth discussion of hydraulic and tracer testing of fractures in rocks is available in a volume by the Committee on Fracture Characterization and Fluid Flow (1996), which also discussed the application of multiple borehole-testing techniques to determine the hydraulic characteristics of a rock mass. Readers are recommended to consult these sources to obtain detailed and authoritative information on modern practice. We therefore provide only a brief discussion pertinent to karst here.

In karstified aquifers borehole data are usually difficult to relate to aquifer structure and behaviour (Bakalowicz *et al.* 1995). However, the distribution of porosity and hydraulic conductivity from the surface to the phreatic zone can be investigated by borehole analysis and, where the number of boreholes is large, borehole tests can provide valuable information on aquifer behaviour, especially in the more porous aquifers such as coral and chalk. Only rarely do boreholes intersect major active karst drains, because the areal coverage of cave passages is usually less than 1% of an aquifer and only exceptionally above 2.5% (Worthington 1999), but when they do their hydraulic behaviour may be compared to that of a spring. More often boreholes intersect small voids with only indirect and inefficient connection to a major drainage line, in which case hydraulic behaviour in the bore is very sluggish compared with that in neighbouring conduits. Nevertheless, borehole analysis permits data to be obtained on (i) the location of relatively permeable zones within the karstified rock, (ii) the hydraulic conductivity of a relatively small volume of rock near the measurement point and (iii) the specific storage of part of the aquifer. Despite not knowing how representative any particular borehole may be of the entire groundwater system, drilling is often the only source of hydrogeological information in many karst areas.

Using the results from 146 borehole tests in the Dinaric karst, Milanović (1981) substantiated previous less formal evidence that karstification decreases exponentially with depth (Figure 5.3). The borehole data can be interpreted as indices of permeability (and consequently of hydraulic conductivity). Thus he inferred that in general the permeability at 300 m below the ground surface is only about one-tenth of that at 100 m and one-thirtieth of that at 10 m.

Castany (1984b) summarized appropriate field techniques for the determination of hydraulic conductivity in karst. Since the permeability of a fissured rock is a function of scale (Figure 5.2) and the geometry of the fissure network, a detailed structural field study is essential prior to the evaluation of permeability results. Fissures are commonly oriented in one, two or three sets of directions. If the spatial distribution can be defined, the permeability K_1, K_2 and K_3 associated with each set may be measured, and a weighted estimate of the hydraulic conductivity of the field area can be derived. Inclined boreholes may be used to intercept vertical joints.

The average hydraulic conductivity K of a fissured rock mass, assuming linear laminar flow conditions, is given by Castany (1984b) as

$$K = \frac{w}{n} \cdot (K_f + K_m) \qquad (6.5)$$

where w is the average width of opening of the fissure set under consideration, n is the average spacing per unit distance of fissures in the same set, K_f is the fissure set hydraulic conductivity and K_m is the hydraulic conductivity of the intervening unfissured rock (the rock matrix). In crystalline karstified limestones, K_m is negligible compared to K_f. Hence a reasonable approximation is

$$K = \frac{w}{n} \cdot K_f \qquad (6.6)$$

Snow (1968) showed that for a parallel joint set with N fissures per unit distance and a fracture porosity $n_f = Nw$ the hydraulic conductivity of the set can be evaluated from

$$K_f = \frac{\rho g}{\mu} \cdot \frac{Nw^3}{12} \qquad (6.7)$$

provided that it is applied to a volume of rock of sufficient size that it acts as a Darcian continuum, otherwise it is invalid. Snow also concluded that a cubic system of similar fractures creates an isotropic net with porosity $n_f = 3Nw$ and permeability k that is double that of any one of its fissure sets. Thus in the cubic system

$$k = \frac{Nw^3}{6} \qquad (6.8)$$

whereas in an array of parallel joints comprising one set

$$k = \frac{Nw^3}{12} \qquad (6.9)$$

The basic procedure in hydraulic testing of boreholes is to inject or withdraw fluid from the hole (or a test interval within it) while measuring the hydraulic head. In the field, hydraulic conductivity is usually determined by borehole **pumping tests** or **recharge tests** (sometimes

called Lugeon tests). The appropriate technique depends upon the purpose and scale of the investigation (Castany 1984b). **Slug tests** involve measuring the recovery of head in a well after near-instantaneous change in head at that well (Butler 1998). This is imposed by suddenly removing or adding a given volume (slug) of water; thus slug tests are the instantaneous equivalent of pump or recharge tests respectively (Barker and Black 1983). An instructive discussion of the slug technique when applied to karst is provided by Sauter (1992). Recharge and pumping tests both distort the initial potentiometric surface into cones of recharge or abstraction, depending on the direction of induced water movement. The resulting flow pattern can be represented by an orthogonal network of equipotentials and streamlines (Figure 6.13). Hydraulic conductivity may also be determined from in-hole tracer dilution (see section 6.10).

6.4.1 Borehole recharge tests

Borehole recharge tests for permeability are of three types: pressure tests, constant head gravity tests and falling head gravity tests. Castany (1984b) noted that three kinds of pressure injection tests are in common

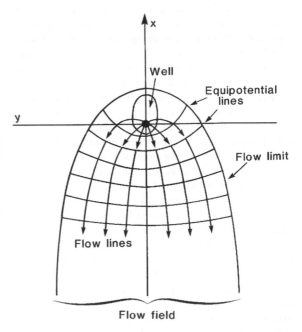

Figure 6.13 Equipotentials and flowlines in a borehole undergoing a recharge test. Reproduced from Castany, G., Determination of aquifer characteristics. In P. E. LaMoreaux, B. M. Wilson and B. A. Memon (eds.), Guide to the Hydrology of Carbonate Rock', P. E. LaMoreaux, B. M. Wilson & B. A. Memon (eds.). Studies and Reports in Hydrology No 41, Paris: UNESCO, 210–237, 1984.

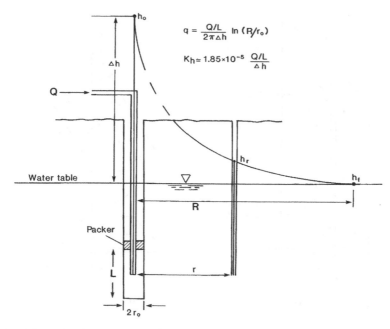

Figure 6.14 Definitions required to interpret the results of a standard borehole recharge test. Reproduced from Castany, G., Determination of aquifer characteristics. In P. E. LaMoreaux, B. M. Wilson and B. A. Memon (eds.), Guide to the Hydrology of Carbonate Rock', P. E. LaMoreaux, B. M. Wilson & B. A. Memon (eds). Studies and Reports in Hydrology No 41, Paris: UNESCO, 210–237, 1984.

use in karst terrains. They are performed in uncased boreholes.

1. The standard recharge (Lugeon) test, which yields an average regional horizontal hydraulic conductivity without taking into account the anisotropy of the rock formation (Figure 6.14).
2. The modified recharge test by which a directional hydraulic conductivity on the basis of relative orientation of the test hole to the system of fissures is determined (Figure 6.15).
3. A triple hydraulic rig test, which gives directional hydraulic conductivities directly.

In the standard recharge test, water is pumped into the borehole to form a recharge cone. It is performed for a selected zone of length L above the bottom of a borehole (Figure 6.14). Injection is organized in fixed steps at prescribed time intervals. The water pressure in the bore is first increased gradually in steps such as 2, 4, 6, 8 and 10 bars (1 bar = 10.2 m head of water), and then is allowed to decrease in a reverse series of similar steps. Test lengths within a borehole are characteristically 5 m, but may be reduced to 1 m, using isolating packers to identify highly permeable zones (Figure 6.16). An acceptable test for analysis is one that produces a matching reverse cycle.

The permeability of the rock is determined from the pressures and measured water quantities injected. It can be expressed as **specific permeability**, defined as the quantity of water that can percolate through the karstified rock per unit length of borehole under a pressure of 1 m of water (0.1 atm) over a duration of 1 min. Milanović (1981) used seven categories of specific permeability for describing permeability in the Dinaric karst (Table 6.3).

The specific permeability q can be derived as follows

$$q = \frac{Q/L}{2\pi\Delta h} \cdot \ln\frac{R}{r_0} \tag{6.10}$$

where r_0 is the diameter of the borehole, Q/L is a volume rate of flow per unit test length and Δh is the difference between the head in the test hole h_0 and that of the natural rest level of the water table h_f beyond the radius of influence R of the test (Figure 6.14).

Castany (1984b) suggested that since $\ln(R/r_0)/2\pi$ is roughly constant and because R/r_0 varies very slightly and can be assumed close to 7, then for numerical computations

$$q* = 1.85 \times 10^{-5} \frac{Q/L}{\Delta h} \tag{6.11}$$

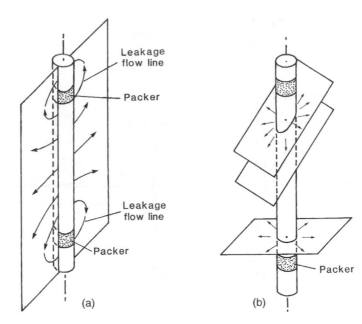

Figure 6.15 Influence of the direction of fissures on flow during a recharge test: (a) fissures parallel to bore and (b) fissures transverse to bore. Reproduced from Castany, G., Determination of aquifer characteristics. In P. E. LaMoreaux, B. M. Wilson and B. A. Memon (eds.), Guide to the Hydrology of Carbonate Rock', P. E. LaMoreaux, B. M. Wilson & B. A. Memon (eds.). Studies and Reports in Hydrology No 41, Paris: UNESCO, 210–237, 1984.

Here the quantity q_* is in $m\,s^{-1}$ and thus has the same dimensions as hydraulic conductivity.

Milanović (1981) discussed the relationships between specific permeability and hydraulic conductivity. If a linear relationship between the two can be assumed then

$$K = jq \tag{6.12}$$

For three differing conditions (the unsaturated zone, the saturated zone and a combination of them), j was calculated for a bore of radius 0.02 m and a test length of 5 m and found to vary from 1.2 to 2.3, being least in the saturated model and greatest in the mixed case. A reasonable approximation is

$$K = 1.7 \times 10^{-5} q \tag{6.13}$$

where K is in $m\,s^{-1}$ and q is in $L\,min^{-1}\,m^{-1}$ under 0.1 atm.

6.4.2 Borehole pumping tests

The boundary conditions of aquifers vary. Some can be assumed for practical purposes to have infinite areal extent. Others are limited horizontally, e.g. by impermeable rocks, a fault, or by a marked reduction in aquifer thickness. A constant potential recharge boundary can be provided by rivers or lakes, whereas the ocean may provide a constant potential discharge boundary. Impervious confining beds in artesian aquifers preclude vertical recharge and discharge, but semi-permeable beds permit leakage. The simplest aquifer configuration for analysis is that which is:

- horizontal and of infinite extent
- confined between impermeable beds
- of constant thickness
- homogeneous and isotropic.

In such a case, where the well fully penetrates the aquifer and at a constant pumping rate, Q, there is no drawdown in hydraulic head at the infinite boundary.

Pumping tests are the most common way of determining water well yields and average hydraulic conductivity over large areas in granular aquifers, but the validity of the results is limited in anisotropic, heterogeneous karst aquifers. The removal of water from a well by pumping produces a cone-shaped zone of depression in the water table, which is unique in shape and lateral extent, depending on the hydraulic characteristics of the aquifer and the rate and duration of

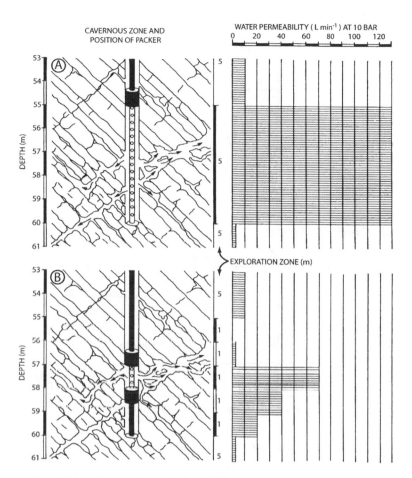

Figure 6.16 The use of variable packer spacing to isolate highly permeable zones within a borehole. Reproduced from Milanović, P.T., Karst Hydrogeology, Water Resources Publication, Colorado, p. 434, 1981.

pumping (Figure 6.17). Two general types of analyses are available for these tests:

Table 6.3 Classification of rocks according to their specific permeability (From Ford, D.C. and Williams, P.W. (1989) Karst Geomorphology and Hydrology; after Milanović 1981)

Category	Specific permeability (L min^{-1})	Rock category
1	0.001	Impermeable
2	0.001–0.01	Low permeability
3	0.01–0.1	Permeable
4	0.1–1	Medium permeability
5	1–10	High permeability
6	10–100	Very high permeability
7	100–1000	Exceptionally high permeability

1. steady-state or equilibrium methods, which yield values of transmissivity T and storativity S;
2. transient or non-equilibrium methods (such as slug tests) from which can also be derived values of storativity and information on boundary conditions.

The relationship of drawdown in the well to the aquifer properties was first elucidated by Theis (1935). Castany (1984b) expressed it as

$$\Delta h = \frac{0.183\,Q}{T} \cdot \log \frac{2.25\,Tt}{r^2 S} \tag{6.14}$$

where Δh is the drawdown (m) of the potentiometric surface in an observation well, T is the transmissivity (m^2 s^{-1}), S is storativity (dimensionless), Q is the

(a) **Plan**

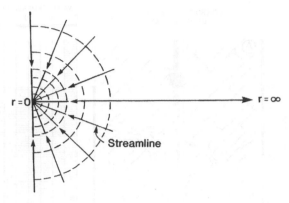

(b) **Confined Aquifer**

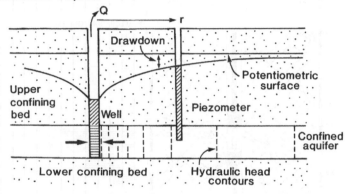

(c) **Unconfined Aquifer**

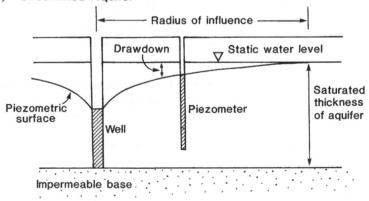

Figure 6.17   Development of drawdown cone and streamlines in confined and unconfined aquifers. Reproduced from Freeze, R.A. and J.A. Cherry, Groundwater, © 1979 Prentice Hall, Inc.

constant rate of pumping ($m^3\,s^{-1}$), t is the time elapsed in seconds since the start of pumping and r is the distance from the pumped well to the observation well. The formula is considered valid to within 5% when $t \geq 10\,r^2S/4T$.

The recovery to the original level of the potentiometric surface is given by

$$\Delta h = \frac{0.183\,Q}{T} \cdot \log\left(\frac{t+t'}{t'}\right) \qquad (6.15)$$

where t' is the time elapsed in seconds since pumping stopped.

The drawdown values at certain times during the pumping test (or recovery test) are recorded as in Figure 6.18. A straight line drawn through the plotted points after 1 to 2 h of pumping is regarded as representative of well behaviour. The slope of the line C is termed Jacob's logarithmic approximation

$$\frac{0.183\,Q}{T}\,C \qquad (6.16)$$

It corresponds to the change of drawdown in one logarithmic cycle on the time scale. Extrapolation of the fitted line to intersect the time axis yields t_0, the time at zero drawdown.

Simple rearrangement of equation (6.16) permits evaluation of transmissivity and, if the thickness of the aquifer b is known, the hydraulic conductivity K may be calculated by rearrangement of equation (5.16). In addition, storativity can be derived from

$$S = \frac{2.25\,T t_0}{r^2} \qquad (6.17)$$

However, the generalized value of K may not be very meaningful in karst given the characteristic variability of permeability mentioned earlier. We emphasize that all of the hydraulic tests for boreholes discussed above were developed for porous media in which Darcian conditions hold. Experience shows that the techniques can sometimes produce acceptable results in karst, provided the constraints of scale are not overlooked (Figures 5.2 and 6.19).

6.4.3 Borehole logging

Further information can be derived on aquifer characteristics by various measurement techniques, mainly geophysical, down boreholes. General details are provided by US Department of the Interior (1981), Robinson and Oliver (1981) and Chapellier (1992); application to karst is discussed by Astier (1984) and Milanović (2004).

Borehole logging identifies variations in physical or chemical parameters with depth down a well and provides *in situ* measurements of their characteristics. Resistivity, spontaneous potential, nuclear, sonic, calliper, geothermal, video and stereophoto techniques are available for well logging and have been used with varying

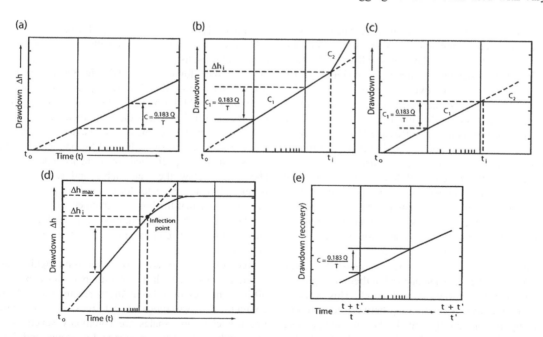

Figure 6.18 Semi-logarithmic plots of drawdown versus time for well pumping and recovering tests in confined aquifers with various boundary conditions: A, infinite areal extent; B, neighbouring lateral barrier; C, neighbouring constant potential recharge boundary; D, leaky confining beds with time lag in recharge transfer; E, recovery straight line plot. C, C1 or C2 defines the slope of the curves. Pumping is at a constant rate. Reproduced from Castany, G., Determination of aquifer characteristics. In P. E. LaMoreaux, B. M. Wilson and B. A. Memon (eds), Guide to the Hydrology of Carbonate Rock', P. E. LaMoreaux, B. M. Wilson & B. A. Memon (eds.). Studies and Reports in Hydrology No 41, Paris: UNESCO, 210–237. 1984.

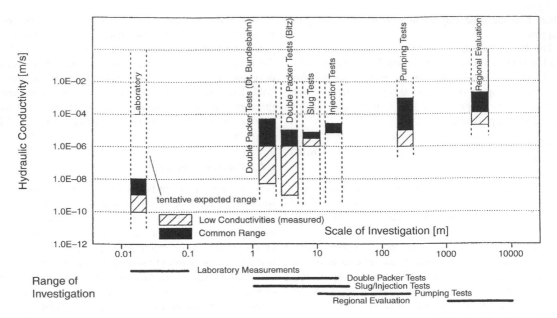

Figure 6.19 Relationship between hydraulic conductivity and scale of investigation in the karst of the Swabian Alb, southern Germany. Reproduced from Sauter, M., Quantification and forecasting of regional groundwater flow and transport in a karst aquifer. Tübinger Geowissenschaftlichen Abhandlungen, Reihe C, 13, 150p, 1992.

success. Well logging is most productive when different methods are used in appropriate combination with each other, as illustrated by Maclay and Small's (1983) study of the Edwards Limestone aquifer in Texas (Figure 6.20).

Electrical logging is of two sorts: spontaneous potential (or self potential) and resistivity. An electrode is lowered down a borehole and in the first case the naturally occurring potential difference at various depths between a surface electrode and the borehole electrode is measured. In the second case, a source of current is connected and the potential difference is measured at different depths for a given current strength. This leads to a log of apparent resistivity versus depth. Both properties are measured in uncased wells, and can be interpreted to distinguish rock-unit thickness and the stratigraphical sequence, i.e. formation characteristics.

Temperature logs are often run in association with electrical logs. They use a sonde with a thermocouple and record temperature variations with depth. This is useful because of the known relationship between temperature and electrical conductivity and also because temperature variations often indicate discrete groundwater bodies sometimes related to water movement from different source areas (e.g. Jeannin *et al.* 1997).

Calliper logging in uncased boreholes measures the variation of well diameter with depth, which helps to identify and correlate solution openings, bedding planes

and zones that may need grouting. However, Milanović (1981) warned that the use of a calliper will not produce meaningful results if caverns of $>10\,m^3$ are encountered – as commonly found in the Dinaric Karst – and that 'experience shows that lowering probes into uncased boreholes should be avoided' because rock fragments may be detached and jam the instrument. Despite such risks to lowered instruments, we have found that the video camera is a most useful tool for boreholes in karst rocks. Modern specialized instruments operate both above and below the water table and can be rotated through 360° to peer into, for example, a partly opened bedding plane that the bore has intercepted. A good deal can be learned about the condition of the rock and nature and location of preferential flow routes through it from a good moving picture.

Recording natural radioactivity (**gamma logging**) uses gamma radiation, with a penetration distance of about 30 cm. The radiation flux is measured by a scintillation counter; flux variations detect rock-unit boundaries because clays and shales are commonly several times more radioactive than sandstones, limestones and dolostones. **Gamma–gamma** and **neutron–gamma** logging techniques require artificial radiation sources and detectors, and hence are used less frequently. Gamma–gamma logging has successfully measured (Milanović 1981, Maclay and Small 1983) rock-density variations down

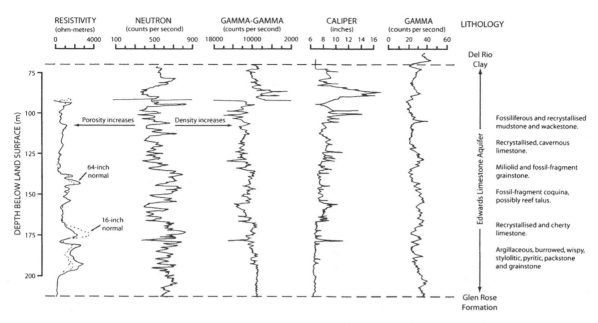

Figure 6.20 Hydrostratigraphical subdivision of the Edwards aquifer, Texas, using borehole logging techniques. Reproduced from Maclay, R.W. and T.A. Small, Hydrostratigraphic subdivisions and fault barriers of the Edwards aquifer, south central Texas, U.S.A. Journal of Hydrology 61, 127–146 © 1983 Elsevier.

boreholes (Figure 6.20). Neutron–gamma logging is a standard technique for soil moisture measurement, but is less frequently used in boreholes. It provides a measure of the hydrogen abundance per unit volume of rock, and so is related to the abundance of water and to porosity.

6.4.4 Borehole hydrograph analysis

Following intense rainfall, the level of standing water in boreholes usually rises. Rapid rises in water level are expected in well-karstified aquifers with high secondary porosity, but only slow responses in porous aquifers with Darcian characteristics. By the same token, drainage and fall of water levels in wells will be quicker in karstified aquifers than in porous media. Provided there is no further recharge, the recession limb of the well hydrograph will show an exponential decrease until the pre-storm water level has been attained. However, just as the storm hydrographs of surface streams show quick flow and delayed flow segments in their recession limbs (Hewlett and Hibbert 1967), so the recessions of well hydrographs may also show discrete line segments of different slope. Shevenell (1996) analysed well hydrographs from a karst aquifer and found up to three straight-line segments (Figure 6.21). The steepest she ascribed to the dominant effect of drainage from the larger karstified conduits (although encompassing the effects of all

three), the intermediate segment to drainage from well-connected fissures, and the third segment with lowest slope to drainage from the matrix portion of the aquifer. Values for transmissivity and specific yield were calculated from the data, and transmissivity was found to show reasonable agreement with values determined from standard slug injection tests. However, the rate of recession of borehole water levels in an unconfined karst aquifer is likely to be reduced by continuous (though diminishing) recharge from epikarstic storage. This was overlooked in Shevenell's study, and could be the explanation of some segments, because Smart and Friederich (1987) identified up to three components in flow recessions of discharges from the subcutaneous zone (the topic of hydrograph recession analysis is dealt with more comprehensively in section 6.5).

6.4.5 Borehole tracer tests

Tracer dilution in boreholes is used to determine transport properties such as kinematic porosity and dispersivity, as well as to establish connectivity in the aquifer, including borehole-to-borehole and borehole-to-spring linkages. The estimation of hydraulic conductivity by tracer dilution is well established in groundwater hydrology in granular aquifers. Fluorometric tracers are now preferred in karst and other rocks, having the advantages of ease

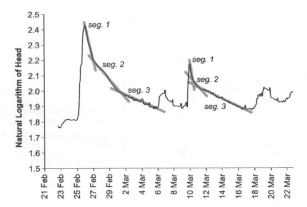

Figure 6.21 The natural logarithm of water level in a bore versus time. The bore is in Palaeozoic Maynardville Limestone in Oak Ridge, Tennessee. The water level culminations are responses to percolation after rainfall and the recessions show the subsequent decline in water level after the recharge event. The differing slopes of the segments 1, 2 and 3 were interpreted as water released dominantly from storage in conduits, fissures and matrix respectively. Reproduced from Shevenell, L., Analysis of well hydrographs in a karst aquifer: estimates of specific yields and continuum transmissivities. Journal of Hydrology, 174, 331–55. © 1996 Elsevier.

and accuracy of application and low toxicity (see section 6.10). Salt dilution, radioisotope tracers and the monitoring of isotope activity by scintillometry down the borehole have also been used (Brown *et al.* 1972, Moser 1998b) Ward *et al.* (1997) and Atkinson *et al.* (2000) provide details of fluorometric tracing borehole-to-borehole and borehole-to-spring connections in Cretaceous Chalk in Britain. Velocities up to 475 m day^{-1} were measured and connections established over several kilometres.

Groundwater flow direction and velocity can be determined either by injection of a tracer up the hydraulic gradient from an observation well (or wells) or from measuring the rate of dye dilution in a given bore. The relationships between tracer travel time t, well spacing R, effective rock porosity n_e, hydraulic conductivity K and hydraulic gradient $dh/dl = i$ may be approximated as follows:

$$t = \frac{Rn_e}{Ki} \qquad (6.18)$$

In karstic rocks an injection quantity of fluorescein dye of 2–10 g 10 m^{-1} of flow path was recommended by Brown *et al.* (1972), who took travel time to be the time to peak concentration in the observation well, rather than the time to first arrival of the dye.

We saw in Chapter 5 that under Darcian conditions groundwater movement can be described by $u = -K \cdot dh/$ $dl = Q/a$ (equation 5.4) and is variously termed **specific discharge**, **macroscopic velocity** or **filtration velocity**. The actual microscopic velocity $u*$ of the water through the pores of the aquifer is obtained by dividing the macroscopic velocity by the effective porosity of the rock

$$u* = \frac{u}{n_e} = \frac{Ki}{n_e} \qquad (6.19)$$

Lewis *et al.* (1966) considered derivation of u from in-hole dye dilution. To achieve valid results there should be steady-state conditions with uniform groundwater flow and tracer distribution; diminution of tracer concentration over time should be due only to dilution by horizontal groundwater flow. Results may be satisfactorily evaluated by plotting against time the logarithm of the ratio C_0/C. The slope of the semi-logarithmic plot is then evaluated and a value for u is determined

$$\log \frac{C_0}{C} = 1.106 \frac{ut}{2r} \qquad (6.20)$$

where C_0 is the initial tracer concentration, C is its concentration at time t and r is the internal radius of the borehole. In bores with well screens, Drost and Klotz (1983) suggested velocity is best derived from

$$u = \frac{\pi r}{2\alpha t} \ln \frac{C_0}{C} \qquad (6.21)$$

The term α corrects for borehole disturbance of the flow field. It usually lies in the range 0.5–4.0, and 2.0 is most often assumed.

If Darcy's Law can be assumed to apply, then the hydraulic conductivity can be calculated from

$$K = \frac{u}{i} \qquad (6.22)$$

In comparing such a result with that obtained from a pumping test, it should be recognized that tracer dilution involves a much smaller rock volume. Estimation of K by this method was considered by Lewis *et al.* (1966) to be more economical from the standpoint of time, cost and repeatability than conventional pumping tests, but we consider it unlikely to be valid in well-karstified aquifers.

6.4.6 Diagnostic tests for karstic conditions

A number of diagnostic tests using borehole data have been identified by Ford and Worthington (1995) and Worthington (2002) to determine if there are conduits (channels) close to the borehole, and therefore if the aquifer is likely to be karstified. These and other tests include the following:

1. Bit drops during drilling – large voids can be identified by bit drops and loss of drilling fluid, but significant voids < 10 cm can be missed.
2. Voids identified by borehole logging – geophysical and down-bore video logging identifies cavities > 2 cm.
3. Well-to-well (or spring) tracer tests – successful long-distance (> 100 m) traces over a short time (few days) provides evidence of conduit flow.
4. Scale effect in packer, slug and pump tests – if hydraulic conductivity increases with the scale of the test (there may be order of magnitude differences), then the implication is that more permeable fissures and channels are encountered.
5. Segmented recession curve of well hydrograph – recession segments of different slope imply drainage of water from an aquifer with dual or triple porosity characteristics, including an epikarst.
6. Non-linear pumping-rate/drawdown response – if the drawdown in an observation well is proportional to the pumping rate, then Darcy's Law is valid within the cone of depression, but if the response is non-linear then it is not and conduits may connect to the pumping well.
7. Irregular cone of depression – in a homogeneous porous medium the zone of depression around a pumping well is symmetrical, but if it is irregular then channelling may be present.
8. Rapid water quantity and quality response to recharge – prompt (hours) water-level and solute concentration responses to charge are expected in bores that are well-connected to channel networks, whereas responses in poorly connected wells are sluggish.
9. Troughs in the water table – water table troughs correspond to flow in channels and they terminate in a downstream direction in springs. In such situations hydraulic gradients also decrease in a downflow direction along the water table trough.
10. Groundwater age distribution – in a porous medium there is an increase in water age with depth, but should channels provide rapid recharge to the subsurface, then younger water in conduits will underlie older water in overlying fractures and the matrix.

None of these tests alone is likely to prove the presence of karst conduits, but the more that apply in a carbonate aquifer, the more likely it is to be karstified. The further the aquifer deviates from the behaviour of an ideal porous or high-density-fracture medium, the less likely a given borehole will intercept conditions appropriate for Darcian techniques of analysis. Hence to understand karst aquifers it is often more appropriate and productive to focus attention on springs rather than bores, because these **must** be integrating significant hydrological conditions.

6.5 SPRING HYDROGRAPH ANALYSIS

Karst-spring hydrograph analysis is important, first, because the form of the output discharge provides an insight into the characteristics of the aquifer from which it flows and, second, because prediction of spring flow is essential for careful water resources management. However, although the different shapes of outflow hydrographs reflect the variable responses of aquifers to recharge, Jeannin and Sauter (1998) expressed the opinion that inferences about the structure of karst systems and classification of their aquifers is not efficiently accomplished by hydrograph analysis because hydrograph form is too strongly related to the frequency of rainfall events. If a long time-series of such records is represented as a curve showing the cumulative percentage of time occupied by flows of different magnitude, then abrupt changes of slope are sometimes revealed in the curve, which have been interpreted by Iurkiewicz and Mangin (1993), in the case of Romanian springs, as representing water withdrawn from different parts of the karst system under different states of flow. Similar information can be obtained from the form and rate of recession of individual hydrographs: they provide evidence of the storage and structural characteristics of the aquifer system sustaining the spring. For these reasons, analysis of the recession limbs of spring hydrographs offers considerable potential insight into the nature and operation of karst drainage systems (Bonacci 1993), as well as providing information on the volume of water held in storage. The potential information that can be obtained is considerably extended if water quantity variations (**hydrographs**) are analysed alongside corresponding water quality variations (**chemographs**). Important recent reviews of karst-spring hydrograph and chemograph analyses are provided by Sauter (1992), Jeannin and Sauter (1998) and Dewandel *et al.* (2003).

The principal influence on the shape of the output hydrograph of karst springs is precipitation. Rain of a particular intensity and duration provides a unique template of an input signal of a given strength and pattern that is transmitted in a form modified by the aquifer to the spring. The frequency of rainstorms, their volume and the storage in the system, determines whether or not recharge waves have time to pass completely through the system or start to accumulate. Antecedent conditions of storage strongly influence the proportion of the rainfall input that runs off and the lag between the input event and the output response. The output pattern of spring hydrographs is, however, moderated by the effect of basin characteristics such as size and slope, style of recharge, drainage network density, geological variability, vegetation and soil. As a consequence of

all the above, flood hydrograph form and recession characteristics show considerable variety (Figure 6.22). The flood hydrographs of vadose cave streams tend to be sharply peaked and similar to surface rivers. But should cave streams flow into a phreatic zone before later emerging at a spring, then the influence of their inflow hydrographs on the composite outflow hydrograph of a spring is similar to that of tributaries flowing into a lake:

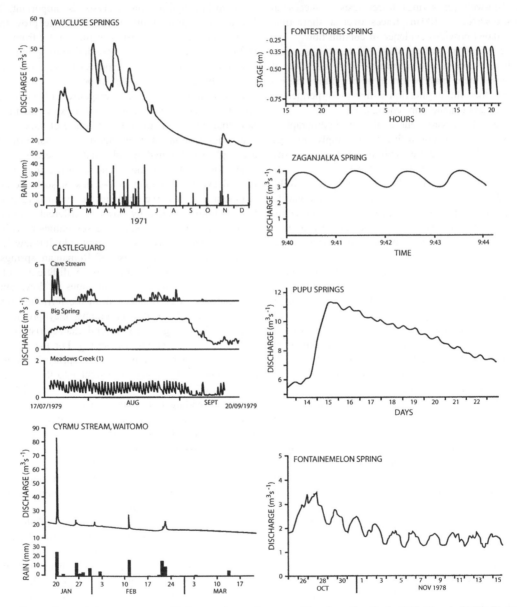

Figure 6.22 Hydrograph forms encountered in karst springs and streams. (Data from Gavrilovic 1970, Mangin 1969b, Durozoy and Paloc 1973, Williams 1977, Gunn 1978, C. Smart 1983b, Siegenthaler *et al.* 1984.)

the overflow or output response is a delayed, muted reflection of the input. Some outflow hydrographs are highly peaked, others are oscillatory (Bonacci and Bojanić 1991), and many are broad and relatively flat. Sometimes a flat-topped, apparently truncated hydrograph is an indication that it is an underflow spring in Smart's (1983a,b) sense, the 'missing' peaked remainder of the hydrograph emerging at an intermittent high-water overflow spring draining the same aquifer (Figure 6.23). Thus in interpreting hydrograph form, it is important to know if the whole output is being dealt with or only part of it. The discussion that follows assumes that the entire output from a karst drainage system is being monitored.

6.5.1 Hydrograph recessions

Given widespread recharge from a precipitation event over a karst basin, the output spring will show important discharge responses, characterized by:

1. a lag time before response occurs;
2. a rate of rise to peak output (the 'rising limb');
3. a rate of recession as spring discharge returns towards its pre-storm outflow (the 'falling limb');
4. small perturbations or 'bumps' on either limb although best seen on the recession.

When the hydrograph is at its peak, storage in the karst system is at its maximum, and after a long period of recession storage is at its minimum. The rate of withdrawal of water from storage is indicated by the slope of the subsequent recession curve. The characterization of the rate of recession and its prediction during drought are necessary for determining storage and reserves of water that might be exploited.

Jeannin and Sauter (1998) pointed out that karst hydrological systems are neither linear nor stationary, especially when considering total precipitation as input and discharge as output. They also supported the conclusion of Hobbs and Smart (1986) that they cannot be considered as a single system with two parallel subsystems (storage and drainage), but should be considered as a cascade of at least three subsystems each with their own storage and transmission characteristics (Figure 5.30). Therefore unlike Darcian porous media, the karst system cannot be characterized by one single transfer function.

Quantitative analysis of hydrograph recession derives from the work of Boussinesq (1903, 1904) and Maillet (1905), who proposed that the discharge of a spring is a function of the volume of water held in storage. Maillet analysed hydrographs of springs near Paris, and described their flow recession using the simple exponential relation

$$Q_t = Q_0 e^{-\alpha t} \tag{6.23}$$

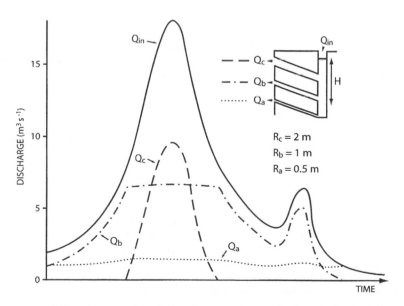

Figure 6.23   Static system model (no storage considered) for a three-component underflow–overflow spring system responding to an arbitrary input hydrograph: *Q*, discharge; *R*, conduit radius. Reproduced from Smart, C.C., Hydrology of a Glacierised Alpine Karst. Ph.D. thesis, McMaster University. p.343, 1983.

where Q_t is the discharge $(m^3\,s^{-1})$ at time t, Q_0 is the initial discharge at time zero, t is the time elapsed (usually expressed in days) between Q_t and Q_0, e is the base of the Napierian logarithms and α of dimension (T^{-1}) is termed the **recession coefficient**. By contrast, Boussinesq expressed recession in quadratic form as

$$Q_t = \frac{Q_0}{(1 + \alpha t)^2} \tag{6.24}$$

where

$$Q_0 = \frac{1.724\,Kh_m^2 l}{L} \tag{6.25}$$

and

$$\alpha = \frac{1.115\,Kh_m}{n_e L^2} \tag{6.26}$$

in which K is aquifer hydraulic conductivity, n_e is its effective porosity and l is aquifer width perpendicular to its length L. Initial hydraulic head h_m is at distance L, and Kh_m represents transmissivity.

Dewandel *et al.* (2003) pointed out that Maillet's exponential equation provides an approximate analytical solution for the diffusion equation that describes flow in a porous medium, whereas the depiction of baseflow recession by Boussinesq's quadratic equation provides an exact analytical solution capable of yielding quantitative data on aquifer characteristics. Both Boussinesq and Maillet assumed the aquifer to be porous, homogeneous, isotropic and unconfined.

Their two equations depict the variety of recessions found in nature with different efficiencies. In a case study from Oman, Dewandel *et al.* (2003) found the quadratic form to provide a correct fit from day 2 of recession, whereas the fit of the Maillet method was valid only from day 22. They also used numerical simulations of aquifer recessions, focusing on porous aquifers and fissured aquifers (which at large scale could be considered as equivalent to porous aquifers). Simulations of shallow aquifers with an impermeable floor at the level of the spring showed hydrograph recession curves to have a quadratic form. Simulations of more realistic aquifers with an impermeable floor much deeper than the outlet also revealed the quadratic form of recession to be valid regardless of the thickness of the aquifer under the outlet, and therefore indicated the relative robustness of the Boussinesq formula. It also provided good estimates of the aquifer's hydrodynamic parameters, such as perme-

ability and storage coefficient. Nevertheless, Dewandel *et al.* (2003) found that aquifers with a very deep floor yield exponential recessions that could be closely modelled by the Maillet formula, even if parameter estimation remained poor. They concluded that the form of the spring recession curve appears to be closer to exponential when flow has a very important vertical component, but is closer to quadratic when horizontal flow is dominant (as it often is in karst). This is somewhat at variance with the emphatic position of Jeannin and Sauter (1998), who concluded that Maillet's formula is only adequate for describing karst systems during low flow or if they are poorly karstified. Nevertheless, the expression has been used frequently in the analysis of karst hydrographs and so merits further discussion.

Maillet's exponent implies that there is a linear relation between hydraulic head and flow rate (commonly found in karst at baseflow), and the curve can be represented as a straight line with slope $-\alpha$ if plotted as a semi-logarithmic graph. It can be represented in logarithmic form as

$$\log Q_t = \log Q_0 - 0.4343 t\alpha \tag{6.27}$$

from which α may be evaluated as

$$\alpha = \frac{\log Q_1 - \log Q_2}{0.4343(t_2 - t_1)} \tag{6.28}$$

Since $e^{-\alpha t}$ in equation (6.23) is a constant, it is sometimes replaced by a term β, the **recession constant**, and the equation is written

$$Q_t = Q_0 \beta^t \tag{6.29}$$

The recession constant can be evaluated from

$$\log \beta = \frac{\log Q_t - \log Q_0}{t} \tag{6.30}$$

A **half-flow period** $t_{0.5}$ is defined as the time required for the baseflow to halve. Then by substitution into equation (6.29)

$$Q_{t_{0.5}} = 2\,Q_{t_{0.5}} \beta^{t_{0.5}} \tag{6.31}$$

hence

$$\tfrac{1}{2} = \beta^{t_{0.5}} \tag{6.32}$$

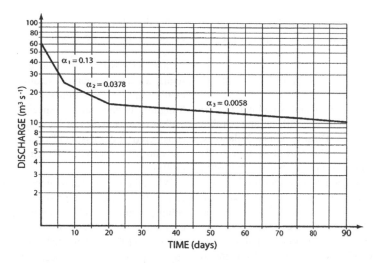

Figure 6.24 Composite hydrograph recession of Ombla spring, Croatia. Note that the three recession coefficients are of different orders of magnitude. Reproduced from Milanović, P., Water regime in deep karst. Case study of the Ombla spring drainage area. In V. Yevjevich (ed.) Karst Hydrology and Water Resources: vol. 1 Karst Hydrology, Water Resources Publications, Colorado, 165–191, 1976.

and

$$t_{0.5} = \frac{\text{constant}}{\log \beta} \tag{6.33}$$

The parameter $t_{0.5}$ has the following properties:

1. it is independent of Q_0 and Q_t and of the time elapsed between them;
2. it is sensitive to change and can take values in the range zero to infinity;
3. it can be easily evaluated from equation (6.33) and is simply related to β by means of that equation;
4. it is a direct measure of the rate of recession and therefore can be used as a means of characterizing exponential baseflow recessions.

6.5.2 Composite hydrograph recessions

Semi-logarithmic plots of karst spring recession data often reveal two or more segments, at least one of which is usually linear (Figures 6.24 and 6.25). In these cases the data can be described by using separate expressions for the different segments. Jeannin and Sauter (1998) and Dewandel *et al.* (2003) explain the various models that have been used to try to conceptualize the structure of the karst drainage system that has given rise to the hydrograph form observed and the means by which its recession might be analysed.

If the karst system is represented as consisting of several parallel reservoirs all contributing to the discharge of the spring and each with its individual

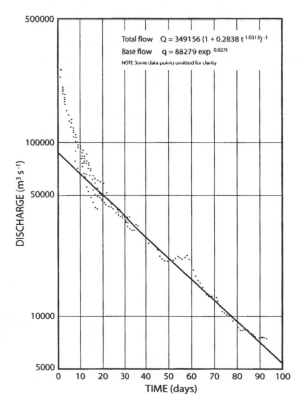

Figure 6.25 Master flow recession at Cheddar spring, England. Note the linear and non-linear segments. Reproduced from Atkinson, T.C., Diffuse flow and conduit flow in limestone terrain in the Mendip Hills, Somerset (Great Britain). Journal of Hydrology, 35, 93–110. © 1977 Elsevier.

hydraulic characteristics, then the complex recession of two or more linear segments can be expressed by a multiple exponential reservoir model:

$$Q_t = Q_{0_1} e^{-\alpha_1 t} + Q_{0_2} e^{-\alpha_2 t} + \ldots + Q_{0_n} e^{-\alpha_n t} \qquad (6.34)$$

This was used by Torbarov (1976) and Milanović (1976) in the analysis of flow from the Bileca and Ombla springs, respectively, in the Dinaric Karst. Milanović interpreted the data for the Ombla regime (Figure 6.24) in Croatia as indicating flow from three types of porosity, represented by the three recession coefficients of successive orders of magnitude. He suggested that α_1 is a reflection of rapid outflow from caves and channels, the large volume of water that filled these conduits emptying in about 7 days. Coefficient α_2 was interpreted as characterizing the outflow of a system of well-integrated karstified fissures, the drainage of which lasts about 13 days; and α_3 was considered to be a response to the drainage of water from pores and narrow fissures including that in rocks, the epikarst and soils above the water table, as well as from sand and clay deposits in caves. However, the effect of emptying of poljes in the catchment was not taken into account, and we now know that the Trebišnjica River is influent when crossing the Ombla catchment and that three overflow springs operate at high flows (Bonacci 1995). Thus it is doubtful if the attributions of linear segments of the hydrograph are entirely correct. Further, one must question whether it is realistic to consider different reservoirs (e.g. conduit, fissured and matrix) to be hydraulically isolated from each other, because under flood surcharge for example conduit water is forced under pressure into fissures, only to be withdrawn as the pressure head falls. Thus although it may be possible to achieve good statistical fits to segments of hydrographs, the attribution of the segments to different stores within the catchment and aquifer is fraught with uncertainty, especially when the basic cascading characteristics of karst systems (Figure 5.31) are considered. A discussion of various causes for changes in the value of recession coefficients is provided by Bonacci (1993).

When the various subregimes yielding flow to a spring cannot be adequately represented by separate semi-logarithmic segments, other approaches must be used. For example, in an analysis of the Cheddar spring, England, Atkinson (1977) constructed a master recession curve by superimposing recession segments of all the winter and spring floods of the 1969–1970 water year (Figure 6.25). He found that this overall recession could be described well by an equation of the form

$$Q_t = Q_0 (1 + xt^y)^{-1} \qquad (6.35)$$

but after 25 days of runoff of the highest flows, the best fit for the remainder of the recession was provided by a simple exponential relation of the type discussed earlier. This experience reaffirmed the important conclusion reached by Mangin (1975), who found that the baseflow phase in all the cases he studied is accurately represented by Maillet's expression (equation 6.23), although it does not adequately represent flood flow.

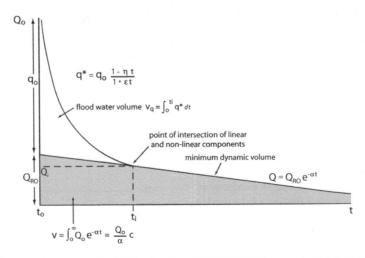

Figure 6.26   Karst spring recession curve analysis following Mangin's (1975, 1998) approach. The shaded area represents baseflow runoff and the unshaded area flood water runoff.

Consequently, Mangin (1975, 1998) considered that a distinction should be made between two basic hydrological entities in the interior of a karst drainage system: the somewhat delayed flow from the saturated zone with a linear baseflow recession represented by the function Φ_t; and the rapid flood flow from the unsaturated zone with a non-linear recession represented with the function ψ_t. Thus

$$Q_t = \Phi_t + \psi_t \tag{6.36}$$

which is a two-reservoir cascade model (Dewandel *et al.* 2003). By comparison to the flood hydrographs of surface rivers, ψ_t corresponds to the quickflow part of the recession and Φ_t to the baseflow segment (Padilla *et al.* 1994). In the case of karst, Φ_t relates to the saturated zone, whereas ψ_t translates the effect of surface recharge through the unsaturated zone to the spring: it is an infiltration function modulated by its transfer through the phreatic zone. Function Φ_t can be described by Maillet's formula (equation 6.23), whereas ψ_t is an empirical function that Mangin considered to be best expressed as

$$\psi_t = q_0 \cdot \frac{1 - \eta t}{1 + \varepsilon t} \tag{6.37}$$

where η characterizes the mean velocity of infiltration (close to 1 when fast infiltration is dominant) and q_0 corresponds to the difference between the total outflow Q_0 at the spring at $t = 0$ and a baseflow component termed Q_{R0} (Figure 6.26). The function is defined between $t = 0$ and $t = 1/\eta$, which is the duration of the flood recession. The coefficient ε characterizes the importance of the concavity of the flood recession curve and is small (<0.01) when infiltration is very slow. A high value for the coefficient corresponds to a recession that falls steeply at first but rapidly levels out. It can be evaluated from the expression

$$\varepsilon = \frac{q_0 - \psi_t}{\psi_t t} - \frac{\eta q_0}{\psi_t} \tag{6.38}$$

Mangin noted that the form of the flood recession can be described by the function

$$y = \frac{1 - \eta t}{1 + \varepsilon t} \tag{6.39}$$

which being independent of discharge permits comparison of different karst systems. After commencement of

the flood recession, the time required for the initial flow to diminish by 50% can be calculated from

$$t_{0.5} = (\varepsilon + 2\eta)^{-1} \tag{6.40}$$

Recalling from equation (6.36) that $Q_t = \Phi_t + \psi_t$ and with reference to Figure 6.26, the entire recession curve may be represented by the expression

$$Q_t = Q_{R0}\, e^{-\alpha t} + Q_0 \left(\frac{1 - \eta t}{1 + \varepsilon t} \right) \tag{6.41}$$

In practice, the baseflow segment is first calculated using equation (6.18) to determine the recession coefficient. Points Q_1 and Q_2 should be as distant from each other as possible, but on a straight line segment. Their choice is critical for the accuracy of α and the reliable estimation of Q_{R0}. The point of intersection of the linear and non-linear components determines t_i, when the flood recession is assumed essentially zero, i.e. $\psi_t = 0$.

The volume of water in storage in the saturated zone above the level of the outflow spring (Figure 5.12) is termed the **dynamic volume** V. Under baseflow conditions this can be found by integrating equation (6.18). Hence

$$V = \int_0^\infty Q_{R0}\, e^{-\alpha t} = c\frac{Q_{R0}}{\alpha} \tag{6.42}$$

where c is a constant that takes the units used into account (when Q_{R0} is in $m^3\,s^{-1}$ and α in days then $c = 86\,400$). In a complex recession of the type shown in Figure 6.24, where there are several α values, the dynamic volume is the sum of the component volumes.

The difference between baseflow reserves and the total volume discharged during a flood recession yields the flood-water volume V_q. It can be obtained by subtracting the baseflow component from the total runoff volume. This was the method used by Atkinson (1977b) in the case illustrated in Figure 6.25. Mangin's (1975) technique was to integrate equation (6.37), with further details and applications provided in Padilla *et al.* (1994).

A valuable insight into establishing whether or not hydrograph recession constants have been correctly attributed is provided by Eisenlohr *et al.* (1997c), who achieved this indirectly through numerical simulation with a finite-element model. Their results showed that the different exponential components do not necessarily correspond to aquifer volumes with different hydraulic conductivities; non-exponential parts of the recession hydrograph do not always yield information about the infiltration process; and the recession coefficient of baseflow depends not only on the hydraulic properties of the

Figure 6.27 (a) Planinsko Polje in the 'classic' karst of Slovenia, flooded in the wet season. (b) Popovo Polje in the Dinaric karst, with the Trebisnjica River incised into its planed rock floor. The residual hill in the background is beside the hamlet of Hum. (c) Polje in the cone karst of Guizhou, China.

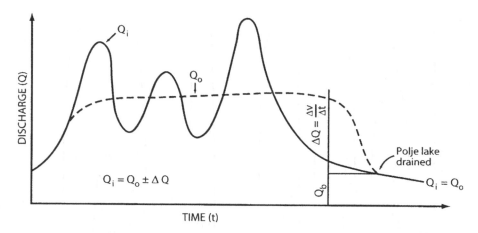

Figure 6.28 Schematic inflow (Q_i) and outflow (Q_o) hydrographs and changes in storage volume (V) in a flooded karst polje. Reproduced from Ristic, D.M., Water regime of flooded poljes. In V. Yevjevich (ed.) Karst Hydrology and Water Resources, Water Res. Pubs., Colorado, B787 301–318, 1976.

low hydraulic conductivity volume but also on the global configuration of the whole karst aquifer. Using the same modelling approach, Eisenlohr *et al.* (1997b) found similar problems with interpretation of aquifer characteristics from the form of discharge–discharge and precipitation–discharge correlograms, which also yielded equivocal results. It is therefore abundantly clear that we must guard against too simplistic an interpretation of the characteristics of karst-spring hydrographs and borehole hydrographs. What is required to complement the information on recession characteristics is corresponding information that identifies the provenance of the water concerned. This is available from water-quality data and the nature of chemographs (section 6.7).

6.6 POLJE HYDROGRAPH ANALYSIS

Poljes are very large, flat-floored depressions with interior drainage (Figure 6.27). They are usually the most extensive of the enclosed basins in a karst region and may measure many square kilometres in area. Their geomorphology is discussed in detail in section 9.9. A very important hydrological feature common to most poljes is their periodic inundation. In the context of their water regime, Ristic (1976) classified them into four simple types: wholly enclosed, open upstream, open downstream, and open both upstream and downstream; and their flooding characteristics range from periodic lakes at one extreme to inundated floodplains at the other. Since this does not emphasize the essential quality of poljes – that they are enclosed internally drained basins – the hydrologically more restrictive but geomorphologically more acceptable classification of section 9.9 (Figure 9.34) is preferred here.

When the sum of the various inflows Q_i into a karst polje exceeds the outflow Q_0, then the excess volume $+\Delta V$ for a time interval Δt is stored in the polje; but when $Q_i < Q_0$ water is withdrawn from storage and flooding recedes. The process is described by a simple water-budget equation

$$Q_i - Q_0 = \pm \frac{\Delta V}{\Delta t} \qquad (6.43)$$

Figure 6.28 illustrates the idealized flooding process in a polje and Figure 6.29 is an example of the more complex real situation.

Inflows into a polje may come from surface streams, springs and estavelles. However, measurement of the output of flooded springs and estavelles is particularly difficult. Estimation of the total swallow-hole capacity is also problematic, but Zibret and Simunic (1976) showed how an order of magnitude value may be obtained from the final dewatering phase of an inundation, provided that the reduced inflow Q_i has attained a nearly constant low discharge; the sum of the inflow and the volume reduction $\Delta V/\Delta t$ during the emptying phase then corresponds to the total stream-sink capacity Q_0 for a given inundation stage in the polje (Figure 6.28). As the lake level reduces the hydraulic head diminishes and the swallow holes may absorb less water.

Although the volume that can be absorbed by swallow holes varies with the depth of flooding, it also depends on underground piezometric pressures. The deeper the water table is below the polje floor, the longer the stream-sink will be able to operate at full capacity. Hence antecedent rainfall conditions are influential in

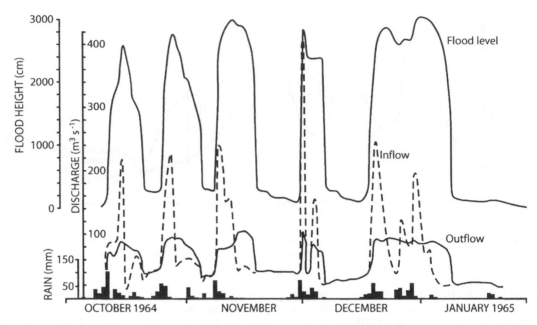

Figure 6.29 Rainfall, hydrographs and flood levels during the inundation of Nevesinjsko polje, Dinaric karst. Reproduced from Zibret, Z. and Z. Simunic, A rapid method for determining water budget of enclosed and flooded karst plains. In V. Yevjevich (ed.) Karst Hydrology and Water Resources. Water Res. Pubs., Colorado, 319–339, 1976.

determining quickflow runoff and the magnitude of Q_i, while seasonal effects influence subterranean storage and the height of the water table. Antecedent rainfall, through its influence on the generation of quickflow runoff, is a particularly important variable in the flooding of piedmont and border poljes, whereas seasonal phreatic zone storage is a critical factor in the inundation of baselevel poljes (Figure 9.34).

The individuality of flooding behaviour in different poljes is well illustrated by a number of excellent case studies: Popovo polje (Milanović 1981), Cerknisko polje (Gospodaric and Habic 1978, Kranjc 1985), Planina polje (Gams 1980), Kocevsko polje (Kranjc and Lovrencak 1981) and Zafarraya polje (López-Chicano *et al.* 2002). The flooding and emptying of poljes has a strong influence on the hydrographs of springs to which they drain, and may be manifest in abrupt changes in recession characteristics. This is evident in the case of Buna spring, for example, which drains Nevesinjsko polje in the Dinaric karst (Bonacci 1993).

Poljes are best seen as subsystems within a broader karst drainage system, the storage and overflow from one polje often affecting the inundation behaviour of others in the series (Figure 6.30). Hydrological interconnections between poljes can often be elucidated by water-tracing experiments (section 6.10).

6.7 SPRING CHEMOGRAPH INTERPRETATION

Discharge variations at a spring are often accompanied by changes in water quality. Drake and Harmon (1973) and Bakalowicz (1979, 1984) applied rigorous statistical methods (stepwise linear discriminant analysis and principal components analysis, respectively) to describe the quality of karst waters emerging at springs and to discriminate objectively between different types. Characteristics that may vary include ions in solution, electrical conductivity, environmental isotopes, pH, suspended sediment and temperature. Although some of these are physical rather than chemical attributes of water, it is convenient to consider plots of any of these water-quality aspects against time as 'chemographs'.

Jakucs (1959) showed that the chemical quality of karst spring water may vary with time (as well as with discharge), and appears to have been the first to recognize that the response of a spring to rainfall depends on the nature of the recharge, whether diffuse via karstified joints or concentrated via stream-sinks (Figure 6.31). Thus the transfer function of the system was perceived as conditioned by the type of recharge and, by implication, by the kind of flow network, since well-developed cave systems are particularly associated with point recharge of sinking streams. This is supported by an

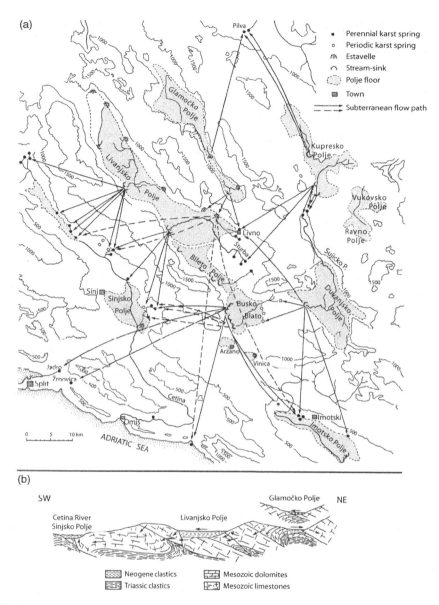

Figure 6.30 The hydrogeological system of poljes, springs and their traced interconnections in southern Dinaric karst. Reproduced from Herak, M. (1972) Karst of Yugoslavia, in Karst: Important Karst Regions of the Northern Hemisphere (eds M. Herak and V.T. Stringfield), Elsevier, Amsterdam, pp. 25–83; and Zötl, J. (1974) Karsthydrogeologie, Springer-Verlag, Vienna.

investigation by Worthington *et al.* (1992), who analysed hardness data from 39 springs in six different countries and demonstrated that more than 75% of hardness variation is explained by recharge type.

An insight into the response to recharge of a conduit-fed spring was also provided by Jakucs (1959), who presented details of a dye-tracing experiment in which the introduction of the dye at a stream-sink was closely fol-

lowed by a heavy rain storm. Analytical data for samples taken at the resurgence are reproduced in Figure 6.32. They clearly show that the recharge pulse from the rain arrived at the spring well before the dye, despite the latter having been introduced first. These data were interpreted by Ashton (1966) in terms of pulses of water from different input points passing through essentially discrete sections of the phreas. He explained the increase in Ca

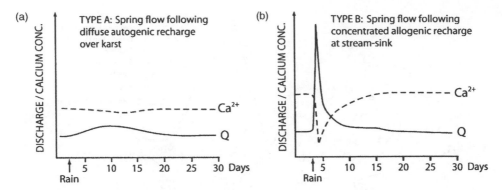

Figure 6.31 Hydrographs and chemographs of karst springs following (a) diffuse autogenic recharge and (b) concentrated allogenic recharge. Reproduced from Jakucs, L., Neue Methoden der Hohlenforschung in Ungarn und ihre Ergebnisse. Hohle 10(4), 88–98, 1959.

concentration at the start of the flood as due to the flushing out of water with a long residence time in the deeper phreatic zone; recharge forces water out by a piston-like process. Thus spring chemographs have a pattern composed of sequential and sometimes superimposed pulses of water of different quality and quantity from different stores and tributary inputs. Hence, if water quality and quantity data are combined, it becomes possible to calculate the volume of stores flushed by the recharge. This enables the hydrograph to be separated into different components.

The reality of 'old' water in storage being pushed out by new recharge water has been confirmed by numerous other workers; see Bakalowicz and Mangin (1980). Using ^{18}O variations in three karst springs, they showed that a precipitation event can cause water in different stores in a karst system to be flushed out and mixed in different proportions, thus giving rise to a fluctuating pattern of $\delta^{18}O$ at the spring that is different to the pattern of the storm event. This has been confirmed in Germany by Sauter (1997) and in the USA by Lakey and Krothe

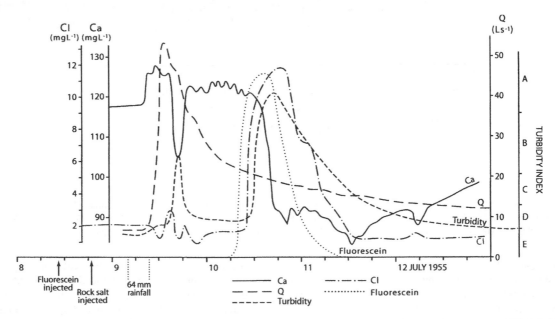

Figure 6.32 Hydrograph and chemograph of the Komlos spring, Hungary, during a water tracing experiment punctuated by rainfall. Note (a) that the discharge pulse arrived before the introduced chemical tracers, in spite of the rain occurring after their injection, (b) that the Ca concentration increased prior to its dilution by floodwaters, and (c) that the turbidity peak lagged the discharge peak. Reproduced from Jakucs, L., Neue Methoden der Hohlenforschung in Ungarn und ihre Ergebnisse. Hohle 10(4), 88–98, 1959.

(1996), whose case study of the Orangeville Rise in Indiana provided a particularly clear example of the flushing effect (Figure 6.33). Isotopic hydrograph separation of storm flow events revealed that over 4–6 days the maximum rainwater contribution to karst-spring discharge was 20–25% and arrived 18–24 h after peak flow, whereas pre-storm waters pushed out from storage contributed 77–80%. In an investigation of a karst groundwater system in Kentucky, Meiman and Ryan (1999) found that 14 h elapsed after a storm event before storm water appeared at the Big Spring beside the Green River, during which time 17 million litres of pre-storm water were displaced from the conduit network.

The separation of karst-spring hydrograph and chemograph data is necessarily guided by the conceptual model of the karst system being used. Thus prior to the mid-1970s the epikarst was not recognized as a store and the variety and efficiencies of flow paths through it were not appreciated. Consequently storage in the vadose zone was overlooked and the volume ascribed to the phreatic zone was too large, whereas now we understand field data better, can identify subcutaneous water and can sometimes distinguish between water displaced rapidly from the subcutaneous zone (fast-event water) as compared with that delayed (slow-event water) by resistance in the fissure system (Sauter 1997, Pinault *et al.* 2001). The model of stores and linkages in a karst system adopted here is presented in Figure 5.31 and an idealized separation of spring data is shown in Figure 6.34. However, we need to recognize that in reality each component separated is a mixture in which the component is the dominant but not the only source of the water. Sauter's (1992) study illustrates the kind of complexity that can result and so provides a warning against adoption of too simple a piston-flow model.

At any given instant, the mass-balance of constituents in spring water can be expressed as the sum of the discharge components and their respective concentrations

$$Q_r C_r = Q_s C_s + Q_e C_e + Q_p C_p \qquad (6.44)$$

where Q_r is the discharge of the resurgence and Q_s, Q_e and Q_p represent the contributions to it made by water

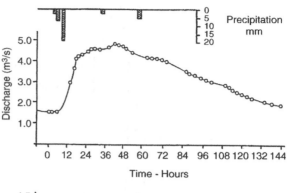

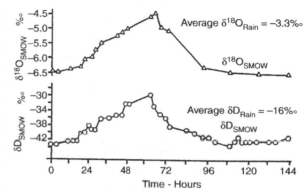

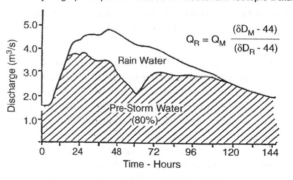

Hydrograph Separation based on Deuterium Isotopic Data

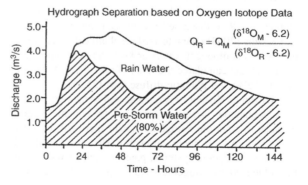

Hydrograph Separation based on Oxygen Isotope Data

Figure 6.33 Rainfall, stable isotopic data and spring flow separation for a 1991 storm event at Orangeville Rise, Indiana. Pre-storm spring water had isotopic values of 44‰ δD and 6.5‰ δO, whereas the rainwater averaged 16‰δD and 3.3‰δO. Reproduced from Lakey, B. and Krothe N. C., Stable isotopic variation of storm discharge from a perennial karst spring, Indiana. *Water Resources Research*, 32(3), 721–731 © 1996 American Geophysical Union.

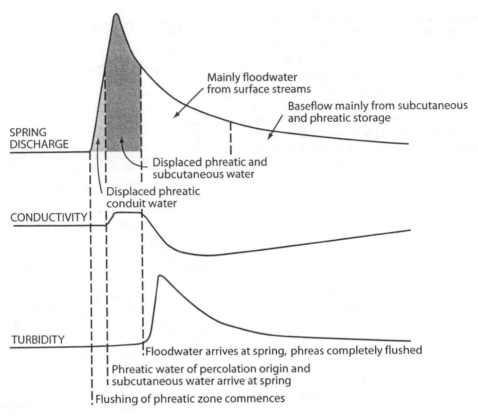

Figure 6.34 Idealized interpretation of a karst spring hydrograph and chemograph. Reproduced from Williams, P.W, The role of the subcutaneous zone in karst hydrology. J. Hydrol. 61, 45–67 © 1983 Elsevier.

from sinking surface streams (allogenic), epikarstic storage and phreatic storage respectively. The concentration of a given water-quality characteristic at the resurgence is C_r, and C_s, C_e and C_p are the contributions made to it by the various components. In practice it is difficult to obtain all the component values for Q, although chemical concentration data are relatively easily acquired; hence a combined approach using graphical hydrograph separation and calculation is necessary to derive a realistic assessment of the proportions involved. Thus in the case of the Orangeville Rise discussed previously (Figure 6.33), Lee and Krothe (2003) characterized the chemical signature of the various component waters and with a three-component mixing model calculated the ratios of rain (16.5%), vadose flow (58.5%) and phreatic diffuse flow (25.0%).

A particular difficulty is encountered in the recognition of subcutaneous water as opposed to phreatic water, because in a simple autogenic system there will be little distinction in their chemical quality. However, in mixed autogenic–allogenic systems, phreatic conduit water will be influenced by allogenic stream inputs and hence will

normally have lower concentrations of Ca than water derived solely from the epikarst. In addition, evapotranspiration losses from the soil in summer tends to elevate Cl concentrations in the soil-water store and hence in the immediately underlying subcutaneous zone. A storm which breaks a summer drought will thus tend to displace the Cl peak. Under such circumstances, the soil-water plus epikarstic component can be recognized. Lee and Krothe (2003) found SO_4 concentration to be a convenient way of distinguishing component waters (rain $1.8\,mg\,L^{-1}$, epikarst water $19\,mg\,L^{-1}$ and phreatic water $210\,mg\,L^{-1}$) in Indiana. In an earlier attempt at a four-component mixing model, Lee and Krothe (2001) were less successful in separating epikarst and soil water using HCO_3 and $\delta^{13}C$ signatures, because with a soil mantle up to 10 m thick, water in the subcutaneous zone will be dominated by the quality of its soil-water recharge.

Chemically complex springs that discharge water from a variety of aquifers, e.g. bicarbonate, sulphate, chloride, thermal, etc., require a modification of this approach. If we assume as an illustration that a spring has two main

Table 6.4 Flood dilution mixing calculations for four karst springs in Mexico (From Ford, D.C. and Williams, P.W. (1989) Karst Geomorphology and Hydrology; after Fish 1977)

Spring	Date	$Q_{r(max)}$ $(m^3 s^{-1})$	$\dfrac{Q_B}{Q_{r(max)}} \times 1000$	SO$_4$ (mg L^{-1}) Predicted	Measured
Choy	7 July 1972	56	1.16	32	30
	15 June 1972	48	1.35	37	35
Mante	17 August 1971	~28	10.7	220	180
	11 July 1972	~23	13.0	225	240
Frio	27 December 1971	13.2+	4.17	90	92
Coy	20 June 1972	<118	>5.93	>112	25

water-quality components, termed *A* and *B*, perhaps from limestone and gypsum respectively, then a two-component mixing equation has the form

$$Q_r C_r = Q_A C_A + Q_B C_B \qquad (6.45)$$

Component *B* may have a deep, inaccessible source, and we may wish to estimate its proportional contribution to the discharge at the resurgence. Dividing through by Q_r, the expression becomes

$$C_r = Q'_A C_A + Q'_B C_B \qquad (6.46)$$

where $Q'_A + Q'_B = 1$. The value of C_A can be assessed from measured values of monominerallic spring waters elsewhere in the region. Knowing the water balance of these other springs, it may also be possible to obtain a reasonable figure for Q_A. Model compositions for C_B are estimated and then tested in high-stage and low-stage conditions where C_A, C_r and Q_r are known, to arrive at an acceptable estimate of the true values of C_B and Q_A. Table 6.4 and Figure 6.35 illustrate results from an application of this method.

6.8 STORAGE VOLUMES AND FLOW ROUTING UNDER DIFFERENT STATES OF THE HYDROGRAPH

We have emphasized that porosity is usually low in karst. As a consequence, the volume of water in storage can vary widely at different stages of the hydrological year and under different states of the hydrograph. The huge range of possible variation may be imagined when the piezometric surface in some places in the Dinaric karst can vary by as much as 312 m in 183 days or 90 m in 24 h (Milanović 1993). Consequently we must consider methods that are applicable to non-Darcian conditions. There are four principal approaches to estimating storage volumes and available water resources in karst aquifers,

the most appropriate depending on local hydrogeological conditions and available data bases: (i) water balance (section 6.2), (ii) multiple-borehole analysis (section 6.4), (iii) spring flow recession (section 6.5) and (iv) residence-time–discharge calculation.

Prestor and Veselic (1993) presented case studies from the alpine and Mediterranean zones of the Slovenian karst and concluded that, with appropriate statistical analysis of time series, a high level of accuracy can be achieved in total discharge estimation from just two hydrological years of observation of precipitation data, with close to 70% accuracy in overall discharge and dynamic reserves estimation, although to raise the level of accuracy complementary information is required.

Multiple-borehole testing is a traditional technique of aquifer evaluation that involves pumping from one well and observing drawdown in nearby observation wells (Figure 6.17). Multiple-borehole testing samples a larger volume of rock mass than single-borehole tests, and so the calculation of storativity (equation 6.17) is subject to less uncertainty. This is acceptable if the rock mass can be treated as a porous medium but, where there is well developed secondary porosity and heterogeneity, the more transmissive conduit porosity of karst may not be sampled; so except in the most well connected, porous carbonate aquifers this technique is unlikely to yield representative estimates of available water resources from karst. Nevertheless, results of sustainable water yield from individual boreholes are still valuable at a local level, and patterns of well yields over large areas provide useful regional pictures of variability of water supplies (e.g. karst in western Ireland in Figure 6.36).

Where reserves are estimated from spring-flow recession data, a master recession curve for a wide range of flows is first constructed from numerous partial recessions, and then an estimate of reserves is made by integrating the area under the master curve. The result is usually interpreted as representing the dynamic volume (Figure 5.12). A particularly good example of the

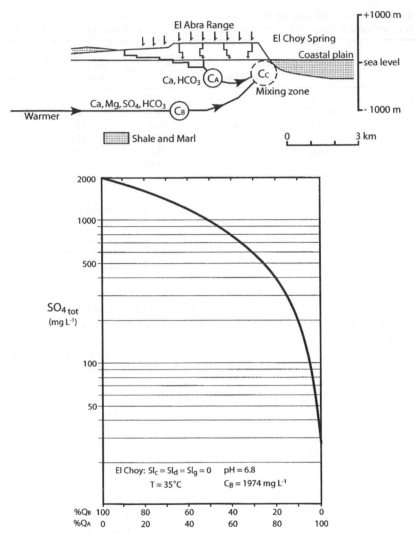

Figure 6.35 Two-source mixing model for the Choy spring, Mexico. Reproduced from Fish, J.E.. Karst hydrogeology and geomorphology of the Sierra de El Abra and the Valles San Luis Potosi Region, Mexico. Ph.D. thesis, 469 p, McMaster Univ., Canada, 1977.

application of the technique in an unconfined karst aquifer (Mendip Hills in England) was provided by Atkinson (1977), who used a combination of an empirical (equation 6.35) and Maillet's equation (equation 6.23). The Mendip aquifer has a deep base below the spring.

Where the impermeable aquifer floor is close to the level of the outflow spring, Dewandel *et al.* (2003) showed that the Maillet formula underestimates the dynamic volume of the aquifer, and considered the Boussinesq equation (equation 6.24) to provide good estimates of dynamic reserves regardless of the thickness of the aquifer under the outlet.

The fourth approach to reserves estimation is to measure and calculate the average amount of water that passes through the aquifer in a year. This should yield a figure that is similar to (but more accurate than) the value estimated from the water-balance approach. Where springs are gauged, this requires a calculation of the outflow over at least 1 yr; the average over many years provides the most reliable long-term estimates and yields information on interannual variability. However, the average throughput in 1 year does not tell us the total capacity of the aquifer. This can be estimated from the product of the mean discharge and the residence time of

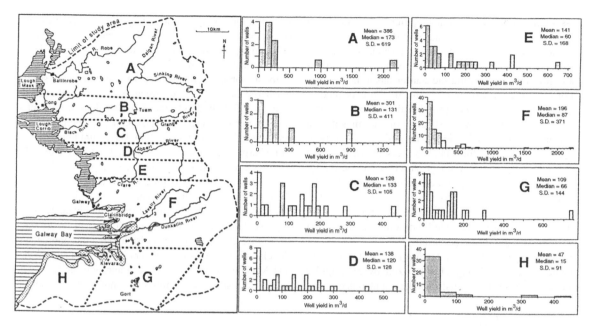

Figure 6.36   Areal variation in borehole yields in unconfined karstified Carboniferous limestones in south Mayo, mid-Galway and north Clare counties, western Ireland. Reproduced from Drew, D.P. and Daly, D., Groundwater and Karstification in mid-Galway, South Mayo and North Clare, Report Series 93/3 (Groundwater),© 1993 Geological Survey of Ireland.

the water. As an illustration, the partly confined Takaka marble aquifer in New Zealand discharges from the Waikoropupu Springs an average flow of $15\,m^3\,s^{-1}$, which is equivalent to $0.47\,km^3\,yr^{-1}$. But the waters emerging have a mixture of ages, estimated by tritium to be mainly in the range of 3–8 yr. Hence the volume of underground storage is in the order of $1.4–3.8\,km^3$ (Williams 1992).

Under different states of the hydrograph there is variation in the relative volumes stored in different parts of the aquifer (Figure 5.31), as well as variation in principal routes followed by water, with some flow paths drying up, especially in the vadose zone. Water resources become especially important as supplies are depleted at the end of a long dry period, especially in the epikarst, but just because there may be plenty of water in a deep phreatic store does not mean that it is available. Abstraction of water by bores with their intake below the level of the outflow can lead to cessation of spring flow and result in serious ecological damage downstream. Hence useable water will depend on decisions regarding ecologically acceptable minimum flows. 'Quarrying' of water is not an option in a world where sustainable use of resources and simultaneous habitat protection are taken seriously. Sadly, it is widely practiced.

6.9 INTERPRETING THE ORGANIZATION OF A KARST AQUIFER

The hydraulic behaviour of karst systems is dependent first of all on the origin of the karstification: whether it is derived from normal surface karst processes with meteoric water, from deep hydrothermal karst flow without ready access to the surface, or influenced by freshwater – saltwater mixing and seawater intrusion near the coast. Here we focus our comment on the first (meteoric water) case, which is quantitatively predominant.

Bakalowicz and Mangin (1980) showed that although a karst aquifer appears very heterogeneous, this is not the result of a random juxtaposition of different types of voids because it stems from the ordered distribution of voids around a drainage axis, according to a certain hierarchy. The principal drain receives water from tributaries, the epikarst, and ancient flooded voids and networks of fissures connected to the main drain. Thus the aquifer in a karst basin can be considered structured with reference to the drainage. This has certain consequences.

1. Water flows are increasingly organized as the structure develops. The point of departure for this evolution is a porous or fissured aquifer with a diffuse flow system, while the end-point is an aquifer with flow limited

essentially to conduits. Karst aquifers can include all possible combinations and stages of drainage evolution.

2. Because of the existence of a particular drainage structure, the scale appropriate to the resolution of any problem concerning the aquifer has to be precise. For example, at a scale of $1-100\,m^3$ the existence of randomness can be recognized in the aquifer – a drill hole can go through a karst cavity or can miss it by a few metres, without being able to predict it one way or the other. On the other hand, at the scale of the karst massif an organized drainage structure can be recognized. Thus at one scale the aquifer appears random, whereas at the other it is structured. Our interest is usually in the regional scale, where the reference unit that encompasses the whole structure is the karst system (i.e. the drainage basin).

Bakalowicz and Mangin (1980) regarded the karst drainage system as being characterized by an impulse response function that transforms the input pulses from precipitation to hydrograph responses at springs, and contended that by analysis of this function it is possible to identify particular features of the aquifer and its degree of organization (sections 6.5 and 6.7).

In a subsequent analysis of rainfall and spring-discharge data using autocorrelation and spectral analysis, Mangin (1981a,b, 1984) added further depth to his interpretation of the Pyreneean karst aquifers. The contrasting reactions of three systems (Aliou, Baget and Fontestorbes) to recharge is illustrated in Figure 6.37, which shows that the speed of decline of the three correlogram is different. The memory effect was interpreted to be very great in the Fontestorbes system, but short in the case of Aliou. The spectral density curves translate the same phenomena. He suggested that karst aquifers may be classified according to these characteristics and later also applied this approach to the classification of karst aquifers in Romania (Iurkiewicz and Mangin 1993). As noted, Jeannin and Sauter (1998) cautioned that these methods are not very efficient in

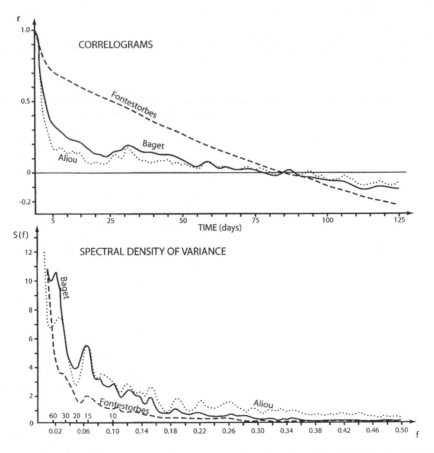

Figure 6.37   Time-series analysis of the discharge of three karst springs. Reproduced from Mangin, A. Utilisation des analysis correlatoire et spectrale dans l'approche des systemes hydrologiques. C. R. Acad. Sci. Paris 293, 401–404, 1981.

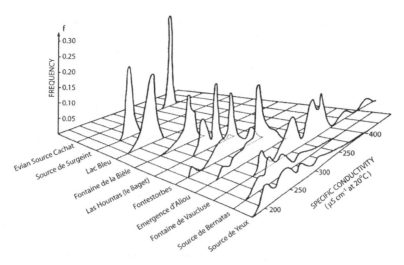

Figure 6.38 Frequency distribution of electrical conductivity of waters from different springs in France. Reproduced from Bakalowicz, M. and Mangin, A., aquifere karstique. Sa definition, ses characteristiques et son identification. Mem. L. sr. Soc. Geol. France 11, 71–79, 1980.

inferring the structure of karst systems or in classifying them because they are too strongly related to the frequency of rainfall events. Recognizing that conventional statistical analyses are based on time-invariance and assumptions of stationarity, Labat *et al.* (2001) have commenced further exploration of the rainfall–runoff relationships of the Pyreneean catchments by means of wavelet analysis and multifractal techniques.

Chemical data reinforce and complement the information derived from the spring hydrographs, although Bakalowicz (1977) argued that the structure of a karst aquifer cannot be defined from the coefficient of variation of chemical variables in spring water, as was suggested by Shuster and White (1971), because the distribution of the values is usually multimodal rather than normal. Atkinson (1977b) also noted that the very small range of $CaCO_3$ in resurgences from the Mendip Hills, England, would suggest that they are diffuse flow springs according to Shuster and White's criteria, yet they are known to be fed largely by conduit flow. This reinforces the conclusion of Jakucs (1959) and Worthington *et al.* (1992) that recharge type explains most of the variation in hardness of springs in carbonate aquifers.

Bakalowicz and Mangin (1980) demonstrated from French examples that the frequency distribution curve of a hydrochemical variable such as conductivity provides a good description of the aquifer and suggested the following interpretation on the curves shown in Figure 6.38:

porous aquifers – unimodal, relatively high electrical conductivity

fissured aquifers – unimodal, relatively low electrical conductivity

karstified aquifers – multimodal, wide range of electrical conductivities.

The development of a preferred conduit drainage axis through the saturated zone permits easy transfer of water and therefore transmits a not too distorted image of the recharge process. But if such a hydrological organization is not developed, then homogenization of infiltration waters occurs.

6.10 WATER-TRACING TECHNIQUES

Water tracing is now a powerful, well-developed tool of the karst hydrologist. It enables catchment boundaries to be delimited, groundwater flow velocities to be estimated, areas of recharge to be determined and sources of pollution to be identified. Dyes and other substances have been used to trace underground water for more than 100 yr. The green dye fluorescein ($C_{20}H_{12}O_5$) was discovered in 1871, and was first used as a karst water tracer at the streamsinks of the upper Danube River in 1877. Its disodium salt under the trade name 'uranine' ($C_{20}H_{10}O_5Na_2$) was introduced a few years later. Early methods were sometimes heavy-handed. A spectacular example was tracing the Doubs River to the Loue Spring in the French Jura Mountains, a distance of about 10 km (Käss 1998). In 1911 about 100 kg of fluorescein was injected into the Doubs and resulted in the Loue Spring and river being green for 3½ days over a distance of 60 km. This added insult to injury, because just 10 year earlier in the same catchment, a fire in the Pernod distillery had cause a vat of absinthe to burst and drain into the Doubs, the resulting alcohol content of water from the Loue indicating that about 1 million litres of absinthe had been lost.

Gaspar (1987), Käss (1998) and contributors review the history and full range of modern techniques from naturally occurring labels to artificially introduced radioisotopes; Dassargues (2000) and contributors give numerous examples of their application; and US EPA (2002) provides a QTRACER2 program for tracer breakthrough analysis. Each of the *Proceedings of the International Symposium on Underground Water Tracing* (SUWT) from 1966 to 1997 enables advances made in water tracing to be tracked. The *proceedings of 7th symposium* (Kranjc 1997a) and the associated regional case study of karst hydrogeology in southwest Slovenia are particularly informative, with valuable commentary on methodological aspects of their experiments. Other important case studies include Batsche *et al.* (1970) for the Danube–Aach region in Germany and Quinlan and Ewers (1981) for the Sinkhole Plain–Mammoth Cave region in the USA.

Four classes of water tracing agents are available.

1. Artificial labels
   (a) dyes
   (b) salts
2. Particulates
   (a) spores
   (b) fluorescent microspheres
   (c) bacteriophages
3. Natural labels
   (a) microorganisms
   (b) ions in solution
   (c) environmental isotopes
4. Pulses
   (a) natural pulses of discharge, solutes and sediment
   (b) artificially generated pulses

6.10.1 Artificial labels

To be suitable for water tracing, an artificial tracer should be absent in the water to be traced and as far as possible be:

1. non-toxic to the handlers, to the karstic ecosystem and to potential consumers of the labelled water;
2. soluble in water, in the case of chemical substances, with the resulting solution having approximately the same density as water;
3. neutral in buoyancy and in the case of particulate tracers, sufficiently fine to avoid significant losses by natural filtration;

4. unambiguously detectable in very small concentrations;
5. resistant to adsorptive loss, cation exchange and photochemical decay, and to quenching by natural effects such as pH change and temperature variation;
6. susceptible to quantitative analysis;
7. quick to administer and technologically simple to detect;
8. inexpensive and readily obtainable.

Unfortunately, no artificial tracer satisfies all the above criteria, but several do so to a significant extent. Non-toxicity is the most important characteristic. Details of warm-blooded and cold-blooded toxicities of dyes used in hydrological testing are provided by Leibundgut and Hadi (1997, tables 1 to 4) and Käss (1998, tables 6 and 7). Leibundgut and Hadi (1997) concluded that most of the fluorescent tracers can be considered harmless in the concentrations usually found in tracer experiments, especially considering the rapid dilution that follows injection under field conditions. Although the usual maximum tracer concentration of $10 \, \text{mg m}^{-3}$ for ground water is below toxic level, dye concentrations should be maintained as far below that level as possible and the mass used should be minimized so that the duration of exposure is limited. Many different formulae have been suggested for calculating the quantity of dye required for a tracer (Käss 1998). Worthington and Smart (2003) took data from 203 tracer tests and found that two equations gave excellent fits to the data:

$$M = 19(LQC)^{0.95} \tag{6.47}$$

and

$$M = 0.73(TQC)^{0.97} \tag{6.48}$$

where M is mass of tracer (g), L is distance between input and output (m), Q is spring discharge ($\text{m}^3 \, \text{s}^{-1}$), C is concentration (g m^{-3}) and T is time (s). These equations facilitate accurate estimation of the amount of dye needed for tracer tests in carbonate rocks under a wide range of conditions.

Dyes

Artificial dyes are the most successful water tracers at the present time, with many properties that are desirable for water tracing in karst systems (Field *et al.* 1995). Although it was known in 1904 that uranine could be adsorbed by charcoal and eluted later (Käss 1998), this convenient property was forgotten, and so for many years detection depended on visual observation. Dunn (1957) revived the method of fluoresein dye adsorption onto charcoal with subsequent elution using potassium hydroxide (5%) in ethanol. The field requirement is simply to suspend small mesh bags containing a few grams of granular activated carbon in a moderate current in the monitored springs. Rhodamine WT is also adsorbed and may be eluted with a warm solution of 10% ammonium hydroxide in 50% aqueous 1-propanol (Smart and Brown 1973). The great advantage of this technique is that detector bags can be changed when convenient and many sites may be easily monitored. There are no time constraints on when analysis must be undertaken, although air drying of detector bags is required if more than a few days will elapse between collection and examination. Charcoal bag detectors also work effectively in the sea should resurgences be submarine or intertidal. A recent evaluation of the use of charcoal for detection is available in Smart and Simpson (2001).

The most significant recent development in dye-tracing techniques has been the advent of quantitative fluorometric procedures. This followed from the work of Käss (1967) in Europe and Wilson (1968) in North America. The present status of these procedures is explained in publications by Käss (1998), Dassargues (2000) and US EPA (2002). Important points about the techniques used are that several dyes can be distinguished from each other when mixed, thus permitting simultaneous dye tracing in the same basin, detection is possible well below visible levels, and quantitative analysis has become routine. Fluorescent dyes can be detected with fluorometers and spectrophotometers. Continuous flow-through fluorometers are available for installation in the field.

Fluorescent substances emit light immediately upon irradiation from an external source. The emitted or fluoresced energy usually has longer wavelengths and lower frequencies than that absorbed during irradiation. This property of dual spectra is utilized to make fluorometry an accurate and sensitive analytical tool, because each fluorescent substance has a different combination of excitation and emission spectra. Since some naturally occurring substances such as plant leachates, phytoplankton and some algae also fluoresce, it is important to know the background fluorescence of the water being traced before any experiments are conducted. Industrial and domestic wastes also introduce background problems. Natural substances tend to fluoresce in the green waveband; thus the use of dyes with an orange emission overcomes problems of possible misidentification.

Numerous fluorescent dyes are commercially available and have been used in groundwater tracing. They are all organic substances and Käss (1998) provides details. Because different manufacturers often assign different product names to what in fact is the same dye, confusion

sometimes occurs in comparing results of traces by different people. Hence publications reporting the use of dyes should record the Colour Index generic name and number, as well as the manufacturer's name for the dye (Table 6.5).

Smart and Laidlaw (1977) assessed the relative merits of eight fluorescent dyes. They considered their sensitivity, minimum detectability, effects of water chemistry, photochemical and biological decay rates, adsorption losses, toxicity and cost. They concluded that the orange dyes are more useful than the others because of lower background fluorescence, which permits higher sensitivities to be obtained. Although on toxicity grounds we recommend that, with the possible exception of Rhodamine WT, the use of the rhodamine group should be limited and restricted to exceptional cases. Smart and Laidlaw (1997) also showed the fluorescence of some dyes, such as pyranine, to be strongly affected by pH of the water which precludes its quantitative use. However, another dye used with spectacular success in water tracing is Direct Yellow 96 (Table 6.6: Diphenyl Brilliant

Table 6.5 Fluorescent dyes used in water tracing, their Colour Index and Chemical Abstract Service (CAS) number (Reproduced from US EPA, The QTRACER2 Program for Tracer-Breakthrough Curve Analysis for Tracer Tests in Karst Aquifers and Other Hydrologic Systems, EPA/600/R-02/001, U.S. Environmental Protection Agency, 2002)

Dye type and common name	Colour index generic name	CAS Number
Xanthenes:		
sodium fluorescein	Acid Yellow 73	518-47-8
eosine	Acid red 87	17372-87-1
Rhodamines:		
Rhodamine B	Basic Violet 10	81-88-9
Rhodamine WT	Acid Red 388	37299-86-8
Sulpho Rhodamine G	Acid Red 50	5873-16-5
Sulpho Rhodamine B	Acid Red 52	3520-42-1
Stilbenes:		
Tinopal CBS-X	Fluorescent Brightner 351	54351-85-8
Tinopal 5BM GX	Fluorescent Brightner 22	12224-01-0
Phorwite BBH Pure	Fluorescent Brightner 28	4404-43-7
Diphenyl Brilliant Flavene 7GFF	Direct Yellow 96	61725-08-4
Functionalized polycyclic aromatic hydrocarbons:		
Lissamine Flavine FF	Acid Yellow 7	2391-30-2
Pyranine	Solvent Green 7	6358-69-6
Amino G acid	—	86-65-7

Table 6.6 Data on some common fluorescent tracer dyes (Reproduced from US EPA, The QTRACER2 Program for Tracer-Breakthrough Curve Analysis for Tracer Tests in Karst Aquifers and Other Hydrologic Systems, EPA/600/R-02/001, U.S. Environmental Protection Agency, 2002)

Dye name	Maximum excitation λ (nm)	Maximum emission* λ (nm)	Fluorescence intensity (%)	Detection limit[†] ($\mu g\,L^{-1}$)	Sorption tendency
Sodium fluorescein	492	513	100	0.002	Very low
Eosine	515	535	18	0.01	Low
Rhodamine B	555	582	60	0.006	Strong
Rhodamine WT	558	583	25	0.006	Moderate
Sulpho Rhodamine G	535	555	14	0.005	Moderate
Sulpho Rhodamine B	560	584	30	0.007	Moderate
Tinopal CBS-X	355	435	60	0.01	Moderate
Phorwite BBH Pure	349	439	2	?	?
Diphenyl Brilliant Flavene 7GFF	415	489	?	?	?
Lissamine Flavine FF	422	512	1.6	?	?
Pyranine	460[‡]	512	18	?	?
	407[§]	512	6	?	?
Amino G acid	359	459	1.0	?	?
Sodium napthionate	325	420	18	0.07	Low

*Different instruments will yield slightly different results.
[†]Assumes clean water and spectrofluorometric instrumentation.
[‡]pH $\geq$ 10.
[§]pH $\leq$ 4.5.

Flavine 7GFF). It was introduced by Quinlan (1976) in the Sinkhole Plain–Mammoth Cave area of Kentucky (Figure 6.39) where hundreds of dye traces have since been conducted (Quinlan *et al.* 1983). The results are available as 1:100 000 scale maps (Ray and Currens 1998a,b).

More than one fluorescent dye may be injected simultaneously into a karst drainage basin and yet be mutually distinguishable at the resurgence. Although the absorbance and emission spectra of many dyes overlap, three groups of them have sufficiently minimal interference to permit ready separation: i.e. those that fluoresce in the blue, green and orange wavelengths. The most suitable dye from each group depends on other characteristics, although Rhodamine WT (orange), uranine/fluorescein (green) and amino G acid (blue) are of proven value. Their spectral properties are shown in Figure 6.40 and other details in Table 6.6.

Käss (1998) provides detail on techniques of dye separation, but warns that if two dyes with fluorescence maxima within about 50 nm of each other are used simultaneously in a tracing experiment, then they will interfere with each other if both are present in the same sample. The easiest method to recognize two or more dyes in one solution is to produce a synchroscan spectrum, by running a synchronized wavelength change of the excitation and fluorescence monochromators through the spectrum of interest with a constant wavelength distance. This is determined by the spectral position of the excitation and fluorescence maxima (between 20 and 25 nm for most xanthene dyes). Other techniques can be used for dye separation, including pH adjustment, thin-layer chromatography and enrichment.

The most useful quantitative water tracing at the time of writing is obtained by a combination of continuous fluorometry, i.e. passing water from a spring or other sampling point directly and continuously through a portable field fluorometer, and tracer breakthrough curve analysis. The shape of the dye breakthrough curve as it emerges at the spring can be analysed and accurately compared with other continuously monitored variables such as discharge, conductivity and temperature. The QTRACER2 program (US EPA 2002) provides a powerful method of analysing tracer breakthrough curves, and permits the calculation of many hydraulic parameters, including tracer mass recovery, mean residence time, mean flow velocity, longitudinal dispersion and characteristics of flow-channel geometry and volume. However, Werner *et al.* (1997) explain difficulties of interpreting tracer tests under unsteady flow conditions. The origin of dispersion and retardation of tracers in karst conduits as revealed by breakthrough curves was explained by Hauns *et al.* (2001) after comparing breakthrough curves obtained by various approaches (field experiments, laboratory experiments and numerical simulations). Dispersion is produced by small-scale turbulent eddies and retardation is mainly the consequence of longitudinal changes in conduit geometry.

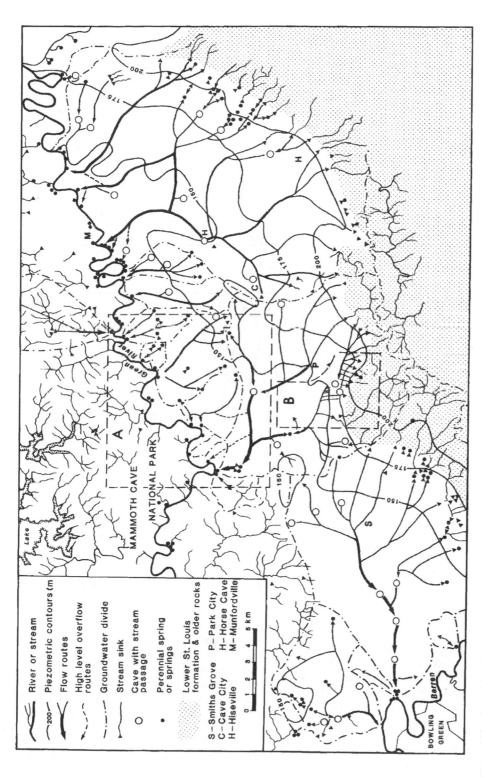

Figure 6.39 Groundwater basins, piezometric surface, subsurface flow routes and surface drainage in the Sinkhole Plain–Mammoth Cave region, Kentucky. Boxes A and B show the locations of Figures 7.11 and 9.67. Adapted from Quinlan, J.F. and Ewers, R.O., Hydrogeology of the Mammoth Cave Region, Kentucky, in Geological Society of America Cincinnati 1981 Field Trip Guidebooks, Vol. 3, (ed.T.G. Roberts), pp. 457–506, 1981.

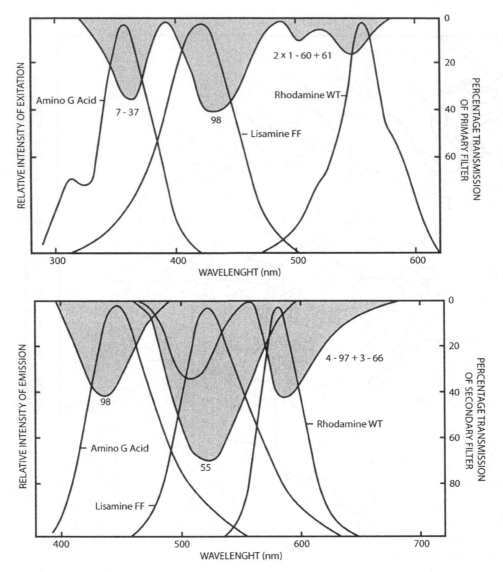

Figure 6.40 Absorption and emission spectra of blue (amino G acid), green (lissamine FF) and orange (Rhodamine WT) fluorescent dyes. Reproduced from Smart, P.L. and Laidlaw, I.M.S, An evaluation of some fluorescent dyes for water tracing. Water Resources Res. 13, 15–23 © 1977, American Geophysical Union.

Salts

Inorganic compounds that dissociate in water into cations and ions have been used frequently to trace water. For example, sodium chloride was one of the earliest artificial water tracers, three tons of it being used in 1899 to trace water sinking at Malham Tarn in Yorkshire, England (Carter and Dwerryhouse 2004). It also featured amongst the 13 different substances used to trace the waters of the upper River Danube to the Aach spring in the catchment of the River Rhine in Germany. In that case 50 t were used (Batsche *et al.* 1970). Lithium and potassium chlorides have also been widely used in water tracing, as have non-fluorescent dyes such as Congo Red and Methylene Blue (Brown *et al.* 1972). Lithium, for example, is relatively rare in nature and so with low background values present is a useful tracer, although uncertainty exists regarding its acceptability on toxicity grounds. An example of the results of karst water tracing with lithium chloride is provided in Kranjc (1997b). An informative review of the use of salts in water tracing is provided by Käss (1998).

6.10.2 Particulates

Spores

Spores of the club moss *Lycopodium clavatum* have a diameter of 30–35 μm and buoyancy only slightly greater than water. This permits them to drift through karstified conduits without excessive loss by filtration. They can be dyed different colours, thus permitting simultaneous traces from different injection sites. They are recovered at resurgences by 25 μm mesh conical plankton nets suspended in the current and are identified and counted under a microscope.

The use of spores to trace water was introduced by Mayr (1953, 1954) and was developed by Maurin and Zotl (1959). The method is fully described by Drew and Smith (1966), Zotl (1974) and Käss (1998). Smart and Smith (1976) review tracing work in tropical waters using spores and fluorescent dyes, and summarize the relative merits of the methods. However, the technique is of little value for quantitative work and so has been largely superseded since the early 1970s by superior fluorometric techniques.

Fluorescent microspheres

Plastic microspheres coloured with fluorescent dye are utilized widely in biological and medical applications and have been used in groundwater tracing since the late 1980s. They are available in a variety of sizes from 0.05 to 90 μm diameter, in different colours and in suspension in deionized water. Their density is 1.055 g mL^{-1}. Käss (1998) reviewed techniques and other applications and provided results of simultaneous comparative tests with conventional fluorescent dyes. Polystyrene microspheres have an electrically neutral surface. Their transport behaviour is similar to bacteria, and so given that the size of most bacteria is around 1 μm, the optimal size of microspheres for a drift tracing test is about that diameter. One of the great advantages of fluorescent microspheres is that many tests can be conducted simultaneously because of the range of colours available and they do not interfere with other tracers such as dyes; however, inadvertent cross-contamination is a serious problem. Ward *et al.* (1997a) gave an interesting example of their use in an investigation of the migration of pathogens through an area of unconfined Chalk and discussed sampling and detection techniques (epifluorescence microscopy and fluorescence spectrophotometry). Injection took place on the Chalk surface beneath the topsoil and monitoring of arrival was by bore water sample drawn from the water table at a depth of 20 m below the surface. The site was irrigated before the test and again lightly after the injection. Microspheres of 6 μm arrived at the water table in 18.5 h, showing that particles of 6 μm or less can migrate rapidly through the unsaturated zone.

Bacteriophages

Bacterial viruses or bacteriophages (or simply phages) have recently been used as water tracers, as reviewed by Käss (1998). They are cultivated in the laboratory, are environmentally harmless, being non-pathogenic, non-toxic and invisible regardless of concentration. However, microbiological laboratories are required to conduct such tests, and considerable work is required both in the preparation stage and in the examination of the samples. Consequently, the use of phages in karst water tracing is uncommon. Nevertheless, examples of their application can be found in Bricelj (1997), Drew *et al.* (1997) and Formentin *et al.* (1997).

6.10.3 Natural labels

Microorganisms

Bacteriological and virological examination of karst spring waters can be undertaken to establish the hygenic quality of the water and, if contaminated, to help trace possible sources of pollution. Karst aquifers are notoriously bad water filters; thus transmission of microorganisms is to be expected. There is an extensive literature on the movement of bacteria through porous media, reviewed by Romero (1970), Gerba *et al.* (1975) and contributors to Gospodarič and Habic (1976). Bacterial species used for groundwater tracing include *Serratia marcescens*, *Chromobacterium violaceum* and *Bacillis subtilis*. These and others are discussed by Käss (1998), who also provides information on the requirements for using bacteria as water tracers. He reviewed tests in karst, although these are rare.

Ions in solution

Naturally occurring chloride has long been used as a tracer to determine the interface of seawater with fresh water in coastal aquifers and to detect possible intrusion of seawater into bore water supplies. Variations in concentration of the calcium ion have also been used to determine flow-through times (Pitty 1968b) and flow-through routes (Gunn 1981b, 1983) especially in the vadose zone. Hydrochemistry of spring waters has also been used to characterize carbonate aquifers, as outlined in section 6.7 (Bakalowicz 1977, 2001, Andreo *et al.* 2002). The provenance of water emerging at springs can also be determined by concentrations of different water-quality characteristics, as discussed by Swarzenski *et al.* (2001) in the case of Crescent Beach Spring in Florida, although environmental isotopes are particularly valuable in that regard.

Environmental Isotopes

The use of stable and radioactive isotopes in hydrology became well established in the second half of the twentieth century. The main applications of environmental isotopes to groundwater hydrology are:

1. to provide a signature of a particular groundwater type that can be related to its area of origin;
2. to identify mixing of waters of different provenance;
3. to provide information about throughflow velocities and directions;
4. to provide data on the underground residence time (age or age spectrum) of the water.

In order to be able to use natural isotopes effectively, the reasons for their distribution in natural waters and processes governing it must be understood. These points are discussed in publications by Fritz and Fontes (1980), Clark and Fritz (1997), Kendall and McDonnell (1998) and Moser (1998a).

The term **environmental isotopes** refers to the naturally occurring isotopes of elements that are abundant in our environment, although the production of some may be increased by anthropogenic activity. These include isotopes of H, C, N, O and S, which are principal elements found in geological, hydrological and biological systems. Hydrogen possesses ^1H, ^2H (also called deuterium, D) and ^3H (termed tritium, T), the last being radioactive. Oxygen isotopes include two of particular interest in hydrology: ^{16}O and ^{18}O, both being stable. These occur in the water molecule, so are the best conceivable tracers of water. They are found in various combinations, such as $^1H_2^{16}O$, $^1H_2^{18}O$, $^2H_2^{16}O$ and $^3H_2^{16}O$. Other important environmental isotopes commonly used in groundwater research are ^{12}C, ^{13}C and ^{14}C. Less frequently used are isotopes of argon, chlorine, helium, krypton, nitrogen, radium, radon, silicon, thorium and uranium (IAEA 1983, 1984, Moser 1998).

Both ^3H and ^{14}C are produced naturally in the atmosphere, but the production of some isotopes is increased by human activity, thus for example ^3H and ^{85}Kr are emitted from nuclear power stations. Very large emissions of ^3H were associated with atmospheric thermonuclear explosions during the 1950s and 1960s, but the subsequent moratorium on atmospheric thermonuclear testing resulted in considerable diminution of this isotope in rainfall to the extent that ^3H levels have returned close to natural background concentrations. Tritium concentrations are expressed in tritium units (TU), defined as one atom of ^3H per 10^{18} atoms of ^1H. Natural background levels are 4–25 TU, whereas ^3H concentration in rainfall

attained a peak of 8000 TU in the Northern Hemisphere in 1963. The Southern Hemisphere experienced a lower peak a year or two later.

The amount of ^3H in precipitation also shows some important natural variations. For example there is a marked latitude dependence, with values lower by a factor of five or so in the tropics (Gat 1980). Tritium values also slowly increase inland, doubling over a distance of 1000 km. Superimposed upon this is a seasonal variation with a maximum in late spring and summer that is 2.5–6 times higher than in winter. These seasonal variations are potentially important in karst water investigations, because they may provide a signature for the season of recharge.

Radioactive isotopes

Radioactive isotopes are unstable and undergo nuclear transformation, emitting radioactivity. Their decay is spontaneous and unaffected by external influence. It occurs at a unique rate for each radioisotope, defined by its half-life $T_{1/2}$, which is the time required for one half of the radioactive atoms to decay following an exponential law

$$N = N_0 e^{-\lambda t} \tag{6.49}$$

where N is the number of radioactive atoms present at time t, N_0 was the number present at the commencement of decay and λ is the half-life or decay constant.

The half-life for ^3H is 12.43 yr, ^{39}Ar is 269 year and ^{14}C is 5730 yr. Consequently, ^3H can be used to date waters up to about 50 year old, ^{39}Ar is useful for dating waters from 100 to 1000 year old and ^{14}C has an upper limit of approximately 35 000 yr. Fontes (1983), Moser (1998) and Solomon *et al.* (1998) explain how groundwaters can be dated and discuss environmental isotopes suitable for that purpose (the ^{14}C production rate is now established back about 50 000 yr, so the range of the method is taken to extend to 45 000 year or so).

Interpretation of data on the abundance of radioisotopes in spring water depends very much on the model of flow and mixing that is adopted (Yurtsever 1983, Moser 1998). Alternatives include the **piston flow** model at one extreme and the **completely mixed reservoir** model (sometimes termed the exponential or one-box model) at the other. The first is similar to the conduit flow or 'perfect pipe' end-member commonly applied to karst (sections 6.7 and 6.9). Piston flow assumes that recharge occurs as a point injection and that a discrete slug of tracer moves through the groundwater system without mixing. In nature this never occurs perfectly because of tributary mixing and dispersion effects but, nevertheless, the flow behaviour of well-developed karsts is sufficiently similar to it (with pre-storm water flushed out

Table 6.7 Interpretation of tritium data for the Waikoropupu Springs, New Zealand (From Ford, D.C. and Williams, P.W. (1989) Karst Geomorphology and Hydrology; after Stewart and Downes 1982)

	Tritium content (TU)			Flow times (yr)	Turnover times (yr)
Date	Annual weighted mean rainfall	Local runoff recharging aquifer	Main spring	Piston-flow model	One-box model
27 May 1966	34		14 ± 0.9	3–4	7–8
29 July 1972	15	20.1 ± 1.2	15.3 ± 1.9	<1 or 8–10	10–12
4 May 1976	8	11.9 ± 2.0	11.2 ± 1.2	2–4 or 12–14	0–20

first) for such a model often to be taken as an acceptable first approximation. By contrast, the completely mixed reservoir model assumes uniformity at all times, each episode of recharge mixing almost instantaneously. It also assumes stationary conditions with respect to reservoir volume, discharge and infiltration rate.

Actual tracer behaviour in most groundwater systems lies between these two extreme cases. **Dispersive** models have been proposed to describe intermediate situations, taking account of mixing and dispersion within the system, but assuming that pulse variations in tracer output can be related back to concentration variations in recharge events. A cascade of mixing compartments connected by linear channels has been used to approximate this situation, being termed a **finite state mixing cell** model. Individual cells are assumed to behave like well-mixed reservoirs but may be of different volume and concentration. Fontes (1983), Yurtsever (1983), (Gaspar 1987) and Moser (1998) discuss applications and theoretical considerations.

Stewart and Downes (1982) provide an instructive illustration of the different interpretation that emerges depending on the model chosen. They dealt with a karstic artesian aquifer in the Takaka valley, New Zealand, which drains to the Waikoropupu Springs (Williams 1977, Stewart and Williams 1981, Williams 2004). Tritium measurements were made at the springs (mean discharge $15\,m^3\,s^{-1}$, Table 5.10) in 1966, 1972 and 1976 and were compared with tritium values in local allogenic recharge and rainfall (Table 6.7). Tritium increased rapidly in local rainfall to about 40 TU in 1964–65. The result for the Main Spring in 1966 showed the input of this young tritium-enriched water, whereas the 1972 figures revealed the presence of some low-tritium (older) water. Values in rainfall declined steadily after 1971, the 1976 results indicating that not much high-tritium water remained. Stewart and Downes (1982) thus concluded that the springs' outflow has a spread of ages, but that young and old components predominate, the best estimate of throughput time depending on the model adopted and the year of sampling (Table 6.7). Geological circumstances in an aquifer that is capped by impermeable rocks at its downstream end suggest that the intermediate dispersion model would be more realistic than either a piston-flow or a one-box model, but detailed stable isotope data demonstrate the aquifer to be very well mixed (Stewart and Williams 1981). It is likely that the mean age of the outflow increases during low discharge conditions.

Our conclusion is that in many karst aquifers the water may have components of different residence times and so to assign one mean age may be misleading. This point was well made by Siegenthaler *et al.* (1984) for a spring in the Swiss Jura. Strong evidence showed that in periods of baseflow relatively homogeneous old water sustains the spring, whereas after a storm or meltwater event rapid runoff of new water is also important. 'Consequently, it would not make sense to talk of **one** mean age of the spring water.' The main interest is in the average residence time of the older reservoir water. Repeat sampling for ^3H at a spring over several years permits definition of the bomb spike as it passes through the aquifer and so gives a clear indication of mean residence time that can be compared to model interpretations.

Tritium concentration has been largely thought of as a means of dating water, but in view of the currently decreasing ^3H values in karst reservoirs it is becoming less important for that purpose. As a consequence, Rozanski and Florkowski (1979) and Salvamoser (1984) suggested the use of ^{85}Kr ($T_{1/2} = 10.8$ yr) for dating, but this requires very large samples sizes (Clark and Fritz 1997, Moser 1998).

Manufactured radioactive and activable isotopes have also been used as tracers in hydrology. Behrens (1998) pointed out that they are used with reservation because of expense in measurement and radiation protection considerations. The International Atomic Energy Agency (IAEA 1984) has given guidance on the choice of artificial isotopes for water tracing.

1. The isotope should have a life comparable to the presumed duration of the observations. Unnecessarily long-lived isotopes will create pollution, a health hazard and interference if the experiment needs repeating.

2. The isotope should be resistant to adsorption by soils and rock.
3. For most problems, it is desirable to be able to measure the radioactivity in the field; thus γ emitters are preferred in general, although β emitters are also suitable.
4. The isotope must be readily available at reasonable cost.

The public health disadvantages of using artificial radioactive isotopes may be overcome for some species by the post-sampling activation analysis technique, although this has the severe drawback of requiring specialized facilities. Buchtela (1970), Schmotzer *et al.* (1973) and Burin *et al.* (1976) have applied the technique to karst. The method involves injection of an initially non-radioactive tracer (e.g. ^{115}In, ^{32}S, ^{6}Li) that is later sampled and irradiated with neutrons in a reactor. If the tracer is present in the sample it is detectable by its activity. Principles of the method are explained by Behrens (1998).

Nevertheless, despite the numerous advances of the past 40 yr, Burdon and Papakis's (1963) recommendation remains true, that before turning to artificial radioactive tracers, which are costly, hazardous and need skilled handling by personnel supported by atomic laboratories, it is better to try to solve the problem with coloured or chemical tracers. Although tritiated (^{3}H) water would seem a perfect tracer, because it is part of the water molecule itself and is identifiable to $1:10^{15}$ or less, using the first criterion above, ^{3}H is immediately eliminated. *In situ* detection (criterion 3) is also not possible.

Stable isotopes

Stable isotopes undergo no radioactive decay. Deuterium, ^{18}O and ^{16}O occur in the oceans in concentrations of about 320 mg L^{-1} HDO, 2000 mg L^{-1} H$_2{}^{18}$O and 997 680 mg L^{-1} H$_2{}^{16}$O. Variations in these proportions are measured by mass spectrometry and compared with the composition of a standard termed VSMOW (Vienna Standard Mean Ocean Water). The ratios D/H and ^{18}O/^{16}O are expressed in delta units, which are parts per thousand (per mil) deviations of the isotopic ratio from the standard

$$\delta\text{‰} = \left(\frac{R_{\text{sample}} - R_{\text{standard}}}{R_{\text{standard}}} \right) \cdot 1000 \qquad (6.50)$$

where R is the isotopic ratio of interest. Thus a sample with δ^{18}O $= +10$‰ is enriched in ^{18}O by 10‰ (i.e. 1%) relative to VSMOW. Values for D usually fall in the range -50 to -300‰ and for δ^{18}O in the range $+5$‰ to -50‰. The accuracy of measurement is approximately ± 2‰ for δD and ± 0.2‰ for δ^{18}O.

Differences in the isotopic composition of water samples reflect fractionation processes that occur in the hydrological cycle. The heavy isotope molecules (HDO, H$_2{}^{18}$O) have slightly lower saturation vapour pressures than the ordinary water molecule (H$_2{}^{16}$O); hence when changes of state occur during evaporation and condensation, a slight fractionation takes place (discussed in detail by Clark and Fritz (1997) and Kendall and Caldwell (1998)). For example, when evaporation occurs from an open water surface the vapour is depleted in heavy isotopes compared with the remaining unevaporated water; whereas when condensation occurs, the initial precipitation is slightly enriched so that later precipitation becomes increasingly depleted with respect to VSMOW. Ingraham (1998) explained the processes that cause isotopic variations in precipitation. From a worldwide analysis of oxygen isotopes in precipitation, Bowen and Wilkinson (2002) defined a negative correlation between δ^{18}O values and station latitude, which they considered to result from the cooling and distillation of water vapour during transport in a polewards direction. Temperature is an important factor in fractionation: the lower the temperature the greater the depletion in heavy isotopes. This influence is reflected in both latitude and altitude differences in heavy isotopic composition of precipitation. Generalizing global data, a linear relationship (the 'meteoric water line') emerges (Craig 1961, Fontes 1980, Rozanski *et al.* 1993):

$$\delta\text{D} = 8.17(\pm 0.07)\delta^{18}\text{O} + 11.27(\pm 0.65)\text{‰} \qquad (6.51)$$

Regional variants can also be identified – for example a Mediterranean curve (Gat and Carmi 1987).

It is evident that four general rules apply to stable isotope fractionation in precipitation:

1. precipitation becomes isotopically lighter with increasing latitude and altitude;
2. D and ^{18}O in precipitation show a good linear correlation (equation 6.51);
3. in mid-latitudes precipitation is isotopically lighter in winter than in summer;
4. enrichment of D and ^{18}O can occur in lakes and pools because of evaporation.

The relationships which underlie these rules and result in the natural labelling of water in stable isotopes can be summarized as in Figure 6.41.

The relationships that exist in the natural distribution of stable isotopes in water have been valuable in determining the recharge areas and recharge seasons of groundwaters, especially where allogenic inputs are involved (Figure 6.42). Autogenic recharge is usually so well mixed in the epikarst that seasonal patterns and recharge events are difficult to identify (Yonge *et al.* 1985, Even

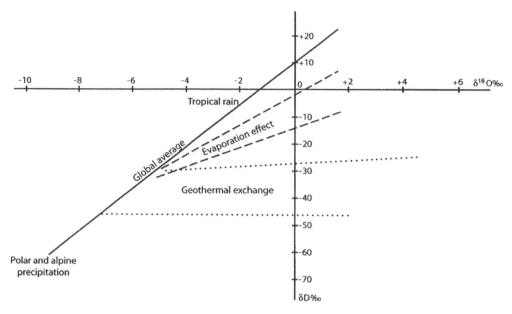

Figure 6.41 Oxygen-18 and deuterium relationships in natural waters. Reproduced from Fontes, J., Environmental isotopes in groundwater hydrology. In P. Fritz & J. Fontes (eds.). Handbook of Environmental Isotope Geochemistry Elsevier, Amsterdam; vol. 1, 75–140 © 1980 Elsevier.

et al. 1986, Williams and Fowler 2002). Mixing of groundwater bodies and leakage of water from lakes and reservoirs can also be discerned (Fontes 1980). Illustrations of the use of these methods in karst aquifers can be found in numerous publications, e.g. by Dincer *et al.* (1972), Bakalowicz *et al.* (1974), Zötl (1974), Moser *et al.* (1976), Stewart and Williams (1981), Celico *et al.* (1984), Stichler *et al.* (1997) and Williams (1992, 2004). However, Margrita *et al.* (1984) pointed out that the concepts of theoreticians do not always appear to be well understood by the users of isotopic tracers, and they identified the kinds of errors that may appear in interpretation of results. A significant limitation applying to the use of stable isotopes in hydrogeology is that they can give information on the general provenance of waters, but do not identify the exact source. Only point-to-point tracers such as dyes can do that.

6.10.4 Pulses as tracers

Natural pulses

A pulse is a significant variation in water quantity or quality. A storm is the usual natural trigger mechanism for its generation, although snow and glacier melt also produce pulse waves (e.g. Meadows Creek in Figure 6.22). Pulses were being used as karst tracers by the mid-19th century (e.g. Tate 1879), although their potential to discriminate details of conduit network geometry was formally

developed by Ashton (1966). He showed how input pulses combine to produce complex output signals and explained how the network geometry might be determined by their interpretation. In practice, it has not yet proved possible to interpret spring hydrographs with confidence in the detail that Ashton's theoretical arguments suggest, but pulse-train analysis as a means of water tracing has proved strikingly successful in a system too difficult for conventional dye tracing (Williams 1977). Further elaboration of pulse-train techniques can be found in papers by Brown (1972, 1973), Christopher (1980), Smart and Hodge (1980) and Wilcock (1997).

Given an input pulse, the related output pulse will vary according to the transfer function of the system, as shown in sections 6.5 and 6.7. Whereas natural and artificial labels such as isotopes and dyes approximately measure the rate of flow of the water being traced, pulses almost always exceed it by a significant amount. A flood wave (discharge pulse) will travel as a **kinematic wave** down an open vadose passage, and as a **pressure pulse** through a phreatic conduit. Large kinematic waves travel faster than smaller ones; both travel more quickly than the water itself, especially through pools. Transmission of a pressure pulse through a flooded tube is almost instantaneous(at the speed of sound). Therefore it is necessary to distinguish between the **pulse through time** (i.e. hydraulic response time) and the **flow through time** of the system. For example, the Komlos spring pulse through

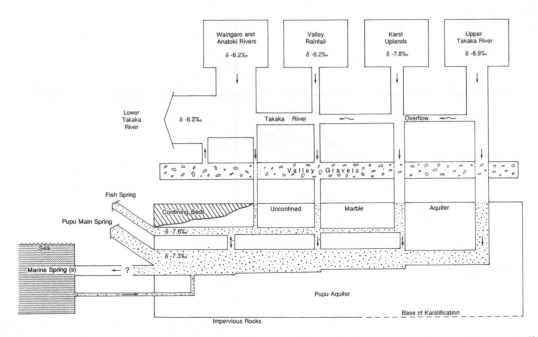

Figure 6.42   A schematic model of the Waikoropupu karst groundwater system, New Zealand, using mass balance of $\delta^{18}O$ to estimate the proportion of recharge from different sectors of the basin. Reproduced from Williams, P.W., Karst hydrology. In M.P. Mosley (Ed.) Waters of New Zealand, Hydrological Society, 187–206, 1992.

time is about 6 h, whereas the flow through time of a tracer dye is roughly 40 h (Figure 6.32).

Brown's (1972, 1973) work on the Maligne basin, Canada, is a particularly fine illustration of the application and development of these techniques. Whereas Ashton (1966) used aperiodic and unique pulses, Brown used time-series analysis on long-term discharge data. The Maligne valley contains Medicine Lake, an intermittent internally draining lake that in winter reduces to small ponds, and overflows about one year in two for a few weeks in summer. The lake is at 1500 m a.s.l. and drains to a line of springs 16 km distant and 410 m lower in altitude. Brown undertook a cross-covariance analysis of lake input and springs output data and identified a negative cross-covariance peak at −20 h. He explained this as the output of daily snowmelt cycles occurring in the lower one-third of the basin (i.e. closer to the springs than the input from the lake upvalley). A subtle secondary 'peaking' at +70 to +124 h was considered to be the response to known flow inputs from the lake, and was later confirmed by a dye trace that appeared +80 to +130 h after injection, peaking at +90 h. Since the travel time of the dye was of the same order as the pulse through time of the river, Brown was able to conclude that open channel (vadose) conditions prevailed underground at the time of the experiments. Later work by Kruse (1980) showed the system's response to vary with discharge conditions.

Artificially generated pulses

Ashton (1966) discussed the use of artificial flood waves produced by collapsing small temporary dams at stream-sinks. Periodic water releases in the course of hydro-electric power generation yield much larger pulses that have been traced for a considerable distance. In the Takaka valley, Williams (1977) showed by cross-correlation analysis that kinematic waves produced by releases from a hydroelectric dam take 5 h to travel about 15 km down an open natural channel to enter a marble aquifer and then 9–11 h to travel a further 20.2 km to the Waikoropupu artesian spring (mean discharge 15 m³ s⁻¹). The karst section consists of a recharge zone where the river loses flow into its gravelly bed for about 10 km and then a confined zone where pressure pulse transmission may be assumed. Whereas the tritium concentration of the spring water suggests a minimum flow through time of 2–4 year (Table 6.7), the pulse through time is less than half a day; and whereas the gradient in the Maligne system analysed by Brown is about 26 m km⁻¹, in the Takaka valley it is a maximum of 2.7 m km⁻¹. Thus pulse-train analysis is an effective means of demonstrating point-to-point connections over long distances with low hydraulic gradients and through large flooded zones (the Takaka marble aquifer probably has a phreatic water volume of 1.5–3.8 km³).

Pulse travel times in these investigations ranged from $3\,km\,h^{-1}$ for kinematic waves travelling down the steep channel of the upper Takaka River to approximately $2\,km\,h^{-1}$ through the marble aquifer of the valley. This compares with 0.2–$1.45\,km\,h^{-1}$ in the much steeper Maligne system under vadose conditions.

Tracer studies in karst aquifers such as we have discussed above have enabled some very detailed models to be constructed of conduit systems leading to springs. Smart's (1983b) proportional geometric model of the Castleguard II conduit system and Crawford's (1994) model of the Cave Creek system are outstanding examples of the use of combined water-tracing technology (several fluorescent dyes with repeated traces, continuous fluorometry, natural discharge pulse analysis, isotope and chemograph analysis) for elucidating inaccessible conduit systems in inhospitable terrains (Figure 6.43).

6.11 COMPUTER MODELLING OF KARST AQUIFERS

There are two main reasons for trying to model karst aquifers:

1. to characterize and understand the system (usually as an aid to improving management of groundwater resources);
2. to try to reproduce the evolution of its characteristics.

We know that karst aquifers have features in common: most water is stored within the primary matrix and secondary fissure system (Table 5.2), yet most water is transmitted through tertiary conduits (Table 5.7); and we know that karst systems have attributes that distinguish them from non-carbonate porous and fissured aquifers:

1. a general lack of permanent surface streams;
2. the presence of swallow holes into which surface streams disappear;
3. the existence of underground channels (conduits) with fast flowing water;
4. the existence of large springs where groundwater reappears at the surface.

We know also that the arrangement of equipotentials and streamlines contrasts markedly in well-karstified aquifers from that in typical porous media (Figure 6.44). Thus unlike standard non-carbonate porous and fissured aquifers, karst aquifers cannot be reduced to a simple elementary volume that ignores the whole drainage structure. The karst system connects the recharge area to the outflow spring (or springs), and karst processes direct groundwater towards the spring(s) along flow paths

that have a hierarchical order. The physical system is defined by its structure (organization of flow paths), hydraulic behaviour (response to recharge) and its evolution (stage of development), and the framework of the system is the rock mass in which it is developed. These aquifers are likely to have turbulent flow components, which immediately presents a problem to the modeller, because most numerical groundwater models are based on Darcy's Law, which assumes laminar flow. Although numerical codes for simulating aquifer evolution generally incorporate turbulent flow, it is seldom included in codes used to simulate groundwater flow in karst at a regional scale. An illustration by Worthington (2004) using the case of the Mammoth Cave groundwater basin (Figure 6.39) highlights the difficult problem confronted by the hydrogeologist trying to model a karst aquifer. Numerical simulation of the aquifer using a standard equivalent porous medium model (EPM) predicts the paths followed by tracer dyes from their various points of injection as shown in Figure 6.45a. But the inaccuracy of this simulation is revealed in Figure 6.45b by the actual paths followed by the dyes (see Ray and Currens (1998a,b) for more detail). Uncritical use of the EPM for management purposes could clearly lead to serious errors. Therefore we need to see what modelling tools are available that might better suit our needs.

Kiraly (1998), Teutsch and Sauter (1998), Kovács (2003) and Worthington and Smart (2004) have reviewed attempts to model karst aquifers. From this we see that in the main two different and complementary approaches have been taken:

1. a global, 'black box' (or probabilistic) approach that focuses on input event and output response (such as rainfall recharge and resulting spring outflow), but takes into account local field observations of flow and transport processes;
2. a deterministic (or parametric) approach that takes into account theoretical concepts of aquifer structure, the physics of each mechanism involved in the transfer and flow of water, and from that tries to simulate the hydraulic behaviour of the aquifer.

These are also known as numerical models.

The earlier discussion in section 6.5 on spring hydrograph analysis provides examples of the black-box approach. These models subsume the flow processes within the aquifer into response functions and are non-physical in nature, so often lack predictive power and are limited when spatially variable output phenomena (e.g. well water levels) have to be modelled. Hence distributed parameter models, which are deterministic models based on physical relationships, represent a viable alternative.

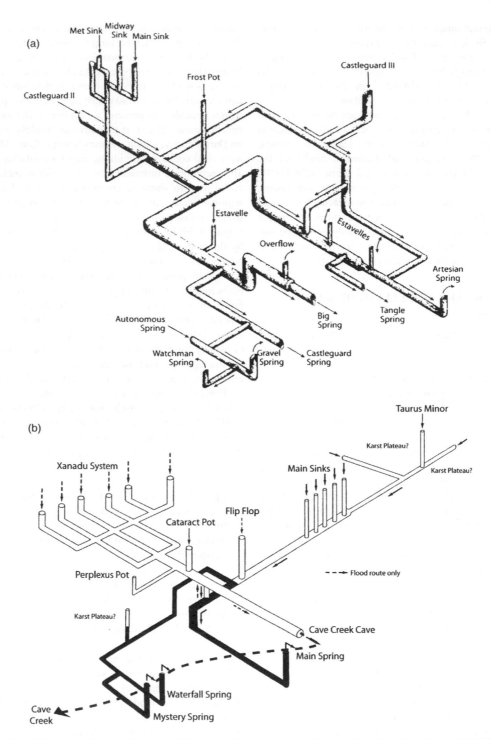

Figure 6.43   (a) Proportional geometric model of Castleguard II conduit system, Canada. Reproduced from Smart, C.C. (1983b) The hydrology of the Castleguard Karst, Columbia Icefields, Alberta, Canada. Arctic and Alpine Research, 15(4), 471–86. (b) Cave Creek floodwater maze system, New Zealand. Reproduced from Crawford, S.J. (1994) Hydrology and geomorphology of the Paparoa Karst, north Westland, New Zealand. Auckland Univ. Unpublished Ph.D. thesis, New Zealand, pp. 240.

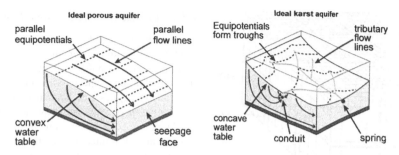

Figure 6.44 (a) Equipotentials and flowlines in an ideal porous aquifer. (b) Equipotentials and flowlines in an ideal karst aquifer. Reproduced from Worthington, S.R.H. (2004) Hydraulic and geologic factors influencing conduit flow depth. Caves and Karst Science, 31(3), 123–34.

However, when considering the adequacy of alternative modelling approaches it is important to be clear about the objective: whether it is to model flow (hydraulic heads, groundwater fluxes and spring discharge) or to model transport (flow direction, velocity and destination). One should also recognize that transport models often are inadequate in karst because of the strong control exerted by fracture and conduit systems of unknown location. These problems and others are confronted in Palmer *et al.* (1999).

Building an appropriate conceptual model of the aquifer is the most important part of the whole modelling process. Figure 6.46 shows an early conceptualization by Drogue (1980). However, the amount of detail required depends on the scale of interest: flow in individual fissures as compared to flow in the entire aquifer. For aquifer-scale modelling, which is our concern here, understanding the pathways and stores involved in water movement to the spring and how the geometry of the system influences flow is central to conceptual modelling. Digital modelling requires us to come to grips with what is happening in the field, and every design step requires an intimate knowledge of the physical system and the processes operating within it.

An important part of a conceptual model is the definition of its boundary conditions, which are along all edges of the domain being simulated. Domenico and Schwartz (1998) explain that there are two commonly used boundary conditions: (i) specified hydraulic head boundaries and (ii) specified flow boundaries. Sometimes these can be readily associated with natural hydrogeological boundaries, such as an underlying aquiclude (impermeable basement), or a no-flow boundary such as a watershed or faulted margin. Inflow and outflow boundaries are also important. An input boundary could be specified as having constant hydraulic head conditions (e.g. an influent stream), although in practice rates of recharge (and discharge) vary over time and space, so are difficult to measure and define. A transient model requires hydraulic head to vary as a function of time.

There are four major conceptual approaches to modelling karst groundwater flow:

1. the **equivalent porous medium** (EPM);
2. the **discrete fracture network**;
3. the **double porosity continuum**;
4. the **triple porosity** approach – matrix, fracture and conduit.

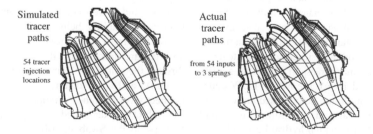

Figure 6.45 (a) Paths followed by tracers injected at 54 locations in the Mammoth Cave groundwater basin as simulated using an equivalent porous medium model. (b) Actual paths followed by tracer dyes determined by field investigation in the Mammoth Cave groundwater basin. These cut across the simulated paths because they follow karst conduits to springs. Reproduced from Worthington, S.R.H. (2004) Hydraulic and geologic factors influencing conduit flow depth. Caves and Karst Science, 31(3), 123–34.

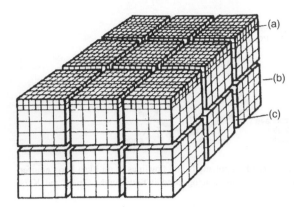

Figure 6.46   Conceptual model of a karst aquifer showing: (a) a highly permeable upper zone (epikarst) with closely spaced fissures; (b) largely unweathered blocks with fissures that permit slow flow; (c) high-permeability karst conduits that permit rapid flow. Reproduced from Drogue, C., Absorption massive d'eau de mer par des aquiferes karstiques cotiers. In Günay, G., Johnson, A.I., and Back, W. (eds). Hydrogeological Processes in Karst Terranes, IAHS Pub 207, 119–128. © 1990 International Association of Hydrological Sciences.

We will briefly discuss each in turn, while recalling that the principal differences between single-, double- and triple-porosity aquifers are outlined in Table 5.11.

6.11.1 Equivalent continuum models

These are appropriate for well-connected porous media or fractured systems at a fairly large scale where information is sought on average behaviour rather than on details of flow paths. They treat large areas as having uniform structural and hydraulic properties and so do not adequately represent the complex flow field in well-karstified heterogeneous and anisotropic conditions. Domenico and Schwartz (1998) provide a description of groundwater flow simulation as applied to standard porous media. They outline methods used to formulate a finite-difference equation for aquifer simulation and, in particular, explain the groundwater industry-standard MODFLOW family of codes, which are digital models designed to calculate head, flow patterns and contaminant transport within an aquifer (Harbaugh and McDonald 1996). The MODFLOW codes are based on a single-continuum porous medium concept and use a three-dimensional grid of columns, rows and layers to identify cells for which hydraulic conditions are simulated. In a karst context these models must be used with great care (Figure 6.45), because the assumptions associated with Darcy's Law are assumed to apply (isotropic, homogeneous medium with laminar flow), but we know that this is seldom applicable in karst (section 5.7).

6.11.2 Discrete fracture (or conduit) network models

Discrete fracture models describe flow within individual fractures or conduits, but ignore matrix characteristics. Examples were provided in section 5.1 when discussing flow through pipes and fissures. For example, equations (5.7) and (5.12) describe laminar flow through pipes and fissures, respectively, and equations (5.8), (5.9) and (5.11) describe turbulent flow through conduits. Such models are of limited use at the scale of the aquifer because of the computational power required to account for every fracture or conduit, but this limitation is disappearing as supercomputers open up the prospect of modelling whole networks of fissures.

6.11.3 Double porosity: matrix and fracture, matrix and conduit, or fracture and conduit

These multiple-continuum models apply to karst aquifers that are considered as being adequately represented by dual matrix–fissure, matrix–conduit or fissure–conduit groundwater systems. They are currently the most useful models for simulating observed aquifer conditions in karst (e.g. Kiraly 1998, 2002, Liedl and Sauter 1998, Adams and Parkin 2002, Kovács 2003). They treat the porous matrix and the fractured medium blocks (or the low-permeability fractured medium and the high-permeability conduit network) as two separate overlapping continua, each with its own characteristic hydraulic and geometric parameters and flow equations (Figure 6.47). The coupling of the two media is handled with a source–link term in each equation, the exchange of flow being controlled by the local difference in potentials. The epikarst can also be treated separately and be linked to the main aquifer (Figure 6.48).

Such models are appropriate in large systems where well-connected fractures provide the dominant flow paths and matrix porosities are high enough to provide significant storage and exchange of water or when conduits provide the dominant flow paths but fracture porosity is high enough to provide significant storage and exchange of water and is well connected to pipe flow, intervening matrix permeability being negligible (Liedl and Sauter 1998). Connectivity controls flow, but when both permeabilities are comparable flow can pass from fractures through the matrix (or from conduits through the fracture system) and so fracture connectivity (or conduit connectivity) becomes unimportant. Sauter (1992) and Teutsch (1993) described the application of such models to karst in the Swabian Alb in southern Germany and Kiraly (1998) and Kovács (2003) provide examples from Switzerland using combined continuum and discrete channel approaches. Kovács (2003) integrated epikarst and saturated zone models and demonstrated that the subcutaneous layer can modify the global hydraulic

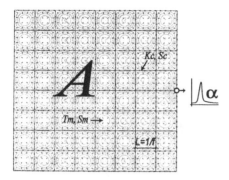

Figure 6.47 Simple conceptual model of a karst aquifer. Characteristic hydraulic and geometric parameters are shown for the low-permeability matrix and intersecting conduit network. T_m and S_m are transmissivity and storativity respectively of the matrix. K_c and S_c represent conductivity and storage coefficient of the karst conduits. The spatial extent of the aquifer is A, the spacing of conduits is L and their frequency f. The recession coefficient of the karst spring is α. Reproduced from Kovács, A. Geometry and hydraulic parameters of karst aquifers: a hydrodynamic modeling approach. Faculté des Sciences, Université de Neuchâtel Thesis, 2003.

response of the entire system by decreasing its recession coefficient. Adams and Parkin (2002) described construction of a physically based model that (i) couples a pipe network to a variably saturated three-dimensional porous medium, (ii) couples surface features such as sinking streams and springs to the conduit system and (iii) includes a bypass flow mechanism to represent fast flow vertical infiltration through the epikarst. This has similarities to the model used by Clemens *et al.* (1999).

6.11.4 Triple porosity: matrix, fracture and conduit

This concept is of particular significance to karst, because storage is often dominant in the rock matrix and fissure system (Table 5.2) whereas flow is achieved mainly through conduits (Table 5.7). In a triple-porosity model aquifer flow is assumed to be laminar in the matrix pores and fissure system, but commonly turbulent in conduits (or macrofissures). The development of realistic triple-porosity models for karst aquifers represents the frontier in karst hydrogeological modelling (e.g. Maloszewski *et al.* 2002). It is the ideal approach for karst, but is currently barely practicable because the data are often unavailable and the level of detail required may be beyond the current computing capability of most computers. An example for dissolutional conduit genesis is shown in Chapter 7.

From this brief review of hydrogeological modelling as applied to karst aquifers, therefore, we can see that for practical purposes we have to deal with less than perfect

modelling alternatives. A valuable discussion of the use of equivalent porous media models in karst has been provided by Scanlon *et al.* (2003). They evaluated two different equivalent porous media approaches (lumped (or black box) and distributed parameter) in a well-karstified, partly confined groundwater basin within the Edwards aquifer in Texas. The MODFLOW code was used for a distributed parameter model, and five cells representing tributary subsystems were used in the lumped parameter model. Transient simulations were conducted using both models for a 10-year period. They found that both the distributed and lumped models simulated fairly accurately the temporal variability in spring discharge, but neither model could simulate the local direction or rate of water flow. The distributed parameter model generally reproduced the potentiometric surface at different times, but more detailed evaluation of the effect of pumping on groundwater levels and spring discharge required a distributed parameter modelling approach. The lumped parameter model has the advantage of simplicity, ease of use and low data requirements, but the distributed parameter model is required to simulate the potentiometric surface, which is necessary to represent regional groundwater flow direction. The distributed parameter model is much more complex, is difficult to parameterize, has large data requirements, and needs longer times to run; however, the MODFLOW code on which it is based is widely used and tested. Key conclusions were that (i) both models could simulate spring flow satisfactorily, but the equivalent porous media models (ii) could not simulate local direction or rate of groundwater flow, because major conduits are not explicitly represented and because turbulent flow is not included, and (iii) could not accomplish objectives such as delineation of protection zones for wells and springs or simulation of point source contamination. The use of these models in karst should therefore be restricted to evaluation of regional groundwater flow issues. Scanlon *et al.* (2003) concluded, like many others, that it is questionable whether any modelling approach can predict direction and rate of groundwater flow from a point source, because many tracer tests demonstrate that directions and rates of flow can be quite variable.

Palmer *et al.* (1999, p. 105) expressed the opinion that 'the heterogeneity of karst aquifers is so severe that it is virtually impossible to acquire sufficient field information to construct a predictive digital model trustworthy enough to allow extrapolation of heads and flow conditions from known to unknown locations, let alone into the future', but 'on the other hand, digital models are well suited to revealing the interactions among water flow, chemistry, and geological setting under idealized conditions'. The purpose of the model is therefore a critical consideration.

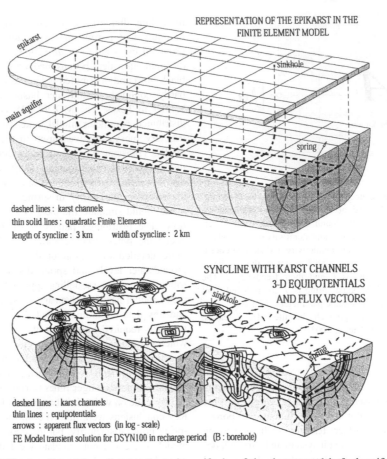

REPRESENTATION OF THE EPIKARST IN THE
FINITE ELEMENT MODEL

epikarst

sinkhole

main aquifer

spring

dashed lines : karst channels
thin solid lines : quadratic Finite Elements
length of syncline : 3 km   width of syncline : 2 km

SYNCLINE WITH KARST CHANNELS
3-D EQUIPOTENTIALS
AND FLUX VECTORS

sinkhole

spring

dashed lines : karst channels
thin lines : equipotentials
arrows : apparent flux vectors (in log - scale)
FE Model transient solution for DSYN100 in recharge period   (B : borehole)

Figure 6.48   (Upper) The coupling of the epikarst to the main aquifer in a finite-element model of a karstified syncline. (Lower) Variation of hydraulic head in a karstified syncline following recharge by concentrated infiltration through the epikarst. Reproduced from Kiraly, L., Karstification and groundwater flow. In Gabrovšek, F. (Ed.) Evolution of Karst: from Prekarst to Cessation. Postojna-Ljubljana: Institut za raziskovanje krasa, ZRC SAZU, p. 155–90, 2002.

We urgently need to improve our capacity to manage karst water resources sustainably. Thus the development of models that adequately represent karst aquifers is vital. Simply because MODFLOW is so widely circulated and used in the hydrogeology world, efforts are being made to adapt the code for karst conduits, trialling what are called 'smeared', 'embedded' and 'dual' approaches. The smeared case drops a high density grid into the basic MODFLOW grid of nodes to provide an accelerated flow-through path under Darcian conditions; the embedded conduit is built of short links between adjoining mesh nodes that have a larger transfer capacity than other routes; and the dual conductivity case has a conduit cutting across the mesh, but with added nodes wherever it crosses that framework, thus permitting easier transfers. We look forward to seeing advances in this critical field of karst research.

Despite the impressive nature of many digital models, we must not forget that they are only representations of reality. They are not susceptible to proof, although their fidelity can be verified by comparing model output with the observed characteristics of the system being modelled, e.g. spring hydrograph or chemograph behaviour, variations in hydraulic head, direction of groundwater flow, and so on. The discrepancy between observed conditions and the ideal model can be a great help in clarifying field conditions (Palmer *et al.* 1999) and provides guidance for further study. Modern modelling techniques also enable us to explore quantitatively the relationships between hydrogeological environment and aquifer properties, and thus enable concepts of conduit development and speleogenesis to be investigated and validated. This is examined further in the next chapter.

7

Speleogenesis: The Development of Cave Systems

7.1 CLASSIFYING CAVE SYSTEMS

In Chapter 5 it was shown that patterns of interconnected conduits created by bedrock dissolution determine the characteristic behaviour of well-developed karst aquifers, differentiating them from the simpler granular (matrix) and fracture types of aquifers. These patterns are systems of caves and proto-caves. Their development is the subject of this chapter.

Dissolutional cave systems are among the most complex of all landforms. This is because they ramify in a great variety of three-dimensional patterns in rock masses that exert many different influences upon their organization, extent and shape. They are further affected by hydrochemical factors dependent upon petrological, tectonic, climatic, biotic and pedologic conditions, and by external baselevel controls. They may survive in the rock as active or relict features after these conditions have ceased to apply, and perhaps be altered under radically different conditions.

Being so varied, there are many different ways in which solution caves are described and interpreted. No single theory of genesis has been able to encompass them all except at a trivial level of explanation and there is no one classification that accommodates all the needs of geomorphologists, hydrogeologists, economic geologists, environmental scientists, etc.

7.1.1 Definitions of caves

The definition adopted by most dictionaries and by the International Speleological Union is that a cave is a natural underground opening in rock that is large enough for human entry. This definition has merit because investigators can obtain direct information only from such caves, but it is not a genetic definition. From Chapter 3, we define a karst cave as an opening enlarged by dissolution to a diameter sufficient for 'breakthrough' kinetic rates to apply if the hydrodynamic setting will permit them. Normally, this means a conduit greater than 5–15 mm in diameter or width, the effective minimum aperture required to cross the threshold from laminar to turbulent flow (section 5.3.2)

Isolated caves are voids that are not and were not connected to any water input or output points by conduits of these minimum dimensions. Such non-integrated caves range from vugs to, possibly, some of the large rooms occasionally encountered in mining and drilling. **Proto-caves** extend from an input or an output point and may connect them, but are not yet enlarged to cave dimensions.

Where a conduit of breakthrough diameter or greater extends continuously between the input points and output points of a karst rock it constitutes an **integrated cave system**. Most enterable caves are portions of such systems. In this chapter we are concerned principally with the building of integrated cave systems, but for brevity will refer to them simply as 'caves'.

7.1.2 Caves and their classification

Tens of thousands of solution caves have been explored, in part at least, and many thousands of them are accurately mapped. Table 7.1 lists some approaches to classifying them. The longest and deepest mapped caves (as at April 2005) are given in Table 7.2 and Figure 7.1. The

Karst Hydrogeology and Geomorphology, Derek Ford and Paul Williams
© 2007 John Wiley & Sons, Ltd

Table 7.1   Some classifications of dissolution caves

By internal characteristics	In relation to external factors
By size: aggregate length or depth or volume	Modes of geological control: rock type (limestone, gypsum, etc.); joint-guided, fault-guided, etc.; horizontal strata, steeply dipping, folded, etc.
By measure of vertical or horizontal dimensions	
By plan form: entrance or niche (abri), chamber (room), linear passage, branchwork, network, anastomosis, spongework, ramiform; multiphase branchworks, rectilinear combinations	By topographic setting: mountain caves, plateau caves, etc.
	By relation to topography: underdrain valley or valley flank, meander cut-off (short cut), connect poljes, foot cave, etc.
By passage cross-section form: circular or elliptical, canyon, breakdown, compound	By role in fluvial system: allogenic river caves, holokarst drains, short-cut caves, combinations, sea cave, etc.
By relation to local or regional water table: vadose, water-table cave, phreatic, compound, relict	By aquifer type: ideal pipe cave → continuum → perfect spongework cave
By categories of deposits: speleothem cave, gypsum (crystal) cave, sand cave, ice cave, archaeological site, etc.	By role in geomorphological and hydrological cycles: active cave → episodic → relict cave (preserved, intercepted, truncated, destroyed)
	By climatic setting: humid tropical, semi-arid, mediterranean, temperate, alpine, arctic, etc.

statistical distributions are highly skewed. This is primarily a function of exploration difficulties. A majority of caves have been explored for less than 1 km and penetrated to less than 100 m in depth; the number of greater examples increases by dozens or hundreds every year, but the form of the distribution curves remains largely the same.

The Mammoth Cave system in Kentucky (see Figure 7.11) has maintained its position as the world's longest set of interconnected passageways for most of the years since the 1840s, when some 25 km were known. It is developed in a thickness of no more than 100 m of limestones, but they are nearly flat lying and the caves extend as multiphase branchworks at several levels beneath sandstone ridges that preserve the older passages. Optimisticeskaya and other great caves in Ukraine (see Figure 7.27) are rectilinear mazes in gypsum strata only 12–30 m thick, but again they are flat lying in low plateaus beneath protective cover strata. Jewel Cave and Wind Cave are three-dimensional thermal water mazes in gently domed limestone hills. Hölloch is a mountain cave in quite steeply dipping limestones, as are caves 8, 11, 12, 14, 15, 18, 19 and 25 in Table 7.2. Lechuguilla and the nearby Carlsbad Caverns (31 km) are ramiform systems in ancient reef rocks overlooking a modern desert. In contrast, cave 9 and 24 in Table 7.2 are anastomosing river caves in flat, very young rocks, now drowned by the post-glacial rise of global sea level. Friar's Hole lies beneath a deep valley in shales and sandstones where the limestone is scarcely exposed at

all. These longest caves thus display a wide range of physical type and setting. They have in common the indefatigability of their explorers.

There is less variety amongst the deepest caverns. It is infeasible to dive below ~300 m in water-filled caves at present; therefore, the deepest known caverns occur in mountain massifs where the drained depth will be greatest. Most are in the alpine ranges of Europe and the Caucasus because these are the most intensively explored. Recently, exploration in the highlands of Mexico has added some tropical contenders. All are alike in being systems of shafts and steeply descending stream canyons that terminate at siphons or breakdown barriers. Only the highly complex Siebenhengste–Hohgant System, Switzerland, occurs in both the longest and deepest lists.

When actually underground most cave scientists categorize caves by the passage shapes – vadose, phreatic or breakdown – that they see about them, or by their deposits. But extensive cave systems may display all of the differing forms and a wide variety of deposits, so that these characteristics are not well suited for general classification.

A majority of karst researchers are concerned with surface landforms or hydrological studies and so classify the unseen plumbing in terms of appropriate external factors, as in Chapter 5 and Table 7.1b. Some geological factors will be stressed later in this chapter. Relationships with topography and with the fluvial systems are of particular relevance to hydrogeologists. Some caves simply underdrain valleys, others abstract water across

Table 7.2 The longest and deepest caves, as at December 2006. (National Speleological Society of America list).

	Number	Cave name	Country	Length (km)	Depth (m)
Longest	1	Mammoth Cave System	U.S.A.	590	116.0
	2	Jewel Cave	U.S.A.	218	193.0
	3	Optimisticeskaja (Optymistychna) (Gypsum)	Ukraine	215	15.0
	4	Wind Cave	U.S.A.	196	202.0
	5	Hoelloch	Switzerland	194	939.0
	6	Lechuguilla Cave	U.S.A.	187	489.0
	7	Fisher Ridge Cave System	U.S.A.	177	109.0
	8	Siebenhengste-hohgant Hoehlensystem	Switzerland	149	1340.0
	9	Sistema Ox Bel Ha (Under Water)	Mexico	144	34.0
	10	Gua Air Jernih (Clearwater Cave-Black Rock-White Rock)	Malaysia	129	355.0
	11	Ozernaja (Gypsum)	Ukraine	124	8.0
	12	Systeme de Ojo Guarena	Spain	110	193.0
	13	Bullita Cave System (Burke's Back Yard)	Australia	110	23.0
	14	Reseau Felix Trombe / Henne-Morte	France	106	975.0
	15	Toca da Boa Vista	Brazil	102	50.0
	16	Shuanghedongqun	China	100	265.0
	17	HirlatzHoehle	Austria	95	1070.0
	18	Sistema Purificacion	Mexico	94	953.0
	19	Zolushka (Gypsum)	Moldova/Ukraine	90	30.0
	20	RaucherkarHoehle	Austria	84	758.0
	21	Sistema Sac Actun (Under Water)	Mexico	80	25.0
	22	Sistema del Alto del Tejuelo	Spain	78	605.0
	23	Friars Hole Cave System	U.S.A.	73	191.0
	24	Easegill System	United Kingdom	71	211.0
	25	Sistema Nohoch Nah Chich (Under Water)	Mexico	68	72
Deepest	1	Krubera (Voronja) Cave	Abkhazia	10	2158.0
	2	Lamprechtsofen Vogelschacht Weg Schacht	Austria	50	1632.0
	3	Gouffre Mirolda / Lucien Bouclier	France	13	1626.0
	4	Reseau Jean Bernard	France	20	1602.0
	5	Torca del Cerro del Cuevon (T.33)-Torca de las Saxifragas	Spain	7	1589.0
	6	Sarma	Abkhazia	6	1543.0
	7	Shakta Vjacheslav Pantjukhina	Abkhazia	6	1508.0
	8	Cehi 2	Slovenia	5	1502.0
	9	Sistema Cheve (Cuicateco)	Mexico	26	1484.0
	10	Sistema Huautla	Mexico	56	1475.0
	11	Sistema del Trave	Spain	9	1441.0
	12	Evren Gunay Dudeni (Sinkhole)	Turkey	0	1429.0
	13	Boj-Bulok	Uzbekistan	14	1415.0
	14	Sima de las Puertas de Illaminako Ateeneko Leizea (BU.56)	Spain	14	1408.0
	15	Kuzgun Cave	Turkey	0	1400.0
	16	Sustav Lukina jama - Trojama (Manual II)	Croatia	1	1392.0
	17	Sniezhnaja-Mezhonnogo (Snezhaya)	Abkhazia	19	1370.0
	18	Sistema Aranonera (Sima S1-S2)(Tendenera connected)	Spain	43	1349.0
	19	Gouffre de la Pierre Saint Martin	France / Spain	54	1342.0
	20	Siebenhengste-hohgant Hoehlensystem	Switzerland	154	1340.0
	21	Sima de la Cornisa	Spain	5	1330.0
	22	Slovacka jama	Croatia	3	1320.0
	23	Poljska jama - Mala Boka System	Slovenia	9	1319.0
	24	Abisso Paolo Roversi	Italy	4	1300.0
	25	Cosanostraloch-Berger-Platteneck Hoehlesystem	Austria		

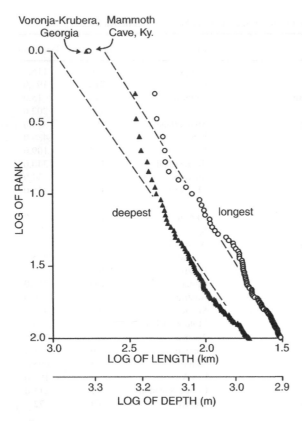

Figure 7.1 Rank/size correlation plots (Zipf diagrams) of the 100 longest and deepest known cave systems. This is a means of predicting size distributions. The log of rank (e.g. No. 1 (log = 0) for Mammoth Cave, Kentucky, the longest known cave, is plotted against the log of its length in kilometres. According to this rank/size rule, Mammoth Cave should be ~250 km in length; it is currently 579 km and gravely distorts the linear relationship, as seen. Lengths of other caves obey the rule quite closely. Cave depth distribution obeyed the rule closely until the summer of 2004, when the Voronja-Krubera system of Georgia was extended from −1710 m to −2160 m, a giant leap downwards.

topographic divides. A very common category is the subterranean piracy that forms a short cut across the neck of an incised river meander or through a spur or bypassing a knickpoint (see Figure 7.6). Some of these are 'ideal pipes', being single straight conduits that neither gain nor lose significant quantities of water during comparatively short passes underground.

Caves are most frequent in the wettest climates. However, large caves do occur in the desert Nullarbor Plain of Australia (section 10.2). There is little relationship between climate and most aspects of cave form

although the largest river cave passages are created by the largest rivers (e.g. Figure 7.2). These are usually allogenic, as in the case of the magnificent Mulu Caves of Sarawak, but Nare Cave and other discoveries in New Britain are largely autogenic. Some very large passages, such as Carlsbad Caverns (see Figure 7.33), display no relation to the modern topography, fluvial channels or groundwater systems at all. They are known because surface erosion has chanced to intercept them.

Classifying caves by these external factors does not explain the structures that the systems adopt or the form of their component parts. Table 7.3 presents a simple classification that is generically based. It is not well balanced. Three quarters or more of caves that have been adequately described and mapped must be placed in the first class. These are caves created by meteoric waters circulating in karst rocks without any unusual constraints such as confinement below aquiclude strata. These we call 'unconfined caves'. Other classes cover those caves formed under unusual geological constraints or where unusual waters are present. They are discussed in the later sections of the chapter.

Many caves display multiple phases of development. One phase ends in a cave and another begins when: (i) a spring position is shifted upwards or downwards sufficiently to compel the creation of extensive new passages; or (ii) there is an externally caused change of water quality or quantity that results in net cave infilling where net erosion prevailed before, or vice versa. Section 7.5 discusses effects of the shift of springs, i.e. local or regional baselevel change. Net erosion–deposition changes are considered in Chapter 8 because they involve cave deposits.

7.1.3 System information

Cave systems are the functional equivalent of river networks in fluvial geomorphology. During the past 60 year morphometric analyses of channel and basin properties have led to major advances in our understanding of them. There has been relatively little of this in speleology because comprehensiveness is lacking in cave-system information in almost every case. Many passages are too small to enter; others are sealed by breakdown, sand, etc. Either it is not known where they originate or where they terminate. In a majority of the great systems much of the passage is water-filled. Cave diving is hazardous, so that only a few active phreatic caves have been well mapped. Most morphometric success so far has been obtained with small features such as dissolutional scallops and alluvial sediment samples, working in local

Figure 7.2　The river cave at Nanxu, Guangxi Province, China. (Photograph by Andy Eavis, with permission.)

sections of caves. These are summarized at the end of this chapter and in the next. There is lacking any substantial body of quantitative results that can link these small-scale findings to the highly generalized system descriptions obtained from the dye-tracing and hydrograph analysis discussed in Chapter 6. That is one of the major challenges for future students of caves.

Table 7.3　Classification of karst solution caves

A Normal meteoric waters	Unconfined circulation in karst rocks (= hypergene caves)	1 Branchwork caves (80% of known caves?)
	Confined circulation in karst rocks, or partial circulation in non-karstic rocks; includes some hypogene caves	2 Maze caves and outlet basal injection caves
		3 Combinations of types 1 and 2
B Deep enriched waters	Enriched by exhalative CO_2 (normally, thermal waters); hypogene caves	4 Hydrothermal caves (~10% of known caves?)
	Enriched in H_2S, etc. (basin waters, connate waters)	5 Carlsbad-type cavities and gypsum replacement caves
C Brackish waters	Chiefly marine and fresh waters mixing	6 Coastal mixing zone cavities
D	Combinations of B or C with A, developing in sequence	7 Hybrid caves

7.1.4 'Speleogenesis' and other sources of information

A major volume entitled *Speleogenesis: Evolution of Karst Aquifers* (Klimchouk *et al.* 2000) has been produced by the International Union of Speleology and the National Speleological Society of America. Dozens of authors from 17 different nations combined their efforts to offer a comprehensive review of cave genesis, in much greater detail and with many more case studies than is possible within the confines of this chapter. Where we describe particular cave genetic features here without giving specific references, it is usually because they can be found in that volume. Following it, *Speleogenesis, a Virtual Scientific Journal* was established by Klimchouk at www.speleogenesis.net. It publishes new papers, abstracts of others, and reviews.

Atlas of the Great Caves of the World (Courbon *et al.* 1989) presents maps, sections and summary overviews of the more prominent caves in 118 nations. In Europe most nations with caves have national and/or regional monographs that provide more details, especially of the geological settings. In the USA there are many state-wide surveys, and the National Speleological Society publishes focused cave symposia from time to time. To our knowledge, Argentina, Australia, Brazil, China, Japan, New Zealand, South Africa, Turkey and Venezuela have also published substantial national and regional reviews. For example, Finlayson and Hamilton-Smith (2003) have edited a book on the natural history of Australian caves.

Speleogenesis (Klimchouk *et al.* 2000) took a very logical approach to the discussion of types of caves, beginning with those that develop in young, newly emerged rocks at the coast, proceeding through the hypogene (deep) cavities that may be formed when these rocks are buried and altered beneath later strata, and concluding with hypergene (unconfined) caves formed when the now diagenetically mature rocks are exposed to meteoric groundwater circulation.

Here we take the reverse approach, dealing with the unconfined caves first, because these are central to understanding much of karst hydrogeology and surface geomorphology. Hypogene and coastal caves are considered more briefly afterwards.

7.2 BUILDING THE PLAN PATTERNS OF UNCONFINED CAVES

The common patterns are single passages that are comparatively short (e.g. cutting off incised river meanders) or the varieties of lengthier branchworks shown in the four left-hand, upper frames of Figure 5.23. Angular patterns predominate where joints or steep faults offer the most penetrable routes for water, and sinuous patterns more like those of river channels where bedding planes are more important. Although some caves may be grid networks or anastomoses in their entirety, more frequently these particular patterns appear as subsidiary components in larger branchworks.

7.2.1 Initial conditions

In section 5.2 we examined the controls on the development of karst aquifers, concluding that the orientation of the maximum hydraulic gradient will reflect a balance between the direction in which resistance to flow is least (i.e. where hydraulic conductivity is maximized) and the direction in which the rate of energy loss is maximized (i.e. the shortest and steepest route). In this section we examine the principles that govern the propagation of solution conduits through the **fissures** (a term encompassing bedding planes and joint and fault fractures) and down the gradient. Before dissolution begins, minimum apertures of connected voids in the fissures are small, $\sim 10\,\mu m$ to $1\,mm$, and their aggregate volume is also small. Available runoff thus is readily able to fill them and the water table is consequently at or near the ground surface. Flow is to dammed springs (Fig. 5.17), the principal type. These are simplified conditions that, because of diagenetic or tectonic history as noted in Chapters 2 and 5, will not always prevail, but they form an acceptable starting point for our discussion.

Figure 7.3 presents a hypothetical situation to provide the hydrogeological framework for this section and the next. It is drawn for the case of steep stratal dip with regional flow in the direction of that dip, because this is easy to illustrate. With the mind's eye, readers may flatten the dip to the horizontal or even reverse it away from the output boundary. Strata may be folded, or the output flow boundary (potential spring line) may be shifted to the strike on one or both sides of the model. The analysis remains the same.

The model and analysis are for bedding planes, because these are areally the most continuous entities permitting groundwater flow in the majority of settings, and normally display greater deviations about the mean aperture than do joints (Figure 2.15). For conduit propagation in normal vertical joints, tip the model on edge.

In this section of the chapter we are concerned only with plan patterns, that is with propagation in the dimensions of length and breadth in one plane (plane A here). Connections with underlying planes such as B, C, D, etc. introduce the depth dimension and are dealt with in the following section. The analysis is based upon a

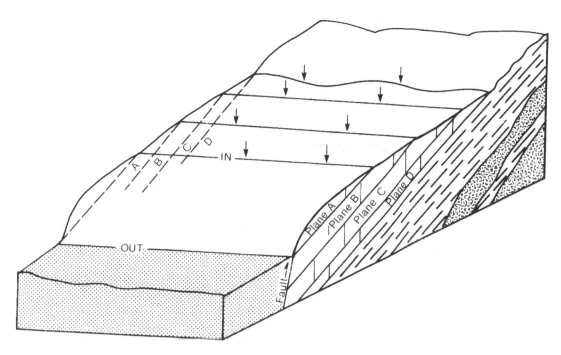

Figure 7.3 Conceptual model structure to explain the development of unconfined cave systems in penetrable bedding planes A to D and joint systems in between them.

systematic sequence of hardware model experiments conducted at McMaster University in 1972–5 by Ewers (1982), and field studies by Ford (1965a, 1968, 1971b). Ewers investigated a variety of initial situations using electrical and sand-model flowfield analogues, and direct dissolutional simulations with Plaster of Paris and salt (see Figure 7.5a). Since 1980, this approach has been replaced by intensive computer modeling, chiefly using finite elements or two-dimensional grids with nodes, and more recently the random aperture convention (Figure 2.15). Dreybrodt (University of Bremen) and his students have been at the forefront of the work (see Bauer 2002, Dreybrodt and Gabrovšek 2002, Dreybrodt 2004, Dreybrodt *et al.* 2005). With great computing capability now available at low cost, there is a bright future for cave genetic modelling: thus far it has largely served to confirm Ewers' hardware findings.

7.2.2 Conduit propagation with a single input

This is the most simple situation (Figures 7.4a and 7.5a and c). Length of the fissure between input and output may be no more than 1 m (in which case we are considering the karren or epikarst scale of aquifer development) or as great as 10 km. The pressure head ranges from the thickness of a single limestone bed to hundreds

of metres. The initial mode of flow is laminar within the plane and can be treated as Darcian.

Distributary patterns of solutional capillary tubes (proto-caves) develop that extend preferentially in the direction of the hydraulic gradient. Their rates of

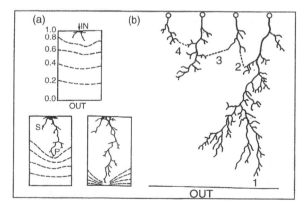

Figure 7.4 (a) The competitive extension of a proto-cave from a single input to an output boundary in plane A. Dashed lines are equipotentials; P, principal (or victor) tube; S, subsidiary tubes. (b) Competitive extension where there are multiple inputs in one rank. Numbers and dashed lines indicate the predicted sequence of breakthrough connections that will occur and their location. (Both figures based on hardware simulations by Ewers 1982.)

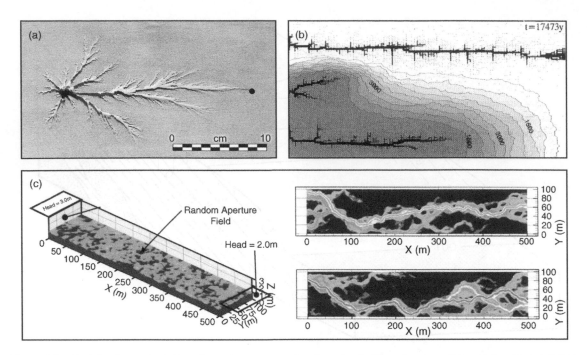

Figure 7.5   (a) A hardware simulation of proto-cave propagation through a bedding-plane type fissure, using Plaster of Paris (gypsum) as the medium and with a constant head applied. Reproduced from Ewers, R.O. (1982) Cavern development in the dimensions of length and breadth. McMaster Univ. PhD thesis, 398 pp. (b) A computer simulation of the multiple inputs in one rank case. Reproduced from Dreybrodt, W. and Gabrovšek, F. (2002) Basic processes and mechanisms governing the evolution of karst, in Evolution of Karst: from Prekarst to Cessation (ed. F. Gabrovšek), Institut za raziskovanje krasa, ZRC SAZU, Postojna-Ljubljana, pp. 115–54. (c) We believe that this is the most elaborate computer simulation of cave development yet attempted. It considers a single input into a Rajaram-type random aperture bedding plane at a depth of 2 m. Flow from the plane into the matrix and out again is incorporated in a three-dimensional mesh 500 m in length, 100 m wide and 3 m deep, that incorporates >400 000 nodes. Two iterations to breakthrough are shown on the right. Using solubility data from Niagaran dolomites, a hydraulic gradient of 0.002 and reasonable P_{CO_2} values, breakthrough is achieved in about 15 000 year. Reproduced from Annable,W.K. (2003) Numerical analysis of conduit evolution in karstic aquifers. Univ. of Waterloo PhD thesis, 139 pp.

extension are determined by solvent penetration distance (section 3.10) and the factors that control it. Actual courses adopted, metre by metre, depend upon geological microfeatures of the plane.

In electrical analogue terms, all tubes are connected in series with the dissolutionally unmodified plane downstream of them. The latter maintains high resistance, so that flow everywhere is slow and small in amount. Variations in cross-section and other properties of the tubes are insignificant while this high resistance element remains.

In every experiment one tube chances to grow ahead of others. It deforms the equipotential field, reducing the gradients at the solution fronts of its competitors ('subsidiaries') and thus slowing their rates of advance. When this 'principal' or 'victor' tube attains the output boundary, three important effects can occur in succession, and rapidly when compared with the slow advance of the proto-caves previously. First, kinetic breakthrough is

quickly achieved, accelerating the rate of enlargement of the tube (Chapter 3). Second, there is hydrodynamic breakthrough if/when flow becomes turbulent, destroying Darcy flow conditions (Chapter 5). Third, when sufficiently large the tube (now a regular karst conduit) can create a trough in the water table. The equipotential field is reoriented onto it, creating yet greater inhomogeneity.

In past writing (e.g. Ford 1971b) these early tubes were termed 'dip tubes', because in steeply dipping strata where they were first studied they are usually oriented close to stratal dip. In reality they are gradient tubes because they are broadly oriented down the hydraulic gradient. Here we name them **primary tubes** because they are the first conduits to develop in the fissure.

The greater the aperture of the fissure before dissolution begins, the straighter and less branched is its pattern of proto-caves. If it is a bedding plane with joints terminating at it or passing through it, distributary proto-caves preferentially extend along the intersections. However, these will become 'victor' tubes only if they are favourably oriented with respect to the hydraulic gradient.

When primary tubes enlarge following dynamic breakthrough, much of the subsidiary branchwork may be swallowed up. However, provided that a pressure head is maintained, surviving elements can continue to extend slowly and may play an important role where later passages develop to create a multiphase cave (section 7.5). In other instances the subsidiary net may be sealed by clay and rendered inert.

Where the output flow boundary is a line or zone, rather than a single spring point, it is common for several downstream distributaries to connect to it and then expand, forming a small distributary network.

This model describes development of the most simple cave, a single passage following a single bedding plane, joint or fault. Many short-cut caves are of this type (e.g. Figure 7.6 left).

7.2.3 Multiple inputs in a single rank

Here we may imagine a number of streams at the input boundary of plane A (Figures 7.4b and 7.5b). This is the common situation in a contact karst, where allogenic streams arrive at the edge of a limestone outcrop, e.g. the Baradla–Domica System, which straddles the Hungary–Slovakia border. Typical competitive development is seen in the figures. Because resistance is not isotropic in the fissure, because input pressure heads will never be precisely equal, or because some inputs are initiated before others, one or more of the input streams extend preferentially. Their flow envelope spaces are increased and those of competitors are diminished.

The closer that parallel inputs such as these are spaced, the greater must be the competition. The steeper the hydraulic gradient, the greater will be the number of inputs with principal tubes that are able to reach the output boundary and so establish separate caves. Very close spacing models the initial conditions for what will become so-called diffuse flow in the epikarst zone described in Chapter 5.

When one proto-cave breaks through, the flow envelopes (or equipotential fields) of its near neighbours are reoriented towards it. Unless already close to the output boundary, the near-neighbour proto-caves will be captured as tributaries. In Figure 7.4b, capture is in the sequence $2 \rightarrow 3 \rightarrow 4$. The principles are further amplified in Figure 7.7. This is drawn for the case of a single point outlet (one spring) that lies to the strike with respect to

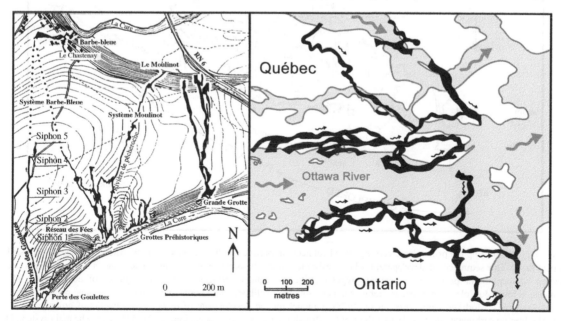

Figure 7.6 (Left) A spectacular example of meander cut-off caves from Arcy-sur-Cure, France. Although now appearing to be an instance of multiple inputs in one rank, development probably has occurred chiefly as a sequence of single inputs, beginning with Grande Grotte, shifting upstream to Systeme Moulinot as a consequence of river entrenchment, and then to Riviere des Goulettes-Barbe-Bleue. (Right) Modern, anastomosing cave passages in two-three master bedding planes under the Ottawa River, Canada. Adapted with permission from Haid, A. (1996) Yonne. *Spelunca*, **62**, 14. © Copyright 1996, la Fédération française de spéléologie.

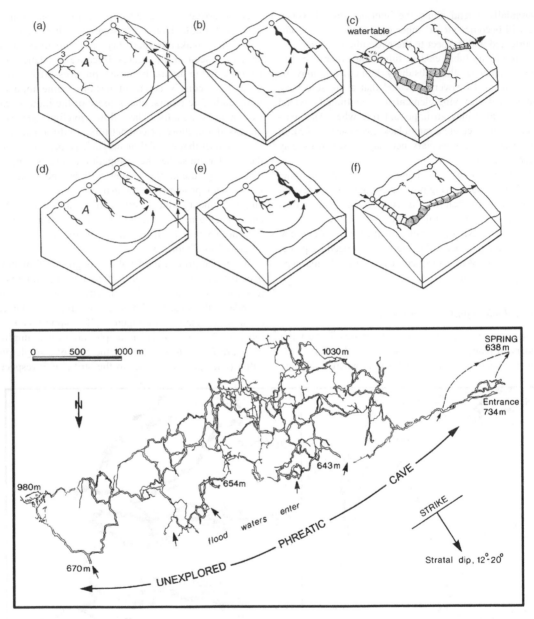

Figure 7.7   Illustrating the principles governing lateral connections between multiple inputs in a rank, where drainage is to the strike. (Upper) (a–c) The sequence of development where resistance in the fissure is high. Single connections at random points occur between adjacent input proto-caves, creating irregular strike subsequent passages. (d–f) The sequence of development where resistance is low. Multiple connections are achieved, increasing the likelihood that a horizontal water-table passage will be created as the cave enlarges. Modified from Ewers, R.O (1978) A model for the development of broadscale networks of groundwater flow insteeply dipping carbonate aquifers. Transactions of the British Cave Research Association, 5, 121–5.(Lower) The main system, Das Hölloch (Hellhole), Switzerland, the fifth longest known cave. The system is largely contained in one major bedding plane dipping at 12–20°. Drainage is to the strike. This is a multiphase system with 400 m of relief. It is seen that principal passages are connected up in the manner of the high-resistance model. Reproduced from Bögli, A. (1970) Le Holloch et son Karst, Editions la Baconniére, Neuchatel.

the near-neighbour inputs. Two different conditions are envisaged. In Figure 7.7a–c the host fissure is tight, with high resistance. The conduit links that are established, first, between victor tube 1 and tube 2, then between tubes 2 and 3 (and so forth) are quite random in their position on the plane. Subsidiaries of adjoining tubes that happen to be nearest each other when the equipotential field is reoriented by newly established connections will be the most likely to link up. In Figure 7.7d–f the fissure offers low resistance. Many alternative linkages can occur at the proto-cave and early cave scales. Those which offer the most direct route to the spring will win, creating a rather straight, near-horizontal cave. In both cases the link-up routes are those of least resistance.

We are now in a position to understand the basic structure of the Hölloch (Switzerland), the fourth longest limestone cave that is known (Figure 7.7 lower). It is a multiphase sequence of trunk conduits draining along the strike in a major, faulted bedding plane that dips at 12–25°. The sinuous, highly irregular patterns on the plan occur because the trunk conduits ascend or descend subsidiaries and angle across the plane between them in the manner of Figure 7.7a–c. Link lengths between subsidiaries of different tubes range from 50 to 250 m. Most of the illustrated passages are now relict, although lower ones may flood during spring thaws, etc. and so may be considered epiphreatic. A similar trunk conveys the modern water through an unexplored phreatic cave at lower elevation in the bedding plane. We may presume that subsidiaries are extending slowly below this active trunk, setting up the situation for a new trunk when external erosion lowers the spring elevation once again.

Strike passages like this are common structural components of caves, although rarely are they as predominant as in Holloch. They are termed **irregular strike subsequents** when displaying the high resistance form of Figure 7.7a–c, and **regular strike subsequents** in the low resistance form. They develop in inclined joint or thrust-fault planes as well as in bedding planes. Figure 7.8 presents a spectacular 'head-on' view of an irregular example that has developed in unloading fractures in marble.

Gabrovšek and Dreybrodt (2000) amplified the single rank situation by modelling some potential effects of mixing corrosion. For example, an input with water equilibrated with P_{CO_2} of 0.05 atm is flanked by inputs with P_{CO_2} merely 0.03 atm. Predictably, the 0.05 atm input breaks through first. Mixing corrosion then accelerates dissolution where the 0.03 atm inputs link up with it, producing more complex patterns of passage connection, with greater enlargement than shown in the simple cases of Figures 7.4 and 7.7. There are no attested field

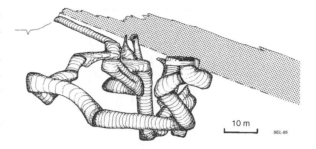

Figure 7.8 A 'head-on' view of the underground outlet of Lake Glomdal, Svartisen, Norway. This shows a complex of 500 m of irregular strike passages developed in steeply dipping marbles. The guiding fissures are low-angle pressure-release fractures, all bedding planes having been sealed during metamorphosis. All except the highest passages remain water-filled today; they were explored and mapped by divers. (Drawing by Stein-Erik Lauritzen, reproduced with permission.)

examples of this complication as yet, but it should be considered when studying cave-passage patterns.

7.2.4 Inputs in multiple ranks

This is the situation most frequently encountered in karst systems. As presented in the highly schematic form of Figure 7.9, it will be best understood if the reader now imagines that plane A is horizontal and that the successive ranks of inputs descend parallel joints to reach it. The analysis, therefore, is appropriate for clint and grike topography, which is small in scale. With somewhat more complex input patterns, it also applies to polygonal karst of intermediate or large scale and to most unconfined caves developed beneath it.

The new element that is introduced in this situation is readily understood. The initial flowfields of further input ranks (rank 2 in Figure 7.9) are obstructed by those closest to the output flow boundary. Proto-caves of further ranks can scarcely connect via joints to plane A until some near inputs have broken through to the output, reducing their resistance and steepening the hydraulic gradient in their rear (Figure 7.9A–C). Lateral connections in near ranks and first connections (principal tubes) from further ranks then proceed simultaneously. High- or low-resistance rules apply as in the single rank situation discussed above. The cave systems will link together headwards and laterally until all of the available karst rock area is incorporated or some minimum hydraulic gradient (for the remaining resistance of the systems) is attained. As systems expand to enterable or greater dimensions their resistance becomes very low so that, given sufficient time and water, most areas of adequately

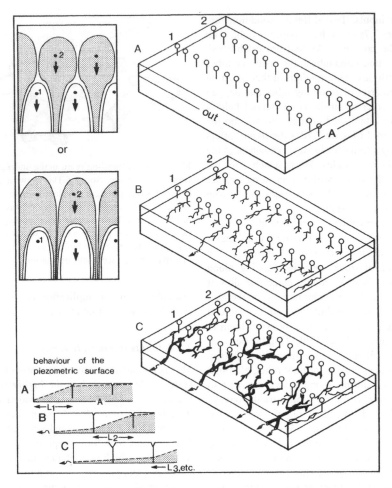

Figure 7.9 The initiation of cave systems with multiple ranks of inputs, drawn for the case of two ranks with constant head conditions. (Upper left) Flowfield configurations limit rear rank access to an output boundary. (Lower left) The cave network is connected up in a headward sequence, with recession of the zone of steep hydraulic gradient.

pure karst rocks will become drained by such cave systems.

Figure 7.10 shows some computer modelling approaches to the multiple input case. Kaufmann and Braun (2000) assume a limestone that is well fractured and also very porous: thus, it is best suited to young, newly emerged limestones or to chalks. Subparallel conduits are initiated very close to the output boundary and extend progressively into the massif. Dreybrodt and Gabrovšek (2002) assume a more normal, low-porosity limestone block 1000 m in length and with a hydraulic gradient of 0.05 (a 5 m head). With reasonable P_{CO_2} and supplies of water, a rank of inputs 500 m from the output boundary breakthrough at it after ~2500 year. Only then does the rear rank (1000 m from the output) begin to extend

significantly. The first hydrodynamic breakthrough to the near rank is achieved at ~5000 year.

Figures 7.9 and 7.10 depict models with uniform conditions (e.g. of pressure head at the inputs) that cannot be expected in reality. Geological, topographic and hydrological variations will always distort it. For example, a larger stream with a higher pressure head in a rear rank may break through first, collecting lesser near-rank streams as its tributaries. Nevertheless, a great many cave systems display patterns of systematic headward and lateral connection such as drawn here. There are obvious similarities to the building of Hortonian stream-channel nets, but Horton's laws have not been successful when tested on caves because of geological distortions and incompleteness of information.

(a)   (b)

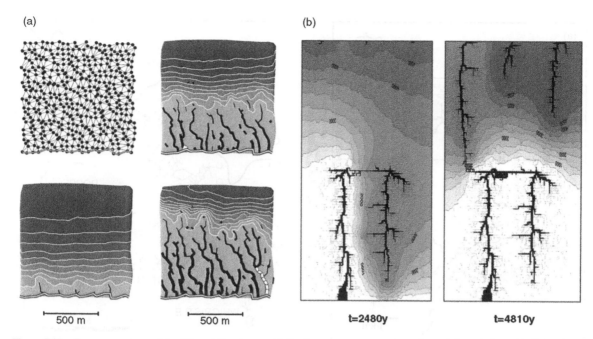

500 m   500 m   **t=2480y**   **t=4810y**

Figure 7.10   Some computer models of the multirank case. (a) For fracture patterns as shown in top left, and also with high primary porosity. Inputs close to the output boundary breakthrough quickly, initiating headward sequences of conduit growth shown in the succeeding frames. Reproduced with permission from Kaufmann, G. and Braun, J. (2000) Karst aquifer evolution in fractured, porous rocks. Water Resources Research, 36, 1381–91. (b) Two successive frames from a model with two near and three rear inputs on a fine fracture mesh. Length is 1000 m and hydraulic gradient is 0.05. Breakthrough from the near rank occurs at 2500 years after initiation. This permits the rear rank to grow; it breaks through to the near rank at 5000 years. Reproduced with permission from Dreybrodt, W., and Gabrovšek, F, Basic processes and mechanisms governing the evolution of karst. In Gabrovsek, F. (ed). Evolution of Karst: from Prekarst to Cessation. Postojna-Lubljana, Carsologica. 115–154, 2002.

Figure 7.11 shows examples of multirank cave patterns below the limestone sink-hole plains and plateaus of the American mid-west. Blue Spring Cave, Indiana (7.11a) is developed in well-jointed rock, whereas Blue Spring Cave, Tennessee (7.11b) is chiefly controlled by bedding planes. Figure 7.11c shows a small part of the Central Kentucky Karst. Spring points have shifted downwards and laterally many times in response to entrenchment by the Green River, and also back up due to its periodic aggradation. As a consequence, the pattern of cave passages is very complicated because much of it belongs to earlier phases and is now relict. However, it is possible to discern incomplete patterns of near and far ranks connecting to the springs. A particularly strong feature is a recent amalgamation along the strike of a far rank (Procter and Roppel Caves) that beheads Mammoth Cave drainage.

7.2.5 The restricted input case

This case completes the range of significantly different karst input–output configurations that can occur. Inputs

are restricted to a line or narrow zone where the karst rock is exposed, most often a valley floor. The spring point is usually at the lower end of the zone. Flow-field geometry constrains the far inputs (Fig. 7.12b), and so the initial cave in plane A is built in a sequence of steps working headward up the valley.

This is a common type of cave. The example in Figure 7.12c is complicated because, once again, the network displays more than a single phase of development and it is incompletely explored. Entrenchment in the Cumberland River has exposed new, lower output points so that, in the latest phase, a new input A in the headward part of the cave has abandoned the subvalley course and propagated beneath the divide toward spring C. These new galleries are not yet enterable in size. Friar's Hole, West Virginia (see Figure 7. 24) is another, lengthier example.

7.2.6 Cave systems and General Systems Theory

In terms of General Systems Theory applied to geomorphology (Chorley and Kennedy 1971), unconfined caves

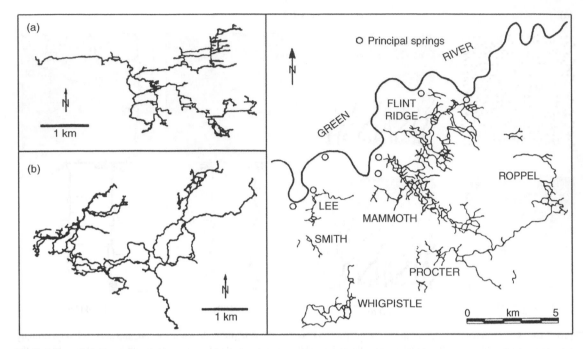

Figure 7.11 The plan patterns of some large multirank cave systems. (Left) (a) Blue Springs, Indiana, a cave dominated by joint control. (b) Blue Springs Cave, Tennesseee, a bedding plane cave. (Right) Mammoth Cave and nearby systems in the Kentucky karst. This is the longest mapped cave system. Only major passages are shown. Although it is a multiphase complex with additional invasion passages the basic structure is of a multirank system draining downdip to the Green River.

are a type of **cascading system**. There is a cascading event each time that a proto-cave connects to a spring or to a pre-existing cave system. In each event, the local flow field and hydraulic gradient are reoriented.

7.3 UNCONFINED CAVE DEVELOPMENT IN LENGTH AND DEPTH

This section completes the formal analysis of the construction of unconfined cave systems by investigating their third dimension, depth, coupled with the length dimension considered above. We are now considering the constraints upon conduits propagating in all planes of the general model (Figure 7.3), or equivalent multifissure sequences in other possible geometric structural arrangements. Sequences of passages may display a predominant vadose or phreatic morphology or appear to have developed along a water table.

7.3.1 The water table controversy

In the first half of the 20th century, the relationship of unconfined cave development to the water table received far more attention than did the genesis of plan patterns

as outlined above. Three conflicting schools of thought can be recognized, as illustrated in the left panel of Figure 7.13. The earliest, associated with Martel (1921) and other European workers, supposed that the locus of most development must be above an already established water table (Figure7.13a). The importance of the soil CO_2 boost was not yet recognized and, because the vadose zone is the first encountered by sinking waters, it was supposed that it would adsorb most of the feeble solvent capacity of sinking streams. It also has the steepest physical gradients, so allowing mechanical corrasion with bedload to take over the enlargement once dissolution had created small openings.

In 1930 the leading American geomorphologist, W.M. Davis, used empirical evidence from cave maps, sections and reports, to argue that many cave passages were developed by ascending groundwater flow. In diametric opposition to the vadose model, he thus proposed that caves developed slowly at random depth beneath regional water tables in the Mature and Old Age stages of his well known Geographical Cycle of Erosion (Davis 1899). Development essentially followed the streamlines of a Darcy groundwater flow net (Figure 7.13b). Bretz (1942) amended the model to include a closing phase of

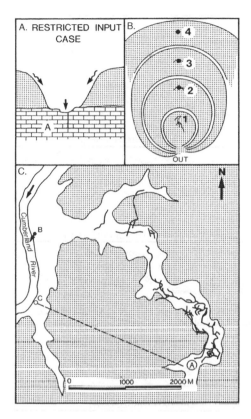

Figure 7.12 The restricted input case of cave system initiation. The example shown is Cave Creek, Kentucky. Cover strata are thick shales. Known caves are a headward sequence that underlie the valley floor and drain to springs at B. A is a late capture that passes beneath the shale interfluve to lower springs at C. Reproduced with permission from Ewers, R.O., Cavern development in the dimensions of length and breadth. McMaster Univ. PhD thesis, 1982.

clay infilling of cave passages beneath the Old Age peneplains because of sluggish circulation due to their very gentle hydraulic gradients.

Swinnerton (1932) was amongst the first to understand the significance of soil CO_2, which weakened one of the principal arguments of the vadose and deep phreatic models. Like them, however, he assumed a stable water table at depth before any cave development. He proposed an opportunistic model in which the likeliest path will be along that water table, generated from the head downstream (Figure 7.13c). Rhoades and Sinacori (1941) recognized that the creation of large dissolutional caves in the rock must affect the water table, lowering it or its gradient in most instances. They proposed an elegant model of development that progressed headwards from

the output spring into the rock, lowering the water table into the conduit as it enlarges. For further discussion of these conflicting ideas see Ford (1998) and Lowe (2000).

7.3.2 Differentiating phreatic and water table types of caves – the Four State Model

The conflict can be resolved by a Four State Model, as shown to the right in Figure 7.13 (Ford 1971b, Ford and Ewers 1978). This is based on the fact that the frequency of connecting initial fissures that are penetrable by water varies from very few in some karst strata to a great many in others. The real variation is as a continuum, but its effect is to permit four distinct phreatic or water table cave geometries to develop, as shown.

At the proto-cave stage the water table is assumed to be close to the surface in the model because there is insufficient effective porosity for it to be lower. As the proto-caves breakthrough, link up and enlarge into caves, the water table is lowered to the stable positions shown, at least under low-stage conditions. Figure 7.14 illustrates the chance nature of the upwards linking process between penetrated bedding planes or other fissures. Location of the link is by chance, where a joint or other fracture gives a short interconnection. This creates a **phreatic loop**, down one bedding plane or low-angle fracture, then up a joint or fault to a higher bedding plane, etc. where the pressure head is lower.

A **bathyphreatic** cave (State 1) makes just a single pass beneath the water table because the frequency of available fissures is too low for any alternative. It has the highest hydraulic resistance. The pass drawn in the figure happens to be quite complex with several looping sections. Successive tubes propagated in planes A, B, C, etc. and then connected as in the multirank input case already discussed. Before any of the others had expanded more than a few centimetres, the tube in plane C (having the headward advantage) captured most available runoff, with the consequence that it alone expanded to enterable size.

Some examples of active and relict bathyphreatic caves are given in Figure 7.15. Note that system information is particularly incomplete. Active (filled) bathyphreatic caves are difficult or impossible to explore. If drained and abandoned, they are most frequently obstructed by detritus towards the base of loops. We do not know the maximum depth attained by bathyphreatic caves but it is certainly greater than 300 m at Vaucluse, or in the Sierra de El Abra, Mexico, where drained higher relicts are particularly well preserved (Ford 2000a). Exploration drilling in many areas has tapped caves filled with young flowing water at depths as great as 3000 m. Some may be bathyphreatic as shown here, but it is likely that

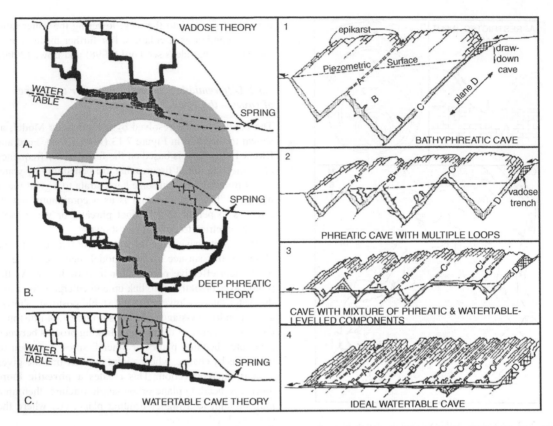

Figure 7.13 (Left) The conflicting theories for the evolution of caves, as they were interpreted in the 1950s. (Right) The Four State Model' showing that vadose, deep phreatic and water-table caves are all possible in unconfined circulation systems. The 'state' depends on the density, penetrability and linkage of fractures and bedding planes, and the orientation of hydraulic gradients. Modified with permission from Ford, D.C. Perspectives in karst hydrogeology and cavern genesis. Bulletin d'Hydrogeologie, 16, 9–29 1998, Université de Neuchâtel Centre d'Hydrogé ologie, Rue Emile-Argand 11, CH-2007 Neuchâtel .

many deep interceptions are of shallower types of caves that have been dropped downwards by tectonic activity.

Multiple-loop phreatic caves (State 2) are created where there is significantly higher fissure frequency. It must be understood that it is the elevation of the tops of higher loops that defines the stable position of the water table and not vice versa. Until the system is quite enlarged, the water table is higher but, as the tubes expand in volume, it falls until arrested locally at the tops of loops. Hölloch is an excellent example where the phreatic loops are built of irregular strike subsequents (Figure 7.15). The vertical amplitude of its looping is ~100 m, rising to 180 m during the greatest recorded floods. In Siebenhengste, Switzerland, amplitude of similar strike loops can exceed 300 m. Other caves loop down the dip and up joints, being built of primary conduits, as in the model of Figure 7.14. Many examples are known where amplitudes range 50–200 m or so (e.g. Czarna Cave,

Poland (Gradziński and Kicinska 2002). In the very gently dipping limestones of the Kentucky Sinkhole Plain the amplitude does not appear to be greater than ~40 m.

Caves that are a mixture of shorter, shallower loops and quasihorizontal canal (i.e. water table) passages represent a third, higher state of fissure frequency and diminishing resistance. The horizontal segments exploit major joints or propagate along the strike in bedding planes as shown in Figure 7.7. Swildon's–Wookey Hole system in the Mendip Hills, England, is an outstanding example. Diving exploration from the upstream end has passed 11 consecutive shallow phreatic loops and is arrested at a twelfth, deeper loop. At least eight loops have been overcome from Wookey Hole, the downstream end (Figure 7.15). At Skalisty Potok, Slovakia, 20 consecutive loops have been dived, to depths up to 25 m, and there are similar examples in Romania (Racoviţă *et al.* 2002). Grottes des Fontanilles, France, has six loops to

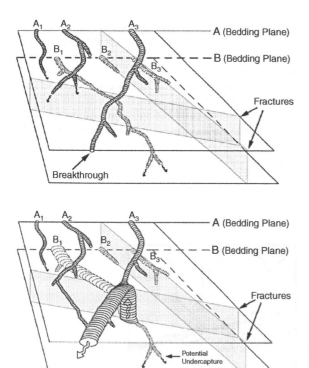

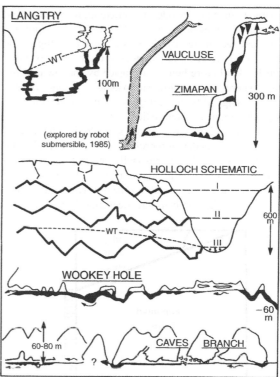

Figure 7.14 A sketch sequence to illustrate the stochastic nature of phreatic loop formation. In the upper frame three proto-caves are extending in each of bedding planes A and B; A_3 has just broken through to the output boundary. In the lower frame, B_1 has connected to A_3 via a lifting shaft where a joint extends between them; B_1 has captured most of the flow and enlarged to explorable size. Its downstream extension will continue to extend slowly unless blocked by silt and clay, and is a prime site for a future undercapture.

Figure 7.15 Examples of phreatic and water table caves. Langtry Cave, Texas, is in horizontally bedded strata After Kastning, E.H. (1983) Relict caves as evidence of landscape and aquifer evolution in a deeply dissected carbonate terrain: southwest Edwards Plateau, Texas, U.S.A. Journal of Hydrology, 61, 89–112. Vaucluse, France is a phreatic lifting passage where water ascends 315 m in a gallery 10–20 m in diameter. It is the outlet end of a deep bathyphreatic system. La Hoya de Zimapan, El Abra, Mexico, is another outlet with at least 300 m of lift but it is now relict. The base is infilled with clay. Upper galleries are 20–30 m in diameter and contain massive speleothem. Das Holloch is an excellent example of a multi-loop phreatic cave, with three relict 'levels'. Reproduced from Bögli, A. (1980) Karst Hydrology and Physical Speleology, Springer-Verlag, Berlin, 284 pp. Wookey Hole, England, is a mixture of phreatic and water tableelements. Caves Branch, Belize, is a great water table cavedeveloped along a polje margin. Reproduced from Miller, T.E. (1982) Hydrochemistry, hydrology and morphology of the Caves Branch karst, Belize. Mcmaster Univ. PhD thesis, 280 pp.

−77 m (Romani *et al.* 1999). Trapped in a short canal section there by rising floods in 2001, cavers experienced air pressure of 2 atm, indicating that the rock above them was airtight, i.e. had no effective primary porosity even at State 3 fissure density.

In State 4, fissure frequency is so high or resistance is so low that low-gradient, most direct, routes are readily constructed through successive ranks of inputs behind the spring points. When sufficiently enlarged these can absorb all runoff; thus the piezometric surface is lowered into them. They become 'ideal' water-table caves.

Ideal water-table caves that are wading or swimming canal passages with low roofs are very common. Hundreds of short examples pass through residual limestone towers on alluvial plains in southern China, Vietnam, Malaysia, Cuba, etc. Large sectors of the Baradla–Domica system, in highly fractured limestones at the Hungarian–

Slovak border, are of this type, as is much of the Caves Branch system, Belize, that is developed in thoroughly brecciated rock. Caves of the Nullarbor Plain, Australia, are other excellent examples of this type, with Cocklebiddy Cave, for example, having been dived just under water table level for more than 6.5 km (Finlayson and Hamilton-Smith 2003).

This Four State Model to differentiate phreatic and water table caves, as shown in Figure 7.13, is simplified and idealized. In reality there is a greater mixing of the types. Ideal water-table passages rarely extend for more than 1–2 km before being interrupted by shallow phreatic loops where local geological conditions change. A cave may be assigned to one category or another according to its predominant characteristics.

Since Cvijić (1918), many speleologists have contended that caves develop preferentially in the epiphreatic zone that is inundated seasonally or in storms by fast-flowing (and chemically aggressive) flood waters. As we have explained it here, such a zone is only of significance to cave genesis where fissure frequency is State 3 or 4.

From careful studies of some major systems in the Alps and Pyrenees, Audra (1994) and Palmer and Audra (2003) proposed the patterns shown in Figure 7.16. Juvenile and perched caves (Figure 7.16 A and B) are discussed as vadose types below. States 2, 3 and 4 cave systems are all assigned to the epiphreatic zone (Figure 7.16 C and D), while State 1 is considered to be an accident caused by aggradation or tectonic lowering (i.e. drowned State 2, 3 or 4). We do not question that both epiphreatic development and deep aggradation can occur; Vaucluse (Figure 7.15) is possibly attributable to aggradation following the Messinian low sea level, for example. However, the El Abra deep systems (State 1) cannot be explained in this manner. The particular examples of

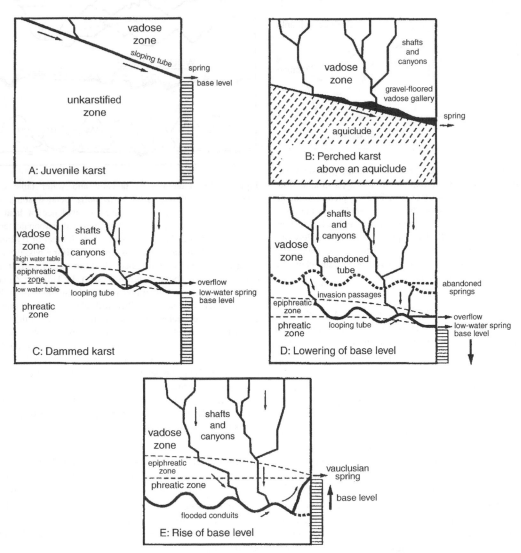

Figure 7.16 The form of vadose, phreatic and epiphreatic caves, as proposed by Palmer and Audra (2004).

State 2 and 3 caves noted above all have large proportions of their known length below the low-stage water table, i.e. are permanently phreatic, not epiphreatic. Castleguard Cave, a well-studied alpine system in Canada, cannot be reconciled with Audra's model, despite the fact that the modern flood amplitude there is as great as 370 m. However, in multiphase caves (discussed below) floods may greatly enlarge one or more upper levels after these have been abandoned by the low-stage flow, i.e. cave enlargement can proceed simultaneously at two different levels (e.g. Ogof Ffynon Ddu, Wales; Smart and Christopher, 1989).

In contrast to Audra, Worthington (2001, 2004) noted that deeply circulating waters may be warmed by geothermal heat, creating lower viscosity that will permit longer solvent penetration before higher order dissolution kinetics begin to inhibit the proto-cave propagation. He argues that deep, State 1 development may be favoured where flow-path lengths are greater than ~3 km and, from analysis of 19 well-mapped sample caves, proposed the empirical relationship presented in equation (5.19). This is a valid possibility but, in perhaps a majority of known cases where the flow path is > 3 km, there are shorter loops between the output and its most distant input points; these tend to breakthrough first despite their possibly higher viscosity because they are significantly shorter in length.

7.3.3 Some effects of geological structure

Structural effects are certainly important in determining the patterns of cave long-sections. Where strata dip quite steeply (2–5° or more) the bedding planes tend to entrain groundwater to greater depths. Their gradients are steeper than most initial hydraulic gradients and so there is a quasi-artesian trapping effect, as Glennie (1954) noted. As a consequence, the deeper types of phreatic caves are favoured.

Where strata are quasihorizontal, bedding planes (the most continuous form of fissuring) may extend to the perimeter of the karst, where they crop out to offer many potential spring points. It is only joints or faults that can guide water to great depth. In addition, conditions are most favourable for perching groundwater streams upon aquitards such as dolomite beds, thick shale bands, etc. States 3 and 4 caves are particularly favoured as a consequence. However, deep phreatic systems can develop where there is strong jointing, e.g. in the Yorkshire Dales, England (Waltham 1970).

In tightly folded rocks (where cave systems extend over several or many folds) fissure frequency is generally high because of the high stressing. Water-table caves are consequently common, as in the folded Appalachians, USA.

Frequency of penetrable fissures may increase where strata are unloaded. This helps to explain why many caves draining plateaus ramify into complex distributary networks where the overburden is reduced close to the springs (Renault 1968), and why State 3 or 4 caves can be well developed in valley sides (Droppa 1966).

Hydration effects in anhydrite inhibit deep phreatic loops; hence gypsum caves tend to be States 3 or 4 where the strata are not confined.

7.3.4 Measure of fissure frequency

There can be no simple assignment of fissure frequencies to the four states because of differing resistances within individual fissures. In a situation of low frequency of penetrable fissuring there may also be low resistance so that State 3 systems evolve, or even State 4 where caves are short, e.g. meander cut-offs. High frequency but high resistance may yield State 2, as in parts of Vancouver Island, Canada (Mills, 1981). Fissure frequency that is measured at the natural surface or exposed in quarries is a poor guide to the effective frequency below the epikarst zone.

In the Mendip Hills, England, all four states of system geometry have developed where the effective porosity is < 1%; hence this measure is not sufficiently discriminating for our purposes.

7.3.5 Increase of fissure frequency with time

At the onset of karstification, initial frequency of penetrable fissures varies within and between formations. With passage of time (and solvent waters) after that it tends to be increased, as pointed out in Chapter 5. As a consequence, later caves in a multiphase complex may display a higher state. For example, in the Mendip Hills over Pliocene–Pleistocene time, system geometry changed from initial State 2 phreatic systems composed of few loops with great vertical amplitude to systems with many loops of lesser amplitude when lowering of spring elevations created second-generation caves underneath them. Repetition yielded systems where gradational processes (outlined in the next section) could produce a State 3 system. In one example the amplitude of early loops was > 50 m, diminishing to ~15 m in a second phase and to < 10 m in a third phase.

This pattern of development can be recognized in the cave fragments that are preserved in many tower karsts. The massive bedding that is necessary to sustain the verticality of the tower walls also favours low fissure frequency. Higher, ancient caves tend to display State 2 features; the modern caves of the floodplain are often State 4. The great tropical river cave systems of

Selminum Tem, Papua New Guinea, and Mulu, Sarawak, have similar histories (Waltham and Brook 1980). However, many alpine systems such as Holloch conserve their high-amplitude, State 2 form through three or more successive phases, so this generalization does not always apply.

7.3.6 Differentiating vadose caves

Primary vadose caves (Bögli 1980) will be the first type to develop where the water table is initially deep in the rock, thus permitting rain or meltwater to invade dry rock. As discussed in section 5.4, this may be because effective porosity is very high, or resistance in fissures is low due to many initial apertures being >1 mm, or the climate is arid. Phreatic elements in the cave morphology are either very limited and rough (**paraphreatic**), or absent. Deep and open fissuring like this can be expected in particular where there is rapid tectonic uplift accompanied by some deformation, i.e. in young mountain systems. The majority of the deepest caves of Table 7.2 thus are simple systems of vertical shafts down fractures, drained by short basal meanders leading to the next shaft. Short sections may exhibit water ponding (e.g. siphons) due to local lithological perching effects. Figure 7.17a shows the vertical profile of the Voronja–Krubera System, Arabika

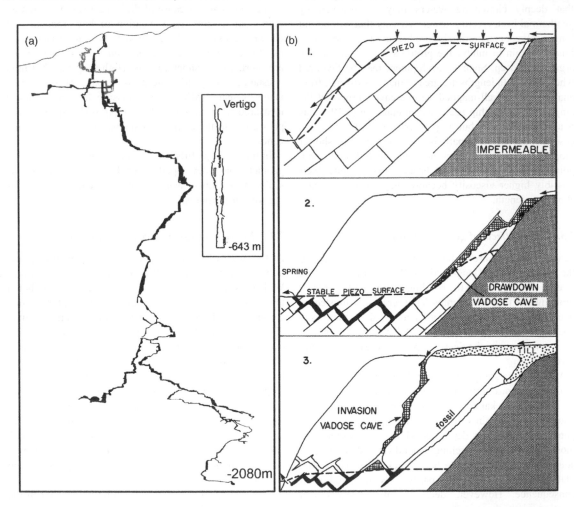

Figure 7.17   (a) The vertical profiles of Krubera-Voronja Cave, Georgia, and Vrtiglavica Shaft (Slovenia) drawn at their scale elevations above sea level (the entrances are at 2250 m and 1910 m a.s.l. respectively). Krubera is currently the deepest explored cave; Vrtiglavica ('Vertigo') is the deepest single explored shaft. Both appear to be largely of primary vadose origin. (From the Ukrainian Speleological Association (Krubera) and R. Stopar (Vrtiglavica), with permission.) (b) The development of the drawdown and invasion types of vadose caves.

Massif, Caucasus, currently the deepest known example at −2160 m. Much of it appears to display primary vadose form. The aptly named Vrtiglavica ('Vertigo') Shaft on Mount Canin in Slovenia is a straight drop of 643 m, lip to floor, the deepest individual pit yet discovered. Many others are known that are > 300 m.

Drawdown vadose caves develop where the water table initially is close to the surface (Figure 7.17b). This is common in low plateaus and hilly regions or where the karst rock is progressively exposed by stripping of insoluble cover strata. The caves are guided by the early networks of phreatic primary tubes. As these break through, connect and then enlarge, the reservoir capacity of the rock increases. The water table then must be drawn down until it finally stabilizes at some minimum gradient to the springs (section 5.5). Although > 90% of the cave volume may be created by later erosion under vadose conditions, the passage plan pattern and skeleton is largely phreatic. If preserved at all, the phreatic morphology is usually in the roof where it may be difficult to inspect. Initial drawdown vadose conditions have probably applied in the majority of the world's karst regions. Lower portions of many of the deep shaft caves can exhibit this control.

Invasion vadose caves develop where new streams invade rock already drained during one or more previous phases of drawdown speleogenesis (Figure 7.17b). The invasion is often a result of karstification of surface streams as they are lowered from an eroding caprock, or it follows from the clogging and sealing of former sink-points, the most potent agency of such infilling being regional glaciation, as hinted in the figure. The effect of glaciation on karst is discussed further in section 10.3.

Juvenile caves that feed free-draining springs (Figure 7.16A) may be of drawdown or invasion vadose type. They are comparatively rare because deeper circulation to and below the spring baselevel is the norm. However, Muruk-Berenice Cave, New Britain, appears to be a good example of simple drawdown with one or two short perched siphons. It is ∼11 km in length and 1178 m deep. Vadose caves perched on an aquiclude (Figure 7.16B) are very common indeed. The basal gallery usually has a drawdown vadose origin. The overlying shafts may be of either type, or a mixture of both.

7.3.7 Extent and magnitude of vadose cave development

The extent of vadose caves (either type) is a function of the depth of the vadose zone and of the many lithological and other effects tending to divert groundwater from a simple, vertical descent. Topographic relief above the springs is most important. The deepest vadose caves occur in mountain areas, as noted. Variations of water-table gradient are also significant. Where vadose caves drain to State 1 or 2 systems variation of resistance in the phreatic passages may cause local variations of tens of metres in the depth of the vadose zone, i.e. in the water-table topography.

In general hydrological modelling, water in the vadose zone is presumed to drain vertically downwards. In many karst regions this is true only in the sense that it does not flow upwards in the zone. Thousands of kilometres of lateral vadose passages are known. The magnitude of vadose caves is a product of the size of their streams and the duration of erosion. The largest occur where large allogenic streams have flowed underground for a long while. Often, simple enlargement plus the gradation processes described below (section 7.4) have converted State 3 or 4 systems into true river caves that rarely or never flood to the roof. In contrast, in many autogenic systems individual streams are tiny because recharge is relatively diffuse and doline frequency is high. Underground, all of the vadose zone (and perhaps more) may be required to effect the amalgamation of streams that is necessary to generate caves of enterable dimensions. Such areas, therefore, may give the impression of lacking vadose caves, but that is only because they cannot be entered.

7.3.8 Review

The following conclusions may be drawn from the models developed above.

1. In the first phase of development an unconfined cave system may be composed entirely of one type between its inputs and its outputs, e.g. bathyphreatic, drawdown vadose, etc.
2. More frequently, in the first phase it will display a combination of one or both vadose types supplying water to one of the four phreatic or water table types (Figure 7.15).
3. In multiphase caves, two or more of the phreatic/water-table types may be present in addition to the vadose types, the later caves often tending to be of higher state.
4. Many systems display some admixture of these types that has developed within a single phase.

7.3.9 Generalizing the Four State Model

In terms of speleogenesis it follows that **six** states may be distinguished in karst rocks. State 0 is that where fissure frequency or effective porosity is too low for unconfined cave systems to propagate in the available geological time. This is true of some marble formations in which karst, therefore, is restricted to the pit and gutter types of karren that do not require cavernous drainage. States 1 to 4 are as specified. State 5 occurs where fissure frequency is too high; the solutional attack is dispersed in innumerable proto-caves or tiny cave channels. Amalgamation to create cave systems of enterable dimensions does not occur. This is true of many thin-bedded formations, some chalks, some corallian and aeolian limestones and some dolomites. It is the ideal diffuse-flow end of this spectrum of karst hydrological states.

7.3.10 Caves in quartzites and siliceous sandstones

Caves also develop in quartzites and siliceous sandstones in situations where the silica cement can be removed by dissolution, and the less soluble residual grains are then efficiently swept away by flowing waters. The grains (usually sand-sized but up to small boulders in some conglomerates) contribute to cave enlargement by mechanical corrasion. It should be noted that without a high hydraulic gradient and low fissure resistance, the mechanical removal of the less soluble particles may not be possible and thus no cave development can ensue. Consequently, caves in siliceous rocks are usually polygenetic rather than strictly karstic in origin, i.e. they cannot develop by dissolution alone. This mode of development actually is much like that envisaged in the early vadose theory of limestone cave genesis (Figure 7.13a). The hydrological settings and cave gross morphology are largely restricted to the simple juvenile and perched types of Figure 7.16.

Caves are quite common in calcareous sandstones, but there are relatively few examples where the cement is silica. Day (2001) describes short and small joint maze and bedding plane caves in sandstone beneath a dolomite aquifer in Wisconsin, i.e. they have developed in the converse of the diffuse-recharge-through-sandstone setting given in Figure 5.23.

Spectacular caves can develop in strong quartzites because, despite the low solubility of these rocks, they are also very resistant to mechanical weathering. Examples are found chiefly above escarpments in wet or seasonally wet climates, where high hydraulic gradients and abundant recharge operate in combination, e.g. the 'Saxon Alps' of Germany and Poland, the High Veldt of South Africa, or Monserrat, Spain. The most celebrated are caves are in the Roraima Formation, a massive Precambrian unit at the Brazil–Guyana–Venezuela border, and in similar strata further south in Brazil (Urbani 1994, Correa Neto 2000). Extensive plateaus there are bounded by vertical cliffs 100–300 m high (section 9.14). Major joints are long, widely spaced, and typically 50–200 m deep before

terminating at a bedding plane. The caves develop as primary vadose shafts down the joints, then drain via smaller floodwater galleries perched on the basal plane. Figure 7.18a shows a typical example. Sistema Roraima Sur, Venezuela, is the largest, having a trunk gallery often ~50 m wide (Šmida *et al.* 2005), i.e. it is on a scale with the great limestone river caves. Many shafts later enlarge along the joints, forming corridor karst (**giant grikes** or **bogaz**) open to the sky. A few caves are more simple juvenile features without shafts, initiated in dipping bedding planes and later entrenched below them, e.g. Lapao Cave, Brazil. Cueva Ojos de Cristal, Venezuela (Šmida *et al.* 2003) is unusual in displaying four separate inlet streams that converge in a spectacular anastomosis on two or more bedding planes. The longest mapped quartzite caves have 4000–11,000 m of passages. The deepest are 300–500 m.

7.3.11 Caves in glacier ice

Caves that are morphologically identical to sequences of shafts and meandering basal drains in limestone can develop where meltwater streams sink into glaciers. However, these are also not strictly karst features because they are formed by melting rather than dissolution. The shafts (**moulins** or glacier mills) are often located on old crevasses annealed by flow of the ice. Caves in shallow glaciers may terminate in river cave passages at the glacier bed, creating systems directly analogous to the perched karst caves of Figure 7.16B. In deeper glaciers, the passages tend to pinch out in the zone of plastic flow below 50–100 m in temperate ice. Badino and Romeo (2005) reports on unusually shallow galleries (at 5–20 m depth) that could be followed for a distance of 1150 m downstream of the entrance moulin in a Patagonian glacier, a record. Glacier caves are also enlarged by sublimation by air currents caused by the cascading water; spectacular patterns of sublimation scalloping (section 7.9) are seen on many walls and ceilings. In polar ice masses glacier snouts may freeze to the bed in winters, impounding late meltwaters which then back up and drown the lower shafts (Schroeder 1999).

Glacier caves develop and close again very quickly if the ice is flowing. Figure 7.18b shows the progress of an example on the Gornergletscher, Switzerland, that was only 6 m deep when first explored on 15 July 1999. It was 60 m deep 3 months later (Piccini *et al.* 1999). Some snout caves in ice at the pressure melting point (temperate ice) can survive for a decade or more and attain diameters of 10 m or more. Caves in stagnant ice or in **firn** (densified snow) can be maintained for several decades if allogenic streams flow into them, e.g. the

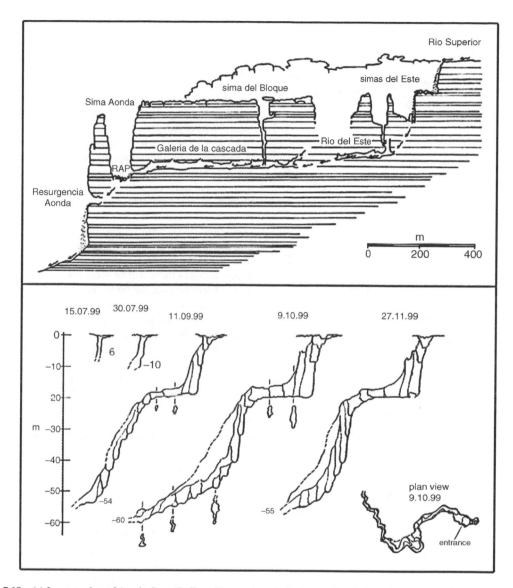

Figure 7.18 (a) Long section of Aonda Cave, Bolivar, Venezuela, a typical example of the profiles of quartzite escarpment caves. Reproduced from Mecchia, M. and Piccini, L. (1999) Hydrogeology and SiO$_2$ geochemistry of the Aonda Cave System, Auyan-Tepui, Bolivar, Venezuela. Bollettino Sociedad Venezolana Espeleologia, 33, 1–18. (b) The growth of vadose shafts and drains in ice of the Gornergletscher, Switzerland, between 15 July and 27 November 1999. reproduced from Piccini, L., Romeo, A. and Badino, G. (1999) Moulins and marginal contact caves in the Gornergletscher, Switzerland. Nimbus, 23/24, 94–9.

Paradise Ice Caves on Mount Rainier, Washington, USA (Halliday and Anderson 1970).

7.4 SYSTEM MODIFICATIONS OCCURRING WITHIN A SINGLE PHASE

Here we are concerned with two sets of effects that can significantly change the cave patterns described above

within the phase that saw their creation. The more drastic pattern changes brought about by multiple phases are discussed in the following section.

7.4.1 Gradational features in phreatic systems

The three different effects are shown in Figure 7.19. All serve to reduce the irregular looping profiles of State 2

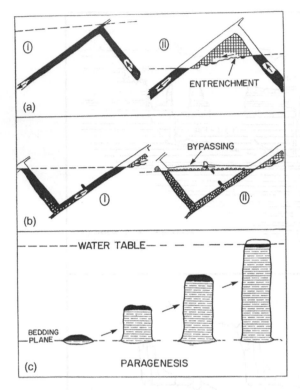

Figure 7.19 Gradational features common in phreatic caves: (a) isolated vadose entrenchment of the upward apex of a phreatic loop; (b) development of a bypass tube above a loop clogged with detritus; (c) development of a paragenetic gallery upwards to the water table.

and 3 caves. **Isolated vadose trenches** develop when, as a result of water table drawdown, apices of upward loops become aerated. They contrast with the **continuous trenches** that extend downstream of a sink point, or upstream from a spring. Isolated entrenchments 50 m deep and hundreds of metres long are known, e.g. the great river passage of Skocjanske Cave, Slovenia, appears to be at least partly of this origin.

Bypass passages or **tubes en raccord** (Ford 1965, Renault 1968) develop where sharp downward loops below a water table become obstructed by alluvial detritus. Normally this will be in upstream parts of a system. When rapid flooding occurs a large head of water may build above the obstruction, greatly steepening the hydraulic gradient across the loop. Local fissures that are impenetrable under normal gradients are then exploited to open one or a series of subparallel bypasses over the loop. The latter becomes fully silted and inert; new clastic load is then carried through the bypass to clog the next loop. The process propagates downstream and is also transitional to the floodwater maze (below).

As originally used by Renault (1968), a **paragenetic passage** is any phreatic or water-table passage where the erosional cross-section is partly attributable to effects of accumulations of fluvial detritus. As applied by Ford (1971b) and other English-speaking authors, it has meant only the type of passage that has a steadily rising solutional ceiling (Figure 7.19c); this is the extreme or 'ideal' type of paragenesis. Pasini (1975) perhaps showed more flair in naming it '*erosione antigravitativa*'!

Paragenetic passages originate as enlarged principal and subsequent tubes. With enlargement, groundwater velocity may be reduced, permitting permanent deposition of a portion of any insoluble suspended or bed load. This protects the bed and lower walls so that dissolution proceeds upwards on a thickening column of fill. The vertical amplitude of such paragenesis can exceed 50 m. Remarkably flat roofs can be bevelled across dipping strata. In many instances there will be some dissolutional development of half tubes at the buried contacts between fill and wall, as well. The process terminates at the water table.

Paragenetic enlargement that is succeeded by partial wash out of the fill in later phases explains much of the complicated plan forms, wall and ceiling shapes, seen in many caves close to geological contacts with sandstones or shales that can supply abundant detritus, e.g. Endless Caverns, in the Ridge and Valley Province of Virginia, USA. Osborne (1999) stresses its role in Jenolan Caves, NSW, Australia, which are in a narrow band of steeply dipping, recessive limestones flanked by sandstones and conglomerates that supply clastic debris. They have a long and complex history of palaeokarst, infilling and hydrothermal speleogenesis succeeded by major paragenetic re-excavation and enlargement.

7.4.2 Floodwater maze passages

Floodwater mazes are evaluated in detail by Palmer (1975). They develop in State 3 or 4 systems or low-gradient vadose caves, where fissure frequency is high. Maze development occurs where allogenic flash flooding invades karst and/or where a trunk passage with large carrying capacity becomes obstructed by clastic or organic debris. Flood heads of many metres can be applied quickly, generating very steep hydraulic gradients, as in bypass passages. Interlinked fissures around the obstruction may be penetrated rapidly to relieve the pressure. This creates a maze that is hydraulically inefficient compared with the trunk, but which may function for the remainder of the active life of the cave.

Floodwater maze development is most significant where caves drain large and rugged allogenic catchments (i.e. where large floods are applied rapidly to one point in the karst) and is most prominent at the upstream end

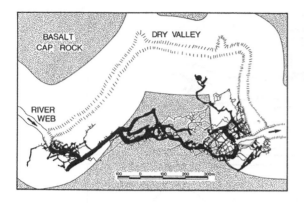

Figure 7.20 Sof Omar Cave, Ethiopia, a spectacular example of a floodwater maze developed as three successive levels in horizontally bedded limestones beneath a basalt caprock. It is a meander cut-off system that floods frequently during the rainy season. (From cartography by Steve Worthington, with permission.)

of systems, e.g. Bullock Creek Caves, New Zealand (Williams 1992). It may extend throughout the length of comparatively short systems such as meander cut-off caves: the spectacular example of Sof Omar Cave, Ethiopia, is shown in Figure 7.20.

7.5 MULTIPHASE CAVE SYSTEMS

Most extensive unconfined caves display multiphase features due to the negative shift of springs (Figure 5.17). Positive shifts flood or fill them with sediment, making them less likely to be explored: at the extreme, the cave system may be decoupled from hydrological throughput to become an inert, buried palaeokarst system. Oil-well drilling penetrates examples down to depths of 3 km or more, e.g. in Romania.

Because of net continental erosion, negative shifts predominate in most karst areas and generally create lower series of passages. Most authors describe the two or more series as cave 'levels'. This raises difficulties because the actual passages are frequently not level (horizontal) at all, e.g. State 2 caves. 'Storey' has similar connotations, and 'stage' has both discharge and evolutionary implications that create misunderstandings. Here we use **phase** and **level** interchangeably, but readers should bear in mind that actual cave passages of a phase need not be level *sensu stricto*.

7.5.1 Types of network reconstruction

The type and scale of reconstruction that follows the lowering of a spring is varied. At its most simple, the spring shifts downwards only, relocating at a lower point on the same vertical fissure. If it follows this fissure, the cave may

display simple vadose entrenchment to the new level, e.g. Grotte St Anne de Tilff (Ek 1961; see Figure 7.39D). Normally, the entrenchment is retrogressive, i.e. a knick-point. There is little or no change of the plan pattern of the cave. This type of reconstruction occurs particularly where the initial cave is one of the vadose types but has a gentle gradient. However, it is comparatively rare.

In many more instances new series of passages are constructed. They propagate headwards through the system and are similar to the connections of further ranks in a first phase. This is because many primary tube systems of previous phases have survived cave enlargement and, although carrying an insignificant proportion of the discharge, have slowly extended and can be exploited when hydraulic gradients are steepened by downward shift of the spring. They serve as 'diversion' or 'tap-off' channels (terms of Ford (1971b) and Mylroie (1984 respectively) but are better described as **undercaptures**, a direct translation of the French *soutirages* (see Häuselmann *et al.* 2003). In addition, in a majority of cases the spring is displaced laterally as well as vertically, which compels the development of at least some distinct new passages. At Hölloch, the lateral shift was of the most simple kind, i.e. by small increments down the edge of a single, inclined plane. Nevertheless, this created several new series of State 2 subsequent passages across that plane (Figures 7.7 and 7.15). At Mammoth Cave, because

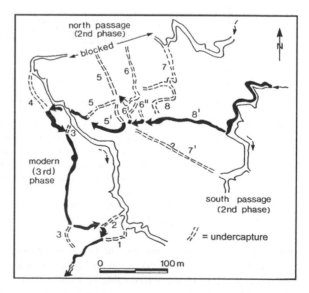

Figure 7.21 The central galleries of Swildon's Hole, Mendip Hills, England. Shown in white solid are north and south passages of a second genetic phase. In black are passages of the third (modern) phase. Numbered dashed lines are the headward sequence of undercapture diversion passages by which the modern route was built beneath that of the second phase.

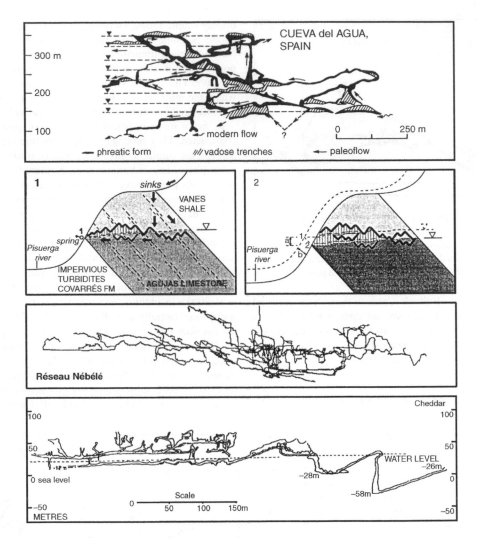

Figure 7.22 Examples of multiphase cave systems, emphasizing the variety that occurs. Réseau Nébélé displays vadose inlet passages, dip tubes, two phases of State 3 loops and water-table passages, plus undercaptures. Cheddar Caves, England, display four levels of outlet passages that successively drained a State 2 looping system with an initial amplitude of ≥ 150 m; the lower three levels are shown here. Cueva del Agua, Spain, after Smart, P.L. (1986) Origin and development of glacio-karst closed depressions in the Picos de Europa, Spain. Zeitschrift für Geomorphologie, NF, 30(4), 423–43; Cueva del Cobre Cave, Spain, after Rossi, C., Munoz, A. and Cortel, A. (1997) Cave development along the water table in the Cobre System (Sierra de Penalabra, Cantabrian Mountains, Spain, Proceedings of the 12th International Congress of Speleology, Vol. 1, pp. 179–85; Réseau Nébélé, France, reproduced from Vanara, N. (2000) Le fonctionnement actuel du réseau Nébélé. Spelunca, 77, 35–8. See text for discussion.

stratal dip is very gentle, each time the Green River (on the downdip, output boundary) entrenched its channel a few metres it might create a favourable new outlet position more than 1 km from previous springs.

An excellent example of the passage complexity that can be created by multiphase undercapture reconstructions is given in Figure 7.21. Swildon's Hole displays just three major erosion phases. The Phase 2 passages were irregular strike subsequents of State 2 type. The stable Phase 3 passage is a river passage 25–30 m lower. The 1000 m of its course that is shown was built piecemeal from at least eight, consecutive and headward, under-captures from the northern Phase 2 tributary, plus two or more from the independent, southern Phase 2 tributary. The earlier undercaptures were wholly phreatic; later ones were vadose at their upper ends, signifying the

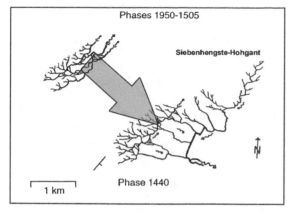

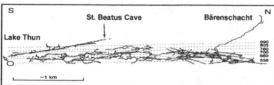

Figure 7.23 The Siebenhengste–Hohgant–St Beatus–Barenschacht multiphase complex cave system near Lake Thun, Switzerland. The upper frame is a map showing the older phases preserved in the Siebenhengste–Hohgant sector. The lower frame is a long section showing the younger phases in the St Beatus and Bärenschacht caves.

lowering of the water table that resulted from creation and expansion of this new cave. Parts of each under-capture were incorporated into the final Phase 3 river passage. Other parts were abandoned entirely and, in some cases, remain too small to enter

Figure 7.22 displays some other multiphase systems chosen to emphasize the wide variety that occurs. Cueva del Agua, Spain, is shown in the outstanding morphological reconstruction by Smart (1986). It is a State 2 multiphase complex containing steep-to-vertical phreatic lifting segments that climb 120 m or more. There are many isolated vadose entrenchments that are taken to represent downcutting to eight different levels within a range of 200 m. Cobre Cave, Spain (Rossi *et al.* 1997) is an instance of two phases of State 2 looping and subsequent vadose entrenchment that, unusually, is against the trend of the stratal dip. Réseau Nébèlè, France (Vanara 2000) displays vadose inlet passages, dip tubes, two phases of State 3 loops and water-table passages, plus undercaptures. The famous Cheddar Caves, England, display four levels of outlet passages that successively drained a State 2 looping system with an initial amplitude of ≥ 150 m; the lower three levels are shown in Figure 7.22.

North of Lake Thun, Switzerland, the Siebenhengste–Hohgant–St Beatus–Barenschacht caves are a multiphase complex containing ~260 km of explored passages beneath ~40 km^2 of dissected escarpments dipping southeast at 15–30° (Figure 7.23). The caves are developed in 150–200 m of well-bedded Cretaceous limestones, capped by sandstones and resting on impervious marls. They have a total relief > 1500 m. Faults have acted as barriers rather than preferred routes for flow, so that explorers have not yet been able to physically interconnect all portions of the caves. The multiphase development can be attributed to Alpine uplift coupled with cover stripping and glacial and fluvial episodes of entrenchment. No fewer than 12 distinct phases are currently recognized (Jeannin *et al.* 2000, Häuselmann *et al.* 2003). In the earlier four there was State 2 irregular strike subsequent development with drainage to the northeast (Figure 7.23, phases at 1950–1505 m a.s.l.), creating a comparatively simple pattern similar to that of Hölloch; the amplitude of deeper phreatic loops was at least 220 m. The direction of flow was then reversed southwest, to the St Beatus and Barenschacht caves that drain into Lake Thun. They display eight successive phases of system reconstruction, descending from initial springs at 1440 m a.s.l. down to a modern low-stage level at 558 m asl. Phreatic looping with amplitudes up to 250 m is well established by exploration, and early looping to 400 m is hypothesized. The younger loops have lesser amplitudes, but epiphreatic flooding can reactivate five relict loops in a downstream succession with an amplitude of many tens of metres (Häuselmann *et al.* 2003, figure 7).

Friars Hole System, West Virginia (Figure 7.24) offers one final contrast. It is an instance of the restricted case of cave development. Limestone crops out in only a few narrow strips in the floor of a steep valley of shale and sandstone, yet 68 km of passages are mapped. Stratal dip is ~3° westerly and the springs lie along strike to the south. The system displays at least six distinct State 4 channels that follow faults or selected bedding planes to the strike. The highest channel is only 125 m above the modern springs. Later ones shifted downdip as at Hölloch. The plan pattern is extremely complex because stream-sinks have been randomly blocked and sealed for long time-spans by copious debris from the surrounding slopes. New sinks were created both upstream and downstream of old ones and many underground diversions have appeared. At present three separate rivers flow through different parts of the explored system, at differing elevations and in different stratigraphical positions. They join somewhere before reaching the springs. The lowest explored river enters a siphon that is 13 km distant from the spring yet only 24 m higher.

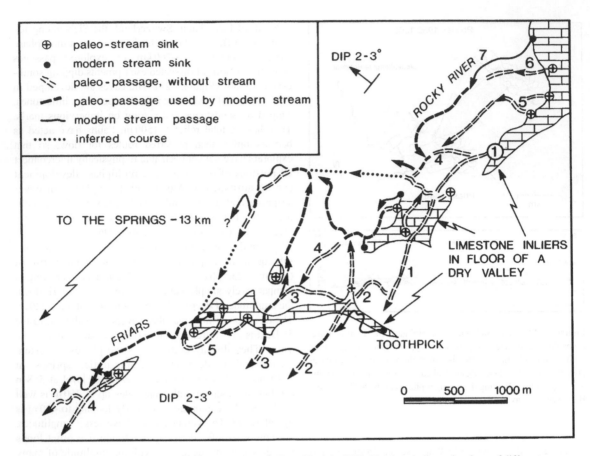

Figure 7.24 The multiphase pattern of Friars Hole System, West Virginia, USA. Numbers indicate the phase of different passages. Reproduced from Worthington, S.R.H., The paleodrainage of an Appalachian Fluviokarst: Friars Hole, West Virginia. M.Sc. Thesis, McMaster University, p. 218, 1984.

7.6 METEORIC WATER CAVES DEVELOPED WHERE THERE IS CONFINED CIRCULATION OR BASAL INJECTION OF WATER

7.6.1 Two-dimensional and three-dimensional maze caves

These caves are encountered quite frequently (Table 7.3, type A2), either as separate entities or as anomalous parts of normal, unconfined caves. Their form is everywhere similar. They are reticulate mazes of comparatively small passages that follow joints confined to one or a few rock beds. Passages are phreatic in form. Parallel passages (i.e. in one joint set) tend to be of similar dimensions, indicating that initial joint apertures were similar (Clemens *et al.* 1997). Characteristically, the morphology is that created by slowly flowing waters. They lack the high-velocity erosional scalloping and sharp cuspate morphology of typical .

floodwater mazes, although transitional examples do occur.

In essence such mazes develop where strata possessing a high fissure frequency are geologically confined. The trapping may be artesian, e.g. in a syncline beneath a shale aquiclude, or it may be a local 'sandwich' situation where one or a few densely jointed limestone beds are trapped between massive limestones with few joints. Artesian maze caves are widespread. Where small areas of reticulate maze occur within normal branchwork caves it is most frequently because a 'sandwich' situation exists.

The highest density of passage per square kilometre is found in confined maze caves. Klimchouk (2003) cites a range of 50–400 $km\,km^{-2}$ for a global selection, which contrasts with 12–24 $km\,km^{-2}$ in such well-known unconfined systems as Mammoth Cave and Friar's Hole, USA. The porosity of the sample maze caves amounts to 1–10% of the rock volume.

7.6.2 Maze formation by diffusion through an overlying sandstone aquifer

Palmer (1975) investigated a sample of maze caves in the USA and found that 86% of them had developed directly beneath permeable sandstones (Figure 7.25a and b). Often, passage roofs were in the sandstone. An equidimensional maze is created because water is introduced to the soluble rock by way of an insoluble but homogeneous diffusing medium, i.e. sandstone with well-developed primary permeability. In Stage 1 the hydraulic conductivity of the sandstone is greater than that of the unkarstified limestone beneath, but the situation is reversed by Stage 2. Crossroads Cave, Virginia (Figure 7.25c) is an example.

More exotic is the case of Cserszegtomaj Well Cave, Hungary (Figure 7.25d). Clint-and-grike dissolutional topography (see section 9.2) developed on exposed dolomites, which were then buried by sands that became consolidated, creating a buried palaeokarst (Bolner-Takacs 1999). Groundwater has selectively removed the limestone clint blocks, leaving walls and ceilings of sandstone resting on a dolomite grus floor. It is uncertain whether the groundwater descended through the sandstone or ascended from below.

7.6.3 Outlet caves created by basal injection of meteoric waters into karst rocks

This category includes simple, single shaft caves, broadly dendritic (branchwork) systems, and reticulate maze caves. They ascend through soluble strata to outlets at the surface or into overlying aquifers. Maze caves are the most frequent and important, the other types apparently being quite rare.

Brod (1964) described a series of small solutional pit and fissure features scattered for 60 km along the crest of a denuded anticline in eastern Missouri, USA. These extend downwards through 30 m of limestone with basal shale, and then through 30 m of dolomite where the form may change from shaft to maze. At the base of the dolomite the underlying sandstone is exposed in some cave floors, including detached sandstone blocks partly rounded *in situ*. The sandstone is 40 m thick and very porous. In Brod's interpretation, its groundwaters were discharged upwards and out through the impure carbonate aquitard where this was fractured at the anticline (Figure 7.26A). Localization of the fracturing allowed sufficient concentration of waters to dissolve caves of enterable size.

Botovskaya Cave, Russia (Filippov 2000) is an excellent example of a two-dimensional maze cave. It is in 6–12 m of limestones confined between argillaceous sandstones; enlargement is attributed to mixing of cool artesian waters. Frumkin (personal communication 1999) describes similar artesian waters rising to the crests of anticlines in the Judean Mountains of Israel–Palestine, where reticulate mazes developed beneath (curiously) chalks functioning as aquitards. The caves drained along the strike to local breaches of the aquitard. There are many similar instances recorded elsewhere.

Grotte de Rouffignac, France, is a highly unusual cave (Figure 7.26B). It is developed in rather soft chalks beneath clays. The chalk has prominent layers of flint nodules, but bedding planes and joints are poorly developed. These strata functioned as an aquitard that confined a productive aquifer below. In our interpretation, diffuse source waters from the aquifer were able to enlarge small inlet passages that converged within the chalk to form a dendritic array that discharged along the flint-rich horizons into an adjoining river valley. The inlet passages are paragenetic, rising on thick residual clays. The modern cave has three levels and displays subtle ceiling bevels and corrosion notching. It is a very famous site of cave-bear habitation and of palaeolithic art.

7.6.4 The gypsum maze caves of western Ukraine

The Ukrainian gypsum caves are amongst the best known reticulate mazes (Figure 7.27). They are created by meteoric waters flowing upwards through them and thus are hypogene (*per ascensum*) in origin. They have been studied intensively (Klimchouk 2000b, 2003): this research provides the basis of much of our understanding of hypogene mazes everywhere, including such hydrothermal giants as Jewel Cave and Wind Cave, South Dakota.

The caves occur in the flanks and spurs of the Dniester and Prut valleys and their tributaries, and across gentle interfluves. They are developed in Tertiary gypsums only 10–30 m in thickness that are underlain and overlain by thin limestones, sands and marls that are efficient aquifers. This sequence is confined beneath upper aquiclude clays. As Figure 7.28 emphasizes, the comparatively massive, homogeneous gypsum functions initially as an aquitard. The aquifers are recharged where streams at higher elevations breach the aquiclude, and then drained by breaches at lower elevation. The caves develop preferentially towards and under the breaches.

Although the gypsum formations are comparatively thin, they display different joint patterns or densities in successive beds, or there may be thin clay layers separating some units. Some of the caves are mazes contained within only one joint set at one level, but the majority of them have two to four distinct levels created by the change of joint patterns or by the clays, etc. As these levels lie almost directly above one another (like the successive storeys in a building) and developed at essentially the same time, they may be termed **storeys** (Ford

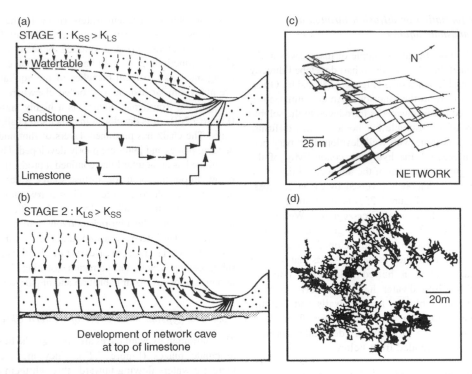

Figure 7.25   (a and b) Development of a two-dimensional maze cave in limestone as a result of diffusion of groundwater through an overlying sandstone aquifer. Reproduced from Palmer, A.N. (1975) The origin of maze caves. Bulletin of the National Speleological Society, 37(3), 56–76. (c) Crossroads Cave, Virginia, USA, an example of this mode of development. (d) Cserszegtomaji Well Cave, Hungary, a highly unusual two-dimensional maze developed from buried palaeokarst.

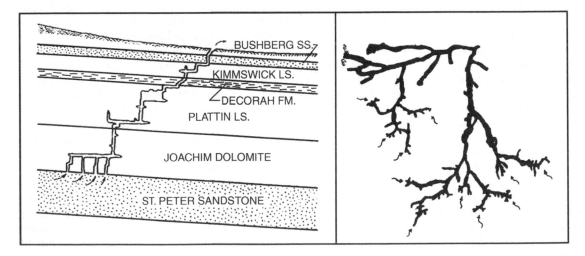

Figure 7.26   (Left) Model for the development of meteoric water caves discharging the St Peter sandstone aquifer in the eastern Missouri Ozark Hills. Reproduced with permission from Brod, L.G., Artesian origin of fissure caves in Missouri. Bulletin of the National Speleological Society, 26(3), 83–112 © 1964 National Speleological Society. (Right) Grotte de Rouffignac, France, displays a pattern of basal inputs from a diffuse flow aquifer that converge into a dendritic network and ascend gently through chalky strata; about 10 km of passages are known.

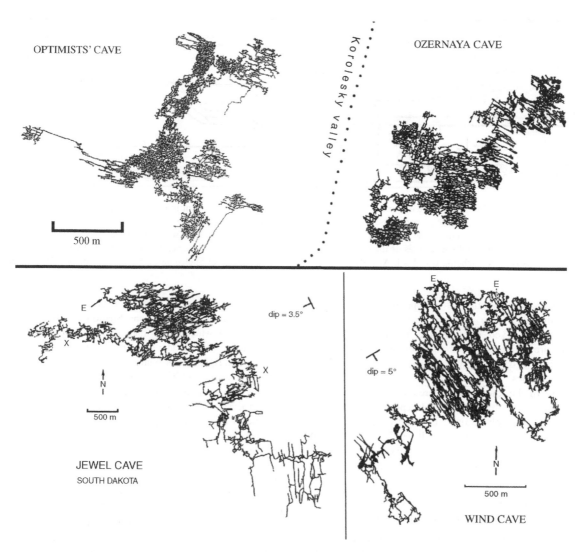

Figure 7.27 The four lengthiest multistorey maze caves currently known. Optimists' and Ozernaya, Ukraine, are gypsum caves with basal inflow of meteoric water. Jewel Cave and Wind Cave, South Dakota, USA, are limestone caves with a complex history of early palaeokarst, and infilling, followed by enlargement (with partial exhumation) by basal inflow of thermal waters.

1991) to differentiate them from the 'levels' of multiphase caves that may be separated by many metres and are of different ages. Typically, the higher storeys are less extensive and are offset towards the discharge points (breaches in the aquiclude).

Figure 7.28b shows details of the construction of storeys and their varying passage morphology. The groundwater circulation is very slow compared with that in most unconfined situations. This permits small thermal or solute density gradients to become significant (**natural convection flow** in Figure 7.28b); cooler or solute-laden water sinks, setting up local convection cells that carry less saturated water to upper parts of passages, preferentially widening them. Distinct cupolas may develop at more soluble places in the ceilings (sections 7.7 and 7.10). **Forced convection flow** is introduced to a given area of the maze if the overlying aquiclude is breached close to it. Flow rates then accelerate a little, which reduces the differential sculpting effects of natural convection. Recently, these processes have been carefully modelled (Birk *et al.* 2000).

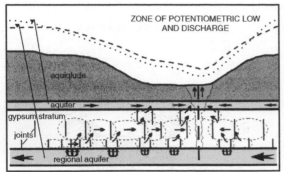 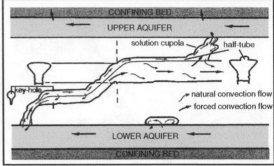

Figure 7.28   (Left) The initial stage of per *ascensum* multistorey maze cave development in the Ukraine gypsum caves. Note that three potentially separate storeys of maze passages may develop in different gypsum members because of differing joint patterns and densities. (Right) The general mode of per *ascensum* dissolution by convective flow; see text for details. Reproduced from Klimchouk, A.B. (2000b) Speleogenesis of the Great Gypsum Mazes in the Western Ukraine, in Speleogenesis; Evolution of Karst Aquifers (eds A.V. Klimchouk, D.C. Ford, A.N. Palmer and W. Dreybrodt), National Speleological Society of America, Huntsville, AL, pp. 261–73.

Where the gypsum is exceptionally massive and impermeable, large and irregular galleries or rooms may be dissolved in it by the underflow. Where the caves are progressively drained by river entrenchment, prominent dissolution notches (section 7.10) can form at the waterlines.

Most of the known caves are now drained and relict. Zolushka (Cinderella) Cave is exceptional because it has been progressively drained by nearby quarrying operations during the past 50 year. It has been a remarkable natural laboratory in which to observe dewatering processes and effects within the span of a human lifetime.

7.7 HYPOGENE CAVES: HYDROTHERMAL CAVES ASSOCIATED CHIEFLY WITH CO_2

This section considers solutional caves created by deeper waters that are usually heated, perhaps enriched in a wide range of dissolved gases and minerals, and flow out to the surface in most instances. By definition, thermal waters are those which, at their springs, have a mean annual temperature $> 4°C$ higher than that of the region (Schoeller 1962). Temperatures at hot springs in karst rocks are most frequently in the range 20–80°C (T.D. Ford 1995). Much higher temperatures can occur at depth. Standard geothermal gradients are $\sim 30°C\,km^{-1}$, increasing to $100°C\,km^{-1}$ around active volcanic areas. In the Transdanubian Mountains of Hungary the gradient is only $\sim 16°C\,km^{-1}$ in limestones but rises to $40–60°C\,km^{-1}$ in shales above them, which emphasizes the role of karst waters in removing heat. In origin, the waters may be juvenile (i.e. volcanic) or connate (trapped depositional waters that are being expelled from sedimentary basins), deeply circulat-

ing meteoric waters (e.g. beneath a syncline), or mixtures of these types in any proportions. Shallow meteoric waters (normal karst waters) may then mix with them as they approach the surface. The gases most commonly reported are CO_2 and H_2S. One or both may be important in a given water. Here we simplify by supposing that one class of cavities (the more abundant) is created predominantly by CO_2 processes, whereas in the second class (section 7.8) the role of H_2S is more significant.

When hot, carbonated water rises and cools in carbonate rocks (with and without mixing with shallow meteoric water), there may be net dissolution, incongruent dissolution or net deposition. A cave may straddle several of the different process zones (e.g. dissolving at its base, receiving subaqueous precipitates at the top) or be entirely within one (section 3.10, equation 3.78). The zones may migrate through it once (erosion $\rightarrow$ deposition) or oscillate across it several times. Many suspected hydrothermal caves experience only dissolution by the hot waters. In others the bedrock walls can scarcely be seen at all, being covered by later hydrothermal precipitates.

Andre and Rajaram (2005) model thermal waters initially penetrating limestone. For an example of 60°C water penetrating a horizontal fracture 500 m in length with an aperture of 50 μm and P_{CO_2} of 0.03 atm, breakthrough occurs in about 6600 year. This suggests that the timescales for thermal cave evolution are similar to those in meteoric waters.

7.7.1 Criteria for hydrothermal origin or part origin

The most certain criterion is hot water flowing from the cave! However, some meteoric caves have been invaded by hot waters because these will migrate to pre-existing conduits if possible. They offer the easiest exits. Most explorable hydrothermal caves are drained and relict. Presence of hot springs in their vicinity offers a strong hint concerning their origin, but cannot be conclusive.

Certain erosional features are highly suggestive but, once again, no one of them may be conclusive. Strong index features are deep, rounded and often multicuspate, solution pockets (cupolas) that are attributed to convectional dissolution (see section 7.10). In a few instances there are highly corroded patches or rills on walls where steam has condensed above hot pools. Dolomitization of limestone walls has also been reported. In some Russian caves, cherts in the limestone are strongly corroded, which is good evidence of hot alkaline waters. Perhaps as significant is the lack of evidence for moderate to fast flowing water that is found in most meteoric caves. Better indices are provided by the exotic deposits. Most important are uncommon forms and abundances of calcite and aragonite, but a wide range of rarer minerals may also be present, especially barite, fluorite, quartz and the sulphide minerals. Crusts of scalenohedral ('dogtooth') or rhombohedral ('nailhead' or Iceland) spar calcite are common, with thicknesses up to 2 m. These may cover all surfaces (floors, walls and ceilings), indicating subaqueous precipitation. Individual crystals are large, indicating slow deposition under uniform conditions. Quite frequently, fine-grained breccia with or without barite or silica is reported beneath the spars. Large quartz crystals are sometimes found that suggest cooling from a hot environment.

Palaeowaterlines tend to be marked by thick botryoidal or coralloid growths or by fin-like accretions of calcite. There may be abundant debris from the most fragile of all speleothems, rafts of calcite that precipitate about dust grains on the pool surfaces. Calcite geyser stalagmites may signify late, feeble effusions of the hot waters through drained cavern floors. In some instances there is further precipitation overhead from the expelled vapour. Some hollow stalagmites may be subaqueous (**white smokers!**). Gypsum as large crystals or extruded **flowers** and **whiskers** is common in freshly drained hydrothermal cavities.

All of these calcite, aragonite and gypsum forms can be found in meteoric water caves. What is distinctive in the hot-water caves is their unusual abundance and their association together (Dubljansky and Dubljansky 1997, Hill and Forti 1997). Strong supporting evidence of precipitation from past thermal waters is obtained from oxygen isotope ratios in the calcites, which will usually be more depleted in ^{18}O than in any meteoric precipitates nearby (section 8.6). Deep-seated hydrothermal calcites luminesce with a brief but strong orange-red glow that is absent in cool-water precipitates (Shopov 1997).

7.7.2 The form of hydrothermal caves

The form of the majority of known hydrothermal caves is similar to that of the confined and basal injection meteoric water types of caves already discussed, because the hydrogeological settings are identical in broad terms, i.e. they are rising water shafts as in Figure 7.26, or single storey or multistorey mazes as in Figure 7.28.

One further type of cave should be considered first, however, because of its comparative simplicity. This is a roughly spherical or cupola-form void created, it is believed, by convection currents within the very slowly flowing water. Individual voids may be no more than football-sized or range up to chambers of several hundred metres in diameter. Clusters of small ones create spongeworks, as in Figure 5.23. Large examples may be complex, with multiple cupolas branching from one another. In thinner strata or where over-enlarged, these grade to deep solution breccias (Figure 7.29a) that, in S-rich waters, may host precipitates of calcite and dolomite with sulphide minerals such as pyrite, galena and sphalerite. Very large, more irregular, cavities full of thermal water are sometimes encountered by exploratory drilling. The largest recorded is the Madan Chamber, in Archaean and Proterozoic marbles in a tin–zinc ore zone in the Rhodope Mountains, Bulgaria (Dubljansky 2000a). The cavity has a volume estimated at $240 \, 10^6 \, m^3$ (more than ten times greater than the largest mapped cave chamber) and at one point apparently measures $> 1340 \, m$ from roof to floor. Water temperatures range 90–130°C in the chamber; pressure increases from 37 to 170 atm, roof to floor.

Simple rising water shafts following major joints, faults or steeply dipping bedding planes are perhaps the most widespread type of thermal cave. The deepest known examples of shaft caves active today are Pozzo de Merro, Italy (−392 m) and El Zacaton, Mexico (> 329 m). Both are close to volcanic centres. El Zacaton (Figure 7.29c) and nearby shafts have thick travertine deposits and rich biomats. Water temperatures are only 5–10°C above regional means, suggesting mixing of thermal and meteoric water. A robot underwater sonar mapper and biosampler, DEPTHX, is being developed to explore this site (Gary *et al.* 2002). In contrast to such mixing, 150 km southwest of El Zacaton in the Sierra de El Abra, two great meteoric springs (Coy and Choy) discharge waters at 25–27°C from depths > 200 m but, between them and from the same strata, Taninul Springs discharge distinct, H_2S-rich water at 40°C.

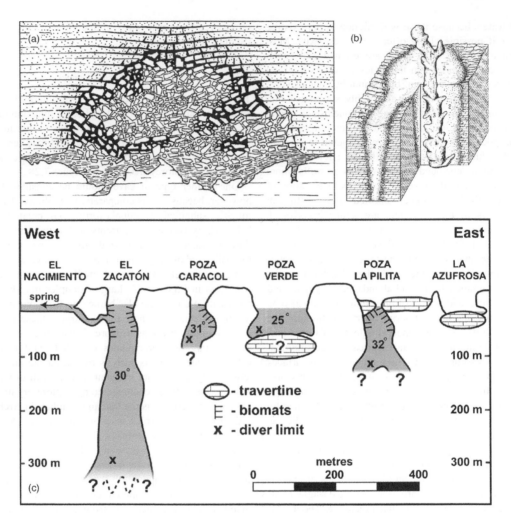

Figure 7.29 (a) Typical features of a deep cave chamber that has collapsed to form a solution breccia. Sagging, slightly displaced strata are **crackle breccia**. The clast-supported central pile is **pack breccia**. Basal fragments dispersed in insoluble residues accumulated before collapse are **float breccia**. Reproduced with permission from Dżulynski , S. and Sass-Gutkiewicz, M. (1989) Pb-Zn ores, in Paleokarst – a Systematic and Regional Review (eds P. Bosźk, D.C. Ford, J. Glazek and I. Horáček), Academia Praha/ Elsevier, Prague/Amsterdam, pp. 377–97. (b) Csiki Shaft, a remarkable palaeothermal water shaft that was exposed in a quarry near Budapest, Hungary: 1, dolomite; 2, grus (dolomite pulverized by thermal water); 3, shaft walls are firmly cemented by precipitated silica. Reproduced from Jakucs, L. (1977) Morphogenetics of Karst Regions: Variants of Karst Evolution, Akademiai Kiado, Budapest, 284 pp. (c) El Zacaton and neighbouring thermal water shafts at Rancho La Azufrosa, Mexico. Adapted from Gary, M.O., Sharp, J.M., Havens, R.S. and Stone, W.C. (2002) Sistema Zacatón: identifying the connection between volcanic activity and hypogenic karst in a hydrothermal phreatic cave system. Geo2, 29(3–4), 1–14.

Most known thermal shafts are now drained and relict. Historically, many were explored by miners exploiting their secondary mineral deposits. The celebrated Fersman Cave, Kirghizstan, was mined to −220 m. It contains an exotic uranium mineral, **tyuyamuyunite**, that was precipitated at 40–60°C, and succeeded by barites and calcites at ∼30°C (Dubljansky 2000). Stari Trg Mine, Croatia, is a complex of shafts with small mazes, single galleries and chambers, 600 m in depth, discovered during zinc–lead mining. Its lower 400 m was filled with warm water until pumped out. The Csiki Shaft (Figure 7.29b) illustrates some other striking effects of thermal karstification. Dolomite is often pulverized (reduced to grus) in the thermal aureole. At Csiki, the walls of the central shaft were then reinforced by silica cementation at high temperature: preferential erosion of the surrounding rock and grus has left the ancient shaft protruding several metres like a hollowed tree trunk!

7.7.3 *Rózsadomb, Hungary*

Budapest was founded upon hotsprings along the River Danube. Almost the full range of thermal cave types can be found there under Rozsadomb (Rose Hill, the diplomatic quarter), including relict shafts, linear passages, two- and three-dimensional mazes, and modern phreatic caves discharging thermal waters at the river level. Cherty Triassic limestones with some palaeokarst are overlain unconformably by 40–60 m of Eocene limestones covered by marls that form a largely impermeable cap. There is a complex block-fault geological structure that may incorporate buried karst towers (Figure 7.30). Uplift has progressively drained the older, higher caves, and left extensive deposits of spring travertines stranded up to 150 m above the modern river. Mean temperatures of meteoric waters invading the caves are 8–13°C. Different thermal springs have mean temperatures between 20 and 60°C, indicating that differing amounts of mixing takes place.

The Mátyás–Pálvölgy and Ferenc-hegy caves are multistorey mazes following joint-guided alteration zones in which the limestone is partly disintegrated and silicified, as in the Csiki shaft. Alteration is attributed to Miocene fore-Alp volcanism (Muller and Sarvary 1977). Calcite raft debris indicates many past water levels. Szemlöhegy Cave is a more simple outlet rift on joints; it has abundant subaqueous calcite crusts. Molnár János is the modern outlet; its warm waters have been dived to −70 m, discovering ~4 km of phreatic maze. Jószefhegy Cave is a shaft descending to ramiform chambers with abundant secondary gypsum deposits. It is transitional to the H_2S class of hypogene caves (section 7.8), indicating how these two classes may mingle within a small geographical area.

7.7.4 *Jewel Cave and Wind Cave, South Dakota*

Jewel Cave and Wind Cave (Figure 7.27) are the greatest known multistorey thermal mazes. They developed in 90–140 m of well-bedded limestones and dolomites uplifted and fractured by Eocene intrusion. There are modern hot springs nearby, but at lower elevation. The caves have a rich and complex history of Mississipian palaeokarst genesis that was succeeded by infilling, deep burial and mineralization, before uplift and modern development with some exhumation began in the Tertiary (Palmer and Palmer 1989, 1995).

Jewel Cave is now entirely relict. It has no relation to the surface topography and is entered where intersected by canyons. Its great features are large passages (up to 20 m in height) mostly or entirely covered by nailhead spar. In the highest places the spar has been partly removed by condensation corrosion. Elsewhere, it is generally 6–15 cm thick and often has a thin silica overgrowth.

Wind Cave is lower in elevation and nearer to the hot springs. The morphology is similar to that at Jewel Cave but there is little spar encrustation. It, too, is a hydrological relict but its lowest accessible parts are filled by semi-static waters fed from below. Calcite rafts form upon them. They are cooling backwaters of mixed thermal–meteoric composition. They have been lowered at a mean rate of ~40 cm 1000 year during the past 400 000 year as a consequence of evolution at the hotsprings (Ford *et al.* 1993).

Boxwork is uniquely well developed in these caves. It occurs in microfractured dolomites where the fractures are filled with calcite. Under incongruent dissolution conditions (section 3.5) the dolomite was dissolved, but the calcite veins were preserved. Later, further calcite precipitated onto the veins, which now protrude as much as 100 cm from roofs and walls.

7.8 HYPOGENE CAVES: CAVES FORMED BY WATERS CONTAINING H_2S

Formation of $H^+ + HS$ and of H_2SO_4 were summarized in Chapter 3. Despite their potency, the quantitative significance in most karst regions is nearly negligible because the quantities of H_2S obtainable are small. However, recent studies strongly suggest that H^+ ion or sulphuric acid from basinal fluids or other sources (equations 3.70–3.75; Figure 3.11) are important in some very large caves, including the famous Carlsbad Caverns and Lechuguilla Cave, New Mexico.

The basic principles of simple, purely H_2SO_4, cave genesis are indicated in Figure 7.31 (left). The H_2S rising from depth mixes with O_2-rich meteoric waters, creating large cavities at and around the water table. The most simple form is a cave with short, horizontal vadose outlet passages in limestone that taper in dimension inwards, where they terminate in springs rising from narrow fissures. The waters are warm and rich in H_2S, which is liberated into the air. Walls and roofs are deeply encrusted with gypsum. The H_2S is oxidized to sulphate and hydrogen ion, which reacts to replace the limestone with gypsum (Figure 7.32; equations 3.56 and 3.57). The gypsum replacement crusts fail because of their weight plus expansion forces, exposing fresh limestone to alteration. Fallen crust is dissolved by the vadose streams. The

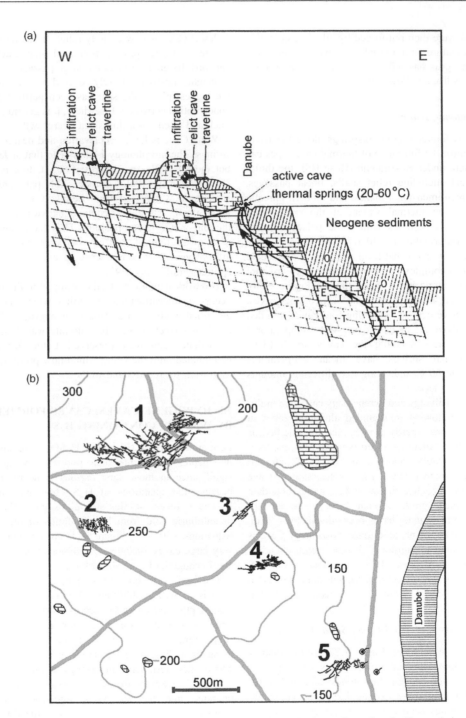

Figure 7.30 (a) Geological section displaying the thermal water circulation at Budapest. Reproduced with permission from Muller, P. and I. Sarvary, Some aspects of developments in Hungarian speleology theories during the last ten years, 53 59 © 1977 Karszt es Barlang. (b) Spring travertines and caves of Ro'zsadomb, Budapest: 1, Mátyáshegy-Pálvölgy Cave; 2, Ferenchegy Cave; 3, Szemlöhegy Cave; 4, Jöszefhegy Cave; 5, Molnár János hotsprings and cave.

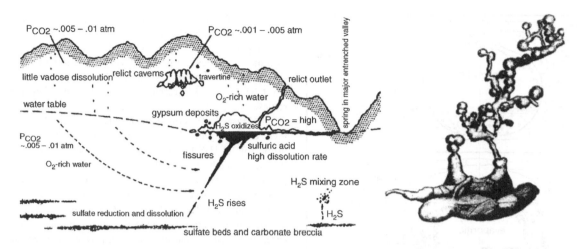

Figure 7.31 (Left) Model for the development of ramiform caves around the water table by oxidation of H_2S. Reproduced from Palmer, A.N. (1995) Geochemical models for the origin of macroscopic solution porosity in carbonate rocks. in Unconformities and Porosity in Carbonate Strata (eds D.A. Budd, P.M. Harris and A. Saller), Memoir 63, American Association of Petroleum Geologists, Tulsa, OK, pp. 77–101. (Right) Sátorköpuszta Cave, Hungary. Spheroidal dissoloution cupolas sprout from a basal source chamber where the limestone walls have been converted to gypsum. Reproduced from Muller, P. and I. Sarvary, Some aspects of developments in Hungarian speleology theories during the last ten years. Karszt es Barlang, 53–59, 1977.

caves enlarge headwards from original valley floor springs, into the rock. Egemeier (1981) described examples in Big Horn Basin, Wyoming; the largest is 420 m in length and has an average diameter of 14 m. Cueva de Villa Luz, Mexico, is an example in which both H_2S and CO_2 concentrations can attain lethal levels, partly generated by chemoautotrophic bacteria (Hose 2004).

7.8.1 Sátorköpuszta Cave and Bátori Cave, Hungary

These remarkable limestone caves consist of basal chambers (that may be likened to magma chambers) formed at a palaeowater-table, from which a branching pattern of rising passages have grown like the trunk and branches of a tree. The chamber walls are largely converted to gypsum. The rising passages are sequences of partly interlocked, semi-spherical convectional condensation cupolas. Sátorköpuszta, the larger cave, is shown in Figure 7.31; it extends for 100 m above the basal chamber. It is possible that these caves are monogenetic H_2S systems; alternatively, they are major enlargements by H_2S processes of earlier rising-water shafts of more conventional origin.

7.8.2 Ramiform caves of the Guadalupe Mountains, New Mexico–Texas, and other regions

The southern Guadalupe Mountains are the type region for very large H_2S caves. The Mountains are built of massive Permian reef, fore-reef and back-reef strata forming an escarpment ∼50 km in length (Figure 3.11). This is dissected by many canyons, which have intercepted and exposed more than 30 major caves, most of which occupy anticlinal trap positions. The region is semi-arid, lacking the large volumes of water needed to develop caves of enormous size by conventional unconfined meteoric waters.

The caves have maze and ramiform patterns, with great rooms linked to higher passages or shafts, and underlain by blind pits. The largest have three or more distinct levels many metres or tens of metres apart in elevation, like typical levels in unconfined meteoric caves. Many contain blocks of layered gypsum, up to 7 m in thickness, resting on thinner residual clay deposits. The Big Room of Carlsbad is representative. It is up to 80 m in height and contains enormous calcite stalagmites and columns. Rooms are interconnected by breakdown or spongework mazes. There are few indicators of the palaeoflow. The caves are relict, without any relation to modern topography or surface hydrology.

Sulphur in the gypsum is very depleted in the heavy isotope [34]S, indicating that it derived from H_2S produced by biogenic reduction in the adjoining oil and gas fields. The H_2S migrated updip with expelled basin fluids that were discharged at a sequence of springs. As the reef complex was uplifted and exposed by differential erosion during the Tertiary, the spring points shifted eastwards to progressively lower elevations. The highest caves are at

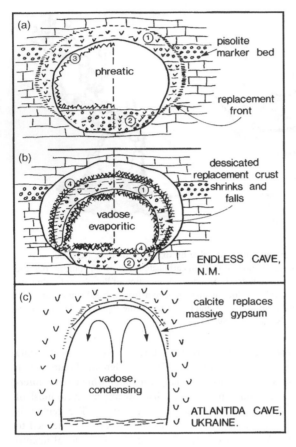

Figure 7.32 Gypsum formation under phreatic conditions in a New Mexico cave. (a) Gypsum replaces limestone wall rock (1) to depths up to 30 cm. Layered gypsum with residual clay and pisolites accumulates on the floor (2). Possible late stage of slow deposition of large gypsum crystals on all exposed surfaces (3). (b) The cave drains and dries. Gypsum replacement crusts and floor deposits are dessicated and shrink. An overgrowth evaporitic crust of gypsum (4) forms on the dessicated deposits. (From observations by M. Buck.) (c) In contrast, at Atlantida Cave, Ukraine, standard atmospheric condensation water with dissolved CO_2 is replacing massive gypsum wall rock with ~5 cm of calcite. (From observations by A. Klimchouk.)

~2200 m a.s.l. Dating of alunite by $^{40}Ar/^{39}Ar$ (section 8.6) places their oldest outlets at ~12 million years. Carlsbad Caverns and Lechuguilla Cave (Figure 7.33) have three or four principal levels between 1100 and 1370 m a.s.l., the higher being ~6 million years in age and the lower (Big Room–Lefthand Tunnel at 1100 m in Carlsbad Caverns) ~3.5 million years. The lowest explored points (950–1000 m a.s.l.), are at or a few tens of metres above the modern water table; studies around 1000 m a.s.l. in Carlsbad Cavern reveal episodes of

protracted flooding there within the past two million years that are probably consequences of the Quaternary colder periods. Lechuguilla Cave, in particular, has an extremely complex morphology that appears to be related in part to exhumation of some much older palaeokarst. Du Chene and Hill (2000) present a comprehensive review.

The Frasassi caves of Ancona, Italy, are a smaller two-level complex with similar morphology and origin. The prime role of H_2S has been recognized there since 1978 (Galdenzi and Menichetti 1990). Auler and Smart (2003) proposed it for maze caves in Bahia Province, Brazil, and it is recognized in caves of the Grand Canyon, Arizona. At Baiyun Cave, Hebei Province, China, the Guadalupe geological structure is reproduced in miniature but, in this case, with basinal coal measures supplying the H_2S; the oldest karst remnants are calcite flowstones preserved on the highest peaks of the denuded anticline. Akhali Atoni Cave, Georgia, is another system of large chambers and galleries created by waters rising through them (Tintilozov 1983). It is in the footslope of the Caucasus Mountains at the edge of the Black Sea. A mixture of cool springs and warm springs with CO_2 and H_2S emerge a few hundred metres distant. When they are in flood the lower cave is also back-flooded. It appears to be essentially of the Carlsbad type, but with deep CO_2 and warm–cold water mixing corrosion processes perhaps playing more of a role than in the New Mexican examples.

7.9 SEA-COAST EOGENETIC CAVES

Here we are considering development of caves in young, newly emergent limestones that have not been buried and compacted beneath later sediments to a significant extent. The term 'eogenetic' refers to the sequence of dissolution and cementation events that occurs during the subaerial diagenesis of young limestones such as calcareous dune sands (Vacher and Quinn 1997). These conditions occur chiefly in those tropical to mid-latitude coastal regions where the rocks are later Tertiary or Quaternary in age. Effective porosity is high so that, although penetrable fissures are less important than in more mature rocks, water-table levels are determined primarily by sea level. If there is fresh water, it is usually as a lens resting on saline water underneath (section 5.8). All development, diagenetic and karstic, is complicated by the fact that sea levels have been oscillating quite rapidly between +10 and −130 m or more due to the Quaternary glaciations. Open cavities on coasts are also subject to concurrent enlargement or destruction by mechanical wave action.

Syngenetic caves are formed simultaneously with lithification (Jennings 1968) and are the most fundamen-

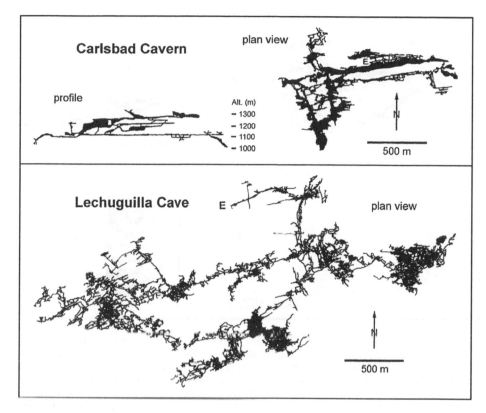

Figure 7.33   (Upper) The long section and map of Carlsbad Caverns. (Lower) The map of Lechuguilla Cave. Both caves are in the Guadalupe Mountains, New Mexico.

tal class of eogenetic cavities. While carbonate sand dunes and beach ridges are accumulating, their surfaces may become cemented to depths of centimetres or tens of centimetres by dissolution and precipitation following rain, spray or dew fall. This can form calcrete that is strong enough to support a cave roof. At their most simple, syngenetic caves are produced by the mechanical washout of unconsolidated sands beneath such thin calcrete layers. Washout occurs where streams flowing from hinterlands, tidal swamps or small shafts, or wave action, can breach the calcrete. Cave chambers > 10 m in diameter are known (White 1994). More often, localized washout is coupled with both dissolution and cementation at the water table and elsewhere in the dune to create shallow spongework caves of greater extent (Figure 7.34).

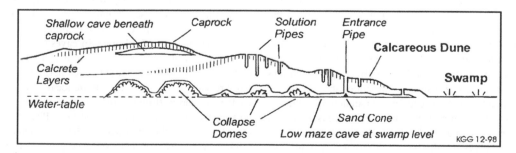

Figure 7.34   A model for syngenetic cave and karst development in accumulating calcareous sand dunes. Reproduced with permission from Grimes, K.G., Mott, K. and White, S., The Gambier Karst Province. Proceedings of the 13th Australian Conference on Cave and Karst Management. 1–7, 1999.

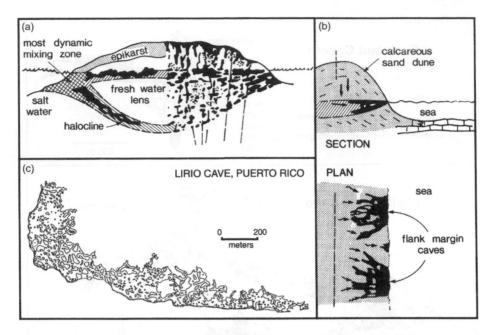

Figure 7.35 (a) General model for zones of enhanced solubility in a young carbonate island. (Left) Development of caves at the water table and along the halocline is stressed. (Right) Model distribution of cavities in the Yates oilfield buried karst, Texas. Adapted from Craig, D.H. (1987) Caves and other features of Permian karst in San Andres dolomites, Yates field reservoir, West Texas, in Paleokarst (eds N.P. James and P.W. Choquette), Springer-Verlag, New York, pp. 342–63.(b) Model for the development of flank margin caves. Reproduced from Mylroie, J.E. and Carew, J.L. (1990) The flank margin model for dissolution cave development in carbonate platforms. Earth Surface Processes and Landforms, 15, 413–24. (c) Map of Lirio Cave, Isla de Mona, Puerto Rico, an outstanding example of a flank margin cave. (By J.E. Mylroie, with permission.)

Where the initial cementation is strong enough to prevent large amounts of washout, small primary vadose caves (rounded shafts, often termed **pit caves** or **soil pipes**) develop in the vadose zone. Conduit caves may be created along the water table where there are allogenic streams or where the autogenic catchment is sufficiently extensive. For example, Mylroie and Carew (2000) note that many conduits developed under the Bahamas when falls of only 10 m below modern sea level increased the dry land catchment area of the Bahamas Banks by ten times or more; these caves are now drowned and inert.

In principle, caves may develop on limestone coasts anywhere along the halocline as a consequence of mixing corrosion, although in practice the effects are much less obvious in ancient crystalline limestones than in young porous limestones. As already noted (section 3.7), intense dissolution is common in many large coastal caves, creating microspongeworks there. But in such cases the caves themselves usually have a different origin, and the halocline entered them as a result of post-glacial sea-level rise. Gunn and Lowe (2000) describe short, inclined caves in the Tongan Archipelago that may have evolved along the length of haloclines beneath fringing reefs,

but little other evidence is reported. Rapid Quaternary sea-level changes have tended to keep the elevation of the halocline in a state of flux, creating small cavities over a substantial vertical range rather than a few larger caves; this pattern is suggested by the empirical model given on the right-hand side of Figure 7.35a.

Mylroie and Carew (1990) have shown convincingly that caves can develop very quickly around the coast where the water table in diagenetically young limestones meets the sea. In these circumstances, fresh- and salt-water mixing is coupled with oxidation/reduction processes there to create **flank margin caves** (Figure 7.35b). A mixing dissolution front recedes irregularly into the rock, leaving wide, low-roofed caves between pillars of more resistant rock in its rear. In the ideal case, the front itself is an abrupt (blind) termination of the cavity, with no suggestion of a proto-cave tube in it. Physically, this model is close to that envisaged by Rhoades and Sinacori (1941; section 7.2) for meteoric cave genesis. The original studies on San Salvador Island, Bahamas, focused on development in the margins of calcareous dunes, where it was shown that the caves were able to extend back into cemented dunes of the previous interglacial stage at mean rates up to 0.5–$1.0\,\mathrm{m\,kyr}^{-1}$. It applies equally well to uplifted reef and platform strata. Figure 7.35c shows Lirio Cave, Isla de Mona, Puerto Rico (Mylroie and Jenson 2002). This exceptional example is an uplifted, relict marginal cave at a contact between platform limestones and dolomites underlying a plateau. The cave extends as much as 250 m into the rock. However, that was no more than 5% of the width of the plateau; a freshwater lens could be sustained under the remainder of it, so maintaining an effective mixing front. The ratio of width to height in the many chambers between the pillars in this cave is always $> 10{:}1$

7.10 PASSAGE CROSS-SECTIONS AND SMALLER FEATURES OF EROSIONAL MORPHOLOGY

The erosional form of a cave passage may be attributable entirely to dissolution in phreatic (pressure flow) conditions or in vadose (free-flow) conditions or alternating (floodwater) conditions. Many passages are compound forms displaying, first, phreatic erosion and then vadose erosion. Dissolutional forms of every kind may be modified or destroyed by collapse of walls or roof. Breakdown processes are reviewed in section 7.12.

There is a great variety of smaller forms (**speleogens**) eroded into cave ceilings, walls and floors, including standard and variant channels plus many types of karren,

most of which also occur above ground. Lauritzen and Lundberg (2000) gave a comprehensive review in which passage cross-sections are classified as *mesoforms*, and lesser speleogens within them as *microforms*. Slabe (1995) illustrated the latter in detail, using many simulations with plaster-of-Paris. Bini (1978) and Zhu (1988) also presented general classifications.

7.10.1 *Phreatic passage cross-sections*

When first connected, phreatic passages (Fr. *conduites forcés*) have a subcircular cross-section or (if resistance is very low) are elongated along the fissure (Figure 7.36a and b). Diameter or width is no more than a few centimetres. The dissolutional attack is delivered to all parts of the perimeter.

The form that develops as the passage enlarges further is a function of the interaction between passive variables (lithological and structural) and active mass transfer variables (fluid velocity, dissolution potential, type and abundance of clastic load). Figure 7.36c illustrates the case where geological properties are isotropic or the active variables are much more significant (normally, flow is very fast). The minimum friction cross-section (a circular pipe) is maintained (Figure 7.37). Circular cross-sections are common, even at great size, e.g. at La Hoya de Zimapan (Figure 7.15) the diameter is $\sim$30 m, both where flow is vertically upwards and where it is horizontal.

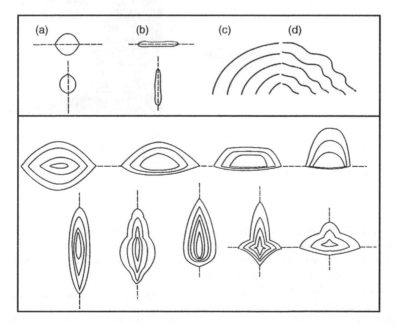

Figure 7.36 The evolution of phreatic passage cross-sections. In the lower frame, effects of differing solubility in beds above and below the exploited bedding plane and along the joint are illustrated.

Figure 7.37   Cross-sections of some phreatic passages. (Top left and middle) Passages centred on bedding planes; residual flood clays mask the floors. (Top right) A passage centred on a small, inclined fault. (These views in Castleguard Cave, Canada). (Bottom right) A joint-centred passage, Grotte Valerie, Canada. (Bottom left) A bedding plane passage in Mammoth Cave, Kentucky. (Photographs by Ford and (bottom left) A.N. Palmer.)

Isotropic geology with slow mass transfer will similarly permit the simple fissure to enlarge uniformly to great size. More often, however, passive variables are anisotropic normal to bedding planes (i.e. properties change significantly from bed to bed). Irregular profiles must then evolve (Figures 7.36d, 7.36 lower and 7.37). The extent of irregularity may be a function of size, and thus of the aggregate amount of erosion time.

The variety of phreatic cross-sections is enormous, but all are variations on these themes. Lange (1968) and Šusteršic (1979) have attempted formal geometric analyses.

7.10.2 Anastomoses, pendants and half-tubes

Anastomoses may be independent forms or, with pendants or half-tubes, can constitute a gradational set of

Figure 7.38 (Top) Dissolution pockets in passage ceilings. Since the cave drained, the pocket at right has begun to fill calcite speleothems fed from the joint that guided its dissolution. (Photographs by Ford.) (Bottom left) Anastomoses in the roof of a passage in Mammoth Cave, Kentucky. (Bottom right) 'Boneyard' – intense pocketing associated with H_2SO_4 formation in oxidizing waters, shown here in reefal rock at Carlsbad Caverns, New Mexico; also characteristic of coastal mixing caves. (Photographs by A.N. Palmer.)

features (Figure 7.38). Independent anastomoses are the subsidiaries of primary tubes. They may continue to extend throughout the phreatic history of a cave. The frequency of their divergence and convergence is a function of properties of the fissure and of its gradient.

Vertical joints, steeply dipping bedding planes, etc. show relatively little anastomosing.

Excellent anastomoses can originate late in the history of a cave where water from an established large passage opens up hitherto impenetrable bedding planes or fractures (i.e. effective fissure frequency is increased). This is best seen where stratal dip is low. Often it appears that the penetration is by floodwater, e.g. into bedding planes overlying the original ceiling in many places in Mammoth Cave, Kentucky.

Pendants (Ger. *deckenkarren*) are residual pillars of rock between anastomosing channels. They can develop in bedding planes and joints, where they are gradational from anastomoses. They also appear on unfissured erosional surfaces such as cave walls where they developed at the contact with rather impermeable clastic fill. They may be as much as 1 m long and can cover tens of square metres of cave wall. They can be carved by water draining up, down or along the contact. Very diverse and complex patterns can be created.

Ceiling half-tubes (Fr. *chenaux de voute*, Ger. *wirbelkanal*) are equally common but more controversial. Some develop where a bed is more resistant. We consider that the great majority develop to carry the surviving flow through a passage that has become choked with sediment, i.e. they are a stage beyond the ideal paragenetic passage of Figure 7.19. Although normally found in the apex of a passage, horizontal examples are known in walls, while others climb from wall to apex. In the apex, they often transform into a broader pattern of pendants (and back again), especially where ceilings are horizontal. Lauritzen (1981) has reproduced them experimentally.

If half-tubes are in the apex, they are the last places to fill during a flood. This led Bögli (1980) to propose that they develop because the CO_2 of trapped air is dissolved there, enhancing the local solvent capability. They are found in deep phreatic conditions (and in gypsum caves), so this explanation can be only partial.

7.10.3 Solution pockets and cupolas

Solution pockets are one of the most attractive features of phreatic caves and one which most surprises geomorphologists who are not cave specialists. They may occur in floors and walls but are best developed in the roofs (Figure 7.38). They are varieties of blind pockets that extend as much as 30–40 m upwards. Many terminate in a tight joint or microjoint that they have followed during expansion. They may be single or multiple features, rounded like conventional cupolas, or elongated along the guiding fracture. Some are multicuspate and transitional to honeycomb structure; these develop particularly well in vuggy rock, e.g. reefs. Some pockets are complexly multicuspate but have neither vugs nor joints to guide them. Osborne (2004) gives a comprehensive review.

It is now well established that pockets can be created by condensation corrosion in vadose conditions. These

are considered in section 7.11. In relict phreatic caves, however, the majority of solution pockets will generally be of phreatic origin except, possibly, in periodically wet–dry entrance zones or climatic chimney sites (section 7.11). Several different mechanisms have been proposed to explain these. Most widely accepted is dissolution by cellular convection in near-static waters, driven either by thermal gradients or by solute density gradients. Bögli (1980) firmly believed that mixing corrosion explained them in meteoric water caves, where enriched subsoil solutions descending the guiding joints mixed with bulk water flowing in the cave passage to create mixing points of accelerated dissolution; Veress *et al.* (1992) presented a comprehensive analysis of the geometry produced by such mixing. However, the likely mixing ratios (one volume from the joint per 100–1000 volumes or more from the passage) seem to preclude this as anything more than a local trigger mechanism in limestone caves; also, pockets of identical form are seen in gypsum caves where mixing corrosion is not a factor and, as noted, many pockets lack a source joint. Pocketing certainly occurs where salt and fresh waters mix at the halocline in coastal caves in young limestones, but it tends to be small and rough, unlike the typical smoothly rounded or edged forms illustrated in Figure 7.38. It has been suggested that pocketing may be due to flooding (i.e. an epiphreatic phenomenon), or to compression of air during floods. In many phreatic caves, however, it is found that the taller pockets terminate upwards at one given elevation that was determined by a palaeowater-table, i.e. they were always submerged. Some of these display a corrosion notch just below the top that marks the low-stage water level, e.g. the Cheddar caves shown in Figure 7.22.

7.10.4 Vadose passages, corrasion and potholes

The form of vadose passages is that of entrenchment, with or without widening. At its minimum development, the entrenchment is an underfit in the floor of a phreatic passage (Figure 7.39a). This suggests that regional phreatic flow has been rerouted and the passage now handles only local epikarst drainage. T-form or keyhole compound passages as in Figure 7.39b suggest that the same river has always occupied them, switching from phreatic to vadose conditions. Figures 7.39c and e are characteristic of drawdown vadose morphology; entrenchment predominates but the phreatic part was first and fixes the position of the trench on the map. There may be multiple entrenchments, where each floor represents a former baselevel, as in the example from Belgium that is shown Figure 7.39d. Single trenches as deep as 100 m are known; they are canyons with a roof on.

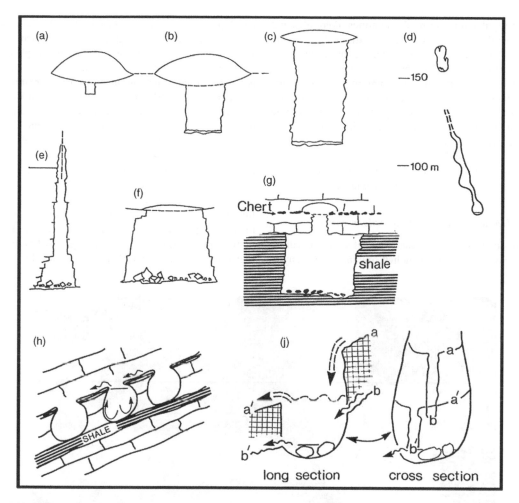

Figure 7.39 Vadose entrenchments: A–F, see text; cross-section D from Grotte Ste. Anne de Tilff, Belgium (Ek 1961); G, McClung's Cave, West Virginia, a typical example of entrenchment from an initial phreatic passage in limestones down into mechanically weaker shales; H, purely dissolutional stream potholing in limestone arrested by a basal shale band; J, a large pothole eliminated by underfit stream entrenchment, which destroys its plunge. Reproduced from Ford, D.C. (1965b) Stream potholes as indicators of erosion phases in caves. Bulletin of the National Speleogical Society, 27(1), 27–32.

Many entrenching streams are able to establish long profiles that in fluvial geomorphology would suggest rough equilibrium. The channel becomes armoured by bedload moved only during floods. This encourages channel widening. Cave walls are undermined, and large blocks collapse from them into the channel which deflects flow to the opposite wall, where the process is repeated. By the mixture of undercutting and breakdown a trapezoid cross-section of stable width is achieved (Figure 7.39f). This is a very common form.

Although all categories of vadose channels can be created by dissolution alone, mechanical erosion by bedload (corrasion) is often important. This is attested by innumerable instances where the entrenchment is largely or entirely in underlying insoluble rocks. Gouffre Pierre St Martin, one of the world's deepest systems (Table 7.2 and Figure 2.17) is a celebrated instance. Figure 7.39g shows a more extreme example; in the 'Contact' caves of Greenbrier County, West Virginia, passages extend for tens of kilometres in insoluble but mechanically weak shales beneath initial, much smaller, limestone solution caves.

Where channel gradients are steep and the rock is hard, stream potholes (Fr. *marmites*) may develop. These occur in all strong rocks, being drilled when grinder boulders are trapped and spun (rock mills) in any small hole in the channel bed. However, they are most frequent and display the most regular form in hard limestones, dolomites and (pre-eminently) in marbles, because dissolution in the swirling water reinforces the grinding there and may replace it entirely (Ford 1965b; Figure 7.39h). To maintain and deepen a pothole a minimum plunging force (height times discharge) is necessary. In some caves this is lost, a narrow entrenchment being cut through the pothole so that it is left abandoned in the wall (Figure 7.39j). There are two successive flights of eliminated potholes in Swildon's Hole, England, and a third flight in the active channel is in process of elimination. It is apparently an effect of changes in the dominant discharge.

7.10.5 Meandering channels in vadose caves

Three distinct types of meandering channels occur in vadose caves. The **entrenched** meandering canyon develops where a waterfall (a knickpoint) recedes along a meandering channel (Figure 7.41a). The canyon has simple vertical walls and meanders only in plan view. This type is common where strata are flat-lying and there is a deep vadose zone with many invasion streams.

Ingrowing meandering canyons are created by meandering entrenchment down into bedrock. There is no waterfall recession. As the channel is cut down it migrates forward (Figure 7.41a) and it may also do so as it rises (Figure 7.41b). Ingrown meanders are common in surface canyons (e.g. the famous 'goosenecks' of the San Juan River, Colorado), but they achieve their supreme morphological development in well-bedded limestones in caves. Many examples are several tens of metres deep and some kilometers in length, but too narrow for an explorer to pass through most parts. Deep ingrowing meanders will often wander laterally far from

Figure 7.40   (Top) Vadose entrenchments; note dissolutional scallops in left and centre frames. (Bottom) Basal undercutting to create trapezoid or breakdown forms. (Photographs by Ford, A.N. Palmer and A.C. Waltham.)

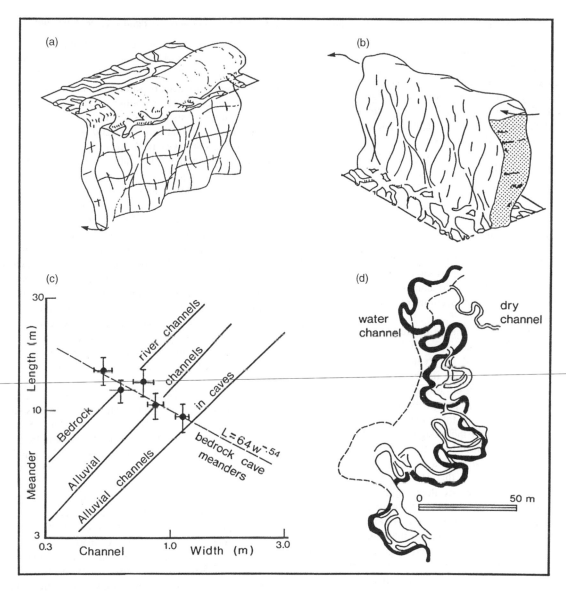

Figure 7.41   Meandering channels in caves. (a) Bedrock meanders ingrowing below an initial phreatic passage. (b) Paragenetic passage meandering above an initial phreatic passage. Reproduced from Ewers, R.O. (1982) Cavern development in the dimensions of length and breadth. McMaster Univ. PhD thesis, 398 pp. (c) Ingrowing bedrock meanders in caves compared with alluvial channels and surficial bedrock channels. (Modified from Smart and Brown 1981.) Bedrock cave meander measurements in Ireland and New Zealand trend counter to the norm found in surface rivers. (d) Part of the composite (bedrock and alluvial) channel pattern in Hurricane River Cave, Arkansas, USA. (Adapted from a map by N.W. and J.O. Youngsteadt, with permission.)

the original passage above if they encounter a particularly penetrable bedding plane.

Smart and Brown (1981) studied ingrowing meanders in the Burren, Ireland, and at Waitomo in New Zealand. The most striking result is that the relationship between meander wavelength and channel width is the converse of that found in other meanders, alluvial or bedrock (Figure 7.41c). In the examples they measured, channel width increases downstream (the standard response to increased discharge from addition of tributaries) but canyon height diminishes as the water table is approached; this may have determined the unusual behaviour.

Alluvial meanders develop where the channel is formed in sand, gravel, etc. on a broad passage floor. Deike and White (1969) showed that examples in Missouri caves behaved morphometrically in the same manner as all surface channel meanders (Figure 7.41d). There is one major distinction, however; because the rock is soluble the alluvial channel may pass smoothly into the wall (becoming a bedrock channel) and back out again. Where there is slow downcutting (or paragenetic rise) of the channel floor this creates tapered fins of rock (bedrock point-bars) projecting into the passage. They are another channel form that is rarely or never found at the surface.

7.10.6 Vadose shafts

Shafts created by falling water are known up to 640 m in depth. Many of the deepest caves are of the primary or invasion vadose types and consist of spacious vertical shafts linked by short sections of constricted meander canyon (Figure 7.17).

The form of shafts ranges between two extremes. The first is that created by a powerful waterfall. The water mass itself will tend to create a simple circular or elliptical cross-section for its fall but this is often modified. Breccia is swept out of any guiding fault, to form a parallel wall shaft. A plunge pool undercuts the walls which, therefore, tend to display irregular taper rather than parallelism near the base. Spray at all levels attacks weaknesses to produce local block fall. Many shafts are highly irregular in form as a result of these effects.

The other extreme is that of the **domepit** created by a relatively slow and steady flow (Figure 7.42). This may occur as leakages at the base of the epikarst (Figure 5.28) or below point-recharge depressions (Figure 5.16a), and is able to attack drained joints in the vadose zone. In the ideal condition the flow is never large enough to detach and fall free in the vertical plane. Instead, it is retained against the rock by surface tension. It disperses radially from an input point and carves a set of dissolutional flutings down the walls. The pit has a symmetric dome at the top (where the first dispersion occurs) and is circular below. This kind of pit is best developed where strata are flat-lying and joints are few and with high resistance. This is the case in the Mammoth Cave area of Kentucky, where the form was first analysed (e.g. Merrill 1960). There is a sandstone caprock there which functions as an additional regulator to maintain a steady, filming flow of aggressive water.

Many shafts show a mixture of the two forms, with waterfall features down the fall line and fluting of farther parts of the perimeter.

7.10.7 Salt caves – rapid vadose cave development

Rock salt is ~1000 times more soluble than limestone. As a consequence, cave development is very rapid. Many caves of enterable size have been created where allogenic streams were introduced by salt mining operations during the past few centuries, including some large ones with tourist cave appeal such as Forat Mico in the Cardona salt diapir, Spain (Cardona and Viver 2002).

Explorable salt caves are largely limited to a primary vadose zone and the water table. Where streams can leak through a caprock, shafts are quickly drilled by (essentially) dissolution mining straight down following intergrain boundaries. Vertical joints or other fractures are not required. The fast, largely first-order, kinetics prohibit lengthy penetration beneath any water table unless it is along an interformation contact such as an underlying dolomite. In most situations, the cave below the entrance shafts is a gallery perched on an insoluble base above the water table, or it has the juvenile type of profile of Figure 7.17. Where there is ponding in perched situations, deep corrosion notches (below) are soon cut at the water line. Similarly, juvenile profiles are quickly lowered to the minimum gradient required to transport whatever bedload is supplied from the caprock.

The greatest known salt caves are 2–6 km in length. Greatest depths are 150–200 m. The leading studies have been by Frumkin (1995, 2000a,b) at the Mount Sedom diapir which overlooks the Dead Sea, a very arid locality. The diaper is capped with anhydrite. It is rising rapidly, with active faulting at the margins where the salt also has the tight concertina folding typical of diapiric deformation. By ^{14}C-dating remnant twigs stranded in upper trenches, Frumkin was able to show that ~25 m of entrenchment had occurred in 3300 year at Mishqafaim Cave, keeping pace with the jerky, aperiodic rise of the diapir (Figure 7.43). Using erosion pins, instantaneous entrenchment rates up to $0.2\,\text{mm s}^{-1}$ were measured, but only for a few minutes of storms during the 5 year observation period. Mean annual entrenchment-rate estimates range 5–$25\,\text{mm a}^{-1}$, being greatest where allogenic catchments are largest. Salt caves in the Zagros Mountains, Iran, have also been studied intensively in recent years (see Bosak *et al.* 1999).

7.10.8 Scalloping by dissolution, sublimation and condensation

Dissolutional **scallops** are spoon-shaped scoops (Figure 7.44). They occur in packed patterns so that individuals are usually overlapping and incomplete. They are common

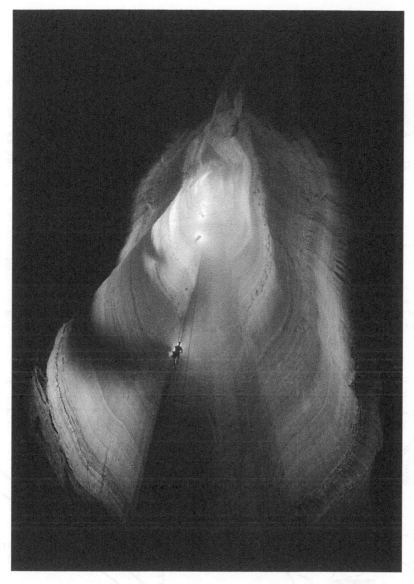

Figure 7.42 Fantastic Pit, Ellison's Cave, Georgia, USA. This is a beautiful example of a fluted domepit. It is 190 m deep, developed in thick to massively bedded limestones that are flat-lying. (Photograph by A.N. Palmer.)

on walls, floors and ceilings in caves. Inspection shows that they are smallest where flow is fastest, e.g. in a venturi. Measurement reveals that their length is log-normally distributed, usually with relatively little statistical dispersion. Most scallops are 0.5–20 cm in length but they range up to 2 m. Width is ~50% of length. In the right conditions it is evident that patterns of scallops of characteristic length extend to colonize all available surfaces. They are the stable form of these surfaces in the prevailing conditions.

It has long been recognized that many scallops are strongly asymmetrical in the direction of flow. The perimeter is steeper at the upstream end, i.e. facing downstream (Bretz 1942). **Flutes** (Curl 1966) show the same asymmetry and are really scallops of infinite width; they are rare. In relict caves, scallops and flutes are

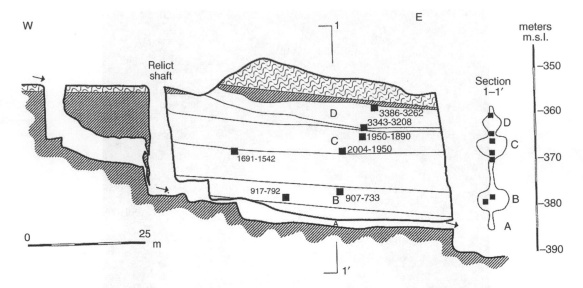

Figure 7.43 Long-section and cross-section through Mishqafaim Cave, a salt cave in the Mount Sedom diapir, Israel. It is an excellent example of 'episodic vadose entrenchment. Numbers' are ^{14}C ages in years of relict wood fragments stranded in the upper trenches. Reproduced with permission from Frumkin, A. and Ford, D.C., Rapid entrenchment of stream profiles in the salt caves of Mount Sedom, Israel. Earth Surface Processes and Landforms 20, 139–152 © 1995 John Wiley and Sons.

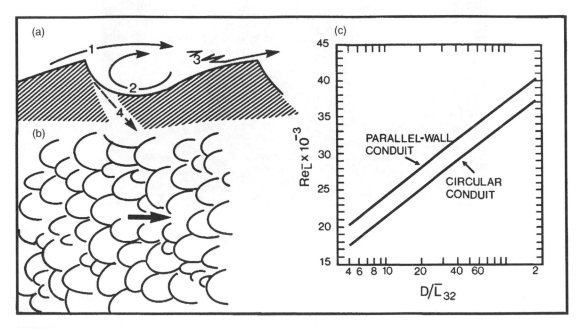

Figure 7.44 Dissolutional scalloping. (a) Section through a scallop: 1, the saturated boundary layer detaches; 2, turbulent eddy (locus of maximum dissolution); 3, diffusion, mixing and reattachment; 4, course of the steepest cusp with further dissolution (the scallop migrates downstream). (b) The characteristic appearance of a fully developed scallop pattern on a surface; the entire surface is occupied and individuals overlap. (c) The predicted relation between Reynolds' number and the ratio, conduit width or diameter d to Sauter mean scallop length, $\check{L}$ 32. Reproduced from Curl, R.L., Deducing flow velocity in cave conduits from scallops. Nat. Speleo. Soc. Am. Bull. 36(2), pp. 1–5 © 1974 The Geological Society of America.

potentially important indicators of both the direction and velocity of palaeoflow.

Scalloping is common in limestone, gypsum and salt caves. However, not all limestone regions display it and it is quite rare in dolomite caves. This is because uniform rock grain size is necessary for good scalloping, and a lack of heterogeneities such as insoluble fragments or open pores. Many limestones and dolomites are too heterogeneous to develop scalloping, e.g. reef rocks.

Scalloping also develops where wind blows over old, well-densified snow or through ice in a glacier cave. It even develops (but without the steeper upstream cusps) on infalling iron meteorites! In these cases the process is mass transfer by sublimation of homogeneous crystalline material. In the karst rocks it is the same, but with dissolution substituting for sublimation. Curl (1966) proposed an elegant theory of scallop and flute formation in which there is detachment of the saturated boundary layer at a specified Reynolds' Number (equation 5.6), occurring in the subcritical turbulent flow regime. Detachment permits aggressive bulk fluids to erode the solid rock directly (Figure 7.44). The frequency of detachment increases as velocity increases, thus reducing the erosion length available to each individual scallop. Scallop length will be inversely proportional to fluid velocity and to fluid viscosity. Air has a much lower viscosity than water; for a given velocity, scallops are longer (larger) in ice and snow than on limestone, etc.

Curl's (1966) theory has been quite exhaustively tested and confirmed with laboratory simulations and in the field (see Lauritzen *et al.* 1986). Mean palaeovelocity (v) in a channel may be computed by obtaining Re_L from Curl's graph for circular and parallel wall conduits (Figure 7.44c) and substituting the value in the equation

$$v = v \frac{\bar{Re_L}}{\bar{L}_{32}} \qquad (7.1)$$

where $\bar{Re_L}$ is a Reynolds number for scallops, based on the fluid velocity at a distance from the wall equal to $\bar{L}_{32}$ and v is the kinematic viscosity (Table 5.4); $\bar{L}_{32}$ is the Sauter mean length of scallops

$$\bar{L}_{32} = \frac{\Sigma L_i^3}{\Sigma L_i^2} \qquad (7.2)$$

being the greatest length (in the direction of flow) of the *i*th scallop. This device is used to suppress the statistical significance of the subpopulation of very short scallops that occurs in many scallop distributions because of bedrock inhomogeneities.

An alternative, direct calculation for a compromise between circular and parallel wall cases is given by Curl (1966) and is accurate to ±15%

$$v = \frac{v}{\bar{L}_{32}[55 \ln(D_h \bar{L}_{32}) + 81]} \qquad (7.3)$$

where D_h is the hydraulic diameter (four times cross-section area divided by length of the wetted perimeter).

The concept of dominant discharge is important throughout hydrology and fluvial geomorphology because this is the flow assumed to be responsible for channel width in alluvial rivers. It is difficult to establish and may not apply to bedrock channels. In well-scalloped cave passages, however, we may think in terms of one or several 'scallop dominant discharges', i.e. those discharges that create the one or several Sauter mean scallop lengths measured. In examples from some Norwegian caves, Lauritzen *et al.* (1982) have shown that the modern scallop dominant discharge is approximately equivalent to the annual snow-melt flood, i.e. to the upper 5% of the flow regime, and some three times larger than mean annual discharge.

At very high velocities sand may be entrained as suspended load (Figure 8.2). Its abrasive effects then may override dissolutional effects so that scallops become highly elongated and polished, resembling 'flute markings' obtained experimentally by Allen (1972). This will only occur where there is an abundant supply of hard (i.e. silica) sand and thus is rare in caves. Excellent examples are seen at narrow points in the great gallery of Niaux Cave, Pyrenees, which was subjected to violent glacial meltwater floods. Simms (2004) compared the mean veolocities required to entrain particles and to produce solutional scallops. He found that the velocities necessary to entrain sand particles of 1 mm would produce scallops of about 10 cm length, and noted that scallop formation continues at flow velocities two orders of magnitude slower than the minimum threshold velocity for entrainment of any particle.

Air scalloping also develops from corrosion by condensation waters, where turbulent air movement near cave entrances introduces humid air that becomes saturated on contact with cooler cave walls, condenses and dissolves the surface (section 7.11).

7.10.9 Corrosion notches, bevels and facets

Scallops are created by an accelerated erosion mechanism requiring turbulent fluids. This survey of solutional forms in caves is concluded by going to the other extreme and considering a special family of features developed because the water is nearly static, so that the slightest fluid density gradients may establish sharply delimited zones of accelerated erosion.

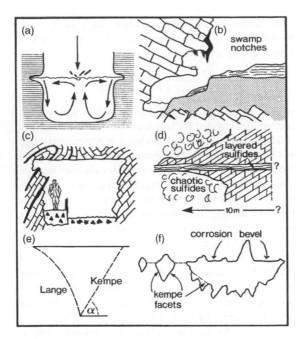

Figure 7.45 (a) Corrosion notching at the surface of a pool. (b) Foot caves or swamp notches in tower karst. (c) Corrosion bevelled ceiling in Fung Kui Water Cave, Guangdong, China. This shows the level of a previous water table. (d) Corrosion notch filled with layered pyrite, Nanisivik zinc–lead mine, Baffin Island, Canada. (e) Corrosional facets; the models of Lange (1968) and Kempe *et al.* (1975). (f) Na Spicaku Cave, Moravia, Czech Republic; corrosion notching is sharpest where passages are largest.

Corrosion notching is illustrated in Figure 7.45. In a standing pool heavy solute ions and ion pairs sink, driving a cellular convection that carries fresh H^+ ions to the walls at the water surface. A sharp notch is dissolved there, tapering off very steeply below the waterline. This is the normal form. It is widespread in limestone caves, e.g. at siphon pools where it develops in low-stage conditions. We have seen examples where such notching is as much as 1 m deep, although this is exceptional. It can signify palaeowater-table levels very precisely. Some times there are several notches one above another.

This notching becomes much larger in the foot caves of karst towers abutting alluvial floodplains (Figures 7.45b and 9.43). Notching at the seasonal flood level and extending several metres into the rock is common. Because such notches are more extensive and create a flat roof regardless of geological structure, they may be termed **corrosion bevels** (Ger. *laugdecke* of Kempe *et al.*

1975). In examples in some karst towers of southern China, the bevelling extends throughout the length of State 4 (water-table cave) passages that pierce through them (Figure 7.45c). These caves have lost their prime hydrological function, because most water now discharges around the towers on the floodplain.

Although bevelling at the level of the water table is extremely common in humid tropical caves, the greatest corrosion notching we know occurs in the hyperacid conditions of some massive sulphide emplacements. For example, at Nanisivik Mine, Baffin Island, the greatest notch is > 400 m wide, but only 1 m deep. It is horizontal, slashing across dolomites dipping at ~15°, and is filled with syngenetic layered pyrite (Figure 7.45d). Stable isotopic evidence suggests that the pyrite was deposited at temperatures between 80° and 150°C (Ghazban *et al.* 1992), so this was a case of immense stability in fluid layering far below the water table.

The first theoretical analyses of density gradient corrosion notching in meteoric water were made by Lange (1968), who supposed that beneath the waterline the corrosional cut would taper away in a smooth exponential function. Kempe *et al.* (1975) suggested on the basis of physical simulations and examples in gypsum caves of southern Germany, that the taper should be linear. They measured density currents 1–3 mm thick descending the natural gypsum walls, and calculated velocities of ~0.5 cm s^{-1}. These linear corrosion surfaces were termed **facets** by Kempe *et al.* (1975). They are comparatively rare in limestone caves in our experience because the surfaces become armoured with silt or clay residuum that inhibits the linear dissolution. They are better seen in maze caves in pure gypsum, such as many of the Ukrainian giants (Klimchouk 1996). Retreat of a Lange or Kempe facet under stable water-table conditions yields a corrosion bevel.

7.10.10 Notch caves

Some striking caves can be created entirely by initial slow dissolutional enlargement of vertical or inclined fissures that is succeeded by notching within them following a fall of the water table. The notching (with bevels and facets, plus breakdown that may be induced by the undercutting) dominates the enlargement. These we term **notch caves**.

Notch caves require an isolation control that permits only limited supply and draining of water. In many instances, the isolation is strictly an effect of local geology. Ochtina Cave, a UNESCO World Heritage cavern in Slovakia, is an excellent example. It is the largest of 16 or more small maze and chamber caves

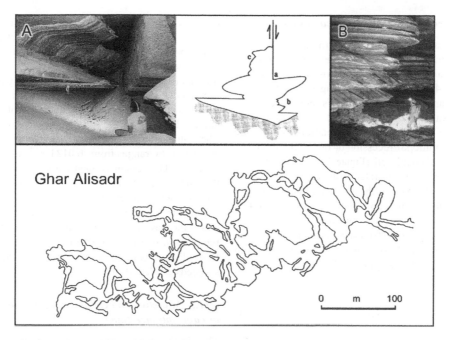

Figure 7.46 (Upper left) A syntectonic passage in Malham Cave, Mount Sedom, Israel–Palestine, showing waterline corrosion notches and an active fault plane in a salt diapir. (Photograph by A. Frumkin, with permission.) (Upper right) Seven corrosion notches can be seen in this view in Cova del Riu, a human-induced cave in the Cardona salt dome, Spain. Adapted from Cardona, F. and Viver, J. (2002) Sota la sal de Cardona, Espeleo Club de Gracia, Barcelona, 128 pp. (Bottom) Ghar Alisadr Cave, a large corrosion notch cave in Iran.

perched far above valleys in the humid temperate hills of the Slovak Raj (Paradise). Water is trapped in lenses of metasomatic limestone, enclosed by phyllites and partly altered to ankerite and siderite by Mg- and Fe-bearing hydrothermal solutions in the past. There is sporadic flooding via fractures in overlying phyllites, the water ponding or draining very slowly out through lower phyllites until it is refreshed by another flood. Convectional dissolution pockets, notches, bevels and facets are all well developed (Bosak *et al.* 2003).

Aridity is another effective isolation control. Rare but violent flooding and consequent notching are prominent in Kartchner Caverns in the Arizona desert (Graf 1999), in Pofaddergat, Thabazimbi Cave, and others in arid regions of southern Africa. The greatest notch cave we have seen is Ghar Alisadr, Hamadan, Iran. This is developed in a narrow ridge of steeply dipping metamorphosed limestones that protrude above weaker schists in a semi-arid basin. The limestones are shaly and impure, probably supplying an H_2S supplement to normal CO_2 dissolution. The only recharge is occasional rain onto the limestone. Narrow dissolution fissures on bedding planes and cross-joints have been greatly enlarged by notching in a large, semi-static pool. Eleven kilometres of passages are

mapped (Figure 7.46 lower). Calcite shelfstones at, above and below the modern waterline indicate 3 m of oscillation of the pool surface over the past 25 000 year or so (Kaufmann 2002); this may be partly related to cool-climate wetter conditions in the Pleistocene and partly to human regulation of a scarce water resource.

7.11 CONDENSATION, CONDENSATION CORROSION AND WEATHERING IN CAVES

In recent years there has been growing interest in the role that processes of aqueous condensation may play in caves. Thus in a New Zealand cave, de Freitas and Schmekal (2003) monitored and numerically simulated for over 1 year the microclimate processes of vapour transfer and condensation. Much of the water is acidic and may attack the rock. Earlier cavities are enlarged, and their shape and proportions may be radically changed. At the least there is slow weathering of the walls and ceilings of relict passages, chiefly in cave-entrance zones.

To simplify, we assume that condensation and its corrosion takes place in the three distinct settings presented below; in practice, there may be a mixture of them occurring in a given cave.

7.11.1 Closed cellular convectional corrosion

This was the first corrosional situation to be considered in detail. Convection is driven by the heat from a pool of thermal water in a chamber that is closed to outside air flow (Figure 7.47a). Vapour (including any H_2CO_3 and H_2SO_4, etc. carried by the water) is condensed onto cooler cave walls. Szunyogh (1989) analysed this in detail, deriving a quite complex equation to estimate the rate of dissolutional recession of the walls about the centre of a hypothetical cell (Figure 7.47b). The film of condensed water (now saturated with respect to calcite) thickens towards the base of the cell, shielding it a little to produce the onion-form chamber shown in Figure 7.47c. For a reasonable range of pCO_2 and thermal gradients in the rock, Szunyogh (1989) estimated wall recession rates of 50–200 μm a^{-1} for water at 60°C, reducing to 4–30 μm a^{-1} at 20°C. Dreybrodt (2003) simplified the model

$$\frac{dM}{dt} = 0.26 \frac{T_{air} - T_{rock}}{t^{\frac{1}{2}}} (\mathrm{g\,m^{-2}\,s^{-1}}) \tag{7.4}$$

and Lismonde (2003) refines this to suggest recession rates around $0.9 \,\mathrm{m}\, 10^4 \,\mathrm{a}^{-1}$. Typical spheroidal pockets attributed to condensation corrosion range 1–5 m or so in diameter, so such rates seem reasonable. Where dissolution by H_2CO_3 is replaced or augmented by H_2S, they may be faster.

7.11.2 Lateral advective corrosion (climatic chimney processes)

Here the ideal situation may be thought of as an inclined cylindrical cave passing from a plateau to a valley below. It is open to the outside atmosphere at both ends. This is the 'chimney' model of cave climatologists. If there are strong contrasts in the external temperature (daily where the cave is short, seasonally where it is long), cave walls are chilled by descending air flow when it is cold outside. Moisture may then be condensed on to them if there is ascending air flow due to subsequent warm conditions.

This setting has been studied meticulously, both in theory and in the field, by V.N. Dubljansky (see Dubljansky and Dubljansky, 2000). At the scale of large caves and extensive blocks of karstified rock around them he proposes

$$A = V\epsilon(e_s - e_u)tJ \tag{7.5}$$

where A is the mass of condensate (g), V is the volume of cave and surrounding karst with active air circulation (m^3), ϵ is the estimated porosity (0–1), $(e_s - e_u)$ is the difference in absolute humidity, surface and underground (g m^{-3}), t is in days and J is the frequency of air exchange (replacement) underground, estimated from cave air-flow data. Field studies were conducted in the Caucasus Mountains and the Crimean Plateau, locations with high relief, cold winters and hot, dry summers. In summer strong correlations between the discharge of karst springs and the humidity of the external air were often detected. The volumes of condensation water in the springs were estimated to range from 0.012 L s^{-1} to as much as 8.6 L s^{-1}. Dry-season condensation water yields ranged 0.3–10 L s^{-1} km^{-2}, or 0.1 to 9% of the mean annual precipitation at different locations.

In contrast to the cellular model, the air flow is linear, along the passage in the ideal case. Dissolved rock is removed in film flow down the walls, or laterally in the form of aerosols when the cycle reverses and evaporation succeeds condensation.

7.11.3 Advective and cellular condensation corrosion in cave entrance zones

This is the most common and widespread setting for condensation corrosion. A little can be seen in all cave entrances except those that are very constricted or thoroughly screened by vegetation. It applies in caves that have only one effective climatic entrance as well as those (such as chimneys) with two or more. However, the quantitative effects are generally more muted than in thermal or chimney settings with strong air circulation.

Figure 7.47d shows a model for diurnal corrosion derived from micrometeorological and condensation field studies in caves of Cayman Brac, Cayman Islands (Tarhule-Lips and Ford 1998a). The caves are short, tapering solution cavities of the flank margin marine type, opening in coastal cliff faces. The climate is tropical, with a diurnal temperature cycle typically ~2–5°C. During the day the cave is comparatively cool and air flows out along the floor, drawing warmer replacement air in at the ceiling. At night this breaks down into weaker advective flow at the floor and cellular flow in any chambers above it. Moisture condenses from the warm daytime air.

In the large majority of caves there is a distinctive climatic entrance zone in which both temperature and relative humidity vary. Condensation and potential corrosion rapidly diminish as the amount of variation (daily or seasonal or both) reduces to nearly zero in the deeper interior. Wigley and Brown (1976) showed that the extent of the zone can be modelled as a relaxation (or 'e-folding') length

$$x_0 = 36.44a^{1.2}V^{0.2} \tag{7.6}$$

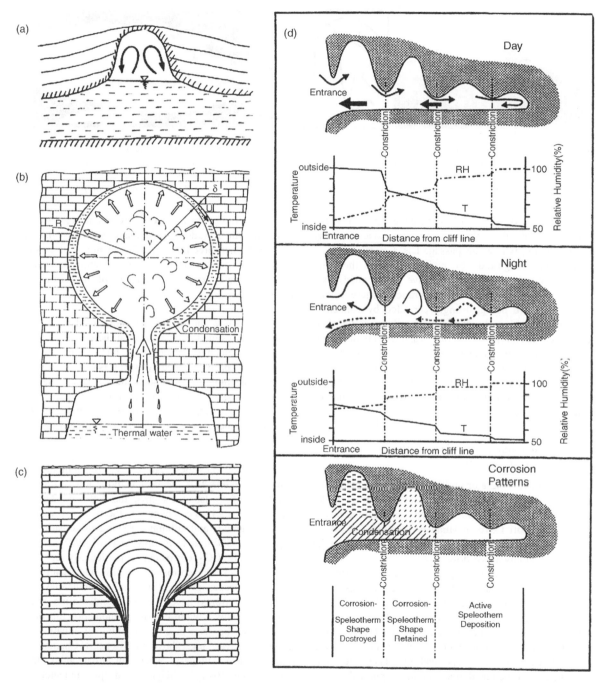

Figure 7.47 (Upper left) A closed convection cell above thermal waters. Reproduced fromMuller, P. and I. Sarvary (1977) Some aspects of developments in Hungarian speleology theories during the last ten years. Karszte's Barlang, Special Issue, pp. 53–59. (Centre left) The model of G. Szunyogh (1989) for condensation in a closed convection cell. (Lower left) The cupola form produced by Szunyogh's closed convection model. (Right) A model of cave-entrance-zone daily microclimate behaviour and resulting condensation corrosion for a tropical island cave. Reproduced from Tarhule-Lips, R. and Ford, D.C. (1998a) Condensation corrosion in caves on Cayman Brac and Isla de Mona, P.R. Journal of Caves and Karst Studies, 60(2), 84–95.

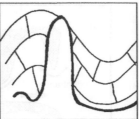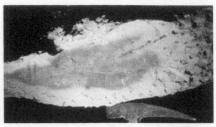

Figure 7.48 (a) Condensation corrosion has bevelled a cave ceiling and severely corroded the stalactite below it in this scene in a cave in northeast Brazil. (Photograph by A. Auler, with permission.) (b) Bellhole morphology. Reproduced from Tarhule-Lips, R. and Ford, D.C. (1998b) Morphometric studies of bellhole development on Cayman Brac. Cave and Karst Science, 25(3), 119–30. (c) The weathering of a limestone pendant in Martinska Cave, Slovenia. (Photograph by N. Zupan Hajna, with permission.)

dependent only on a (the radius of the passage) and V (the velocity of air flow through it). The zone typically extends 5–$6 x_0$ where the passage has a constant cross-section. The magnitude of x_0 is reduced if there are constrictions. At one prominent cave on Cayman Brac, only 30 m separates the cave entrance, where calcite speleothems are being strongly corroded by condensation, from interior chambers, where there is vigorous speleothem accumulation. From measurements of solutes in condensation water and from the dissolution of gypsum tablets suspended in the air, mean corrosion rates around $20\,\mathrm{mm\,ka^{-1}}$ were obtained; from theoretical considerations, Dreybrodt (2003) suggests that these may be somewhat too great.

7.11.4 Condensation corrosional features on bedrocks and speleothems

As noted above, single or multiple spheroidal pockets are produced where cellular convectional air flow is predominant. The best examples, perhaps, are Sátorköpuszta Cave and Bátori Cave in Hungary (section 7.8). Well-formed pockets are also seen where there is local cellular convection in advective flow caves.

Where advective flow predominates, the principal forms on bedrock walls and ceilings are varieties of **air scallops** and **flutes**. These have the same basic form as scallops in stream channels described above, including the flow asymmetry, but are much larger (0.5–2.5 m or more in length) as a consequence of the lower viscosity of air, and are generally very shallow. They are more prominent and better defined in and approaching air flow constrictions. Where denser air may be ponded behind a natural dam in a cave, corrosion bevels that trim off pendants or trim the entire ceiling can be seen (Figure 7.48a).

The impact of condensation corrosion is most apparent where speleothems such as stalactites, stalagmites and columns protrude into a moist air stream and are corroded by it. Growth of vadose speleothems indicates that net calcite (or other) precipitation is dominant at a given site. Reduction by corrosion thus signifies a significant change in the balance of forces. Three levels of impact may be recognized:

1. minor, where the stalagmite (etc.) is damaged by small corrosional rills, usually on the upwind side;
2. where as much as 75% of the volume of the deposits may have been corroded away but its original form is still discernible (e.g. the stalactite in Figure 7.48a);
3. major, where the original form is lost and an air scallop form cuts smoothly across its stump and the adjoining bedrock wall or ceiling (Tarhule-Lips and Ford 1998a).

Such damage is widespread in caves in subhumid and semi-arid regions. It is spectacular in the lowest (youngest) level of Carlsbad Caverns, the Lefthand Tunnel, where there is a H_2S component in the corrosive vapour; most stalagmites and columns growing in constrictions display 50–75% loss from upwind (Hill 1987). Where there is only CO_2 corrosion, a particularly common feature is pocketing corrosion upwards underneath protruding carapaces in large stalagmites, indicating that rising air currents have focused condensation there. Although chiefly a warm-climate phenomenon this occasionally can be seen in climatic chimney situations in temperate caves, e.g. above the Concert Chamber in Baradla Cave, Hungary, where an ornamented stalagmite some 10 m in height is deeply indented by pockets.

In the case of speleothems it is often apparent that the condensation corrosion has been episodic. Periods of net corrosion are succeeded by hiatuses with very slow weathering, or by renewed calcite deposition. Evidently there is climatic or microclimatic control, but attempts to firmly correlate episodes of corrosion to, for example, global glacial conditions or lower sea levels, have so far failed because of local factors such as erosion or infilling disrupting the cave air-flow patterns.

7.11.5 Bell holes

Bell holes (Wilford 1966) are a geomorphological peculiarity, strictly vertical cylinders of nearly perfect circular cross-section, seen in the ceilings of a small minority of caves in the tropics. Their verticality is maintained regardless of lithological variation, stratal dip or the regularity or gradient of the ceiling (Figure 7.48b). There may be shallower **bell pits** (Lauritzen and Lundberg 2000) in the floor beneath them. Careful study in Cayman Brac determined that they occur only in the illuminated or partially illuminated entrance zones (Tarhule-Lips and Ford 1998b). Height ranges 0.5 to 5.7 m in four sample caves in dolomitic limestones; their diameters are proportional, 0.6–1.3 m. It is proposed that these bell holes were formed by condensation corrosion strongly abetted by microbial activity. Colonies of microorganisms (biofilms) establish themselves in patches on ceilings in the entrance zones, where they obtain moisture from condensing water. Dissolution of the bedrock by these microorganisms initiates a depression in the ceiling. When deep enough, this traps the warmest air, inducing extra condensation in a positive feedback mechanism. The verticality of the holes is attributed to the fact that only the most soluble intergrain calcite cements are dissolved by the microorganisms and condensation corrosion; the less soluble grains themselves are removed by gravity, which is most effective when operating straight down.

In some very large tropical cave entrances we have seen apparent bell holes that are as great as ~8 m in height and 1.5 m diameter, exposed to view by later condensation cutting across them; thus the maximum sizes of bell holes are not yet established. Densely packed, small, bell-hole-like features occur in limestone and dolomite in the zone of water-level oscillation of some freshwater lakes in Canada and Ireland; they are weakly illuminated and may also be mixed bio- and condensation corrosional in origin.

7.11.6 Weathering in relict caves

Where limestone and dolomite caves have been drained and relict for long periods it is common to find that the walls and ceilings are weathered. Condensation is the source of the moisture. However, its rate of deposition is too low to permit effective film flow that can remove the solutes and so sculpt the walls as it does in spheroidal pockets, for example. Velocities of air flow across the

surfaces are similarly too low for rapid removal of solutes as aerosols, so that air scallop and flute corrosion does not occur either. Instead, the rock surfaces are subject to a very delicate, preferential dissolution of all tiny weaknesses. Microboxwork may develop where calcite veins are frequent. Most common are patterns of dense, irregular pits 0.3–1.0 cm in diameter and a millimetre or two deep; in Bone Cave, West Virginia, we have found that they follow silty streaks in the limestone. Rims are soft and can be scraped away by fingernail, or even the claws of mice! Zupan Hajna (2003) has studied such weathering in detail in caves of Slovenia (Figure 7.48c). The weathering rinds generally extended to depths of 1–5 cm there, and were $\geq 90\%$ calcite or dolomite. The water content ranged 24–40%.

The weathering is greatest in the cave-entrance zones defined above, although it often extends throughout chimney caves. It can also be found deep in the interiors of some relict caves without vigorous airflow, where it is probably the result of lengthy wetting and drying cycles, and proceeds very slowly indeed.

Palmer and Palmer (2003) describe a different mode of weathering in Mammoth Cave, Kentucky. Diffuse seepage loses its CO_2 passing through sandstone above the cave; it becomes acidified when it is wicked out to the cave walls and encounters the atmosphere, contributing to weathering that is 0.2–2.5 cm deep. Precipitated in the rind there may be trace amounts of secondary silica derived from the sandstone.

7.12 BREAKDOWN IN CAVES

Jagged surfaces of rupture in walls and roofs and piles of angular rocks that have fallen from them are found in a majority of caves (Figure 7.49). They constitute the third basic morphology seen underground, modifying or replacing previous phreatic or vadose solutional forms. This morphology is termed **breakdown** or **collapse** by English-speaking speleologists, **incasion** by many European specialists.

The cause of all breakdown is mechanical failure within or between rock beds or joint-bounded masses. The summary of cave breakdown that follows is also applicable to cliff faces exposed in surface karst, e.g. at stream sinks, pinnacles, towers, etc.

The load on a point in a rock mass may be expressed simply as

$$p = \rho g h \tag{7.7}$$

where ρ is rock density and h the thickness of the rock mass or height of the cliff overhead. The distribution of

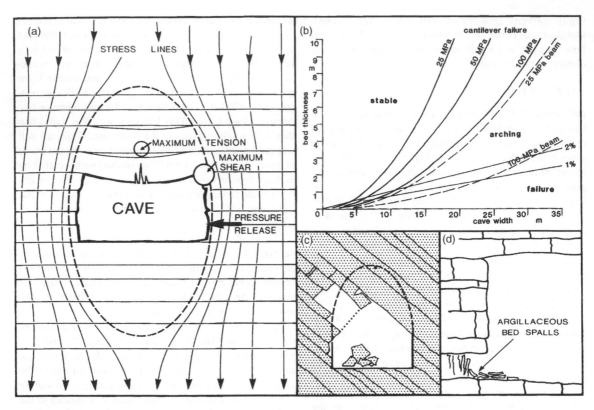

Figure 7.49   (a) Distribution of stress lines around a cave and tension dome. (b) Critical spans (passage widths) or breaking thicknesses for beams and cantilevers in limestones with horizontal bedding. Reproduced from Waltham, A.C. (1996) Ground subsidence over underground cavities. Journal of the Geological Society of China, 39(4), 605–26. (c) Deformation of the theoretical breakout dome where stratal dip is steep. (d) Pressure-release spalling where load is maximum, e.g. at a pillar or the foot of a wall.

this load, as a stress field about a cave cross-section, is given in Figure 7.49a. A tension dome is created in the rock above the passage. Its height is determined principally by passage width. Rock in the dome is subject to sagging and the overlying weight is transferred to the adjoining passage walls, greatly increasing the stress there. There is also tension in the floor, but in natural caves (as opposed to mined cavities) this is of little significance.

The largest cave breakdowns are failures in the tension dome. These are most regular in form where strata are well-bedded and horizontal. As a consequence, this is the situation used in basic analysis. Bonding across bedding planes is presumed to be much weaker than strength within the beds; therefore, the beds sag elastically away from each other. Each bed then can be considered to function as a separate **beam** if it extends the full width of the passage, or as a **cantilever** where it is fractured through (e.g. by a wide central joint) or does not extend the full width. Fractured spans are usually much stronger

than simple cantilevers. Cliffs at the surface most often fail as cantilevers.

Mechanical rupture and fall of the bed will occur where a critical span is exceeded for a given thickness and strength, and vice versa. A simple equation for a beam is

$$t_{\text{crit}} = \frac{\rho \lambda^2}{2S} \qquad (7.8)$$

where t_{crit} is critical thickness, λ the span of the beam (passage width) and S the bending stress. For a cantilever:

$$t_{\text{crit}} = \frac{3\rho \lambda^2}{2S} \qquad (7.9)$$

i.e. cantilevers are much weaker. Rock density is used here as an approximation for strength: in mining practice, triaxial compressive test values are substituted for it (Brady and Brown 1985).

Figure 7.50 Ceiling breakdown occurring at a passage junction. (Photograph by A.N. Palmer.)

The bending stress is defined as:

$$S = \frac{Mc}{I} \tag{7.10}$$

where M, the maximum moment of the beam, is given by

$$M = \beta t \rho \lambda^2 \tag{7.11}$$

with β the width of the beam, i.e. normal to t and λ

$$c = \frac{t}{2} \tag{7.12}$$

and I, the moment of inertia of the beam, is given by

$$I = \beta t^3 \tag{7.13}$$

Figure 7.49b models critical spans (passage widths) or breaking thicknesses for beams and cantilevers in limestones with horizontal bedding (Waltham 1996). The unconfined compressive strengths (UCS) range from a weak 25 MPa ($\rho = 18\,\text{kPa m}^{-3}$) that is found in many chalks up to 100 + MPa ($\rho = 26\,\text{kPa m}^{-3}$) found in the strongest cavernous limestones. Note the critical importance of bed thickness. Thick-bedded carbonates of medium strength can support roofs up to 20 m wide before failure where they are unbroken beams, and massive beds ($> 1.0\,\text{m}$ thick) may extend 35 m or more.

This model uses only flexural strength, assuming that the rock breaks like a piece of elastic that is suddenly overstretched. This does not allow for plastic creep or the development of cracks in highly stressed areas. Failure usually occurs where open cracks allow rotational movement of the bending beds. This is considered in a crack propagation model by Tharp (1995):

$$v = c \left(\frac{K_I}{K_{Ic}} \right)^n \tag{7.14}$$

where v is the velocity of crack propagation, K_I is the stress intensity (MPa m$^{0.5}$), K_{Ic} is the 'fracture toughness' of the given rock and c is an empirical constant (Tharp 1995). In practice it will be difficult to estimate K_{Ic}, especially in soluble rocks where dissolution within the widening cracks may disturb simple, progressive, physical failure; Tharp suggests that this may reduce the rock strength to about one half within 100 year. In Figure 7.49b the values 1% and 2% estimate the reduction in strength of a cantilever in which crack width is 1% and 2% of the cantilever length.

Failure in the tension dome tends to progress upwards, one or a few successive beds falling at one time. As the form becomes increasingly arched it is strengthened and the rate of bed fall may reduce. Breakdown finally stabilizes, however, when either (i) the span of the bed newly exposed is less than the critical width for its thickness, or (ii) the dome below becomes fully clogged with fallen breakdown which supports the remaining span. Clogging must occur unless there is concurrent removal of the breakdown by dissolution, settling, stream action, etc. at and towards the base of the pile because the volume occupied by breakdown will always exceed that of the previously unbroken rock. The stabilizing height where there is no concurrent removal is given by

$$h = h_0 \frac{k}{k-1} \qquad (7.15)$$

where h_0 is the initial height of the cave (before there is any collapse into it) and k the volume increase created by piling the breakdown (Andrejchuk 1999). For example, where $h_0 = 10\,\mathrm{m}$ and the void volume in the pile (its bulk porosity) is 10%, clogging occurs 100 m above the original ceiling; for 20% porosity the limit is 50 m and for 50% it is only 20 m.

The proportion of failure in the tension dome (partial, or completed to either upwards limit) and its geographical extent varies greatly within and between caves. There may be no breakdown. For example, it is negligible in much of the Big Room at Carlsbad Caverns (Figure 7.51) although the room is often wider than 50 m. This is because it is located in massive reefal rock. Where the

wide Carlsbad passages extend into adjoining backreef beds, breakdown promptly occurs.

Where cavernous strata are generally horizontal and medium- to thick-bedded, it is common to find collapse domes at particular places only. They tend to develop at passage junctions or where a prominent joint intersection creates a cantilever. At the extreme there is partial or complete failure throughout the length of a passage. This is characteristic of most large passages in gypsum because the rock is weakened by hydration stresses. Klimchouk and Andreichuk (2003) give a detailed account of breakdown in the Ukraine gypsum maze caves, recording 1800 collapses per square kilometre in Zolushka Cave as a consequence of its dewatering for quarrying purposes.

Where strata dip steeply the same tension dome with a vertical orientation exists in theory, but it is evident that updip walls and roofs are the more unstable, creating asymmetric breakdown (Figure 7.49c). Here, the adhesion across bedding planes and joint faces (neglected in the analyses given above) becomes an important variable. In mines, adhesion will be measured by laboratory or *in situ* shear-box testing.

Tension dome breakdown that is completed to stability at a given time may not be stable over the thousands to millions of years that the cave may then survive, especially if it is in a vadose state. Infiltrating waters drain preferentially towards domes and, if chemically aggressive, will attack the rock, reducing S values and converting beams to cantilevers, etc. This renews the process of upwards failure of rock, termed **stoping**. Many karst terrains contain deep breccia pipes or geological organs

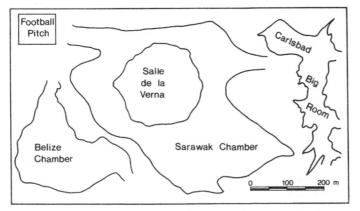

Figure 7.51 Examples of the largest cave chambers or rooms. Sarawak Chamber, Lubang Nasib Bagus (Good Luck Cave), Mulu National Park, Sarawak; Belize Chamber, Actun Tun Kul (Tun Kul Cave), Belize. Big Room, Carlsbad Caverns, USA and Salle de la Verna, Gouffre Pierre St Martin, France. The great shaft of El Sotano, Mexico, is compared with the 430 m high Empire State Building, New York.

(section 9.13; Quinlan 1978) that are the debris-filled products of long sequences of tension dome formation and collapse. Frequently, they extend into non-karstic cover strata. Where the latter are mechanically weak but elastic (e.g. clays, many shales), it is common for the progressive fragmentation to be replaced by cylindrical downfaulting *en masse*. Stoping upwards has propagated through more than 1000 m or more of cover rocks at sites in Canada and Russia. In mines, rock bursts (explosive release of stress) sometimes shatter walls and floors. This is unknown in natural caves, because the slow rate of their excavation permits less violent adjustments to the stresses being imposed. However, it is common to see pressure-release spalling of particular beds exposed in cave walls (especially at pillar sites such as passage junctions), or at or near the bases of limestone or dolomite cliffs (Figure 7.50d). The spalling beds are usually the most argillaceous and/or fine-grained in the sequence The spalled fragments display platy or conchoidal fracture. Such basal spalling undermines upper walls, widening the tension dome, and so may induce a general collapse at the site.

White and White (1969) divide breakdown fragments into three categories.

1. Block breakdown: rock fragments consisting of more than one bed remaining as a coherent unit.
2. Slab breakdown: rock fragments consisting of single beds.
3. Chip breakdown: rock fragments derived from the fragmentation of a bed.

Single blocks with volumes greater than 25 000 m^3 are known, e.g. from a reef–backreef junction at Carlsbad Caverns. Slab breakdown is predominant in thin- to thick-bedded, horizontal strata. There are many forms of chip breakdown because here petrological properties will determine the shape; splinters, flakes, plates, arrowheads, cubes and saucer-shaped conchoidal fragments are amongst the more common forms.

Mechanical failure is the proximate cause of all breakdown and cave genesis is the ultimate cause. Between these extremes, other factors determine the sites and timing of the breakdown events. Probably, the three most important factors are: (i) draining of phreatic caves (e.g. Zolushka), which removes the buoyant force of water, effectively increasing the load by 30% to 50% in different karst rocks; (ii) vadose streams, which over-widen a passage beyond its beam or cantilever width limits, either locally at meander undercuts and passage junctions, or generally along a passage that becomes alluviated; (iii) aggressive vadose seepage waters weak-

ening the roofs, as noted above. In very well-aerated caves dissolution by water condensed onto walls may produce much slab and chip breakdown.

In many caves it is suspected that major falls may have been triggered by earthquakes. In the headwater passages at Castleguard Cave, Canada, it is evident that the addition of 200–400 m of glacier ice overhead induced pressure-release spalling (Schroeder and Ford 1983): probably, this was a factor in caves in many formerly glaciated terrains. Gypsum precipitating within limestone from SO_4-enriched seepage waters causes widespread chip and slab breakdown at Mammoth Cave (White and White 2003) and in many other caves around the world; salt plays the same role in some Nullarbor caves, Australia (section 10.1).

7.12.1 Frost shattering and frost pockets

Frost shattering is an effective agent of breakdown in limestone and dolomite caves in cool regions. In caves where there is only one entrance that admits or discharges thermally significant quantities of air or where such entrances are far apart, frost penetration is limited to the entrance zone of vigorous temperature fluctuation, as discussed above (equation 7.6). Effective frost action is usually limited to the first one or two relaxation lengths. The shattering zone will be longer or frost action more severe where the cave is a chimney with two or more large entrances at different elevations, as is the case of many relict caves in mountainous regions. The frost debris produced are normally varieties of chips. Slab breakdown is common close to entrances.

Frost shattering on cliffs may create shallow, shelter-type caves termed **frost pockets**. These develop in all stronger rocks where there are wetted fractures, but they appear to be most abundant in the carbonate rocks because they are wetted via solutionally enlarged openings. Solution is the trigger process for larger scale frost action about the point of groundwater emergence (Schroeder 1979). The pockets often expand upwards into the tension dome that they create, forming a rounded arch that may be metres to tens of metres in height. At a distance these cannot be distinguished from the frost-modified entrances to large caves, which has led to many exaggerated reports of dissolution cave frequency in mountainous regions.

7.12.2 The largest known voids

Great open spaces in caves are termed **rooms** by American cavers and **chambers** by British cavers. Some of the largest that are known are shown in Figure 7.51. Sarawak

Chamber, discovered in 1980 in the Mulu karst, Sarawak, is the largest. It has a volume of approximately $20\ 10^6\ m^3$. Miaos Chamber, Gebihe Cave, China has $\sim\!10\ 10^6\ m^3$. Belize Chamber, Salle de la Verna and the Carlsbad Big Room are each in excess of $1\ 10^6\ m^3$. Hundreds of rooms are known with volumes of 100 000 to 500 000 m.

A majority of these rooms are centred at vadose river passage junctions where undercutting has widened tension domes to produce repeated breakdown. The rivers are able to remove much of the debris in solution, and so maintain open voids rather than breccia pipes. Carlsbad Big Room and some large chambers occurring in hydrothermal caves probably owe most of their excavation to the exotic corrosion effects discussed above and display relatively little breakdown morphology.

Impressive as they are, these greatest cave rooms are minor in scale when compared with the debris-filled volumes of many breccia pipes.

8

Cave Interior Deposits

8.1 INTRODUCTION

Caves function as giant sediment traps, accumulating samples of all clastic, chemical and organic debris mobile in the local environment during the life of the cave. They are the most richly varied deposits that form in continental environments and tend to be preserved for greater spans of time than most others do. It is not by chance that a majority of important Lower Palaeolithic archaeological sites have been caves.

Cave sediments have received much more study than the cave erosional genesis discussed in Chapter 7, because of their archaeological and palaeoenvironmental significance and because they can be sampled and analysed like other sedimentary deposits. They may be divided into cave entrance plus rock shelter (**abris**) deposits or facies, and cave interior deposits (Figure 8.1). This chapter focuses on the latter. Entrance deposits, because of their archaeological importance, can be considered a partly separate subject.

Table 8.1 presents a comprehensive classification of cave interior sediments. There is a basic division into clastic, organic and precipitated types and into sediment formed outside of a cave and transported into it (allogenic or allochthonous) and that created within a cave (autogenic or autochthonous).

This introduction closes with a warning. Cave sediments can be most complex. The law of superposition (that an upper deposit is younger than a lower deposit on which it rests) is often violated because of shrinkage, slumping, flowstone intrusion, burrowing and other effects. Many facies are diachronous (i.e. they differ in age laterally). Variation in rates of deposition can be extreme. Downstream movement of sediment may be much obstructed because breakdown barriers sieve it to varying degrees or dam it entirely for long periods. Undoubtedly, there is much reworking and redeposition.

8.2 CLASTIC SEDIMENTS

8.2.1 Mechanics of fluvial sediment transport

Local breakdown (section 7.12) and allochthonous fluvial sediments are the predominant categories of clastic deposits. Sediment transport in open alluvial channels such as rivers has been studied intensively (e.g. Middleton 2003). However, vadose caves are usually rock-bound channels, hence width is fixed during an event, channels are particularly rough and they usually lack floodplains. Phreatic and floodwater caves are approximately equivalent to manmade pipes, but with much higher friction due to their irregularity. Transport of solids in slurries through filled pipes has been the subject of much engineering experiment. In particular, a study by Newitt *et al.* (1955) was particularly important, so its findings are followed in our discussion.

Sediment begins to be moved when the shear stress at the bed (boundary) exerted by the specific stream power exceeds a critical value (equations 5.17 and 5.18). A general expression for boundary shear stress τ_0 ($kg\,m^{-2}$) is:

$$\tau_0 = \rho g \theta \frac{a}{2d + w} \tag{8.1}$$

where ρ is fluid density, θ is channel slope, a is cross-sectional area, d is depth and w is width. The critical value τ_{crit} increases with particle diameter and density but is also dependent on its shape, packing among other particles, etc. An approximation is

$$\tau_{crit} = 0.06(\rho_s - \rho)gD \tag{8.2}$$

where D is particle diameter (mm), and ρ_s is particle density. The greatest particle moved by a given flow is estimated by

$$D_{max} = 65\tau_0^{0.54} \tag{8.3}$$

Karst Hydrogeology and Geomorphology, Derek Ford and Paul Williams
© 2007 John Wiley & Sons, Ltd

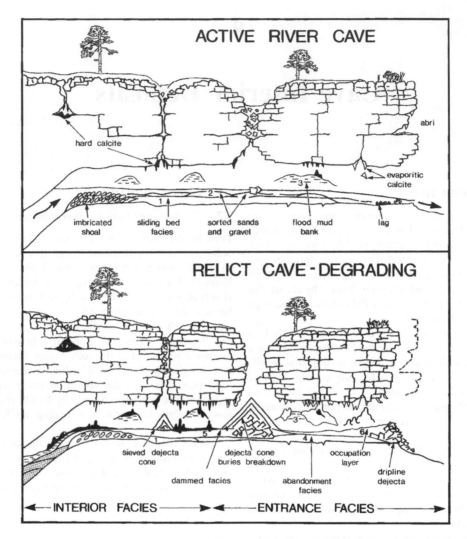

Figure 8.1 A model depicting the principal categories of clastic and precipitate deposits in hydrologically active or relict caves.

Equations estimating bedload transport *en masse* face many problems. However, that of Meyer-Peter and Muller (1948) is widely used

$$\text{rate} = 0.253(\tau_0 - \tau_{\text{crit}})^{3/2} \qquad (8.4)$$

For deposition, the terminal settling velocity v_t for a spherical particle is given by **Stokes' law** or similar equations

$$v_t = \frac{1}{18}\frac{(\rho_s - \rho)}{\mu}gD^2 \qquad (8.5)$$

The effects of these relationships in open channels are summarized in Figure 8.2. Dogwiler and Wicks (2004)

measured grain sizes at 59 stations in two stream caves in Kentucky and Missouri and used shear-stress analysis to determine that 85% of grains could be transported at the underground equivalent of river 'bankfull' stage, which in these regions had a recurrence interval of $\sim$1.7 year.

For pipefull flow Newitt *et al.* (1955) noted five modes of transport:

1. a slowest mode of rolling grains producing ripples on a stationary bed;
2. saltation of individual grains above the bed;
3. a sliding bed involving, at first, the upper part of the bed load but extending to all of it with further increase of velocity;

Table 8.1 Cave interior deposits

Source	Deposit type	Origin	Comments
Allochthonous or allogenic	Clastic	1 Fluvial	Many kinds – dominant allochthone
		2 Filtrates	From seepage – minor
		3 Lacustrine	Rare
		4 Marine	Beach facies
		5 Aeolian	Normally minor except at entrances
		6 Glacial and glaciofluvial injecta	Common in glaciated regions
		7 Dejecta, colluvium and mudflows	Normally restricted to entrance areas
		8 Tephric	Volcanic areas; inwashed ash and pumice
	Organic	9 Waterborne, windborne, etc.	Scale ranges from spores to tree trunks
		10 Exterior fauna	Sometimes cave-using species Bones, nests and middens, faeces
Autochthonous or autogenic	Clastic	11 Breakdown	Mainly by failure; minor thermoclastic
		12 Fluvial	Derived from breakdown or erosion of karst rock
		13 Weathering earths and rinds	
		14 Aeolian	Derivatives of 11, 13
Precipitates and evaporites*		15 Ice	As water ice, glacières, frost and glacial injecta
		16 Calcite	Most significant autochthone
		17 Other carbonates and hydrated carbonates	
		18 Sulphates and hydrated sulphates	
		19 Halides	
		20 Nitrates and Phosphates	
		21 Silica and silicates	
		22 Manganese and hydrated iron oxides	
		23 Ore-associated and miscellaneous minerals	

*More than 250 different minerals are known to be generated in caves. The dominant ones are listed.

4. heterogeneous suspension;
5. homogeneous suspension at the highest velocities (Figure 8.3).

Their experiments were limited to horizontal pipes of circular cross-section, and to grain sizes up to 10 mm. Later experimenters have obtained similar results from larger pipes.

There is no doubt that at least the first four of these transport modes occur in phreatic or flooded caves. There are many instances of pebbles and cobbles being carried straight up vertical shafts and expelled at the top. This represents heterogeneous suspension, at the least. An example from Castleguard Cave is plotted on Figure 8.3; 40–100 mm diameter pebbles of dense limestone ($\rho = 2.85$) are carried > 7.5 m up a floodwater shaft there (Schroeder and Ford 1983). Minimum velocities > 0.8–1.0 m s^{-1} are calculated from the settling laws, in good agreement with mean velocities of 1.2–1.3 m s^{-1} obtained from measured maximum flood discharges.

Note that the Castleguard grains plot in the 'stationary bed' domain of Newitt *et al.*'s (1955) experiments. This does not invalidate those experiments, but does emphasize that grain shape and other factors create great variability in real behaviour.

Figure 8.4 from Gillieson (1986) presents a summary of the relationships between cave hydraulic conditions and common sedimentary structures.

8.2.2 Deposits of gravel to boulder-sized material

This is typical bedload in many hilly karst regions if there are extensive allogenic catchments. It is important in most glaciated terrains, where it is reworked from till and outwash. It can also be generated from autogenic breakdown.

The extreme of transport is seen in steep floodwater caves such as many alpine systems, where the maximum permitted grain size may be decided by the minimum passage diameter. Allogenic boulders of 1t or more firmly wedged between floor and roof deep in the cave are

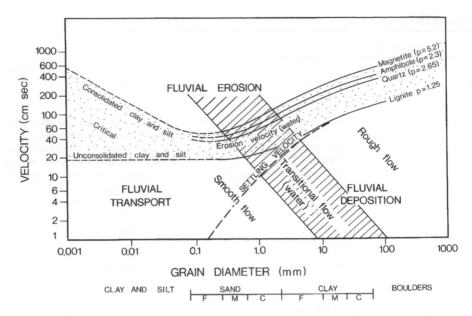

Figure 8.2 Curves showing the relationships of grain size to critical erosion velocities for flowing water with material of differing densities. Fluvial erosion curves refer to velocity at 1 m above the channel bed. Erosion, transport and deposition regimes are defined. Reproduced with permission from Sundborg, A, The River Klaralven, a study in fluvial processes. Geografiska Annaler, 38, 125–316. © 1956 Blackwell Publishing.

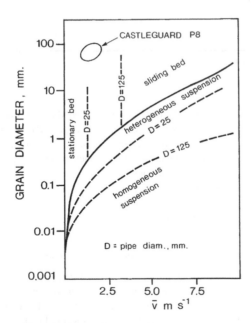

Figure 8.3 Clastic sediment transport regimes under conditions of pipefull water flow (Reproduced from Newitt, D.M., Richardson, J.F., Abbott, M. and Turtle, R.B. (1955) Hydraulic conveying of solids in horizontal pipes. Transactions Institute of Chemical Engineers, 33, 93–110.; with data for Castleguard Cave added.)

common. The dynamics can be likened to those of a vigorous toilet flush.

A facies that is frequently encountered is a poorly sorted to chaotic mixture of all sizes up to boulders, with patches that are grain-supported and others that are matrix-supported. There is little preferred orientation or other structure. Such deposits can have varying origins and, in general sedimentology, are known as **diamictons**. The top of the deposit is usually an abrupt contact that is often succeeded by well sorted gravel or sand. If in a sequence of deposits it is usually the basal unit. Such deposits in caves we attribute to the pipefull, sliding bed mode, i.e. all the mass was in motion, and then deposited simultaneously with sorting only by the dispersive pressure of colliding particles (McDonald and Vincent 1972). It is the equivalent of the antidune mode in alluvial channels. Gillieson (1986) presented the first detailed descriptions from caves, in the very wet highlands of Papua New Guinea. Sliding bed deposits are also recognized in subglacial esker deposits, i.e. deposited in floodwater caves formed in glaciers (Saunderson 1977). Figure 8.5 displays examples of Gillieson's results (curves 17 and 18) and the grain-size envelope for six esker samples (envelope 2).

In open cave channels or where pipefull velocities are too low, gravels and boulders move by rolling and sliding. They are deposited as shoals or bars of medium-

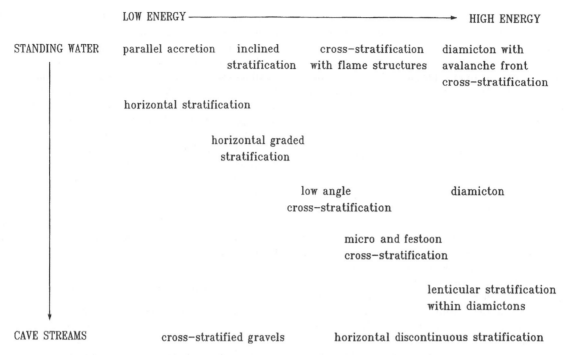

LOW ENERGY ⟶ HIGH ENERGY

STANDING WATER   parallel accretion   inclined   cross-stratification   diamicton with
                                        stratification   with flame structures   avalanche front
                                                                                  cross-stratification

                  horizontal stratification

                          horizontal graded
                          stratification

                                  low angle   diamicton
                              cross-stratification

                                      micro and festoon
                                      cross-stratification

                                                      lenticular stratification
                                                      within diamictons

CAVE STREAMS   cross-stratified gravels   horizontal discontinuous stratification

Figure 8.4 Relationships between water flow and depositional energy as expressed in cave clastic structures. Reproduced with permission from Gillieson, D., Cave sedimentation in the New Guinea highlands. Earth Surface Proc. and Landforms, 11, 533–543 © 1986 John Wiley and Sons.

to well-sorted grains, exhibiting lateral, downstream and upwards fining trends. Bar form is best downstream of constrictions such as short siphons. These deposits very often show channel **cut-and-fill** structures that include

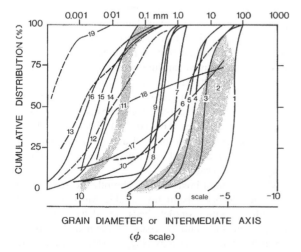

Figure 8.5 Cumulative distribution of grain size curves for 18 different cave deposits plus samples from six eskers. Data from many sources.

sands and fines. The gravels may be well imbricated. Wolfe (1973) described a palaeodeposit in West Virginia where the imbrication indicated flow in the direction opposite to that indicated by solutional scallops on adjoining walls: it was a consequence of an alluvial stream invading an abandoned phreatic cave.

A distinct facies may occur where there is a choice of routes for floodwater, so that any given passage can be progressively infilled. As its flood cross-sectional area is decreased by filling, velocities increase and thus the deposit coarsens upwards, perhaps terminating with a lag of boulders or cobbles jammed into the roof.

8.2.3 Sands and their sedimentary structures

Allochthonous sand is an abundant constituent of cave sediments because of the common occurrence of limestone and sandstone together in basins. The typical sand structures of flumes and alluvial channels (Middleton 2003) have been reported many times in caves. Sorting ranges from moderate to very good. Fining upwards, laterally or downstream is common, as is coarsening upwards. Sands are often interbedded with silts and clays.

In Figure 8.5, curves 6, 7 and 8 are representative of coarse and medium sands measured in the temperate river caves of Slovenia (Gospodarič 1976, Kranjc1981); curve 9 is the mean for well-sorted sands from the Niaux System (Pyrenees), where the deposits are proglacial outwash into giant passages (Sorriaux 1982); and curve 10 is sand ejected from a phreatic tube only 67 mm in diameter (Gale 1984). More recent studies have produced grain-size curves that are very similar in form and range (e.g. Valen *et al.* 1997).

8.2.4 Silts and clays deposited from suspension

Silts and clays are the most widespread clastic deposits in caves. Because they are transported in suspension they may coat walls and even ceilings, although most accumulation is on the floors.

Their sources are the most diversified. Allogenic sources include eroded soils, reworked fluvial and lacustrine deposits, and windborne dust and tephra (Figure 8.5, curves 14 and 16). There may be filtrates from soils overhead, and there is often a significant autogenic component from the weathering of walls (curve 15) or the winnowing or decomposition of older sediments. *Terra rossa* at the surface in southern France (curve 12) is winnowed to curve 13 at a depth of 30 m. Mixed allogenic and autogenic fines are often the final deposit in relict caves and are referred to as **cave earths**.

The clays and silts usually accrete as laminae that are parallel to the depositional surface, which may be horizontal, inclined, vertical or inverted. There may be marked fining upwards within a lamina, which creates a colour change that gives it the appearance of a couplet, e.g. from buff (silt dominant) to grey (fine silt–clay) in many glaciated regions. Thickness of individual laminae are reported to range from ~0.2 to >50 mm, but 1–10 mm is most common. Aggregate sections can total many metres.

Some clays lack lamination or fining structure, which suggests that they were decanted from a homogeneous suspension that was steadily renewed. However, this is rare. Lamination and fining (sometimes coarsening) upwards are normal and represent deposition from waning floodwaters or other pulsed flow, termed **slackwater** deposits (e.g. Gillieson 1996).

Typically, silts and clays grade laterally from centres of passages and are finest grained in deep recesses (curve 19). Floods can repeatedly scour and fill along the main passages but there may be only accretion in the recesses which thus can build up great banks with the most complete sedimentary records. 'Mud mountains' > 20 m in height are known in protected places.

Where frequent flooding and draining occur silt and clay banks steeper than ~ 40° may exhibit patterns of

rilling; rills are dendritic on slopes between 40° and 70°, becoming parallel where they are steeper. These 'surge-marks' are depositional, at least in part (Bull 1982).

Thin deposits onto vadose cave roofs are often reorganized into stringers and subcircular clusters of clay when the cave dries. They have the appearance of worm trails and are termed **vermiculations** (Bögli 1980). The patterns appear to be produced by a combination of droplet distribution and electrical charge attraction.

In extraglacial regions clays and silts are normally composed of clay minerals, quartz fines and some dolomite and calcite silts. Where there are allogenic catchments the clay minerals are most often dominated by the suites in their soils. In glaciated areas the proportion of carbonate grains typically increases to 20–80% or more of the fines. They are reworked 'glacier flour' produced by basal ice erosion.

8.2.5 Characteristic sedimentary facies and suites

All categories of sediments noted above can be found in common phreatic caves. Allogenic gravels and even boulders may be transported through a surprising number of deep loops. At Wookey Hole, the downstream end of a classic 'multiloop' cave system in the Mendip Hills, England (Figure 7.15; Farrant and Ford 2004) divers have found that medium gravels are in constant agitation even at low stages of flow, being driven up a very steep slope at −90 m by the force of the cave river, and rolling back again. Where there is such large and coarse clastic load, paragenetic deposition with roof dissolution and wall pendant formation is particularly common. The facies can include sands, granules and gravels, usually in fining upwards sequences. However, laminated silts and clays are more usually dominant in the phreatic environment, and paragenetic dissolution is more muted. Laminated fines in small amounts are the chief deposits in thermal water caves and artesian mazes.

The standard suite in gentle vadose or shallow phreatic caves is a series of cut-and-fill deposits with marked lateral fining (Figures 8.6 and 8.7). There may be a sliding bed facies at the base of any given sequence or this facies may cross-cut remnants of earlier sequences along a central channel.

Where a dam of breakdown is abruptly created in such caves the effects can be highly variable. Most often, there is infilling with parallel beds or delta facies on the upstream side, initially coarse, becoming finer as the barrier becomes increasingly clogged. Downstream, there is reworking or complete scouring of older deposits, plus a varying addition of finer grain sizes passed through the dam or winnowed from its components. If much of the dam

Figure 8.6 Some representative cave sediments. (Upper left) A basal chaotic (probably sliding bed) layer is terminated by clay at the level of the hand. A sequence of pebble–silt–clay overlies this and fills the passage to the roof. A later, imbricated fill of pebbles is emplaced in the upper clay by cut-and-fill processes. (Photograph by Ford.) (Upper right). Banks (dunes) of flood silt and clay, Selminum Tem Cave, New Guinea. (Photograph by A.S. White, by permission.) (Lower left) An 'abandonment suite' of pebbles fining upwards to clays succeeded by silts fining upwards to clays. (Photograph by Ford.) (Lower right) Typical flood clays ('slackwater deposits') in a backwater passage. Note the dendritic surge marks. (Photograph by A.N. Palmer, by permission.)

is then swept away in a flood, a particularly complex 'pond-and-sieve, winnow, cut-and-fill' suite is left.

Where upper passages are progressively abandoned by flowing water distinctive 'abandonment suites' may occur. These are cut-and-fill sequences with marked lateral fining, that also tend to become finer upwards as the floods become weaker and weaker in their capability. The final deposits tend to be finest at the upstream end (the overspill flood entry point where only suspended load can now enter) and coarsen downstream as a result of local rework-

ing. In Figure 8.5, grain size curves 3, 4 and 5 generalize the progression from perennial channel sediments to the overspill channel and then to the completely abandoned channel in river caves of West Virginia (Wolfe 1973).

Varved clays are rhythmic sequences of laminated silt and clay each appearing as a couplet that is darker and coarser below, finer and lighter above. They form in proglacial lakes, the finer fraction only settling out when the lake is frozen over in winter, i.e. ideally, each couplet represents one year's accumulation. Thick sequences

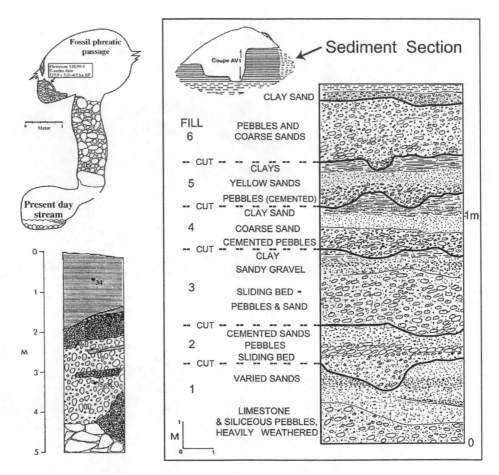

Figure 8.7   Representative drawings of sedimentary sections in caves. (Top left) Coarse fill in a compound (phreatic → vadose) conduit in Sirijordgrotta, Norway, with a modern stream below. Note that, despite the small size of the passage, the filling is capped with a calcite flowstone that has survived there for ∼128 000 yr. Reproduced from Valen, V., Lauritzen, S-E. and Løvlie, R. (1997) Sedimentation in a high-latitude karst cave: Sirijordgrotta, Nordland, Norway. Norsk Geologisk Tidsskrift, 77, 233–50. (Lower left) D. Gillieson's (1986) drawing of a colluvial diamicton (mud flow), with clay above, in a New Guinea cave. Reproduced from Gillieson, D. (1986) Cave sedimentation in the New Guinea highlands. Earth Surface Processes and Landforms, 11, 533–543. (Right) The author's interpretation of a complex sedimentary section in Grotte de l'Entre de Vénus, France. There are six channel cut-and-fill sequences, including two with partial sliding bed facies, within a section only 1.5 m in height. The passage cross-section (above) shows that this sequence was then cut through in a seventh erosional event. Reproduced from Delannoy, J-J. and Caillault, S. (1998) Les apports de l'endokarst dans la reconstitution morphogénique d'un karst: exemple de l'Antre de Vénus. Karstologia, 31, 27–41.

of varve-like rhythmites are common in caves in gla- ciated regions. Figure 8.5 curve 11 is the envelope for 14 samples from Castleguard Cave. They may be the predominant deposits, with later channels carrying bed- load being cut through them. They represent deposition from pondings beneath or along the flanks of Quaternary glaciers. They have a high carbonate content. It is not yet established whether each couplet is an annual (true varve) deposit or represents some other event; see Schroeder and

Ford (1983), Maire (1990), Campy (1990), Lignier and Desmets (2002) for discussions. Reams (1968) attributed couplet laminae in caves of Missouri to individual floods generated by thunderstorms.

Where surface rivers impound and backflood the downstream ends of caves, any sediments they are able to introduce tend to be those typical of the local flood- plains, chiefly fine sands, silts, clays, floating and water- logged organic debris, etc. (e.g. Springer and Kite 1997).

8.2.6 *Other types of clastic deposits*

Caves everywhere display cones of dejecta accumulated by a mixture of piecemeal falls and small slurries from steep or vertical openings to the surface (Figure 8.1). Passages that slope more gently from the surface may channel tongues of colluvium for many tens of metres. This is unsorted debris, usually matrix-supported and with a high organic content. Sorriaux (1982) described instances where fluvial sediments in Pyrenean caves have been remobilized as mass waste flows. Gillieson (1986) showed that fluidized polymodal mudflow deposits (diamictons) were able to penetrate up to 3 km into caves of the Papua New Guinea highlands.

Glaciers may discharge meltwater and clastic load directly into caves. Typical here is an injection facies; the passage entrance is blocked by wedged boulders and a tail of gravels and sands extends inwards. There are also simpler injecta where an advancing glacier merely bulldozed till or outwash into entrances, filling them entirely.

Caves that terminate in the coastal breaker zone may contain beach sand and shingle deposits. These take the form of berms (barriers) with steep fronts facing the waves and gentle backslopes that can extend inwards for several tens of metres. Williams (1982b) described a marine gravel deposit injected into a limestone cave graded to a 60 m emerged coastal terrace in New Zealand. Cosmogenic burial dating has since shown the quartz gravel to have been emplaced 1.25 ± 0.43 million years ago (D. Fabel, personal. communication).

8.2.7 *Particle shape and changes of shape and size in caves*

The shape of a particle is described in terms of form (platy, bladed, etc.), sphericity and the roundedness of corners (e.g. Sneed and Folk 1958, Cailleux and Tricart 1963). Most studies in caves have been concerned with pebble and cobble sizes, and have taken care to differentiate autogenic clasts (usually limestone or dolomite) from allogenic, non-carbonate clasts. Bull (1978) found no systematic downstream changes of autogenic bedload in river caves in Wales and attributed variations to original lithological properties. In general, this is not surprising. Caves that offer sufficient accessible lengths of alluvial channel for substantial changes of shape to be expected will be low gradient and likely to contain breakdown barriers that screen out upstream particles and substitute fresh stocks of angular fragments. Excessive roughness of channels also causes much breaking into angular pieces rather than steady attrition over all the surfaces of grains. Major rounding or shingling can occur

in situ where clasts of local origin are trapped against a barrier that the flowing water must ascend. Curve 1, Figure 8.5 presents an extreme illustration of this, depicting pebbles of the dense, particularly resistant, limestone that were trapped and rounded at the bottom of the 7.5 m floodwater shaft mentioned above in Castleguard Cave. Each pebble will have swept up the shingle beach there hundreds of times, and rolled back down; the net downstream movement of most was zero. However, in comprehensive studies in the Slovene river caves Kranjc (1989b) found that autogenic gravels could become very well rounded by 5–10 km of travel downstream.

The sizes of allochthonous pebbles, cobbles and boulders, and their proportion in the total bedload is certainly reduced during transport through caves. In large karst basins in West Virginia, Wolfe (1973) found that mean bedload grain size was weakly related to channel length and gradient (Table 8.2), but filtering in breakdown destroyed any trend at the outlet of the longest basin there. Kranjc (1982) noted a mean diameter loss of 4 mm km^{-1} for sandstones transported at least 11.5 km through Kacne Jama, but random samples of all clasts (allochthonous plus autochthonous) yielded no trend there. The sandstone roundedness generally increased to 350–550 on the Cailleux and Tricart (1963) scale (where $\sim$550 is close to spherical), but more resistant chert pebbles remained quite angular.

There have been many electron microscope studies of the shape of sand and clay size quartz grains in caves. These grains suffer comparatively little damage during underground transport, so the studies determine their origin or provenance. Quartz silts from local soils are most often detected. In a pioneer study at Agen Allwedd, Wales, Bull (1977b) recognized also a suite of glacigenic sands modified by fluvial transport and a later suite of diagenetic quartz sands from the local caprock, plus distinct glacial, fluvial or aeolian features in the silt and clay grains. He was able to correlate particular groups of silt and clay laminae over distances up to 5 km in Agen Allwedd, and later to recognize a fini-glacial loessic phase (component) in capping muds in caves throughout southern Britain.

8.2.8 *Provenance studies*

Many studies have used pebbles and cobbles of some distinctive lithology as an easy but accurate means of tracing sediment sources and flow paths. Normally they will be allogenic.

In fine fractions heavy mineral abundances may differentiate the sources. Other studies have emphasized clay minerals and their ratios. Although there is much variation, abundant kaolinite tends to indicate warm

Table 8.2 The principal minerals deposited in caves

Group	Mineral	Formula
Carbonates	Calcite	$CaCO_3$
	Aragonite	$CaCO_3$
	Magnesite	$MgCO_3$
	Huntite	$CaMg_3(CO_3)_4$
Hydrated carbonates	Mono-hydrocalcite	$CaCO_3.H_2O$
	Tri-hydrocalcite	$CaCO_3.H_2O$
	Nesquehonite	$MgCO_3.H_2O$
	Hydromagnesite and others	$4MgCO_3.Mg(OH)_2.4H_2O$
Sulphates and hydrated sulphates: halides	Anhydrite	$CaSo_4$
	Barite	$BaSO_4$
	Celestite	$SrSO_4$
	Thenardite	Na_2SO_4
	Halite	$NaCl$
	Bassanite	$CaSO_4.H_2O$
	Gypsum	$CaSO_4.2H_2O$
	Epsomite	$MgSO_4.7H_2O$
	Hexahydrite	$MgSO_4.6H_2O$
	Mirabilite	$Na_2SO_4.10H_2O$
	Bloedite and others	$Mg_2SO_4.Na_2SO_4.4H_2O$
Sulphides	Pyrite	FeS_2
	Marcasite	FeS_2
	Galena	PbS
	Sphalerite	ZnS
	plus fluorite (fluorspar)	CaF_2
Phosphates and nitrates	Whitlockite	$Ca_3(PO_4)_2$
	Monetite	$CaHPO_4$
	Hydroxyapatite	$Ca_5(PO_4)_3.OH$
	Carbonate-apatite	$Ca_{10}(PO_4)6CO_3.H_2O$
	Crandallite	$CaAl_3.(PO_4)_2.(OH)_5.H_2O$
	Taranakite	$(K.NH_4)Al_3(PO_4)_3(OH).9H_2O$
	Potassium nitre (Saltpetre)	KNO_3
	Soda nitre	$NaNO_3$
	Nitrocalcite and others	$Ca(NO_3)_3.4H_2O$
Iron, manganese and aluminum oxides	Geothite	$Fe\,OOH$
	Hematite	Fe_2O_3
	Limonite	$Fe_2O_3.nH_2O$
	Birnessite	MnO_2
	Hollandite	BaM_8O_{16}
	Psilomelane	$(Ba.H_2O)2Mn_5O_{10}$
	Todorokite	$(Na.Ca.K.Mn.Mg)_6O_{12}.3H_2O$
	Bauxite	$Al_2O_3.3H_2O$
	Boehmite	$AlO(OH)$
	Gibbsite and others	$Al(OH)_3$
Silica	Quartz, chalcedony and cristobalite (opal)	SiO_2
Ice		H_2O

conditions either in the source rock or during its weathering. Illite and chlorite are usually the most prominent clay minerals in glaciated regions, while increasing montmorillonite suggests a drier climate, etc. Wolfe (1973) found a kaolinite:illite ratio of $\sim 3:1$ in Carboniferous source rocks. This reduced to 1:1 in modern sediments entering the caves and towards 0.6:1 as they were traced 8–12 km underground. The ratio in abandoned passages also decreased downstream, from 0.6 to < 0.4, indicating that there was further alteration of kaolinite as the ancient deposits weathered a little in the caves. Lynch *et al.* (2004) showed that most clay discharged from a major Texan spring that is an important water resource was allochthonous, deriving from distant upland soils and prone to convey bacteria and other contaminants from there.

8.2.9 Diagenesis of cave sediments

The nature and amount of diagenesis occurring in cave sediments is usually much less significant than that in original limestones (Chapter 2) and has been little studied. Physical processes include effects such as loading (load casts or deformation), drying out, abrupt rewetting, etc. Shrinking away from walls is common; it permits later sediment to intrude. Polygonal dessication cracking may extend to depths greater than 1 m.

Burrowing by animals (**bioturbation**) is a second physical category. Cave-specific (**hypogean**) fauna in the interior are small and weak. They can do little damage but disturb the uppermost laminae of soft sediments which, however, may suffice to ruin their palaeomagnetic signals (see below). Burrowing by large mammals such as badgers (and humans) is important in many entrance deposits.

Drying permits oxidation, which extends throughout most well-drained sediments. Periodic wetting encourages further hydrolysis of aluminosilicate minerals and the alteration of clay minerals. Such chemical weathering takes place in most cave sediments, but normally at very low rates.

Osborne (2001) has investigated such diagenesis in Australia, where many caves are unusually old and so provide more time for the slow changes that occur. Chemical precipitation within sediments is the most important process. In vadose settings, meniscus calcite cement is first deposited, succeeded by void-filling spar that can convert the loose sediment into a solid rock of low porosity. Between these extremes, patchy, layered or graded cementation is common. In phreatic settings there is acicular cementation. Because of their low permeability, allogenic clays (e.g. kaolinite, illite, quartz) resist cementation and can survive for millions of years in a soft, wet condition.

Other precipitates include surface veneers of gypsum and other crystals where enriched solutions in the original sediment have been drawn out by capillary suction as it dried. In parts of Mammoth Cave silts and clays have suffered **evapoturbation** to depths of 1 m or more, i.e. gypsum precipitated between the laminae breaks them up. In quartz sand deposits amorphous silica (crystallites) may precipitate onto grain surfaces, although this will rarely cement them. Nitrate and phosphate minerals also accumulate in cave earths, as discussed below.

8.3 CALCITE, ARAGONITE AND OTHER CARBONATE PRECIPITATES

Calcite is the principal secondary precipitate in caves. General terms for it include **speleothem, sinter, travertine** and **tufa**. The first two terms are generally used to describe dense crystalline deposits found in cave interiors, although both sinter and travertine are also used to describe predominantly chemical precipitates deposited at springs. Tufa refers to softer porous material found near cave entrances where rapid deposition is encouraged by evaporation. In most karst regions the mass of secondary calcite will far exceed that of all other secondary minerals combined. Aragonite is more widespread than has generally been supposed and is possibly the second mineral in abundance, with gypsum being the third. The other carbonates and hydrated carbonates are much less significant; they form small but attractive decorations in a minority of caves.

Speleothems (stalactites, stalagmites, helictites and flowstones, etc.) are the chief attraction in the majority of the world's tourist caves (e.g. Figure 8.8). It was their visual aesthetic appeal that also attracted many scientists to study caves. As a consequence, speleothems are the subject of many detailed studies, general reviews and lavishly illustrated popular books. A review by Hill and Forti (1997) with other contributing specialists presents the most complete account; see also Cabrol (1978), Bögli (1980) and Cabrol and Mangin (2000). Railsback (2000) has prepared an online atlas of sample microfabrics. Here we can give only a broad summary of the many interesting features.

8.3.1 Calcite and aragonite crystal growth

In the evaporitic conditions of most cave entrances and seasonally dry caves, calcite deposition is by the rapid formation of separate crystallites. This creates a soft calcite of globular or irregular clumps of tiny crystals, earthy in texture, with high porosity and often pasty to the touch, at least on first deposition. Hard calcite is precipitated by slow exsolution of CO_2 from solution in humid cave interiors (Figure 8.1). Some speleothems are composed of sequences of hard and soft calcite, or have soft calcite on upwind sides.

Most hard calcite grows by the competitive enlargement and coalescence of crystallites deposited as syntaxial (axially aligned) overgrowths on previous crystals (Figure 8.9). In most speleothems it is present in

Figure 8.8 Stalactites, stalagmites and columns growing in a cave. J.W. Valvasor's engraving of AD 1689 showing a decorated chamber in Postojna (Postumia or Adelsberg) Cave in Slovenia. Cave lighting then was not as revealing as it is today but visitors clearly did not lack in imagination. Reproduced from Valvasor, J.W., Die Ehre des Hertzogthums Crain. Lubljana, Endter. 4 vols, 1687.

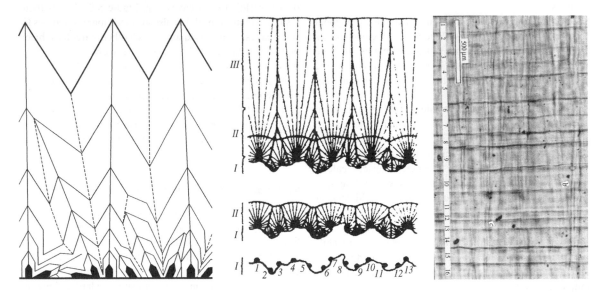

Figure 8.9 Ilustrating calcite crystal ontogeny. (Left) Competitive growth and selection from an initial substrate that is flat; favourably oriented crystals succeed in a length-fast mode. (Centre) Length-fast growth from a more realistic, irregular, substrate; initial crystallites are numbered. The gently undulating surfaces of stage II growth in the figure are very common features in stalagmites and flowstones. Adapted with permission from Self, C.A. and Hill, C.A., How speleothems grow: a guide to the ontogeny of cave minerals. Journal of Cave and Karst Studies, 65(2), 130–51. [BCRA] (Right) A photomicrograph of length-fast growth upwards in a stalagmite from China. It is in the equivalent of stage III in the centre illustration. Vertical extension is clearly seen in the crystal fabric but transverse banding is more prominent. The 16 numbered bands are considered to be successive annual accumulations. The scale bar is 500 mm in length. Reproduced with permission from Ming, T., Tungsheng, L., Xiaoguang, Q., and Xianfeng, W. 1998. Signification chrono-climatique de spéléothèmes laminés de Chine du Nord. Karstologia, 32(2) ; 1–6, 2003.

microcrystalline form with the *c* axis oriented roughly across the direction of growth (**length-slow** or 'coconut meat' texture of Folk and Assereto 1976) or as coarser palisade, columnar or equant crystal aggregates with the *c* axes oriented to growth (**length-fast** – see Railsback 2000). Large monocrystalline speleothems occur but are rare. A final crystalline type is an irregular mosaic, created by inversion from aragonite. Russian mineralogists adopt an ontogenetic approach that aims to link a particular crystal fabric with the specific environmental conditions in a cave; see Onac (1997) and Self and Hill (2003) for reviews. Frisia *et al.* (2000) correlate columnar and fibrous fabrics with low supersaturation and continuously wet conditions. Where discharge varies or there are abundant growth inhibitors, the defects increase. They are at a maximum in evaporative conditions, as many have noted.

Spaces left between the growing individual crystals are sealed off by later deposition and constitute inclusions, normally filled with fluid. Variation in the abundance and scale of **fluid inclusions** creates the smallest features that can be distinguished by eye or hand lens, which are faint, parallel growth layers. They reflect variation in rates of deposition. Stronger layering is created by temporary cessation and drying, by periods of dissolving at the growth front (**re-solution**) or other erosion, or by deposition of foreign particulates. Syntaxial crystal growth may extend through such breaks, or new generations of crystals may start upon them with greater or lesser bonding across the break. At the extreme, there is no bonding; the speleothem can fall apart at the hiatus.

Aragonite in caves displays a principal habit as radiating clusters of needles, termed 'whiskers', 'flowers', 'anthodites', etc. They grow from bedrock or calcite speleothems. There is also massive, acicular aragonite in regular stalagmites and flowstones. It may be interlaminated with calcite or display a patchy appearance where inversion to calcite is occurring.

There has been much discussion of why aragonite is found in some caves (or parts of them) and not others. Figure 8.10 indicates that it is metastable in the domain of calcite and dolomite. The latter is almost unknown as a primary precipitate in caves, so it is widely accepted that depletion of Ca ions in Mg-rich solutions is the principal factor (Hill and Forti 1997, p. 238), usually by prior deposition of calcite. The 'popcorn → aragonite → hydromagnesite globule' model (see Figure 8.14) thus is a potent example of the process at work. Other suggestions include ion substitution or 'poisoning' (e.g. by Sr) or unspecified effects of organics or other seed nucleii when growth is initiated.

8.3.2 Stalactites and draperies

The principal forms of speleothems are shown in Figures 8.11 and 8.12. The fundamental form is the

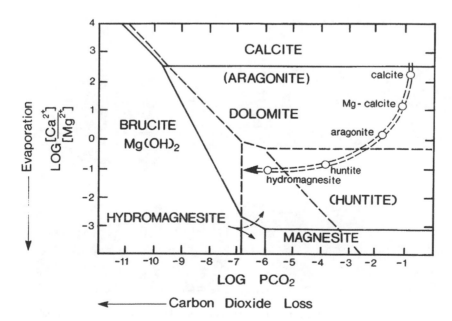

Figure 8.10 Stable and metastable (dashed lines) phase relationships in the system Ca–Mg–CO₂–H₂O, showing a model evolutionary path for a hypothetical water.

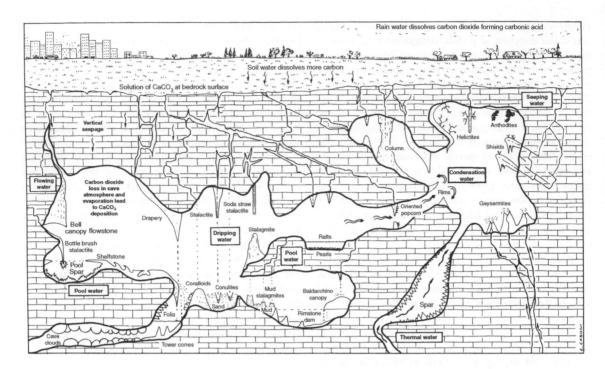

Figure 8.11   Model depicting the many different forms of aragonite and calcite precipitates found in caves. Reproduced from Hill, C.A. and Forti, P., Cave Minerals of the World (2nd Edition). Huntsville, AL, p. 463 © 1997 National Speleological Society of America.

Figure 8.12   (Upper left) Soda straw stalactites, stalagmites and flowstone. (Lower left) Stalactites and stalagmites aligned along joints. (Photographs by A.C. Waltham, with permission.) (Right) Giant stalagmites and columns in Ogle Cave, New Mexico. The large scale and highly ornamented form are typical of warmer caves with some evaporation. (Photograph by A.N. Palmer, with permission.)

soda straw stalactite (Fr. *fistuleuse*), a single sheath of crystal enclosing a feedwater canal. Growth occurs at the tip and the *c* axis is oriented down the straw. Andrieux (1963) suggested that straw stalactites require a slow, steady feedwater without suspended fines or organics to block the canal. Some straws grow to 3–6 m in length before breaking under their own weight.

Leakage through the walls of straws may overplate and thicken the sheath. Partial or complete blockage of the canal compels leakage, usually at the stem, thus forming the common tapered or carrot-shaped stalactite. *C* axes can be subparallel to the canal, radiate from it, or be randomly oriented.

Curl (1973a) suggested that because of gravitational forces and surface tension effects the minimum diameter of straw stalactites must be ~5 mm. Hill and Forti (1997) cite diameters of 2–9 mm. Very fragile forms of submi-nimum diameter can grow as sheaths on hanging tree roots in caves where there is little air movement.

The more complex forms of tapered stalactites have been analysed by Short *et al* (2005). Because of acceler-ated deposition at points of water extrusion they can grow protuberances in a myriad of forms, including crenulations and corbels with curtains (drapes) or subsidiary stalactites. Maximum dimensions are determined by the strength of attachment to the roof, strength of roof rock, or dimensions of the cave. It is rare to see free-hanging stalactites longer than ~10 m and greater than 1–2 m in diameter.

Aragonite soda straws are rare. Somewhat more com-mon is the clumped 'spathite' or 'chapelets de boules' form (Figure 8.11). The narrow necks may represent temporary halts in growth.

Draperies or curtains develop where the feedwater trickles down an inclined wall or beneath a tapering stalactite. Deposition is along the trickle course, with the *c* axis perpendicular to the growth edges so that scalenohedral crystal patterns normal to it are well dis-played. The form may be sinuous. Overgrowth may not occur because the feedwater is limited to the lower edges; thus, curtains tend to have the width of a few crystals and to be translucent. Colour-banded varieties are common and are known as 'bacon rashers'.

8.3.3 Stalagmites and flowstones

The forms of stalagmites are similarly varied, and they grade into general coverings spread across floors and walls termed **flowstones**. Stalagmites were divided into three categories by Franke (1965). Those of uniform diameter add almost all new growth at the top (not down the sides) and are considered to be created by a uniform supply of feedwater with uniform solute concentrations. Curl

(1973b) specified their diameter by the relation

$$D = \frac{(C_0 q)^{\frac{1}{2}}}{\pi, z} \tag{8.6}$$

where C_0 is $CaCO_3$ available for deposition ($C - C_{eq}$ of equation 3.76), q is flow rate and z is the rate of growth in height. The minimum diameter is about 3 cm. Increase in diameter is broadly proportional to q (i.e. drip rate). Kaufmann (2003) developed this relationship to model the hypothetical variation of diameter of a speleothem growing throughout one or more glacial cycles in a temperate region: the results are strikingly like the forms of many ice stalagmites that grow and disappear annually (see Figures 8.17 and 8.18). Gams (1981) showed that diameter also increases with drip fall height, a first com-plicating factor. 'Candlestick' stalagmites of this type are quite common.

Stalagmites with terraced or corbelled thickening are Franke's (1965) second type. At the extreme they possess leaf-like protuberances and are likened to piled plates or palm tree trunks (e.g. in Aven Armand, France, and Ludi Cave, Guilin). In between these growths the stem has a uniform diameter. Franke attributed them to periodic variations in growth rate and to greater fall height causing greater splash.

The third, most common, type is the conical or tapered form. Franke (1965) attributed it to a decreasing growth rate, Gams (1981) to increasing fall height. It is evidently due to there being significant calcite deposition from film flow down the sides. There is a transition to shallow bosses and flowstone sheets. A rare subtype is the 'logo-mite', a Christmas-tree-like tapering pile of calcite plate-lets accumulated below drip points in cave ceilings subjected to condensation corrosion (i.e. the corrosion solutes are precipitated immediately below); examples found in Jewel Cave, South Dakota, where a subaqueous crystal lining is being dissolved, are up to 1.5 m in height and decorated with evaporitic overgrowth. An analogous form is the 'conulite' (Figure 8.11) where calcite rafts on pool surfaces are sunk and pile up below fixed drip points.

The greatest reported heights of broomstick and corbelled stalagmites are ~30 m. Tapered stalagmites can be higher. Some bosses attain diameters approach-ing 50 m.

Flowstones are deposited from a rather uniform flow and accrete roughly parallel to the host surface. They are most common on floors or gentle slopes but occur on vertical walls where they are transitional to stalactites, drapery, etc. Maximum thicknesses up to a few metres are known underground. Flowstones are the dominant

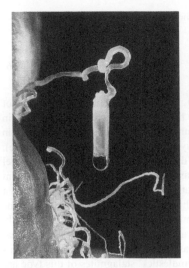  

Figure 8.13 Excentric speleothems. (Left) A straw stalactite grows from the tip of a vermiform helictite in Lechuguilla Cave, New Mexico. (Photograph by N.R. Thompson, with permission.) (Centre) Cave pearls, Castleguard Cave, Canada. (Photograph by Ford.) (Right) A 'heligmite' bush that is probably of subaqueous origin, draped by sunken calcite rafts, Wind Cave, South Dakota. (Photograph by A.N. Palmer, with permission.)

deposits at karst springs where, because of rapid growth due to evaporation and incorporation of vegetation, thicknesses of tens of metres can occur (section 9.11). Flowstones may cover a cave floor for tens or hundreds of metres downstream of a single source of feedwater.

Growth layering and hiatuses achieve their maximum development in flowstones. For example, Gradziński *et al.* (1997) described three strongly contrasted layers in a deposit 2 m thick in a Slovak cave: a basal, porous layer with acicular fabric and much detritus was abruptly succeeded by dense calcite with varve-like microlaminations; above this was a very dense, darker flowstone containing many internal corrosion surfaces and dated to the later stage of the previous interglacial. Many sediment sections in caves everywhere display flowstones interbedded with fluvial deposits, etc.

8.3.4 Excentric speleothems

Excentric or erratic deposits occur in most well decorated caves. They are termed **excentric** because they grow outwards from a wall or earlier formation, seeming to defy gravity (Figure 8.11). 'The form develops when crystal growth forces are dominant over the hydraulic forces of vertically moving water. Specifically, this implies that the flow to the excentric must be sufficiently slow to prevent drops forming' (White 1976). If drops form, a regular straw stalactite will grow downwards from the excentric; Figure 8.13 (left) shows it all!

There are innumerable varieties. Most excentrics are small, individuals rarely being > 1 m and commonly < 10 cm in length. These can be divided into (i) helictites or linear forms and (ii) globular or semi-spherical forms. In addition, there is one large calcite erratic form, the **shield** or **palette**.

Hill and Forti (1997) recognize four types of helictites:

1. **filiform** or thread-like growths of ~0.2–1 mm in diameter, in aragonite or calcite;
2. **vermiform**, sinuous to twisting, often bifurcating calcite sheaths of 1–10 mm diameter, is the most common type – in a tangled mass on a wall it looks like the hair of the Medusa;
3. an aragonite, **beaded** variety, which also curves and branches;
4. twig-like helictites which develop more angular branching.

All these types possess a tiny central capillary, with the water evaporating from its tip. The cause of the curving is attributed to trace impurities coprecipitated at the evaporation front twisting the *c* axis, coupled with regular rotation of crystal axes (Sletov 1985, Ghergari *et al.* 1998). Branching is due to blockage by crystal growth during drier periods, compelling diversion or bifurcation of flow. From the preferred distributions of helictites, very stable microclimate and hydraulic conditions also appear to be important in many caves, but not all. Kempe and Spaeth (1977) showed that sample capillaries widen and narrow, appearing like pearls on a string, and suggested that a seasonal effect might also drive the deformation.

Cser and Maucha (1966) noted two further types:

5. straight monocrystals, 0.2–3.0 mm diameter, which lack a capillary and are believed to form from aerosol $CaCO_3$ accreting at the tip;
6. straight or tapered monocrystals 2.0 mm or more in diameter and often lacking a capillary – most often this is a wind flag (**anemolite**), growing into oncoming air from an extrusion in a stalactite.

The globular family contains many varieties. Names given to them include 'cave coral', 'popcorn', 'globulite', 'botryoidal calcite'; 'boules' in French, etc. Occurrence ranges from single individuals growing from a wall or regular speleothem, to lines or patches one globule deep and, finally, great clusters like bunches of grapes. Clusters may be tightly packed together and many layers deep.

Figure 8.14 gives a comprehensive model for the subaerial type. Feedwater is extruded from microfissures and evaporated slowly from the globule surface, i.e. it is an optimum evaporation form. Depletion of Ca ions may permit acicular aragonite to grow outwards from the globule, tipped by hydromagnesite (Figure 8.10). In most caves, only the calcite is present. Calcite popcorn also accretes by condensation around nucleii on the lower parts of condensation corrosion cells above thermal waters, and globular deposits form underwater in saturated pools.

Shields or palettes are composed of an upper and a lower plate, both displaying concentric layering, jutting from a wall. Kunsky (1950) showed that they form where water is discharged under hydrostatic pressure from a fissure inclined upwards. The water diffuses radially outwards between the plates and new calcite accretes to their edges. There is usually deposition of draperies

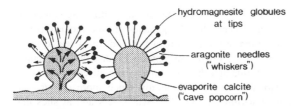

Figure 8.14 Development of the 'popcorn–whiskers–globules' or 'calcite–aragonite–hydromagnesite' sequence common in many caves where there is slow evaporation.

underneath. Shields are up to 5 m in diameter and 4–10 cm thick.

8.3.5 Pisoliths or cave pearls

Cave pearls are regular accretions of radial calcite about a foreign nucleus such as a sand grain (Figure 8.13). They grow in groups of a few to thousands in shallow pools agitated by falling drops of feedwater. Homann (1969) showed that gentle oscillation (rather than repeated rolling) is all that is needed to cause regular spherical accretion. Ideal cave pearls are spheres that range ~0.2 to 10–15 mm in diameter. Those in a given nest are similar in size. If the nucleus is elongated, then bladed or roller shapes develop. Irregular shapes form about larger granules. There are a few known instances of edge-rounded cubical pearls (e.g. at Castleguard Cave; Roberge and Caron, 1983) and one example of nested hexagonal pearls, the optimum packing pattern.

Growth of cave pearls calls for a delicate balance in the supply of water. If too little, the pearls cement together to form the core of a new stalagmite; if too much, they are displaced from the pool. Nevertheless, the balance is attained often enough. Nests of cave pearls are known in caves everywhere between the arctic and the tropics. Irregular examples develop rapidly in old mines (e.g. to 2 cm diameter in 5 months in highly saturated waters in a Cu-porphyry mine in Hungary (L. Lennart, personal communication).

8.3.6 Rimstone dams or gours

These features build up in stream channels or on flowstones. The height ranges from millimetres (microgours, often on the downstream faces of larger dams) to many metres. They may be single, or interlocking to create a staircase of pools. Rims are straight, curved or crenulated. The form is also common in rivers (e.g. Plitvice Lakes, Croatia; Figure 9.53) and at tufa springs and hotsprings, such as those of Yellowstone Park, USA, and Pamukkale in Turkey (Figure 9.54).

Rimstone dams originate at any irregularities that will thin a film of flow. They are the best example of positive feed-back operating in speleothems. Calcite precipitation is fastest at the rim because the water layer is thinnest there. Transitions to turbulent or to supercritical flow may help accelerate the process. Cavitation (detachment of flow from a bed, to form temporarily free sheets or drops) may explain dense patterns of microgours on steep flowstone slopes.

8.3.7 Moonmilk

This term is applied to white, amorphous masses of crystals that are pasty or plastic when wet, powdery when dry. Moonmilk has been reported from caves in all climatic regions, although it is more prominent in cold or dry caves because there is often a lack of other types of speleothems in them.

Moonmilk occurs on rock and clastic sediment surfaces or as overgrowths on earlier speleothems, in irregular patches varying from centimetres to many metres in area. Patches are often bulbous, like cauliflower heads. Thickness may be several tens of centimetres, though that is exceptional.

Moonmilk is composed of loose, fibrous crystals with an average thickness of only 1 µm or so. They display needle, branched or helicoidal forms. In very humid regions the crystals are entirely or predominantly of calcite or aragonite, plus hydromagnesite in dolomite caves. Where there is effective evaporation, the hydrated carbonate minerals and gypsum become more important, even huntite, dolomite, some phosphates and silicates.

Moonmilk has been studied intensively (Hill and Forti 1997). The presence of loose crystals of evaporite minerals is readily understood as being due to late-stage deposition from minute quantities of concentrated solutions. But this habit is quite anomalous for calcite. In many instances, it is believed that bacteria break down ordinary calcite crystals and redeposit them as microfibres; however, some wholly inorganic deposits are known.

8.3.8 Subaqueous deposits

A characteristic suite of precipitates develops in and on the surfaces of pools of supersaturated meteoric water, normally rimstone pools. The basic form is a wall encrustation of scalenohedral calcite (dogtooth spar) with projecting faces. Sometimes there are botryoidal crusts instead. At the water surface calcite may precipitate on dust particles and float as **calcite ice** or **rafts** supported by surface tension until too thick, when it sinks (e.g. Figure 8.13). Raft material and laminated flowstone grow outwards from the pool edges at the water surface. It may also grow radially from any projecting rocks in the centre, creating lily pad patterns. Calcite even encrusts bubbles on pool surfaces, a testimony to the extreme physical stability that can pertain in some cave environments.

Rising pools (e.g. as a rimstone dam rises) may inundate stalactites. Their submerged parts then accrete calcite or aragonite encrustations and appear as **dipsticks**.

The greatest subaqueous deposits are associated with the degassing of thermal waters enriched in exhalative CO_2. At the extreme, all surfaces become coated with dogtooth or nailhead (rhombohedral) calcite spar. These are **crystal caves**. In Jewel Cave, South Dakota, all surfaces are covered with 6–15 cm of nailhead spar in sheets and bulbous masses (Figure 8.15), except the highest places where it has been removed by condensation corrosion. Other caves nearby contain sheets > 1 m thick.

Continuous calcite precipitation over a vertical range of > 130 m as at Jewel Cave is exceptional. It is more common in thermal caves to see strong outward growth of calcite fins or curving folia at and immediately below the water surface, with botryoidal or dogtooth encrustations below them that rapidly diminish with depth. In some parts of the Budapest thermal caves there are no deposits below −6 m except sunken raft detritus. Wind Cave, South Dakota, displays thin, discontinuous crusts to −70 m or more. This deeper precipitation is associated with very slow renewal of water and slow, steady degassing (Ford *et al.* 1993). Renaut and Jones (2003) survey hot-springs calcites.

8.3.9 Deposits of other carbonate minerals

These are the Mg-rich carbonates plus hydrated carbonates (Table 8.3). They are all tiny in amount, and products of concentrated solutions created by evaporation or by previous deposition of calcite and aragonite. They are difficult to identify in the field and so have been reported from comparatively few caves. Most have been recognized only as components in moonmilk. For example, in the cold and humid centre of Castleguard Cave (+3°C and relative humidity > 95%) small nodular growths of aragonite or of calcite plus huntite are surrounded by aureoles of pure hydromagnesite moonmilk (Harmon *et al.* 1983). Brucite (Fig. 8.10) has not been identified in caves.

8.3.10 Colour and luminescence in speleothems

Pure, inclusion-free calcites and aragonites are colourless and translucent. High frequencies of fluid inclusions will make them white and opaque. Flood muds sealed in by calcite overgrowth can appear to discolour an entire deposit although they are only minor, discrete layers.

However, there is generally much more colour contained **within** individual growth layers and at sites that never flood. The common range is from reds through redbrowns, browns, buffs and ochres, to yellows and creams, with brightness (intensity) varying widely in all. This is

Figure 8.15   Typical passage in Jewel Cave, South Dakota. Walls are covered by 6–15 cm or more of nailhead spar precipitated from thermal phreatic waters. (Photograph by A.N. Palmer, by permission.)

Table 8.3 Conditions of calcite speleothem growth, destruction or decay

Condition	Process	State
The speleothem feedwater is the dominant control	1 Continuous deposition	Net deposition
	2 Periodic deposition	Net deposition
	3 Periodic deposition, periodic erosion	Net deposition
	4 Periodic erosion, periodic deposition	Net erosion
	5 Periodic erosion, no deposition	Net erosion
	6 Continuous erosion	Net erosion
Other waters are the dominant control (usually floodwater)	7 One of 1–6 may pertain but body (floodwater) re-solution predominates	Net erosion
	8 One of 1–6 may pertain but mechanical erosion predominates	Net erosion
	9 One of 1–6 may pertain but burial by clastic sediments predominates	Burial
	10 Permanently inundated by fresh water or seawater	No change, or net erosion, or marine boring and overgrowths
All effective water flow has ceased	11 Very slow weathering with minor accumulation of dust and aerosols; loss of lustre	Weathering
	12 Corrosion by acidic waters condensed from flowing, CO_{2-} or H_2S-rich air	Net erosion
	13 Speleothem freezes	Shattering
	14 Speleothem buried by weathering earth, organic matter, etc.	Burial

one of the major draws in tourist caves. It was long supposed that such colour must be due to metal oxides (chiefly iron and manganese) being distributed as particles of pigment amongst the carbonate crystals. These do play a role in some cases, but it is now firmly established that most or all colour in this range derives from humic and fulvic acids that were dissolved in the feedwater and then co-precipitated with the calcite or aragonite. These are complex organic molecules with masses ranging from hundreds to a few thousand daltons (the fulvics) up to tens of thousands of daltons (the humics): 1 dalton = 1/12 the mass of ^{12}C, the common isotope of carbon. These substances absorb strongly at the blue end of the spectrum, so producing the red-cream range of colours. In a comprehensive review, White (1976) suggested that higher proportions of heavier humics create the darker red and brown hues, with lighter humics and fulvics produce the yellows and creams. Van Beynen *et al.* (2001) confirmed this with detailed studies and also showed that undissolved particulate organic matter ($> 0.7\,\mu m$ in diameter) contributes to the very dark brown-black shades seen in many speleothems.

Transition metal ions in the crystal lattice can also produce spectacular colouring: Cu^{2+} gives green in calcite, blue in aragonite; Cr^{2+} and Co^{2+} give blue and pink to blue respectively; Ni^{2+} gives green to yellow (White 1997b). All of these are very rare, however, and in most instances are found only in speleothems that are small.

As cave photographers have long known, many calcite speleothems luminesce (emit light) when excited by a U/V flash. X-rays, electron beams, heating or crushing also produce luminescent responses. If the light vanishes immediately after the flash (excitation) it is termed 'fluorescence'; if it fades slowly it is 'phosphorescence'. The brilliance of the emission is generally proportional to the concentration of luminescing centres in the crystal lattice. Shopov (1997b) gives a comprehensive review. Although some trace elements in calcite will fluoresce, in ordinary vadose speleothems it is now known that most emission is from the fulvic and humic acid contents. The study by van Beynen *et al.* (2001) encompassed speleothems from subarctic and alpine tundra, temperate and tropical forest, grassland and desert-margin environments, and all colours from translucent white to nearly black. Fulvic acids yielded the higher fluorescence (per ppm) and, because of their greater abundance in most samples, were usually more important emission sources than the humic acids. Larger particulate organic matter was also present in many samples and tended to quench the luminescence.

Shopov (1987) was the first to recognize that, in many speleothems, the fluorescence signals appear as stronger–weaker couplets a few micrometres or tens of micrometres in thickness and parallel to any growth banding visible to the eye. He suggested that in many instances this must represent annual banding. Annual or seasonal banding of the fluorophores has now been established by several field studies in Europe and North America that extracted the humics, fulvics and particulate organics from speleothem drip waters. For example, van Beynen *et al.* (2000) found that the strongest concentrations of fluorophores occurred during the annual spring thaw flushing of the soil above a cave in Indiana.

Luminescence may also discriminate hydrothermal calcite environments. Those precipitated from very hot solutions ($\gg100°C$) yield a brief orange-red phosphorescence; lengthy afterglow indicates temperatures down to 60°C or less.

8.3.11 Rates of growth of calcite speleothems

Most visitors to caves are interested to know how fast the speleothems grow. Although the theory of their growth is now quite well understood (section 3.10), this remains a difficult question because rate-control environments are complex. Table 8.3 cites 14 different conditions that a given deposit may experience. Some re-solution (net dissolution by waters from the original depositional source) or by floodwaters, condensation corrosion, etc. is seen in most well-decorated caves.

Rates of growth are usually quoted in terms of the extension of a given form rather than its accumulation of mass. Straw stalactites 'grow' fastest because they have the greatest extension per unit of mass deposited. Rates of many millimetres per year are reported beneath bridges, in cellars, etc. where dissolved Portland cement is being re-precipitated in evaporative conditions, e.g. $3000\,mm\,a^{-1}$ at Caracas, Venezuela (de Bellard-Pietri 1981). Other drip-stones and flowstones grow especially rapidly where they are tufaceous, i.e. built upon vegetal matter.

There have been many attempts to measure current rates of extension of stalactites in show caves. Rates between 0.2 and $2\,mm\,a^{-1}$ are quoted for straw stalactites, between <0.1 and $3\,mm\,a^{-1}$ for carrot stalactites, and between <0.005 and $7\,mm\,a^{-1}$ for stalagmites. The sampling is often biased in favour of deposits that appear to be growing the fastest.

There may be significant seasonal variations in rate of deposition even where deposition is continuous. Gams and Kogovšek (1998) summarize long-term field observations in the great caves around Postojna, Slovenia. At Postojna Cave itself, rates of deposition approximately doubled when summer soil water reached the speleothems in the autumn months; the range was 18–$40\,g\,m^{-3}$ of dripwater. Rates of deposition on the instrumented flowstones appeared to be inversely proportional to the volumes of flow across them.

A different approach has been to calculate long-term mean rates in stalagmites and flowstones that have been dated by radiometric methods (section 8.6). At present, the range that has been detected is about three to four orders of magnitude, even though the samples are restricted in their morphological type and largely from humid temperate regions. This is to emphasize that growth rates are site-specific to a considerable degree; they can vary a lot within one cave.

In summary, in humid cave interiors, carrot stalactites probably grow up to $10\,mm\,a^{-1}$ and straw stalactites may grow several times faster. Stalagmites rarely extend more than $1\,mm\,a^{-1}$ and mean rates as low as $0.001\,mm\,a^{-1}$ have been calculated. Flowstone thickening rates are generally less than stalagmite extension rates, perhaps by as much as one order of magnitude.

8.3.12 Patterns of distribution of calcite speleothems

This subject has attracted comparatively little discussion, which is surprising because distribution patterns can reveal much concerning effective fissuration, the epikarst and environmental controls above a cave or in its region. The two extremes are readily identified: many caves have no speleothems; others are largely or entirely filled with them.

Absence of speleothems in a given region or cave or particular passage is due to four principal causes.

1. Although vadose, all of the passage is in a net erosional state. This is typical of young caves in any region, of the 'active' (usually lowest) passages in a multiphase system, etc. The first speleothems become established in sheltered recesses in the roof when the flood frequency begins to drop.
2. There is insufficient soil CO_2 so that infiltrating waters are not supersaturated with respect to the cave air. This is the major factor in explaining the small number and size of speleothems in many arctic and alpine caves and some semi-arid caves. However, lack of well-developed soils does not always imply their absence. Castleguard Cave is covered by glacier ice yet some sections of it are profusely decorated with modern deposits, as a consequence of suspected exotic acid effects (Atkinson 1983).
3. A non-carbonate caprock prohibits infiltration of feedwaters, or changes their ionic composition so

that calcite precipitation cannot occur. This is true of many middle and upper level passages in the Mammoth Cave. Underneath its caprocks there is some precipitation of gypsum (section 7.10), but calcite appears in profusion only if the galleries pass under valleys where the capping has been removed. Low-level passages in the cave generally lack speleothems because they are in the net erosional condition.

4. There is no hydraulic connection between the epikarst zone and the passage, or connections are intercepted by overlying galleries which capture all precipitates. It is rare to find an entire, extensive cave that does not possess effective linkages to the epikarst, but particular passages remain unconnected or intercepted in many systems. Partial interception is common, reducing the number of feedwaters entering lower passages. These effects are most important where fissure frequency is low.

Where speleothems are abundant their distribution varies with characteristics of the fissures transmitting the feedwater. At one extreme is limestone possessing high primary permeability. Feedwater emerges at points a few millimetres to centimetres apart, tending to create a dense and uniform cover of straw or carrot stalactites in a roof. This is characteristic of shallow caves in Quaternary limestones, but particular beds or lesser patches of rock in older strata also display it. At the other extreme, feedwater supply is limited to a single fissure and the stalactites, etc. are aligned along it. In most grottoes the supply is from a variety of alignments.

Often the type and scale of precipitation are functions of the hydraulic efficiency of the final fissures that supply the water. The largest fissures support the largest stalactites, stalagmites, etc. Small fissures support only straw stalactites and a few tapered examples plus thin calcite crusts, while the smallest may yield only excentrics and evaporite minerals.

Microclimatic factors may be locally significant in determining the type, scale or density of speleothems. This is most obvious in the cases of evaporites and anemolites; they grow best where air currents are strongest. Many hard calcite speleothems have soft calcite overgrowths on the upwind side, especially near entrances. When there is a sequence of hard and soft calcite layers oscillation of the cave climate about some critical value of relative humidity may be implied. Some caves display strong vertical zonation of temperature or humidity, with the consequence that excentrics (particularly the cave coral types) are seen to terminate abruptly at some upper or lower limit on the walls.

8.4 OTHER CAVE MINERALS

Hill and Forti (1997) report more than 180 non-carbonate minerals in caves. Some are alteration products such as the standard clay minerals. However, a majority are precipitated from solutions. Some of these are rarely found except in caves (i.e. they can almost be described as 'cave specific'). Many others attain their greatest or purest development in the stable, unchanging environments of deep but inactive caves. Here, we summarize the most commonly encountered types.

8.4.1 The sulphates and halides

Gypsum is believed to be the third cave mineral by volume and frequency. Its source may be gypsum strata, inclusions, etc. in the feedwater course, or it may be complexed from oxidized sulphides and dissolved calcite as explained in equation (3.58).

Gypsum is deposited in three principal modes in caves. The first is as evaporite inclusions in rock or clastic sediments. The inclusions are typically coarse, tubular or fibrous crystals in lenses up to a few centimetres thick. They disrupt the rock or sediment (evapoturbation).

The second mode is as speleothems growing from rock, sediment or calcite speleothems. Hill and Forti (1997) list 18 forms including dripstones, flowstones and crusts. The Chandeliers in Lechuguilla Cave are perhaps the most celebrated example of a dripstone. However, more abundant are **gypsum flowers** ('oulopholites'), **needles** and **whiskers** or **hair**. Flowers are curving, branching bundles of fibrous crystals as long as 50 cm. Epsomite and mirabilite, the most common Mg and Na cave sulphates, also display this form. Gypsum needles usually grow in dense clusters from sediment, like needle ice in freezing soils. Whiskers and hair are the finest, monocrystalline strands, readily swayed by a puff of air. Massive, sword-like crystals of gypsum that may exceed 1 m are also known.

The third mode of deposition is as regularly bedded deposits or wall encrustations from evaporating lakes or pools, etc. (Figure 7.32).

Gypsum, plus the minor but frequent epsomite and mirabilite, is most abundant in gypsum caves and in temperate to tropical limestone caves that are quite dry, at least at some seasons. They may suffer significant condensation corrosion in such caves. They form in lesser amounts in cold, humid caves such as Castleguard and in arctic caves where temperatures are -1 to $-3°C$.

The other sulphate and halide minerals in Table 8.2 are much rarer and are known only in warm, dry-to-arid caves. Epsomite loses one water of hydration to form hexahydrite. Mirabilite dries to powdery thenardite. Anhydrite, bassanite and bloedite are known in polymineral crusts or inclusions on or in drying cave earths. Halite forms crusts, single crystals or flowers in desert limestone and gypsum caves and is abundant in caves in salt. Celestite is occasionally reported as crusts or distinctively blue needle clusters on gypsum, and barite is reported as inclusions in cave earths. Both minerals occur in hydrothermal precipitates.

The sulphide minerals and fluorite (fluorspar) are primarily deposited from metalliferous thermal or other waters invading caves, with or without contemporaneous dissolution (section 7.7). Occasionally, there is dissolution and redeposition of these minerals as small stalactites in later caves.

8.4.2 Phosphates, nitrates and vanadates

Most of these minerals are produced by urine, faeces and other decaying animal matter reacting with cave rock, speleothems or clastic sediment. Bat guano is the principal source of reactants. The minerals therefore are most common in temperate and tropical cave areas with abundant bat populations. They attain significant bulk as microcrystalline aggregates within cave earths and also appear as thin, discontinuous alteration crusts on walls and speleothems.

The cave phosphate system is outlined in Figure 8.16. Reactions between guano, etc. and calcite normally produce hydroxyapatite. Other calcium and magnesium phosphates are rare. Alumino-phosphate minerals such as crandallite and taranakite form in permeable silts and clays on cave floors. Onac and Vereş (2003) discuss a richly complex example in a Romanian cave.

Nitrate minerals are of unusual interest because they have been extracted from cave earths for millenia, to use as medicines and in ceramics. There was widespread mining in USA caves during wars of the 18th and 19th centuries to manufacture saltpetre (KNO_3) by addition of potash (wood ash) mixed with water. River caves with deep clastic fills in their upper levels were therefore

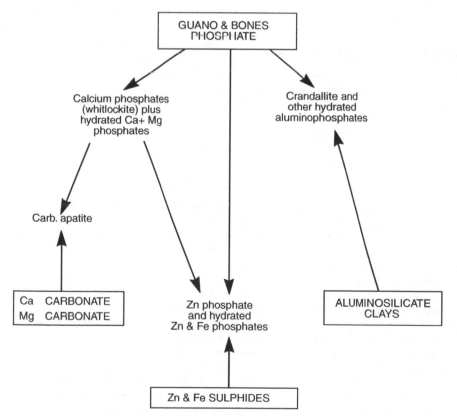

Figure 8.16 Phosphate minerals derived from the reaction of guano, bone, etc. decay products with wall rocks, clay fillings or sulphide ores in caves.

favoured and in the Appalachian region few of these escaped attention. The excavations, tools and vats are still well preserved in many of them; e.g. in the Historic section, Mammoth Cave.

The subject is comprehensively reviewed by Hill (1981). Where bats are abundant, the nitrates derive from nitrous guano. There are few bats in most Appalachian 'saltpetre caves', however, and their typical nitrate earths are alkaline and low in organic matter. They contain 0.01–4% nitrates by weight, mainly in the top few metres. These are soil nitrates carried down in solution in seepage waters.

Nitrates crystallize only where caves are dry and warm ($> 12°C$). KNO_3 and $NaNO_3$ occur as thin wall crusts and small stalactites, and darapskite (Na_3 $(NO_3)(SO_4).H_2O$) occurs as small flowers, hair, dripstones and flowstones.

Trace amounts of vanadium leached from clays or intrusive ores in vadose caves may complex with Ca or other elements and compounds to form vanadates. Tyuyamyunite ($Ca(UO_2)_2V_2O_8.5–8H_2O$), named for Tyuya Muyun Cave in Khirghizstan, is best known. It and other vanadates can appear as bright yellow crusts on calcite and other precipitates. All are rare.

8.4.3 Aluminium, iron and manganese coatings

Most aluminium and iron oxides are carried into caves in suspension but some form *in situ*. Alumina (in bauxite, boehmite or gibbsite; Table 8.3) mostly occurs as thin encrustations in caverns directly beneath bauxite sediments filling karst or palaeokarst depressions (Bárdossy 1982). Iron oxides occur chiefly as limonite coatings on walls and sediments. Some thick limonite within sediments is reported, and a few instances of limonitic stalactites. In thermal caves or very dry caves haematite is formed. Tiny crystals of magnetite precipitate within the calcite of many ordinary speleothems, where they carry a chemical remanent magnetism (Latham *et al.* 1979).

In many caves, dark brown-black stains or coatings are prominent on cobbles and boulders in stream channels and may extend up walls or speleothems as obvious deposits from ponded floodwaters. They are usually shiny and soft. These are manganese oxides, chiefly birnessite but also psilomelane, hollandite and todorokite. They are deposited from highly oxidized (e.g. turbulent vadose) waters carrying Mn, Ba, etc. ions from the solution of impure limestones. They are associated also with hydrated iron oxides, for example in some caves in North Island, New Zealand, where the iron is derived from volcanic ash soils. Thick deposits of pure manganese in Jewel Cave, South Dakota, appear to be residuals

of local limestone solution also, in this case settled out of slowly circulating thermal waters (Hill 1982).

8.4.4 Silica

Most silica precipitated in caves occurs as thin coatings of euhedral quartz in thermal waters. Fibrous, length-slow chalcedony replaces gypsum on some speleothems.

Opal (cristobalite) is a very fine-grained silica that occurs as thin interlayers in some calcite and gypsum speleothems. It is precipitated by high silica supersaturation attributable to evaporation.

8.5 ICE IN CAVES

Ice caves are those containing seasonal or perennial ice or both. Ice caves may be 'static' or 'dynamic' in their meterological behaviour (Figure 8.17). The static cave is a simple deep hole with no air drainage out of the bottom; thus it traps cold, dense air and excludes warmer air. Such static caves extend permanent ice to lower latitudes and altitudes than other natural cold traps. Most dynamic caves have two or more effective entrances for airflow and distinct entrance zones where the temperature varies $> 1°C$ over the year, as explained in section 7.11 Racoviţă and Onac (2000) present the most comprehensive discussion in English of the theory of ice-cave climates, ice growth and stability, hoarfrost deposition and sublimation. It is based on long records of field studies at Scărişoara Ice Cave, at 1160 m a.s.l. in Romania: see also Maire (1990, p. 507 *et seq*) and Mavlyudov (2005).

8.5.1 Types of ice

At least seven different types of ice in caves are known. The most common consists of dripstones and flowstones formed by freezing of infiltrating waters (Figure 8.18). These appear every winter in many cave entrances in temperate regions. They are often predominant in perennial ice caves. The morphologies are similar to those of calcite dripstones and flowstones but with less variety and ornamentation. The ice is clear to opaque and polycrystalline. Eisriesenwelt, at 1660 m in the Austrian Alps, is a perennial ice cave with outstanding ice of this type. It includes an entrance zone of ice flowstone 650 m in length and up to 12 m thick.

The second type is the **glacière** (Balch 1900), a large ice body created by the densification and recrystallization of old snow (**firn** or **névé**) accumulated in a static cave entrance trap site (Figure 8.17a). It is perennial and always contains a substantial proportion of infiltrating

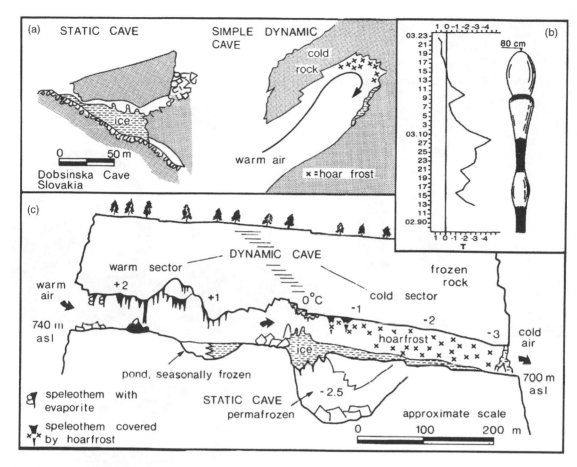

Figure 8.17 (a) The structure of static and simple (one entrance) dynamic ice caves. (b) The growth of an ice stalagmite in Scărişoara Glacier Cave, Romania, over 49 days in February–March 1965. The width is strongly positively correlated with ambient temperature. Reproduced with permission from Racoviţă, G. and Onac, B.P., Scărişoara Glacier Cave, Editura Carpatica, Cluj-Napoca, 2000 (c) Features of Grotte Valerie, a complex dynamic ice cave in the subarctic Nahanni karst, Canada. Vertical scale is much exaggerated.

water ice. The ice is opaque and bluish in section. It occurs as irregular masses conforming to the topography plus snow or water supply patterns of the host cave. It usually displays distinct dirt layers formed by concentration during superficial melt in the summer. At Scărişoara there are > 1000 layers in a mass ~15 m thick; a wood fragment 4 m above the base gave a ^{14}C age of AD 870. Dobšina Ice Cave, at 970 m in Slovakia, is another excellent example (Figure 8.17a). The ice mass is 145 000 m^3, with a mean depth ~13 m. It is true (flowing) glacier ice in many respects because plastic deformation at annual rates of 10–27 mm has been measured in it (Lalkovič 1995). The ice temperatures range between −0.2 and −1.0°C. This is delicately balanced; melting problems were encountered in the early years of tourist display, and were solved by modifying the natural air circulation. The mean annual external air temperature at the site is +6°C.

The third type is formed by freezing of ponded, static water. It may be seasonal or perennial. Freezing takes place from the top downwards and tends to occur at steady rates once begun. This creates coarse polycrystalline ice with horizontal c axes (Marshall and Brown 1974). Normally, it is the most transparent of all types of ice, although containing spectacular bubble patterns. At Coulthard Cave, Rocky Mountains, water-filled phreatic passages are now completely frozen up (Figure 8.18). It is possible to view along them through the clear ice until the passages turn away at an apparent optical depth of many metres. The invariant air temperature is approximately −2.5°C and the ice is sublimating at about 3 mm a^{-1}, forming large scallops.

Figure 8.18   Ice in caves. (Upper left) Ice stalagmites in the entrance area of an Italian cave. (Photograph by P. Forti, with permission.) (Upper right) Spectacular growth of hexagonal hoarfrost crystals in a cave in the Rocky Mountains, Alberta, Canada. (Lower left) A cold trap. The ice in the foreground traps a lake of very cold, dry air in the rear. Hoarfrost is seen above the lake; it is due to turbulent air drainage down the walls. Grotte Valerie, Nahanni, Canada. (Lower right) Pond water frozen in Coulthard cave, Alberta, and now ablating slowly in cold dry air. Ablating surfaces are large scallops. (Photographs by Ford.)

Hoarfrost is the fourth type, deposited from water vapour onto rock or ice surfaces below 0°C. There are at least four different kinds.

1. Mixtures of small needles, rosettes and hexagonal crystals that develop rapidly at temperatures close to 0°C. They form densely packed crusts of clear to opaque ice growing into oncoming drafts.
2. The most spectacular are larger, hexagonal plate crystals of transparent ice that grow by accretion at the edges. Individual plate diameters as great as 50 cm have been measured and many interlocked, sequential plates may extend up to 2 m as stalactitic masses (Figure 8.18). These deposits require formation temperatures of −1 to −5°C and very slow, gentle air circulation to supply vapour.
3. Fused, regelation masses derived from type 1 or 2 hoarfrost. They are opaque and bulbous and signify an increase of temperature. In Arctic Canada many relict cave passages are completely filled with such ice at short distances from the entrance (Lauriol and Clark 1993).
4. Regularly tapering, pyramidal crystals with hexagonal cross-sections. These are small and rare.

Extruded ice forms curving fibrous crystals similar to gypsum flowers. Lengths up to 20 cm are recorded in Canadian caves. This ice probably forms where supercooled water is forced under pressure through microfissures in frozen rock. It freezes where pressure is relaxed at the point of emergence. It is rare.

Intrusive ice is true glacier ice at the pressure melting point, intruded by plastic flow into open caves beneath glaciers. The only known examples fill passages at the head of Castleguard Cave, which is situated beneath ∼280 m of temperate glacier ice. The face of the principal intrusion remained constant in its position, neither advancing nor receding by melt or sublimation, between 1974 and 1993.

A final category is ground ice in clastic sediments. Where there is rapid freezing it forms thick, irregular masses. Slow freezing creates segregated ice, consisting of thin, scattered lenses. At intermediate rates **needle ice (pipkrake)** is formed, and may extrude as it does in ordinary soils; it is perhaps the most common type reported in caves. Pulinowa and Pulina (1972) and Schroeder (1979) discuss Polish and Canadian examples in detail, and also describe their effects on the sediments.

8.5.2 Distribution patterns of cave ice

In static caves and simple dynamic examples the ice may extend to fill all of the cold trap, seasonal or perennial.

This is rarely more than 100 m in length (e.g. Dobšina) and for hoarfrost it is normally only a few tens of metres.

In multi-entrance dynamic caves patterns become more complex, especially where entrances are at widely differing elevations. However, most ice will form in the entrance dynamic zones, as defined in section 7.12. The greatest quoted extent of ice into cave interiors is 1300 m at the Eisriesenwelt, but this is something of an overestimate because subparallel galleries are being added together. The usual maximum is ∼400 m. Perhaps the most complex and unexpected patterns occur in permafrost regions. The example of Grotte Valerie, Nahanni, Canada, is given in Figures 8.17c and 8.18. The site is at 61.5°N and the mean annual external temperature is about −7°C. The ice zone is a composite of dripstone and flowstone ice plus hoarfrost. It extends throughout what is, in the summer season, the interior or downwind half of the dynamic cave. This is because the upwind half (i.e. the conventional dynamic entrance zone) is thawed by warm exterior air that is drawn in. A static cave below the dynamic cave remains a cold trap at all times; it is permafrozen, arid and dusty, without ice.

8.5.3 Seasonal patterns of ice accumulation and ablation

In a majority of caves, the ice accumulates in the autumn and early winter and melts in the spring and early summer. In colder mountain regions there may be more complex sequences. In alpine Canada dripstone ice accumulates during the autumn, at a decreasing rate as soil-water sources freeze up. Hoarfrost accretes to the dripstone in late autumn to early winter. It may continue to form all winter where there are flowing allogenic streams to provide some humidity; otherwise, the balance swings to net sublimation in the depth of winter if cave entrances remain open to permit cold, dry air to enter. A brief burst of renewed hoarfrost formation in early spring is succeeded by general melting that normally proceeds from the entrance inwards. At Ciemniak Ice Cave in the High Tatra of Poland, Rachlewicz and Szczuciński (2004) found that sublimation was predominant when temperatures fell frequently below −8°C, hoarfrost or ice stalagmites grew when they oscillated between −8 and −1°C, and melting dominated when they were warmer.

Racoviţă (1972) studied rates of net accumulation of perennial dripstone and glacière ice at Scărişoara Cave, over periods ranging from a few years to approximately the past 370 year. He correlated them to variations in the severity of central European winters and concluded that a period of net ice build-up began around AD 1700 and terminated about AD 1920. Using oxygen isotope evidence, Marshall and Brown (1974) concluded that pond ice in Coulthard Cave may be ~4000 year in age, accumulating at the close of the Boreal Optimum climate phase. In colder regions it is possible that some ice survives from the last (Würm/Wisconsin) glaciation.

8.6 DATING OF CALCITE SPELEOTHEMS AND OTHER CAVE DEPOSITS

This section briefly reviews the principal methods used in dating and analysing cave deposits. It is an area of research where new analytic and interpretive methods are developing at a rapid pace (Onac *et al.* 2006). General reviews of Quaternary dating methods are available in Bradley (1999) and Noller *et al.* 2000. Bosak (2002) summarizes methods for all of karst research, while Ford (1997) and White (2004) focus on speleothems. We begin with the absolute (radiometric clock) methods. These rest upon the decay of natural radio isotopes in a statistically random manner such that there is a fixed decay constant (λ) for a given isotopic species, or on the filling of traps (holes) in the crystal lattice by emitted decay products. Thus the decay is exponential

$$N = N_0 e^{-\lambda t} \tag{8.7}$$

where N is the number of radioactive atoms present at time t and N_0 is the number present at the beginning of the decay.

8.6.1 Absolute dating

Cosmogenic isotopes

Carbon-14

Carbon-14 is created in the atmosphere when cosmic radiation interacts with ^{14}N. The half-life ($t_{1/2}$) is 5730 ± 40 year. This new carbon is then available for take up in all organic and inorganic systems, e.g. in HCO_3^-. This gives it very broad dating potential. The cosmic ray flux has varied over time, however, causing some deviations in the production rate, a first source of errors. Although age estimates up to 75 000 year ago (75 a) are occasionally published, most workers place the limits of the method around 50 a (i.e. eight or nine half-lives), and beyond ~30 year uncertainties in cosmic ray flux rates may render most results too young. Carbon-14 ages are now usually determined by atom counting in an accelerator mass spectrometer. Samples need be no larger than 1 mg, permitting careful selection of the points to be dated.

In carbonate regions ^{14}C dating is further complicated because of the abundance of stable ^{12}C and ^{13}C in rocks and soil (= 'dead carbon'). This may distort the age calculation, which assumes standard proportions between the three C isotopes, ^{12}C, ^{13}C and ^{14}C. From the simple calcite and dolomite dissolution reactions (equations 3.32 and 3.35) it might appear that half the carbon in, for example, a stalagmite will derive from soil air enriched in ^{14}C and half from rocks which are without it because they are too old. Soil CO_2 and HCO_3^- in water are very reactive, however, which usually increases the proportion of carbon from 'live' sources that is taken up by organisms or in precipitates. Several studies have found that 80–90% 'live' carbon is usually present in young speleothems (e.g. Genty *et al.* 2001a) but even in caves close to the soil it can drop to ~60%. Where there is little CO_2-enriched soil air and the groundwater flow path through the dead carbon of the rocks is long, this may fall to as little as 35%. Conversely, bone, shell, even wood fragments, buried in cave deposits may take up 'new' ^{14}C from HCO_3 in groundwater passing through them, which will result in too young an age. Carbon-14 dating is quite widely applied in caves but the results should always be treated with caution, and calibrated by comparison with results from other methods wherever possible.

$^{10}Be/^{26}Al$

In addition to manufacturing new isotopes in the upper atmosphere, cosmic rays strike exposed rocks and soils to create them at the Earth's surface, chiefly radioactive ^{10}Be, ^{14}C, ^{26}Al and ^{36}Cl. During the past two decades there has been substantial exploration of this phenomenon for geomorphological dating. Harbor (1999) presents a general review. For karst studies, ^{10}Be ($t_{1/2} = 1.5$ Ma (million years ago)) and ^{26}Al ($t_{1/2} = 0.71$ Ma) are most important because they are created in the fixed ratio of 1:6 in quartz, which is common amongst pebbles and sand in cave fills. When quartz rocks are exposed, cosmic rays may penetrate them effectively to depths of 1 m or more (e.g., at −0.6 m the flux is ~1/e of the surface value). Regardless of the total amount of exposure time any given rock or soil receives, if the host quartz should be swept into a cave where it is out of reach of further cosmic radiation the fixed ratio will generate a steady decay curve for the two isotopes considered together. At present, the effective dating range of the method is 0–5 Ma or a little more (see Granger and Muzikar (2001) for full discussion). Unfortunately, the abundance of these isotopes can be measured only with a cyclotron or similar expensive equipment.

At the beginning of this chapter we emphasized that there can be much reworking of clastic sediments within a cave. In addition, they may have been buried below cosmic-ray penetration depths in surficial channel or terrace deposits before being transported underground. Therefore, results must be interpreted with caution. The ^{10}Be/^{26}Al age for a particular sediment section in a cave is the **maximum** residence time of the quartz there; it may be considerably less. Nevertheless, the work undertaken by Granger and colleagues has been most impressive for its care and the coherence of the results. These include dating river gorge entrenchment in Virginia (Granger *et al.* 1997), confirming that the previously speculative pre-Quaternary history of Mammoth Cave and caves of the Cumberland Plateau, Kentucky and Tennessee begins at ~5 Ma (Granger *et al.* 2001, Anthony and Granger 2004), and that older sediments in the complex multilevel caves of the Siebenhengste–Bärenschach system of Switzerland described in section 7.5 may be 4.4 Ma in age or greater (Häuselmann and Granger 2004).

Uranium series methods

Disequilibria in the U species are widely used for dating, for the study of weathering systems and groundwater flow. They are the principal methods of dating speleothems at the present time. Comprehensive general reviews are given in Ivanovich and Harmon (1992) and Bourdon *et al.* (2003); see details for speleothems in Richards and Dorale (2003) and Dorale *et al.* (2004).

There are two natural parent isotopes, ^{238}U($t_{1/2} = 4.47 \times 10^9$ year) and ^{235}U ($t_{1/2} = 7.04 \times 10^8$ year). With such long half-lives they and their daughter isotopes survive as common trace elements in igneous and derived rocks, especially black shales. They decay by emission of α particles (^{4}He nucleii), electrons (β) and photons (γ) to produce stable ^{206}Pb and ^{207}Pb respectively. The heavier of the intermediate daughters, ^{234}U($t_{1/2} = 2.45 \times 10^5$ year), ^{230}Th($t_{1/2} = 7.57 \times 10^4$ year), ^{226}Ra($t_{1/2} = 1602$ year) and ^{231}Pa($t_{1/2} = 3.27 \times 10^4$ year) are also suitable for dating because of their comparatively long half-lives.

When a rock containing U is weathered, a higher proportion of ^{234}U atoms is mobilized than of ^{238}U or ^{235}U atoms, i.e. there is 'daughter excess' (Figure 8.19). This is because many of the ^{234}U atoms became loosened in their crystal lattice position when emitting the α particle. All three species are readily oxidized and transported in solution in bicarbonate waters as the complexed ions UO$_2$ (CO$_3$)$_2^{2-}$ and UO$_2$(CO$_3$)$_3^{4-}$. They may then be coprecipitated in calcite or aragonite, the latter normally accepting up to ten times more U atoms because of its larger lattice. The long-lived daughters, ^{231}Pa and ^{230}Th, are essentially insoluble. When detached by weathering these bond to clay or other particles. Therefore, they are not precipitated in the calcite. In an ideal closed system they will accumulate there only as a function of the decay of the parent U species. One gram of calcite with a trace U content of 1.0 ppm contains 10^{15} atoms of uranium available for spontaneous decay.

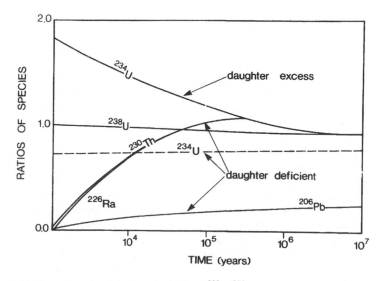

Figure 8.19 Evolution of the longer lived radioisotopes in the series ^{238}U–^{206}Pb to display the various dating schemes. The diagram is illustrative only; gradients are not true to scale.

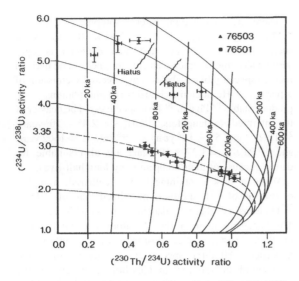

Figure 8.20 Graphical illustration of the ^{230}Th:^{234}U:^{238}U dating method, the principal method used for carbonate speleothems today. Most clean speleothems are deposited with an initial ^{234}U:^{238}U activity ratio greater than 1.0 and a ^{230}Th:^{234}U ratio of 0.0. With the passage of time, ratios evolve to the right. Sample 76501 grew between 250 a and 50 a (with one hiatus) with an initial ^{234}U:^{238}U ratio always close to 3.35. The initial ratio in sample 76503 varied between 5.3 and 6.3. The examples are of α-dated samples: modern mass spectrometric methods have reduced the errors by an order of magnitude.

The chief dating method uses the decay of excess ^{234}U to ^{230}Th, with allowance for the parent ^{238}U. Figure 8.20 shows the graph of the dating equation with results from two different stalagmites plotted on it. Until the late 1980s the abundance of the different isotopes was estimated by counting α disintegrations in a scintillometer. Counting took approximately 1 week before there were sufficient numbers for reliable statistics, and one standard deviation (σ) errors in the age were typically 10%. This has been replaced by direct counting of isotopes by mass spectrometry: for corals first by Edwards *et al.* (1986–7), for speleothems first by Li *et al.* (1989). In thermal ionization mass spectrometry (TIMS) separate extracts of U and Th are burned off of filaments at 1800–2200°C, with manual control required for the Th measurements. In induction-coupled mass spectrometry (ICPMS) liquid extracts are heated to a plasma at 8000–10,000°C and measured automatically. The two methods give very similar results. Calcite sample sizes need be no more than 0.5–2.0 g and 2σ age errors are reduced to ~1% or better: Richards and Dorale (2003) cite $2\sigma = \pm40$ year for a sample 10 ka in age, ±200 year for 50 ka, $\pm15\,000$ year for 500 ka. The current limit of the method is a little over 600 a (Figure 8.20). *In situ* U series dating by laser-ablation multicollector ICPMS is currently being investigated (Eggins *et al.* 2005) with the prospect of dating at a spatial resolution of 100 μm or better, although currently with less precision and accuracy than TIMS methods.

The flow of groundwater through U source rocks or soils that have already been highly leached may lead to deposition of calcite with a deficiency of ^{234}U rather than an excess with respect to ^{238}U. This offers an absolute dating method if the initial deficiency can be determined, and establishes the duration of growth of a deficient calcite where it cannot. Daughter-deficiency is encountered in a few per cent of calcite speleothems.

The ^{231}Pa/^{235}U method permits dating only to ~250 ka because the half-life of ^{231}Pa is shorter than that of ^{230}Th. A major problem is that the natural abundance of ^{235}U is low; ^{238}U:^{235}U = 137.9:1. As a consequence this method has been used most often as a check on ^{230}Th/^{234}U results. However, ^{231}Pa dating of carbonates by TIMS has increased its precision tenfold and so improved its value for dating in its own right (Edwards *et al.* 1997).

Decay of ^{230}Th to ^{226}Ra permits dating to ~10 ka. Because of this short timespan the method is rarely used. Latham *et al.* (1986) applied it when studying the magnetic record of a Mexican stalagmite that had apparently grown to a height of 72 cm in the past 2000 year. Decay of ^{226}Ra to ^{210}Pb will date for just 100 year; Tanahara *et al.* (1998) have used it to date straw stalactites in Japan and Condomines *et al.* (1999) used it to study some quickly growing thermal calcites in France.

Decay of ^{238}U to ^{234}U may be differentiated for five to six half-lives of the daughter, i.e. to 1.25–1.50 Ma. Unfortunately, the initial ^{234}U/^{238}U ratio (U/U$_0$) cannot be determined analytically unless ^{230}Th is also in disequilibrium, i.e. unless the calcite is younger than ~600 a. Where it is older, ^{234}U/^{238}U > 1.0 only indicates that a sample is younger than 1.25–1.50 Ma. However, where U/U$_0$ has been measured at many points in a single speleothem or in a very small geographical area over the range, 0–400+ ka, the mean ratio may be applied to older samples if the standard deviation is low. This is RUBE dating, for 'Regional Uranium Best Estimate' (Gascoyne *et al.* 1983). In Figure 8.20 it is seen that sample 76501 has a rather steady U/U$_0$ ratio and might be suitable, whereas sample 76503 (from the same cave) is emphatically not. Such large variations are common in vadose speleothems. Some display apparent systematic change over the course of a glacial cycle (~100 ka), the ratio decreasing as the duration of the cold conditions increases, then reversing as a warmer climate accelerates weathering rates again. ^{234}U/^{238}U thus has been most successful for dating thermal water calcites in large aquifers where U/U_0 variations are substantially damped by groundwater mixing, for example in Devil's Hole, Nevada (Ludwig *et al.* 1992), and Wind Cave, South Dakota (Ford *et al.* 1993).

Finally, the decay of parent uranium to stable lead in principle permits dating of calcite back to the origin of the Earth itself. There is a great deal of background lead in the natural environment, however, that also accumulates in trace amounts in most calcites (Jahn and Cuvellier 1994). Discriminating radiogenic ^{206}Pb or ^{207}Pb from the background is a major problem because there will be little of the radiogenic lead until a speleothem is at least several million years old. Lead-204 is non-radiogenic: the parent/daughter ratios, $^{238}U/^{204}Pb$–$^{206}Pb/^{204}Pb$ and $^{235}U/^{204}Pb$–$^{207}Pb/^{204}Pb$, thus may build isochron graphs that give ages by approximation. For statistical reasons, however, this requires that the U content must vary substantially within a sample: in practice, that rarely occurs, at least within the limits of current technology. Using TIMS, we have failed in attempts to date ancient vadose speleothems from the oldest caves surviving in the Austrian Alps and the Canadian Rockies, the thermal subaqueous calcites from Jewel Cave (Figure 8.15), and a very thick and extensive subaerial tufa (Pliocene?) on a hilltop in Spain. In every case there was insufficient variation in the U content. We have succeeded with a calcite filling in the earliest palaeokarst phase in the Guadalupe Mountains, New Mexico (90–96±7 Ma; Lundberg *et al.* 2000). Richards *et al.* (1996) have succeeded with much younger speleothems from sites exhibiting exceptionally high U contents. Using MC-ICPMS, Woodhead *et al.* (2006) have obtained a reasonable late Tertiary age for speleothems from caves in the Nullarbor Plain, Australia (section 10.2) and ages around 1 Ma for samples from Antro del Corchia Cave, Italy: poor resolution of ^{204}Pb by ICP methods is overcome by alternative isochron techniques but it is acknowledged that the estimation of initial $^{234}U/^{238}U$ ratios remains a considerable problem where ages are less than a few million years. Because multiple analyses are required, U/Pb dating is more expensive than other U series.

There are three basic requirements for all U series methods.

1. The calcite or aragonite (or gypsum) speleothem must contain sufficient U. Measured concentrations range $\ll 0.01\,ppm$ to $>300\,ppm$. $0.01\,ppm$ is the current reasonable minimum for $^{230}Th/^{234}U$ dating. $>80\%$ of assayed calcite speleothems and all aragonite speleothems contain more than this minimum concentration.
2. The system must be closed after coprecipitation of U and calcite. Often this will be violated. Many speleothems are partly or wholly recrystallized. Others are porous so that water flows freely through them and may preferentially leach ^{234}U. This results in too great an age being calculated. For this reason stalactites (with their central feedwater canals) and porous tufa deposits are to be avoided, if possible.
3. The most important requirement is that no ^{230}Th or ^{231}Pa be deposited in the calcite. In fact, most calcite contains a proportion of these species and of $^{232}Th(t_{1/2} = 1.39 \times 10^{10}\,year)$ bonded to clay or other particulate detritus deposited in the speleothem. They are contaminants. As they increase, reliability of calculated dates deteriorates. In practice, where the ratio, $^{230}Th/^{232}Th$, is >20 it is presumed that radiogenic ^{230}Th completely predominates and that contamination is insignificant; most dating computer programs now correct for it. For highly contaminated deposits $(^{230}Th/^{232}Th < 5.0)$ multiple determinations are recommended in order to calculate the isochron, $^{230}Th/^{232}Th$ versus $^{234}U/^{232}Th$. Multiple leachings of samples have also been tried. These approaches often fail, however. If possible, samples that are visibly dirty should be avoided. Unfortunately, those are often the most interesting, e.g. flowstones in cave entrances, or spring travertines.

Uranium series methods are also applied to bones and shells. Because of *post mortem* uptake of ^{234}U and other factors the analytical difficulties are considerable (Schwarcz 1980).

Several thousand U series dates of speleothems have now been published and their number increases by hundreds every year. The coverage is worldwide. Dates are used to determine mean rates of growth of stalagmites and flowstones, as noted in section 8.3. More fundamentally, vadose speleothems give the minimum ages for the cutting of the vadose trenches or draining of the phreatic passages that they now occupy. By extension, mean maximum rates of channel entrenchment can be computed (e.g. section 4.4) and the rates at which river valleys or glacial valleys have been entrenched below palaeospring positions; the principles are set out in Ford *et al.* (1981). In early work, Ford (1973) used the dated drainage of phreatic caves on a rising anticline to estimate the age of antecedent canyons along the South Nahanni River, Canada. Williams (1982b) dated the tectonic uplift of coastal terraces in New Zealand by the same means. A feature of most work of this kind is that the caves or their draining have proved to be much older than was previously supposed.

Speleothem ages can be used to date episodes of clastic sedimentation and erosion in the cave interior facies

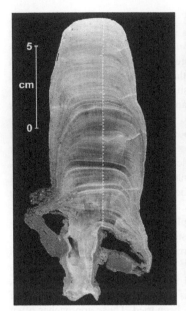 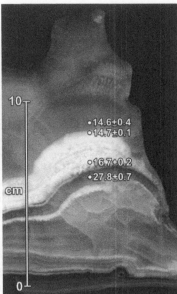 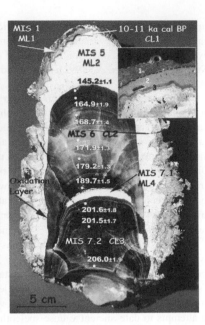

Figure 8.21   Polished sections through three small stalagmites to show something of the depositional variety that occurs. (Left) A typical stalagmite from a quarried cave beneath forest in the eastern USA. Continuous growth banding is clearly seen from the base upwards but it lacks strong contrasts because this sample grew under quite uniform environmental conditions in the later Holocene. Note its deep root into underlying clays. (Centre). A stalagmite from Cueva del Cobre, Spain, close to the modern alpine treeline. It exhibits stop-and-go growth. Calcite (grey layers) grows today. The sharp white layers are aragonite that grew under cooler, drier conditions. Uranium series TIMS ages are in thousands of years, with two standard deviation errors cited. Calcite deposition ceased shortly after 27 ka and there was slow weathering in glacial climatic conditions until aragonite accumulation could begin at 17 ka; it switched to calcite again 14.7 ka. (From C. Rossi, with permission.) (Right) Stalagmite I from 18.5 m below sea level in a drowned cave on the west coast of Italy. ML1, 2, 4 are marine deposits of serpulid worm casts and calcite overgrowth. The hard dark calcite layers grew when the cave was above sea level, between ~206 and 145 ka, with one marine interruption ~200-190 ka. Reproduced from Antonioli, F., Bard, E., Potter, E.-K., et al, 215-ka history of sea level oscillations from marine and continental layers in Argenterola Cave speleothems, Italy. Global and Planetary Change, 43(1–2), 57–78 © 2004 Elsevier.

discussed above. More obviously, the methods are applied to date spring travertines or speleothems in entrance facies containing bone, artefacts or other remains of early humans or fossil fauna, even fossil footprints (Onac *et al.* 2005).

Speleothems in caves of the Bahamas, Bermuda and elsewhere that are now submerged supplied the first absolute dates for Quaternary global low sea levels (Harmon *et al.* 1978, 1983; Figures 8.21 and 8.22b). Speleothem growth in glaciated regions and their peripheries tends to cease during periods of greatest cold. Harmon *et al.* (1977), Baker *et al.* (1993a) and others have used this feature to broadly date interglacial and interstadial periods (Figure 8.22a).

Finally, U-series dating fixes the chronology of stable isotope and other palaeoenvironmental records recovered from the speleothems more accurately than is possible for the other long Quaternary records such as marine cores or ice cores (section 8.7).

$^{40}Ar/^{39}Ar$ dating of alunite formed in H_2S caves

During the formation of caves by the H_2S processes described in section 7.8 common clay minerals such as montmorillonite, illite and kaolinite that are present in small quantities in the limestone may react with the acids to form alunite, natroalunite and other hydrated sulphate minerals that then accumulate in small quantities in solution pockets or on passage floors. Alunite (KAl_3 $(SO_4)_2(OH)_6$) has very small, tight crystals that function as closed systems after precipitation. The K content decays to ^{40}Ar and thus may be dated by the $^{40}Ar/^{39}Ar$ method that is well established for dating lavas and other K-rich rocks. Polyak *et al.* (1998) achieved great success when they made the first applications of the method in Carlsbad Caverns and other H_2S caves of the Guadalupe Mountains, New Mexico, obtaining ages ranging from ~12 to 3.5 Ma (section 7.8). From the perspective of cave genetic chronology this application is most significant

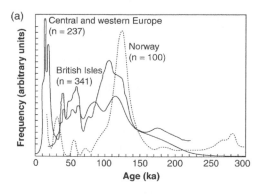

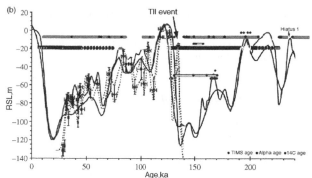

Figure 8.22   (a) The probability density function of compilations of spectrometry U series ages from central and western Europe (Hercman 2000), British Isles (Baker et al. 1993a) and Norway (Lauritzen 1995). Reproduced from Richards, D.A. and Dorale, J.A. (2003) Uranium-series chronology and environmental applications of speleothems, in Uranium-series Geochemistry (eds B. Bourdon, G.M. Henderson, C.C. Lundstrom and S.P. Turner). Reviews in Mineralogy and Geochemistry, 52, 407–61. (b) Global sealevel behaviour during the past 240 ka as recorded by corals and drowned vadose speleothems dated by U series. Reproduced from Antonioli, F., Bard, E., Potter, E.-K., et al. (2004) 215-ka history of sea level oscillations from marine and continental layers in Argenterola Cave speleothems, Italy. Global and Planetary Change, 43(1–2), 57–78.

because it dates a process occurring **during the actual dissolutional excavation of a cavity**, rather than its later filling with the clastic or precipitate sediments used by the other methods. Alunite has been reported from caves in other regions around the world, but we are not aware of $^{40}Ar/^{39}Ar$ dating of them.

Electron spin resonance decay, thermoluminescence and optically stimulated luminescence

These methods rest on the principle that electrons released during radioactive decay or by solar or cosmic radiation become trapped at charge defects ('traps') in crystals. They accumulate at a rate proportional to the annual dose rate of radiation until all traps are filled, when the crystals are said to be 'saturated'. In an unsaturated sample, electron spin resonance (ESR) age may be determined from

$$age = accumulated\ dose\ (AD)/$$
$$environmental\ dose\ rate\ (DR) \qquad (8.8)$$

Because of the natural variation in the response of natural substances, the accumulated dose is determined by an additive technique, i.e. samples are given stepwise additional radiation (usually from a γ source). Their response is measured, extrapolated back to zero, and thus the pre-irradiation dose (AD) is determined. In the thermoluminescence (TL) method a sample is heated to 450°C and the luminescent glow curve produced is measured, whereas in ESR dating microwave absorption

of the additional radiation is determined by spectroscopy. In the optically stimulated luminescence (OSL) method samples are stimulated by argon laser at a wavelength of 514 nm, which releases only the AD in light-sensitive traps, allowing for greater precision than the other methods. The environmental dose rate is estimated from concentrations of radio elements (U, Th, K) in the environs of the sample and by γ-ray measurements at the site (Hennig and Grün 1983).

These methods are inherently less reliable than U-series methods because of the assumption that internal and external (environmental) contributions to annual doses will be constant. In fact, considerable differences of dose rate have been recorded a few centimetres apart in cave deposits (Debenham and Aitken 1984). Great uncertainty still exists in the reliable determination of both AD and DR, with the consequence that meaningful error limits cannot yet be quoted for estimated ages. In studies of speleothems that compare ESR results with U series ages agreement is sometimes excellent, but too often poor (Smart *et al.* 1988; Hercman 2000). There have been few TL and ESR studies of speleothems recently.

The TL and ESR methods have been applied chiefly to shells, teeth, bones (including samples in cave deposits), plus loess, sand and tephra (Schwarcz 2000), and OSL only to quartz and feldspar grains. There have been only a few dating studies of clastic deposits in caves. One interesting case is that of Victoria Cave at Naracoorte in Australia, where sedimentary deposits containing fossil bones and teeth are interbedded with speleothems. Speleothems were dated by TIMS from 41 ka to > 500 ka

(Moriarty *et al.* 2000) and revealed a pattern of wet and dry phases. The fossil teeth were dated by ESR from 125 ka to 500 ka (Grün *et al.* 2001) and all results agreed within the constraints given by the U series ages.

8.6.2 Comparative dating methods

Palaeomagnetism

The Earth's magnetic field displays secular variations in its polar declination and inclination and in its intensity. Most variations are generally small in amount, irregular and only of regional extent. Greater, complete reversals of the field occur at intervals of 10^5–10^6 year, termed magnetic epochs or 'chrons'. The present Brunhes epoch commenced 780 ka ago and is defined as 'normal'. It was preceded by the Matuyama 'reversed' epoch, 0.78–2.5 Ma, which contained some prolonged normal episodes, termed 'events' or 'subchrons'. These were worldwide in their extent. Use of records of ancient variations or reversals as a dating tool relies on matching the curves of declination and inclination (and perhaps intensity) in a given deposit with established curves that have been dated by independent methods, chiefly lavas dated by the K/Ar method.

In caves palaeomagnetic studies have been applied principally to deposits of laminated clays and silts in which detrital grains of magnetite or haematite retain the magnetic declination and inclination of their time of deposition (detrital remanent magnetism – DRM). In many long depositional sequences of such fines it is found that the concentration of ferromagnetic minerals (the magnetic susceptibility) also varies over time, as a presumed consequence of weathering intensity that is climatically driven. This can permit some local correlation of sedimentary sections. Sands may be used where they have remained moist and stable (show no obvious deformation structures), and Williams *et al.* (1986) reported some success with calcite-cemented clays and even with layers of cemented cave pearls.

A principal problem with cave clastic sediments is that their deposition was rarely continuous so that, in comparison with, for example, lake-bottom deposits, the magnetic records are much interrupted. They may also suffer post-depositional alteration of the D and I signals, especially if they have drained, and bioturbation can be a major problem, e.g. at the Mulu Caves, Sarawak (Noel and Bull 1982). As a consequence, cave sediments are now studied chiefly to establish whether their declinations are normal or reversed, the latter implying that they are probably > 780 ka in age, and to correlate any older reversals of subchron or greater magnitude.

Schmidt (1982) published pioneer studies from the deposits in Mammoth Cave, Kentucky, finding that the earliest silts and clays were older than 1.7 Ma. With others he used clays in relict caves in river valleys of the Cumberland Plateau, Tennessee, to derive maximum river channel incision rates of $\sim 60\,m\,Ma^{-1}$ (Sasowsky *et al.* 1995). Auler *et al.* (2002) obtained incision rates of 25–34 m Ma^{-1} for the eastern Brazilian craton by the same means. Audra *et al.* (2001) studied fragmentary records preserved on ledges in a 100-m-deep vadose trench in a cave above the Ardeche River in the French Alps, correlating them with bedrock terraces in the river valley and obtaining a Late Pliocene age (2.2–2.5 Ma) for the beginning of entrenchment. The most detailed and comprehensive studies and analysis recently have been by Bosak and colleagues in caves of the Czech Republic, Slovakia and Slovenia, as illustrated in Figure 8.23 (Bosak *et al.* 1998, 2004, Bosak 2002).

Post-depositional alteration cannot occur if the magnetic grains are cemented inside calcite, and their preserved records of magnetic variation then may be dated independently by U series methods, etc. Latham *et al.* (1979) recognized this possibility and showed that many stalagmites and flowstones carry natural remanent magnetism either as a chemical precipitate (CRM) or as floodwater or filtrate detrital grains (DRM) or both. Magnetite is the carrier. The signal is weak, requiring comparatively large amounts of speleothem calcite and a high sensitivity magnetometer to measure it. Full technical details are given in Latham and Ford (1993). Palaeomagnetism of speleothems has been used to test for normality or reversal where a sample is known to be older than $\sim 500\,a$ ($^{230}Th^{234}U$ method) or older than ~ 1.5 Ma ($^{234}U/238U$ method), and also to obtain a few dated, high-resolution curves of recent secular variations of the Earth's magnetic field, e.g. in Mexico (Latham *et al.* 1986).

Biostratigraphical methods using fauna, flora and pollen

Deposits may be used for approximate dating by correlation with external type sections, and also in the reconstruction of past environments such as successive ecological assemblages above a cave.

Troglobitic flora and fauna (i.e. living only in caves) are too small in number and volume to be significant in most instances (although even they can disturb the palaeomagnetic signals in soft sediments!). Animals that roost or nest in caves but forage outside (trogloxenes)

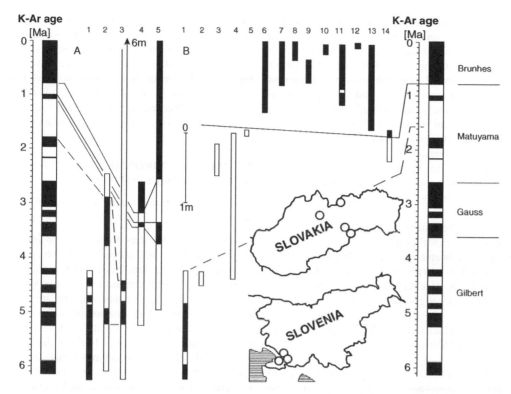

Figure 8.23 Fourteen magnetostratigraphical profiles of clastic sediments measured in some Slovak and Slovene caves (located by dots). On left the scale shows the global magnetic normal (black) and reversed (white) chrons with their K/Ar radiometric dating; on right, the names of the chrons. Correlations between the cave records and the chrons are suggested by the linking lines. Profile 1 cannot be correlated yet; it is said to be 'floating'. Slovakia and Slovenia are offset from their correct geographical locations in order to minimize space. Reproduced with permission from Bosak, P, 'Karst processes from the beginning to the end: how can they be dated? 'In Grabrovsek, F. (ed.) Evolution of Karst: from Prekarst to Cessation. Postojna-Lubljana, ZRC SAZU; 191–223, 2002.

are more important. Rodent nests, bones and faeces are often found in the furthest interior parts of shallow cave systems. More striking in many caves of middle Europe are remains of an extinct bear, *Ursus spelaeus*, that is conventionally dated to the last interglacial and the lower half of the Würm glacial period. For example, Abel and Kyrle (1931) estimated that Drachenhohle, Austria, contained the skeletal remains of $3–5 \times 10^5$ individuals, accumulated over approximately 40 ka. Little dating use has been made of other trogloxenes. Extinct fauna are common in cave entrance facies, where they have been studied intensively.

Flora carried into cave interiors are generally small in amount and prone to rapid decay. Most attention has focused on pollen and spores. Many clastic deposits are barren or contain only degraded grains because of oxidation. There have been recoveries of well-preserved pollen from laminated silts and clays in some Belgian and

French caves, in Kentucky and a few other places (Damblon 1974, Peterson 1976).

Geurts (1976) and Bastin (1979, 1990) pioneered the extraction of pollen from spring travertines, stalagmites and flowstones. McGarry and Caseldine (2004) have reviewed its use and potential value. Comparatively large volumes (e.g. 100–200 g) of sample are usually needed to obtain significant pollen counts because pollen is usually preserved at concentrations of <1 to ~ 10 grains per gram of calcite, depending on the site in the cave. This means that the time resolution will be poor unless the host deposit grew rapidly. However, this is offset by the ability to date the calcite independently by the absolute methods. Bastin (1990) reported on 241 pollen spectra from 45 speleothems in nine Belgian caves. They were mostly of Holocene age but some fragmentary records from the previous interglacial were detected. Brook *et al.* (1990) emphasized that pollen from dated speleothems may be particularly useful in desert areas, where palaeovegetational records are sparse on the surface; they were able to recognize wetter past phases in samples from the Chihuahuan, Kalahari and Somali deserts, and also found savannah grassland pollen in speleothems from a cave now covered by tropical rainforest in Zaire. Similarly, in glaciated regions the conventional palynological sources, pond and lake-floor silts and clays, are usually post-glacial in age; caves may preserve older samplings.

Caution is needed in interpreting pollen assemblages from cave deposits because there are three potential, distinct sources of supply:

1. aeolian, which presumably gives a valid sample of the contemporary regional pollen;
2. speleothem feedwater or other infiltration – pollen grains range 0.5–100 µm in diameter so that where infiltration is an important source of them, species represented by larger grains will probably be screened out;
3. floodwaters, in which much or all of the pollen may be reworked from older deposits.

Bastin (1990) emphasized that the greatest densities of pollen grains in speleothems are usually found in the detrital layers, which may be of flood origin. Burney and Burney (1993) set air traps for 2 year in two American caves and showed that the modern aeolian pollen was representative of the exterior spectra; caves with larger entrances and stronger air flow were the most productive, as would be expected. Genty *et al.* (2002) carefully sampled speleothem feedwaters in caves in southwest France and found that they were without pollen. It appears that most is screened out.

Amino acid racemization

Any organic matter transported into caves ultimately decomposes. After death the protein amino acids in organic matter slowly convert from an L to a D configuration, at a rate controlled chiefly by the ambient temperature. This is the basis of the **amino acid racemization** dating technique. The principles and common applications are summarized by Williams and Smith (1977) and Miller (1980). As noted in section 8.3 a large proportion of vadose speleothems will contain humic and fulvic acids and some larger organic fines. Cave interior temperatures also may be essentially constant over the year, although most will vary over the course of a glacial climatic cycle. Lauritzen *et al.* (1994) extracted nine different amino acids from a calcite flowstone in an arctic Norwegian

cave. They investigated the L: D ratio in isoleucine, the chief acid used for dating. A linear correlation with U series α-spectrometric ages back to 350 ka was obtained and used to calculate amino acid ages of 420 and 505 ka for two basal layers of the flowstone. We are not aware of further applications to cave deposits, although it has potential value because in principle amino acid racemization dating can extend for one million years or more.

8.7 PALAEOENVIRONMENTAL ANALYSIS OF CALCITE SPELEOTHEMS

8.7.1 'Trees of stone' – stable isotope studies of speleothems

Caves are excellent sites from which to obtain paleoenvironmental data because their deep interiors tend to be sheltered, protected repositories with extremely stable climates. Away from entrances and major streams their temperatures vary by less than 1°C and are close to the mean annual external air temperature; this is the homothermic zone of Figure 5.20. Relative humidity is similarly invariant, often being close to 100% so that evaporation is negligible. Evidence of environmental changes at the Earth's surface is recorded in the speleothems of the homothermic zone. It can be accessed by analysing variations in the record of isotopes, trace elements, luminescence, pollen and other larger organic matter noted above (Harmon *et al.* 2004, McDermott 2004, McGarry and Caseldine 2004, White 2004, Fairchild *et al.* 2006).

Analysis of the stable isotopes, chiefly ^{16}O, ^{18}O, ^{12}C and ^{13}C, is the most widely used approach, and its applicability is shown in Figure 8.24. The Devil's Hole (Nevada) speleothem is a layered, subaqueous calcite precipitated from the well-mixed, mildly thermal waters of a large regional aquifer in a desert setting. It has been precisely dated by U series methods. It is seen that the O and C isotope behaviour tracks the climatic cycles of most of the past 500 000 year, with a distinct anticorrelation. However, in the lower frame this impressive record of climate change is shrunk within just one small envelope on a general plot of ^{18}O versus ^{13}C speleothem variations along a transect from the Caribbean to mid-USA.

Oxygen is more readily fractionated than carbon because C is held at the centres of CO_3 groups in calcite and aragonite. As a consequence, most study has focused on the enrichment or depletion of ^{18}O with respect to ^{16}O. Speleothem stable isotope data are expressed in per mil or 'delta' (δ) notation as explained in section 6.10. Fractionations in water are expressed against standard mean ocean water as defined by the International Atomic Energy Agency in Vienna (VSMOW), and those in

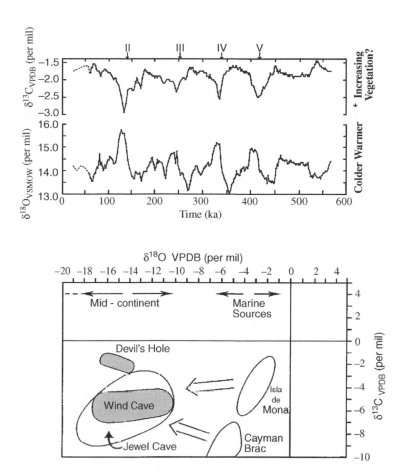

Figure 8.24 (Upper) The $\delta^{18}O_c$ and $\delta^{13}C_c$ records of DH11, a thermal subaqueous calcite deposit 34 cm in thickness on the walls of Devil's Hole, Nevada. It is precisely dated by TIMS $^{230}Th/^{234}U$ and $^{234}U/^{238}U$ methods (from Coplen *et al.* 1994). This is the longest continuous published speleothem O and C isotopic record at time of writing. (Lower) The $\delta^{18}O_c$ and $\delta^{13}C_c$ isotopic range of DH11 (shaded envelope) placed in a broader North American speleothem isotope context. Devil's Hole and Wind Cave (South Dakota) represent large and medium sized thermal water aquifers, respectively. The Jewel Cave $\delta^{18}O_c$ and $\delta^{13}C_c$ envelope is from vadose stalagmites and flowstones in a cave close to Wind Cave; over the many glacial cycles recorded here vegetation oscillated between C_3 and C_4 types. Most precipitation in South Dakota comes from the Caribbean, where the Cayman Brac envelope is from speleothems beneath C_3 vegetation (rainforest) and the Isla de Mona envelope is from speleothems beneath C_4 vegetation (xerophytic scrub). NB: the $\delta^{18}O_c$ values of the Devil's Hole and Wind Cave calcites have been increased by 2‰ to compensate the thermal effect in them.

calcite are expressed against a standard based on a fossil belemnite (VPDB). The relationship between the two standards is (Clark and Fritz 1997):

$$\delta^{18}O_{VSMOW} = 1.03091 \cdot \delta^{18}O_{VPDB} + 30.91‰ \quad (8.9)$$

and

$$\delta^{18}O_{VPDB} = 0.97002 \cdot \delta^{18}O_{VSMOW} - 29.98‰ \quad (8.10)$$

Carbon isotopes from calcite ($\delta^{13}C$) are also measured against VPDB. Precision of VPDB analysis is about $\pm 0.05‰$ for $\delta^{18}O$ and $\pm 0.02‰$ for $\delta^{13}C$, whereas results relative to VSMOW have an error of $\pm 0.2‰$.

Sampling is conducted along the growth axis of the speleothem. Because of their internal structure and means of growth, stalagmites and flowstones are preferred to stalactites. X-ray tomography (a non-destructive technique) reveals the best locations for sampling (Mickler *et al.* 2004b). Depending on the growth rate, close sampling with a drill bit giving ~10 mg of powder can yield resolutions of about one determination per 10 to 100 year. Laser ablation (McDermott *et al.* 2001) may reduce this to a few years and micromilling at 20 to

100 µm increments or high spatial resolution ion microprobe analysis can yield average resolution of less than 1 year in some instances (Frappier *et al.* 2002). Trace element analysis by laser ablation (Treble *et al.* 2003) and ion microprobe (Baldini *et al.* 2002) yields similar temporal resolution.

By providing a terrestrial palaeoenvironmental record, speleothem sequences complement the marine and polar records from deep-sea cores and glacier-ice cores. In marine sediments, stable isotopic data are obtained from the calcite tests of Foraminifera. They provide superb continuous records of environmental change (as recorded in the oceans) that extend back in time for more than one million years. However, deep-sea cores have limitations. They are not well dated; absolute ages between independently dated horizons are subject to significant error, and resolution is seldom better than 3 ka due to bioturbation. Being marine deposits, they offer only a limited and uncertain insight into environmental changes on the continents, although this is enhanced when continental deposits such as pollen and dust are incorporated in the core sediments. Records of comparable length have been obtained from polar ice cores. These yield continuous data of stable isotopes, gases, dust and trace elements for as long as 650 ka (e.g. Petit *et al.* 1999). They are especially valuable for the information they provide on environmental change in the atmosphere. But ice cores also have limitations, because dating errors increase with depth sometimes to as much as ±15 ka at ~150 ka, and their polar location permits only a distant insight into environmental change on land at lower latitudes.

Many of these shortcomings are overcome by speleothems, because they can be closely and precisely dated back to ~0.5 a using Th ages and with the prospect of millions of years as U/Pb dating improves. Their environmental data relate directly to the land above the cave. They are distributed in carbonate karst over about 10% of the planet (Figure 1.2). Therefore they provide a coverage that neatly complements deep-sea cores and ice cores. However, speleothems also have limitations because their records are often discontinuous and relatively short and the relationship of speleothem variables to climatic factors is sometimes ambiguous and difficult to quantify (Fairchild *et al.* 2006). Nevertheless, considerable progress has been made in addressing these problems in recent years. McDermott (2004) emphasized the importance of speleothems in providing precise estimates for the timing and duration of major O-isotope-defined climatic events.

A basic principle when interpreting palaeoenvironments from speleothems is that the heavier isotope is preferentially concentrated or retained in the denser phase (gas–liquid–solid). When water evaporates, for example, relatively more ^{16}O moves into the vapour phase and the remaining water becomes preferentially enriched in ^{18}O. The amount of such fractionation when calcite is deposited may be determined by the ambient temperature alone (termed **equilibrium fractionation**) or by a combination of temperature plus evaporation (**kinetic fractionation**). Criteria used to determine if a speleothem was deposited in isotopic equilibrium are: (i) there is no correlation between paired values of $\delta^{18}O$ and $\delta^{13}C$ measured along a growth layer and (ii) the value of $\delta^{18}O$ along a growth layer shows no enrichment from the axis of deposition towards the edges of the growth layer (Hendy 1971, Gascoyne 1992). However, the growth layers of some speleothems are difficult to differentiate and individual δ values may average hundreds of years; thus obtaining unequivocal results using the second criterion is not always easy (Harmon *et al.* 2004). The degree of isotopic equilibrium can also be assessed by directly comparing the $\delta^{18}O$ and $\delta^{13}C$ values of modern speleothems and their corresponding drip waters (Mickler *et al.* 2004a).

It is now recognized that speleothems often do not grow entirely under equilibrium. This is unlikely if there has been significant evaporation or rapid degassing. Thus samples that have grown in isotopic equilibrium are most likely to be found in recesses or closed chambers where relative humidity is always high, air flow is minimal and drip waters fall only short distances. Speleothems that have grown continuously over long spans of time sometimes display an alternation of equilibrium and kinetic fractionation conditions. This information has palaeoenvironmental value because it can be interpreted as representing changes between no evaporation and effective evaporation that may be correlated with Pleistocene climatic oscillations. If there is doubt concerning the suitability of a speleothem's isotopic record for palaeoclimatic reconstruction, then the issue can often be resolved by comparing the record with coeval speleothems from other parts of the cave or from another cave nearby. If the isotopic variations displayed in both speleothems are similar, then they are likely to be reflecting the same regional environmental signal and so equilibrium deposition is implied.

The equilibrium fractionation factor, calcite–water, is expressed as

$$1000 \ln \alpha_{c-w} = \delta^{18}O_c - \delta^{18}O_w \qquad (8.11)$$

(O'Neil *et al.* 1969), and if a speleothem grows in oxygen isotope equilibrium with drip waters then variations in $\delta^{18}O$ of the calcite $(\delta^{18}O_c)$ are governed by the equation

$$\delta^{18}O_c = \frac{d\alpha_{c-w}}{dT}T + \frac{d(\delta^{18}O_p)}{dT}T + (\delta^{18}O_{sw}) \qquad (8.12)$$

(Serefiddin *et al.* 2004). This relates changes in $\delta^{18}O_c$ to the temperature dependent changes in fractionation between calcite and water (α_{c-w}) and precipitation $(\delta^{18}O_p)$ and to changes in global ice volume as reflected in $\delta^{18}O$ of seawater $(\delta^{18}O_{sw})$. An assumption here is that the value of $\delta^{18}O_p$ is approximately the same as the drip water $\delta^{18}O_w$.

The temperature dependence of the calcite–water fractionation factor $(d\alpha_{c-w}/dt)$ ‰ $°C^{-1}$ decreases from 0.27 at 5°C to 0.21 at 25°C (Harmon *et al.* 2004). It is this inverse relationship between speleothem O-isotope composition and depositional temperature that forms the basis of a palaeothermometer.

The average effect of the temperature during precipitation on $\delta^{18}O_p$ is positive

$$\frac{d(\delta^{18}O_p)}{dT} = 0.55‰\ °C^{-1} \qquad (8.13)$$

(Kohn and Welker 2005), although there is considerable site-specific variability. The positive association is reflected in the relationship of cave drip water $\delta^{18}O_w$ to cave temperature, although it is non-linear. However, the effect cannot be assumed to have remained stable if climatic boundary conditions changed.

The average glacial–interglacial change in $\delta^{18}O_{sw}$ of ocean water was ~1‰ (Shackleton 2000), but showed significant spatial variation. Ocean bottom-water values were often less than the mean and surface waters (from which precipitation is derived) rather more. From an investigation of isotopic values in the equatorial Pacific (Lea *et al.* 2000), we may estimate the change from glacial to interglacial conditions in surface waters to be ~1.2‰. The importance of this is that for every 1‰ change in the value of the seawater, there will be an equal change in the $\delta^{18}O_p$ value of precipitation derived from it.

The net result of the above competing influences on $\delta^{18}O_c$ varies significantly between sites from which speleothems are obtained. Nevertheless, in equilibrium calcite, if the value of $\delta^{18}O_c$ varies along the growth axis of a speleothem, then change in environmental conditions is indicated that usually involves change of temperature.

Expressions used to estimate the temperature (T) of deposition of speleothem calcite (or of a cave atmosphere at the time when a speleothem was deposited) are based on the results by O'Neil *et al.* (1969) and have been developed by Hays and Grossman (1991)

$$T(°C) = 15.7 - 4.36\,(\delta^{18}O_c - \delta^{18}O_w)$$
$$+\,0.12(\delta^{18}O_c - \delta^{18}O_w)^2 \qquad (8.14)$$

and Genty *et al.* (2002)

$$T(K) = \left\{ 2780 \Big/ \left[\ln\left(\frac{1 + 10^{-3} \times \delta^{18}O_c}{1 + 10^{-3} \times \delta^{18}O_w}\right) \right.\right.$$
$$\left.\left. +\,0.00289 \right] \right\}^{-2} \qquad (8.15)$$

To determine temperature using these equations it is necessary to measure $\delta^{18}O_w$ of the formation water as well as $\delta^{18}O_c$. This can be achieved for modern conditions by determining the $\delta^{18}O_w$ of cave seepage water and the $\delta^{18}O_c$ of actively growing straw tips. It can also be achieved for past conditions, because one of the great advantages of the dense, crystalline speleothem calcite typical of humid cave interiors is that it may retain the formation water in fluid inclusions (section 8.3). However, the present $\delta^{18}O_w$ value of an inclusion may not be a valid reflection of the original water, because oxygen can exchange with that in the calcite lattice if there is a change of temperature. This uncertainty can be overcome because the ^2H:^1H ratio (D/H ratio) in the inclusion is stable, and thus the $\delta^{18}O_w$ value of the formation water can be estimated from the meteoric water line (equation 6.51) or its local variants. This approach has been shown to apply in modern karst waters ranging from subarctic to tropical locales (Schwarcz *et al.* 1976). For a given site, $\delta^{18}O_w$ values can be expected to remain relatively constant provided climatic boundary conditions do not change.

From the use of fluid inclusions it has been suggested that cave temperatures shifted as much as 8°C between glacial and interglacial times at sites south of the ice limits in the interior USA (Harmon *et al.* 1978). The method has seen comparatively little use, however, because of problems in obtaining valid fluid extractions (Yonge 1982). The technical difficulties are now being overcome (Dennis *et al.* 2001, Serefiddin *et al.* 2005), so greater application of this technique can be anticipated. Fleitmann *et al.* (2003), for example, successfully analysed fluid inclusions in speleothems from a cave in Oman and were able to identify five pluvial periods over the past 330 ka during which δD and $\delta^{18}O$ values were much more negative than in modern rainfall, suggesting a southern (Indian Ocean) moisture source at these times. Genty *et al.* (2002) identified macroscopic fluid inclusions in speleothems from southern France that were large enough to permit direct injection of the water into the spectrometer, thus eliminating the problems associated with extraction from microscopic inclusions. However, in such macro-inclusions there may be post-depositional exchange of water.

Before $\delta^{18}O_c$ and $\delta^{13}C_c$ records from speleothems can be used reliably to interpret past environmental conditions, we must understand the processes that determine these data (Figure 8.25). Meteoric precipitation is the source of the feedwaters that sustain speleothem growth. Percolation through the soil and epikarst (section 6.3) results in mixing and storage of recharge waters. As a result, the wide variation in isotopic values found in precipitation is suppressed and drip waters often have $\delta^{18}O_w$ values that are close to the mean annual value of precipitation in the region above the cave, as shown, for example, by investigations in Tasmania (Goede *et al.* 1982), across North America (Yonge *et al.* 1985) and in New Zealand (Figure 8.26a). However, in warm semi-arid regions evapotranspiration may result in some fractionation in the soil and epikarst with the result that drip waters become isotopically heavier than rainfall (Bar-Matthews *et al.* 1996). A tabulation of worldwide examples by Harmon *et al.* (2004) shows that the difference between observed seepage water and average precipitation $\delta^{18}O$ values can range from at least $+1.1$ to $-1.9‰$.

Because there can be great variability in epikarstic flow routes and storage (Figures 6.10 and 6.12), homogenization of recharge may not be perfect, and so drip waters in the same cave may have $\delta^{18}O$ values that vary considerably. Individual drips have been shown in Yorkshire caves to have a monthly range of $\delta^{18}O_w$ variation of up to 4.9‰, even though the overall mean of all sites measured is close to that of precipitation (Harmon *et al.* 2004). Where such variability exists, then the isotopic composition of calcite precipitated on different speleothems in the same cave will also vary. For example, Serefiddin *et al.* (2004) measured differences of up to 4‰ $\delta^{18}O_{VPDB}$ in coeval speleothems just a few metres apart. However, although the $\delta^{18}O_c$ values of such neighbouring speleothems that grow in isotopic equilibrium may differ (because of differences in drip water residence times and pathways), the changes in δ values along their length can still be responsive to (and record) the same major environmental changes.

From equation (8.12) and Figure 8.25a we see that the relationships between regional environmental change and the value of $\delta^{18}O$ found in equilibrium calcite are complex. When a cave interior cools in response to a fall in the external temperature the $\delta^{18}O$ value of deposited calcite will increase; i.e. it becomes isotopically 'heavier'. This is the **cave temperature effect**. However, the ultimate source of water from which the calcite is precipitated is the ocean, so if the isotopic characteristics of that source water vary, so should the $\delta^{18}O$ value of deposited calcite. Water that evaporates is relatively enriched in ^{16}O compared with its source. Consequently, when ice sheets accumulated during the Pleistocene, the glaciers had abundant ^{16}O (hence very low $\delta^{18}O$ values) and the remaining ocean water became progressively more enriched in ^{18}O. As a result, during glacial phases, precipitation and cave seepage waters had high $\delta^{18}O$ values compared with interglacial intervals – termed the **ice volume effect**. However, the isotopic value of precipitation is also affected by the temperatures at which evaporation and condensation take place, the average effect of temperature on precipitation being about $0.55‰\,°C^{-1}$ (equation 8.13). Thus if seawater temperatures change in the source areas of precipitation, so will the $\delta^{18}O$ values of precipitation. During the Last Glacial Maximum around 20 kyear BP, cooling of the tropical oceans has been estimated as up to $4°C$ in the Indian Ocean (Barrows and Juggins 2005) and $2.8 \pm 0.7°C$ in the equatorial Pacific, with glacial–interglacial differences being as great as $5°C$ over the past 450 000 year (Lea *et al.* 2000). Thus $\delta^{18}O_p$ and $\delta^{18}O_w$ values during glacial stages reflected both ^{18}O-enriched oceans and cooler water temperatures.

The $\delta^{18}O_c$ value of speleothems growing under isotopic equilibrium therefore can be determined by two sets of characteristics: the first represents the thermodynamic fractionation between calcite and water, the **cave temperature effect**, and the second relates to the combination of factors that influence the isotopic composition of the feed water, termed the **drip water function** by Lauritzen and Lundberg (1999b). These characteristics have different temperature sensitivities. The former always has a negative response to temperature, whereas the latter may respond negatively or positively, depending on regional meteorology and the scale of climatic change. The overall net outcome depends on their relative magnitudes.

Because the **cave temperature effect** and the **drip water function** may oppose each other, the isotopic characteristics in each karst region should be studied carefully before $\delta^{18}O_c$ trends are interpreted in terms of palaeotemperature changes. It is likely that in many continental sites the cave temperature effect predominates, i.e. the sign of $d(\delta^{18}O_c)/dT$ is negative, e.g. in China and Austria (Wang *et al.* 2001, Yuan *et al.* 2004, Mangini *et al.* 2005). But in mid-latitude oceanic settings and some continental sites the resultant can be positive

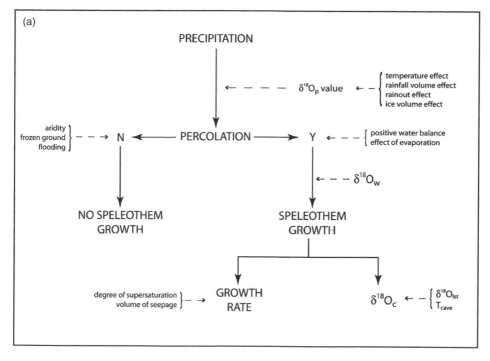

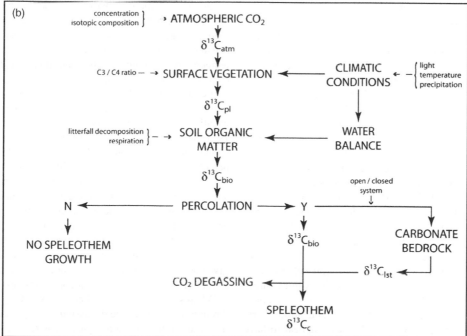

Figure 8.25 (a) Factors determining the $\delta^{18}O_c$ composition of speleothem calcite (assuming equilibrium deposition) and speleothem growth rate. (b) Factors determining the $\delta^{13}C_c$ composition of speleothem calcite (assuming equilibrium deposition). Reproduced from Williams, P.W., King, D.N.T., Zhao, J.-X. and Collerson, K.D, Speleothem master chronologies: combined Holocene $\delta^{18}O$ and $\delta^{13}C$ records from the North Island of New Zealand and their palaeo-environmental interpretation. The Holocene 14(2), 194–208, Sage, 2004.

(Goede *et al.* 1986, Dorale *et al.* 1992, Gascoyne 1992, McDermott *et al.* 1999, Xia *et al.* 2001, Paulsen *et al.* 2003, Williams *et al.* 2005), because the cave temperature effect is dominated by the influence of temperature on the $\delta^{18}O$ of precipitation. Different speleothems in some places have been found to display opposite $d(\delta^{18}O_c)/dT$ relationships (e.g. in Norway, cf. Lauritzen 1995, Linge *et al.* 2001, Berstad *et al.* 2002), and this can even occur in the same cave (e.g. in Reed's Cave, South Dakota; Serefiddin *et al.* 2004). Therefore, no assumptions should be made about the direction of a T relationship; the polarity of each speleothem should be checked individually. Furthermore, variations in $\delta^{18}O_c$ cannot always be assumed to be determined by temperature changes, because changes in rainfall source and amount can sometimes be more important (Bar-Matthews and Ayalon 1997, Cruz *et al.* 2005a,b, Treble *et al.* 2005). For example, the $\delta^{18}O$ of rainfall in low and middle latitudes can be strongly influenced by rainfall amount, the relationship being negative.

Factors influencing speleothem $\delta^{13}C_c$ values are illustrated in Figure 8.25b. Whereas it was once common to interpret $\delta^{13}C_c$ variations mainly in terms of vegetation changes, it is now appreciated that such variations are a result of a complex interplay of several variables. Five factors are especially important: the concentration and isotopic composition of CO_2 in the atmosphere; production of biogenic CO_2 by plant and soil processes; carbon sourced from karst bedrock and ageing soil; the ratio of open to closed system dissolution; and degassing of CO_2 (Baskaran and Krishnamurthy 1993, Baker *et al.* 1997, Denniston *et al.* 2000, Genty *et al.* 2001, Williams *et al.* 2004). The latter factor is well illustrated by the effect of dynamic seasonal ventilation changes in a chimney cave in the Austrian Alps (Spötl *et al.* 2005).

During glacial–interglacial cycles there are changes in the concentration and isotopic composition of atmospheric CO_2 (Petit *et al.* 1999). Concentrations varied from 180 to 200 ppm by volume during the Last Glacial Maximum ($\sim$20 kyear BP) to around 260 ppm by 8 kyear BP (and 375 ppm by AD 2003). The 60–80 ppm pre-industrial glacial-interglacial shift is the **glacial atmosphere effect**, the increase in atmospheric CO_2 being partly responsible for reducing $\delta^{13}C_c$ values during glacial–interglacial transitions. There is a weak anticorrelation between CO_2 concentration and atmospheric $\delta^{13}C$ ($\delta^{13}C_{atm}$) values. Values of $\delta^{13}C_{atm}$ appear to have been around $-6.7\permil$ during the Last Glacial Maximum and about $-6.3\permil$ by 8 a (Indermühle *et al.* 1999).

Photosynthesis preferentially removes $^{12}CO_2$ from the atmosphere; so the atmosphere becomes enriched in $^{13}CO_2$ under interglacial conditions. This has a direct effect on plant $\delta^{13}C$ values. For every 100 ppm increase in CO_2 there is a $-2.0 \pm 0.1\permil$ change in the $\delta^{13}C$ of plants (Feng and Epstein 1995). The photosynthetic pathway that plants follow affects the $\delta^{13}C$ values of their respired CO_2, which is in the range of -26 to $-20\permil$ for C_3 plants and -16 to $-10\permil$ for C_4 plants (Cerling 1984) and this in turn affects the $\delta^{13}C$ value of soil CO_2. During periods of low plant activity soil CO_2 has high values of $\delta^{13}C$, probably due to a greater admixture of atmospheric CO_2. The water balance of the soil also affects $\delta^{13}C$ of plants with δ values becoming higher (less negative) as aridity increases (Stewart *et al.* 1995).

The principal sources of speleothem carbon are the atmosphere, biological activity via the soil and the bedrock. Carbon-14 dating indicates that there is substantial 'dead' carbon in speleothems, up to about 65% having been measured, although it is usually $< 20\%$. Ageing soil organic matter is responsible for a high proportion of this, but dissolution of limestone bedrock typically yields 5–15%. This shows that substantial dilution of modern carbon can occur during carbonate dissolution. Delta ^{13}C values of limestone bedrock are relatively high, usually in the range -5 to $+5\permil$ (e.g. -1.72 to $-0.98\permil$ in the case of Oligocene limestones from New Zealand). Thus the greater the addition of the old carbon component from the bedrock, the higher the resulting $\delta^{13}C$ value of the seepage water as it approaches the cave. Thermal water calcites are comparatively enriched in ^{13}C due to high-temperature leaching of limestone along the flow paths (Bakalowicz *et al.* 1987).

If carbonate dissolution occurs under open-system conditions, resulting $\delta^{13}C$ values will be considerably more negative than under closed-system conditions (Hendy 1971, Salomons and Mook 1980). The relative significance of the bedrock component also varies with the time that percolating water is in contact with the rock. In periods when rainfall is relatively high, flushing rate is increased and residence time of water in the epikarst is decreased; so the opportunity for inorganic $\delta^{13}C$ enrichment also decreases (Shopov *et al.* 1997). The effect on $\delta^{13}C_c$ values operates in the same direction as the effect of high rainfall on plants, leading to a decrease in $\delta^{13}C$. So wet conditions give rise to relatively low $\delta^{13}C_c$ values, and particularly so should open-system dissolution conditions also prevail.

When seepage water enriched with soil CO_2 percolates through the aerated zone degassing of CO_2 occurs. This is important in the cave atmosphere, where P_{CO_2} is close to that found in the open atmosphere. It is also sometimes important in aerated fissures *en route* to the cave, where it leads to the critical supersaturation that results in calcite deposition on speleothems. This process leads to enrichment of the solution in ^{13}C. Dulinski and Rosanski (1990) modelled the processes that lead to the formation of $^{13}C/^{12}C$ isotope ratios in speleothems, and identified the importance of time since first deposition of calcite in accounting for variations of $\delta^{13}C_c$.

Where special effects (e.g. geothermal) may be discounted and assuming climatic boundary conditions have remained relatively stable, a change of several parts per thousand over time in an equilibrium speleothem could be interpreted either as a major change in the type (C_3/C_4) of vegetation cover or biotic activity at the feedwater source or as a change in precipitation (or both). In some countries with almost exclusively C_3 vegetation (such as New Zealand), changes in $\delta^{13}C_c$ values are mainly ascribable to changes in vegetation density and water balance conditions.

Given the foregoing, it is evident that integration of O and C isotopic information from speleothems will permit reconstruction of past environmental conditions. The usual first step is to establish the polarity of the $d\delta^{18}O_c/dT$ relationship, as we have stressed above. In many cases, and especially continental areas, the relationship will be a negative one. This is the case for caves in Israel (Frumkin *et al.* 1999, Bar-Matthews *et al.* 1996) and is especially well exemplified by the composite 185 ka series compiled from 21 overlapping speleothem records from Soreq Cave shown in Figure 8.26b (Ayalon *et al.* 2002), where low values of $\delta^{18}O_c$ are associated with interglacial periods (e.g. at 8 ka and 125 ka) and high values occur during the glacial maximum around 18–25 ka. The strong negative polarity in these cases arises from the cave temperature effect being reinforced by the rainfall amount effect and the $\delta^{18}O$ values of the source waters of the eastern Mediterranean Sea. In sharp contrast is the long $\delta^{18}O_c$ record with positive polarity from the Devil's Hole calcite from Nevada (Figure 8.24; Coplen *et al.* 1994). This continuous 60–566 ka record was deposited in warm supersaturated water in a well-mixed desert aquifer ~400 km from the Pacific coast. It reflects the composition of regional meteoric water and has been convincingly correlated with sea-surface palaeotemperatures along the California margin derived from alkenone palaeothermometry (Herbert *et al.* 2001). The Devil's Hole data can be interpreted as being in phase with oceanic temperatures, but leading oceanic ice-volume records as determined from foraminiferal $\delta^{18}O$; the explanation being that sea-surface temperature changes in the eastern Pacific are transmitted via the moisture carried by westerly winds to the western interior of North America (Lea *et al.* 2000).

Once the polarity of records has been established, the issue of representativeness of samples becomes important. Whereas the Devil's Hole calcite provides an integrated but muted isotopic signal from a large groundwater catchment, individual vadose zone stalagmites record the signal transmitted via epikarstic flow paths from a limited area immediately above the cave. Consequently, stable isotope records from particular stalagmites can differ in detail from neighbours, even from the same cave (Figure 8.24; Dorale *et al.* 1998, Denniston *et al.* 2000, Serefiddin *et al.* 2004). Thus to obtain an isotopic series that is robust and representative of the region, records are needed from several coeval speleothems. The issue is similar to that faced in dendrochronology, because tree-rings record site factors as well as regional influences.

Regionally representative stable isotope time series can be obtained by merging the individual records of several speleothems, having first determined that they were deposited in isotopic equilibrium and that all have the same $d\delta^{18}O_c/dT$ polarity. This yields a composite curve. Williams *et al.* (2004, 2005) show how this can be done, merging the $\delta^{18}O_c$ and $\delta^{13}C_c$ records of eight different speleothems from six different caves, suppressing local effects and highlighting general trends by smoothing the data with a running mean (Figure 8.27a).The main source of error in the process is from dating, because ages of individual δ values are usually estimated by linear interpolation between dated points. To minimize errors of interpolated ages, dates should be close together and consideration should be given to possible advantages of curvilinear interpolation.

Before an estimate can be made (even qualitatively) of the temperature changes implied by $\delta^{18}O_c$ variation in a record that extends into the Pleistocene, adjustment of $\delta^{18}O_c$ values will be required to compensate for the ice-volume effect. The enrichment of ocean surface source waters was ~1.2‰ at the LGM, with a lowering of sea level by approximately 130 m. Thus ice volume adjustment at a rate of 0.009‰ m^{-1} of sea level change is of the correct order. The effect of such adjustment is illustrated in Figure 8.27b. From this we see that the difference in the average adjusted $\delta^{18}O_c$ values between the LGM (~20 a) and the Holocene in this example is about 0.55‰. Most of this probably can be ascribed to temperature change because the rainfall amount effect is not strong in the region concerned.

Reconstruction of palaeotemperatures can be achieved by estimating $\delta^{18}O_w$ from δD determined from speleothem fluid inclusions. Thus last interglacial (17–22°C) and LGM (8°C) temperatures in Israel were derived in this way by McGarry *et al.* (2004). A quite different approach is to calibrate the $\delta^{18}O_c$ signal against other independent records of temperature and then to apply the calibration curve to the $\delta^{18}O_c$ series. This method was used by Lauritzen and Lundberg (1999b) and Mangini *et al.*

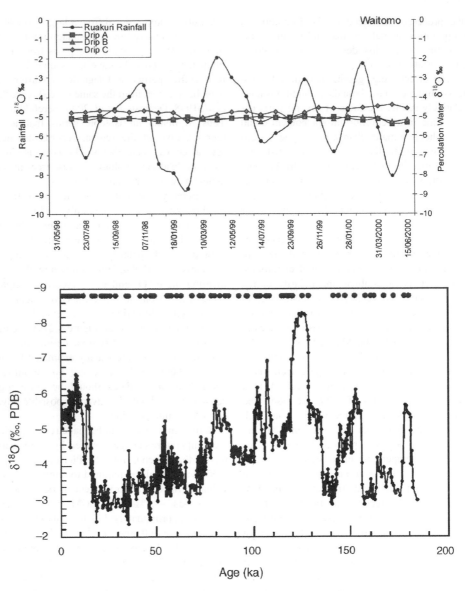

Figure 8.26 (Upper) Relationship between $\delta^{18}O_{VSMOW}$ in rainfall above Aranui Cave, New Zealand, and $\delta^{18}O_{VSMOW}$ in seepage waters in the underlying cave 40 m below the surface. Reproduced from Williams, P.W. and Fowler, A. (2002) Relationship between oxygen isotopes in rainfall, cave percolation waters and speleothem calcite at Waitomo, New Zealand. New Zealand Journal of Hydrology, 41(1), 53–70. (Lower) A 185 a record of $\delta^{18}O_c$ from Soreq Cave, Israel, assembled from 21 overlapping speleothem data sets. It involves more than 2000 isotopic analyses and is constrained by 95 TIMS ages, the locations indicated by dots along the top of the graph. Reproduced from Ayalon, A., Bar-Matthews, M. and Kaufman, A. (2002) Climatic conditions during marine oxygen isotope stage 6 in the eastern Mediterranean region from the isotopic composition of speleothems of Soreq Cave, Israel. Geology, 30(4), 303–6.

(2005) in the case of speleothems from Norway and the central Alps. However, the success of this approach depends on the availability and reliability of palaeotemperature indicators from the past.

One of the interesting attributes of speleothem isotope records is the periodicity often shown by the data. It is assumed to reflect cyclic changes in the terrestrial environment. One technical problem associated with

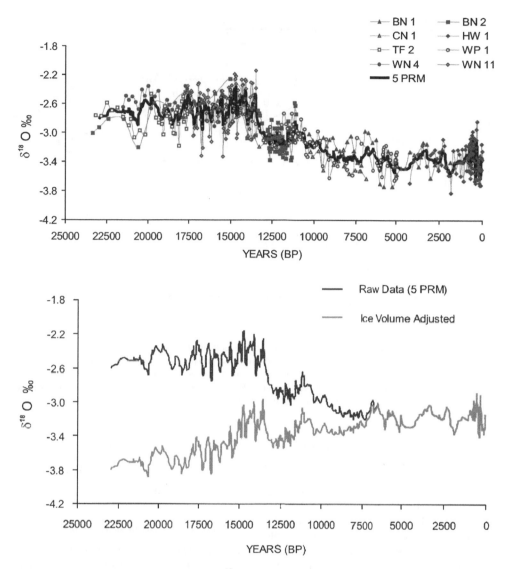

Figure 8.27 (Upper) Derivation of a composite regional $\delta^{18}O_c$ curve from the merger of records from eight different speleothems from six caves in New Zealand. A five-point running mean has been used to smooth the composite curve and reveal its principal features. (Lower) The composite regional curve adjusted for the ice volume effect (lower curve). The adjustment is greatest (1.2%) at the Last Glacial Maximum and tends to zero when sea level attained its present level about 6.5 a. Reproduced from Williams, P.W., King, D.N.T., Zhao, J.-X. and Collerson, K.D, Late Pleistocene to Holocene composite speleothem chronologies from South Island, New Zealand – did a global Younger Dryas really exist? Earth and Planetary Science Letters 230 (3–4), 301–317 © 2005 Elsevier.

identifying periodicity in records from speleothems is that their time series are unevenly spaced.

Fortunately this can be overcome by using the programs SPECTRUM (Schulz and Statteger 1997) or REDFIT (Schultz and Mudelsee (2002) that were specially created for unevenly spaced palaeoclimatic time series. Serefiddin *et al.* (2004) applied the routine to data from speleothems

from Reed's Cave in South Dakota and demonstrated periodicity at 1000 to 2000 year, which is similar to the millennial-scale variability seen in North Atlantic sediments and Greenland ice-cores. In Europe, McDermott *et al.* (2001), Spötl and Mangini (2002) and Genty *et al.* (2003) found centennial to millennial-scale oscillations in $\delta^{18}O_c$ in speleothems from Ireland, Austria and

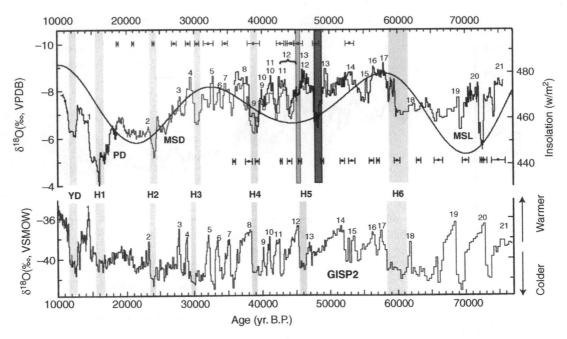

Figure 8.28 Relationship between linked $\delta^{18}O$ records from three speleothems from Hulu Cave in China, the GISP2 ice core from Greenland, and summer insolation at 33° N. The shaded vertical bars show possible correlations of Heinrich events (H1–H6) and the Younger Dryas (YD) cold reversal in the GISP2 ice core with the Hulu Cave record (two possible correlations are shown for H5.) Reproduced with permission from Wang, Y.J., Chen, H., Edwards, R.L. An, Z.S., Wu, J.Y., Shen, C.-C., and Dorale, J.A,. A high-resolution absolute-dated late Pleistocene monsoon record from Hulu Cave, China. Science 294, 2345–2348. © 2001 American Association for the Advancement of Science.

France, respectively, that they considered coincident with Dansgaard–Oeschger events of the Greenland ice-core records. In China, Wang *et al.* (2001) found a correspondence with Heinrich events (massive iceberg discharges in the North Atlantic) (Figure 8.28). These results are important, because TIMS-dated speleothems define the chronology of palaeolimate episodes, such as D–O events, more accurately than can be achieved from ice cores. Frappier *et al.* (2002) also demonstrated approximate correspondence of $\delta^{13}C_c$ from a stalagmite from Belize with the Southern Oscillation Index, and Dykoski *et al.* (2005) found indications of solar forcing of $\delta^{18}O_c$ in a speleothem record from Dongge Cave, Guizhou, China.

Wavelet analysis has also been used to identify variations in $\delta^{18}O_c$ and $\delta^{13}C_c$. For instance, Paulsen *et al.* (2003) detected cycles of 33, 22, 11, 9.6 and 7.2 year in a 1270 year high-resolution record from China; Holmgren *et al.* (2003) identified millennial and centennial scale oscillations in a 24.4 ka record from South Africa; Qian and Zhu (2002) found quasi-70 year climatic oscillations in the East Asia monsoon; and Cruz *et al.* (2005) identified a 23 ka cycle equivalent to the Earth's precessional cycle in a 116 ka speleothem record from Brazil. It

is most intriguing that evidence for solar forcing of climate change can be found in lightless caves; it underlines their value as sensitive natural archives of global environmental change.

8.7.2 Trace elements in speleothems

Many different elements are present in trace amounts in calcite and aragonite speleothems. The most abundant are usually Mg and Sr, readily substituting for Ca in the crystal lattice. Also widely detected are Na, K, Ba, Cu, Fe, Mn, P, Pb, U and Zn, and Cd, Co, Cr, Ni, Ti and rare earth elements (REE) have been recorded, all studied for their possible contributions to speleothem colour (James 1997). The potential use of trace abundances and ratios for palaeoenvironmental reconstructions has been appreciated and is currently a focus of much study, both in the field and in the laboratory.

In early work at McMaster University speleothems with differing U concentrations were irradiated to map the fission tracks, i.e. the distribution of U atoms in the lattice. Some displayed very sharp periodic banding, others random or weakly varying distribution patterns.

Gascoyne (1977) followed up with analysis of the stable trace elements and found wide differences between individual samples. He turned to the field, instrumenting speleothem drip sites in Vancouver Island caves. The Mg/Ca ratio in calcite should be temperature dependent (*ceteris paribus*), while the Sr/Ca ratio is not; comparison of the two thus might yield an index of temperature changes over the history of a stalagmite, etc. However, during 12 months of drip sampling little coherent variation could be found.

Subsequent research by many others has confirmed that trace-element behaviour in speleothems is variable and complex. It may differ between samples, within samples and over time at a site.

Where growth banding can be seen either optically or in the luminescence spectra or both, there is usually correlative periodicity in some of the trace-element distributions. There are several reports of Sr enrichment in the lighter bands, while Fe, P and Zn are more abundant in the dark bands (e.g. Huang *et al.* 2001). Attention has focused on the Mg/Ca and Sr/Ca ratios. Roberts *et al.* (1998) found the very strong negative covariance in part of a Holocene stalagmite from a shallow cave in the north of Scotland that is shown in Figure 8.29. That was an exceptionally clear result, however. Fairchild *et al.* (2000) studied drip and pool water in four caves between southwest Ireland and northern Italy, and found much variation in behaviour which they attributed chiefly to varying feed-water paths between soil and cave, open versus closed dissolution conditions, and to varying residence times. Where there was statistically good positive

covariance between the two ratios ($R^2 > 0.8$), this was attributed to precipitation of calcite along the path during drought periods, a finding which has important implications for the stable isotope interpretations discussed above. Similarly, in two eastern Australian caves McDonald *et al.* (2004) obtained relatively low average Mg/Ca ratios in drips that responded rapidly to rains compared with higher values in relatively unresponsive (i.e. long resident) waters. Both Mg/Ca and Sr/Ca ratios showed a systematic increase through drought conditions, with peak values immediately before drought was broken by rain. Tooth and Fairchild (2003) have presented a range of fast versus slow, wet versus dry, soil → drip physical plumbing models to explain such differing patterns of behaviour. They are now being intensively explored to further our understanding of the complex environment of vadose speleothem growth.

8.7.3 *Optical and luminescent banding in speleothems*

When sectioned many vadose and phreatic speleothems display banding in their calcite texture, or depositional terminations, colour and/or colour density that is easily visible to the naked eye. Figures 8.9 and 8.21 show good examples. If they are visible, however, these individual layers usually represent hundreds to thousands of years of accumulation. Under the optical microscope, much banding can be resolved to layers that are only 1–100 μm in thickness. Causes of such layering where the net deposition is periodic are set out in Table 8.3. Where it is continuous (Condition 1, Table 8.3), layering may be due

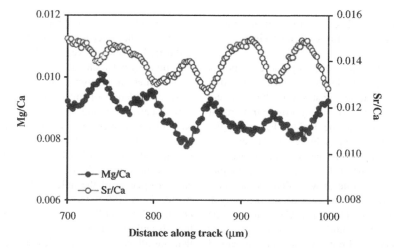

Figure 8.29 Comparison of the Mg/Ca and Sr/Ca ratios across 300 mm of calcite growth bands in a 17 cm high stalagmite from Uanh am Tartair Cave, Scotland. Reproduced with permission from Roberts, M.S., Smart, P.L. and Baker, A, Annual trace element variations in a Holocene speleothem. Earth and Planetary Science Letters, 154; 237–246, 1998.

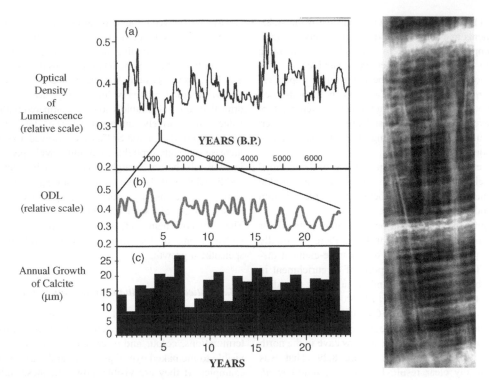

Figure 8.30 (Left) (a) Variation of luminescence intensity measured along a stalagmite from Coldwater Cave, Iowa. The sample is 16.4 cm in height and began growing 7000 yr ago (U series TIMS dates). (b) Blow up of a segment from just before AD 1000. Annual variation of luminescence is strongly developed, which permits (c) measurement of the thickness of each annual band of calcite. The range of thickness is ×4, from ~7 μm to ~28 μm, reflecting wetter and drier hydrological years. (Right) Periodic microbanding in a stalagmite from the northern USA revealed by photography of the UV luminescence. (Photograph by Y.Y. Shopov.) The true height of the frame is ~1.0 mm. The bands are probably annual.

to abrupt changes in crystal texture, rates of accumulation, or to varying concentrations of trace elements or organic compounds.

As described in section 8.3, where optical evidence is lacking, the periodicity in vadose speleothems can often be detected by investigating variations in the luminescence intensity attributable to varying concentrations of fulvic and humic acids in the lattice. For example, Figure 8.30 shows approximately 40 lighter–darker luminescence couplets within ~0.4 mm of calcite from a speleothem in the northeastern USA, i.e. average couplet thickness is ~10 μm. The line drawing depicts the varying luminescence intensity measured along the growth axis of a stalagmite from Coldwater Cave, Iowa. The sample is 16.4 cm in height and began growing 7000 year ago (U series TIMS dates). Frame B is an enlargement of a segment of it from just before 1000 year ago. Annual variation of luminescence is strongly developed, which permits (Frame C) measurement of the thickness of each

annual band of calcite. The range of thickness is ×4, from ~7 to ~28 μm, over a span of 25 year, reflecting wetter and drier hydrological years at this site, which is close to the boundary between natural forest and tall grass prairie in the American mid-west (Shopov *et al*. 1994).

The correlation of much of this fine-scale optical or luminescence banding with the hydrological year at particular cave sites has been firmly established in recent years by the study of banded calcites accumulating in mines and other artificial excavations. Bands can be counted back from the date of collection to the known date of opening of the hole or the cessation of work in it, etc. Genty (1992), Genty and Quinif (1996) and Genty *et al*. (1995) published the early analyses from both tunnels and natural caves. It should be stressed, however, that not all stalagmites and flowstones will display continuous banding. Baker *et al*. (1993) reported it in only a few of more than 40 older samples studied with care. Often the banding in a speleothem is interspersed with sections that are without it. Other samples (even in the same cave) may display no detectable banding at all.

The 25 year record in Figure 8.30 (C) shows that there is scope for multiyear hydrological reconstructions using speleothem periodic banding and so evidence of longer, more general, climatic cycles is being studied. For example, for speleothems from northwest Scotland, Proctor *et al.* (2002) resolved 50–70 year and 72–94 year spectral periodicities in the North Atlantic climate record and Fleitmann *et al.* (2004) identified a 780 year record of the Indian monsoon from speleothems from Southern Oman. However, even when obvious banding is not present, there can still be scope for hydrological reconstructions as shown by Treble *et al.* (2005) for a modern speleothem from Western Australia. Sometimes temperature and irradiance rather than precipitation is recorded. Thus the 11- or 22-year sunspot cycle has been reported (see Shopov (1987, 1997) for details) in many analyses around the world. In an interesting case of a stalagmite with a 2650 year record from Beijing, China, Tan *et al.* (2004) have correlated annual banding with warm season temperature and total solar irradiance.

8.7.4 Seismospeleology

In many caves a proportion of the calcite speleothems have been fractured by natural causes and have fallen from their growth position. Straw stalactites break under their own weight and larger carrot stalactites fall when their weight overcomes the bonding to the ceiling. Stalagmites and flowstone masses built upon unconsolidated sediments may topple because the sediment foundation fails under the increasing load. However, there are instances where columnar stalagmites are broken off

above a base that appears not to have shifted. More than 100 year ago European researchers suggested that such breakage might be caused by earthquakes (see Quinif (1996) and Forti (1997) for reviews). Using U series methods, the date of an earthquake might then be determined: for example, using α-spectrometric dating Agostini *et al.* (1994) studied a cave near the epicentre of the strongest earthquake recorded in Italy in modern times and suggested that there had been distinct older events there at 0.5 ka, between 30 and 40, 90 and 100, and > 350 ka.

Early attention focused on the position and orientation of the broken columns. A strong earthquake shock propagated from due east might be expected to place the broken section on the west side of its base, with a westerly orientation. Directions to epicentres of some past earthquakes have been estimated by this means, and Moser and Geyer (1979) used the relationship between basal diameter and length of the broken fragment in an attempt to determine the magnitude of the vibration in Austrian alpine caves. However, the orientations of fallen fragments vary chiefly due to compaction irregularities in sediments beneath columns or to the topography of the cave floor. More promising is the observation that a stalactite–stalagmite pair function as a recording pendulum (Schillat 1977). A slight tilt of the cave due to earthquake activity thus will offset the point of accretion at the tip of the stalagmite (as is seen in Figure 8.21). The same effect can be achieved by compaction irregularities at the base (as noted) or by an independent shift of the drip point in the ceiling but, where these two possibilities can be eliminated, the pendulum concept is valid. Forti and Postpischl (1985) investigated the changes in long-axis orientation of some Italian

Table 8.4 Estimated dimensions and mass fluxes through the Friar's Hole Cave system, West Virginia (Adapted from computations by Worthington, from data of Worthington 1984)

Physical characteristics		Mass fluxes	Total (10^3 m^3)	In cave now (10^3 m^3)
Contributing drainage basin	85.7 km^2	Dissolved host limestone:		
(including 2.6 Km^2 of		from the input surface	57 300	Trace
Host limestone exposed at		from the known cave	2400	Trace
surface)		Cave breakdown	1000	280
Friars Hole cave system		Authigenic fluvial detritus	400	20
Explored length	68.12 km	(mostly from breakdown)		
Total volume of known cave	2700×10^3 m^3	Allogenic fluvial detritus	3 000 000	600
Volume now open	1800×10^3 m^3	(mostly from siliciclastic rocks)		
		Aeolian deposits	< 1?	< 0.001
Volume now infilled with	900×10^3 m^3	Organic matter – all sources	100	0.001
detritus		Calcite speleothems	~1	0.15
Age of the oldest passages	> 4.0 Ma	Gypsum and other precipitates	0.001	0.001

stalagmites and showed that the directions of shifts fitted the main tectonic trends in the areas. In one example, 21 small shifts of the axis were measured in a specimen only 36 cm in height.

8.8 MASS FLUX THROUGH A CAVE SYSTEM: THE EXAMPLE OF FRIAR'S HOLE, WEST VIRGINIA

It is instructive to close this chapter with an estimate of the flux of all matter through a cave system during its lifetime. Few such estimates have been attempted because of the evident difficulties in devising them. Many quantities may be in error by at least one order of magnitude.

One set of estimates (Table 8.4) has been prepared by Worthington from his 1984 study in one of the world's most extensive caves, Friar's Hole (Table 7.2). This is something of an extreme example because only 3% of the surface catchment basin consists of the host limestones (Figure 7.24). These crop out as inliers in narrow valley floors, into which flow the detritus from steep slopes of shale, sandstone and argillaceous limestone, i.e. it is a situation favouring the maximum flux of allogenic debris through a system. Extrapolating from U series, RUBE and palaeomagnetic results, the earliest passages are 4×10^6 year or a little greater in age. The mean solute

mass flux from the host limestone has approximated $15\,000\,\text{m}^3\,\text{ka}^{-1}$ over that period. Note that only 4% of this flux is represented by dissolution in the known cave. Generous estimates of the additional volume of unknown cave passages in the system will increase this value to 15–25%, i.e. over 75% of net dissolution has occurred in the epikarst developed in the small windows or inliers.

Thirty-seven per cent of the volume of the known cave was created by mechanical breakdown. Seventy per cent of the clasts produced have been removed, chiefly in solution.

Clastic rocks crop out over 97% of the area of the basin and are estimated to have furnished 95-98% of the total mass flux in the system during its history. This component overwhelms all others. Only 0.2% of the aggregate detrital flux is in transit through the caves at the present time, yet this suffices to infill about 22% of their volume. The mean underground transit time of a clast is ~80,000 year. Flow path lengths (sink to spring) will be between 15 and 60 km. Effective hydraulic gradients have been 0.006 or lower.

Although it has a few grottoes with large and abundant speleothems, Friars Hole is not a well decorated cave by world standards. It is estimated that only the tiny proportion of 0.0016% of solutes in transit from the epikarst have been intercepted and precipitated as cave calcite during its history.

9

Karst Landform Development in Humid Regions

9.1 COUPLED HYDROLOGICAL AND GEOCHEMICAL SYSTEMS

It was explained in Chapter 4 that most dissolution is expended near the surface, in the epikarst. We now consider the landforms and assemblages of landforms that are created there. These vary from small features such as karren to large-scale landforms measured in kilometres such as poljes. Within the dynamic karst system they can also be classified as input, output or residual features (Figure 1.2). This chapter is organized to discuss input forms first, proceeding from smaller to larger, and then output and residual features. We also review the special solution features associated with evaporite and quartzose rocks. The chapter concludes with discussions of landform sequences in carbonate terrains and considers the extent to which karst landscape evolution can be simulated by computer modelling.

We introduced karst hydrogeology (Chapters 5 and 6) and cave development (Chapter 7) before discussing surface landforms because the essence of karst is that its drainage is subterranean and the initiation of karst plumbing is an essential pre-condition for the early development of medium- to large-scale surface landforms. Karst landforms result from processes operating in coupled hydrological and geochemical systems. Essentially the same processes can operate over a very wide range of environments, but limiting conditions are provided by aridity and extreme cold. Karst is therefore characteristic of humid regions, where water normally occurs in its liquid phase. In this section we examine 'normal' karst development in humid areas, leaving consideration of karstification under extreme climatic conditions until the next chapter.

In endeavouring to understand the relationship between processes, karst rocks and resultant landforms the following points are important.

1. Hydrological processes determine the general location of erosion within karst lithologies and hence are usually the principal control on landform development. In particular, the nature of hydrological recharge, whether autogenic, allogenic or mixed, has considerable morphological significance because of its influence on the horizontal and vertical distribution of corrosion and corrasion.
2. Lithology and structure can be so important as to dominate landform development, although in general geology influences karst development through its control of (i) the provision of solute pathways, (ii) rock strength and (iii) susceptibility to corrosion and corrasion.
3. Different amounts of runoff in various humid regions influence annual karst erosion and hence the rate of landscape evolution, but not necessarily the morphological style of karst topography that is developed.
4. Temperature variation is significant to morphological development mainly through its influence on (i) the water balance (via evapotranspiration), (ii) the rate of chemical reactions and hence the vertical distribution of dissolution, and (iii) biochemical processes leading to the acidification of infiltrating water. Depositional landforms are also influenced by temperature via evaporation and biological processes.

9.2 SMALL-SCALE SOLUTION SCULPTURE – MICROKARREN AND KARREN

The German term 'Karren' and the French term 'lapiés' are widely used to describe small-scale dissolution pit,

Karst Hydrogeology and Geomorphology, Derek Ford and Paul Williams
© 2007 John Wiley & Sons, Ltd

groove and channel forms at the surface and underground. Here we anglicize the German and define **microkarren** as features with a greatest dimension or characteristic dimension (length, width, diameter, depth, etc.) that is normally less than 1 cm. **Karren** range from 1 cm to 10 m in greatest dimension in most instances, although kluft-karren and some solution channels can be longer. Assemblages of many individual karren, termed **Karrenfeld**, may cover much larger areas (Figure 9.1).

Karren develop upon the carbonate and sulphate rocks and dominate all outcrops of salt. They are also the dissolutional landforms encountered most frequently on other rocks such as sandstone, quartzite and granite. Lithological properties are of great importance; many specific karren forms develop only where rocks are homogeneous and fine grained.

There is a vast range of karren features. Bögli (1980) wrote 'The multiplicity of possible karren forms makes a morphological system endless, while a genetic one allows a meaningful collection'. In 1960 he proposed a classification based primarily upon whether the host rock was bare ('free karren'), partly covered ('half free'), or entirely covered by soil or dense vegetation ('covered karren'). We sympathize with the principle that a genetic classification is to be preferred to a morphological one, but believe that the genesis of many karren is not sufficiently understood to support a wholly genetic basis at this time. In particular, much of the variety in karren occurs because two or more differing processes combine to produce a polygenetic form. The classification adopted here (Table 9.1) is based on morphology, with subdivisions that incorporate genetic factors. Fornos and Gines (1996), Gines 2005), Macaluso and Sauro (1996) and Veress (2004) adopt similar ones. Bögli's (1960) nomenclature is retained wherever possible; see Perna and Sauro

Figure 9.1   Clint-and-grike topographies or 'limestone pavement'. (Upper left) Glaciated limestone pavement with drumlin hill in the background, Co. Clare, Ireland. (Upper right) A pavement staircase (Schichttreppen) on the Burren in western Ireland (see also Figure 6.7 upper). (Lower left) 'Classic' clint-and-grike at Malham Cove, Yorkshire, UK. Rundkarren revealed by stripping of soil dominate the foreground, with simpler grike forms at scarp edge in the rear. (Lower right) Staircase pavement on dolomite, arctic Canada.

Table 9.1 Classification of karren forms

A Circular plan forms

Micropits and etched surfaces – wide variety of pitting and differential etching forms commonly less than 1.0 cm in characteristic dimension.

Pits – circular, oval, irregular plan forms, with rounded or tapering floors, > 1.0 cm in diameter.

Pans – rounded, elliptical, to highly irregular plan forms: planar, usually horizontal floors in bedrock or fill, > 1.0 cm in diameter.

Heelprints or trittkarren – arcuate headwall, flat floor, open in downslope direction. Normally 10–30 cm diameter.

Shafts or wells – connected at bottom to proto caves/small caves draining into epikarst. Great range of form.

B Linear forms – fracture controlled

Microfissures – microjoint guided, normally tapering with depth. May be several centimetres long but rarely more than 1.0 cm deep. Transitional to

Splitkarren – joint-, stylolite- or vein-guided solution fissures. Taper with depth unless adapted for channel flow. From centimetres to several metres in length, centimetres deep. Closed type terminates on fracture at both ends. Open type terminates in other karren at one or both ends.

Grikes or kluftkarren. Major joint- or fault-guided solutional clefts. Normally 1–10 m in length. Master features in most karren assemblages, segregating *clint* blocks (Flachkarren) between them. Scale up to karst bogaz, corridors, streets, etc. Subsoil forms are termed cutters.

C Linear forms – hydrodynamically controlled

Microrills – as on *rillenstein.* Rill width is ∼ 1.0 mm. Flow is controlled by capillary forces and/or gravity and/or wind.

Gravitomorphic solution channels

   1 *Rillenkarren* – packed channels commencing at crest of slope; 1–3 cm wide. Extinguish downslope. Rainfall-generated, no decantation.

   2 *Solution runnels* – Hortonian channels commencing below a belt of no channelled erosion. Sharp-rimmed on bare rock (Rinnenkarren), rounded if subsoil (Rundkarren). Channels enlarge downslope. Normally, 3–30 cm wide, 1–10 m long. Linear, dendritic or centripetal channel patterns.

   3 *Decantation runnels.* Solvent is released from an upslope, point-located store. Channels reduce in size downslope. Many varieties and scales up to 100 m in length, e.g. wall karren (Wandkarren), Maanderkarren.

   4 *Decantation flutings* – solvent is released from a diffuse source upslope. Channels are packed; may reduce downslope. 1–50 cm wide.

   5 *Fluted scallops or solution ripples* – ripple-like flutes oriented normal to direction of flow. A variety of scallop. Prominent as a component of *cockling patterns* on steep, bare slopes.

D Polygenetic forms

Mixtures of solution channels with pits, pans, wells and splitkarren. Subsequent development of *Hohlkarren, Spitzkarren* and subsoil *pinnacles.* Superimposition of small forms (microrills, Rillenkarren, small pits) upon larger forms.

Assemblages of karren

Karrenfeld – general term for exposed tracts of karren.

Limestone pavement – a type of karrenfeld dominated by regular clints (flachkarren) and grikes (Kluftkarren). Stepped pavements (*Schichttreppenkarst*) when benched.

Pinnacle karst – pinnacle topography on karst rocks, sometimes exposed by soil erosion.

Arete-and pinnacle, stone forest, etc., with pinnacles to 45 m high and 20 m wide at base.

Ruiniform karst – wide grike and degrading clint assemblage exposed by soil erosion. Transitional to *tors*.

Corridor karst – (or *labyrinth karst, giant grikeland*); scaled-up clint-and-grike terrains with grikes several metres or more in width and up to 1 km in length.

Coastal karren – distinctive coastal and lucustrine solutional topography on limestone or dolomite. Boring and grazing marine organisms may contribute. Includes intertidal and subtidal notches, and dense development of pits, pains and micropits.

(1978) for the equivalent names in other European languages. It is emphasized that our classification stresses the comparatively simple end-member forms. In reality there is a great mixture of forms created by factors of lithological variation and polygenesis combining together.

9.2.1 Microkarren

All the microforms distinguished in Table 9.1 are considered together here. Dissolutional topography can be recognized at a scale of a few microns under the electron microscope, but where the relief is less than 1 mm or so it

Figure 9.2   Small-scale dissolution forms on limestone and marble. (Upper left) Microrills on the walls of a shallow solution pit in micrite. Coin is 2 cm in diameter. (Upper right) Dissolution pan. (Lower left) Meandering microrills from stem flow into rundkarren, freshly deforested. (Lower right) A pattern of multiple trittkarren (heel prints) in marble, with rillenkarren developing upon residual ridges. Can is 12 cm in length.

is convenient to consider the surface to be smooth. Exposed karst rock surfaces generally display relief greater than 1 mm unless they are being subjected to vigorous scouring or polishing action. This relief can develop upon limestones within a few decades.

Many apparently bare carbonate surfaces are partly or entirely covered by bacteria, fungi, green algae, blue-green algae or lichens. These may contribute to the preferential etching of weaker grains and to the development of micropits (sections 3.8 and 4.4). Most attention has focused on the activity of blue-green algae (cyanophytes) since Folk *et al.* (1973) suggested that they produce much of the relief of coastal phytokarst. Most species are surface dwellers (epilithic), but in ecologically stressful environments some cyanophytes bore into rocks to depths of ~ 1.0 mm while others dwell in vacated borings or other microcavities. Borers create pits directly; other species may contribute to their creation or enlargement by way of the organic acids or CO_2 that they excrete. Once established, small pits and fissures may be preferentially deepened if fungi, lichens or mosses can establish in them and excrete CO_2. Cyanobacteria-induced pits have been measured to 14 mm (Figure 4.13).

Microrills are typically 1 mm wide, round-bottomed dissolutional channels that are tightly packed together (Figure 9.2). They are sinuous or anastomosing on gentle slopes, becoming straighter on steep slopes. Lengths are up to a few centimetres. Most reports of them are upon fine grained to aphanitic limestone but they also occur on gypsum. They are known in most climates. Clasts with microrills are *rillensteine* (Laudermilk and Woodford 1932). Some microrills are created by waters flowing down surfaces, e.g. from acid stemflow over clasts. In other instances the rilling takes place when waters move upwards, drawn by capillary tension exerted at an evaporating front. Capillary flow is believed to explain much of their characteristic sinuosity.

9.2.2 Solution pits, pans, heel-prints (tritt), shafts or wells, cavernous weathering

Solution pits are round-bottomed or tapering forms that are circular, elliptical or irregular in plan view. Diameters greater than 1 or 2 m are rare; the form merges to a pan as that scale is approached. Pits can occur singly, aligned, clustered or packed. They may drain by evaporation, by overspill and/or by basal seepage via primary pores or tight microfissures. Together with shafts, they are the most widespread karren form globally, both on bare rocks and beneath soil. They are predominant where the rocks are very heterogeneous (e.g. many limestone and most dolomite reefs).

Many pits are located along small joints, taper down into them and are transitional to shafts or to fissure karren (below). Others have developed at a cluster of primary pores or a vug, or where an insoluble fossil has fallen out. Deeper pits are often colonized by moss that appears to have accelerated the deepening by algae, etc. Some pits display raised rims, where water has precipitated calcite upon evaporation, armouring them. At an experimentally cleared limestone site in Yorkshire, England, Sweeting (1966) noted that pits 3–5 cm deep developed in 10 year. The water was peaty, i.e. enriched by organic acids.

Solution pans display a flat or nearly flat bottom that is usually horizontal (Figure 9.2). This may be created by an organic or clastic filling in a round-bottomed pit, but most often it is a dissolutional bevel in the bedrock with an organic or other veneer on the floor below. Walls are steepened by undercutting and may display a basal corrosion notch. Overflow channels are common. Individual pans attain diameters of several metres and depths greater than 1 m. Amalgamation of adjoining pans is common, creating larger features of cuspate or irregular form. Pans develop well on limestone, dolomite, gypsum, quartzites, granites and well-cemented sandstones. They are termed **solution basins**, **kamenitze** and **tinajitas** by other authors.

Solution pans occur on bare or lightly vegetated rock. They appear to be rare or absent beneath a soil cover, whereas solution pits are abundant there. This emphasizes that pans develop where a pool may form with a floor that becomes partially armoured by detritus, focusing dissolution around the perimeter. Pans cease to function when the floors, lowered by dissolution, intercept a penetrable bedding plane or other fissure.

Trittkarren or **heelprints** (Figure 9.2) are comparatively rare. They occur on bare limestone and dolomite surfaces that are gently inclined or shallowly stepped. Each tritt comprises a planar corrosion bevel (the heel) open in the downslope direction. The bevel is usually horizontal, 10 to 30 cm in diameter. Upslope it is enclosed by a steep, cirque-like backwall a few centimetres in height. This may be indented by rillenkarren. The contact between backwall and bevel is sharp but not undercut. Tritt occur singly, adjoining one another where they indent a step, or in a sequence down a slope.

Some trittkarren are modified solution pans but the majority appear to be of different, though related origin.

Bögli (1960) ascribed their development to the accelerated dissolution that might occur where a film of flowing water is thinned upon descending a pre-existing step. Possibly the early process is a boundary layer detachment as in dissolution scallops, though this cannot be true once rillenkarren are established.

Trittkarren we have seen have been limited to homogeneous, fine-grained to aphanitic, limestone, dolomite or marble. They are also limited to surfaces where scouring agencies (chiefly glaciers, but also waves and flood waters) generate microscarps such as chatter marks.

Karren shafts or **wells** are very short caves draining into the epikarst. Most are guided by joints, bedding planes or calcite veins. More sinuous examples in porous rocks follow primary porosity. They may be vertical, horizontal or inclined. Length (depth) ranges from a few centimetres to 2–3 m. Cross-sections tend to be circular or elliptical and up to 1 m in diameter, but there is great variety.

These features develop from proto-caves as described in section 7.2. In addition, pits and pans are converted into shafts where their floors intersect bedding planes. Many grikes are initiated by shaft development down to an underlying bedding plane. Karren shaft forms and assemblages can be complex and variable where they develop beneath a deep, periodically saturated soil cover because the dominant condition is epiphreatic; it becomes a 'boneyard' morphology that is much favoured for ornamental rockeries throughout the world (Figure 9.3).

Cavernous weathering refers to this boneyard type of cave and pit morphology. It can also describe individual pockets (**tafoni**) or clusters of hollows produced by weathering back into steep faces. The latter are common on some dolomites, as well as on sandstones, quartzites, conglomerates and granitic rocks where salt weathering and/or hydrolysis may play a role. In this kind of weathering, the water does not pass via the cave into the epikarst; it is a superficial phenomenon.

9.2.3 Fracture-controlled linear karren

Many linear karren are elongated along minor joints, veins, stylolites or microfractures, such as may develop normal to stylolite seams (Pluhar and Ford 1970). Many such karren range in length from one or a few centimetres to several metres. Length:maximum-width ratios are greater than ~ 3:1 and depth is usually much less than length. Unless adapted by a channelled flow, the features taper sharply with depth. Thus they appear to be splitting the rock. Closed linear karren terminate on the host fissure, open ones terminate at one or both ends in other karren (e.g. a grike). Linear karren may be transitional to pits, karren

Figure 9.3 Subsoil pit and shaft development in a limestone block set up as an ornament in the garden of the Imperial Winter Palace, Beijing, China.

(flachkarren) and host the smaller forms of karren. In bedded rocks most grikes terminate at penetrable bedding planes at depths of one half to a few metres. A small minority may extend down to deeper bedding planes and receive the drainage of the shallower members.

The length of grikes is inversely proportional to the density of major joints. In most karsts it ranges from 0.3 m to a few tens of metres. At a given site grikes tend to be longest, widest and deepest near escarpment edges where jointing is expanded by tension unloading. This is where they will develop first. In many young reef rocks, grikes are the only linear karren form that develop because the others are prohibited by the textural heterogeneity.

Grike walls may be parallel or taper with increasing depth. They are often indented with cavernous weathering, rillenkarren or cockling, or dissected by linear karren, rinnen or rundkarren. Many grikes have been created by the amalgamation of earlier shafts developed at intervals along the joint, which creates a pattern of widenings and narrowings. Beneath a deep soil, grikes become much widened at the top and taper with depth. These forms are termed **cutters** (Figure 5.25) by American authors (e.g. Howard 1963). Intervening clint tops are reduced in area and sharpened by runnel cutting to form **subsoil pinnacles**.

9.2.4 Hydraulically controlled linear forms – dissolution channels

Channel karren have received more study than the other types because of their similarities to erosional channels and, therefore, their supposed amenability to an hydraulic explanation.

Rillenkarren are perhaps the most striking, because they appear to be the antithesis of the normal (or Hortonian) erosional rills that are generated by runoff on soil (Figure 9.4). Rillenkarren head at the crest of a bare slope, where they are uniformly packed together and, at a given site, display only one or two characteristic widths. They diminish in depth lower downslope (Figure 9.5) until they are replaced by a planar solution surface or **Ausgleichflache** (Bögli 1960). In contrast, Hortonian dissolutional **runnels** head below such a belt of no channelled erosion and, on a simple surface, are uniformly separated by interfluves.

Rillenkarren do not develop on gentle slopes and on the steepest slopes degenerate into **cockling patterns** – a mixture of scallops, fluted scallops or ripples (below) and discontinuous rills. Rillenkarren must be the product of direct rainfall because there is no other feasible source of water. Channels that also head at crests but are notably

shafts or grikes. On slopes they are often intermingled with the larger, gravity-controlled karren. Where a rock is densely fractured they can display a bewildering variety of orientations and intersections; development of other karren types then is prohibited by their density, with the exception of grikes and some shafts.

Grikes or **Kluftkarren** are the master features in most karren assemblages (Figure 9.1). They are the principal drains, either to the deeper epikarst, or to dolines or to surface discharge such as river channels. They develop along the major joint sets or systems and thus tend to intersect at angles of 60°, 90° and 120° (tension and shear systems). Blocks isolated between them are termed **clints**

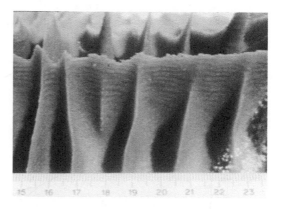

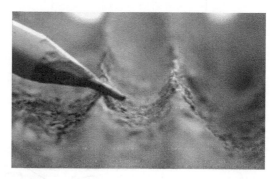

Figure 9.4   (Left) Typical rillenkarren (fluting) on limestone. (Upper right) Rillenkarren crests on an experimental salt block. (Lower right) Cross-sections of a few of the rillen shown in the left-hand frame.

wider (> 4–5 cm) are varieties of the decantation flutings discussed below. Admixture of the two types occurs in nature, so they are readily confused.

Rillenkarren develop well upon fine-grained, homogeneous limestones and marbles. On dolomite and other more heterogeneous carbonates rilling is only partial or is absent entirely. They develop well upon gypsum and are the predominant karren form on salt outcrops. In a worldwide study, Mottershead *et al.* (2000) found median lengths and widths to be 300 mm and 18 mm respectively on limestone, 120 mm and 11 mm on gypsum, and 210 mm and 17 mm on salt.

At many sites it appears that, setting aside textural factors, length of rillenkarren increases with an increase of gradient. Glew and Ford (1980) investigated this question by hardware simulation, exposing texturally uniform plaster of Paris slabs at differing inclinations to constant rainfall at 25°C (Figure 9.6). It was found that rillenkarren propagate from the crest downslope until a stable length is reached. They are produced by a hydraulic 'rim effect'. At the crest of a slope raindrops penetrate the fluid boundary layer (section 3.10), permitting turbu-

lent reaction at the mineral surface. Depth of flow increases downslope to some critical value (0.15 mm in the experiments) where drops cannot impact the surface directly. Uniform mass transfer then creates the ausgleichflache, after which both rillen and ausgleichflache are removed by parallel retreat. Formation begins as many short, shallow rills that deepen and lengthen, coalescing laterally to achieve the characteristic width. Short rills appear between 5 and 10° (depending upon texture). Mean rill lengths increase with slope, being 250–300 mm at 60° in the experiments. The rill cross-section approaches the parabolic, which focuses rain splash into the centre (Glew and Ford 1980, Crowther 1998). Fiol and Gines (1996) suggested that, on limestone, mechanical processes may also play a role because raindrop impact can remove small particles loosened by algal corrosion of the surface; however, this is not fundamental because it does not apply to gypsum or salt rillen.

Solution runnels are normal Hortonian channels, heading where sheetflow or wash on a slope breaks down into linear threads (Figure 9.5). On steeper slopes the channels are parallel; on gentle slopes there may be dendritic

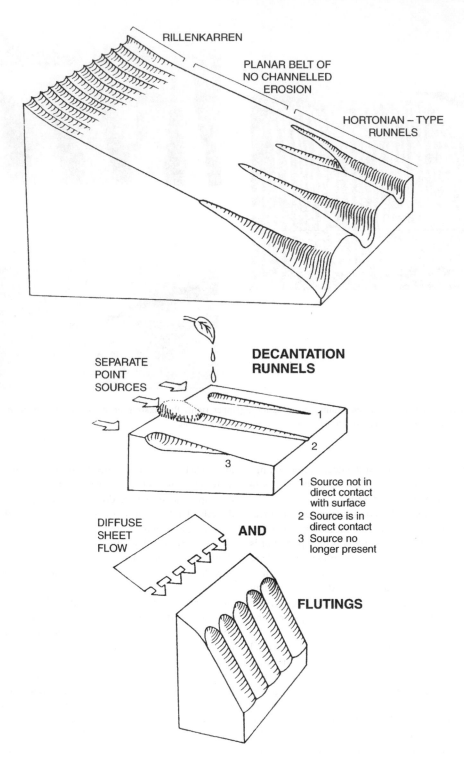

Figure 9.5   Rillenkarren, Hortonian-type dissolution channels and decantation channels, as defined in Table 9.1. (Drawings by J. Lundberg.)

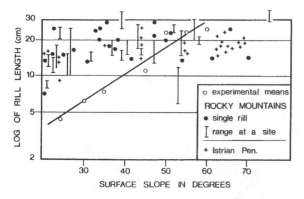

Figure 9.6 Relationships between the length of rillenkarren and the angle of slope. Open circles and the linear relation are from the controlled experiments of Glew and Ford (1980) with plaster of Paris and a rain machine. Rocky Mountain data are from different blocks in a Jasper Park landslide pile. This was an ideal site to study rillenkarren because there is no inheritance effect; nevertheless, it is seen that correlation with the experimental results is poor.

confluence or centripetal orientation into a karren shaft or grike. **Rinnenkarren** display sharp rims and flat or rounded bottoms (Figure 9.7). They develop on bare slopes. **Rundkarren** have more rounded cross-sections because they develop beneath vegetation or soil; with pits, they are the predominant form on soil-covered clint surfaces. However, when exposed they may be sharpened into rinnenkarren (Figure 9.8).

These are conventional stream channels that gain discharge downstream; thus they normally widen and deepen downstream. For example, rinnen and rundkarren on 35–40° dip slopes on Vancouver Island, Canada, increased in mean width from 4 to 8 cm along lengths of 3–5 m and maintained width:depth ratios between 1.5 and 2.5 (Gladysz 1987). After 3 to 5 m all examples amalgamated with decantation forms (i.e. became composite or polygenetic) or were intercepted by linears, shafts or grikes. This is the normal pattern elsewhere, as well, but in the spectacular exception in Patagonia noted below rinnen can be 1–4 m wide and several hundred metres long.

Rinnenkarren occur on slopes as low as 3°. Their courses may be sinuous but full meandering (sinuosity ratio > 1.5) is rare. They have been measured on slopes as steep as 60°. Rundkarren can propagate on slopes that are nearly vertical because cavitation rarely occurs beneath soil. Rinnen and rundkarren often accumulate moss or excess soil at places along their channels. These may be widened to create a segment with overhanging walls (**hohlkarren**), or deepened to create a locally

reversed gradient or even a flat pan bottom. Overdeepened segments occur in most rundkarren we have seen, making them more complex forms. Rinnenkarren and rundkarren develop on most carbonate rocks but are best formed where they are homogeneous and medium to fine grained. They develop well on gypsum, basalt, granites and sandstones, but are not known on salt.

Decantation runnels and **flutings** are classes proposed by Ford and Lundberg (1987) to differentiate channels created where water is released steadily from an upslope store, from the rillenkarren and Hortonian types that are created during episodes of storm-generated runoff. Decantation runnels occur principally upon bare and partly covered surfaces, but also beneath continuous soil or vegetation covers if small surfaces are temporarily freed by soil piping or root decay. The diagnostic characteristic of pure decantation forms is that, because they do not collect additional acidic water downslope, their cross-sections are largest at or close to the input point and may diminish downstream (Figure 9.8).

Decantation flutings are adjoining, shallow channels formed where water is released from a linear source such as a bedding plane or a soil mat at the top of a cliff. They develop best where slopes are steep to overhanging, as on grike walls or in vadose shafts in caves (section 7.10). Channel widths typically are 5–25 cm, and lengths are up to 25 m. Depths and depth:width ratios have not been reported, but the features appear to be shallow in proportion to their width when compared with rillen, rinnen and decantation runnels. On steep surfaces fluting requires that the film flow be thin enough to be retained on the rock by surface tension rather than detaching as happens when cockling is formed. Such films develop 'parting lineations' oriented in the direction of flow (Allen 1972). These probably establish the flutings in the rock; their separation (flute width) is inversely proportional to the velocity of flow.

Decantation runnels are approximately equivalent to the **wandkarren** (wall karren) type of Bögli (1960). Each individual is supplied from a point store such as a patch of moss or the stem of a tree. The dimensions of the runnel are in proportion to the volume of water released from storage and its acidity. At the two extremes, tiny pits along the crests of rillenkarren slopes overspill to enlarge particular rills below them, while perennial snow banks may support runnels 50–80 cm wide and deep that are up to 100 m in length. A majority of wandkarren have widths and depths in the range 1–10 cm and are distinguished or have amalgamated with other karren types within 10 m of their source.

Decantation stores accumulate within earlier linear karren, rinnen and rundkarren to produce composite or

Figure 9.7   (Upper) Rinnenkarren on steep limestone face in the Pyrenees. Note the figure for scale on rock to left of center. (Lower) Rinnenkarren showing a tendency to meander on a steep marble slope, Mount Owen, New Zealand.

Figure 9.8 (Upper left) Rundkarren exposed by modern deforestation, Vancouver Island. (Upper right) Long denuded rundkarren are sharpened into rinnenkarren, Pyrenees. (Lower) A meandering decantation runnel with its source in a broad solution pan that supports blue-green algae, Burren, western Ireland.

polygenetic forms. These are more abundant than the pure forms at many sites; e.g. at a Vancouver Island site, Gladysz (1987) measured 66 rinnenkerren, 27 decantation runnels and 423 composite runnels.

The slow release of water from storage may permit meandering, creating striking meander incisions into bare rock surfaces (**mäanderkarren**). Veress (2000a,b,c) and Veress and Nacsa (2000) have analysed complex examples from the Austrian Alps, finding that many are polycyclic and polygenetic. 'True' meanders in the bedrock have asymmetric cross-sections (cutbank and slip-off slopes), as they do in alluvial river channels; they are initiated from the head. 'False' meanders have symmetric cross-sections and are created by knickpoint recession. This is the same behaviour as is seen in many vadose channels in caves (section 7.10). Cut-off is sometimes seen in true karren meanders. Larger troughs may exhibit multiple sinuous entrenchments.

Fluted scallops resemble transverse ripples in sand. They are oriented normal to the direction of flow and may

have an asymmetric cross-section, being slightly steeper on the upstream side. Where fully developed they adjoin one another and extend across a wall or a cave roof. Curl (1966) defined them as an ideal end-member of the class of dissolutional scallops (section 7.10) and termed them 'flutes'. Jennings (1985) termed them 'solution ripples'. They develop only partially on many steep walls such as grikes, to form a prominent element in cockling patterns.

The foremost modelling of hydraulically controlled karren in recent times has been by Szunyogh (2000), Veress (2000c), Veress and Zentai (2004) and colleagues in Hungary. Their combination of formal geometric analysis, direct simulation and careful field measurement set the stage for future karren computer modelling studies of the type presented for large-scale karst at the close of this chapter.

The 'glaciers de marbre' of Patagonia (Chile) are the most spectacular example of hydraulically controlled karrenfeld yet reported (Figure 9.9). They occur on

Figure 9.9 Hydraulically controlled karren from the 'glaciers de marbre' of Patagonia. (Photographs by R.Maire and L.Pernette, with permission.) (Left) Large-scale picture. Note the figure for scale. (Right) Megawandkarren.

steeply sloping surfaces of resistant marble scoured by glacial action. The region has ~5000–7000 mm a^{-1} of precipitation, and strong winds. There are few penetrable joints or bedding planes to abstract the water underground and little vegetation, so that surface dissolution by intense runoff predominates. Rinnen and wall karren of giant size, tritt, pans, flachkarren and cockling are the principal forms. Maire (1999), Hoblea *et al.* (2001) and Veress *et al.* (2003) give details and many illustrations.

9.2.5 The composite nature of karren

We re-emphasize that there is an immense variety of karren forms. Simple or monogenetic end-members have been stressed but many karren are composite, being both polygenetic and also varying with lithology. Chief among the lithological factors are chemical purity, grain size, textural homogeneity (including pores), bedding thickness and joint frequency. Greatest density of karren occur where strata are

thin, closely jointed and heterogeneous in composition but types will be limited largely to grikes, linear karren, pits and shafts. The greatest variety of karren and their best-developed form is associated with massively bedded, fine-grained and homogeneous limestones and marbles.

Some authors have sought to relate particular karren types or scale of development to specific climatic conditions, but this encounters many problems. The **rate** of their development is greatest where it is wettest (section 4.1) and types that are limited to bare rock (principally tritt and rillenkarren) are less common where the climate can support forest. For the other types it appears to us that variations in lithology, hydraulic gradient and the duration of the karst denudation outweigh climatic control.

9.2.6 Assemblages of karren and giant karren

Great areas of karren evolve beneath complete soil and vegetation cover. They are sometimes termed

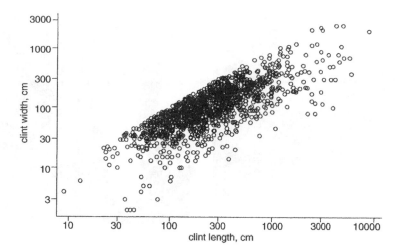

Figure 9.10   Scatter plot of clint widths versus clint length for limestone pavements in Ireland, Britain and Switzerland. Reproduced from Goldie, H.S. and Cox, N.J., Comparative morphometry of limestone pavements in Switzerland, Britain and Ireland. Zeitschrift für Geomorphologie, N.F. Supplement-Band, 122, 85–112.E.© 2000 Schweizerbart'sche Verlagsbuchhandlung, Science Publishers.

cryptokarst, e.g. Salomon *et al.* (1995) and Knez *et al.* (2003). By contrast, bare or partly bare exposures of karren are widespread above the treeline in alpine areas, on desert slopes, or where there has been deforestation and soil erosion. There are contiguous areas of many square kilometres of karrenfeld in the Adriatic karst and elsewhere. Subaerial dissolution of newly emerged coral reefs leaves the surfaces sharply serrated and extremely pitted. They are eogenetic karren, known as **makatea** in the tropical Pacific (Taboroši *et al.* 2004).

Where strata are flat-lying or gently dipping, karrenfeld are termed **limestone pavements** (Williams 1966a, Vincent 2004) if they are dominated by patterns of regular clints and grikes so that they appear like artificial paving (Figure 9.1). These are the most studied type of karren assemblage. Figure 9.10 shows the range of clint lengths and widths from a sample of nine areas in Britain, Ireland and Switzerland (Goldie and Cox 2000); median lengths and widths were 235 cm and 99 cm respectively, but the largest dimensions (from the Burren in western Ireland) were 88 m and 25 m. Where bare or lightly vegetated the clints are indented by pits, pans with decantation runnels, shafts and linear karren. Beneath dense vegetation or an acidic soil rundkarren tend to become predominant. In addition to limestone, they develop well on dolomite and on some well-bedded sandstones and quartzites.

Pavements develop best upon thick to massively bedded strata. Variations in rock strength associated with cyclothemic deposition with limestones and dolomites (section 2.6) favours stripping at particular horizons

(e.g. Vincent 2004), which explains the terraced nature of many pavements (**Schichttreppenkarst** of Bögli 1980). But where beds are thinner, clints are readily broken up by mechanical processes such as frost wedging, root wedging and fire. Massive beds eventually break up along subsidiary (previously impenetrable) sedimentation planes also, degrading the clint surfaces to rubble (or **shillow**). As a consequence, optimum pavement development occurs where an agency periodically can scour off the shillow and dissected upper beds to restore pristine surfaces. Globally, the chief scouring agents have been Quaternary glaciers. Extensive pavements are most common in areas subject to the last (Wisconsin–Würm) glaciation and so have developed largely during the 10 000–15 000 year of postglacial time. However, they also occur where the scour is by wave action, river floods or sheetfloods on pediments; the Katherine pavements of the Northern Territory, Australia (Karp 2002), are an outstanding instance of stripping by sheetflood action.

Where soils are deep but acidic, dissolution occurs on the limestone surface beneath. Unless they are very wide, clints then become tapered by rundkarren entrenchment. This creates subsoil pinnacles. If their tops are then exposed by soil erosion, they become sharpened (depending on the extent of vegetation cover) by rillen, rinnen, wandkarren and decantation flutings, especially the latter. As they begin to emerge they have the appearance of teeth ('dragons' teeth' in China), transforming to a pinnacle karst when the tops are several metres or more above the soil (e.g. Knez *et al.* 2003).

Figure 9.11   The Stone Forest of Yunnan, China. (Upper left) Rounded stone teeth emerging from beneath eroding soil. (Upper right) Stone teeth sharpened by exposure into pinnacled terrain. (Lower left) Fully developed stone forest. (Lower right) Stone forest landscape with isolated pinnacles in an advanced stage of development. See also Figure 6.6b.

The most celebrated pinnacle karst of this type is the **Shilin** ('Stone Forest') in Yunnan, China (e.g. Chen *et al.* 1986, Ford *et al.* 1996, Song *et al.* 1997). This is a rugged tor-and-pediment topography that is the product of a long and complex history, involving mid-Permian burial of karren by continental basalts; exhumation, re-karstification and then reburial by Eocene continental deposits; and finally late Tertiary and Quaternary re-exhumation and re-karstification. Pinnacle development occurred beneath these clastic deposits and the old topography is being exhumed and sharpened, exposing tracts of pinnacles in large patches over an area of 30 000 ha (Figure 9.11). Characteristic heights range 1 to 35 m and diameters, 1 to 20 m.

However, even more spectacular, but with a less complex geomorphological history, are the **tsingy** terrains of Madagascar (Figure 9.12a), which in the Bemaraha region alone cover an almost continuous area of 152 000 ha (Rossi 1986, Middleton 2004). The bare tsingy blocks are dissected by joint-aligned canyons up to 80 m deep and are surmounted by razor-sharp pinnacles.

Spectacular arête-and-pinnacle terrains also occur on Mount Kaijende in Papua New Guinea (Figure 6.6a) (Williams 1971, 2004) and on Mount Api in Sarawak (Osmaston and Sweeting 1982). Although covering relatively small areas, they have spire forms in excess of 45 m high, penetrating and rising well above tropical rain forests (Figure 9.12b). The pinnacle karsts from Madagascar, Papua New Guinea and Sarawak appear to have developed directly from dissolution of the limestone without previous burial and exhumation phases. Similar features occur in the semi-arid karsts of northern Australia (Jennings and Sweeting 1963) and Minas Gerais, Brazil. Essential prerequisites are dense rocks, massive bedding and widely spaced major joints.

(a)

(b)

Figure 9.12 (a) Oblique aerial view of the **tsingy** of Bemaraha, Madagascar. Plateau tops, which are dissected by great joint corridors up to 80 m deep, are sharpened by rillen and rinnen karren into inaccessible spires. Forest trees occupy the joint corridors. (Photograph by A. Clarke.) (b) Spire-like pinnacle karren of Gunung Api in the Gunung Mulu World Heritage Park, Sarawak. The spires are of the order of 45 m high and penetrate through humid tropical rain forest.

Ruiniform (Perna and Sauro 1978) describes terrains where soil has been removed from exceptionally deep and wide grikes, but where the clints are not sharpened into pinnacles (Figure 9.1). Instead, they stand out like miniature city blocks in a ruined townscape; Lessini (Italy) and Torcal de Antequera (Spain) are well-known examples. Ruiniform tracts are common on gentle hillslopes that have suffered deforestation and substantial

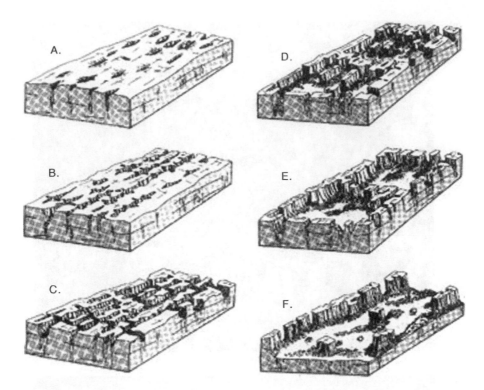

Figure 9.13   Model for the development of the Nahanni labyrinth karst or giant grikeland, Mackenzie Mountains, NWT, Canada. Grikes are created by shaft formation and elision along master joints (A, B). A combination of further dissolution and frost shattering widens them into 'streets' (C, D), reducing clints to residual towers (E) and, finally, creates large closed depressions or 'platea' (F). Grikes are up to 1000 m in length and 50 m deep. Reproduced with permission from Brook, G.A. and D.C. Ford, The origin of labyrinth and tower karst and the climatic conditions necessary for their development. Nature 275, 493–496.© 1978 Nature Publishing Group.

soil erosion. On hill crests they are transitional to *tors*, which develop particularly well on massive, coarsely crystalline dolomites.

Grikes may be expanded to larger scales to create aligned or intersecting corridor topographies (Figure 9.12). Types of large grikes were termed **bogaz** by Cvijić (1893), **corridors** (Jennings and Sweeting 1963), **zanjones** (Monroe 1968) and **streets** (Brook and Ford 1978). Squared valleys or closed depressions formed by grike-wall recession are **box valleys** and **platea**, respectively. The assemblages have been termed **giant grikelands**, **corridor karst** or **labyrinth karst**. Examples are known in tropical and temperate rain forests, and desert and semi-desert areas (section 10.2). Grikelands develop on sandstone crests in dry mountain ranges of the Australian and USA deserts, also. Small grikelands are common in limestones, dolomites and sandstones along crestlines in the subpolar, periglacial Mackenzie Mountains of Canada and are expanded to a spectacular labyrinth with individual streets longer than 1 km and deeper than 50 m in the

Nahanni limestone karst in that region (Figure 9.13). The prime requirements for such assemblages appear to be very major joint sets (often, small faults with lateral displacements), massive strata, climatic or topographic conditions that permit a deep water table, and a long duration of sustained dissolution.

9.2.7 Littoral karren

A variety of karren forms and assemblages occur on limestone and dolomite sea coasts and around lakes. The coasts may be sculpted by wave action, wetting and drying, salt weathering and hydration, but in addition to physico-chemical dissolution and bioerosion, there can also be bioconstruction (Figure 9.14). The relative effectiveness of these competing processes is determined by many different factors, chief of which appear to be (i) wave energy, (ii) tidal range and (iii) variations of lithology and structure. Combining the range of processes and factors, two extreme (or end-member) types of

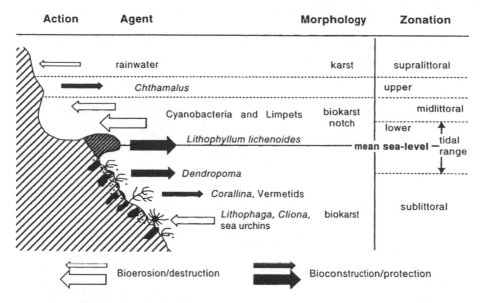

Action	Agent	Morphology	Zonation

rainwater — karst — supralittoral

Chthamalus — upper

Cyanobacteria and Limpets — biokarst notch — midlittoral — lower

Lithophyllum lichenoides — mean sea-level — tidal range

Dendropoma

Corallina, Vermetids

Lithophaga, Cliona, sea urchins — biokarst — sublittoral

⇐ Bioerosion/destruction   ⮕ Bioconstruction/protection

Figure 9.14 Zonation of bioerosive and bioconstructive activity on carbonate shorelines. (From Spencer and Viles 2002.)

Figure 9.15 (Left) Karst tower being undercut by intertidal zone solution notches, Bay of Halong, Vietnam. (Upper right) Coastal dissolution pans on northwest coast of South Island, New Zealand. (Lower right) Freshwater micropitting, Lough Mask, western Ireland.

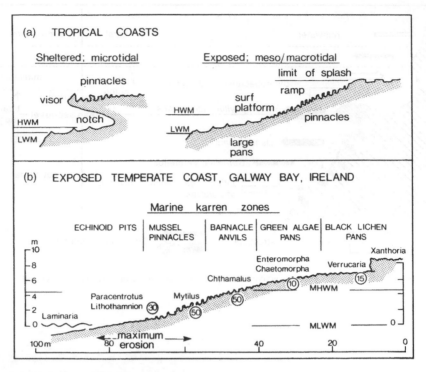

Figure 9.16   (a) Zonal features of tropical limestone coasts; generalized models to display the contrasts between sheltered and exposed, micro- and macrotidal settings. (b) Transect of an exposed limestone coast on Inishmore in Galway Bay, Ireland. Latin names indicate principal colonizing species. Circled numbers are local relief in centimetres.

eroding carbonate coasts can be recognized. The first is that dominated by mechanical erosion; dissolution forms are few or absent. This will tend to be a high wave energy (or exposed) coast with low tidal range and weak strata. At the other extreme is the low wave energy coast where a high tidal range exposes wide intertidal flats upon mechanically resistant rocks. Karst studies have favoured this latter type for obvious reasons; even there, structural factors such as dipping strata often blur other karst zonation (Ley 1979).

Karst sea coasts display two distinctive features, **notches** (or **nips**) and a high density of pits and pans (Figure 9.15). On protected coasts gypsum cliffs also display sharp intertidal notching. Notches in limestone or dolomite are most prominent in the tropical and warm temperate regions, but are known along the cold shores of Newfoundland and Patagonia. A prominent notch develops in the intertidal zone around mean sea level. It is cut partly by boring organisms (algae and sponges) and molluscs that graze upon them, rasping the limestone to obtain their prey (Neumann 1968). On Aldabra Atoll, Trudgill (1976) measured notch recession rates as high as 1.0–1.25 mm a^{-1}, with grazing contributing 0.45–0.60 mm. The sharpest notches occur on the most

protected, microtidal coasts, where the depth of the undercut may be 2 m or more. On exposed coasts the intertidal notch is usually replaced by an eroding ramp (Figure 9.16).

Pitting attributable to boring and grazing organisms may be densely packed or coalescing (Figure 9.17) and can reduce divides between pits to a lacework of sharp crests and pinnacles. Pit depths range from < 1 cm to ~1 m, and diameters can be as great as several metres. Width:depth ratios tend to fall between 1:2 and 6:1.

The interaction of marine, biological and chemical processes may produce a karren zonation on limestone shores that varies with climate and exposure. It is best displayed where unbroken beds dip gently towards the sea in areas of high tidal range, e.g. on the Aran Islands, western Ireland (Figure 9.16b). In contrast, the west coast of Newfoundland has many steep dip sites, is microtidal and has sea-ice fast to the shore for three or more months per year; bioerosion is much reduced but there is a clear trend to progressively larger pitting inshore between the mean low-water mark and the inland limit of swash (Malis 1997).

There has been little study of the freshwater littoral karren. Usually development appears to be restricted to formation of small pits and pans (Figure 9.15) in a narrow

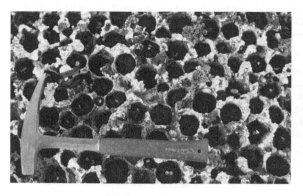 

Figure 9.17   (Left) Carboniferous limestone surface exposed at low tide with tightly packed echinoids (sea urchins, *Paracentrotus* sp.) in pits surrounded by white calcareous encrustation deposited by calcareous algae (*Lithothamnion* sp.), Aran Islands, western Ireland. (Right) Surface similar to that illustrated on the left but scoured of biota, showing pits of 7–10 cm diameter that had been incised by echinoids.

zone about the waterline (e.g. Perna (1990) at Lago di Loppio, Italy). They may develop to high densities, and coalesce. Many are occupied by blue-green algae, which possibly initiate them. Narrow cylinder-like bell holes (section 7.11) extend vertically **upwards** from opened bedding planes in a zone of seasonal lake-level oscillation round Lough Mask and some other Irish lakes. On dolomite shores of Georgian Bay, Canada, Vajocki and Ford (2000) undertook sample traverses between the swash zone and -25 m. There was a strong positive relationship between pit depth and water depth ($r^2 = 0.93$), which was considered to be due to greater exposure age in the deeper water. Lithological factors obscured any other correlations (e.g. with pit width) and there was no evidence of bioerosion.

9.3 DOLINES – THE 'DIAGNOSTIC' KARST LANDFORM?

Karst geomorphologists have always accorded special importance to dolines since Cvijić (1893) identified them as giving karst topography its particular character. The term is derived from *dolina*, an everyday Slovenian expression for a valley (Gams 1994); Cvijić also introduced the term **vrtača** (Šušteršič 1994) which is widely used locally for medium-sized enclosed depressions (e.g. Gams 2003). Sweeting (1972) discussed other local terms and Roglić (1972) reviewed early ideas on their origin. Dolines are called **sinkholes** in engineering and North American literature (e.g. Sowers 1996, Beck 2003, Fookes 2003, Waltham *et al.* 2005), but in this chapter we use **doline** for any small to intermediate enclosed karst depression regardless of genesis or climatic context. Karst is always developed when dolines are found and so

they can be considered index landforms of karst; indeed Grund (1914) considered their place in the karst landscape to be similar to that of valleys in fluvial terrain. However, we emphasize that their absence in carbonate terrain should not be taken to mean that karst is not developed, because karst groundwater systems can evolve without doline formation at the surface.

Dolines are usually circular to subcircular in plan form, and vary in diameter from a few metres to $\sim$1 km. Their sides range from gently sloping to vertical and they vary from a few to several hundred metres deep. They form by various processes including dissolution, collapse and subsidence (Table 9.2). This yields a spectrum of features from saucer-shaped hollows, to funnels to cylindrical pits. In the landscape they may occur as isolated individuals or as densely packed groups that totally pock the terrain (Figure 9.18).

Cvijić (1893) recognized that dissolution and collapse account for the formation of most dolines, although he considered the majority to have a predominantly solutional origin. He was one of the first geomorphologists to apply morphometry to landform description when, on the basis of depth:diameter ratios from numerous field measurements, he distinguished (i) shallow trough or bowl-shaped basins with flattish floors from (ii) steeper, deeper funnel shaped dolines and (iii) shaft-like dolines in which the breadth is usually less than the depth.

Thus early ideas on the nature of dolines were drawn mainly from field experience in Europe, although Daneš (1908, 1910) also investigated humid tropical karsts in Jamaica and Java. He and Grund (1914) were of the opinion that tropical dolines, or **cockpits** as they are known in Jamaica, develop mainly by solution in much the same way as dolines in the temperate zone. Later work by

Table 9.2   Doline/sinkhole English language nomenclature as used by various authors (Reproduced from Williams, P.W., Dolines, in Encyclopedia of Caves and Karst Science (ed. J. Gunn), pp. 304–10. © 2004, Fitzroy Dearborn, Taylor and Francis)

Doline-forming processes	Ford and Williams 1989	White 1988	Jennings 1985	Bogli 1980	Sweeting 1972	Culshaw and Waltham 1987	Beck and Sinclair 1986	Other terms in use
Dissolution	Solution	Solution	Solution	Solution	Solution	Solution	Solution	
Collapse	Collapse	Collapse	Collapse	Collapse (fast) or subsidence (slow)	Collapse	Collapse	Collapse	
Caprock collapse		—	Subjacent collapse		Solution subsidence	—		Interstratal collapse
Dropout	Subsidence	Cover collapse	Subsidence			Subsidence	Cover collapse	
Suffosion	Suffosion	Cover subsidence		Alluvial	Alluvial	—	Cover subsidence	Ravelled, shakehole
Burial	—	—	—	—	—	—	—	Filled, palaeo

Figure 9.18   Examples of solution dolines in limestone (Upper left) An isolated solution doline, Causses Mejean, France. The floor of the basin has been infilled by slopewash and periglacial debris and levelled for agriculture. (Upper right) Cross-section of a solution doline in Verd quarry, Slovenia. (Photograph by A. Mihevc.) (Lower left) Solution dolines in the Waitomo karst, North Island, New Zealand. (Lower right) Aerial view of polygonal karst depressions around 2000 m altitude on Treskavica Mountain, Hercegovina. (Photograph by I. Gams.)

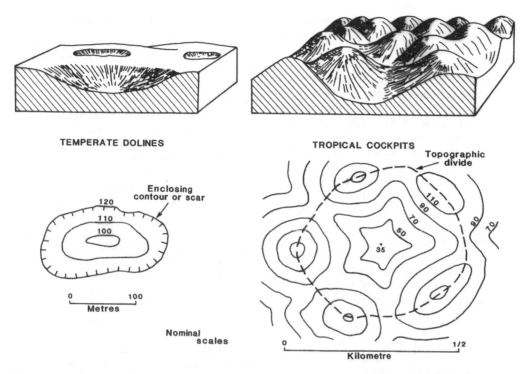

Figure 9.19   (Left) Delimitation of a doline on a plain (conventional temperate doline). (Right) Delimitation of a doline in hilly terrain (conventional humid tropical cockpit). Reproduced from Williams, P.W, The geomorphic effects of groundwater. In R. J. Chorley (ed.) Water, Earth and Man, Methuen, London, 269–284, 1969.

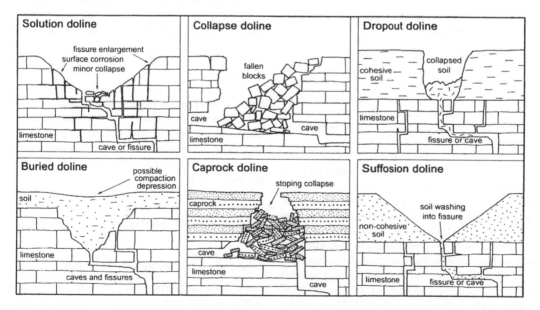

Figure 9.20   Six main types of dolines. Reproduced from Williams, P.W. (ed.), Karst Terrains; Environmental Changes and Human Impacts. Catena Supplement, 25 © 1993 Elsevier; modified from Waltham, A.C. and Fookes, P.G. (2003) Engineering classification of karst ground conditions. Quaterly Journal of Engineering Geology and Hydrogeology, 36, 101–18.

Lehmann (1936) indicated tropical cockpits to be morphologically distinct from most temperate dolines (Figure 9.19), although he supported an origin by solution processes.

The basic genetic distinction between solution and collapse dolines was elaborated by Cramer (1941) in an extremely thorough paper on their development and morphology. He examined topographic maps of karst regions from many parts of the world and made a comprehensive morphometric description of the doline terrains concerned. His study revealed a picture of variations in the density of doline fields in different areas.

From this early work emerged an awareness that there are several types of dolines developed in different ways and in different materials, but with a convergence of form. The end-members are illustrated in Figure 9.20. We distinguish **collapse**, which involves fracturing and rupture of rock and soil, from **subsidence**, which is a more gradual process involving sagging or settling of the surface without any obvious breaking of the soil. Natural dolines of a purely subsidence origin are rare, being found where there is interstratal solution of evaporite rocks at shallow depth (artificial subsidence dolines are often created during salt mining). The actual form of most dolines is of polygenetic origin, plotting within a ternary diagram of end-members, but it is strongly biased towards one or another of the end-points; thus we discuss them under those headings.

9.4 THE ORIGIN AND DEVELOPMENT OF SOLUTION DOLINES

The bowl-shaped form of dolines indicates that a greater mass of rock has been removed from their centres than from around their sides. This implies there to be a common natural process that focuses dissolution. Since mass transport in solution is the product of solute concentration and runoff, variations in either or both factors could explain the focused corrosion within dolines. If local variability in solute concentration alone were sufficient to explain them, then they would be found on every type of limestone in a given climatic zone, but this is not the case, as is illustrated by comparison of the Cretaceous, Jurassic, Carboniferous and Devonian limestones of England, for example. If the solute concentration factor is eliminated, it follows that sufficient inequalities in corrosion to form dolines must be created principally by local spatial variations in water flow, but by what mechanism can the flow of water through the rock be focused? Here we must refer back to our discussion of the initiation of proto-conduits in Chapter 7, specifically to Figures 7.4, 7.5 and 7.9, and to our explanation of epikarstic processes in Chapter 5.

9.4.1 Initiation of dolines by point recharge

In an essentially horizontal sequence of limestones interbedded between clastic rocks, fluvial incision of the cover beds will eventually reveal inliers of limestone. These provide an input boundary for a developing karst; whereas the valley of a main stream cut deeply into the limestones will provide an output boundary at a lower elevation. Subterranean connections from input to output are established as explained in section 7.2. Dolines can only commence to form when proto-caves extending from recharge points breakthrough to a spring or earlier cave conduit. Until such links are established, resistance to flow through the rock is too high to permit removal of enough limestone to create a doline-sized depression, but once connected the focus of drainage and solute removal is a surface funnel and the karst pipe that drains it.

Dolines formed in this way can be termed **point-recharge depressions** in recognition of the process responsible for focusing corrosional attack (Figure 5.16(1)). In the early stages of the removal of a caprock, only the bottom of the enclosed basin is composed of limestone, the flanks being relatively impermeable cover beds. These dolines act as centripetal drainage points for allogenic runoff and, clearly, the larger their contributing catchments the bigger these depressions may become.

As more clastic rocks are eroded, so more limestone inliers are exposed. A second generation of point-recharge sites may be envisaged, not necessarily in the stylized ranks of Figure 7.9, but nevertheless with the same implications for subsurface conduit development and its feed-back effects permitting doline growth. Continued caprock erosion and more doline initiation further increase the frequency of point-recharge sites. The drainage areas of individual dolines are reduced as a result. Where caprock has long been entirely removed and soils are thin or discontinuous, the high density fissuring of weathered bedrock, sometimes exposed as karren fields, may begin to substitute for dolines as drainage routes. Under such circumstances recharge has moved to the diffuse end of the spectrum (section 5.3) and accompanying corrosion is also spatially diffuse; thus unless flow is focused lower down in the epikarst the form of any dolines that have been initiated may not be maintained.

9.4.2 Initiation of dolines by drawdown

Some karst landscapes have been developed in limestones without caprocks, either because none were deposited or because morphological development followed the uplift of a baselevelled erosion surface. Solution dolines are often found in such terrains, although corrosion is unlikely to have been focused by point recharge. In other karsts where caprocks have long gone, there are apparently active, rather than degenerate, doline landscapes.

Any uplifted erosion surface subjected to renewed karstic attack will possess a vestigial conduit network developed beneath it by dissolution in the phreatic zone in the previous phase. It can be assumed that some input to output connections already exist, although they will certainly develop further. Recharge from rainfall will be relatively diffuse, depending on soil cover, and percolating water will accomplish 50–80% of its solutional work within about 10 m of the surface (Chapter 4). Hence fissures that are considerably widened by corrosion beneath the soil close rapidly with depth (Figure 5.24). As a result, infiltration into the top of this highly corroded subcutaneous zone is much easier than drainage out of it (Williams 1983). The bottleneck effect results in significant storage of water in this zone after heavy rain, and the stored water constitutes a suspended epikarstic aquifer with a base that is essentially a leaky capillary barrier (Figure 5.28). Because of initial spatial variability in fissure frequency and permeability arising from tectonic and lithological influences, preferred (i.e. low resistance) vertical leakage paths develop down connected pipes at the base of this suspended aquifer. These paths are enlarged by dissolution with the result that a depression develops in the overlying subcutaneous water table similar to the cone of depression around a pumped well. Flow paths then adjust in the epikarstic aquifer to converge on the dominant leakage route. The extra flow encourages more solution and with it the enhancement of vertical permeability. The zone of influence of the leakage route(s) widens according to the radius of the cone of depression in the subcutaneous water table. The dimension of this radius depends on the hydraulic conductivity of the epikarst and the rate of water loss down the leakage path at its base.

These processes, then, can explain the focusing of corrosion where caprock point recharge is not important. As the surface lowers the more intensely corroded zones begin to obtain topographic expression as solution dolines, their diameters being controlled by the radius of the subcutaneous drawdown cone. The theoretical relationships between surface doline topography, underlying relief on the subcutaneous water table and vertical hydraulic conductivity near the base of the epikarstic aquifer are illustrated in Figure 5.16.

Klimchouk (1995, 2000) and Klimchouk *et al.* (1996) have pointed out that flow concentration at the base of the epikarst can sometimes lead to the development of vertical shafts that have no open entrance to the surface, and that a karren field may develop at the surface with a diameter

corresponding to the cone of depression of the epikarstic water table. The progressive lowering of the surface and widening of the shaft eventually leads to collapse and an open entrance at the surface. Degradation of its slopes leads to the development of a cone-shaped depression. Such features are especially evident in the alpine zone where they are not masked by vegetation and thick soils.

From this discussion we see the need to distinguish (i) between doline initiation where there has been no proto-cave development from that where a ready-made, permeable, vadose zone is inherited from an earlier phase of karstification, and (ii) between doline corrosion focused by point recharge as opposed to that focused by epikarstic drawdown. Although dissolution is the initiating and dominant process, other factors such as collapse may also contribute to doline form. We can also appreciate that different generations of dolines can develop either from successive caprock stripping or from secondary drawdown within a major drawdown cone (Drake and Ford 1972, Kemmerly 1982).

9.4.3 The occurrence and enlargement of solution dolines

Once a solution doline is established, positive feed-back will encourage its further development because of the centripetal focusing of flow and hence corrosion (Figure 4.16 and 9.21). The aggressiveness of the water may be enhanced by greater biogenic CO_2 production in the thick soils that tend to accumulate in the bottoms of depressions. Such soils may also be damper longer because of drainage accumulation and lingering snow melt; thus the duration of active corrosion may also increase (Figure 4.16; Zámbó and Ford 1997). Further, with efficient vertical drainage encouraged by the corrosional enlargement of shafts, the increasing average velocity of water flow leads to a corresponding growth in mechanical transport of downwashed soil and rock, evacuating them underground. The freer vertical drainage permits much greater leakage from the basin and so steepens the epikarstic hydraulic gradient, stimulating further drawdown in the subcutaneous water table, and encouraging expansion of the radius of influence of the centripetal drainage system.

Although the general tendency during solution doline development is for self-reinforcement (Figure 9.21), some effects have negative feed-back influences. For instance, some soils covering karst may have a lower saturated hydraulic conductivity than the underlying rock. Depending on its composition it could range from 10^{-2} to 10^{-5} m day^{-1}, whereas in karstified limestone conductivity may be 10^{-1} to 10^{3} m day^{-1} or more. The soil therefore commonly acts as an infiltration regulator

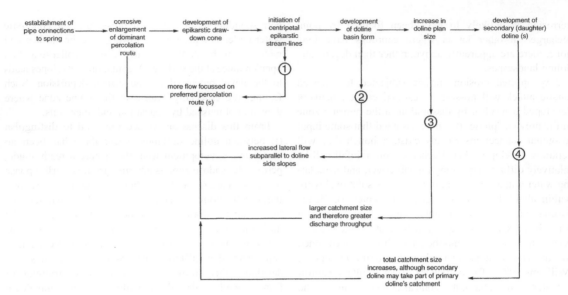

Figure 9.21   Positive feed-back loops in the development of solution dolines. Modified from Williams, P.W. (1985) Subcutaneous hydrology and the development of doline and cockpit karst. Zeitschrift für Geomorphologie, 29(4), 463–82.

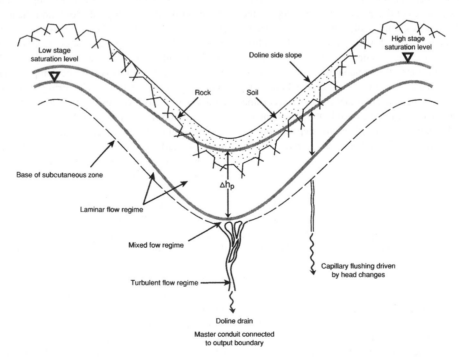

Figure 9.22   Flow regimes, saturation levels and pressure head changes Δh_p within the epikarst of a solution doline. Reproduced from Williams, P.W. (ed.), Karst Terrains; Environmental Changes and Human Impacts. Catena Supplement, 25 © 1993 Elsevier.

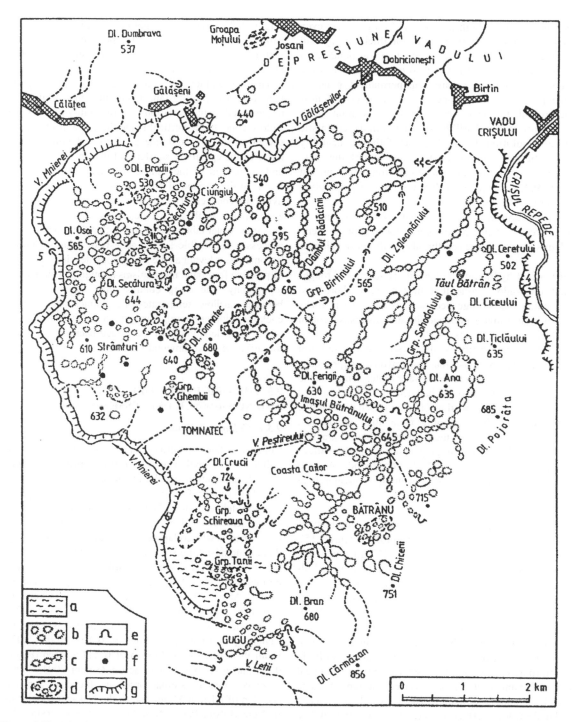

Figure 9.23 Morphological map of Zece Hotare karst plateau in Romania, showing the alignment of numerous dolines along dry valleys: a, karrenfeld; b, doline; c, doline-valley; d, uvala; e, cave; f, pothole shaft; g, plateau limits. Reproduced from Racoviţă, G., Moldovan, O and Onac. B, Monografia carstului din Munţtii Pădurea Craiului. Cluj-Napoc, 264 p. © 2002 Cluj University Press.

and, where it has a high proportion of clay, it may even be an infiltration inhibitor. Hence a thick clayey soil mantling a doline bottom can result in the ponding of runoff and in the reduction of peak throughput following storms. It can even create semi-permanent ponds – doline ponds – that reduce storm surcharges of flow underground to zero. However, even without thick soils stormwater will still be stored within the epikarstic zone of a solution doline, because of the permeability contrast between the epikarst and the underlying rock. This gives rise to hydraulic head variation in the epikarst (Figure 9.22) and to head-driven surges of percolation throughput into the vadose zone.

Solution dolines are a common feature of karst in humid areas, yet not all carbonate rocks support them. Williams (1985) suggested that they will not develop if:

1. if vertical hydraulic conductivity is so great throughout the vadose zone that little or only short-lived subcutaneous storage occurs, as is the case in rocks with high primary permeability such as many aeolian calcarenites (Jennings 1968);
2. vertical permeability is always so spatially uniform and sufficiently dense that no drawdown depressions develop in the subcutaneous water table, as may be the case within Cretaceous Chalk terrain of England and northern France as well as on some raised coral atolls;
3. on steep hillsides, sloping more than about 20°, where the dominant subcutaneous hydraulic gradient is subparallel to the topographic slope – in dry valleys on the Mendip Hills, UK, Ford (1964) found dolines to be absent when the valley gradient is greater than about 0.04 (~2°).

Factors 1 and 2 that exclude dolines are products of primary rock control on hydraulic variables (Figure 5.15). With respect to factor 3, Williams (1972a) noted that closed depressions incised into moderately sloping hillsides in Papua New Guinea are asymmetrical, being elongated downslope and having their deepest point off-centre close to the downslope margin. Salomon (2000) illustrates asymmetrical dolines as found down slopes and along dry valleys in France. It is interesting that the shape of drawdown cones around pumped wells in sloping (phreatic) groundwater surfaces have an identical asymmetry; and so it is to be expected that rapid leakage beneath a sloping epikarstic aquifer would produce a similar drawdown effect and create the observed doline asymmetry. Dolines in dry valleys can sometimes be so closely spaced as to justify the term **doline valleys**. Racoviţă *et al.* (2002) provide an informative illustration of this from the Zece Hotare karst plateau in Romania

(Figure 9.23), where many dolines are aligned along dry valleys and doline clusters on the plateau surface are set within larger enclosed basins (thus constituting **uvalas**, i.e. compound enclosed depressions).

The development of solution dolines leads both to increased permeability (and hydraulic conductivity) and to increased spatial variation in permeability in the upper vadose zone. In the vertical plane there is a difference of several orders of magnitude between hydraulic conductivity in the epikarstic aquifer and that in the remaining underlying part of the vadose zone (see Figure 5.3 and 5.15). In the horizontal plane, major variations occur in hydraulic conductivity near the base of the subcutaneous zone, and it is these that determine the relief on the epikarstic water table. It is probable that the subcutaneous processes responsible for the genesis of solution dolines become less relevant as the topography develops, because of positive feed-back mechanisms strongly enhancing the topography. The sharp contrast that sometimes exists between shallow temperate-zone dolines and tropical cockpits with larger relief may be a direct consequence of the variable strength of those feed-back factors under the differing environmental conditions.

9.5 THE ORIGIN OF COLLAPSE AND SUBSIDENCE DEPRESSIONS

9.5.1 Collapse dolines

Collapse dolines are usually steeper sided than solution dolines and of smaller plan area. But as their side slopes degrade and bottoms infill with slopewash or other detritus, their surface form may assume the bowl shape of solution dolines, with which they can then be confused. Only excavation will reveal their true origin. Nevertheless, when newly collapsed or still actively developing, there is no mistaking some collapse dolines. Collectively the most impressive are the **tiankeng** ('large collapse dolines') of the Dashiwei group in Leye, Guangxi, China. About 50 tiangkeng are known in China, of which three are more than 500 m deep and over 500 m in entrance diameter. The deepest reported is Xiaozhai Tiankeng in Chongqing Province (Chen *et al.* 2003). From its lowest rim at 1180 m, it is 511 m deep (Figure 9.24) and it is about 600 m across (Senior 2004). Other large collapses of similar order are found in Sarawak (Garden of Eden at Mulu) and in the Nakanai Mountains in New Britain, Papua New Guinea (Maire 1981a). All are in polygonal karst terrain and overlie large underground rivers. Another immensely impressive collapse contains Crveno Jezero (Red Lake) in Croatia. It is 528 m deep from its lowest rim, the bottom of the collapse extending

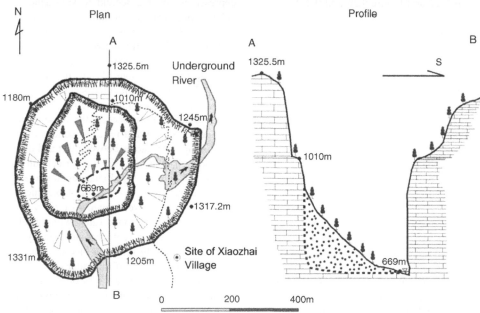

Figure 9.24 (Upper) Oblique aerial view from the northwest of Xiaozhai tiankeng, Chongqing, China. This great collapse doline is in fencong karst terrain. At the rim, the collapse is over 0.5 km wide. The location of the winding footpath is shown on the plan below (Photograph by Chen Lixin.) (Lower) Plan and profile of Xiaozhai tiankeng. Reproduced from Chen, W., Zhu D. and Zhu X. (2003) Study on Karst Landscape and World Natural Heritage Value of Tiankeng-Difeng Tourist Site, Fengjie, Chongqing. Geological Printing House, Beijing, 133 pp.

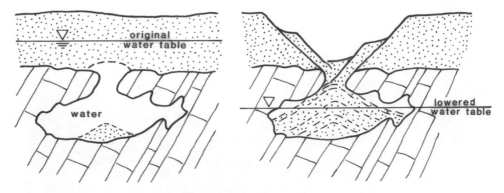

Figure 9.25   Development of collapse doline by removal of buoyant support.

10 m below modern sea level (Bonacci and Roje-Bonacci 2000). The lake averages ∼285 m deep, but fluctuates by ±35 m in level. The collapse diameter at the surface is about 350 m and at lake level is about 200 m. Some filled breccia pipes above deep evaporites may be larger still, but lack prominent topographic expression (section 9.13).

Collapse can occur in compact bedrock, in caprock over karstified beds, and in superficial unconsolidated sediment cover. It can occur suddenly or progress gradually. Three main mechanisms are responsible for collapse dolines (section 7.12 and Figure 9.20 and 9.25):

1. solution from above that weakens the span of a cave roof;
2. collapse from below that widens and progressively weakens the span of a cave roof (in rock or unconsolidated sediments);
3. removal of buoyant support by water table lowering, which increases the effective weight on the span so that its strength is exceeded.

In practice these mechanisms often interact. For example, corrosion by water percolating down a fissure may combine with upwards stoping of a cave roof along the same plane of weakness. Failure then occurs because of weakening both from above and below. Illustrations of depressions formed in this way are the tiankeng dolines of China and the huge collapses of the Nakanai Mountains in New Britain, mentioned above. They are in polygonal karst terrain and overlie large underground rivers with low flow discharges of many cubic metres per second. Numerous examples of this kind of polygenetic doline are found above river caves running through cockpit karsts. They are frequently encountered just upstream of spring heads and just downstream of stream-sinks.

The nature of rock strength and the stresses to which it is subjected when associated with underground openings such as caves are discussed in sections 2.8 and 7.12. It suffices to note here that no single criterion defines rock mass strength, but unconfined compressive strength, shearing resistance and tensile strength are important elements (Selby and Hodder 1993), with compressive stresses being the most significant in underground environments. If stress exceeds the rock's strength then the cave roof will fail. This will be by fracturing in the case of competent brittle rock or by deforming in the case of incompetent lithologies. The first gives rise to collapse depressions and the second to subsidence depressions. When consolidated non-karst caprocks are breached (Figure 9.20), the depressions formed are termed **caprock collapse** or **subjacent collapse** dolines (Table 9.2). Examples of this type in gritstones above carbonate rocks are described from South Wales by Thomas (1974) and Bull (1977). They frequently occur in clastic beds above evaporite rocks (see Figure 9.56).

As karstification proceeds, the regional water table is gradually lowered and phreatic conduits are dewatered. Oscillating water tables between wet and dry seasons and temporary backing up of water in conduits due to flash floods also produce rapid changes in stress patterns in a karstified massif. Seasonal water-table fluctuations of up to 100 m or more are known, so large volumes of rock can be affected. Groundwater withdrawal can be as abrupt where water resources are heavily exploited by pumping. In a fully saturated medium, the buoyant force of water is $1\,t\,m^{-3}$ and if the water table is lowered 30 m, the increase in effective stress on the rocks is $30\,t\,m^{-3}$ (Hunt 1984). If unconsolidated overlying materials are affected by such dewatering, compression occurs and the surface subsides. Compression in sands is essentially immediate, but cohesive materials exhibit a time delay as they drain. Thus the amount of subsidence is a function of the magnitude of the decrease in the water table, which determines the increase in overburden pressures,

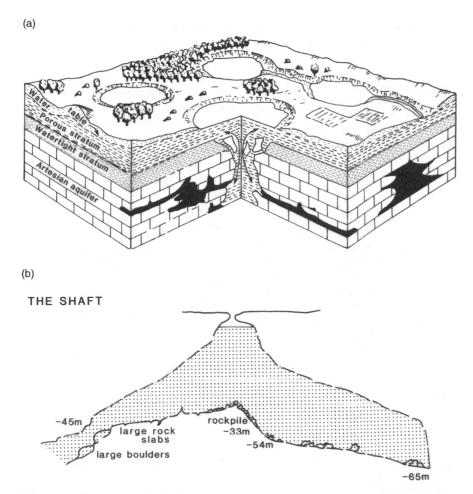

(a)

(b)

THE SHAFT

Figure 9.26 (a) Drowned collapse dolines in Florida formed by a combination of undermining from below and periodic loss of buoyant support. Reproduced with permission from, Davies, W.E., and LeGrand, H, Karst of the United States. In Herak, M. & V.T. Stringfield, (eds.), 1972. Karst: Important Karst Regions of the Northern Hemisphere. Elsevier, Amsterdam © 1972 Elsevier. (b) Flooded, cenote-style collapse doline in southeast Australia formed by a combination of undermining from below and periodic loss of buoyant support. Reproduced from Lewis, I. and Stace, P. (1980) Cave Diving in Australia, I. Lewis. Adelaide.

and the strength and compressibility of the strata. The abruptness of the process is a function of the rapidity of dewatering and the nature of the materials (e.g. Zolushka Cave, Ukraine, in section 7.12). The propagation of **dropout** (or **cover collapse** or **soil arch**) sinkholes/dolines (Figure 9.26a) by rapid dewatering following groundwater pumping is discussed in section 13.3. However, rapid doline formation can also arise from natural processes such as flooding. At least 312 cover-collapse depressions in alluvium developed within 48 h following flooding by a tropical storm in the Flint River valley, Georgia, USA (Hyatt and Jacobs 1996), and even more quickly in terrace alluvium over gypsiferous beds in northeast Spain (Benito *et al.* 1995), many being exacer-

bated by human activity. However, even without human intervention, the rapid rate of dissolution of gypsum can still yield major problems. For example, natural ground collapse over gypsum in northeast England resulted in about $1.5M worth of damage in the interval 1984–1994 (Cooper 1995).

Water-table fluctuations have also occurred more gradually as a result of repeated glacio-eustatic sea-level changes during the Quaternary Period. This has caused numerous major fluctuations as much as 130 m in the level of the water table. Consequently cover beds have been penetrated by collapse in many locations. The present high stand of sea level results in most of the depressions being water filled, giving a landscape of doline ponds.

Many of these in Florida have been successfully investigated using high-resolution single-channel seismic profiling (Kindinger *et al.* 1999). Glacio-eustatic oscillations probably also made a major contribution to the development of flooded collapse depressions in Yucatán (Salomon 2003a, Beddows 2004), where they are known as **cenotes**, and southeast Australia (Marker 1976, Grimes 1994). Diving and sounding has shown them to be steep-walled with large chambers extending below the water table. Some are the top of enormous flooded bell-shaped cavities with disconcertingly thin roofs (Figure 9.26b). Shaft-like depressions deeply flooded by seawater in low-lying coral islands are known as **blue holes**. They are common features in the Caribbean and are also found on the Great Barrier Reef (Backshall *et al.* 1979, Gascoyne *et al.* 1979, Hopley 1982, Smart 1983, Mylroie *et al.* 1995). They resemble cenotes at the surface because of their usually circular opening. However, some are produced along massive shear fractures formed by steep-reef-edge foundering following removal of buoyant ocean support during low glacial sea levels. Thus not all blue holes are the product of karst processes, although the cavities are highly decorated with drowned speleothems (Palmer and Heath 1985). When a thick microbial layer occurs near the surface, the upper part of the water column is coloured black and the features have been termed **black holes** (Schwabe and Herbert 2004).

9.5.2 Subsidence and suffosion depressions

In the ideal subsidence depression the ground and underlying beds gradually sag downwards with no significant faulting of the rocks, although folding always occurs. This may be contrasted with the collapse depression where rupture is a characteristic feature. In practice there is a continuum in nature, many predominantly subsidence depressions having an element of fault displacement in them, depending largely on the competence of the lithologies involved.

Trough subsidence above underground cavities can take place at different scales. At a small scale and at shallow depth it can create subsidence dolines where the cavity is at the rock–soil interface. Artificial subsidence depressions up to a few hundred hectares in extent have been created by injection mining of shallow salt deposits. At much larger scales and sometimes involving considerable depths are processes leading to development of depressions by solution sag. These are limited to the evaporite rocks and are discussed separately in section 9.13.

Seepage through thick unconsolidated regolith or allogenic detritus over karst rocks can generate **suffosion** depressions. Suffosion is the process whereby fines are evacuated through underlying pipes in the karstified bedrock by a combination of physical downwashing and chemical solution. Infiltrating water beneath regolith creates subsoil karren and widened joints connected with

Figure 9.27   A field of partly snow-filled suffosion dolines in glacial moraine on Mount Owen, New Zealand.

deeper cavities. Suffosion then causes a dimpling of the surface with a multitude of small dolines (Figure 9.27). Very often inheritance plays a part, the veneer of alluvium, loess or glacial till, etc. having been deposited onto a well-developed karst. Dolines formed in the thick alluvial veneer that commonly floors blind valleys and poljes were termed 'alluvial dolines' by Cvijić (1893).

Kemmerly and Towe (1978) studied change in suffosional/cover-collapse dolines with time, based on growth of 18 dolines over 35 year. The dolines were mainly of suffosional origin, being developed in drift deposits up to about 10 m thick, with no carbonate rock outcrops showing. They used sets of aerial photographs flown in 1937 and 1972, supplemented by field survey. Their conclusions were that:

1. depression length and width axes enlarged over the 35 year period, length usually growing more rapidly than width;
2. doline growth is a function of surficial geological setting, areal growth rates averaging 40, 70 and 100 m^2 100 a^{-1} for loessic, clayey residual and silty colluvial surface material respectively;
3. estimates of doline age on the three surficial materials, based on linear growth rates, varied from 25 a (silty colluvium), through 38 a (clayey residuum) to 65 a (loess) – these ages were found compatible with independent biological evidence.

Magdalene and Alexander (1995) re-examined sinkholes mapped in 1984 with those found 10 year later. The area studied was a till- and loess-mantled fluviokarst in Minnesota, where dolines are mainly formed by cover subsidence and cover collapse. They found 34 new dolines to have been formed in the interval, all within high probability sinkhole zones as defined in the earlier survey. The dolines were strongly clustered, with the majority of the new dolines clustered around other freshly developed sinkholes. This conclusion was confirmed by Gao *et al.* (2002).

9.6 POLYGONAL KARST

Dolines can occur as scattered isolated individuals, as scattered clusters of individuals, as densely packed groups, or as irregularly spaced chains along allogenic recharge margins and dry-valley systems. In humid tropical and temperate zones dolines sometimes totally pock the land surface, occupying all the available space. Viewed from the air the landscape resembles an egg-box-like topography with the divides between adjoining depressions forming a cellular mesh pattern

(Figure 9.18 (bottom right)). This style of karst landscape was first recognized in Papua New Guinea where it was called **polygonal karst** (Williams 1971, 1972), but it occurs in many other carbonate karsts including parts of the Caribbean, USA, New Zealand, Tasmania, China, Herzegovina and Turkey. It is also found in evaporite terrains, where it displays its greatest density and uniformity.

Polygonal karst mainly comprises solutional depressions in bedrock, as in the 'cockpit' karst of Jamaica, although polygenetic depressions also feature when the surface collapses into underlying caves. When the surface is thickly blanketed in superficial deposits, many of the depressions involve cover subsidence and are in the process of exhuming buried depressions.

The polygonal pattern is interesting, because it constitutes a geometrically efficient way of subdividing space, minimizing the distance from the divide to the centre of the cell. Rainfall is rapidly transmitted from the surface down to the bottom of individual depressions (mainly as throughflow in the epikarst). Consequently polygonal karst is one of nature's most efficient drainage systems – possibly the most efficient.

Limestone polygonal karsts have different mesh sizes (Figure 9.28). For example, in the Guilin region of China polygons have an average area of 0.51 km^2, which is almost 30 times larger than the 0.018 km^2 mesh size in the Waitomo region of New Zealand (Williams 1993). Average depression radii vary from 400 m to 80 m, respectively, and internal depression relief from 200 m to 64 m in the two cases. The 'egg-boxes' are of very different dimensions! This is significant, because the contrast in polygonal karst geometry is almost certainly a reflection of the marked difference in hydraulic conductivity of the epikarsts in the two regions (high in New Zealand, low in China), as well as a reflection of past and present rainfall regimes (currently around 2000 mm a^{-1} in both areas, although with much stronger seasonal distribution (62% from April to July) near Guilin). The geometry of the surface karst reflects the balance between the recharge rate and the throughput regime, the latter being controlled by the vertical hydraulic conductivity of the bedrock. In the Chinese case this is relatively low because of the massive nature of the bedrock with low joint frequency and widely spaced drainage shafts, but in the New Zealand case it is very high because of thin bedding and closely spaced joints (Williams 2004c), although thick volcanic soils make direct comparison difficult.

The divides between the inosculating basins of polygonal karst are surmounted by residual hills that assume a variety of shapes across the range of hemispherical–conical–pyramidal (Balazs 1973, Day 1978) that in

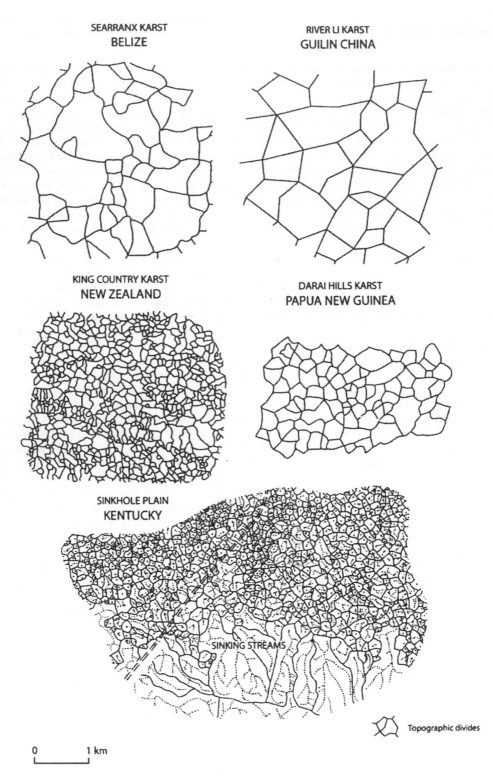

Figure 9.28 Comparison of polygonal karst textures from different karst landscapes. Reproduced from Williams, P.W. (2004a) Polygonal karst and palaeokarst of the King Country, North Island, New Zealand. Zeitschrift für Geomorphologie, Suppl.-Vol 136, 45–67.

particular reflect the structural attributes of the bedrock. The differing styles and sizes of dolines and residual hills impart a unique character to the terrains in which they are found. It has been claimed that morphologically distinct assemblages develop in different climatic zones (Lehmann 1954, Jakucs 1977), but it is also evident that there is a considerable range of karst landscapes even within individual climatic zones (see section 10.1). Therefore questions arise as to whether the perceived topographic styles can be verified objectively, and whether the landscape differences between climatic zones are greater than those within them.

Morphometry is a technique that has been used to help to resolve such questions. Its aim is the objective and quantitative description of landforms. The application of morphometry to karst in fact commenced long before it became widely used in geomorphology as a whole. Cvijić (1893) and Cramer (1941) in particular were pioneers in this field. Their work and later contributions by many others is discussed by Williams (1972b), Jennings (1985), Ford and Williams (1989), Bondesan *et al.* (1992) and Day (2004).

9.7 MORPHOMETRIC ANALYSIS OF SOLUTION DOLINES

9.7.1 Objective description of doline styles and patterns

The most accurate data for morphometric analysis are obtained from field survey (e.g. Jennings 1975, Šušteršič 1994). However, this is time consuming and permits coverage of only a small area. The most practical medium for morphometric analysis of karst is generally found to be large scale (say 1:15 000) aerial photographs viewed stereoscopically under magnification. These are preferable to topographic maps, because even when maps are of large scale and small contour interval, significant terrain information is lost, especially when dolines are shallow. In Barbados, Day (1983) used maps as large as 1:10 000, but found their representation of dolines to be arbitrary, underestimating depression numbers by up to 54% when compared with field surveys. The major problem with maps is their variable quality and density of information, e.g. Troester *et al.* (1984) endeavoured to compare results from 1:50 000 maps of 20 m contour interval with those of 1:24 000 with ~1.5 m interval. Problems are not entirely absent when using aerial photographs either, because heavy forest or shadows mask hydrographic and topographic detail. However, handling large quantities of morphometric information from maps has become easier with the advent of geographical information systems (GIS) and digital elevation models (DEM),

e.g. Gao *et al.* (2002, 2005a), Florea *et al.* (2002) and Denizman (2003).

On detailed maps, the contour pattern of deep and large dolines (cockpits) in the humid tropics is often star shaped (Figure 9.19). This arises from the small and usually dry gullies that incise the depression slopes and focus on the bottom. Aerial photographs reveal this even more clearly and also show that a similar internal dissection often characterizes the commonly smaller dolines of the temperate zone. The spatial location of a doline is therefore usually represented by its drainage focus, which is the lowest point and not necessarily coincident with the geometric centre of the depression (Figure 9.29a). The basin is normally delimited by its topographic drainage divide, which is the surrounding high land in the case of polygonal karst or the break of slope at the basin perimeter in the case of isolated depressions. The plan geometry of the individual features thus defined can be expressed by various length:width ratios and basins can be ranked hierarchically according to the order of channel networks draining their sides (Williams 1971), although other ranking systems (e.g. basin size) are possible.

Relief attributes of dolines have received less attention than plan form because sufficiently detailed height data are not readily obtainable from maps and field measurement is time consuming, although the increasing availability of digital terrain data is making this easier. Doline depth is usually taken to be the difference in elevation between the lowest point in the depression and the lowest point on the basin rim or divide. Depth information is important because as Troester *et al.* (1984) remarked, the most obvious difference between temperate and tropical karst is often the amount of relief in the landforms. From map analysis they found an average depth of the order of 20 m in Caribbean examples compared with < 10 m in the Appalachians, Kentucky and Missouri.

Šušteršič (2006) has considerably advanced methodology for analysing solution doline form by using Fourier techniques to consider doline geometry in its entirety. He investigated which rotated power function best fits the geometry of 38 dolines in the 'classic' karst of Slovenia and examined how geometry changes with volume. He concluded that the doline radius:depth ratio increases as doline volume increases. Doline volumes are very close to those of regular cones of the same dimensions, and grow relatively deeper as the volume increase.

Terrain roughness is an associated attribute of interest, because it also can be used to help discriminate between karst landscape styles (Day 1977a). 'Smooth' and 'rough' topography (Figure 9.29b) can be defined by vector orientation, strength and dispersion (Day 1979b). Mathematical models simulating tropical karst

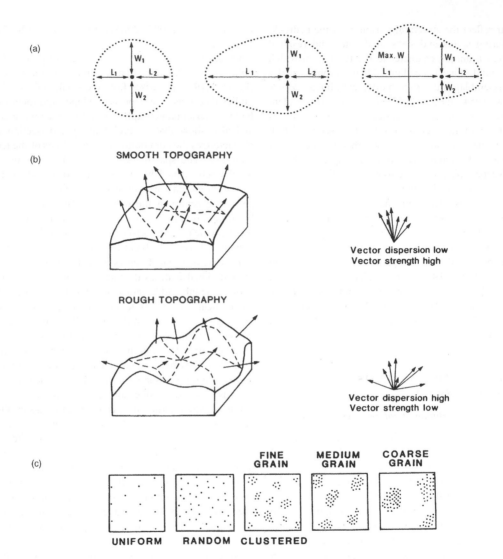

Figure 9.29 (a) Measurements made to estimate the two-dimensional geometry of karst depressions. Reproduced from Williams, P.W. (1971) Illustrating morphometric analysis of karst with examples from New Guinea. Zeitschrift für Geomorphologie, 15, 40–61. (b) Definition of 'smooth' and 'rough' topography using the criteria of vector dispersion and strength. Reproduced from Day, M.J, Surface roughness as a discriminator of tropical karst styles. Zeitschrift für Geomorphologie, Supplement-band, 32, 1–8 © 1979 E. Schweizerbart'sche Verlagsbuchhandlung, Science Publishers. (c) Dispersion patterns with different intensities and grains. Reproduced from Jarvis, R.S. (1981) Specific geomorphometry, in Geomorphological Techniques (ed. A. Goudie), Allen & Unwin, London, pp. 42–6; after Pielou, E.C. (1969) An Introduction to Mathematical Ecology, J. Wiley & Sons, New York.

terrain with different degrees of roughness have been developed by Brook (1981). The analysis by Brook and Hanson (1991) was particularly successful in showing the strong structural control of cockpit karst relief in Jamaica and explaining its 'grain'.

The spatial patterns of dolines can have two distinct aspects, termed 'intensity' and 'grain'. Intensity is the extent to which density varies from place to place, whereas grain is independent of intensity and concerns the spacing and areas of high- and low-density patches in a dispersion pattern (Figure 9.29c). Most geomorphological interest in pattern analysis focuses on intensity, and the question of particular interest in the investigation of doline fields is whether the pattern is best described as random, clustered or regular. Two methods have been used to assess this, quadrat analysis and nearest neighbour analysis, e.g. Drake and Ford (1972), McConnell and Horn (1972), Williams (1972a, b) and Gao *et al.* (2002, 2005b).

Quadrat analysis compares the number of dolines occurring per cell with expectation according to a model distribution. Thus, for example, if the actual occurrence is not significantly different from that described by the negative binomial frequency distribution, then the pattern can be described as clustered.

Nearest neighbour analysis has frequently used Clark and Evans' (1954) test in which pattern is assessed with a nearest neighbour index that compares the average actual distance (L_a) between points in a spatial distribution with the average expected distance (L_e) if the points were randomly disposed:

$$L_e = 1/2\,D^{-1} \qquad (9.1)$$

where D is the doline density. The nearest neighbour index $R = L_a/L_e$ varies from 0 for a dispersion with maximum aggregation or clustering, through 1 for a random case, to 2.149 for a regular pattern that is as evenly and widely spaced as possible. A summary of nearest neighbour statistics for various karsts is provided in Table 9.3. Vincent (1987) pointed out that reliance on a single measure based on nearest neighbour distance is a rather superficial way of investigating spatial dispersion, and showed that an improvement is obtained by

calculating an empirical distribution function of nearest neighbour distances. Complete spatial randomness can then be tested for by using the Poisson distribution. He illustrated the method using Williams' (1972a) data from New Guinea, broadly confirming previous conclusions for seven of the eight sites.

9.7.2 Evolution of spatial patterns

Morphometric techniques have been used to address the question of karstic evolution. Three approaches have been adopted: one analyses growth patterns, the second compares change over time and the third substitutes space for time.

By means of quadrat analysis Drake and Ford (1972) showed that in the Mendip Hills there was a constant cluster consisting of (on average) four smaller, younger 'daughter' dolines about one 'mother' doline, the mothers themselves being distributed in random clusters. We may thus think in terms of two 'generations' of dolines. There is a definite spatial and causal relationship between them; the daughters grow in the drawdown cones of their mothers.

Substitution of space for time to obtain information on landform evolution is a well-known strategy in geomorphology. Its validity rests on the assumption that evolution has followed a similar pattern in the places being compared; hence a weakness is that process rates and combinations may have varied significantly through

Table 9.3 Depression density and nearest neighbour statistics for various polygonal karsts (Reproduced from Ford, D.C. and Williams, P.W. (1989) Karst Geomorphology and Hydrology)

Area	Number of depressions	Depression density (km^{-2})	Nearest neighbour index	Pattern	Author
Papua New Guinea (8 areas)	1228	10–22.1	1.091–1.404	Near random* to approaching uniform	Williams 1972a
Waitomo, New Zealand	1930	55.3	1.1236	Near random*	Pringle 1973
Yucatan (Carrillo Puerto Fm)	100	3.52	1.362	Approaching uniform	Day 1978
Yucatan (Chichen Itza Fm)	25	3.15	0.987	Near random	Day 1978
Barbados	360	3.5–13.9	0.874	Tending to cluster	Day 1978
Antigua	45	0.39	0.533	Clustered	Day 1978
Guatemala	524	13.1	1.217	Approaching uniform	Day 1978
Belize	203	9.7	1.193	Approaching uniform	Day 1978
Guadelope	123	11.2	1.154	Near random*	Day 1978
Jamaica (Browns Town-Walderston Fm)	301	12.5	1.246	Approaching uniform	Day 1978
Puerto Rico (Lares Fm)	459	15.3	1.141	Near random*	Day 1978
Puerto Rico (Aguada Fm)	122	8.7	1.124	Near random*	Day 1978
Guangxi, China (3 areas)	566	1.96–6.51	1.60–1.67	Approaching uniform	Fang 1984
Spain, Sierra de Segura	817	18–80	1.66–2.14	Near uniform	Lopez Lima 1986

*Significantly different from random expectation at the 0.05 level, i.e. although near random there is a tendency to uniform distribution.

Pleistocene climatic changes. Yet despite this problem useful results have been obtained. For example, Day (1983) studied solution dolines on a series of raised reef terraces in Barbados, the purpose being to measure how doline morphology changes with duration of evolution, i.e. on surfaces of successively greater age. Doline densities were found to increase on terraces up to an elevation of 150 m, but to decline above that, whereas doline dimensions reached a maximum at about 225 m a.s.l. On a suite of rock-cut fluvial terraces in northern Italy, Ferrarese *et al.* (1998) found the volume of solution dolines to increase with terrace height (and relative age). Similarly, Strecker *et al.* (1986) described a time sequence of karst landform development on coral terraces of increasing age in Vanuatu, with cone karst first appearing on surfaces about 500 a.

The karstification of drainage as a normal dendritic stream pattern is lowered from cover beds onto limestone has also been studied morphometrically using this space–time substitution approach. Williams (1982a) described results from Waitomo, New Zealand, where karst is developed in limestones that rest unconformably on an impervious basement and are overlain by relatively impermeable clastic beds. Every degree of stripping of the caprock and limestone is represented. The doline pattern progresses from a clustered dispersion when karstification of surface drainage commences to a distribution on the regular side of random when the polygonal karst is at its maximum development (Figure 9.30). Coarse-grained clusters of dolines mark the end stages of the evolutionary sequence as blocks of karst are isolated and reduced by denudation and as surface drainage resumes and expands across the growing exposure of basement rocks. Comparison of this model with that of the development of cave systems with multiple ranks of inputs (Figure 7.9) shows how results of independent investigations of surface morphometry and speleogenesis can converge.

The incision of dolines and valleys often leaves residual hills around their perimeters or along interfluves. Sometimes the hills become isolated and of conical or tower-like form. The morphometric characteristics of areas of cone karst with depressions and cone karst with dry valleys (Figure 9.31) were examined in Guizhou Province, China, by Xiong (1992) and Tan (1992) respectively, using a combination of field measurement, map analysis and photogrammetry. Residual hills were found to be of pyramidal form and to have symmetrical slopes of 45° to 47°. The morphometric evidence suggested that when the deeper depressions reached base-level, depression bottoms widened horizontally and cone slopes retreat parallel to themselves, because the slopes

remained the same angle as the residual hills became smaller. Further information on the morphometric characteristics of karst in China can be found in Zhu (1988), who synthesized findings concerning peak–cluster–depression and peak–forest landscapes around Guilin.

9.7.3 Principal conclusions from karst morphometry

One of the great values of morphometry is that detailed scrutiny of a landscape throws up unexpected observations and stimulates fresh hypotheses. For example, until the 1960s it was supposed that karsts are chaotic, their landforms being a random jumble of collapse and dissolution features. Morphometric research since then has shown this not to be so. Many karsts in widely separated parts of the world have similar spatial organization, especially the polygonal karsts. Furthermore, their dolines are dispersed in patterns that are significantly different from random, tending towards uniformity of distribution. This has been explained as being a result of spatial competition, attainment of perfectly uniform distribution being hindered by distorting factors such as general topographic slope, regional dip and fissure patterns which orient corrosion by directing runoff and infiltration (Williams 1972a) and giving a grain to the relief (Brook and Hanson 1991).

Reviewing several published data sets, White and White (1995) found that the number of closed depressions of given depth or diameter falls off exponentially with increasing size, suggesting that beyond a certain threshold size (different for each region because of variation in structural and climatic conditions) depressions subdivide into smaller basins because of the availability of alternative drainage paths. This may help to explain the development of 'daughter' dolines in the epikarst drawdown cone.

Ever since Penck's (1900) views were expressed on the importance of a fluvial phase before karstification commenced, there has been interest in the effect of lowering a normal drainage pattern from impermeable cover beds onto underlying karst rocks. The results of morphometry show that such karstification does not **disorganize** the stream pattern, as was commonly assumed; it **reorganizes** it (Figure 9.30). Lehmann (1936) postulated a fluvial phase in Java prior to the development of Kegelkarst, although at that site it was not lowered from cover beds but developed from consequent streams on an uplifted erosion surface. Lassere (1954), Monroe (1974) and Miller (1982) kept this idea alive from karsts in Central America and the Caribbean. Miller's morphometric work enabled reconstruction of palaeodrainage lines in Belize and Guatemala, and convincingly showed their imprint on polygonal karst terrains.

1. KARSTIC REORGANISATION OF DRAINAGE COMMENCES
AS STREAMSINK ONE (S1) AND RESURGENCE ONE (R1)
DEVELOP, FOLLOWED LATER BY S2 AND R2.

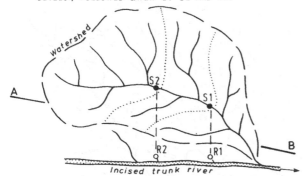

PART OF THE S1 CATCHMENT IS CAPTURED BY S2,
REDUCING THE SUBSEQUENT SIZE OF THE CAVE
TO R1.

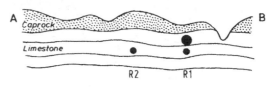

2. CONTINUED REMOVAL OF THE CAPROCK LEADS TO THE
DEVELOPMENT OF FURTHER STREAMSINKS, AND SOME COLLAPSE
SINKS, WHICH SHOW A CLUSTERED PATTERN.

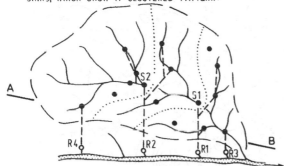

FURTHER DISMEMBERMENT OF THE ORIGINAL SURFACE
STREAM LEADS TO DISTRIBUTION OF ITS FLOW TO
NUMEROUS SMALL CAVE STREAMS.

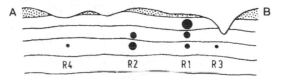

3. THE ORIGINAL LARGE STREAMSINK CATCHMENTS ARE
BROKEN DOWN INTO MANY SMALLER CENTRIPETALLY
DRAINING SINKS, SPATIAL COMPETITION PRODUCING
A MORE UNIFORM DISTRIBUTION PATTERN.

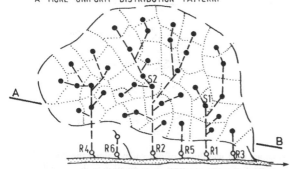

CONTINUED INCISION OF THE TRUNK RIVER PROMOTES
DOWNCUTTING OF CAVE PASSAGES WHICH ADJUST IN
SIZE (CAPACITY), ACCORDING TO THE DISCHARGE
FROM THEIR SURFACE CATCHMENTS.

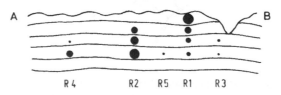

DENDRITIC SUBTERRANEAN DRAINAGE NETWORKS REPLACE
THE INITIAL SURFACE STREAM SYSTEM, WHICH IS REORGANISED
INTO ADJOINING SMALL CENTRIPETAL BASINS FEEDING TO
THE CAVE STREAMS.

Figure 9.30 A model of the reorganization of drainage by karstification. Reproduced from Williams, P.W. (1982a) Karst landforms in New Zealand, in Landforms of New Zealand (eds J. Soons and M.J. Selby), Longman Paul, Auckland, pp. 105–25.

Figure 9.31 Morphological map showing the association of karst cones with interfluves of dry valleys along the floors of which arekarst depressions. This is a case of fengcong–valley karst. Development of cones occurred in association with a fluvial network, which dried up as a consequence of incision of the trunk stream. Reproduced from Tan, M. (1992) Mathematical modelling of catchment morphology in the karst of Guizhou, China. Zeitschrift für Geomorphologie, N.F., 36(1), 37–51.

However, the palaeodrainage (fluvio-karst) impression across polygonal karst in Papua New Guinea is almost certainly ascribable to lowering from a caprock (Williams 1972a).

9.8 LANDFORMS ASSOCIATED WITH ALLOGENIC INPUTS: CONTACT KARST

Section 5.2 discussed input controls on the development of karst aquifers. The landforms associated with allogenic inputs depend upon seven factors:

- input discharge
- hydraulic conductivity
- hydraulic gradient
- location of input (lateral or vertical)
- geological context
- process environment
- time.

9.8.1 Through valleys and gorges

When allogenic discharge into a karst is considerable it may exceed the capacity of the karst to absorb it underground. Such rivers maintain their flow at the surface and may entirely cross the karst to the output boundary. Whether they do so also depends partly upon the hydraulic gradient; the steeper this is, the greater the tendency for drainage loss underground. This is because for a given hydraulic conductivity throughput discharge can increase if hydraulic gradient steepens (section 5.1). The morphological consequence of a large allogenic river entering a karst region with little elevation difference between input and output boundaries is normally that a through valley will be incised across the karst. Where the height difference is greater, a through gorge will still form if discharge remains sufficient to maintain competent surface flow or if uplift occurs at a rate that does not exceed the river's capacity to incise (an antecedent gorge in this case). Gorges formed in either of these two ways should not be confused with those developed by cavern collapse.

Such perennial **effluent** allogenic rivers traversing a karst have an important morphological function, because they act as the regional baselevel. The Green River in Kentucky and the Krka River in Yugoslavia are examples. They gain discharge from tributary karst springs emerging along their banks. Therefore they contrast with **influent** allogenic rivers that gradually lose flow, in part or totally, but still maintain a continuous channel downstream. The upper River Danube in Germany and the Takaka River in New Zealand are good examples. In these cases, their valley floors display no obvious karst morphology, but karst drainage is well developed and spread along many kilometres of channel.

An interesting general review of the styles, contexts and rates of development of gorges in karst was provided by Nicod (1997). The significant points include the following.

1. Gorges that drain karst as effluent water courses should be distinguished from gorges that lose flow into the karst, even though their canyon form continues downstream.

2. The beds of some gorges are close to or have reached an impermeable basement, so they act as effluent systems draining the karst (such as the Tarn in France), whereas gorges entirely in limestone may become influent if there is uplift or baselevel lowering (as in the case of the Reka-Timavo in Slovenia).

3. Some gorges contain large deposits of tufa that dam water flow, produce lakes and so block further incision of the canyon. The Krka and Plitvice lakes in Croatia and Huanglong valley in Sichuan, China, are examples. At the extreme, at Huanguosho, China, the waterfall is now prograding, its vertical face of tufa advancing into the earlier gorge.

4. Sometimes cavern collapse has clearly been the principal process producing a gorge (e.g. the Rak and Skocjan gorges in Slovenia, the Marble Arch gorge of Ulster, the Da Xiao Cao Kou gorge of the Yijiehe River in Guizhou, and the Rio Peruaçú gorge in Minas Gerais, Brazil), but probably the majority of gorges – and especially the larger ones – are attributable to antecedent drainage, where river flow has been maintained at the surface as uplift continued (such as the Three Gorges of the Yangtze River and numerous other cases in China).

Although simple cavern collapse gorges are important features of karst (Figure 9.32), most gorges traversing limestone terrains are fluviokarstic landforms and polygenetic in origin. They often have a complex morphoclimatic heritage that is a consequence of their long evolution, especially through the changing climatic conditions of the Quaternary Period, when discharge conditions also varied widely. Some evolved as pro-glacial meltwater canyons. Nicod (1997) assembled evidence for the ages and rates of incision of numerous gorges in karst up to 40 km long and 1.2 km deep. There is evidence that some commenced their evolution in the late Miocene or Pliocene, whereas others are mainly of Quaternary origin. Average rates of incision range from 0.1 to 4.3 m ka^{-1}, the faster rates normally involving shorter timespans.

9.8.2 Blind valleys

As hydraulic gradient steepens and/or the average flows of incoming allogenic streams reduce, sink points become more localized and morphologically more distinct. Such valleys are characterized by one or more permanent stream-sink depressions that sometimes overflow, the channel downstream being interrupted by further sinkholes. Given time, and especially where drainage has been

Figure 9.32   (Left) A natural bridge in a gorge produced by cavern collapse, Guizhou, China (Photograph by A. Milhevc.) (Right) View through a natural arch in the Rio Peruaçu valley, Brazil, to a gorge of cavern collapse upstream. The figure standing left of centre on the rock pile beneath the arch provides scale. During floods, water level rises to the figure.

lowered from above, stream-sinks can incise to the extent that all overflow ceases and the former channel downstream becomes dissected by dolines. We then have a blind valley with an abandoned higher level valley continuing downstream. There are many instances in Appalachian America, e.g. Friar's Hole dry valley (Figure 9.33), which overlies the cave of Figure 7.24. A modification of this kind of situation is encountered as an 'island' of caprock is stripped back. Upstream retreat of a sink point may leave a line of abandoned stream-sinks along a normally dry valley downstream, some of which may be periodically reactivated during floods.

Karstification of drainage lowered from above can lead to the development of innumerable small stream-engulfments. Flow is directed into the karst either laterally or vertically, depending largely on whether runoff is from an inlier or a caprock (Figure 5.16). Blind valleys consequently vary in form from normal elongate valleys with a cliff-foot sink point at the downstream 'blind' end to those which are circular in plan with drainage focused centripetally into a stream-sink where the underlying karst rock is exposed. The latter features are often scattered in clusters across a surface, whereas the elongate forms, the conventional blind valleys, line up along a lithological contact as **contact karst** features (Gams 1994).

As caprock is eroded, centripetal blind valleys subdivide into smaller features best regarded as point-recharge dolines (section 9.4). Eventually all the caprock is removed and only the dolines remain. In contrast, contact-karst blind valleys may be deepened, widened or both. Incision is most likely when the hydraulic gradient is steep, otherwise lateral enlargement occurs. This may lead to the coalescence of neighbouring blind valleys and so to the production of a larger enclosed depression. If this develops an alluviated floor (the coalesced floodplains of allogenic streams), it can then be termed a **border polje** (see section 9.9).

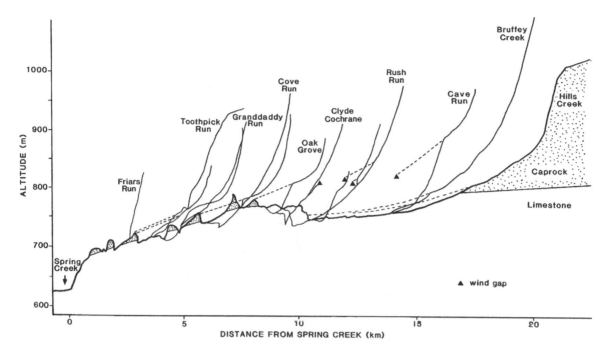

Figure 9.33 Dissected valley above Friars Hole cave system, West Virginia. The multiple blind valley has developed by headwater stripping of the caprock. Wind gaps and caprock residuals mark the level of the former active valley. Numerous stream-sink points pock the valley floor. Modified from Worthington, The paleodrainage of an Appalachian Fluviokarst: Friars Hole, West Virginia. McMaster Univ. MSc. Thesis, 1984.

Sometimes aggradation buries stream-sinks and reduces their absorbance capacity. More frequent overflow then occurs and the valley is considered semi-blind. The semi-blind condition can therefore arise both in the early stages of evolution before subterranean conduits are fully developed and at a later stage if aggradation and fill reduces conduit throughput capacity.

9.8.3 Recharge response

Recharge is a process, valley development is the morphological response. Other things being equal, the greater the lateral point recharge the larger is the valley penetration into the karst, to the extent that very large recharge permits complete traversing of the karst by the allogenic river. This general principle is illustrated in Figure 9.34, where rivers in Belize flow from an inlier of resistant insoluble rocks across a faulted boundary into a polygonal karst in brecciated limestones. Lubul Ha blind valley drains 3.3 km^2 of impervious rocks and penetrates merely 0.5 km into the karst; Actun Chek catchment covers 20 km^2 and penetrates 2 km; Caves Branch has an 88 km^2 allogenic basin and penetrates 9.2 km when in flood; whereas the Sibun River draws its large discharge from 250 km^2 of metasediments and completely traverses the karst in a through valley.

9.8.4 Response to volcanism

An interesting case of allogenic inputs and contact karst arises from the intrusion of volcanic features into and onto karst rocks. Salomon (2003b) reviewed volcanic influences on karst and showed that the response of karst systems to volcanism is mainly determined by the consequences for water movement. Volcanism distorts surface strata, with associated swelling, folding and fracturing, and this guides erosion processes through its direction of karstification (Figure 9.35). However, very large lava flows and deposition of welded ignimbrite can totally overwhelm karst and disconnect it from the existing hydrological system (Williams 2004c).

9.9 KARST POLJES

9.9.1 Definition

Poljes are large, flat-floored, enclosed depressions in karst terrains (Figure 6.27). They are landforms

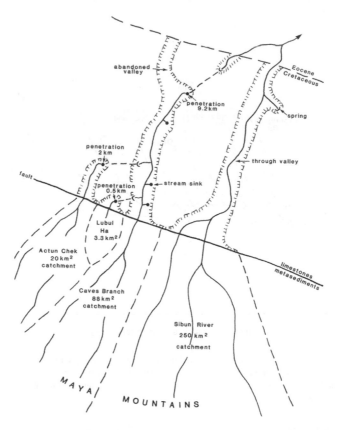

Figure 9.34   Variation in penetration distance of allogenic streams according to catchment size, Belize, central America. Modified from Miller, T.E. (1982) Hydrochemistry, hydrology and morphology of the Caves Branch karst, Belize. Mcmaster Univ. PhD thesis, 280 pp.

associated with the input and throughput of water and in many respects can be considered inliers of a normal fluvial landscape (Figure 6.30). The word *polje* signifies a field or a large plain, and it is still widely used in Slav languages without particular reference to karstic terrain (Sweeting 1972, Roglić 1974), although poljes in the karst of Slovenia are usually called *dolina* or *dol* (Gams 1994). Nevertheless, in technical karst literature ''polje'' has acquired a special usage, particularly since the writings of Cvijić (1893, 1901) and Grund (1903). Similar landforms are termed *plans* in France, *campo* in Italy and Spain, *wangs* in Malaysia and *hojos* in Cuba. Although they are written about mainly in tropical and temperate contexts, poljes are also reported in the subarctic zone (Fig. 10.14; Brook and Ford 1980). The most thoroughly studied are in the Dinaric karst, where about 130 have been recognized (Gams 1978, 1994; Mijatović 1984b; Bonacci 2004; Milanović 2004). The hydrological behaviour of poljes is discussed in section 6.6.

Gams (1978) considered the many geomorphological definitions of polje that have been published and found several elements to be common to most definitions. He identified three criteria that must be met for a depression to be classified as a polje:

1. flat floor in rock (which can also be terraced) or in unconsolidated sediments such as alluvium;
2. a closed basin with a steeply rising marginal slope at least on one side;
3. karstic drainage.

He also suggested that the flat floor should be at least 400 m wide, but this is arbitrary because Cvijić (1893) took 1 km as a lower limit. In fact, poljes vary considerably in size. The floors of reported poljes range from ~ 1 to $> 470\,\mathrm{km^2}$ in area (Lika Polje is the largest at $474\,\mathrm{km^2}$), but even in the Dinaric karst most are less than $50\,\mathrm{km^2}$, and elsewhere in the world a majority are less than $10\,\mathrm{km^2}$.

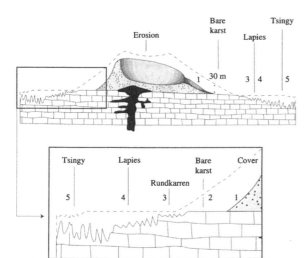

The strombolian cone of the Ankopatra (Ankarana, North Madagascar.)

Figure 9.35 Influence of volcanic cone on karstification. Reproduced from Salomon, J.-N., Précis de karstologie. p. 250 © 2000 Bordeaux, Presses Universitaires.

Regardless of individual particularities, all poljes have a common hydrological factor in their history: their development occurred close to the local water table, even though it may be perched in some cases and even though subsequent events (such as uplift and karstification) may since have separated the polje floor from the contemporary position of the water table. Where a low-gradient water table is close to the surface, lateral fluvial planation (corrosion and corrasion) and deposition processes are more important than incision; hence plains are created rather than deep valleys. The relative importance for polje development of tectonic processes and planation by karst dissolution varies, but in most poljes there is evidence for both being important. An interesting case is that of Jiloca polje in Spain, which has developed in an active half-graben (Gracia *et al.* 2003). Although having clear tectonic origins, a sequence of eight stepped corrosion terraces truncating limestones in the basin gives clear evidence of the importance of dissolution in the evolution of its morphology. Cerkniško Polje, Slovenia, is being deformed by movement along a transverse fault. There has been > 120 m displacement since the last interglacial stage, and outlet drainage has been rerouted into a previously infilled cave (Šušteršič and Šušteršič 2003).

Five types of polje were recognized by Gams (1973b, 1978) and examples were mapped in the vicinity of Postojna (Gams 1994). These are: (i) **border polje** located at a geological contact across which it receives allogenic surface runoff; (ii) **piedmont polje** in an alluviated valley

usually located downslope of a glaciated terrain; (iii) **peripheral polje** that receives surface runoff from a large (often downfaulted) internal area of impermeable rock; (iv) **overflow polje** which is a depression underlain by relatively impermeable rock that acts as a dam and forces groundwater to flow on the surface to stream-sinks on the other side of the basin; and (v) **baselevel polje** in which the floor is cut entirely across karst rock and is located in the epiphreatic zone.

We consider that the above categories of polje can be reduced to the three basic types illustrated in Figure 9.36.

9.9.2 Border poljes

Border poljes (German **Randpolje**) are dominated by allogenic input controls. They develop where the zone of water table fluctuation in non-karst rocks extends onto the limestone. This ensures that allogenic fluvial activity is kept at the surface and that lateral planation and alluviation dominate valley incision; otherwise blind valleys will form. Floodplain deposits may partly seal the underlying limestone and encourage water to stay near the surface, although leakage may be widespread upstream of the final point of engulfment. This kind of polje is very common along contacts with highlands that can supply abundant clastic load to rivers.

9.9.3 Structural poljes

Structural poljes are dominated by bedrock geological controls. They are often associated with graben or fault-angle depressions and with inliers of impervious rocks or even of relatively less permeable rocks such as dolomite. Their depression form is elongated with the structural grain, although their tectonic boundaries may be modified by extensive planation across karst rocks. Structural poljes are very important features, giving rise to the largest karst depressions in the world. They are the dominant class of polje in the Dinaric karst (Figure 6.30) and other active tectonic regions such as the Taurus Mountains of Turkey. They are inliers of a normal fluvial landscape (a floodplain, often with terraces) within an otherwise karstic terrain, the local water table being near the surface because of the low permeability of the enclave. Water escapes from the basin where the hydraulic gradient steepens, usually on the karst rock side of a bounding fault, where there may be literally hundreds of ponors (stream-sinks). For example, the total sinking capacity of ponors in Popovo Polje is more than $300 \, m^3 \, s^{-1}$ (Milanović 2004). Numerous boreholes through poljes of this type have shown their flat floors to be mainly the result of aggradation of Neogene terrestrial and lacustrine deposits that frequently bury an irregular

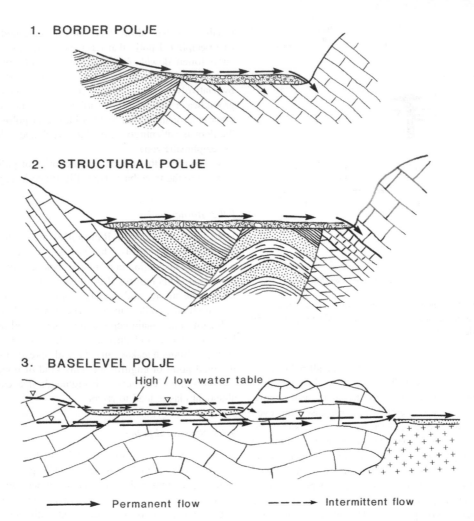

Figure 9.36   Basic types of polje.

topography (Mijatovic 1984b). In Duvanjsko Polje, for example, such sediments reach 2500 m in thickness.

9.9.4 Baselevel poljes

These are water-table dominated, occurring where dissolution has lowered the karst surface to the regional epiphreatic zone; i.e. they are windows on the water table. Typically, they develop within or on the outflow side of karst systems. Because they do not depend on allogenic inputs or geological control, they can be considered the purest kind of polje. They can develop in an entirely autogenic setting. Water-table control extends inland from the output boundary, where the sea or an impermeable formation may act as a dam or threshold.

Where long-term denudation finally erodes part of the terrain down to the piezometric level, subterranean streams must be exposed and flow across the surface. Their lateral rather than vertical activity expands the window on the water table, creating an interior fluvial plain that is cut entirely across carbonate rocks, although often veneered with alluvium.

Seasonal or major storm inundation of the floor is characteristic of many poljes; in certain respects this tendency to flood has defined them classically. In many tropical poljes and in most temperate poljes there is evidence of effects of Quaternary climatic change. It takes the form of episodes of accelerated filling, detrital fan formation (e.g. Cerknisko), or channel entrenchment, rerouting of surface flow and/or sink-points underground.

Both alluvial and corrosion terraces are common as a consequence. In the temperate zone, many poljes that are dry or only seasonally inundated today had permanent lakes occupying part or all of their floors during cold wet phases of the Pleistocene. A spectacular example is Zafarraya Polje, Andalucia, Spain (Duran Valsero and Lopez Martinez 1999). This is a highland tectonic polje with 60 m of clastic fill on a marl bottom. It has a polycyclic overflow outlet in which more than 100 m of travertine has accumulated, most of which was eroded during the latest overspill episode. This polje has only a small area of seasonal flooding today.

9.10 CORROSIONAL PLAINS AND SHIFTS IN BASELEVEL

9.10.1 Steepheads and pocket valleys

Springs occur on the output margin of karst regions. They are controlled in elevation by the regional or local water table which may be determined by lakes or the sea, by an impermeable rock threshold on the outflow side (German **Vorfluter**) or by aggradation. Some springhead sapping occurs where water issues from a major penetrable bedding plane or contact within carbonate rocks that may be a little above the true allogenic baselevel, i.e. it is not strictly baselevel, but the next penetrable level above the base.

Headwards retreat of the spring produces an amphitheatre-like **steephead** or **pocket valley**. Springs can retreat either by gravitational undermining and slumping of a slope, a process termed 'spring sapping', or by irregular collapse of a cave roof above a subterranean river, the collapse depressions ultimately coalescing to yield a gorge with a spring at its upstream end. A particularly good example of a pocket valley is Malham Cove in England (Sweeting 1972). Salomon (2000) provides interesting examples from France, where these landforms are known as **reculées** (Figure 9.37). Sapping canyons of similar length and length:width proportions are well developed in the massive, resistant sandstones of the Colorado Plateau, where dissolution is considered only a minor contributor (Howard *et al.* 1988), but perhaps is a critical one. Still larger features, ascribed to the emergence of groundwaters and measuring kilometres across, are termed **makhteshim** (erosion cirques) in the Negev and Sinai deserts of Israel and Egypt (Issar 1983). Long-term denudation results in the merger of neighbouring valleys and in the production of a corrosion plain at the level of the water table. Such features are known as **karst margin plains** (German **Karstrandebene**). However, karst margin plains also include those created by exterior

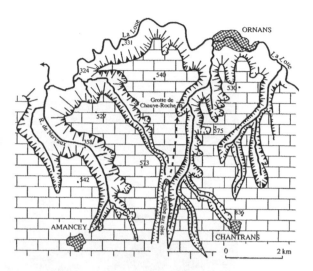

Figure 9.37 Steephead valleys on the outflow side of a karst plateau in France. Reproduced from Salomon, J.-N., Précis de karstologie. p250 © 2000 Bordeaux, Presses Universitaires.

river overbank flooding, the water being partly allogenic and (perhaps) partly from karst springs.

9.10.2 Baselevelled corrosion plains

Corrosion plains have been described in many parts of the world, e.g. by Blanc (1992), Pfeffer (1993), Sunartadirdja and Lehmann (1960), Sweeting (1995) and Williams (1970). Although most often associated with the output side of a karst terrain and (on a smaller scale) as corrosion terraces within poljes, they can also be produced after a long period of denudation on the input margin. When a border polje at the input boundary becomes connected at the surface to a corrosion plain on the outflow side, very extensive corrosion surfaces of low relief are formed. The Gort Lowland in western Ireland is an especially good example (Figure 9.38) of a corrosion plain that extends continuously from input to output margins.

Because of their low elevation, corrosion plains are commonly veneered with alluvium. Beneath this is a surface cut across the karst rock regardless of structure. When uplifted, glaciated or strip mined for placer deposits (as in the Kinta valley in Malaysia), removal of the clastic veneer reveals an impressively planar rock floor that in the tropics, in particular, can sometimes be very rugged in detail because of etching down joints, but has little or no perceptible slope and can extend across many square kilometres. Corrosional plains of this type

Figure 9.38   The Gort lowland, an extensive baselevelled corrosion plain cutting across folded Carboniferous limestones in western Ireland. The lowland is the product of Tertiary solution planation followed by scouring by Pleistocene glaciers. The view is to the northwest with the Burren Plateau about 10 km away in the distance.

develop by solutional removal of irregularities down to a surface controlled by the water table. Since mechanical work is not involved, except where there are significant insoluble residues, slopes can be very low indeed ($< 0.1°$). Once the topography is corroded down to the level of frequent inundation (the epiphreatic zone), the plain expands by gradual retreat of adjoining karst uplands and by elimination of residual hills. Even on a perfect plain corrosion can be expected to occur near the water table because of the continuous supply of aggressive water from rain, but most activity will be restricted to a shallow depth because of the low hydraulic gradient.

The complex of processes producing baselevelled plains in karst has been called **lateral solution planation**, but in fact **corrosion planation** is a better term because it involves a combination of (i) vertical dissolution of upstanding remnants by direct rainfall, (ii) lateral under-cutting of hillsides by accelerated corrosion in swampy zones at their base, and (iii) spring head sapping. The relative importance of these three activities depends upon hydrogeological and biogeographical circumstances, which control the aggressiveness of water as discussed in Chapters 3 and 4, and the deposition of alluvium, which influences (im)permeability of the surface. Warm climates are not essential for corrosion plain development. More important is a long stable period under conditions sufficiently humid to permit denudation of the relief down to the epiphreatic zone.

Corrosion plains are not **peneplains** in the sense of W.M. Davis (1899), because first they are not undulating surfaces of low relief and second, they are not roughly the same age everywhere across the surface, as is the assumption with peneplains. Corrosion plains are as near perfectly flat as can be produced by natural denudation processes (Figure 9.38) and, having reached the water table, are worn back rather than worn down. Thus they are youngest at the foot of the retreating uplands which they replace. In this respect they resemble **pediplains**.

9.10.3 Baselevel control

In all the schemes of karst denudation that have been proposed, the ultimate baselevel of erosion recognized is either sea level or underlying impermeable rocks, if they occur above sea level. However, it is well established that karst springs can resurge well below sea level and we saw in section 5.4 that that groundwater circulation can extend to considerable depths. Since the karst hydro-chemical system can operate well below sea level and can corrode deep phreatic caves, sea level is clearly not the ultimate basis for karst dissolution. The question of note then is: what is the depth of the lowest active conduit?

The baselevel issue is best understood by referring to the explanations of controls on aquifer development (Chapter 5) and cave evolution (Chapter 7), which point out that a karst circulation system undergoes a continuous process of self-adjustment, depending in detail on fissure frequency and hydraulic potential. Where fissuring supports a State 3 or 4 conduit system, a corrosion plain will develop near the level of the water table with only shallow groundwater circulation beneath it. Should the fissure frequency initially have been low, the increase of fissuring with age that appears to occur in a majority of States 1 and 2 systems also tends to favour generally shallower circulation, as will the declining hydraulic gradient. However, there will be exceptions, and deep phreatic loops once formed may maintain their flow unless they are sealed by clastic detritus.

Therefore:

1. the baselevel for corrosion plain development is the water table;
2. the base of active corrosion varies (i) with the system, being deepest in States 1 and 2, (ii) with age (if fissure frequency increases it will tend to be shallower) and (iii) with the extent of any clastic infilling.

Freshwater:saltwater interface effects are ignored for the purpose of this discussion, but it should nevertheless be recognized that considerable corrosion is likely in the mixing zone.

The most dramatic known eustatic shift to have affected karst was the Late Miocene Messinian fall of sea level in the Mediterranean around 6 to 5.3 a, when sea level dropped about 2000 m. Most of the Mediterranean Sea dried up and the Black Sea also fell considerably (Hsü *et al.* 1973, Gargani 2004). Water subsequently poured in through the Strait of Gibraltar in the Pliocene and refilled these basins. All karst systems active in the Mio-Pliocene with these seas as their baselevel would have been affected by these dramatic eustatic changes, except perhaps those in the headwaters of the longest drainage basins. Thus karsts between the Cevennes Mountains and the Gulf of Lions in southern France drained towards the deep canyon that was incised along the course of the River Rhône. With the Black Sea 200 m lower than at present, water was captured underground from the lower Danube River and discharged through deep subterranean passages that seem to be still active (Lascu 2004). Other seas such as the Persian Gulf and Red Sea have been similarly affected, as were the karsts that drained to them.

9.10.4 Rejuvenation

A negative shift in baselevel resulting from tectonic uplift or sea level fall is a common cause of rejuvenation of

erosional activity. Uplift of a well-developed karst has three principal effects: (i) major allogenic through-rivers entrench to produce deep valleys; (ii) the vadose zone expands but may be perched above valley bottoms, depending on the rate of valley entrenchment; and (iii) the phreatic zone moves downwards into less karstified rock. Once the water table drops beneath an uplifted corrosion plain, percolating rainwater can escape vertically again instead of running off laterally. Solution dolines can incise, and suffosion and collapse dolines appear on the alluvial plains. Any residual hills become incorporated into divides between developing dolines. Figure 9.39 schematically represents rejuvenation of a tower karst that had reached water-table level. A wave of morphogenesis of this sort could be expected to progress inland from the coast or from an incised river. Rejuvenation underground (i.e. development of a multiphase cave) is a prerequisite for rejuvenation of the surface topography.

Figure 9.40 shows an uplifted corrosion plain with residual hills in the Guizhou Plateau, in an area still unaffected by the current wave of rejuvenation which is working headwards up neighbouring gorges. Superb examples of uplifted, stripped karst margin corrosion plains are found along the Hongshui River (Guizhou, China), which drains the Dahua and Qibailong areas, the most intensely rugged karst topography we have seen; Song and Junshu (2004) give details. The lower courses of the Cetina and Krka rivers in Croatia have similar entrenched plains (Figure 9.41).

9.10.5 Submergence

Quaternary sea-level history reveals that the level of the ocean has been well beneath its present height for most of the past several hundred thousand years (Chappell 2004; Figure 10.17). Karsts near the sea have therefore generally evolved in response to lower baselevels than at present. Consequently, the positive shifts of baselevel during the post-glacial transgressions of the multiple glaciations have given rise to:

1. contracting vadose zones;
2. expanding phreatic zones, possibly with deeper stagnant sections being removed from the layer of active groundwater circulation;
3. upward displacement of the saltwater–freshwater interface on the coast and hence of the associated zone of enhanced corrosion;
4. drowned coastal springs, some of which are too deeply submerged to operate;
5. aggradation of coastal lowlands and infilling of caves by waterborne sediments.

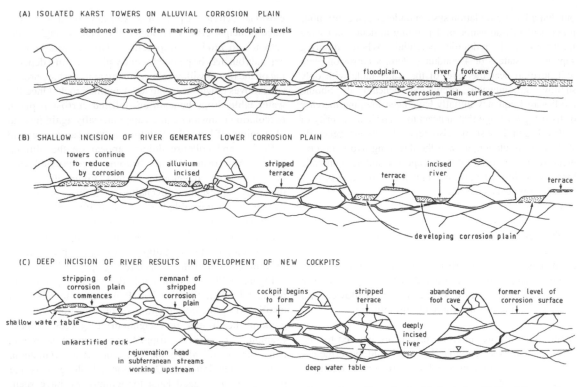

Figure 9.39   Redevelopment of polygonal karst from rejuvenation of tower karst. Note that in (C) the isolated hills of the previous phase may become the tops of the highest tier of cones in the modern phase.

Figure 9.40   An ancient corrosion plain with residual hills on the Guizhou plateau, China, at 1200 m above sea level.

Figure 9.41 An uplifted corrosion plain in the Krka valley, Croatia. The plain truncates the structure and has been incised to form a rock terrace above the river.

In addition, tectonic subsidence has also contributed to the submergence of some karsts. The Florida (Land and Paull 2000) and Yucatan (Beddows 2004) peninsulas and the Bay of Halong in Vietnam (Khang 1991, Waltham and Hamilton-Smith 2004) provide outstanding examples of drowned karsts (Figure 9.42). A detailed survey of drowned reefs and antecedent karst topography has been undertaken in the southeast Hawaiian Islands using multibeam high-resolution bathymetric survey coupled with videotransects and observations from submersibles (Grigg *et al.* 2002). Flat-floored enclosed depressions with steep walls and cliffs undercut by notches provide evidence for the operation of subaerial karstic processes when sea levels were at least 120 m lower.

Figure 9.42 Partly submerged tower karst topography, Bay of Halong, Vietnam.

9.11 RESIDUAL HILLS ON KARST PLAINS

9.11.1 Tower karst landscapes

A landscape of residual carbonate hills scattered across a plain is referred to as **tower karst**. The residual hills display a variety of shapes from tall sheer-sided towers (Figure 9.43) to cones or even hemispheres (Figure 9.44). Some are symmetrical in form, while others are asymmetric, reflecting the influence of dip or erosional processes. Although some residual hills rise

Figure 9.43 The sheer-sided karst tower in Guilin, Guangxi, at the base of which are the swamp notches shown in Figure 9.48.

Figure 9.44 Isolated cones in the Thousand Hills area of Bohol Island, Philippines. (Photograph by R. Wasson.)

directly from the plain, many surmount pedestals. Some towers are isolated and others are in groups rising from a common base. The word 'tower' therefore subsumes a myriad of forms, and a variety of terms in different languages has been used to describe them, including *turm*, *mogote*, *cone*, *piton*, *hum* and *pepino*.

Tower karst can be visually spectacular. That of southern China must be said to be one of the world's great natural landscapes (Figure 9.45). Its evolution in the Guilin region of Guangxi Province has been explained by Lu (1986), Zhu (1988), Yuan (1991) and Sweeting (1995). For Guizhou Province, Song *et al.* (1993) and Chen *et al.* (1998) have published substantial international collaborative studies.

Chinese geomorphologists distinguish between isolated towers on a plain (termed **fenglin** or '**peak forest**') and groups of residual hills that emerge from a common bedrock base above the plain (termed **fengcong** or '**peak cluster**'). Fengcong may develop as an accentuated cockpit karst, the towers being residuals around large and deep dolines (= fengcong–depression karst). Or the towers may be separated by dry valley networks (= fengcong–valley karst), in which the valley floors are often incised by closed depressions separated by low cols.

The international literature suggests four main genetic types of tower karst (Figure 9.46):

1. residual hills protruding from a planed carbonate surface veneered by alluvium;

2. residual hills emerging from carbonate inliers in a planed surface cut mainly across non-carbonate rock;
3. carbonate hills protruding through an aggraded surface of clastic sediments that buries the underlying karst topography;
4. isolated carbonate towers rising from steeply sloping pedestal bases of various lithologies.

The plains between the towers thus need not be cut entirely across carbonate lithologies, even though this often appears to be the case. Tower karsts sometimes incorporate planed surfaces in other rocks, those in the Sierra de los Organos in Cuba (Panoš and Stelcl 1968) and at Chillagoe, Queensland, Australia (Jennings 1982) being examples. Where deep residual deposits cover the plain between the towers, as is found at the coast in the Bay of Halong, Vietnam (Pham 1985), or in the blanket sand plains between the towers (locally known as **mogotes**) of Puerto Rico (Monroe 1968), the third type of terrain may exist or perhaps aggraded versions of types 1 or 2. An added complexity in all of these cases is that the intervening plain is frequently terraced. The terraces are often developed in the veneer of superficial deposits, although they can also be in bedrock. Zhu (1988) commented that the bedrock surface of the peak-forest plain near Guilin, which in many respects could be regarded as the global type-site of tower karst, is generally quite smooth, gently sloping towards the centre or towards one side. The overburden of sediment on the

Figure 9.45 (Upper) Tower karst (**fenglin**) along the River Li in Guangxi, China. Note that the shape of 'towers' varies from sheer-sided to conical. (Middle) Peak forest plain landscape beside the Lijiang, Guangxi. (Photograph by A. Waltham.) (Lower) Cone karst (**fengcong**) along the skyline transitional to tower karst in a valley in the Guilin region of Guangxi.

1. RESIDUAL HILLS ON A PLANED LIMESTONE SURFACE

2. RESIDUAL HILLS EMERGING FROM LIMESTONE INLIERS

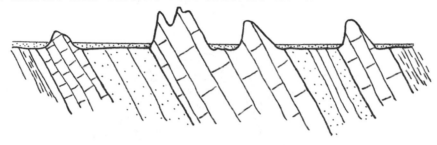

3. RESIDUAL HILLS PROTRUDING THROUGH AN ALLUVIATED SURFACE

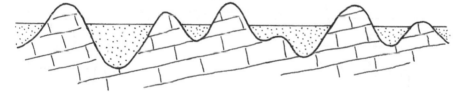

4. TOWERS RISING FROM SLOPING PEDESTALS OF VARIOUS LITHOLOGIES

Figure 9.46 Types of tower karst. Reproduced with permission from Williams, P.W., Geomorphic inheritance and the development of tower karst. Earth Surface Processes and Landforms 12(5), 453–465 © 1987 John Wiley and Sons.

plain is generally quite thin, although in places can attain 20–30 m thickness. Annual water-table fluctuation is of the order of 3–5 m.

9.11.2 History of a karst tower

Williams *et al.* (1986) investigated the evolution of an individual tower at the edge of a fengcong group at Guilin. The hill rises abruptly from the floodplain of the River Li. A reconnaissance examination of the palaeomagnetism of deposits from caves at different levels in the tower suggested that sediments up to 23 m above the floodplain possess normal geomagnetic polarity, but that some deposits above this have a reversed magnetism. Alluvial sediments within lower caves in the tower indicated that its base was buried by fluvial aggradation and then re-exposed following floodplain incision. The combined information indicated the tower

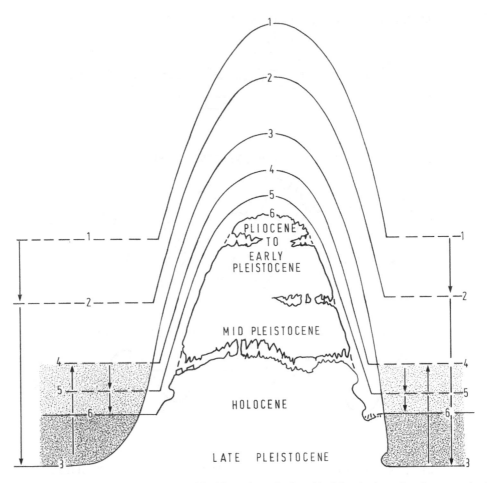

Figure 9.47   Model of karst tower evolution near Guilin, illustrating episodes of burial and exhumation that occur simultaneously with tower reduction by dissolution. Abandoned caves from earlier phases penetrate the tower and newer swamp notches and foot caves mark present and previous floodplain levels. Reproduced with permission from Williams, P.W., Geomorphic inheritance and the development of tower karst. Earth Surface Processes and Landforms 12(5), 453–466, © 1987 John Wiley and Sons.

to have grown by the lowering of its base at a net rate not exceeding $23 \, \mathrm{mm \, ka^{-1}}$ during the past 1 Myear. This evidence reinforced that already long established from the fossil record in other towers in southern China, where Pliocene to mid-Pleistocene vertebrates of the *Stegodon–Ailuropoda* faunal complex had been recovered from high-level caves (Kowalski 1965). The inescapable conclusion is that karst towers in this region are time-transgressive landforms, being considerably older near their summits than at their base, although showing an oscillatory age pattern in detail because of episodes of partial burial and re-excavation as morphological evolution progressed (Figure 9.47).

In Belize, McDonald (1979) demonstrated that the morphogenesis of towers located on low interfluves is governed by the progressive lowering of the alluvial plains by erosion caused by overland runoff. As the interfluvial plains are lowered bedrock slopes are formed around the towers, with angles between 20° and 60°. Undermining of towers only seems to occur where rivers actually flow against their base or foot-caves receive water. Field observation points towards similar conclusions around Guilin.

Evidence often cited for the effectiveness of lateral undercutting is the occurrence of swamp slots or notches (Figure 9.48) and cliff-foot caves. Jennings (1976) and McDonald (1976a) mapped distributions of foot-caves and swamp slots around isolated limestone towers in Selangor, Malaysia, and Sulawesi, Indonesia, respectively. Jennings concluded that their frequency of

Figure 9.48   Swamp notches near the base of the karst tower illustrated in Figure 9.43. The different notches reflect changes in local water-table level.

occurrence was compatible with the supposed importance of lateral solution undercutting. McDonald measured 12 km of hill bases and found 31% to be composed of foot-caves, some being relict, and 59% to be characterized by footslopes that owe their origin to processes other than solutional undermining. He concluded (p. 89) that the 'erosion of hillslope support and the retreat of limestone hillslopes appears not to take place uniformly . . . but takes place in scattered locations along hillsides.' Our observations elsewhere indicate that the relative significance of lateral solution planation depends on the location of the residual hill: if beside a river or on a floodplain, swamp-slot corrosion is very important, but if on a terrace above the reach of contemporary inundation then other slope processes become comparatively more significant.

9.11.3 Hypotheses concerning tower karst evolution

Some controversy exists concerning the way in which tower karst evolves. There are two main hypotheses:

1. tower karst (fenglin) evolves sequentially from a previous cockpit-style polygonal karst landscape (fengcong–depression);
2. it evolves directly.

These alternatives were examined by Williams (1987, 1988) and Zhu (1988), who reached similar conclusions. Williams found that both styles of evolution can occur (Figure 9.49), but that the former appears to be the more common case. Zhu concluded that peak-cluster depression and peak-forest plain styles are two patterns of karst that can develop synchronously, the condition favouring peak-cluster depression development being relatively high relief and a great depth to the water table (see Figure 9.65). The transition from cockpit to tower karst is forced when the water table is reached; it is an important hydrological threshold that controls the operation of solution processes. A major fall in the water table as a consequence of uplift and rejuvenation causes tower karst to revert to cockpit karst, but a minor or gradual lowering of the water table promotes the vertical extension of towers by incision around their base and planation of a lower corrosion plain (Figure 9.39b).

The sequential and direct modes of evolution can occur simultaneously in neighbouring regions, hence giving rise to the two synchronous patterns of karst that Zhu noted, because the course that is followed depends on the depth to the water table (Figure 9.49). In the sequential case karst development starts on an upland with the water table deep beneath the surface. As a polygonal karst dissolves downwards through a thick carbonate mass, minor

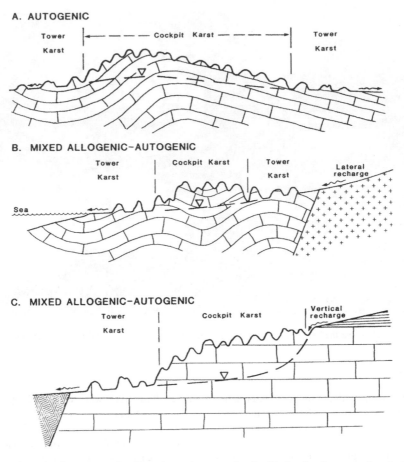

Figure 9.49 Autogenic and mixed autogenic–allogenic settings associated with the development of cockpit and tower karsts. Reproduced with permission from Williams, P.W., Geomorphic inheritance and the development of tower karst. Earth Surface Processes and Landforms 12(5), 453–465 © 1987 John Wiley and Sons.

adjustments to the doline (or cockpit) pattern will occur as beds or zones of different permeability are encountered, but otherwise the network geometry will remain until the water table is reached. When this occurs, the epikarst merges with the phreatic zone and free vertical drainage is no longer possible. Consequently, vertical deepening of depressions ceases and their floors begin to widen in the epiphreatic zone. The location and form of residual hills is inherited from the position and shape of hills around the enclosed basins of the earlier stage. Rainfall reduces these residual hills by corrosion, so that they become isolated on an expanding corrosion plain. Where allogenic rivers cross the karst, they substantially aid the planation process by periodically introducing large volumes of aggressive floodwaters. The tower forms can then be steepened-up at their base by swamp-notch and foot-cave undercutting, resulting in a lower cliffed or concave slope element.

Flooding can also lead to floodplain aggradation that may bury the base of some towers (Figure 9.47).

The usual starting point for direct evolution of tower karst is the slight uplift of a corrosion plain. If uplift is gradual and the water table is at a shallow depth, then surface streams will be maintained during uplift. The streams gradually incise and cones may develop on their interfluves. When uplift ceases or decelerates, floodplain widening then isolates these cones. However, if uplift accelerates the drainage will be captured underground and valleys will become dry, resulting in a fengcong–valley landscape (Figure 9.31), where cones are reduced by parallel slope retreat (Xiong 1992) in the absence of swamp or fluvial processes. Even without surface streams, if the water table is at a shallow depth, as on the flanks of a fold along an inflow or outflow margin, limestone blocks can be isolated by intersecting joint corridors. If these widen at the level of the water table, the residual blocks will become isolated as towers or tower groups on a plain. As in the first case, the influence of allogenic rivers is to accelerate the rate of tower development.

The direct style of tower evolution is also known in the semi-arid zone. For example, in the limestone ranges of northwest Australia, karst towers are isolated on a limestone pediment that evolved by replacement of an upland erosion surface (section 10.2). No previous cockpit-style of evolution was involved (Jennings and Sweeting 1963). This karst also has some similarity to the giant grikeland or labyrinth karst in the Nahanni region of northern Canada (Figure 9.13) described by Brook and Ford (1978). As the labyrinth of widened grikes, closed depressions and small poljes is expanded, steep residual towers to 50 m high are left emerging from uneven karst margin plains.

Both courses of tower karst evolution probably occurred simultaneously in the Guilin karst, although case 1 could be dominant. Two lines of evidence support this. First, where hillslope angles of fengcong–depression karst are similar to slope angles on fenglin karst, the first case of evolution is possible because it supports the inheritance of tower form from an earlier stage. In Guilin, Tang and Day (2000) found mean slope angles of towers to range from 60° to 75° (mean 62.4°), there being no significant difference in the mean slope angle and slope-angle distribution between towers in fenglin or fengcong. Second, as polygonal karst evolves, the water table lowers and multilevel caves form. The occurrence of large-diameter cave remnants at different heights in towers indicates dismemberment of large cave systems and so favours case 1 evolution. However, in case 2 evolution, multilevel foot-caves would also be found if periodic uplift occurred. The well-defined tiers of tower summits in the Guilin–Yangshuo area indicate episodes of uplift and stability. In fact, the entire karst landscape around Guilin has had a long period of evolution involving considerable tectonic and palaeoenvironmental change, as indicated amongst other things by Cretaceous clastic red beds, which are found unconformably on Palaeozoic limestones at many altitudes around Guilin, including on the sides of karst towers (Drogue and Bidaux 1996). Although there was undoubtedly a pre-Cretaceous palaeokarst in the region (Yuan 1991), most of the landforms found in the area today were probably developed during the stripping of these red beds following uplift and climatic change in the Tertiary and Quaternary. Therefore, both in southern China and elsewhere in the world, we must admit different histories and styles of evolution of tower karst, not all of which

are associated with the humid tropics or subtropics. Computer modelling of karst has more to contribute to the issue of how tower karst evolves (section 9.16).

9.12 DEPOSITIONAL AND CONSTRUCTIONAL KARST FEATURES

9.12.1 Case-hardening of residual hills and limestone surfaces

Weathering crusts are often observed on limestone outcrops in tropical to warm temperate environments. When carbonate-rich waters infuse soil, alluvium or weathered rock in regions where potential evaporation exceeds rainfall, chemical precipitation may occur in the profile. The surface itself may become indurated, a process known as **case-hardening** that significantly increases its strength (as discussed briefly in sections 5.5 and 7.9). Or there is dissolution of carbonate clasts near the top of the deposit succeeded by aragonite and calcite precipitation lower down, to produce a calcium carbonate layer termed **calcrete** (or **caliche** or **kunkur**). Its nature, origin and distribution has been explained by Goudie (1983), Wright and Tucker (1991) and Nash (2004). There are three principal modes; as dispersed intergranular cements, as nodules and as consolidated sheets that amount to the insertion of a new calcareous conglomerate bed within, for example, a soil or a terrace gravel. Gradations between these modes are common. Sheets up to 1 m thick are reported and, like laterites, are usually resistant.

The significance of case-hardening for karst was first appreciated in the Caribbean region by Monroe (1964) in Puerto Rico and Panoš and Stelcl (1968) in Cuba. Induration of the rock produces an outer shell of greater strength than the interior material, and so is particularly important in the development of karst landforms on highly porous, mechanically weak and diagenetically immature lithologies. Case-hardening invests the rock with increased strength and resistance to erosion and collapse. It also renders the surface much less permeable, with porosity being reduced by a factor of ten or more. Thus it is a type of vadose diagenesis (section 2.3). The zone of induration closely follows the topography (Figure 9.50) and on average is 1 to 2 m thick, but it can vary from 0.5 to 10 m (Ireland 1979).

In Cuba, Panoš and Stelcl (1968) found case-hardening on practically all bare limestone surfaces, the thickest crusts developing on the most porous rocks, especially where the surfaces are relatively old (pre-Pleistocene). In Puerto Rico, Monroe (1964, 1966, 1968) considered case-hardening to be thickest on the windward sides of mogotes, encouraged by their more frequent wetting

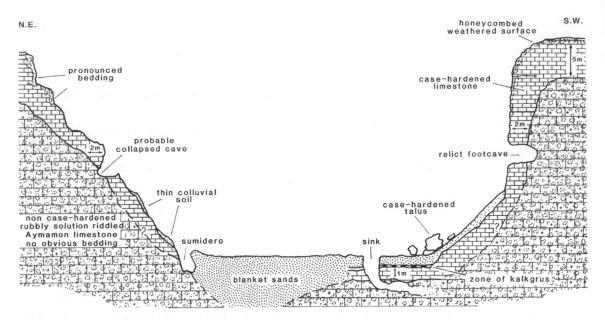

Figure 9.50 Variation in thickness of case-hardening on mogotes from north-central Puerto Rico. Note the extension of casehardening beneath the Blanket Sands, indicating that the sands were deposited after case-hardening had already occurred. Reproduced form Ireland, P., Geomorphological variations of 'case hardening' in Puerto Rico. Zeitschrift für Geomorphologie, Suplement-Band, 32, 9–20, © 1979 E. Schweizerbart'sche Verlagsbuchhandlung, Science Publishers.

and drying. This was offered as an explanation of the apparent asymmetry of mogotes first observed by Thorp (1934), but work by Day (1978) showed that the asymmetry does not have a simple windward–leeward pattern. Ivanovich and Ireland (1984) proposed a model for case-hardening with two main diagenetic processes. In limestones composed of more than 50% fossils, the dominant process is precipitation into solution cavities. In rocks with less than about 20% fossils, the main process is aggrading neomorphism, a wet recrystallization process resulting in the progressive increase of microspar. During these processes total carbonate porosity is reduced from about 30% to 5% or less. They suggested that formation of an indurated layer 1 m thick could occur within 10 000 to 20 000 year, assuming a constant denudation rate of between 50 and 100 mm ka^{-1}.

A more subtle form of case-hardening occurs on emerged coral reefs. These rocks are mechanically much stronger than dune sands or the chalky limestones of Puerto Rico, but can also be extremely porous, depending on the facies. Primary pore spaces within coral tend to be very large, typically centimetres or decimetres across. Dissolution near the surface leads to carbonate precipitation in the voids a few metres down the profile, banded flowstone and silt (from soil) being common deposits. In this way the reef rock in the upper vadose zone is rendered

less permeable as the terrain is denuded. The relief of uplifted atoll reef crests can be accentuated into 'limestone walls' when dissolution is greater beneath a soil-covered former lagoon floor in the interior than around the encircling, exposed, case-hardened main reef. Santo and Minamidaito islands provide good examples (Strecker *et al*. 1986, Urushibara-Yoshino 2003).

Case-hardening is particularly well developed on aeolianites (calcareous dune limestones) as discussed in section 7.9. Jennings (1968) recognized that lithification and karstification are likely to occur simultaneously, for the same agents are responsible for both. Therefore he proposed the concept of **syngenetic** karst development. Rain falling directly onto the dune sands is responsible for their case-hardening, but point recharge by allogenic streams on the inland side of the dune fields is responsible for the numerous caves and collapse dolines. Water flow channelled by the topography of underlying impermeable rocks can direct caves and 'underprint' alignment of collapse dolines at the surface (Twidale and Bourne 2000).

Induration in calcareous aeolianites is often characterized by vertical piping because dissolution and any re-precipitation occurs where percolation is guided by tree roots. The pipes formed are typically 0.3 to 0.6 m diameter and up to 20 m deep in Australian dune calcarenites

(Grimes 2002); similar 'pit caves' are found on young carbonate islands in the Caribbean (Mylroie and Carew 1995). A notable feature of the pipes in dune sands is that they are lined with an indurated skin. When they are truncated and stand in relief, they can be mistaken for petrified forests. Should the soil profile be deeply eroded, intervening solution residuals between the pipes can stand proud and give rise to impressively pinnacled 'tombstone' terrains (Figure 9.51); note the analogy with the thermal water sheaths shown in Figure 7.29b.

9.12.2 Tufa deposits, dams, terraces, waterfalls and mound springs

Springs, waterfalls and outflowing rivers in karst often display the precipitates known as **tufa** and **travertine** (sections 2.2 and 8.3). Overhanging cliffs may also be draped with tufa stalactites of diverse morphology (Figure 9.52). Many of the deposits and forms that are found arise partly through the intervention of biological activity, as do some erosional forms. Such features were termed **biokarst** by Viles (1984) whose suggested typology is presented in Table 9.4. T.D.Ford (1989), Ford and Pedley (1996) and Pentecost (1995) have reviewed the occurrence of travertine deposits in Europe, North America and Asia Minor.

Calcareous tufa is mixed with organic remains and so the relative importance of organic and inorganic processes in its formation is not immediately obvious. The roles of a range of cyanobacteria, algae and higher plants in the accumulation of tufa have been investigated by the Association Francaise de Karstologie (1981), Chafetz and Folk (1984), Viles and Pentecost (1999) and Carthew *et al.* (2006). Drysdale *et al.* (2002) examined hydrochemical factors influencing deposition. These studies demonstrated that both inorganic and organic deposition occur, but that organic processes are much more important than previously assumed. Chafetz and Folk provided convincing evidence that bacterially precipitated calcite forms a large percentage of the carbonate in many tufa and travertine accumulations in Italy and the USA, exceeding 90% of the framework grains comprising some lake-fill deposits. Their investigations found individual accumulations to range up to 85 m thick and to cover hundreds of square kilometres. They concluded that harsh environmental situations (e.g. hot geothermal water) favour inorganic deposits, while increasingly more moderate conditions favour organically precipitated material. Many other researchers have since confirmed these findings. Chafetz and Folk also recognized five morphological variations of surface tufa deposition: (i) waterfall, (ii) lake-fill, (iii) sloping mound or fan, (iv) terraced

mound and (v) fissure ridge. Waterfall or cascade deposits accumulate at the loci of both increased agitation and a place where algae and mosses can readily attach and grow. Tufa accumulation at such sites can produce dams and pond substantial lakes. The 100-m-thick travertines of Zafarraya Polje (section 9.9) are the remains of a great fan downstream of a dam.

The world's best known modern impoundment of water by tufa dams are the Plitvice lakes in Croatia (Bozicevic and Biondic 1999, Bonacci and Roje-Bonacci 2004), a World Heritage site. Sixteen lakes have formed along a 6.5 km reach of the upper Korana valley (Figure 9.53). They lie in a gorge immediately downstream of the confluence of two main tributaries, one flowing from dolomite and already containing tufa, and the other from limestone and without calcareous deposition. The saturation index is over 3 and pH between 8.2 and 8.4. An increase in Mg in already saturated carbonate waters is known to induce supersaturation when its concentration exceeds about 7%, because of the common ion effect (Chapter 3). Thus the mixing of the water from these two streams could be responsible for much of the deposition, although an ecosystem in which bacteria, algae and mosses play an important part is also significant (Chafetz *et al.* 1994). The tufa barriers are up to 30 m high, with compound dams holding back lakes as deep as 46 m. In the largest lake (Kozjak, 0.815 km^2), a drowned dam is found 4.6 m beneath the water surface; presumably a consequence of a downstream tufa dam growing upwards at a greater rate.

An outstanding example of dominantly inorganic deposition of calcite from geothermal water is found at the travertine terraces of Pamukkale (Figure 9.54), a World Heritage site in Turkey (Simsek 1999, Nicod 2002, Dilsiz and Günay 2004). In addition, impressive tufa-dammed lakes also occur in China at the World heritage sites of Jiuzhaigou and Huanglong in Sichuan Province (Sweeting 1995). The tufa deposition rate at Huanglong is estimated to be 0.1 cm a^{-1} (Yoshimura *et al.* 2004). Similar lakes with barriers up to 20 m high extend 15 km down a canyon at Band-i-Amir in Afghanistan (Brooks and Gebauer 2004).

Springs can produce stepped mound deposits, the water flowing through radially disposed pools dammed by tufa barriers similar to rimstone pools in caves. Sulphate and carbonate deposits often occur around artesian springs in arid regions, the spring water emerging at the top of a mound of its own construction; hence the term **mound spring**. The largest known is Solomon's Prison in Iran, which rises 69 m, and other examples occur in Australia. Elongate fissure ridges form where spring waters upwell through fissures running along mound crests. By contrast, tufaceous waterfall deposits can have a tapered dome

Figure 9.51 (Upper) Soil pipes dissolved in aeolian calcarenite (dune limestone), probably closely associated with former tree roots (that exude CO_2). (Lower) The 'Pinnacles' north of Perth, Western Australia. These are calcite cemented residuals developed between vertical soil pipes in indurated dune limestone, the overlying soil having been stripped off (mainly by wind) in the seasonally arid coastal environment.

Figure 9.52 (Upper) Bulbous tufa stalactites on the overhanging edge of a seasonally flooded enclosed depression, Northern Territory, Australia. Water submerges the face by 2–4 m in the wet season. (Photograph by D. Karp.) (Lower) Eccentric tufa stalactites suspended from overhanging cliff in Minas Gerais, Brazil.

Table 9.4 A tentative typology of biokarst forms (Reproduced from Viles, H. (1984) Biokarst: review and prospect. *Progress in Physical Geography*, 8(4), 523–42).

Erosional forms	Mixed erosional and depositional forms	Depositional forms
Phytokarst (Folk *et al.* 1973)	Calcareous crusts	Tufas and travertines
Directed phytokarst (Bull and Laverty 1982)	Case-hardening	Directed speleothems
	Lake crust and furrow systems	Moonmilk
	Degraded tufas	Stromatolites
Coastal biokarst		Reefs
Root grooves		
'Zookarst'		

form acquired from the trajectory of cascading water, up to 40 m high in the case of the Aquidaban Falls in the Bodoquena Plateau in Brazil.

Tufa and travertine deposits are often datable radiometrically (chiefly, by U series methods) or by means of their fossil fauna and flora assemblages. Their occurrence at palaeospring sites in dry regions is important in studies of early humans, because stone tools are quite frequently found within them (e.g. Schwarcz 1993). On terraces they can provide lengthy chronological and palaeoclimatological records, e.g. in central Spain (Martin-Algarra *et al.* 2003; Ordonez *et al.* 2005). Particularly notable is a case near Malaga, Spain, where Delannoy *et al.* (1997) have identified six large phases of travertine accumulation that they consider to represent episodes from the Messinian to the Holocene. Carthew *et al.* (2006) discuss fossil tufa facies in monsoonal Australia and emphasize their significance for palaeoclimatic reconstruction.

9.13 SPECIAL FEATURES OF EVAPORITE TERRAINS

Evaporite deposits are widespread (Figure 1.3) and support a wide range of karst features in all climates. Because of their great solubility (section 4.3) salt rocks are exposed only in the most arid or cold regions such as Death Valley, the Dead Sea, the Qinghai (Tibetan) plateau and the high Arctic islands of Canada. Even there, individual outcrops are limited to a few square kilometres at most. Gypsum is much more stable in outcrop than salt, but its karst is still best expressed where mean annual precipitation is low. Gypsum karst is widespread in the dry midwest–southwest of the USA, in the northern interior of Canada, the Arkhangel'sk, Bashkir and Perm regions of Russia and Ukraine, across

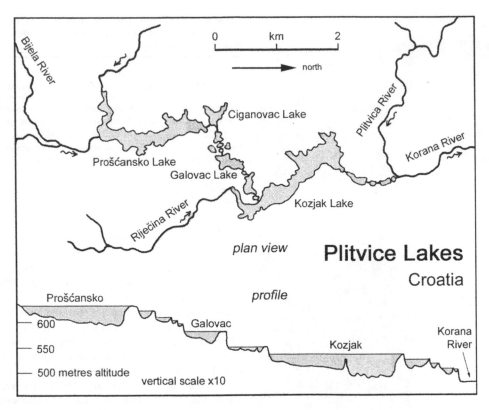

Figure 9.53 The Plitvice Lakes in the Korana valley, Croatia, are impounded by tufa dams. Reproduced from Bonacci, O. and Roje-Bonacci, T. (2004) Plitvice Lakes, Croatia, in *Encyclopedia of Caves and Karst Science* (ed. J. Gunn), Fitzroy Dearborn, New York, pp 597–598.

Figure 9.54 Calcite travertine (or sinter) terraces deposited by geothermal waters at Pamukkale, Turkey. Springs with temperatures up to 59°C emerge along a fault line and yield a flow of $0.39\,\text{m}^3\,\text{s}^{-1}$. Calcite deposition occurs by CO_2 degassing enhanced by turbulent flow.

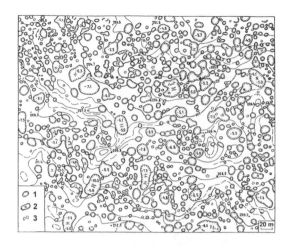

Figure 9.55 Typical gypsum karst doline landscape in the Iren River basin, Pre-Urals, Russia. The area mapped is less than 0.2 km². Reproduced with permission from Klimchouk, A. and Andrejchuk, V., Sulphate rocks as an arena for karst development. International Journal of Speleology 25 (3-4), 9–20 © 1996 International Congress of Speleology.

the Middle East and in northeastern China. Outcrops may amount to thousands of square kilometres.

Many classic gypsum karst studies were conducted in Russia and the Ukraine (Gorbunova 1979, Pechorkin and Bolotov 1983, Pechorkin 1986). Nicod (1976), Forti and Grimandi (1986), Cooper (1996) and Gutiérrez and Gutiérrez 1998) have presented western European work. Quinlan *et al.* (1986) and Johnson (1996) reviewed gypsum karst in the USA, and Ford (1997) in Canada. A global review has been edited by Klimchouk *et al.* (1996).

Evaporite terrains display many of the landforms typical of carbonate karst, including varieties of karren, dolines, blind valleys and poljes. Gorbunova (1979) asserted that dolines are the most widespread features in the gypsum karsts of the former Soviet Republic, and this is true elsewhere (Figure 9.55). Collapse and suffosion processes are more prominent in doline development than in most carbonate karsts (Figure 9.56), the former because interstratal dissolution is of much greater extent and the latter because, in Russia, Ukraine and Canada, great tracts are veneered with glacial debris or loess. Nevertheless, wholly solutional dolines are also common. In parts of the Pecos Valley of Texas and New Mexico and in the Pinega valley of northern Russia they are packed to form a high-density polygonal karst (Figure 9.57). Sauro (1996) provided an excellent aerial illustration of a doline landscape with a honeycomb structure in the Baisun-Tau mountain area between Uzbekistan and Tagikistan; and Dogan and Yesilyurt (2004), Klimchouk (2004) and Waltham *et al.* (2005) illustrated polygonal karst in Turkey. Günay (2002) described associated lakes and springs.

In the gypsum karsts of Perm and Bashkir, Gorbunova (1979) reported doline densities of 32 and 10 km⁻² respectively, although densities up to 1000 km⁻² sometimes occur at the crown of folds or at contacts with other lithologies. Densities of 1100–1500 km⁻² occur in the Italian Alps where hydraulic gradients are high (Belloni *et al.* 1972). Dolines there have a mean diameter of only 5 m. The densely packed schlotten type of doline or large karren shaft have mean diameters of 0.5–1.5 m and depths of 0.3–3.3 m in Newfoundland and Nova Scotia. Nearest-neighbour *R* values for 21

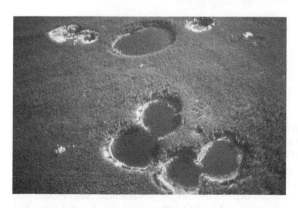

Figure 9.56 (Left) Doline ponds in Wood Buffalo National Park, northern Alberta, Canada. These are collapse dolines propagated through dolomite cover beds as a consequence of interstratal dissolution of gypsum. (Parks Canada photograph) (Right) Vermilion Creek doline, near Norman Wells, Northwest Territories, Canada; Latitude 65°N. This spectacular feature measures 180 × 100 m and is approximately 40 m deep to the waterline. It is a collapse of Holocene age through calcareous shales overlying gypsum. (Photograph by R.O. van Everdingen.)

Figure 9.57   (Upper) Intensely dissected solution forms in gypsum that are transitional between karren field and polygonal karst. For scale, the trees in the background are 5–7 m in height. Pinega River valley, east of Arkhangel'sk, Russia. (Photograph by V.Nikolaev.) (Lower) Gypsum tent produced by expansion following hydration of anhydrite to gypsum, Sorbas, Spain. (Photograph by J. Calafora.)

different sites there mostly fall between 1.5 and 2.0, i.e. there is regular packing of sink-points (Stenson 1990). Hypothetical densities extrapolate to $10\,000\,\text{km}^{-2}$, but here (and in the Pinega gypsum karst near Arkhangel'sk, Russia) they are confined to escarpment edges where hydraulic gradients are greatest (Figure 9.58). Elongated closed depressions along escarpment edges are also referred to as **karst trenches** (Klimchouk and Andrejchuk 1996).

Breccia pipes are common features of gypsum karst that are created by progressive stoping above sites of interstratal dissolution (Figure 9.59). Although well

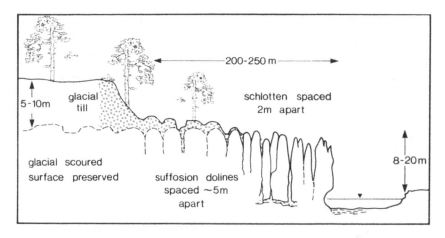

Figure 9.58 Model for the progressive development of schlotten topography in massive gypsum near Windsor, Nova Scotia, Canada. The features are of Holocene age.

developed in carbonate rocks, they are most abundant and largest above gypsum/anhydrite and salt. Breccia pipes may exhibit one of four dynamic/topographic states:

1. active, and propagating upwards towards the surface but not yet expressed there;
2. active or inactive, expressed at the surface as a closed depression or a depression with a surface outflow channel;
3. inactive, and buried by later strata (= palaeokarst);
4. inactive and standing up as a positive relief feature because the breccia (probably cemented) is more resistant than the upper cover strata.

Quinlan (1978) reviewed the nature and distribution of an estimated 5000 breccia pipes over salt or gypsum in the USA. They range in diameter from 1 to 1000 m and in depth up to 500 m. Similar features interrupt the potash mine workings of Saskatchewan (Figure 9.60) where they may propagate from depths as great as 1200 m (i.e. beneath 1000 m or more of cover strata). Numerous examples of deep breccia pipes are also known from China.

Solution subsidence troughs (Olive 1957) are elongated depressions created by interstratal solution. The largest solution-induced depositional basins tend to occur along the margins of the great interstratal halite deposits, creating a solution form that may be represented by a shallow **salt slope** at the surface (Figure 9.59). Dissolution can begin as soon as the salt is buried, and the immediately overlying strata (usually, dolomites, gypsum/anhydrite or redbeds) may be comprehensively

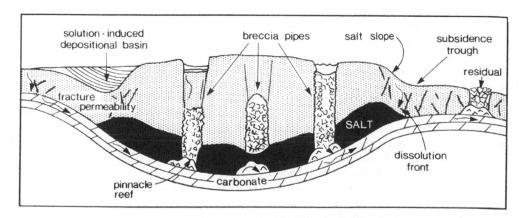

Figure 9.59 Model for the development of breccia pipes and residuals, subsidence troughs and solution-induced subsidence basins.

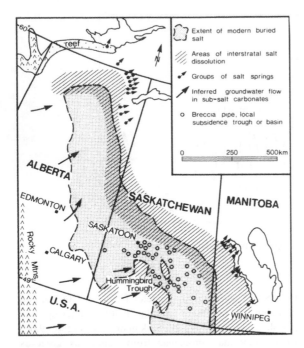

Figure 9.60 Features of the interstratal dissolution of the Elk Point (Devonian) salt deposits beneath the Prairie Provinces of Canada.

brecciated. Figure 9.60 shows the Elk Point Formation of Canada, a sequence of salts with lesser gypsum and redbeds that accumulated to thicknesses of 50–500 m in a lagoon barred by the Presqu'ile Reef (Figure 2.12) during the Devonian Period. It is now at a depth of 200–2400 m beneath later carbonate and clastic rocks. Where the burial is shallow along the eastern edge, the dissolution front has receded an average of 130 km along a distance of 1600 km. The Hummingbird Trough is a re-entrant in the deeply buried southwest side, where 200–300 m of salt has been entirely removed over an area of 25 000 km^2 (De Mille *et al.* 1964). The Trough is now inert, a palaeokarst feature, but the eastern dissolution front continues to recede. Its mean rate of recession, averaged over the 365 Myear since Late Devonian times, is 36 cm a^{-1}.

Many small examples of solution subsidence troughs occur in the gypsum plain south of Carlsbad Caverns, being 0.7–15 km in length, 100–1500 m wide but no more than 5–10 m deep (Quinlan *et al.* 1986). Larger troughs tend to be infilled by terrigeneous or other sediments and so lack topographic expression in most instances. Quinlan (1978) termed them **solution-induced depositional basins**. They are noted in the palaeokarst reports of

many nations. Many continue to be active but do not appear as strong topographic depressions, because the rate of sedimentation approximately equals the rate of subsidence. Intermediate-scale samples (5–100 km long, 5–250 km wide and with 100–500 m of subsidence and sedimentation) are reported in Canada (Tsui and Cruden 1984), New Mexico and Texas (Bachman 1976, 1984), and Spain (Gutiérrez *et al.* 2001). Pleistocene and Holocene rates of subsidence in active examples in these regions are estimated at 5–10 cm ka^{-1}.

9.13.1 Positive relief features created by diapiric or hydration processes

Salt has a low density (2.16 g cm^{-3}). Consequently, when considerable deposits of it are buried by denser strata it deforms in a cellular manner and flows upwards into them (section 2.3; Jackson *et al.* 1995, Alsop *et al.* 1996). The displaced salt approaches or emerges at the surface in the form of stocks (diapirs), dikes and sills (canopies). Diapirs are also known in gypsum and have been described in the Andes (Salomon and Bustos 1992) and Spain (Calaforra and Pulido-Bosch 1999). These are not karst landforms, but they support solution features that are.

Diapirs are initiated beneath 2000 and 10 000 m of cover strata and rise episodically. Calculated rates range from 0.1 to 1.0 mm a^{-1} in the Gulf of Mexico to 4.6 mm a^{-1} in northwest Yemen and 6–7 mm a^{-1} in Israel. Some of the most active modern diapir formation occurs in the Zagros Mountains of Iran, where convergence of European and Arabian crustal plates is squeezing deep salt and extruding it more rapidly. The crests of diapirs in the Great Kavir Province of Iran rise 1.5 km above the desert surface.

Emerged salt diapirs generally range from 2 to 20 km in diameter. In wetter climates they are decapitated by groundwater dissolution, but may still create fractured dome-shaped hills up to 100 m in height because of the displacement of insoluble superficial rocks. In dry and cold regions relief of 500 m or more can be created, with an exposed core of salt (Figure 9.61).

Where the extrusion is slow the salt displays rillenkarren, wallkarren, schlotten and pinnacle karst. Where it is more rapid it flows like macrocrystalline ice to create salt glaciers or **namakiers** (from *namak*, the Farsi for salt). These display standard glacier features such as crevasses, icefalls and ogives (overthrust ridges). Salt-glacier flow rates in the Zagros Mountains average a few metres per year, i.e. one or two orders of magnitude slower than in conventional mountain glaciers formed of ice. The salt flow is episodic, resulting from recrystallization

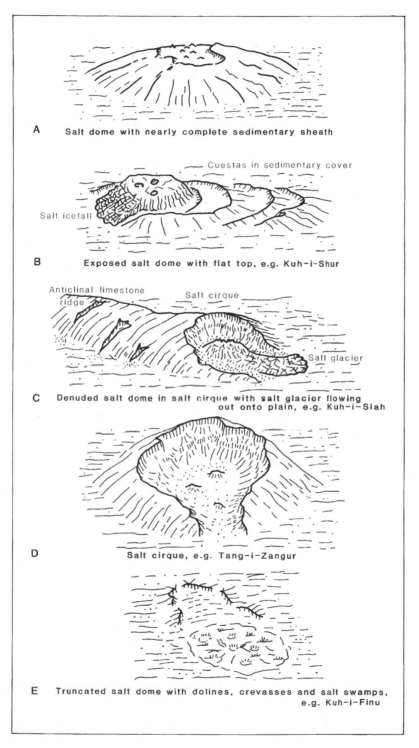

A   Salt dome with nearly complete sedimentary sheath

Cuestas in sedimentary cover

Salt icefall

B   Exposed salt dome with flat top, e.g. Kuh-i-Shur

Anticlinal limestone ridge

Salt cirque

Salt glacier

C   Denuded salt dome in salt cirque with salt glacier flowing out onto plain, e.g. Kuh-i-Siah

D   Salt cirque, e.g. Tang-i-Zangur

E   Truncated salt dome with dolines, crevasses and salt swamps, e.g. Kuh-i-Finu

Figure 9.61   Surface expression of salt domes in the Laristan Desert, Iran. Reproduced from Jennings, J.N., Karst Geomorphology © 1985 Blackwell Publishing.

following the addition of water (Urai *et al.* 1986). Water content as low as 0.1% by weight can induce the process; thus it may operate sporadically in the driest climates.

Anhydrite is hydrated to gypsum by a sequence of dissolution and reprecipitation (section 2.4), increasing the volume by about 63% under open-system conditions at shallow depths where water may enter or leave the system (James 1972). Hydration expansion and/or the viscous (diapiric) flow of the newly formed gypsum can create many landforms. Features such as gypsum tents (below) appear to be attributable to hydration alone, probably also involving weathering crust formation by recrystallization, while undoubted gypsum intrusions are commonplace in highly deformed redbed and dolomite sections. Weathering crust features on gypsum surfaces are described by Macaluso and Sauro (1996).

Gypsum tumuli (bubbles or tents) are hollow domes of freshly formed gypsum that are round or elliptical in plan (Breish and Wefer 1981, Pulido-Bosch 1986, Macaluso and Sauro 1996). Small examples are a few tens of centimetres in diameter, but most described in the literature are 2–10 m in diameter and up to 2.5 m in height (Figure 9.57). These are hydration features created by compressive and shear stresses that separate the gypsum crust from underlying gypsum or anhydrite. Stenson (1990) studied their fresh development in the floor of a gypsum quarry in Nova Scotia that had been abandoned only 35 year earlier. In an area of a few hectare there were 69 new tents, ranging 0.8 to 8.2 m in diameter. Crustal hydration was apparent, but here the process may have been aided by mechanical 'pop-up' due to lithostatic pressure release in an artificial excavation.

Karst domes in the New Mexico gypsum karst are larger features, up to 200 m in diameter and 10 m in height. The cores are of gypsum and insoluble residue, with disturbed dolomite or clastic beds or calcrete crusts draped around their annular rims. Bachman (1987) suggested that they are the surficial remnants of pervasive near-surface salt dissolution that has caused general subsidence with projections remaining above insoluble remnants; the domes being part of those remnants.

More dramatic are domes and anticlines in the gypsum terrains of northern Canada described by van Everdingen (1981) and Tsui and Cruden (1984) and similar features in the Arkhangel'sk gypsum karst of Russia (Korotkov 1974). These features range from 10 to 1000 m or more in length or diameter, and up to 25 m in height in a majority of cases. Many are highly fractured, with individual blocks being displaced by heaving and sliding. At the extreme they become a megabreccia, an upthrust jumble of large blocks. The largest reported Canadian example is

a steep-limbed anticline that extends along the shore of a lake for a distance of 30 km and is up to 175 m in height. Its crest is marked by 'chaotic structure and trench-like lineaments' (Aitken and Cook 1969). These features may be due to post-glacial hydration processes or to diapiric injection of gypsum during times of rapidly changing glacial ice loading (section 10.3). These are also regions of widespread modern permafrost, so that accumulation of ground ice in initial fractures probably contributes to the heaving and other displacement.

A sabkha sequence of 250–300 m of dolomites and anhydrite underlies 50 000 km^2 of the Mackenzie River valley around 66°N in Canada. Where exposed in outcrop in mountain ranges on either side, hydration and dissolution have reduced this to a multiphase dolomite breccia cemented by calcite and with minor residual gypsum (Figure 2.9). It thins progressively from ~140 m at new outcrop to zero higher up the mountains. The top few metres are further case-hardened by evaporitic precipitation. The hardened crust is breached where sinking streams are able to penetrate the ~50 m of permafrost below. This produces a spectacular 'dissolution drape' topography of fragments of the crust tilting and sliding down into the karst depressions (Hamilton and Ford 2001).

The final category of positive relief features associated with evaporite karsts are cemented sinkhole or breccia pipe fillings that come to be exposed as residual hills when strata surrounding them are preferentially eroded, as noted above. Breccia pipe residuals are normally a few tens or hundreds of metres in diameter and rise 5–40 m above the general land surface. More than 1000 such hills are mapped in the plains of western Oklahoma (Fay and Hart 1978). Shallow hills with a doughnut form (a central depression) in that region are believed to be doline fillings, rather than breccia pipe fillings (Myers 1962).

Castiles are steeper, irregular masses of secondary calcite rising 3 to 30 m above the Gypsum Plain south of Carlsbad Caverns, New Mexico (Kirkland and Evans 1980, Hill 1995). These formed to replace gypsum locally at the base of the Castile Formation at the start of the H_2S-generating process described in section 3.6 and Figure 3.11. They are exposed by the preferential dissolution of the remaining gypsum along the margins of the Plain.

9.14 KARSTIC FEATURES OF QUARTZOSE AND OTHER ROCKS

The dissolution of silicate rocks yields a range of landforms that vary in scale from karren to large enclosed depressions, although formation of the larger features is aided by mechanical erosion. Robinson and Williams

(1994) reviewed such features in Europe and a global review has been provided by Wray (1997). Dissolution forms on silicate rocks are found most widely in warm temperate to tropical regions, where the legacy of long periods of chemical denudation has not been destroyed by vigorous mechanical processes such as glaciation, although karren forms on quartz sandstones and quartzites are also well known in colder regions. Joints and other fissures provide the avenues by which water penetrates quartzose rocks. Dissolution then proceeds along crystal boundaries.

Martini (1979) expressed the opinion that karst cannot normally develop on quartzitic rocks, because released quartz grains then require removal by mechanical processes. However, we explained in section 1.2 that landforms developed on these rocks can be considered karst if they are produced predominantly by dissolution, but are fluvio-karst when the intervention of running water is essential for their development. Landscapes on quartzites and siliceous sandstones clearly occupy the transition zone between normal fluvial landscapes and karst. Drainage remains surficial in most cases and over most of the area. There is little dense epikarst development on quartzites, implying that deep solutional attack is restricted to the biggest fractures and bedding planes, so that any karst that develops is comparatively impoverished.

9.14.1 Karren in basalts, granites and quartzites

Although quartz and silicate minerals have low solubility in water at surface pressures and temperatures (see Chapter 3), given enough time and not too vigorous effacing effects from competing processes, such as freeze–thaw or sand abrasion, the result of dissolution will become apparent, and is commonly expressed in dissolution pans and hydraulically controlled linear forms of karren such as rundkarren and rinnenkarren. These can develop quite quickly as is evident from flutings coursing down the sides of neolithic granite menhir in Brittany, France, where rinnenkarren developed at the rate of a few tens of millimetres per thousand years (Lageat *et al.* 1994). See Williams and Robinson (1994) for a review of weathering flutes on siliceous rocks in Europe. Rundkarren are also commonly found on basaltic lava flows in humid warm temperate regions such as Northland, New Zealand.

Martini (2004) has reviewed silicate karst research and pointed out that the most spectacular karren features are developed by weathering under soil; a process that is conducive to progressive formation of sand and clay along joints and bedding planes. By erosion of the soft, deeply weathered material, he explained that unaltered or only moderately weathered quartzite rock is then left standing as irregular pinnacles (Figure 4.10). He noted that in areas with dry seasons the pinnacles are often indurated at the surface by iron oxides or by an opal matrix deposited by evaporation. Fields of pinnacles formed in this way in South Africa were estimated to have evolved in less than two million years.

9.14.2 Caves and closed depression in quartzites

Most caves in quartzose rocks are developed by an essential combination of dissolution and mechanical erosion by running water. They are therefore fluvio-karst features. Although dissolutional widening of joints and bedding planes is critical to their inception, under surface temperatures and pressures mechanical processes are always heavily implicated as the dominant agent of stream passage enlargement. Nevertheless, although caves in quartzite are not normally pure karst features (quartzite caves produced by hydrothermal dissolution are the exception), the silicate speleothems sometimes found in them are, because they are formed primarily from the products of dissolution.

Cave networks develop close to the escarpment edges of quartzitic plateaux, especially on the downdip side, and are usually associated with unloading fractures and gravity tectonics (see section 7.3.10 and Figure 7.18). Martini (2004) noted that these systems do not usually extend more than a few hundred metres from resurgences, although exceptionally they may be several kilometres; e.g. Sistema Roraima Sur in Venezuela is 10.82 km long (Urbani 2005). Most frequently these caves tend to run parallel with or develop close to plateau edges rather than develop under central parts. However, the widened fissures capture surface streams and so enable an invasion style of speleogenesis. The majority of quartzite caves are vadose and active, although abandoned dry levels are known. Fissuring often guides water to the base of the quartzose formation, where caves may develop at the contact with underlying impermeable rocks. Speleogenesis is by piping through zones of deeply weathered quartzite that has been transformed into friable neosandstone by the process illustrated in Figure 4.10. Thus two steps are involved: weathering first, followed by mechanical removal of sand. The weathering stage may be extremely protracted. The onset of the second stage requires triggering by uplift or regional incision, because it requires a hydraulic gradient steep enough to generate turbulent water flow of sufficient velocity to permit the entrainment and transport of sands. Piping and removal of sand starts at the spring and works headwards. Hence most caves in

Figure 9.62 Stylized depiction of landforms and water circulation on a small plateau (tepuy) in the quartzites of Roraima, Venezuela. Surface drainage from the centre of the plateau is absorbed down joints around its margins: 1, remnant of an old erosion surface; 2, tepuy surface; 3, lower platform; 4, ledge; 5, cliffs; 6, large towers in the dissected edge; 7, towers and mushroom rocks on the upper surface; 8, small towers amongst channel systems; 9, streams and ponds; 10, canyons and fissures at plateau edge; 11, fissure system; 12, depressions; 13, depression truncated by cliff retreat; 14, collapsed depression; 15, elongated shaft in fracture zone absorbing surface streams; 16, talus; 17, resurgences from caves; 18, waterfalls; 19, talus at cliff foot; 20, resurgences from bedding planes. Reproduced from Galan, C. and Lagarde, J. (1988) Morphologie et evolution des caverns et formes superficielles dans les quartzites du Roraima (Venezuela). Karstologia, 11–12, 49–60.

quartzose rocks are vadose features. They are preferentially developed in sites where high hydraulic gradients are attainable, and they are excavated mechanically along previously weathered fissures. In most cases passages show rectangular cross-sections, but some have low arched roofs. Martini (2004) observed that a characteristic feature of such caves is extreme variation in the size of a single passage which, in a downstream direction, can narrow from the order of 10 m to much less than 1 m. He explained this as an outcome of variation in the degree of quartz weathering. Caves in quartzose rocks are known in Venezuela (Urbani and Szcerban 1974, Pouyllau and Seurin 1985, Galan and Lagarde 1988, Briceño and Schubert 1990, Šmida *et al.* 2005, Urbani 2005), Brazil (Corrêa Neto 2000), South Africa (Martini 1987), Australia (Jennings 1983, Wray 1997, Young and

Young 1992) and Saharan Africa (Busche and Erbe 1987, Busche and Sponholze 1992).

Most drainage in silicate rock landscapes remains on the surface. However, enclosed depressions can occur particularly near the edges of quartzite plateau. These include swallow holes, rift potholes over open joints and collapse dolines over caves (Figure 9.62). Such dolines in the Precambrian orthoquartzite plateaux of Venezuela can be 300 m in diameter (Pouyllau and Seurin 1985), and they overlie caves of huge dimensions, the deepest in the region descending 383 m. Although solution pans can be large, the occurrence of a significant epikarst in quartzite has not been demonstrated, and so the hydrological circumstances that could give rise to the genesis of solution dolines appear to be absent. However, some enclosed depression in quartzose rocks in Venezuela and the Sahara have been reported up to several square kilometres in area. They are point-recharge depressions similar in morphology to shallow blind valleys, but are entirely autogenic. Dissolution may have played a significant part in their origin.

9.14.3 Residual towers and beehive hills

Deep and prolonged weathering along joints provides avenues that can be exploited by surficial erosion processes that sluice out the fines. Enlargement of intersecting joints by such processes can yield tower form residual hills up to 30 m or more in height (Figure 4.11) that may take tens of millions of years to evolve, although they are difficult to date. These 'ruined cities', which sometimes resemble the ruiniform relief of karst in dolomitic limestones, are a common feature of Upper Proterozoic siliceous cemented sandstones of Arnhem Land in Northern Australia (Jennings 1983, Wray 1997) and orthoquartzite plateaux in Venezuela (Briceño and Schubert 1990). Just as the style of carbonate cone and tower karst varies with the structural attributes of the rock, especially bedding thickness and joint frequency, so this also affects the form of residual quartzose hills which vary from beehive-like to tower-like. However, recalling the discussion about tower karst development in section 9.11, there is no question about quartzose towers being developed from a previous cockpit karst phase. They can only form directly, and mainly by runoff processes exploiting lines of weakness afforded by a network of weathered joints. So although ruined cities and beehive hills occur in quartzose landscapes, and quartz dissolution may have been a critical initial step in their development, such features are not simply karstic, and nor is the groundwater network beneath them. Nevertheless, quartzite landscapes have some morphologically similarities to some arid-zone limestone landscapes with the low solubility of quartzite substituting for lack of water in limestone desertsf.

9.15 SEQUENCES OF CARBONATE KARST EVOLUTION IN HUMID TERRAINS

9.15.1 Early ideas

At the turn of the 20th century the influential ideas of the American geomorphologist W.M. Davis on the cycle of erosion in 'normal' landscapes were reverberating through the world of geomorphology. In Vienna, which is on the doorstep of the Dinaric karst, one of the best known European geomorphologists, Albrecht Penck, had amongst his students Jovan Cvijić (Figure 1.5), who is considered by many to be the father of modern karst geomorphology. Hence, the landforms of what we now regard as the 'classic' karst provided the main source of inspiration from which concepts of karst landscape evolution emerged. Imagine the excitement and stimulation of a fieldtrip in 1899, when Penck and his students accompanied by Davis set out to investigate the karsts of Bosnia and Hercegovina. Roglić (1972) considered this meeting in the karst of two outstanding masters of geomorphology to have been of decisive importance for the further development of geomorphological concepts, although we should not forget that arguably the most influential work ever written on karst had already been published by Cvijić in 1893.

Penck and Davis both contended that karstification is preceded by an episode of fluvial erosion, an idea with which Cvijić concurred. The problem then remained to identify the erosional stages through which the karst landscape progressed. Although Richter (1907), Sawicki (1909) and Beede (1911) first offered solutions, a scheme proposed by Grund (1914) claimed most interest. His theoretical sequence of landscape evolution took into account both the Dinaric region, with which he had first-hand experience, and the humid tropical karsts of Jamaica and Java, which he visualized from the writings of Daneš (1908, 1910). His scheme is shown in Figure 9.63. It depicts a doline karst of several generations in which individuals enlarge, coalesce and gradually consume intervening residual hills. The terrain is ultimately reduced to a corrosional plain.

Cvijić's own thinking evolved significantly on the question of karstic evolution. In his 1893 monograph he suggested a genetic sequence involving amalgamation from dolines to uvulas to poljes, but it was not until 1918 that he published his considered opinion on the morphological evolution of karst and its relationship to subterranean hydrology. In this work he drew attention to hydrographic zones within karst and pointed out that subterranean evolution can proceed without the intervention of baselevel change, because karstification itself leads to the lowering of the hydrographic zones as permeability progressively increases at depth. Whereas Grund attempted to produce a universal model, Cvijić's scheme of karstic evolution (Figure 9.64) was proposed for the Dinaric karst only; and whereas Grund's model assumed an indefinite thickness of limestone, with clear implications for uplift and rejuvenation, Cvijić's focused on a sequential development that terminates on impermeable underlying beds.

The conceptual framework provided by these early geomorphologists persisted essentially unchallenged until the emergence of ideas on the critical influence

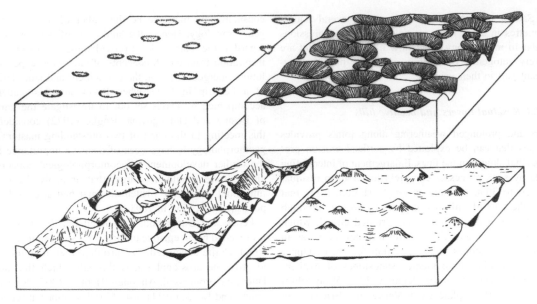

Figure 9.63   The karst cycle according to Grund (1914).

of climate on landform evolution. This commenced in 1936 with publication of H. Lehmann's observations in Java and thereafter followed a series of perceived schemes of karst landscape evolution, each tailored to the specific climatic zone in which they occurred. Zhu's (1988) schematic model of the evolution of humid subtropical fengcong and fenglin karst in the Guilin region of China is a particularly good example (Figure 9.65).

9.15.2 *Alternative conceptual models*

One of the major difficulties with proposed models of karst evolution is that they do not adequately accommodate the range of hydrogeological and geomorphological circumstances that occur. For a comprehensive summary of the possibilities that incorporates eogenetic and mesogenetic karst activity in maturing rocks, see Klimchouk and Ford (2000). Here we focus our attention on surface landform development at the telogenetic stage. Karstic erosion may be envisaged as commencing from one of three main starting points, i.e. cases:

(1) uplifted unkarstified dense rock protected by impervious cover beds;
(2) uplifted unkarstified rock of high primary porosity with no cover beds;

(3) uplifted rock karstified in a previous erosion phase.

There are two important variants of the first case (Figure 9.66):

(1a) stratification is horizontal or dips upstream and the impervious cover beds are stripped down and back from the spring (output) boundary;
(1b) the strata dip downstream and the caprock is stripped down and towards the spring boundary.

Highly tectonized terrain with complex geology can often be subdivided into more simple sectors such as the above two, although continuing uplift and tilting during karstification presents special problems. **Stripe karst** in steeply dipping strata adds the complexity of indefinite thickness.

Case 1a

The important point here is that surface karst landforms greater than superficial karren cannot evolve until subterranean connections have been established from input to output boundaries. Having acquired a hydraulic potential and output boundary by deep incision of an allogenic trunk stream, denudation of the caprock begins to expose the carbonate formation beneath. A

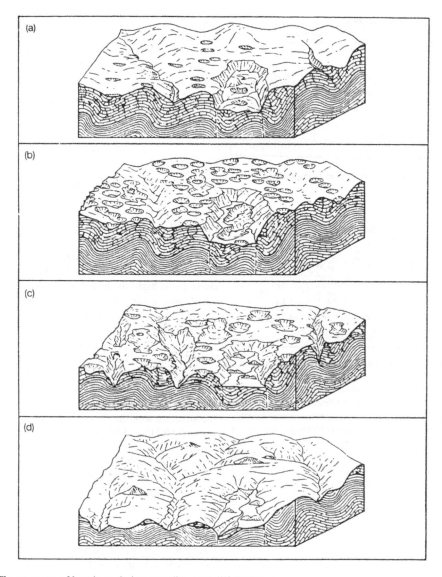

Figure 9.64 The sequence of karstic evolution according to Cvijić (1918).

first rank of point-recharge depressions develops at the input points where the first throughput connections are made. As the caprock retreats upstream, successive ranks of new connections form, each supporting new dolines along the lines depicted in Figure 7.9. By this stage the karst has become one of multiple inputs and multiple ranks, with caves at different levels if local baselevel has also changed. During and following complete stripping of the cover beds, the epikarst will develop. The initial solution dolines will grow in plan size, but may be limited in their extension or subdivided by development of daughter dolines above new leakage routes in the epikarst. Where fissure frequency is high and soil is thin, diffuse autogenic recharge down innumerable fissures coupled with the absence of a significant capillary barrier in the epikarst promotes the development of a karren field morphology rather than a doline karst.

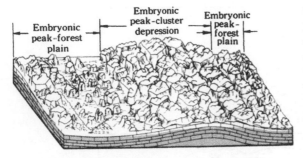

Peak-forest geomorphology system being formed

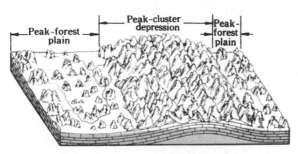

Sequential development of peak-forest system

Figure 9.65 Development of humid subtropical cone (feng-cong) and tower (fenglin) karst in the Guilin district of Guangxi, China. Note that the sequential development accommodates synchronous evolution of peak-cluster depression and peakforest plain styles in neighbouring areas. The peak-cluster depression style develops where the water table remains at depth below the surface; whereas the peak-forest plain case occurs where the water table is at shallow depth and rivers are at the surface. Reproduced from Zhu Xuewen 1988. Guilin Karst, p188© 1968 Shanghai Scientific and Technical Publishers.

Case 1b

The same principles apply, but with some important variations. Because strata dip downstream, the limestone is first exposed at its upstream boundary. Successive ranks of inputs then migrate downdip and downstream with the contracting caprock. Stripping is therefore towards the spring into a previously established deep and well-drained vadose zone. Any underlying impervious beds are also first exposed at the upstream boundary; thus the karst area contracts as the input boundary migrates downdip. In Case 1a the oldest surface landforms are closest to the output margin, whereas in Case 1b they are closest to the input margin.

Field examples of Case 1a are found in Yorkshire and Derbyshire, England; Fermanagh, Ireland; the Dordogne, France; eastern Kentucky, Tennessee and West Virginia,

USA. The most celebrated instance of Case 1b is the Mammoth Cave–Sinkhole Plain region of Kentucky (Figures 6.41 and 7.11). The change in landform texture across part of this area is depicted in Figure 9.67. In these cases, landscape evolution follows a sequence rather than a cycle, because the effective thickness of carbonates is limited and no further karst development is possible once erosion has removed them.

Case 2

Here we may imagine the extensive surfaces exposed on coral reefs during Pleistocene glacio-eustatic low sea levels. This is an eogenetic situation and is discussed more fully in Chapter 10. Essential points are that the rock possesses a high density of openings of all types that provide **ready made** hydrological connections from the recharge zone to the output boundary. Hydraulic conductivity is high, its spatial variability is comparatively small, and water retention in the epikarst is minimal. Most caves form at the water table and at the freshwater:saltwater interface, although glacio-eustatic fluctuations have forced this activity to be located at different levels. Autogenic karst of this type never becomes highly developed because of very high primary porosities, although where denudation exposes an impermeable inlier, then point recharge of allogenic streams transforms the situation with the development of large State 4 caves (Figure 10.21) and blind valleys merge to form interior lowlands (small poljes). Collapse above caves becomes common, but solution dolines remain rare. The Cook Islands in the tropical Pacific provide an interesting example (Stoddard *et al.* 1985, 1990).

Case 3

The course of karstic evolution following uplift of a previously karstified surface is strongly influenced by inheritance. The uplifted surface may be a previously baselevelled plain (Figure 9.38) or a karst with more relief (Figure 9.40), but in each case the drained phreas provides an instant vadose zone with ready made connections from input to output boundaries. Any inherited topography will guide runoff underground and hence guide solution in the epikarst. Residual hills become incorporated into the topographic divides of the emerging rejuvenated karst (Figure 9.39c). Minor uplift results in development of a new corrosion plain with dissected fragments of the former surface left as terraces, whereas major uplift and deep gorge entrenchment by allogenic trunk rivers may lead to incision below the base of karstification in the former phreatic

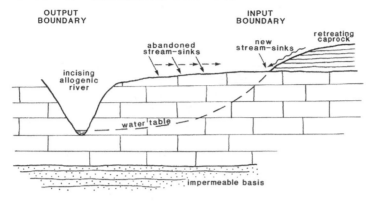

CASE 1A HORIZONTAL STRIPPING

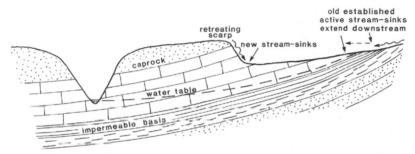

CASE 1B DOWN-DIP STRIPPING

Figure 9.66 Karstic evolution in uplifted dense carbonate rock protected by cover beds: Case 1a horizonal beds; Case 1b strata dip downstream.

zone. This leaves new vadose tributaries suspended above the trunk river. Their long profiles consequently steepen downstream as the gorge is approached, and undercapture (subterranean knick-point recession) works upstream. When the water table eventually lowers upstream, surface incision can commence. Hence underground rejuvenation is an essential precursor of surface rejuvenation. Water-table lowering occurs in two stages: initially by gravity drainage of the former phreatic zone during incision of the trunk river and later as a result of development of secondary permeability in previously unkarstified rock. The margins of the Guizhou plateau, China, offer outstanding examples of this pattern of development (Smart *et al.* 1986, Song 1986, Ahnert and Williams 1997).

9.16 COMPUTER MODELS OF KARST LANDSCAPE EVOLUTION

Readers are referred to Ahnert (1996) for a discussion of the purpose of modelling in geomorphology. Process models were discussed in Chapter 4, hydrological models in 6.11, and morphometric models that describe the two-dimensional statistical characteristics of karst landscapes in section 9.7. Conceptual models as discussed in section 9.15 are based on empirical observations and represent their authors' understanding of karst evolution. They are representations of reality that are intended to encapsulate the essential elements of the landscape; detail assumed irrelevant is ignored. However, because the models are static time-slice representations of landscape development, they do not permit us to explore the relative importance of factors that guide landscape evolution. Therefore in this section we consider theoretical three-dimensional computer models of the landscape that are designed deductively from general principles based on results of prior empirical research. These enable us to investigate variables influencing karst landscape evolution over time.

Topographic profiles across cockpit karst have long been recognized to resemble sinusoidal waves. Using this as a starting point Brook (1981) developed

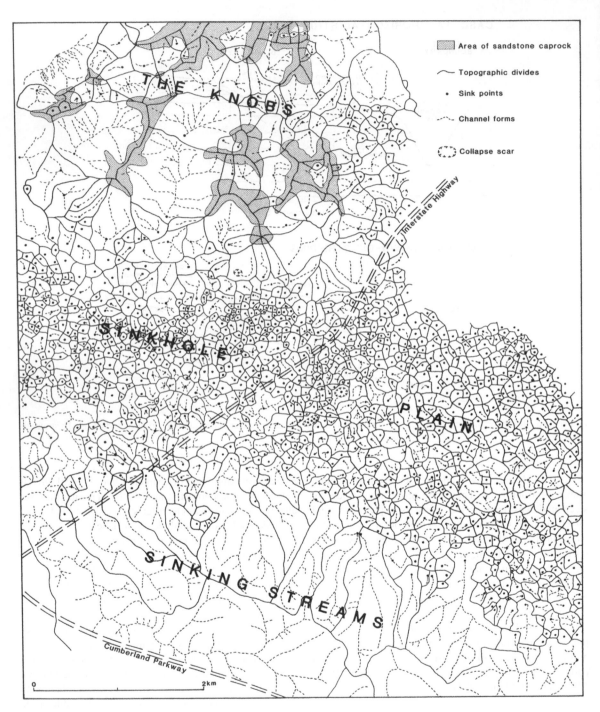

Figure 9.67   Morphological map of part of the Mammoth Cave–Sinkhole Plain district showing the change in closed depression texture from the youngest point-recharge depressions amongst the Knobs, where sandstone caprock is being pierced, to the oldest features at the updip Sinking Streams boundary, where underlying impermeable beds are being exposed. The situation is as depicted in Figure 9.66 (b), with the groundwater hydrology shown on Figure 6.39.

three-dimensional models of such landscapes by representing karst forms along intersecting fracture sets by intersecting sine waves, the wavelength being equal to the spacing of dominant fractures and the amplitude being determined by the ratio between the rate of vertical and horizontal dissolution. The latter is dependent mainly on climate and rock strength, and was taken by Brook to be revealed by the depth:diameter ratio of enclosed depressions. Different styles of karst relief could be replicated, therefore, by varying the depth:diameter ratio. This was an interesting start, but without a built-in process function and feed-back mechanisms could not give insight into the factors influencing the development of the relief over time.

Ahnert and Williams (1997) developed a three-dimensional process–response model to assess (i) the minimum requirements for karst landscape evolution, (ii) the effect of different starting conditions on end-stage landforms, and (iii) whether different landform types are the result of different environments or merely represent successive stages under unchanging conditions. Variables tested that influence the rate and location of dissolution included topographic control, structural control, flow divergence and slope of the initial surface. Locally higher dissolution rates caused by flow convergence were found sufficient to explain the formation of solution dolines and polygonal karst, but not cone or tower karst, which require dissolution rates to be lower at points of flow divergence. In different runs of the model, cones (or towers) were never found to develop directly, but were always derived from interdoline residual hills. This result has an important bearing on our understanding of how tower karst evolves, clearly favouring the sequential hypothesis explained in section 9.11. To explain the model landscapes developed, there was no general need to invoke any climatic factors, except sufficient rainfall to permit dissolution.

The main components of the process–response elements of Ahnert and Williams' model are illustrated in Figure 9.68. This was applied to a simple autogenic karst system represented by an uplifted limestone block with already established input to output connections. All rain passes through the system without overflow and there is no storage. Dissolution rate is proportional to the amount of runoff. The model surface is an X–Y grid with fixed plan coordinates and variable elevation. Runoff follows the steepest slope, rainfall at each input point having eight grid-point neighbours towards which it could flow (Figure 9.69). Water sinks underground at the lowest points of topographic hollows.

Figure 9.70 illustrates the results of a run of the model. In this case the initial surface at time zero ($T = 0$) is horizontal, but has small random irregularities. An arbi-

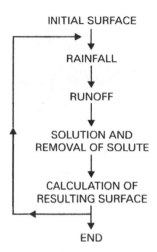

Figure 9.68   Components of a process–response model representing one time unit of land surface development. Reproduced from Ahnert, F., and Williams, P.W., Karst landform development in a three-dimensional theoretical model. Zeitschrift für Geomorphologie, Supplementband 108, 63–80 © 1997 E. Schweizerbart'sche Verlagsbuchhandlung, Science Publishers.

trary baselevel is shown by a dashed line at $Z = 450$. By nine time units of the model ($T = 9$) corrosion by water converging on topographic low points has caused solution depressions to develop. These deepen until $T = 45$ by which time the bottom of the deepest depression has reached baselevel (in reality the water table). By $T = 69$ the floors of some depressions have

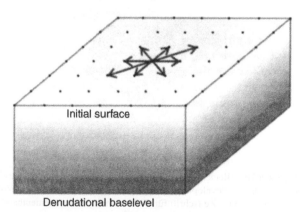

Figure 9.69   Points on a landscape model initial surface showing eight possible directions of runoff. Reproduced from Ahnert, F., and Williams, P.W., Karst landform development in a three-dimensional theoretical model. Zeitschrift für Geomorphologie, Supplementband 108, 63–80 © 1997 E. Schweizerbart'sche Verlagsbuchhandlung, Science Publishers.

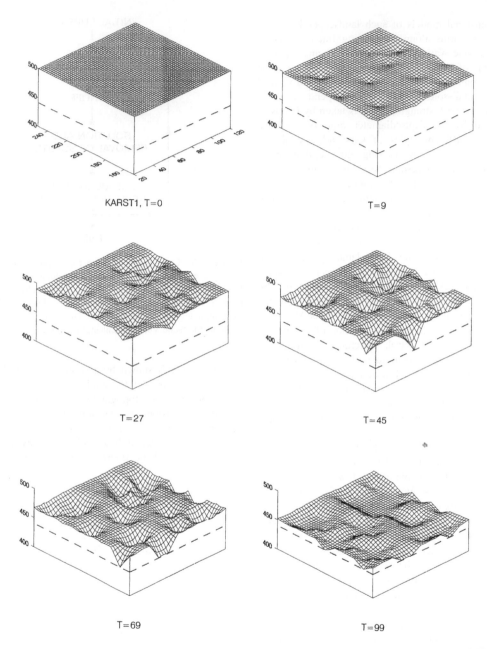

KARST1, T=0

T=9

T=27

T=45

T=69

T=99

Figure 9.70   Block diagrams illustrating selected time unit stages in the course of running a process–response model (KARST1) of karst landscape development. Reproduced from Ahnert, F., and Williams, P.W., Karst landform development in a three-dimensional theoretical model. Zeitschrift für Geomorphologie, Supplementband 108, 63–80 © 1997 E. Schweizerbart'sche Verlagsbuchhandlung, Science Publishers.

widened at the water table. This continued until $T = 99$ by which time the surface had lowered further and neighbouring dolines had coalesced at baselevel. Model results implied that locally higher solution rates caused by flow convergence are sufficient to explain the development of doline and polygonal karst. However, no cones or towers were generated by the conditions imposed on this model.

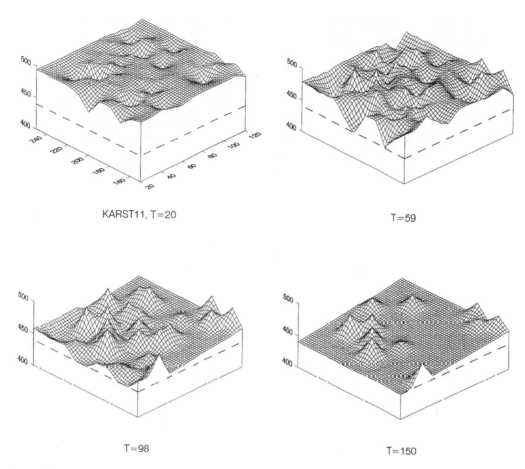

Figure 9.71 Block diagrams illustrating selected time unit stages in the course of running a process–response model (KARST11) of karst landscape development. The sloping corrosion plain at T = 150 follows the hydraulic gradient. Reproduced from Ahnert, F. and Williams, P.W., Karst landform development in a three-dimensional theoretical model. Z. Geomorph NF Suppl.-Vol. 108, 63–80 © 1997.

Figure 9.71 shows stages in the evolution of a more complex case in which baselevel is a sloping water table and a divergence effect is added. Divergence was measured in the model by the number of directions from a given grid point that have downslope gradients (Figure 9.69). On a peak it is 8 and on a planar sloping surface it is 3. The more directions in which an incoming increment of rainfall is dispersed, the less effective will be its solutional denudation at a point. In the model this was programmed as a reduction in local denudation as a function of the number of downslope flow directions. This leads to relatively reduced solution on topographic highs where flow diverges and to increased solution in hollows where it converges. In this run of the model a doline karst at $T = 20$ evolves into a cockpit karst with conical interdepression hills by $T = 59$. Cockpit floors converge at water-table level producing a sloping corrosion plain, the beginning of which is evident by $T = 98$. Further development sees the isolation of cones on the corrosion plain ($T = 150$). The model results clearly point to the importance runoff divergence and convergence effects in generating strong spatial variation in solutional denudation, which gains topographic expression in the evolution of cockpit karst (fengcong–depression). The model also shows that tower karst (fenglin) can develop from a prior polygonal cone karst stage. However, in this model solution is focused at points with unlimited infiltration capacity where flow sinks underground, which produces circular depressions. If instead infiltration were increased along intersecting fractures, then crossing joint corridors with intervening blocks would be produced (Telbisz 2001), which could be clints or plateau depending on scale assumptions.

Ahnert and Williams (1997) also explored the effect of a sudden fall in baselevel (simulating uplift) during the course of model development. When uplift occurs at a stage similar to $T = 150$ in Figure 9.71, the residual hills of the first cycle become the tops of hills of the second cycle. This can be confirmed to occur from field evidence around the head of rejuvenation gorges that penetrate uplifted karst plateau in Guizhou, China.

Karst landscape modelling is in its infancy. Much can be learnt from this approach, because process–response models can be built that can accommodate numerous variables, enable their relative importance to be assessed, and permit changes in their relative importance to be compared over time. Computer models of karst landforms and of karst groundwater system evolution have been developed independently of each other. We look forward to the next generation of models that will link surface and subsurface karst development and so tackle the problem of karst evolution in a fully integrated way.

10

The Influence of Climate, Climatic Change and Other Environmental Factors on Karst Development

10.1 THE PRECEPTS OF CLIMATIC GEOMORPHOLOGY

During a study of karst in Java, Lehmann (1936) recognized the coincidence of Kegelkarst (cone karst) relief with the humid tropics. This reinforced a growing belief amongst geomorphologists generally that landform assemblages are strongly influenced by climate through its control of natural processes. There then began a quest to identify the singular relief styles that were expected theoretically in the various climatic zones, and maps of morphoclimatic regions were produced, e.g. by Tricart and Cailleux (1972) and Büdel (1982). European karst geomorphologists such as Lehmann (1954), Lehmann et al. (1956) and Corbel (1957) were particularly active in this. Descriptions and comparisons of karsts from the tropics to the arctic provided a focus for research for 40 year, the aim being to determine and explain in morphogenetic terms the contrasting landform assemblages that were found. Nevertheless, some detailed work showed lithology and structure to have a larger effect on relief forms than had been appreciated (Verstappen 1964, Panoš and Stelcl 1968).

It is evident from more recent writing, for example in the volume edited by Salomon and Maire (1992), that most geomorphologists now agree that broad landscape differences exist in regions with contrasting climates, but at the same time it is admitted that subtler variations in style had often been claimed than objective scrutiny could justify. A more important weakness of climatic geomorphology is that it has been unable to explain convincingly why many of these landscape contrasts occur. For example, it has not revealed why karstic activity in the humid tropics sometimes results in the development of cockpit karst, whereas in the humid temperate zone doline karst is apparently more typical – even though dolines are also found in the tropics. It appears to us that **qualitative** climato-genetic geomorphology has reached about the limit of its contribution. Nevertheless, the effect of climate is real even if difficult to pin down, Salomon (2000) providing a valuable summary of principal conclusions. We should now pass on, but avoid the mistake of dismissing the broad value of climatic geomorphology. There is a risk of this, as is evident from the content of a recent book about karst evolution from pre-karst to cessation, which completely omits even mention of the importance of climate in the evolution of karst landscapes (Gabrovšek 2002). Perhaps it is now taken for granted?

The availability of water is the key climatic factor in karst development. It is certainly the principal variable controlling total denudation by dissolution (Figure 4.3), although the targeting of corrosion is determined by the controls on water flow, as explained in Chapter 5. In Chapter 9 we explained how 'normal' karst evolves in areas where water is abundant, but remarked that aridity and extreme cold place constraints on development. Both of these climatic conditions lead to a scarcity of water in its liquid state, thereby limiting dissolution and permitting other geomorphological processes to dominate morphological evolution. But if the other processes are themselves not very active, then dissolution effects may persist unaltered for a considerable time. We shall see from field evidence that many karst features (landforms and groundwater circulation systems) in arid and cold

Karst Hydrogeology and Geomorphology, Derek Ford and Paul Williams
© 2007 John Wiley & Sons, Ltd

zones are the legacy of times in the distant past when conditions were much wetter or warmer. Thus it is not always easy to separate the effects of modern processes from those of earlier times. A subtheme of this chapter, therefore, is **inheritance** because, where the contemporary dissolution regime is feeble, palaeofeatures persist and guide modern processes. In this chapter we examine the karsts that result.

10.2 THE HOT ARID EXTREME

We agree with Jennings' (1983) assessment that 'Less is known about karst in deserts and semi-deserts than anywhere else except beneath glaciers and permafrost', but progress has nevertheless been made since the publication of our first review of this subject (Ford and Williams 1989), as can be seen in the latest global survey of karsts in the arid and semi-arid zones by Salomon (2005). Present information indicates that the determinants of karstification are the same in hot arid environments as elsewhere, although there are differences in the relative influence exercised by the various controls. Because soil is usually thin and patchy (or absent altogether) in the arid zone, it is less influential as an infiltration 'governor' and as a moisture store than in more humid regions. Since it can support only a small biomass, it has reduced significance as a CO_2 source. One major consequence is that individual drawdown solution dolines are rare, and no extensive tracts of polygonal doline karst are reported. As a result, collapse dolines assume greater relative importance, although they are still not common.

Precipitation in these regions is delivered typically in short but violent, aperiodic convectional storms. This favours flash-flooding, especially in rugged terrains. For example, Frumkin (1992) studied the hyperarid Mount Sedom salt diapir in the Dead Sea basin, where the meteorological 'mean annual precipitation' is cited as $\sim50\,mm\,a^{-1}$. Instruments in stream caves there measured just two short-lived flow events in 5 years. Gillieson *et al.* (1991) described similar behaviour in the limestone ranges of the Kimberley region of Western Australia, where proxy palaeoenvironmental records can be extended back for 2000 years. The rapid delivery, and then loss by runoff and evaporation, tends to limit the development of epikarst when compared with humid regions. Rillen, rinnen and pitting are common on bare desert surfaces but extensive clint-and-grike pavements draining efficiently to collector conduits in bedding planes are comparatively rare.

The morphological consequences of these process factors vary according to the lithology, but it is possible to offer some generalizations. In many carbonate rock terrains (especially where the strata are chiefly medium- to thick-bedded rather than consistently massive) and in the majority of gypsum terrains we have seen, the morphology is much more fluvial in character than it is in humid karst regions. There are regular Hortonian patterns of dry valleys with rounded interfluves in gentle uplands, or of steeper-walled canyons or reculées dissecting massifs and plateaus. Both types lose most storm runoff into small solutional shafts (ponors) along the thalwegs. This is true of the Judean Hills around Jerusalem, the Edwards Plateau of Texas and, more spectacular, the reefal Guadalupe Mountains (New Mexico–Texas) that contain the great H_2S outlet caves such as Carlsbad Caverns and Lechuguilla Cave discussed in section 7.7. These three limestone examples all extend across the semi-arid to arid climatic transitions in their respective regions. They display little surface karst except the limited kind of epikarst described above, plus relict caves that have been intercepted and drained by entrenchment of the valleys or canyons. Nevertheless, they support regionally important aquifers that usually discharge at just one or a few large springs at the junction between limestone uplands and desert detrital plains. The relict caves tend to be well decorated with speleothems which indicates that, despite the characteristic flash-flood runoff to ponors, there is some diffuse recharge via the epikarst.

This dessicated fluvial morphology is perhaps best developed in gypsum and anhydrite, however. Folded or block-faulted terrains with only narrow ridges, such as the semi-arid hills of Catalyud in the Ebro valley of Spain (Gutierrez and Gutierrez 1998) or the very arid Bir al Ghanam in the Libyan Desert south of Tripoli (Kósa 1981), are dissected by consequent valleys regularly spaced between ranges of well rounded to conical hills. Recharge is via sinks in stream beds (thalweg ponors), with very limited epikarst elsewhere. Where the gypsum forms plateaus and escarpments, however, stream-sink and collapse dolines, uvalas and even small poljes may develop, as in the arid Pecos Valley of New Mexico or in subhumid western Oklahoma (Johnson 1996).

In contrast, in some dense, massively jointed crystalline carbonate plateaus, the relatively 'naked karst' becomes dissected along joint corridors into blocks heavily fluted by karren. Dissection penetrates to the level of neighbouring pediplains, which may truncate rocks of any lithology. The karst surface is then efficiently drained down open networks of grikes or corridors (section 9.2) that spill onto them.

No matter how arid, exposed salt always displays dense karst because of its great solubility. Salt pan deposits 2–3 m deep on the floor of Death Valley, California, are pierced through by solution pits with densely fluted walls. At 150 m below sea level, the summit of the Mount Sedom (Dead Sea) salt diapir displays regular Hortonian valley patterns on its anhydrite caprock, but they terminate abruptly in deep solution shafts at the contact with the bare salt (Frumkin and Ford 1995).

Speleogenesis in hot arid regions follows the principles explained in Chapter 7, but the size and frequency of caves developed is necessarily more limited than in humid regions. There is preferred development where simple cut-off caves can capture flash-flood waters. For example, Sof Omar Cave (Figure 7.20) is a floodwater cut-off maze in a semi-arid region. Short vadose shaft-and-drain systems fed by ephemeral streams are common along escarpment edges. Intensive exploration in the Bir al Ghanam gypsum karst discovered similarly simple patterns of dendritic caves underdraining the dry valleys there, with a few tributaries passing under divides for short distances to join the trunk conduits (Kósa 1981). However, there is also deep phreatic circulation in mountainous deserts; deep interbasin flow is common in carbonate rocks in the Russian deserts and western USA, for example. A case in point is the great groundwater system draining to the Ash Meadows spring line (and Devil's Hole) in the eastern Amargosa Desert in Nevada (Riggs *et al.* 1994). Relict phreatic caves, especially of the maze type, are quite abundant. The system, though still active, may be a legacy of previous wetter conditions.

Russian researchers have emphasized that seasonal condensation waters in otherwise dry areas can play a major role in enlarging relict phreatic caves by processes of corrosion, breakdown and scaling. A valuable review by Dublyansky and Dublyansky (2000) of work in the former Soviet Union on the role of condensation in karst has enabled their research findings to be more accessible to non-Russian speakers. Their data suggest that condensation does not normally exceed 9% of the annual precipitation, but that it occurs in the summer when there may not be much rainfall. An estimate from the western Caucasus indicated that it could be responsible for about 3.7% of the gross annual denudation by dissolution. Domepit-type shafts with small drains may possibly be produced in escarpments entirely from dew. Castellani and Dragoni (1986) calculated that shafts 0.5 m diameter and 10–15 m deep may develop in 500 ka by this mechanism; their field site, the Hammada de Guir plateau and scarp in Morocco, receives only 50–60 mm a^{-1} of conventional precipitation and has a mean annual temperature of 19.6°C. However, the precipitation of condensation water underground is not just a feature of arid karsts because it occurs in caves in humid areas too

(section 7.11), but its relative importance is more significant in the arid zone.

Jennings (1983) recognized that hot arid karsts present us with a dilemma. With little process study accomplished in these areas, it is difficult to know if we should ascribe their landscapes to the cumulative effect of repeated if sporadic modern events, or to more humid periods in the past when conditions for karst development may have been more favourable. Indeed, evidence is mounting that many, if not most, karst features in hot arid zones are largely the products of more humid intervals that may have occurred 10^5 or even 10^7 years ago. For example in the western desert of Egypt, El Aref *et al.* (1987) found evidence for post-Eocene cone karst development; and from cave deposits in the same region, Brook *et al.* (2003) determined that humid intervals and, therefore, further karstification had also occurred there during Marine Isotope Stages 5, 7, 9 and 13. Similarly, in hyperarid northeastern Saudi Arabia, Edgell (1993) attributed weathered karren, ponors and collapse dolines giving entry to multilevel caves, and even some poljes, to Pleistocene pluvial periods. Nevertheless, complicating factors are:

1. the difficulty of distinguishing between karst and desert elements, e.g. in the broken drainage patterns often found in arid landscapes;
2. the difficulty of distinguishing between landforms developed under the present process regime and those inherited from palaeokarst, but modified by modern processes and incorporated into the present landscape.

We illustrate the points made above by reference to the better studied dryland karsts of the world in Australia (Figure 10.1).

10.2.1 Karst of the Nullarbor Plain

The Nullarbor Plain lies immediately inland from the Great Australian Bight (Lowry and Jennings 1974, Gillieson and Spate 1992, Webb *et al.* 2003). It covers an area of about 200 000 km^2 and extends across the Western Australia–South Australia border (Figures 10.1 and 10.2). From 40 to 90 m at the top of its precipitously cliffed and almost 900 km long coastal boundary, the plain rises almost imperceptibly inland to about 240 m over a distance of 350 km. Annual rainfall diminishes inland from up to 400 mm a^{-1} near the southwest coast to only 150 mm a^{-1} inland and is greatly exceeded by annual potential evapotranspiration, which varies from about 1250–2000 mm near the coast to 2500–3000 mm in the

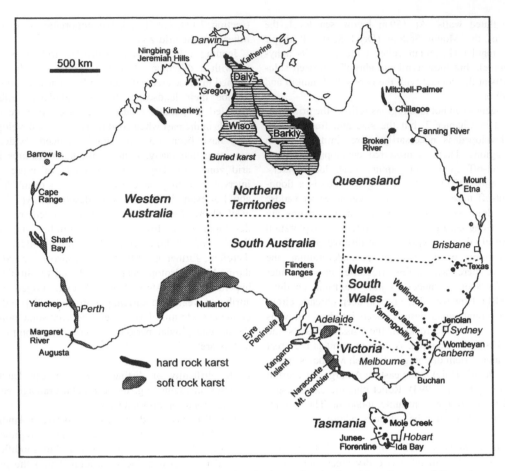

Figure 10.1 Distribution of karsts in Australia. 'Soft rock karst' refers to that developed in carbonates of Cenozoic age and 'hard rock karst' to that in older carbonates. Modified from Lowry, D.C. and Jennings, J.N. 1974. The Nullarbor karst, Australia. Zeitschrift für Geomorphologie, 18, 35–81.© 1974.

north. The mean annual temperature is about 18°C, although mean maximum summer (January) temperature is 35°C. Most of the Nullarbor is treeless, as its name implies, and is covered by bluebush, saltbush and tussock grasses; however, in places near the coast there are small trees.

The low plateau is underlain by an Eocene to Miocene carbonate sequence that extends well below sea level. The lower Wilson Bluff Limestone (up to 300 m thick) is chalky and porous, but although porosity can be high (up to 30%), poor interconnection results in relatively low permeability. The overlying Abrakurrie (100 m thick) and Nullarbor Limestones (maximum of 45 m) both have high porosity (about 40%) and permeability, except the top 15 m which is indurated by secondary reprecipitation of calcite. This calcrete case-hardening reduces surface porosity, especially in the top 1 m or so.

The plateau comprises an almost undisturbed, uplifted sea bed. At its inland margin up to 250 km inland from the present coast, lines of ancient coastal dunes define two distinct shorelines that mark stages in the emergence of the Nullarbor Plain; one is Early Oligocene in age (~35 Ma), when the Wilson Bluff Limestone was exposed to erosion for about 10 Ma before resubmergence, and the other is Mid-Miocene (~14 Ma), when the final emergence occurred. During this period of regression, rivers originating on the surrounding Precambrian basement rocks extended their courses onto the emerging plain and, even though incised no more than 10 m, their now dry meandering channels have been traced across it for up to 130 km. It is probable that the climate was particularly humid 5–3 Ma, when the channels were active (Clarke 1994, Alley *et al.* 1999).

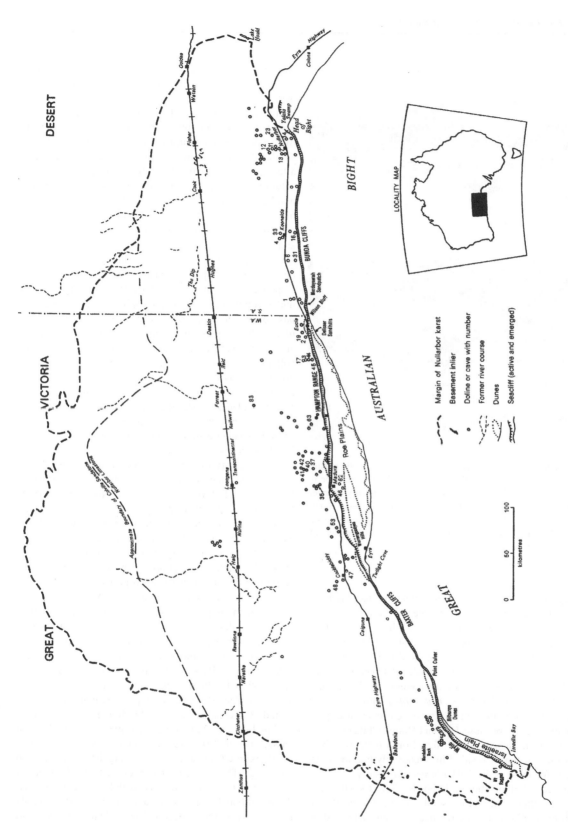

Figure 10.2 The Nullarbor Plain in southern Australia. Note the traces of former river courses. Most of the dolines and cave entrances marked are collapse features and have their greatest density near the coast. The Bunda and Baxter Cliffs have a height of about 75 m. Reproduced from Lowry, D.C. and Jennings, J.N. (1974) The Nullarbor karst, Australia. Zeitschrift für Geomorphologie, 18, 35–81.

Figure 10.3   (Upper) Doline with blowhole (beside the person looking down the shaft) on the Nullarbor Plain. The function is that of a centripetally draining stream-sink. Note the dry channel that leads storm runoff to the blowhole drain. (Lower) Small blowholes on the Nullarbor Plain perforating a case-hardened surface. The area around the blowholes has been washed clear of soil by surface runoff.

Judging by the 30 to 70 m of limestone that has been removed since the final emergence of the plains (more having been removed near the coast where it is wetter), dissolutional lowering has proceeded at an average rate of only 2–5 mm ka^{-1} over the past 14 Ma. This long-term rate compares well with an estimate of Quaternary surface lowering derived from cosmogenic ^{36}Cl measurement of < 5 μm a^{-1} (Stone *et al.* 1994). However, it is possible that the present phase of extreme aridity is relatively recent, because Lee and Bland (2002) provide evidence from desert varnish found on a meteorite from the Nullarbor that it resided in a relatively humid environment for most of its 5.9 ka terrestrial history.

Surface relief on the Nullarbor is always small, generally less than 6 m, and is usually assumed to be karstic (Jennings 1983). In the wetter areas it is characterized by

shallow claypan corridors between low rocky ridges or by a lattice of claypans around bedrock outcrops. In drier parts there is a scatter of shallow, circular depressions locally called '**dongas**'. These basins may be up to 1 km across, but have a depth of only 1.5–6 m. The extensive limestone plains ('**hammadas**') of the Sahara display similar features. Dongas tend to direct surface runoff after occasional heavy rains to **blowholes**, which are solution pipes (Figure 10.3). There are an enormous number of blowholes on the Nullarbor Plain, possibly of the order of 10^5. They are subvertical, smoothed-walled tubes that penetrate the surface indurated layer and appear to connect to complex anastomosing tube and cave systems. Reversing airflow through blowholes can be very strong (up to 70 km h^{-1} has been recorded) in response to air pressure adjustments in large-volume underground networks as atmospheric pressure systems pass overhead. The origin of blowholes remains speculative, but may be similar to solution pipes in Quaternary dune limestones (aeolian calcarenites). Deeper collapse depressions and accessible caves in the Nullarbor are largely confined to a southern belt parallel to the coast roughly 75 km wide, although some are known 150 km inland. More than 150 collapse depressions have been mapped, ranging up to 240 m diameter and 35 m deep (Figure 10.4). Some lead into caves, which can be of impressive dimensions.

More than 100 caves are known (James 1992, Webb *et al.* 2003), many being 50–120 m below the surface of the plain and developed largely in the Wilson Bluff Limestone. Gillieson (2004) pointed out that stream-cut passages are uncommon and that running water is absent except during flash flooding. Nevertheless, the occurrence of a large enclosing arc of Precambrian hinterland from which allogenic runoff has been focused onto the Nullarbor since its emergence has probably been a critical factor in the development of the caves. Most of these had a predominantly phreatic origin, the mixing corrosion of fresh and saline waters probably also assisting dissolution. Old Homestead Cave has 30 km of mapped passages and appears to be the southern extension of a surface palaeochannel. Passage dimensions in many caves are large, often 15–40 m diameter, thus implying considerable throughput discharges compared with present. Cocklebiddy Cave has been shown by diving to be more than 6.5 km long and Mullamullang Cave has some 5 km of passages up to 30 m wide. The largest chamber known is in Abrakurrie Cave and is 300 m long, by 30 m wide and 15 m high. The large cave passages are typically flattened collapse arches with numerous breakdown blocks on their floors, although some still have flat roofs. Presumably much of the collapse occurred when hydrostatic support

Figure 10.4   Collapse doline on the Nullarbor Plain. Note the flat treeless landscape.

was removed during glacio-eustatic low sea-level stages, although considerable contemporary breakdown is caused by salt wedging.

Calcite speleothems provide evidence for past wet phases, although they are largely inactive at present and frequently broken by salt and gypsum wedging. There appears to have been a transition in speleothem deposition from calcite to gypsum to salt that probably reflects the growing aridity of the region. Salt speleothems are known up to 2.7 m long. The calcite speleothems are commonly dark brown to black in colour due to humic compounds and their ages are generally beyond the range of U/Th dating. Some are of Pliocene age as shown by three recent U/Pb dates of 3.28, 3.96 and 3.93 Ma (2σ errors of 1–5%) (J. Woodhead, personal communication), although it is conceivable that some of the younger speleothems may date from late Quaternary wet phases, perhaps of the same stadial and cool interstadial ages as those identified from speleothems at Naracoorte further east (Figure 10.1) (Ayliffe *et al.* 1998, Moriarty *et al.* 2000). Red aeolian quartz sand accumulations in some caves probably reflect arid phases, when the process of salt wedging (**exsudation**) in cave passages also caused much breakdown. Gypsum and salt speleothems have been dated to 185 ka and < 40 ka respectively (Webb *et al.* 2003).

The water table beneath the Nullarbor Plain lies at a depth of about 120 m in the south near the coast and rises to within 30 m of the surface in the north. The present hydraulic gradient is extremely low at about 20 mm km^{-1}. The saturated zone connects directly with the sea, although no submarine springs are known, and the water table has

no doubt been subject to considerable Quaternary oscillation in response to glacio-eustatic shifts. Water in the phreatic zone is exceptionally clear and highly saline; the salinity arising from a mixture of sea spray carried far inland and evaporated rain and percolation water. James (1992) has recognized three mixing zones in the caves and rock mass beneath the Nullarbor: mixing zone 1 occurs at the runoff–phreatic brackish or saline water interface of cave lakes and canals; mixing zone 2 is in the porous bedrock at the vadose–phreatic interface; and mixing zone 3 occurs at a halocline deep beneath cave lakes and flooded passages. Calcite precipitation occurs at the mixing zone 3 interface in close association with microbial activity and results in deposition of 'snow-fields' of subaqueous spindle-shaped calcite microcrystals (Contos *et al.* 2001).

Apart from a few solution pans, karren are entirely absent on the Nullarbor despite induration of the uppermost limestones. Yet solution flutes are developed at the same latitude some 1000 km to the east in the Flinders Ranges (Figure 10.1) near Brachina Gorge, where rainfall is also slight (250 mm a^{-1}), but the rock is a dense Cambrian limestone (Williams 1978).

A feature of the relatively moist southwestern corner of the Nullarbor is the occurrence of numerous inliers of crystalline basement rock. Some scarcely rise above the general level of the plain, but one emerges 450 m. The inliers are commonly ringed by annular depressions 50–150 m across and 3–10 m deep, resembling dry moats. They are clearly formed by the corrosive action of centrifugal allogenic runoff from the impermeable inliers. Jennings (1983) considered these moats to be the only distinctive karst landform at the meso- to macroscale in this climatic zone. However, similar landforms created by the same process surround the volcanic inliers of some coral islands (section 10.5). They are a product of physical juxtaposition that is independent of climatic factors.

10.2.2 Karst of the limestone ranges of Western Australia

The relationship of modern processes to existing landforms is of particular interest when trying to understand the geomorphological history of the area. This is a problem that particularly applies to arid-land karsts, because of the likelihood of significant wetter periods ('pluvials') in the past when some of the morphological features may have formed. On the Nullarbor Plain, for example, it is possible that the region's landscape is mainly a reflection of process environments of late Miocene–Pliocene times and has been subject to only relatively minor modification during the Quaternary Period, the more recent aridity of the region providing little

opportunity for further morphological development. In the Kimberley region of northern Western Australia (Figure 10.1), the imprint of past processes may be even more ancient because as Playford (2002) explains, some features of the present landscape may be ascribable to the exhumation of palaeokarst. We discuss this further in section 10.6.

The limestone ranges of the Kimberley region lie about 1500 km north-northwest of the Nullarbor Plain in northern Western Australia (Figure 10.5). In this area Middle Devonian reef limestones with lesser dolomite are found in the Laidlaw, Lawford, Napier and Oscar Ranges. The Napier Range is formed of a fringing and barrier reef complex approaching 110 km long and up to 5 km wide; the Oscar Plateau is significantly wider. A well defined but stratigraphically complex forereef–main reef–backreef facies sequence is exposed in walls of limestone gorges in the area and supports a well-developed karst (Jennings and Sweeting 1963, Goudie *et al.* 1989, 1990, Allison and Goudie 1990, Playford 2002). Average monthly temperatures range from 22°C to 33°C with air temperatures probably exceeding 38°C on over 100 days in the year.

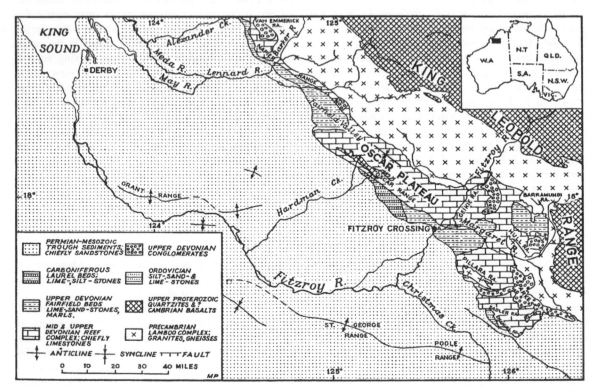

Figure 10.5 The limestone ranges of the Kimberley region, northern Western Australia. (a morphological map at 1:125 000 is available in the original publication). Reproduced from Jennings, J.N. and Sweeting, M.M. (1963) The limestone ranges of the Fitzroy Basin, Western Australia. Bonner Geographische Abhandlungen, 32, 60 pp.

Annual rainfall varies from about 640 to 760 mm, depending on distance inland, and is distributed over 30–80 days. The monsoonal wet season from December to March is short but intense, with rain-day averages of 16–18 mm and 50–year intensities of 80–90 mm h^{-1}. Since annual potential evaporation is around 3400 mm, arid conditions prevail for most of the year. High-intensity rains during the wet season give rise to seasonal floods, however, and the flood history of the Lennard River where it passes through Windjana Gorge in the Napier Range (Figure 10.5) has been investigated by Gillieson *et al.* (1991) using slackwater deposits. The modern dissolution denudation rate has been estimated by the Corbel technique (see Chapter 4) as 6.4–10.4 mm ka^{-1} (Ellaway *et al.* 1990), i.e. in places perhaps up to twice that of the Nullarbor Plain but still small by global standards.

Figure 10.7 Detail of dissolution runnels on towers in the limestone ranges of the Kimberley region, Western Australia. (Photograph by A. Goede.)

Figure 10.6 (Upper) Tower karst rising from a bare limestone pediment near J.K. Yard in the limestone ranges of the Kimberley region, Western Australia. (Photograph by J.N. Jennings.) (Lower) The sharp break of slope as the pediment gives way to the limestone plateau. Note the large, long wall karren that are incised into the scarp face. (Photograph by J.N. Jennings.)

The limestone ranges of the Kimberley region are an international type-site of semi-arid karst. In their study, Jennings and Sweeting (1963) described a distinctive landscape (Figures 10.6 and 10.7) and a particular sequence of landform evolution which has since been found applicable to other seasonally humid karsts in northern Australia. They argued that during Tertiary and Quaternary denudation of clastic coverbeds a drainage pattern was lowered onto the underlying limestones that resulted in the formation of gorges of superimposition. For example, the Lennard River has cut the 4-km-long Windjana Gorge through the Napier Range, and drains runoff from a hilly impermeable catchment of about 500 km^2 through the gorge. Although most allogenic rivers in the area cross through the karst in gorges, some smaller streams pass through the plateau in caves, the largest known being 8 km long.

The exhumed and re-karstified plateau surface stands 80–100 m above a neighbouring lower pediment that rises inland from about 90 m in the northwest to 140 m in the southeast (Figure 10.6). The pediment surface truncates limestones, shales and siltstones and has a gradient of 1–2° or less (Goudie *et al.* 1990). The sequence of evolution proposed by Jennings and Sweeting (1963) envisages the plateau being gradually dissected and reduced to the level of the pediment, mainly by a process of parallel joint-aligned retreat in interfluvial areas. Through-rivers provide regional baselevel control. The main stages in the present phase of evolution of the karst were identified as follows.

1. The plateau surface is stripped of soil and joints are penetrated by corrosion, producing fissure caves which enlarge to become intersecting sets of closed solution corridors or '**giant grikes**' that isolate large bedrock blocks. The block surfaces are densely fluted by rillenkarren and solution pans, and the grike corridors that isolate them are up to 5 m wide, 50 m deep and hundreds of metres long. Vertical wall karren flute their sides. Modern fissure caves prolong the open grikes underground and link up joints with different orientations. Intersection points of solution corridors sometimes widen into steep-walled closed depressions.

2. Solution corridors and infrequent closed depressions amalgamate to form integrated valley systems reflecting the joint geometry in plan. Termed '**box valleys**', they have rectangular cross-sections with steep walls, flat floors and plateau-like divides. Their long profiles grade to the adjoining pediment. Closed depression floors are also adjusted to this level. Significant tufa deposition in some valleys may seal their floors.

3. Plateau remnants are consumed by the widening of box valleys, thereby isolating towers that are scattered across bedrock pediments. In places, the landscape comprises a bare fluted tower karst (Figure 10.7) that is sharp and abrupt, but of comparatively small relief (< 40 m). However, the bounding slopes display a wide range of morphologies from abrupt free faces to gently convexo-concave slopes.

4. Pediplanation results from continued dissolution of the towers and from direct scarp recession into the margins of the plateau. Scarps are fluted by wall runnels up to 2 m deep and to 30–60 m long. Ultimately the upper surface is completely replaced by the lower. Valley floors are the sites of abundant tufa and calcrete deposition.

Just how much this pattern of evolution has been preconditioned and directed by palaeokarst developed between the first late Devonian emergence of the limestones and their Early Permian burial under clastic deposits (Playford 2002) remains an interesting question well worthy of further research, but it seems certain that some karst depressions and therefore probably some caves and fissures have Early Permian precursors. It is also likely that a major part of the present phase of evolution occurred during more humid periods of the Tertiary when, as was the case with the Nullarbor karst, Australia was in more southerly latitudes and more exposed to a strong westerly circulation.

We suggest that the extension of pediments into some karstlands is the one truly distinctive climato-genetic feature so far recognized in desert karst. The frequent occurrence of thick tufa and calcrete crusts is also notable (see section 9.12), with these deposits often sealing valley floors. The patterns of poorly integrated to non-integrated channels, claypans and dongas ('dayas' on the Moroccan hamada) are quite distinctive, but similar patterns occur on humid northern plains where the deranging agency is glaciation rather than aridity. Salomon (2005) provides numerous illustrations of such features around the globe.

We agree with Jennings' (1983) general conclusion that carbonate karst declines with precipitation; quite simply, the richness and diversity of landforms diminishes. We observe also that the mode of development and the scale and form of landscape produced in the limestone ranges is very similar to that of the subarctic Nahanni karsts (sections 9.2 and 10.4) save in one crucial respect. This is that the Nahanni corridor karst is drained via younger generations of caves developed in the floors of the corridors and platea (box valleys), whereas the limestone ranges of Kimberley drain surficially across a pediment.

10.3 THE COLD EXTREME: KARST DEVELOPMENT IN GLACIATED TERRAINS

10.3.1 The Late Tertiary–Quaternary glaciations

The cold extreme is represented by land that is ice-covered or that is bare but permafrozen. At the present time (considered to be 'post-glacial') 10% of the aggregate continental area is occupied by glaciers (**glacierized**) and a further 15% is widely or continuously permafrozen. At the maxima of the Quaternary glaciations ice cover increased to approximately 30%. Much of the terrain that is now permafrozen was then glacierized. Because the glaciers were so extensive, relationships between the karst system and glacial processes will be discussed first. Karst development in permafrozen terrain is considered in section 10.4.

More than 95% of the volume of modern glacier ice is in Antarctica. Ice sheets have existed there for at least the past 8 Ma, though not necessarily at the modern scale. Expansion to ∼30% continental cover during glaciation maxima was accomplished by minor extension of the Antarctic sheets, growth of major ice sheets over Canada and over Scandinavia plus the Russian northwest and

Barents Sea shelf, and of ice caps and valley glaciers in all high mountain regions. Withdrawal of so much water from the oceans lowered global sea level as much as 130 m. This radically affected the world's coasts (section 10.5) and significantly changed the $^{18}O{:}^{16}O$ ratio in the remaining seawater (section 8.7). Study of fluctuations in that ratio, as it is recorded in foraminiferal tests recovered from tropical ocean sediment cores, suggests that there may have been as many as 17 cycles of glacier ice growth and decay within the past 2 Ma, i.e. roughly 100 000 years per cycle (Bradley 1999). These are correlated with changes of net global solar radiation induced by periodic irregularities in the Earth's orbital motion about the Sun. The number of glaciations recognized on the continents is fewer because of destruction of early evidence. In well-studied regions at least three or four glaciations separated by warm interglacial conditions such as the present are known. The first glacial perturbations appear in lowland mid-continent sites such as Nebraska at 2.5–3.0 Ma and in the Mediterranean region at 2.4–2.6 Ma. Local glaciations in high mountain areas probably began earlier, e.g. 8 Ma in Alaska.

The warm climate peak of the last interglacial occurred about 125 ka. This was followed by a worldwide glacial cycle; all glaciers expanded close to their greatest known extent, reaching their maxima in most regions between 26 and 18 ka. Recession to the modern sizes or less was completed by about 7 ka. There have been at least two other glaciations of this magnitude since ∼780 ka.

The effects of such radical, comparatively rapid changes of conditions upon karst in the glaciated regions are fundamental and complex. **Glaciokarst** relationships range from the perfect preservation of 'pre-glacial' karst to its complete destruction and from the prohibition of any post-glacial karst to its most rapid development.

10.3.2 Relevant conditions in glaciers

Glacier ice flows by intracrystalline creep and by sliding on its bed (Drewry 1986, Martini *et al.* 2001). With creep, the velocity of flow is proportional to temperature and to the surface slope of the glacier; it varies from 1–2 m a^{-1} in cold, near-horizontal ice sheets to 100–1000 m a^{-1} or more in steep, 'temperate' glaciers. Sliding over the bed is limited largely to temperate glaciers, where basal ice temperature is at the pressure melting point (0°C or a little below), so that water is present to lubricate the sliding. Velocity is proportional to bed slope. Wet-based glaciers, flowing by both creep and sliding and loaded with rock debris, can scour rock surfaces very effectively. Dry-based glaciers are frozen to the rock, either partially or everywhere. They may protect it completely from effects of creep in the overlying ice, drag it to create

local folding, or wrench it and remove large blocks as integral parts of the basal flowing mass. Blocks of sedimentary bedrocks as great as 1 km^2 in area may be extracted intact and moved downstream for short distances. Where the wrenching is only partial, however (the rock block is not entirely extracted), glaciotectonic cavities can be created. Often, these are best developed in carbonate rocks because earlier dissolution weakened bedding plane and joint contacts there, thus permitting differential slip. Figure 10.8 shows spectacular examples of these processes from the work of Schroeder and colleagues in glaciated Quebec.

Water from two different sources may be present at glacier beds. The first source is of meltstreams that have descended from the glacier surface or from allogenic terrain (e.g. nunataks). It flows in channels melted upwards into the ice (R channels, which may closed by annealing during the winter) or in channels partly protected in the bedrock (N channels). A subglacial limestone cave is an N channel fully protected against annealing, although its entry may become obstructed. The second type of water is a thin pressure melt/regelation film present wherever temperate ice and rock are in direct contact (Hallet 1979). It may function as a wholly closed system, melting against an obstruction and freezing again in its lee, or part may escape into the R or N channels or pass down under pressure into the bedrock (Smart, 1983b, 1984).

Smart (1997) undertook detailed studies at the Small River site in the Canadian Rocky Mountains, where a temperate alpine glacier 2.8 km^2 in area is drained underground into dipping limestones sandwiched between quartzites. Moulins (ablation shafts in glacier ice – section 7.3) develop above bedrock ridges there, channelling supraglacial meltstreams down into the rock. Repeated dye traces at one moulin demonstrated karst groundwater flows of 0.5–1.0 m^3 s^{-1} to springs 1.75 km distant; maximum straight-line velocities ranged 480–650 m h^{-1}. Boreholes to pressure-melt areas of the glacier sole adjoining the moulin found that there was no hydrological connection of the pressure melt to the moulin flow (Ross *et al.* 2001). Lauritzen (1996a) reports on larger glaciers in South Spitsbergen, where much of the ice is dry-based. At the coast, Trollosen Spring discharges 13–15 m^3 s^{-1} of englacial and supraglacial water from the large Vitkovski Glacier. It is mixed with a smaller proportion of thermal water from deep sources.

The extent to which meltwaters from supra-, en- and subglacial sources can accumulate above a glacier base and so raise the water table there is controversial. At the least, such accumulation is inescapable where the ice occupies a closed depression such as an overdeepened

cirque, unless karst channels are open already to drain it. Lauritzen (1984a,b) has deduced glacier aquifer depths of > 100 m at Norwegian cave sites where there is no overdeepening, as shown in Figure 10.9. Schroeder (1999) found that englacial caves flooded to similar depths when the outlets (R channels) annealed during the polar winter in Spitsbergen (section 7.3). Schroeder and Ford (1983) contended that Castleguard Cave (section 8.1) was back-flooded and filled with varved clays when the Columbia Icefield expanded to fully bury it during two or more glacials. Melt-season water tables will then have been ~1000 m above the local glacier base, which was > 50 km upstream of the glacier snout.

Continental glacier conditions exist when all of the topography is buried by ice flowing over it. The ice is normally 500–5000 m deep. These conditions prevailed over Canadian karst areas east of the cordillera, over most karsts of Scandinavia, northwest Russia and the glaciated British Isles. Alpine glacial conditions prevailed over most other glaciated karsts, where the flowing glaciers

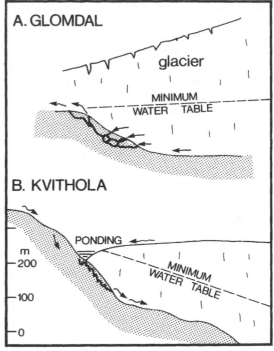

Figure 10.8 (Upper) Pull-apart cavity created by glacial wrenching of horizontally bedded limestones along some major joints. Ste. Foy, Quebec. (Lower) A spectacular instance of subglacial drag-folding exposed in the wall of a limestone cave at Pont Rouge, Quebec. There is glacial till in the cores of the folds. (Photographs by J. Schroeder, with permission).

Figure 10.9 Two examples of glacier-induced phreatic cave development from the Norwegian karst. Adapted with permission from Lauritzen, S.-E. (1984b) Evidence of subglacial karstification in Glomdal, Svartisen. Norsk Geografisk Tidesskrift, 38(3–4), 169–70. Lauritzen, S.-E. (1986) Kvithola at Fauske, northern Norway: an example of ice-contact speleogenesis. Norsk Geografisk Tidesskrift, 66, 153–61.

Table 10.1   Categories of karst landforms in glaciated terrains

Category	Type
Post-glacial	1. Simple forms – independent of preceding glacial forms: many karren, collapse dolines
	2. Suffosion dolines, many solution dolines, limestone pavements, etc., where the location or part of the form is determined by preceding glacial features
Adapted glacial	3. Glacial grooves, scourholes, potholes, S and P forms, kettle holes, cirques, moraine-dammed valleys, etc., adapted to karstic drainage
Subglacial and glacier marginal	4. Normal karst landforms (e.g. solution shafts) occupying anomalous hydrological positions
	5. Subglacial calcite precipitates; some arctic corridor karst?
Glaciated karst	6. Dolines, etc., subject to one apparent episode of glacial erosion and/or deposition
Polygenetic forms	7. Most large closed karst depressions in glaciated terrains; created by repeated glacial and karst episodes
Preglacial karst	8. Some relict caves (surface karst features are modified into types 6 or 7)

were channelled between ice-free ridges and summits that often supplied allogenic streams to the glacier systems. Alpine local relief normally ranges 500–4000 m.

Alpine karst has received more attention than the continental type because it is well developed in the Alps and Pyrenees, close to the classic centres of karst study (Kunaver 1982, Maire 1990). It is important to recognize that two distinct ice–karst relationships have existed in alpine terrains. The first is where glaciers were confined to the highest ground (cirques, summit benches, upper valleys) so that meltwater could discharge underground into lower karst valleys that were ice-free at all times, i.e. the karst water input points are glacierized but not the outputs. This relationship prevailed in most of the Pyrenees and Picos de Europa, much of the western and southern Alps, the Taurus, the Caucasus, the Rocky Mountains (USA), etc. and may be termed the **Pyrenean type**. Many detailed case studies around the world are given in 'La Haute Montagne Calcaire' (Maire 1990); see also Bini *et al.* (1998a,b) and Audra (2000). In the second relationship, glacier ice occupied all valleys and extended far beyond the output spring positions. This condition prevailed in the Canadian Rocky Mountains, parts of the central Alps and Scandinavian mountains and may be termed the **Canadian type**. Opportunities for karst development are more restrictive in these conditions because glacial effects have a direct impact on all of the karst system. Canadian findings are emphasized here; Lauritzen (1996b) gives details from glaciated stripe karst in the mountains of northern Norway. In the Southern Alps of New Zealand, the situation is transitional, being of Canadian type in the south (Fiordland) and Pyrenean type in the north.

10.3.3 Surface karst morphology in glaciated terrains

Where there have been multiple glaciations many different relationships between surface karst landforms and glacial

action are possible. Those that we have noted are listed in Table 10.1. Post-glacial karst forms are freshest and easiest to categorize. In type 1 (Table 10.1) the form, its dimensions and location owe little or nothing to prior glacial effects. Examples are many types of karren, collapse dolines such as that of Vermilion Creek, NWT (Figure 9.56) and the densely packed schlotten in gypsum (Figure 9.58).

In type 2 the location and/or recognizable parts of the morphology and dimensions are determined by inherited glacial features. As discussed in section 9.2, the placement and extent of modern limestone pavements are most often determined by prior glacier scour. Suffosion dolines also develop particularly well because fine-grained glacial detritus is readily piped into surviving epikarst cavities such as grikes. Probably the majority of post-glacial solution dolines also belong in this category. They are located at low points prepared by glacial scour and the form is often irregular because it retains part of the scour morphology. Steep- to vertical-sided dolines on scoured fractures trap and conserve snow, becoming sites of accelerated corrosion termed **'schachtdolinen'**, **'schneedolinen'** or **'kotlic'**. Solution dolines in glaciated terrains rarely compose regular polygonal karst because of the deranging glacial effects. However, suffosion dolines over buried pavement may yield excellent polygonal drainage.

Type 3 are glacial depressions adapted to karst drainage. Small depressions in bedrock include potholes, grooves and irregular, shallow scours of weaker strata or fracture lines that range from centimetres to a few hundred metres in length (P forms and S forms – Shaw 1988, Tinkler and Stenson 1992). Their origin may be by glacier scour, fluvial corrosion, dissolution by meltwater floods, or a combination. Smaller examples may be modified into karren pits and runnels, larger into one or a line of small kotlic dolines.

Innumerable closed depressions of all shapes and sizes are created during the glacial and pro-glacial deposition of clastic detritus. Amongst those of small to intermediate scale, such as kettle holes, four different conditions may arise (Ford 1979):

1. the depressions cannot drain underground, and so fill up as ponds;
2. the depressions drain underground through the glacial detritus without modifying it;
3. the depressions drain underground through drift with a high content of soluble clasts, so that there is dissolution with collapse or piping;
4. the depressions drain underground into bedrock karst – subsidiary suffosion or collapse features may develop.

Only this fourth condition belongs to the karst system, but in many continental glaciated terrains covering thousands of square kilometres there is the greatest difficulty in distinguishing it from the others. Interesting large examples are provided by the **turloughs** (dry lakes) of western Ireland (Williams 1970, Coxon 1987), where more than 100 are known up to 250 ha in area. In the lowlands, they flood as a consequence of seasonal fluctuations of the regional water table, whereas in upland areas such as the Burren plateau their flooding can be a response to prolonged intense rain at any season, the capacity of their swallow holes being exceeded. Canadian examples from permafrozen regions are illustrated in section 10.4.

Moraine-dammed valleys that drain karstically are comparatively rare. Figure 10.10 shows a fine example from the Mackenzie Mountains, NWT, Canada, that

Figure 10.10   (Upper) Post-glacial landforms. An erratic block on a pedestal on a glacier-moulded ridge in marble, Pikhauga, Norway. Stream-sink in medium-bedded limestones, Anticosti Island, Quebec. (Lower) 'Moraine Polje', a karst-adapted glacial feature. The photograph is taken from the crest of a terminal moraine which blocks a valley in dolomites (catchment is $90\,km^2$). The valley floor is seasonally inundated; all waters sink in a cave at the arrow. Mackenzie Mountains, Canada.

Figure 10.11   (Left) Subglacial calcite precipitates (white material) upon an ice-scoured surface. Ice flow was towards the camera. Ice receded approximately 20 year before this picture was taken. (Right) Detail of the calcite deposition around a limestone drumlinoid. A pattern of subglacial microrills is seen at the stoss (upstream or near) end of the drumlinoid. Scale in inches. Castleguard karst, Canada.

functions in the same manner as a seasonally inundated polje.

Karst landforms created beneath the ice include shaft dolines of normal appearance occupying what become anomalous hydrological positions when the ice is gone. Examples are shafts in the crests of subglacial ridges, created because crevasses formed above them and were adapted as moulins. The Small River moulin described above is an example. There are also stream sinkholes stranded high in valley sides where they once swallowed marginal streams from the ice. They are characteristic of alpine karst.

Subglacial calcite precipitates (Figure 10.11) are a class of karst forms unique to the subglacial environment. They are crusts deposited from the basal pressure-melt film as it refreezes (Hallet 1976). They are highly stream-lined in accordance with local ice flow, including ridge-and-furrow forms and even small, horizontal helictites oriented downflow. In some instances it is apparent that the feedwater flowed through the bedrock in a miniature epikarst before reaching the freezing site. Crusts are up to a few centimetres thick and may cover 70–80% of a limestone or dolomite surface if it is comparatively regular and free of basal clastic debris. In the Alps the forefield of the Glacier de Tsanfleuron is a particularly well-studied locality (e.g. Hubbard and Hubbard 1998). There are a few instances of the precipitates being found on non-carbonate rocks (Hillaire-Marcel *et al.* 1979), including in Antarctica.

In southern parts of the continuous permafrost zone of Canada there are patterns of solutional corridors (giant grikelands) that are now draped with glacial till, permafrozen and relict. They occur on lowland and plateau surfaces where, despite their northerly situation, the continental ice flow was probably wet-based for at least parts of each glaciation. It is suggested that they may have been created by subglacial dissolution (Ford 1984). They are not glacial erosion forms but it is evident that the carbonate rock surfaces between corridors were scoured by ice able to remove overlying beds.

Simple glaciated karst (Table 10.1, type 6) describes karst features subject to one significant episode of glacial action which has modified their form. The example of a small doline is shown in Figure 10.12. Such simple forms are comparatively rare. Most karst features subject to glaciation must be placed in type 7 – landforms that have suffered several or many later episodes of glacial action plus, in most instances, significant karst solution when ice was absent. They are polygenetic, multiphase features. There are no examples where the precise number of phases has been reliably determined.

A majority of the largest closed depressions now functioning karstically in glaciated regions belong in this category. It includes hundreds, perhaps thousands, of cirques in the carbonate ranges (Figure 10.13; Ford 1979, Maire 1990). The cirque is the basic alpine glacial landform, normally 0.5 to 5 km in length or diameter. Many are overdeepened by scour into the bedrock, i.e. their bases are closed depressions. In others closure is created, partly or wholly, by a moraine barrier. Some contain seasonal or permanent lakes that may overflow at the surface. Others are always dry, with karren and

Figure 10.12 A simple (i.e. one-event) example of a karst form overridden and scoured by warm-based ice. Ice flow was from left to right across this small doline which displays glacial plucking on the upstream (left-hand) wall, abrasion and sub-glacial precipitation on the downstream wall.

varieties of the post-glacial and adapted dolines in their floors. Many drain to springs at the foot of the **riegel** (bedrock step below the cirque) and thus are local karst hydraulic systems subordinate to the glacial topography. Others contribute to regional aquifers. Together with spreads of schichttreppen karren, kotlic, and suffosion dolines on patches of glacial till, karst-drained cirques are the type features of alpine karst.

High, rugged limestone ranges in southern regions of Italy and Greece, and in Mexico, Guatemala, etc. were never glaciated. Their slopes are often dissected by deep, very asymmetric dolines that may be 1000 m or more in diameter. These are 'prefabricated cirques' should glaciers become established there. They may help to explain the high frequency and regularity of form that is characteristic of cirques in glaciated limestone mountains.

Large depressions also occur in alpine plateaus and valleys (e.g. Figure 10.13; Smart 1986), and even on plateaux and plains subject to continental glaciation. Intensive geophysical exploration for kimberlite pipes is currently discovering closed depressions up to several kilometres in length and 200 m deep scoured into the very extensive limestone plains of northern Ontario. The process is glacial overdeepening of initial subsidence holes created by dissolution of minor underlying gypsum beds. These depressions are now fully infilled with glacial and pro-glacial deposits.

In most glaciated terrains truly pre-glacial karst landforms (i.e. antedating all episodes of glaciation) cannot be recognized. A few exceptions are noted, such as in southern Ireland where numerous enclosed depressions with organic deposits containing early Tertiary pollen are

Figure 10.13 (Upper) Racehorse Cirque and (lower) Surprise Lake II. These are two deep, ice-scoured depression in massive, steeply dipping limestones in the Rocky Mountains of Canada. The waters drain karstically, their water surfaces being 50 m and 80 m respectively below the lowest points in the bedrock rims. Features such as these are probably produced by many successive episodes of karst and glacial action.

clearly pre-glacial in age (Drew and Jones 2000). In most other glaciated regions such features possibly existed but have been modified into the glaciated and polygenetic types. In the literature, 'pre-glacial' is often used where it is established only that a feature is older than the last glaciation.

In the alpine karsts of Europe and North America, landforms of types 1, 2 and 3 (Table 10.1) are numerically predominant. Many authors have recognized strong altitudinal zonation of these features, beginning with a doline zone at the treeline, succeeded by a higher karren zone and then a highest zone where frost shattering is predominant (e.g. Bauer (1962) in the Alps; Jennings and Bik (1962) in Papua New Guinea; Miotke (1968) in the

Picos de Europa). Ford (1979) emphasized that in the Rocky Mountains most modern alpine karst zonation may be more strongly tied to ice extent, erosion and deposition during the last glaciation. In particular, terrain covered by glaciers displays mixed assemblages of karren and differing types of dolines, varying with slope gradients and with the depth of any detrital cover. Terrain that was not ice covered tends to be dominated by frost debris, regardless of altitude, and so does not host karren. It will display kotlic and other dolines if sufficient drainage can be focused.

10.3.4 Effects of glacier action upon karst systems

This section briefly considers the effects that glaciers may have upon the karst system as a whole. The most extensive tracts of formerly glaciated karst rocks occur in Canada where Ford (1983a, 1996b) recognized the nine distinctive effects listed in Table 10.2. There may be others awaiting analysis.

Erasure of shallow karst features by glacial scour and plucking is perhaps the most widespread and frequently recognized effect. Positive forms such as spring mounds, pinnacles and hums may be removed entirely, as are microkarren and the smaller karren such as rillen. Shallow epikarst aquifers such as limestone and dolomite pavements can be stripped away, leaving unbreached, ice-polished surfaces in their stead. However, it is common for the bases of deeper grikes to survive a single glaciation: unless they are rendered hydrologically inert by infilling and shielding (below) they then may guide the post-glacial renewal of karstification. Most dolines are too deep to be removed by the scouring capacity of a single glaciation although, once again, they may be rendered hydrologically inert.

Glacier ice cannot be abstracted into karst aquifers as invading allogenic rivers can. Therefore, extensive aquifers may be **dissected** by means of glacial groove, trough

or cirque cutting. The process is most effective in alpine terrains with temperate glaciers. In perhaps the majority of such regions the consequence is that fragments of large, deep phreatic cave systems that are long drained and relict are preserved high in valley sides and even close to the summits of horn peaks. The famous Eisriesenwelt near Salzburg, Austria, is an example; Ford (1983c) and Lauritzen (1996b) illustrate many others in the Canadian Rockies and north Norway respectively.

Infilling is used to describe the filling of surface karst forms by glacial detritus. In Canada it is known that closed depressions as great as $300 \, \text{km}^2$ in area were filled completely during the last glaciation and ceased to exist as topographic entities. It may be presumed that, at the least, tens of thousands of doline-scale depressions have been filled in the world's glaciated regions during the Quaternary Period.

Filling and burial do not necessarily imply that a doline will be unable to perform its hydrological role after the ice has receded. If there is an adequate hydraulic gradient the clastic fill may function merely as an infiltration regulator as discussed in section 9.4. With time the depression then is partially re-excavated by dissolution or suffosion or, most often, by a combination of them both. The most striking Canadian example is Medicine Lake, where it is estimated that $1.8 \times 10^8 \, \text{m}^3$ of detritus has been removed into the Maligne River conduit aquifer (section 5.3) during the Holocene. This has created a closed depression 6 km in length that is a partial re-excavation of a greater, infilled bedrock depression.

Glacial debris can be **injected** deep into conduit aquifers. At Castleguard Cave, the heads of six inlet passages are sealed by intruded glacier ice itself; this is the only explored cave that extends substantially beneath modern flowing ice and the intrusions are believed to be very short. Normal injecta are clastic detritus of all grain sizes, transported by meltwater (section 8.2). The streams

Table 10.2   Effects of glacier action upon karst systems

Type	Effect
Destructive, deranging:	Erasure – of karren, and residuals
	Dissection – of integrated systems of conduits
	Infilling – of karren, dolines and larger input features; aggradation of springs
	Injection – of clastic detritus into cave systems
Inhibitive:	Shielding – carbonate- or sulphate-rich drift protects bedrock surfaces from post-glacial dissolution
Preservative:	Sealing – clay-rich deposits seal and confine epikarst aquifers
Stimulative:	Focusing inputs, raising hydraulic head – with superimposed glacial streams or aquifers
	Lowering spring elevations – by glacial entrenchment
	Deep injection – of glacial meltwaters/groundwater when bedrocks are being flexed during crustal depression or rebound

may be supraglacial or englacial in position (e.g. at the Kvithola cave, Figure 10.8) but in the majority of cases studied appear to have been subglacial or pro-glacial. Substantial sections of many deep alpine cave systems have been filled, nearly or completely, by injecta of varied grain size. Repeated filling is common. However, other sections escape filling so that, during the deglacial or post-glacial periods the aquifer can be substantially renewed by a combination of re-excavation of some conduits and opening of new bypasses to others.

Where the relief is intermediate or low, so that it is completely buried by ice, and abundant clay is available from nearby shales or other strata, aquifers may be more comprehensively deranged or rendered quite inert by injection of the clay sealant. The Goose Arm Karst, Newfoundland, is an excellent example of severe derangement (Karolyi and Ford 1983). It is a rugged terrain with a local relief of 50–350 m that may have been a cone karst before glaciation. It contained more than 40 large, bedrock closed depressions (100 to > 1000 m diameter), implying the existence of mature conduit aquifers draining to a small number of regional springs. Following clay injection, some depressions were completely blocked and have filled with water to function as normal lakes. Others discharge via short or tiny postglacial conduits to their individual springs. Orientation of the modern flow paths correlates poorly with both geological structure and maximum topographic gradient, the

two factors that control the directions of flow in normal aquifers (section 5.2).

Post-glacial karst development is inhibited where the bedrock is **shielded** from solutional attack because the solvent capacity of the waters is expended on soluble clasts in the cover of glacial detritus. Because they are comparatively weak most carbonate rocks and gypsum yield abundant clasts to scouring glaciers, enriching the local till. Some marbles do not yield significantly and tills immediately downstream of yet weaker rocks such as shale may be deficient in soluble fragments.

At different sites in Ontario and Quebec 1–2 m of till or outwash, or no more than 0.25 m of marl, have served to protect underlying limestones and dolomites entirely during the 10 000–14 000 year since the ice receded (Figure 10.15) despite the fact that an ideally open (i.e. optimal) dissolution system may be operating. At Windsor, Nova Scotia, an ice-scoured surface of gypsum is beautifully preserved beneath 4–6 m of till rich in gypsum fragments.

Much of the terrain underlying sluggish, cold-based glaciers may suffer negligible erosion; e.g. tors created by deep weathering are widely preserved in Swedish Lappland (André 2001) and, more remarkably, spreads of felsenmeer in the Torngat and Kaumajet Mountains of Quebec and Labrador (Marquette *et al.* 2004). Eyles (1983) presents zonal models. In such situations karst features as fragile as densely fissured pavement may be **preserved** more or less intact. The principal Canadian example is pavement in medium- to thick-bedded dolomites that underlies the city of Winnipeg, perhaps extending for as much as 3000 km^2 (Figure 10.15; Ford 1983b). It is overlain by up to 4 m of meltout till that is not injected into it significantly. This **sealed** the pavement and has converted a quickly drained epikarst into a productive confined aquifer. Such extensive preservation appears to be rare, however.

Probably the fastest rates of development of ponors and short sections of new cave in carbonate rocks occur during glacial recession in rugged terrains such as alpine areas. This is because large melt streams can be **focused** on specific input points at a glacier sole or margin for a few tens or hundreds of years, and because the hydraulic head building up in a glacier might be **superimposed** functionally upon that in an underlying karst aquifer. There are many examples of shafts drilled into the crests of subglacial ridges, as noted above, and of others aligned along the outer margins of end-moraine ridges where they were evidently fed by supraglacial streams, e.g. Glazek *et al.* (1977) in Poland. In the modern Columbia Icefield a small melt-lake builds up in a depression in the ice until it discharges abruptly through its base, apparently into the underlying Castleguard aquifer, i.e. it is a karstic

Figure 10.14 A dry cirque in marble on Mount Arthur, Southern Alps, New Zealand. The glaciokarstic depression failed to maintain a lake because it is drained underground resurging 6.3 km distant and 990 m lower. Several stream-sink positions are evident as the ponor point retreated. (Photograph by M. Tooker.)

Figure 10.15 Illustrating two extremes of glacial effects upon epikarst solutional pavements in limestone and dolomite. Both sites were overridden by the (Wisconsinan) Laurentide ice sheet but were close to its margins. (Upper) Scour beneath warm-based ice led to complete removal of upper beds plus infilling of grike bases with injected clay, to create a truncated and inert palaeo-karst. Striae are perfectly preserved beneath 1.3 m of diamict rich in dolomite clasts. Hamilton, Ontario. (Lower) Perfect preserva-tion of open clint-and-grike epikarst beneath meltout tills re-leased from cold-based ice, Winnipeg, Manitoba.

'jokulhaup' (ice-dam burst), an extreme form of super-imposition (Ford 1996a).

Glacial entrenchment also steepens groundwater hydraulic gradients by **lowering** potential spring points. In alpine glacial regions of the Canadian type many springs 'hang' above the floors of limestone valleys because they have not yet adjusted.

The final effect is somewhat more speculative than those summarized above. In Canada and elsewhere there is increasing evidence to suggest that there may be **very deep injection of meltwater** into karst aquifers, into interstratal karst and into palaeokarst features that were

buried and inert before the Quaternary glaciations. Injec-tion may be abetted by the crustal flexure that occurs when land is rebounding isostatically from the release of ice load, especially in extensive lowlands.

The most important evidence is that some deep col-lapse structures have been either initiated or rejuvenated during glaciations (Ford 1983a, Anderson and Hinds 1997, Grasby *et al.* 2002). The principal examples are of giant breccia pipes above salt in the plains of Sas-katchewan and Alberta (Figure 9.59). In Figure 10.16a, Howe Lake is a closed depression in which tills of last glacial age were downfaulted 78 m (Christiansen and Sauer 2001). The depression has been infilling during the Holocene and is now only 10 m deep though it is 300 m in diameter. The 'Saskatoon Low' is more com-plex. In one or a series of early collapses, Cretaceous and older clastic rocks downdropped at least 190 m to create a principal depression measuring 25×40 km (Christiansen 1967). These collapses may be Quaternary in age and glacially triggered but could be earlier. The final collapse occurred at the close of the last (Wisconsin/Wurm) glaciation, when tills of that age were downfaulted 70 m into the older depression (Figure 10.16b). The collapse propagated from salt approximately 1000 m beneath the surface of the Cretaceous cover rocks. The depression was infilled by post-glacial lake sediments and has been inert during the Holocene.

Evidence of glacial rejuvenation of palaeokarst of Palaeozoic age has been revealed at Pine Point, a zinc–lead mining region on the south shore of Great Slave Lake, Canada. The ores fill palaeocaves and collapse dolines in a Devonian barrier reef (Figure 2.11). Certain dolines display cores of Quaternary till (or multiple tills) passing through the centres of the older breccia and sulphide fillings. In a few cases these cores extend below the palaeokarst groundwater circulation base in the carbonates and enter a deeper anhydrite formation. Modern hydraulic gradients are very low and could not support such deep circulation.

This effect operates at large spatial and temporal scales. The Saskatoon Low is as large as the greatest modern poljes. The deep injection concept implies that a proportion of the water in some deep aquifers or emer-ging at regional springs may have been resident under-ground since the last or earlier glaciations.

To conclude, the development of karst landforms and systems in formerly glaciated terrains is particularly com-plex. Glacial action may create a wide range of effects within comparatively small regions. The problem of inheri-tance from previous glacial or interglacial conditions complicates all analysis. In general, destructive or inhibitive effects tend to predominate where the relief is intermediate

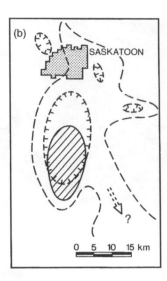

Figure 10.16   (a) The surface expression of the Howe Lake collapse structure, Saskatchewan. (b) The 'Saskatoon Low'. The shaded feature is a Late Wisconsinan collapse of 70 m depth that is infilled by lake sediments. Earlier depressions (unshaded) are probably Quaternary and glacially triggered but could be earlier. Arrow indicates possible glacial spillway. Note the very large scale of these features. Based on figures in Christiansen, E.A. (1967) Collapse structures near Saskatchewan, Canada. Canadian Journal of Earth Sciences, 4, 757–67.

or low. For example, in Midwest USA density and variety of explorable caves and other karst features increases markedly to the south of the glacial limits. Where relief is high and alpine glacial conditions prevailed there is always drainage derangement and complex inheritance present in an extensive karst but individual features or systems may develop rapidly and to a spectacular scale.

10.3.5 Nival karst

Nival karst is a term widely used to describe regions where the snowfall is comparatively heavy and its seasonal melt contributes a large proportion of the groundwater budget. There are two distinct situations. In the first, the terrain has been glaciated so that the modern

nival conditions are superimposed on assemblages of glaciokarst features. In terms of karst form and distribution, the effects of modern snow-patch deepening are subordinate to the effects of glacier and karst interaction already discussed.

The second situation occurs where the terrain was not glaciated and was also too warm for deep permafrost to establish during glacial periods. Examples are found among the lower slopes or foothills of most alpine karst regions of the Pyrenean type, such as the Dolomites and the Julian Alps. The clearest examples of nival effects, however, are found in wholly unglaciated mountain ranges such as the Carpathians of Romania, the Peloponese and Crete (Greece), the Zagros in Iran, etc. that are remote from former glaciations. Frost shattering and

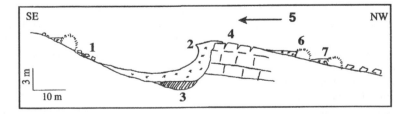

Figure 10.17   Type-section of a 'wave' (vague) on the Ait Abdi Plateau, a nival karst in the High Atlas Mountains, Morocco: 1, gelifracts (frost- shattered clasts); 2, snowdrift; 3, colluvial soil; 4, karren; 5, wind direction; 6, clumps of xerophytic vegetation; 7, snow. Reproduced from Perritaz, L. (1996) Le 'karst en vagues' des Ait Abdi (Haut-Atlas central, Maroc). Karstologia, 28(1), 1–12.

solifluction limit the development of karren in many cases, so that dolines are normally the dominant landforms. The 'karst en vagues' at 2200–3000 m a.s.l. on the Ait Abdi Plateau of the High Atlas Mountains, Morocco, is a striking example from a mediterranean (summer dry) setting. Well-bedded limestones dipping westerly form sequences of small scarps and dipslopes that together appear like waves at sea. The recharge of ~150–250 mm a^{-1} is largely derived from snow drifts against the scarps, and sinks at their feet in shallow dolines or dry valleys (Figure 10.17; Perritaz 1996). Maire (1990) presents many other well-illustrated examples of nival karst.

10.4 THE COLD EXTREME: KARST DEVELOPMENT IN PERMAFROZEN TERRAINS

10.4.1 The nature and distribution of permafrost

Bedrocks or detrital cover are said to be **permafrozen** if their temperature remains below the freezing point for 1 year or longer (French 1996; Smith and Riseborough

2002). A classification for karst in permafrost zones devised for Canada is shown in Figure 10.18. **Glacières** are the distinctive karstic category of permafrost that can form and survive outside of these zones because of the configuration and large size of some cave cold traps, as explained in section 8.5. **Sporadic** permafrost is confined largely to silts and similar detritus at the surface. **Widespread** permafrost extends to all types of rocks and in the **continuous** zone it is present everywhere except beneath lakes, large rivers or in a few special circumstances described below. Pulina (2005) shows the relationship between the limit of permafrost and occurrence of karst in Russia.

Permafrost in alpine regions may be placed in a separate category because of the great local irregularity that it displays in its distribution and depth. Here we treat it under a more general 'rugged terrain' model outlined below.

Groundwater conditions in widespread and continuous permafrost may be most complex. The **active layer** is the

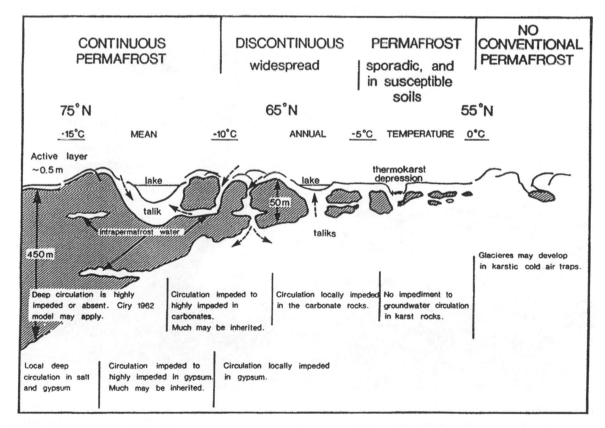

Figure 10.18   Model to depict the general relationships between permafrost and karst activity in terrains of low to intermediate relief. The model is based on conditions in the interior platform and arctic islands of Canada.

top layer, present everywhere, that is thawed and refrozen seasonally. Its depth diminishes from ∼2 m at the warm limits of permafrost to no more than 30 cm at the coldest sites. The **thermoactive layer** (the downward limit of **any** seasonal temperature change) is deeper, ranging −5 m to −30 m at differing sites. **Taliks** are unfrozen areas extending below the active layer. They may reach the base of the permafrost or terminate above it. They are present to some extent beneath all permanent lakes and ponds that are too deep to freeze to their beds in winter. Static or flowing groundwater bodies may be present in taliks or elsewhere within or below the permafrost (Figure 10.18). In karst rocks most intrapermafrost waters are contained within conduit systems or in particular strata with high diagenetic porosity.

The freezing process is weak because of the release of latent heat as water freezes. As a consequence little energy is required to keep conduits open once flow is established below the thermoactive layer. Maximum seasonal discharges into ponors (at the top of that layer) of as little as $5 \, \text{L s}^{-1}$ provide enough heat to keep them perennially open as taliks at the northern limit of the discontinuous permafrost zone in Canada (van Everdin-

gen 1981). Finally, it should be noted that the freezing point of water decreases as its content of dissolved solids increases; thus sulphate or saline groundwaters may flow where bicarbonate waters would freeze. Figure 10.19 shows one of a series of springs at Gypsum Hill at 79°N on Axel Heiberg Island, Canada, where there is perennial discharge of saline groundwater as cold as −4°C: the mean annual air temperature is −15°C (Pollard *et al.* 1999)

The most extensive tracts of karst rocks in the permafrost zones occur in Canada and Russia. Most of the Canadian areas and those of northwest Russia were glaciated. Large tracts of carbonate and evaporite rocks in the Angara–Lena Platform and adjoining Yakutia in Siberia were not (Pinneker and Shenkman 1992, Pulina 1992, 2005, Alexeev and Alexeeva 2002). However, the models presented here from Canadian experience (Figures 10.18 and 10.20) are broadly appropriate to them.

10.4.2 Karst development in permafrozen terrains

Development of karst systems does not appear to be significantly restricted in the sporadic permafrost

© 1999 NRC Canada

Figure 10.19 Gypsum Hill, a sulphate and saltwater karst spring emerging from a talik leak in permafrost under the channel of the Expedition River at 79°N on the west coast of Axel Heiberg Island, Nunavut, Canada. This 'frost blister' is ∼20 m in diameter. (Photograph by Wayne Pollard, with permission.)

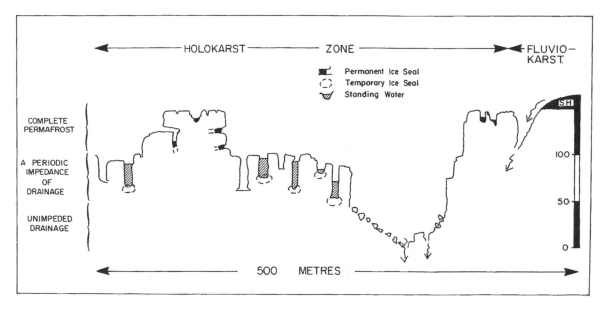

Figure 10.20 A zonal model for the relationships between permafrost and karst drainage in alpine permafrost and for rugged terrain in the widespread-to-continuous permafrost regions. Based upon the doline and corridor karst of Nahanni, Mackenzie Mountains, NWT, Canada. SH, shale caprock.

zone in Canada. Permanent freezing there is confined to patches of susceptible soils such as silts that readily accrete ground ice. In never-glaciated southern Siberia the morphology is of long shallow valleys tributary to the Angara and Lena rivers, with mixtures of carbonate, gypsum and clastic rocks. On the interfluves karst aquifers are recharged by diffuse flow or shallow dolines. In the valley bottoms, however, alluvium is permafrozen and supports surface streams for tens of kilometres until leak points via taliks are encountered; stream-sink dolines may develop there to lengths of several kilometres and be seasonally inundated (local term – **suchodol**; Salomon and Pulina 2005).

For alpine permafrost conditions and intermediate to high relief in the widespread permafrost zone the Nahanni karst may be taken as a model (Ford 1984; Figure 10.20). It extends between 61 and 62°N in the Mackenzie Mountains of Canada. Mean annual temperatures range from −6 to −8°C. There is little snow to insulate the ground and winter temperatures fall below −50°C. Summer temperatures can attain 35°C. As a consequence the thermoactive layer is deep. The region was glaciated at some time or times before the last interglacial but not since then. Topographic and climatic conditions are broadly similar in the central and northern Lena and Angara highlands.

The Nahanni is a mixed doline and corridor karst in 200 m of limestones overlying ~1000 m of dolomites. There are some 'scablands' (proglacial lake outburst channel landforms) superimposed on the karst. Larger depressions are >100 m deep and up to 1000 m in lenght (Figure 10.21). Their walls are frost-shattered and floors filled with talus. Where allogenic streams supply sand or clay these depressions become alluviated and may function as ideal baselevel poljes (Brook and Ford 1980).

There is unimpeded, highly integrated conduit drainage from all larger depressions to springs at the two ends of the karst belt. Mean velocities of ground water flow through conduits in the dolomite can exceed 4 km per day over distances of 20 km or more, as proven by recent dye traces. Between the depressions the highest karst is permafrozen and relict. An intermediate zone displays aperiodic impedance of the recharge. Its catchments are small, a few hectares at the most. Being in the thermoactive layer, the drains of dolines become sealed by ice or detrital plugs that freeze. Over one or a few melt seasons water accumulates above the seals until, it is supposed, the hydrostatic pressure is sufficient to rupture them; then they drain in the space of a few hours or days. Such behaviour tends to create cenote-form point-recharge dolines.

Further north in the Mackenzie Mountains the Bear Rock cemented dolomite breccia karst (Figure 10.22

Figure 10.21   Aerial oblique views of the Nahanni karst. (Upper) Raven Lake, a doline that is 150 m deep to the waterline. It drains dry in winter. (Lower) Flood waters rising in a small polje.

and section 9.13; Hamilton and Ford 2002) is at the widespread–continuous permafrost boundary. It was ice-covered during the last glaciation, so permafrost is only ~50 m in thickness. Many former (fini-glacial?) dolines there retain perched ponds for decades or longer but all allogenic streams are taken underground along the

geological contacts. To the northwest in the Yukon Territory, Tsi-It-Toh-Choh is a rugged limestone karst which escaped all glaciation because of regional aridity (Cinq-Mars and Lauriol 1985). Mean annual temperature is −10 to −11°C. Frost-shattered surfaces (**felsenmeer**) are more extensive than in the Nahanni and Bear Rock

Figure 10.22   (Upper) Solifluction lobes descending into a large doline in the Bear Rock karst at 66°N, Mackenzie Mountains, Canada. (Lower) At the same latitude, a turlough in the lowland karst of the West Arm, Great Bear Lake, Canada, in the widespread permafrost zone. It is ~4 km in length, a seasonally inundated closed depression in shallow glacial deposits underlain by dolomites over gypsum. (Photograph by R.O. van Everdingen, with permission.)

regions. Permafrost to 50–100 m depth is widespread to continuous. Relict caves are well decorated with speleothems which (from U series and palaeomagnetic studies) are almost entirely of Tertiary age (Lauriol *et al*. 1997). This suggests that most epikarst drainage has been eliminated in the Quaternary. Nevertheless, over a large area focused drainage to large dolines or valley bottoms permits unimpeded groundwater circulation to perennial springs that are marked by prominent winter-ice buildups (**'aufeis'**, German: or **'naledi'**, Russian): these may linger long into the summer season. Clark and Lauriol (1999) reported aufeis spreads up to 32 km² in the Yukon ranges, typically being ~1.3% of the areas of the source groundwater basins.

Aperiodic impedance is particularly relevant to the study of temperate highlands that experienced episodes of permafrost but not of glacier cover during the last or earlier glaciations. Examples include the Mendip Hills and Peak District of England, parts of the Ardennes, Jura, Moravia, Carpathians, French fore-alpine regions such as the Vaucluse Plateau, etc. At least the smaller threads of drainage from the epikarst base were often halted because speleothem growth in underlying caves was widely arrested (section 8.6). Blockage of dolines is suggested by deep accumulations of layered clays that are now dissected and drained by ponors. Occasionally, relict overspill channels cross the doline rims (e.g. Ford and Stanton 1968). The clay often has a large component of periglacial loess. Some dolines became infilled entirely with solifucted debris.

East of the mountains in northern Canada, as the southern limit of continuous permafrost is approached the topographic relief also diminishes. The constraint of low hydraulic gradient is combined with the constraints imposed by permafrost so that the latter are often difficult to distinguish. On carbonate rocks modern karst is poorly developed except along escarpment edges where groundwater gradients are high (e.g. Lauriol and Gray 1990). Karst is widespread where gypsum crops out or is covered by comparatively thin dolomites or shales. Van Everdingen (1981) mapped more than 1400 closed depressions functioning karstically and 67 perennial springs in gypsum around the western end of Great Bear Lake. This is at the widespread–continuous permafrost transition. The largest depressions probably antedate the last glaciation, contain much glacial sediment and function as seasonal turloughs (Figure 10.22).

Further north, active dolines become fewer but are recognized until relief becomes very low 200 to 300 km inside the continuous permafrost belt. Seasonal meltwater may accumulate and freeze in small dissolutional cavities to heave up shallow rock blisters a few metres in diameter. There is one known example of an open **pingo** (a larger blister form in which the ice derives from subpermafrost water) at 69°N close to the arctic coast. It is in dolomite, ~60 m in diameter and 22 m high (St-Onge and McMartin 1995). The region was glacierized during the last cold stage; much of the karst does not appear to have expanded during the postglacial period and may actually be shrinking in its extent. This last observation again raises the issue of **inheritance**.

In the modern permafrost regions it is suspected that many functioning karst systems may have been estab-

lished under more favourable combinations of hydrological and thermal conditions. They are maintained today because of the comparatively feeble freezing capacity below the thermoactive layer, but are essentially inherited. In arctic Canada and other glaciated terrains this implies inheritance from conditions of subglacial, or ice-margin deglacial, thawing of the permafrost. In nonglaciated regions of Siberia researchers consider comparable conditions to be warm interglacial episodes or to be the legacy of pre-permafrost times in the Cenozoic or Mesozoic (Pulina 2005).

Where there is groundwater circulation today it is difficult to prove that this has been inherited, although in Siberia mining and other engineering work has revealed numerous buried karst depressions, some filled with Cretaceous and others with Jurassic deposits. Origin under more favourable conditions is better understood where the inheritance could not be passed on, i.e. where permafrost has grown and arrested the circulation during post-glacial times. Examples include shallow corridor karst that, as noted, is quite widely scattered on limestone and dolomite plains between 68 and 73°N in Canada, and some large breccia collapses cemented by ground ice that have been found in arctic mines or on sea-cliff faces. Figure 10.23 shows the example of the Nanisivik zinc–lead mine at 72°30′N on Baffin Island. The ore body is a paragenetic cavity filling of pre-Cambrian age that is 100 m or more in width and >2 km in length. Along the south side, abutting a glacial valley, it is a very open pack breccia that is now firmly cemented by ground ice. The temperature is −13°C. Brecciation is attributed to dissolution following the thaw of permafrost beneath a valley glacier during the last glaciation (Ford 1996b). On west Spitsbergen, Salvigsen and Elgersma (1985) suggest that a doline terrain of low relief in gravels overlying gypsum on the seashore was initiated after it had thawed beneath a marine inundation and then had risen isostatically.

In the extreme conditions prevailing in the northern Canadian islands, northeast Spitsbergen, the Siberian high arctic, and Antarctica, mean annual temperatures are below −12°C and the warmest months are cooler than 5°C. As a consequence the thermoactive layer is shallow. Precipitation generally is <200 mm, in addition. If glaciated most land remains permafrozen. Recent karst development on the carbonate rocks is very limited in these conditions, although there may be effective groundwater circulation with dissolution to the base of the active layer at 30–100 cm. Ciry (1962) suggested that this would favour development of subcutaneous karst such as limestone pavements. Vestigial pavement no more than a few

1. LAST INTERGLACIAL?

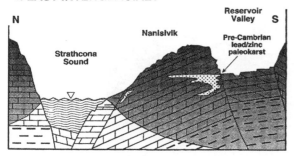

2. LAST GLACIATION

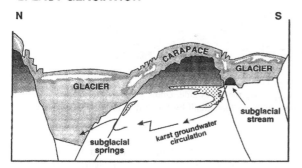

3. HOLOCENE

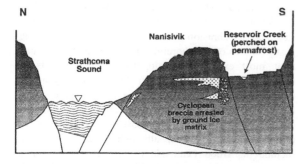

Figure 10.23 Schematic cross-sections through the Nanisivik zinc–lead ore body (Baffin Island, NWT) to model the development of a megabreccia arrested by ground ice. (1) Before brecciation (the last interglacial or earlier). The terrain is ice-free and permafrozen to depths >100 m except beneath Strathcona Sound, an arm of the sea. (2) The brecciation event (the last (Wisconsinan) glaciation?). Permafrost largely or entirely thaws beneath a cover of glacial ice. Subglacial meltwater is diverted through the Nanisivik palaeokarst into the Sound. Dissolution induces collapse of the southern part of the ore body along a frontage of ~200 m. (3) Post-glacial. Permafrost returns. Brecciation is halted by growth of ground ice. Reproduced from Ford, D.C. (1996b) Karst in a cold climate, in Geomorphology sans Frontie'res (eds S.B. McCann and D.C. Ford), John Wiley & Sons, Chichester, pp 153–79.

metres wide is occasionally seen along escarpment edges and small channel karren may develop on steep, bare slopes (e.g. Woo and Marsh 1977; at 73°N on Cornwallis Island). However, in most instances the surface is reduced to felsenmeer by frost shatter (Figure 10.24). This is the predominant aspect of carbonate outcrops in arctic Canada (Bird 1967). Shattering is often more effective on them than on other comparatively strong rocks because solution opens up cracks to permit veins of ice to form more readily (e.g. Goudie 1999). Solvent capacity is then dissipated on the rubble. When the water freezes or is sublimated much of the solute carbonate is redeposited as calcretes or biocrusts on the undersides of clasts or in joints beneath the rubble (Lauriol and Clark 1999). Some groundwater flows through for short distances because seasonal seeps are common along the bases of low escarpments.

Deeper, subpermafrost karst systems may develop in salt and gypsum because of the greater solubility of these rocks and the lower freezing point of strong solutions. The systems are artesian. Most of them discharge where permafrost is dissipated at the sea coast, e.g. the Gypsum Hill springs noted above. There are stream-sink dolines on the flanks of salt diapirs in Axel Heiberg Island, 80°N. The groundwater flow paths are no more than a few hundred metres in length and appear to be shallow supra- or intrapermafrost features. Gypsum solution breccia has been noted in sea cliffs on the north coast of Ellesmere Island (83°N). This may be the most poleward karst, but it is believed to be inactive (arrested by ground-ice cementation) today.

10.5 SEA-LEVEL CHANGES, TECTONIC MOVEMENT AND IMPLICATIONS FOR COASTAL KARST DEVELOPMENT

Mean sea level closely approximates the geoid, i.e. the equipotential surface of the gravitational and rotational potentials. Mörner (2005) has explained the complex interplay of processes that determine its level and produce changes in its position. The interface between this surface and the land has varied considerably over geological time, particularly in the past 2 Ma. This is because there have been worldwide oscillations in ocean level, termed **eustatic** variations, plus regional- to continental-scale vertical crustal adjustments and linear mountain building or trough-forming disturbances attributable to plate tectonics. These movements of both land and sea have forced repeated vertical displacement of the sites of coastal karst development.

The most important factors producing eustatic changes are of glacial, tectonic and geoidal origin; the former

Figure 10.24   Limestone pavement reduced to felsenmeer, Akpatok Island, Ungava Bay, Canada. The island is underlain by permafrost, which is breached here where the hydraulic gradient is steep along the valley side. Drainage down into a clint-and-grike epikarst is clearly indicated by the pattern of microdolines in the frost rubble. (Photograph by Bernard Lauriol, with permission.)

being the most significant over the timescale with which we are concerned. Glacio-eustatic changes stem from the transfer of water mass from the ocean to continental ice sheets during glaciations. Glacial eustasy accounts for sea-level movements of up to 130 ± 20 m in periods of 10^3 to 10^5 years (Figure 10.25); sea-level change through the last glacial cycle having been reviewed by Lambeck and Chappell (2001).

The world's most important data on sea-level variation over time have been obtained from the datable evidence available on limestone coasts. However, Mörner (2005) emphasized that since tectono-eustatic and glacio-eustatic rises and falls cannot occur without accompanying geoidal eustatic changes, we cannot expect to find a perfectly parallel history of absolute sea level for different regions of the globe. Although the major cause of Quaternary sea-level variations must be glacial eustasy, Nunn (1986) stressed the danger of assuming negligible geoidal eustatic changes in an ocean with $\sim$150 m of movable relief, particularly when the above type-sites are in areas of large geoid anomaly, where the probability of change could be correspondingly high.

One of the consequences of variations in sea level is that, even in tectonically relatively stable areas, the present coastline has only been vertically within 30 m of its present position for about 30% of the past 800 000 years (Figure 10.26). So with the baselevel of erosion well below its present position for much of the

Quaternary, the water table inland and the associated freshwater–saltwater interface zone were also lower. Hence there was ample time in the Quaternary to produce karst almost anywhere in the zone exposed by glacial low sea levels down to -130 m or more. On tectonically stable coasts, abrasion platforms and coral reefs marking past positions of the shoreline are presumed to have had their most extensive development during periods of relative stillstand of the ocean. On tectonically active coasts, the corresponding position is at an interval of tangency between uplift and sea-level oscillation. Excellent examples of carbonate coasts where this sort of movement has occurred, although at different rates, include the Huon Peninsula (Figure 10.27) of Papua New Guinea (Bloom 1974, Chappell 1983, Chappell and Shackleton 1986, Ota and Chappell 1999), Barbados (Fairbanks and Matthews 1978, Schellmann *et al.* 2004) and Bermuda (Harmon *et al.* 1978, 1983). In Patagonia there is a spectacular series of horizontal notches in the cliffed marble coast formed by chemical and biological erosion in the intertidal zone during brief stillstands on an otherwise rapidly uplifting coast (Maire *et al.* 1999), but unfortunately they are very difficult to date.

Smart and Richards (1992) analysed over 300 published U series analyses of corals from Quaternary marine terraces and resolved eight high seastands with means and standard deviations of 129.0 ± 33.0, 123.0 ± 13.0, 102.5 ± 2.0, 81.5 ± 5.0, 61.5 ± 6.0, 50.0 ± 1.0, $40.5 \pm$

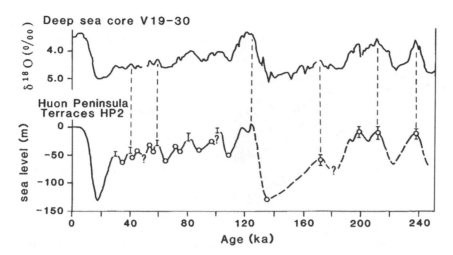

Figure 10.25   Sea-level curve derived from coral terraces on the Huon Peninsula, Papua New Guinea (Figure 10.27), correlated with Pacific Ocean deep-sea core V 19-30. Reproduced from Chappell, J. and Shackleton, N.J. (1986) Oxygen isotopes and sea level. Nature, 324, 137–40.© 1986 Nature Publishing Group.

5.0 and 33.0 ± 2.5 ka. Using both U series and electron spin resonance dating on coral from Barbados, Schellmann *et al.* (2004) differentiate three high seastands during the last interglacial (MIS 5e) at ~132 ka (ESR) to 128 ka (U/Th), at ~128 ka (ESR) and at ~120 ka (U/Th) to 118 ka (ESR). In Western Australia, coral ages indicate that the last interglacial started at ~128 ± 1 ka and terminated at 116 ± 1 a with the main period of coral growth being a short interval from 128 to 122 ka (Stirling *et al.*

1995, 1998). Reviewing coral age data from four sites (Bermuda, Bahamas, Hawaii and Australia), Muhs (2002) concluded that the last interglacial had a sea level at least as high as present from ~128 to 116 ka.

Marine overgrowths on speleothems (Figure 10.28) can also be used to determine the dates of high sea-level stands. Thus for example, Fornós *et al.* (2002) found high MIS 5e sea levels in the Mediterranean to have occurred at 135–130 ka and 125–112 ka. From an analysis

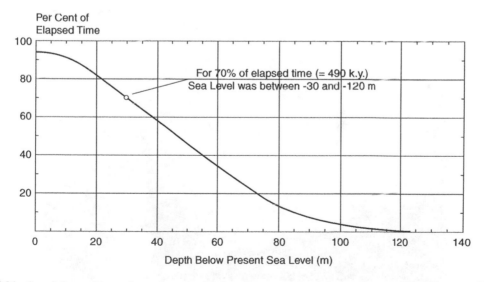

Figure 10.26   Cumulative residence time of sea level below its present level during the past 800 000 years. Reproduced with permission from Purdy, E.G. and Winter, E.L. 2001. Origin of atoll lagoons. Geological Society of America Bulletin 113 (7), 837–854, Geological Society of America.

Figure 10.27   Oblique aerial photograph of the Huon Peninsula showing the raised coral reefs used in the construction of the sea-level curve of Figure 10.25. (Photograph by D. Dunkerley.)

of speleothems drowned by rising sea level, it is also possible to estimate the maximum level attained by the sea at different times in the past, e.g. Richards *et al.* (1994) and Figure 8.22b.

In addition to sea-stand data, information on the ages of emerged coastal terraces and the rate of tectonic tilting may also be obtained from cave deposits. Williams (1982b) constructed uplift-rate curves for different cave locations along the northwest coast of the South Island,

New Zealand, and by comparing the different rates was able to estimate the rate of tilting. Fornós *et al.* (2002) identified the rate of tilting along the coast of Mallorca in the Mediterranean by comparing the elevation of MIS 5e speleothem deposits in caves.

On stable coasts, new features may be built on or cut into a base inherited from the past. Thus post-glacial morphology may be superimposed onto that developed in the last interglacial around 125 000 year ago, when sea level briefly stood up to 6 m above present. Similarly, landforms developed in coastal zones exposed during low sea levels may have been reactivated and resubmerged several times during their history (Figure 10.29); thus drowned karst is normal along all carbonate coasts except those subject to very rapid tectonic uplift. Large cave networks discharging at the coast, in particular, have been drained and reflooded repeatedly with Quaternary rise and fall of sea level. Caves in the Yucatan Peninsula provide spectacularly good examples of this (Beddows *et al.* 2002) (see section 5.8). The conduits feeding many submarine springs would also have been formed in association with lower baselevels.

10.5.1 Karst phenomena in the coastal zone

In the coastal zone a distinction must be made between those features formed at the coast and those that land–sea interface changes have by chance located there. Those

Figure 10.28   Marine molluscs growing part way up a stalactite in a cave on Manus Island, Papua New Guinea. They record precisely the position reached by a former sea level.

Figure 10.29   Dolines on the New Zealand coast drowned by the Holocene rise of sea level. These are solution dolines that were filled by volcanic deposits, partly removed by suffosion prior to drowning.

formed at the coast mainly comprise littoral karren and notches of the intertidal zone, discussed in section 9.2. Here we focus predominantly on the other characteristics.

The coastal zone in karst is defined by the inland limit of marine tidal influences. Controls on the seawater–freshwater interface are explained in section 5.8. In porous, emerged coral islands the sensitive groundwater lens readjusts twice daily to the low-amplitude variations of marine tides. On Niue Island in the Pacific (Figure 5.36), for example, the seawater–freshwater transition at the level of the water-table occurs about 0.5 km inland and tidal effects are measurable to the centre of the island more than 6 km from the shoreline (Jacobson and Hill 1980, Williams 1992). The intrusion of marine influences along coastlines composed of dense crystalline carbonates is more irregular than in porous limestones, and penetration of tidal effects along conduits can be considerable. For instance, in the Gort Lowland of western Ireland (Figure 9.36), tidal Lough Caherglassaun is 5.3 km inland and in summer fills and empties twice daily with a lag of 3 to 4 h behind tides in neighbouring Galway Bay. Nearby Hawkill Lough is 8.75 km inland and is also reputed sometimes to show the effect of tides.

Difficulties exist in the interpretation of some morphological features of carbonate islands. Thus whereas Darwin (1842) believed barrier reefs around volcanic islands to be due to subsidence, Stoddart *et al.* (1985) explained them as principally the result of karstic erosion of the inner margin of a fringing reef by allogenic runoff,

followed by drowning by a rising sea; and Purdy (1974) and Purdy and Winterer (2001) ascribed the origin of atoll lagoons to differential solution by rainwater during glacial low sea levels. Whereas numerous authors, such as McNutt and Menard (1978), have attributed suites of coral island terraces to uplift, Nunn (1986) has warned that some could alternatively be a result of geoidal eustatic regression. It is clear that many carbonate islands have undergone long and complex histories with the alternation of marine and terrestrial influences leaving complex patterns in the landscape. Mylroie and Carew (1995) have provided a valuable review and explanation of these phenomena.

Carbonate islands are like dip-sticks in the sea, so the effects of eustatic variation is well displayed in them (Figure 10.30). Emerged coral atolls record the relative level of the ocean as it has moved in the past, as well as lithospheric flexure (Spencer *et al.* 1987). Negative shifts in sea level around atolls exposed precipitous fore-reef submarine slopes, and with the loss of buoyant support, massive scale cliff failure often followed. Dislocations arising from this are schematically illustrated in Figure 10.30. Long fracture traces subparallel to the coast now mark the inland margins of foundered blocks. Deep fissures sometimes open at the surface have been confused with karst caves because of their vadose canyon-like morphology and speleothem decoration, but many are simply tectonic features. Good examples of such fracture caves occur in the Bahamas (Palmer 1986) and on Niue

Island (Terry and Nunn 2003). Accelerated corrosion at the base of cliffs produces prominent notches in the intertidal zone of tropical seas (Figures 9.15 and 9.16). Notches have been found to $-105\,\mathrm{m}$ in the submerged wall of San Salvador Island in the Bahamas (Carew *et al.* 1984).

Falling sea levels result in the drawdown of the freshwater lens and in deepening the vadose zone. The location of mixture corrosion at the halocline and at brackish springs is correspondingly lowered and with it an important horizon of karst cave development – the flank-margin

caves of Mylroie and Carew (1995). According to Back *et al.* (1984), differential solution generates pocket valleys at spring heads round the coast of Yucatan; their later submergence being responsible for coves and crescentic beaches.

The karstification of atolls and fringing reefs during low sea levels has been recognized for many years, although obvious suites of surface karst landforms other than karren are uncommon. Since hydrogeological prerequisites for the focusing of percolation are generally absent, because of the extreme porosity of coral

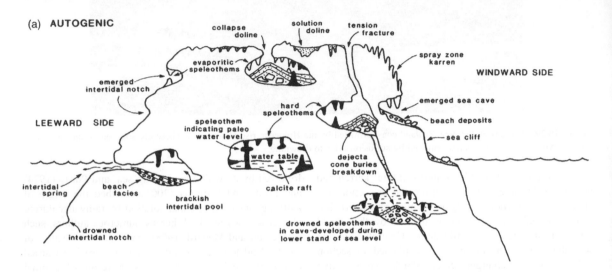

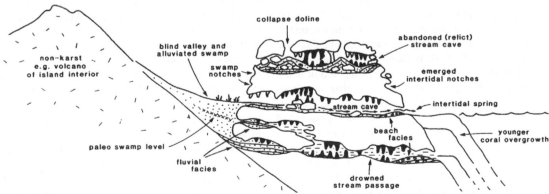

Figure 10.30 Geomorphological and stratigraphical effects of sea-level changes observed in coastal cave systems. (a) Mixing corrosion at the freshwater–saltwater interface promotes cave development near the water table and especially flank-margin caves around the island perimeter. The tension fracture is a non-karst feature resulting from rock failure due to loss of buoyant support during glacio-eustatic low sea levels, although it is exploited by karst circulation. (b) Allogenic runoff causes seawards retreat of the inland-facing reef margin and promotes cave development through the emerged reef. Submarine or intertidal springs occur where outlets of stream passages remain clear of reef overgrowth.

limestones, the generation of solution dolines by draw-down is uncommon and the evolution of cockpit karst relief is unlikely even in the humid tropics. Suffosion dolines can occur in residuum. Very high rock porosity works against the centripetal focusing of runoff (which requires **lateral** movement), and so argues against the central lagoons of atolls being drowned basins formed by focused dissolution. Instead, it is more likely that their basin form is a product of differential corrosion achieved by **vertical** percolation of water acidified in the soil. Evidence for this is provided by Urushibara-Yoshino (2003) on Minamidaito Island, Japan, where she found the dissolution rate of limestone tablets buried in the soil of an emerged lagoon floor to be four times greater than the rate in air. This differential corrosion helps to explain the emergent 'walls' of relatively soil-free limestone that rim the island and provides a mechanism for increased dissolution in the atoll interior. If such islands were later submerged, new coral growth would be favoured on the higher rims rather than in the deeper lagoon interior; thus accentuating the atoll form. These processes are accentuated by repeated exposure and submergence.

The karst that develops on emerged reefs depends on the hydrochemical controls (Figure 4.1). Entirely auto-genic systems such as Niue and the Trobriand Island (Ollier 1975) usually support relatively few karst features; this being more a function of their vadose hydrology than their coastal location. Nevertheless, tropical cyclones ensure the penetration of salt-spray influences well inland. Emerged coral islands often have rough corroded surfaces (**makatea**) with the texture of scoriaceous lava rather than sculpted karrenfeld; these features being developed on a weakly case-hardened crust that commonly occupies the top few metres of the vadose zone (Hopley 1982). Makatea are eogenetic karren, as described by Taboroši *et al.* (2004) on Guam. Across these surfaces are scattered the occasional collapse doline, a few giving access to karst caves. Only rarely, as in Barbados, is doline development comparatively extensive (Table 9.3).

The highly porous nature of coral encourages diffuse infiltration and circulation and hence inhibits karst cave development. But where there is pronounced point recharge from allogenic streams flowing from adjacent non-karst rocks, exposed coral reefs may develop a dis-tinctly karstic drainage with large river caves. For example, streams draining radially from the impervious volcanic inliers of some coral islands (e.g. Mangaia Island) usually terminate in a ring of blind valleys on the inland side of an emerged reef, from where subterra-nean conduits lead runoff to brackish springs on the coast. Neighbouring blind valleys coalesce to yield a circular moat-like depression with a steep inland-facing limestone cliff; the retreating corrosion front (Figure 10.30). During high stands of the sea, the moat becomes a swamp, lake or lagoon behind the barrier. Dry valleys across the emerged reef mark abandoned stream courses or atoll passes, but otherwise the surface topography displays a limited range of karst features, including karren, pit caves and occa-sional collapses. Corrosion plains may form along the allogenic input boundary, but residual towers are not expected unless there was an earlier cockpit phase. In a detailed critical examination of the applicability of the antecedent karst hypothesis to the Great Barrier Reef, Hopley (1982, p. 218) concluded that 'there is no evidence for larger karst features such as marginal plains or towers having evolved. Only the blue holes, and possibly the gorgelike interreefal channels, provide evidence of a major karst process, and both features are very localised'. However, as we saw in section 9.5, some (if not most) blue holes are related to fractures rather than to karst conduits.

The syngenetic karst of Pleistocene calcareous coastal dunes (section 9.12) is in many respects similar to the karst found on exposed reefs. Even though aeolian calcarenite is extremely porous, case-hardening welds its surface together and lateral allogenic recharge from the hinterland generates blind valleys and river caves. Few other karst attributes are evident except solution pipes (or pit caves), occasional collapses and a limited range of karren forms. Some enclosed depressions in aeolian limestones are not karstic, but interdune hollows.

Calcareous beach sands in tropical and warm tempe-rate regions are often indurated between the limits of high and low spring tides, although sometimes cementa-tion extends to the immediate supratidal zone. The resulting coarse-grained limestone is termed **beach-rock** (Figure 10.31). The cemented sands are usually identical to unconsolidated materials nearby and retain the seawards dip of the beach bedding. Induration appears to proceed inland, with the stratigraphically lowest of the dipping beds being the most recently cemented. Cementation has been attributed to micro-organisms, inorganic processes involving fresh water and inorganic processes from seawater (Neumeier 1999, Webb *et al.* 1999). Hopley (1982) noted the frequent occurrence of aragonite cement, as opposed to the calcite spar that would be expected from fresh water. Cementation of beach-rock can be very rapid, with considerable indura-tion having been recognized since World War II.

Where dense crystalline limestones form the coast the features are more clearly karstic, even where drowned, simply because such rocks are more favourable for karst development. For instance, spectacular examples of partly submerged tower karst occur in Ha Long Bay in

Figure 10.31   Sheets of beach-rock accumulation on Vava'u Island, Tonga group.

Vietnam and the Phuket Island of Thailand. Doline karst affected by high sea level is known along the Dinaric coast of Adriatic Sea and elsewhere (Figure 10.29), and often these regions have submarine springs (section 5.8), which along the Dinaric coast are called *vrulje*. It is assumed that their conduits were formed during Quaternary lowstands of the sea, but that their activity has been maintained because of sufficient hydraulic head inland. More unusual are sea swallow holes, the best known case being the Sea Mills of Argostoli in the Ionian Islands (Maurin and Zötl 1963). Seawater is drawn into the karst and passes right through the island to springs on the other side. Although there may be an ejector effect caused by autogenic recharge, differential tide levels on each side of the island aids the head difference. Bögli (1980) noted the occurrence of saltwater swallow holes elsewhere and the relevance of the Bernouilli equation to their explanation. A more common phenomenon is the reversing flow of coastal subterranean rivers with the ebb and flow of tides (sea-estavelles), particularly in regions with a large tidal range. Tidally generated reversing flow through blue holes is reported from the Bahamas.

Brackish karst springs are sometimes located above sea level. The main outflow pool of the Waikoropupu springs of New Zealand, for example, is 14 m above sea level but its water is 0.5% salt. Williams (1977) suggested this to be the result of a venturi mixing process drawing in water from the underlying coastal intrusion into the aquifer. This mechanism is supported by the observation of increasing salinity with spring discharge. Stringfield and

LeGrand (1971) and Milanović (1981) discussed other cases.

Deep submarine karst has also been observed. Maire (1986) briefly reported the discovery of karren-scale grooves and pitting on steep limestone slopes at depths of 1000 m to 3000 m off the Iberian Peninsula. This is far too deep for any eustatic fluctuation of Quaternary age. He suggested that the features may be dissolved by descending cold waters, perhaps with thermal or chemical mixing effects. Possibly they are very ancient relict forms, although this seems unlikely. However, palaeokarsts that have been taken deep below sea level by tectonic subsidence are well known, including the Rospo Mare beneath the Adriatic Sea (Soudet *et al.* 1994).

10.6 POLYCYCLIC, POLYGENETIC AND EXHUMED KARSTS

W.M. Davis (1899) maintained that all uplifted land will ultimately be worn down to a surface of low relief termed a peneplain at an elevation close to baselevel, which is usually sea level. This unidirectional erosion process he called the **cycle of erosion**. As the cycle progressed, he considered landforms to pass through an identifiable sequence of stages termed 'youth', 'maturity' and 'old age'. The occurrence of uplift before the completion of the cycle would lead to its interruption; erosion processes would be rejuvenated and a second cycle would ensue, with development proceeding headwards from the sea. Thus landforms of the present cycle would be found downstream of rejuvenation heads (marked by knickpoints in stream profiles), while the still unconsumed morphology of the previous cycle would persist upstream. Grund (1914), Cvijić (1918) and others incorporated Davisian cyclic ideas into their schemes of karst landscape evolution (section 9.15) and Davis himself proposed a two-cycle hypothesis of cave evolution (Davis 1930). Landscapes showing evidence of more than one cycle of erosion are termed **polycyclic**. The relationship of stepped erosion surfaces to karsts is discussed in many contexts, e.g. in the eastern USA by Palmer and Palmer (1975) and White and White (1983) and in southern China by Song (1981), Zhu (1988) and Sweeting (1995). In a review of karst types in China, Zhang (1980) described numerous cases of knickpoint recession into such surfaces, which probably range in age back to the Cretaceous.

Rejuvenation results from increased erosive energy. This can stem from (i) increased potential energy due to a negative shift in baselevel (following either uplift or sea level fall) or (ii) increased erosive power (chemical as well as mechanical) particularly following a climatic change that results in increased runoff. Uplift, if sufficiently great, can also lead to climatic change, as in the case of the Tibetan plateau. Many variations in climate have occurred during geological time, for example at the height of the Quaternary glaciations both the cold polar and semi-arid mid-latitude zones expanded, with compensating contraction of the intervening mediterranean and humid tropical zones. The direct effects of glaciation were particularly influential on karst, because glacial deepening of valleys lowered the local baselevel of karst springs, and because large volumes of glacial meltwater incised valleys and invaded pre-existing caves. The effect and timing of rejuvenation of karst in Indiana and Kentucky following the entrenchment of the Ohio River (by glacial meltwaters) and its tributary the Green River has been investigated by Pease and Gomez (1997), Granger and Smith (2000) and Granger et al. (2001).

Climatic change does not always stimulate rejuvenation. Often it alters the balance of processes instead, most radically when fluvial erosion gives way to glaciation, as discussed in section 10.3. In numerous cases the climatic shift persisted long enough to have its geomorphological consequences engraved in the landscape, yet was of insufficient duration to obliterate all traces of previous regimes. Landforms moulded under one climato-genetic regime are then remodelled under the next. But if modification is incomplete before the climate changes again, then the landscape will bear the imprint of more than one genetic environment and can be termed **polygenetic**. Good examples are provided by pre-Quaternary

but subsequently glaciated karsts in Ireland (Williams 1970, Drew and Jones 2000) and by karst hills similar to tropical mogotes buried under detrital cover in western Europe (Salomon et al. 1995).

Because changes in baselevel and climate have affected large parts of the world, most karsts can be expected to have both polycyclic and polygenetic elements in their landscapes and groundwater systems. To these, in conclusion, we should add the complication of the **exhumation** of palaeokarst topography and hydrological systems that is occasionally detected contributing to modern karst assemblages. The Australian continent is particularly interesting in this regard because it has so much ancient landscape that has evolved slowly in comparatively low-relief and low-energy (dry) conditions.

10.6.1 The Tindall Plain and Riversleigh karst, Northern Territories and Queensland

The Northern Territories of Australia contain very large areas of buried carbonate rocks in synclinal structures known as the Daly, Wiso and Barkly basins (Figure 10.1). In places around the margins of this region the cover beds have been eroded and karst is exposed over hundreds of square kilometres. This is the case in the north around the town of Katherine where at the edge of the Daly basin there is a broad strip of karst, and also further east on the Barkly tableland, especially in the Lawn Hill–Riversleigh area of Queensland. In both of these districts there is clear evidence for a long complex period of karstification.

In the Katherine region, karst is developed at 100–220 m a.s.l. on Lower Cambrian Tindall Limestone, a platformal rock of 180 m thickness. The limestone is thick to massively bedded with deep regular jointing, and is overlain by thinner, weaker ferruginous Middle Cambrian sandstones and silicified limestones. All these strata remain flat-lying and with little tectonic disturbance. They were dissected to form tableland topography with shallow scarp fronts on the stronger beds at some time or times before they became buried by Lower Cretaceous sandstones (Lees Formation) during the course of a low-energy marine transgression. The unaccounted Middle Cambrian to Early Cretaceous time interval is about 400 a.

The modern climate is tropical semi-arid, with rainfall at Katherine ranging from 800–1020 mm during a 5 month rainy season; the district being near the inland limit of penetration of monsoonal rains. Average annual temperature is about 27°C and potential evapotranspiration approaches 2300 mm.

Local relief on the limestones is up to 60 m, and the Katherine River, which is shallowly entrenched across

Figure 10.32   (Upper) Air oblique photograph showing a clint-and-grike paleokarst surface on the Tindall Plain, Australia, being exhumed and destroyed. (Photographed by D. Karp, with permission.) (Lower) Corroded clint surface at the exhumation margin and the lower surface of the Tindall Plain that replaces it.

the Tindall outcrop, provides the regional baselevel for karstic groundwater discharge. Allogenic streams collect on the Cretaceous cover, dissect and strip it and the Middle Cambrian strata before sinking at the limestone contact. Water- and windborne detritus is carried onto lower areas of the limestones where it forms deep soils: subsoil pinnacle karst formation is active beneath them, with much suffosion doline development (Karp 2002). In the wet season caves fill with water and there is inundation of some very shallow poljes.

In 'The enigma of the Tindal Plain', Twidale (1984) suggested that the limestone morphology developed as a shallow scarpland with mesas and plateaus during the Jurassic before being buried by the Lees Formation sandstones. Burial was then followed by later Cretaceous and early Tertiary deep weathering with lateritization. Beginning in the Miocene, the lateritic surfaces began to be stripped back, exposing limestone plains decorated with pinnacles of the Cambrian limestone *in situ*, plus tors and corestone piles from the Cretaceous sandstone. These stratiform plains thus must be at least of Jurassic age, and are being exhumed intact and scarcely marked. Twidale noted the 'anomaly' that, although the plain appears to be a youthful landform, field evidence points to it being an ancient feature of complex origin perhaps at least 135 Ma; this is the 'enigma' to which he referred.

The complexity in detail is illustrated in a low elevation aerial oblique picture of a Tindall Plain surface and

Figure 10.33 Strongly cemented fill of ferruginous sands and limestone clasts emplaced within solutionally widened joints in Cambrian limestone bedrock (pale *in situ* blocks on the right). Layering of the fill was disturbed by slumping in the centre prior to induration. The fill may be pre-Cretaceous in age and forms part of the wall and ceiling of Cutta Cutta Cave near Katherine, Northern Territory, Australia.

its bounding scarp edge (Figure 10.32). The intact plain is densely dissected by rectilinear grikes (up to 1 m or more in width) that have been uniformly infilled with detritus. In the top right corner of the upper figure some Lees Sandstone remains *in situ*. Modern storm wash keeps grikes clear of filling along the scarp margins, leaving the intervening clint blocks upstanding as pinnacles ~2 m in height that are degrading comparatively quickly. The exhumed surface is replaced by a more recent surface just a few metres lower in elevation (Figure 10.32). Thus the modern Tindall Plain mimics the pre-Cretaceous surface, but stands just a little lower as a consequence of post-exhumation dissolution. The wash detritus is deposited on lower plains (often across surfaces of individual Cambrian

limestone beds) in the form of deepening veneers that will retain some moisture and so induce the modern subsoil pinnacle karst and suffosion doline development.

Several caves up to 3000 m in length have been discovered beneath the plains. They are restricted to one or a few of the massive beds, none being deeper than 30 m or so before reaching the regional water table, although extensive flooded joint networks extend beyond. They have developed along the major, grike-forming joints, with parts of the cave roofs being composed of the filling materials (Figure 10.33). Lauritzen and Karp (1993) found that these consisted chiefly of limestone clasts within a ferruginous sand matrix that often contained fragments of (vadose) stalagmites. The caves lack regular flowing streams today but some can be flooded during modern storms. There is evidence of earlier development in thermal waters (ceiling cupolas, etc.), and there are thermal karst springs on the Katherine River today. The caves therefore appear to belong to the modern (Miocene and later) cycle of erosion but are guided by grikes that may be of palaeokarst origin and thus could be considered at least partial exhumations.

Evidence of conditions that affected the karst during Oligo-Miocene times is available further east at the edge of the Barkly tableland in Queensland (Figure 10.1) where, in the Lawn Hill–Riversleigh area, the karst is also developed on tabular Cambrian carbonates (Thorntonia Limestone). The limestones are dolomitic and cherty, and rest unconformably on a basement of Proterozoic sandstones (Figure 10.34). The O'Shanassy and Gregory Rivers traverse the karst in gorges before emerging onto a pediplain cut across basement clastic deposits. They then flow north to the Gulf of Carpentaria. Of particular interest is the occurrence in the district of linear patches of dense, fossiliferous Tertiary freshwater limestones set within the Cambrian carbonates (Figure 10.34). These occupy basins that may have been small lakes along river valleys, and they indicate that the water table was near the surface. The exceptionally rich fossil fauna that they contain (the area is a World Heritage site for that reason) indicates a rainforest environment in Oligo-Miocene times. At the time, the continent had a more southerly latitude: in the course of drifting further north the region has become drier, the rain forest has disappeared, and so morphogenetic conditions have changed. In the process the region was also uplifted ~150 m with the result that main rivers were rejuvenated and able to entrench their courses, but their tributaries could not respond as vigorously and so became a network of dry valleys. Across the tableland this encouraged the deepening and widening of joints, and has produced a pinnacled terrain of a few metres local relief.

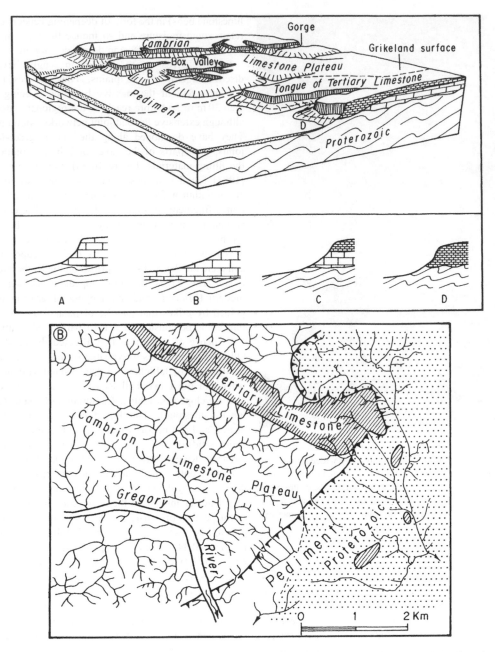

Figure 10.34   (Upper) A sketch of karst landforms in the Riversleigh area of Queensland. (Lower) A belt of Tertiary (Oligo-Miocene) freshwater limestones set within Cambrian limestones in the Riverleigh area. Reproduced from Williams, P.W., Interpretations of Australasian karsts. In J. L. Davies & M.A.J. Williams (eds.), Landform Evolution in Australasia, 259 286 © 1978 ANU Press.

The combined evidence from the Katherine and Riversleigh areas reveals a long and complex history of which we have gleaned only the sketchiest of details. More than one episode of karstification has been involved and there have been considerable shifts in morphogenetic environment. Evidence from Riversleigh indicates early karst development under lowland rain-forest conditions; while seasonally humid tropical to subtropical conditions are suggested by the ferruginous sandy matrix of cemented cave and joint deposits in some caves near Katherine. Some of the filled grikes on the Tindall Plain are even reminiscent of those exposed by glacier-ice stripping of clint-and-grike terrain on the carbonate plains of Canada (Figure 10.15). Interestingly, it is known from numerous other sites in Australia that there was exposure to the Permian ('Gondwana') Glaciation (Twidale 1976), so a subglacial contribution to the evolution of this already genetically complex terrain is possible.

10.6.2 The Kimberley region, Western Australia

The main stages of the present phase of evolution of the karst of the limestone ranges of the Kimberley region in Western Australia (Figure 10.1) were discussed in section 10.2. However, there is a much older dimension. This has been investigated and explained by Playford (2002), who provides compelling evidence that the Devonian reef limestones of the area have been subjected to four periods of karstification. The first followed Late Devonian reef emergence, evidence for which is now seen as sequence boundaries within the carbonate rocks. A second period involved latest Devonian or earliest Carboniferous hydrothermal karst development beside active faults, resulting in caverns that were filled penecontemporaneously with zinc–lead mineralization. The third phase, Mid-Carboniferous to early Permian, occurred when the area was faulted, tilted and mildly folded before being truncated and intensively karstified. Playford observes that a well-exposed glacial pavement provides evidence that a continental ice sheet (part of the Permo-Carboniferous 'Gondwana Glaciation') moved towards the north to northwest, and he suggests that during this interval glacial meltwaters could have been responsible for considerable dissolution. Largely glacigenic claystones and sandstones of the Lower Permian Grant Group were deposited over the karstified and glaciated terrain, and the karst remained buried until Cenozoic times. In the fourth period of karstification Playford argues that as these clastic coverbeds were denuded an Early Permian palaeokarst was progressively exhumed (Figure 10.35) and incorporated into the developing modern topography.

Playford (2002) contends that many features of the present landscape are the legacy of Early Permian subglacial dissolution. Thus the Mimbi Caves system, which is the largest cave and karst corridor network of the area, is suggested as having been formed by subglacial solution along a series of rectilinear joints, which resulted in a maze of narrow passages and corridors 2–5 m wide and several kilometres long, and up to 50 m deep. However, under modern conditions most of the system is inundated

Figure 10.35 Rugged eroded surface in Devonian reef limestones (foreground) being exposed as coverbeds (cliffs in the background) of Lower Permian glacigenic clastic deposits are stripped back. Kimberley region, Western Australia. (Photograph by P. Playford, with permission.)

during times of flood and there are many recent speleothems; so present processes must be seasonally active and could have been so (with variable intensity) throughout Cenozoic times. Section 10.3 discusses the issues associated with dissolution under continental glaciers in more detail.

According to Playford (2002), much of the Permo-Carboniferous subglacial meltwater would have drained through networks of subglacial channels cut in bedrock, but roofed by ice (such features are known as N or Nye channels, see section 10.3.2). He interprets numerous gaps and gorges that cut through the limestone ranges to be examples of N channels that were exhumed by Cenozoic erosion of the relatively soft Permian sedimentary rocks that once filled them. The major Windjana, Geikie, Galeru and Brooking Gorges are considered to have been initiated in this way. However, because subglacial channels are produced under pressure flow, their floors can be undulating and their morphology can change rapidly over very short distances. Such morphological characteristics must therefore be demonstrated before their origin as N channels can be confirmed. An alternative and less demanding hypothesis is that the rivers were superimposed from the overlying clastic beds, the larger rivers having maintained their courses as incision produced the gorges, but the smaller streams having passed entirely underground leaving dry valleys at the surface. In this process, blocks of

Early Permian clastic rocks would have slumped into developing karst features and sediments would have been redistributed and deposited in slackwater areas and hollows.

Playford (2002) demonstrates that Permian sediments occur in certain depressions on the limestone plateau and provides evidence from drilling that some dolines and pipes contain fills of siliceous sandstone and claystone more than 200 m deep, although 50–100 m is more usual. Jennings and Sweeting (1963) commented that the clastic sediments may have been let down by solution from a higher level and 'may represent the fill of former enclosed depressions in the limestone'. Some flat-floored depressions bounded by Devonian limestones in the Lawford Range are up to several hundred metres across, and are floored by clastic deposits that sometimes surround interspersed limestone towers. These depressions are interpreted by Playford as sites of Early Permian subglacial lakes, but confirmation of this requires the discovery of laminated lacustrine deposits.

A key issue in this interesting interpretation is the extent to which the main elements of the landforms we see today (Figures 10.6 and 10.7) could in fact have been developed before and during the Early Permian and subsequently survived erosion beneath a continental ice sheet. It is unlikely that the pediment surface and its bounding cliff line, for example, could have developed under such cold conditions, and given the evidence for wet-based glaciers with intense scouring indicated by glacial striations, it is also unlikely that pinnacles and towers would have survived the glaciation. At the very least they would have been moulded into streamlined moutonée forms.

Playford's (2002) hypothesis concerning landscape evolution of the Kimberley limestone ranges therefore remains to be confirmed. Nonetheless, it brings a searching light to bear on the much less demanding interpretation by Jennings and Sweeting (1963), who essentially considered only Cenozoic evolution and recognized the influence of two episodes of erosion, superimposition of drainage and the importance of climatic change. At the very least one must now conclude that there is strong evidence for several erosion phases in the past and for exhumation of a previously eroded limestone topography from beneath glacigenic deposits: the extent to which the exhumed landforms dominate or have at least directed fluvial and karstic elements in the modern topography remains to be clarified.

The examples that we have discussed in this section show that when the effects of major, repeated climatic changes, burial, uplift and exhumation are superimposed on stage of development and underlying lithological influences, there is clearly considerable potential for great complexity of morphology and landform assemblage. Herein lies the difficulty of interpreting many of the world's great karsts in terms of climato-genetic models.

11

Karst Water Resources Management

Karst water resources have been important for millennia. Karst springs are larger than most others, so in all karst regions many of the earliest villages and towns centred upon them. Such natural 'fountains' were exploited by ancient Greeks, Persians and Romans to supply their cities, sometimes delivering water via aqueducts and qanats. For example, the giant Ras el Ain springs of Lebanon (Burdon and Safadi 1963) were being captured with dams and an aqueduct to convey water to the Phoenician port of Tyre as early as 1200 BC (Figure 11.1). By 453 BC karst spring water was used for irrigation in Shanxi Province, China. The Mayan people of Yucatan (Classical period AD ~317–889) centred their civilization around cenotes penetrating the karst aquifer there (e.g. at Chichen Itza). Thermal karst springs were used at Pamukkale (Turkey) by Greeks and Romans, and at Budapest (Hungary) and Bath (England) by Romans. These and other examples of karst water exploitation over several millennia are reviewed in Drew and Hötzl (1999). Many cities and rural populations still depend heavily on water supplies from karst. In Europe, for example, carbonate terrains occupy about $3 \times 10^6 \, \text{km}^2$ (35% of the land surface) and many important cities are supplied totally or partly with karst waters, including Bristol, London, Paris, Rome and Vienna. In some European countries karst water contributes 50% of the total drinking water supply (COST Action 65 1995), and in many regions it is the only available source of fresh water. Forty per cent of the area of the USA east of the Mississippi River is underlain by karstic aquifers. At a global scale, if one compares the world's population distribution to the distribution of carbonate rocks, then it is possible that 20–25% of the world's population depends on karst water supplies to a greater or lesser extent. In south China alone well in excess of 100 million people live on karst.

11.1 WATER RESOURCES AND SUSTAINABLE YIELDS

Sustainable management implies the use of resources for the benefit of the present generation without limiting the potential use of the same resources by future generations. Effective sustainable management will leave the environment in at least as good a condition as when its use first started. It is a major challenge to achieve sustainable use of groundwater resources and difficult to prove that the management regime implemented is being successful.

Karst groundwater resources are contained in aquifers that involve recharge, storage and discharge. The volume in storage is dependent more on the size and porosity of the aquifer than on amounts of recharge, and if the combined natural and pumped output is greater than the recharge, then the resource will be depleted regardless of the size of the store. The situation is not unlike a lake: the basin may be large but if more water is used than comes into it, the lake level will fall. In karst, water abstraction can take place in allogenic streams, thus reducing the recharge; by pumping the aquifer, thus reducing the storage; and at the outflow spring, thereby depleting streamflow downstream. One of the major problems we face is that exploitation often occurs without any evaluation of the resource until problems of depletion emerge. So a key question is: how can we achieve sustainable use? And is the target moving because climate change may be altering the water balance in some locations? Younger et al. (2002) considered the sensitivity to climate change of four major carbonate aquifers in Europe, and concluded that 'significant reductions in available

Karst Hydrogeology and Geomorphology, Derek Ford and Paul Williams

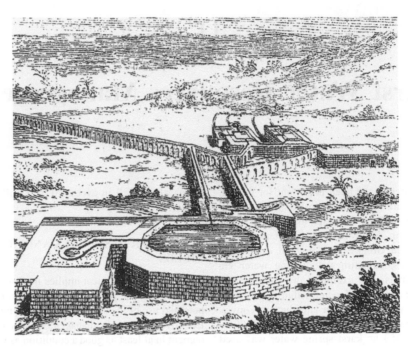

Figure 11.1 An artist's reconstruction of the waterworks at Ras el Ain Spring, Lebanon, as they would have appeared around 1200 SC. The elevation of this conduit spring was raised by a surface enclosure in order to furnish gravity flow to an aqueduct supplying the city of Tyre. Reproduced from Bakalowicz, M., Fleyfel, M. and Hachache, A. (2002) Une histoire ancienne: le captage de la source de Ras el Ain et l'alimentation en eau de la ville de Tyr (Liban). La Houille Blanche, 4(5), 157–60.

resources can be expected in low- and mid-latitude carbonate aquifers in Europe by the middle of the present century'. Thus even without pressures on the resource caused by rising demand on one hand and reduction of useable supplies because of pollution on the other, changes in the water balance caused by climatic change may in some places exacerbate supply shortages in the future.

To achieve sustainable management of karst water resources we need first to assess the magnitude and quality of the resource; then compare the demand to the potential supply; and, finally devise methods of equitable allocation of available water surplus amongst competing users (including ecological instream demands). Long-term monitoring against agreed baselines of quality and quantity can then be used to assess the effectiveness of management.

11.2 DETERMINATION OF AVAILABLE WATER RESOURCES

Chapter 6 discussed how karst drainage systems can be delimited and analysed. A valuable and proven starting point for karst water resource surveys is speleological investigation (i.e. cave exploration and mapping) with associated water tracing. This establishes the basic facts

about catchment limits and drainage network geometry. Other exploration and survey techniques, including methods for estimating the water budget, are described in section 6.2. So suffice to say here that a climatic water balance calculation, when combined with data on the drainage area, will provide an order of magnitude estimate of annually renewed reserves. This approach adopts a reference period, such as a water year (dry season to dry season), and a well-defined reference area, such as a representative basin or a karst district with known boundary conditions. Error terms associated with the calculation of the water surplus in this way are typically quite large though seldom evaluated; probably of the order of ±25% or more. But the estimate can sometimes be cross-checked by a complementary technique based on spring hydrograph analysis (section 6.5), provided that the watershed of the basin feeding the spring is known accurately. The technique usually expresses exploitable resources as a volume in storage or as a mean annual discharge, but can be converted to equivalent depth of runoff if the basin area is known. In the allocation of resources, knowledge of annual and seasonal variations in water surplus is important, as is information on the recurrence intervals of low flows at karst springs. For a recent review that takes into account advances in aquifer modelling, we particularly recommend that of Bakalowicz (2005). He asserted that in the evaluation of karst water resources 'the classical hydrogeological approach is generally inoperative because of the inability to show the existence of conduits' and proposed an investigation approach that:

1. characterizes the geological structure;
2. delineates the karst system;
3. characterizes their **lump** functioning by hydrodynamic methods (spring hydrograph analysis, spectral analysis, etc.) and hydrogeochemical methods (including isotopic techniques);
4. characterizes their **local** functioning by artificial tracing tests and pumping tests.

In this way a karst system can be characterized by its structure and functioning. Armed with this knowledge the hydrogeologist can then (i) make realistic estimates of its water resources and exploitable storage, and (ii) draw up plans for water quality protection and aquifer management.

Various methods of spring hydrograph recession analysis (Chapter 6) permit calculation of the dynamic reserves of karst groundwater systems. Integration of the area under a master spring-flow recession curve provides a measure of the volume of dynamic reserves, which includes water stored in the epikarst as well as in the phreatic zone. This provides an estimate of the total volume of water discharged between floodflow, when storage is at a maximum, and baseflow, when it is at a minimum. However, the volume calculated will depend on the model used (section 6.5) and it does not include the reserves impounded below the level of the spring (which are still in storage after spring-flow has ceased). In field situations where water is extracted directly from the spring, available water resources are usually estimated conservatively from the baseflow component, because floodwaters runoff rapidly and so are not easily captured, but if the total outflow is collected in a reservoir then the floodflow component should be included.

In very large groundwater systems the water can have a long residence time, the duration of which can sometimes be assessed by dating using radioactive isotopes (section 6.10). Using this information, the total water stored in the system can be calculated from the product of the average age of the water and the mean annual discharge of the spring(s). However, estimating the average age of the water can be complex, because the average residence time progressively increases from conduits to fissures and then to matrix. So here we need to be pragmatic and estimate the average age of the mix of water appearing at a spring,

noting that even that average may vary with discharge (usually inversely). Having determined a representative age, the next problem is that average turnover time calculated for water in the aquifer will vary with the model used (Table 6.7), as explained in section 6.10. Nevertheless, potentially useable groundwater reserves estimated in this way can amount to many cubic kilometres. Boreholes are often used to exploit this water, utilizing various forms of borehole investigation (section 6.4) to assess the locally accessible water supply. However, there is a significant practical problem in locating subterranean conduits that would be a suitable target for drilling, as noted in section 6.4: Bakalowicz (2005) expressed the opinion that almost no geophysical method is suitable for locating conduits at depths greater than 40–50 m. The most successful sites are likely to be close to permanent karst springs.

Having identified a suitable site for water resources extraction, the next problem is to decide how much of the resource can be exploited without causing unacceptable environmental impacts. A common reality is that a surprisingly small proportion of the aquifer volume may be useable without unacceptable depletion of spring-flow, because spring discharge is reduced as the water table is drawn down by pumping (Figure 11.2). Excessive pumping not only depletes water resources available at the spring, but also may give rise to other problems such as saltwater encroachment and sink-hole collapse (sections 9.5 and 12.3). Saltwater encroachment in coastal sites (section 5.8), especially following heavy pumping during dry summers, imposes a severe limitation on exploitable resources. Case studies from Croatia and Italy illustrate efforts to manage such problems (Biondic *et al.* 1999, Tulipano and Fidelibus 1999). Because over half the Earth's population lives along the coast, pressure on coastal aquifers is set to increase and so these problems will become severe.

In practice, water resource managers are often forced to adjudicate between competing users only when demands exceed available supplies. Identification of all competing users is essential before determining the priority of their claims. Some require water at the spring or downstream of it; others require it from bores into the aquifer or from allogenic streams that recharge it. These cases can be subdivided into **exploitive users** that deplete or degrade the resource in some respect and **sustainable users** that have relatively little or no adverse impact on the total quantity or quality of the resource. In the first group are demands for domestic, agricultural and industrial water supplies, irrigation, fish farming and wastewater dilution (including factory, farm and domestic waste waters); and in the second group are requirements for ecosystem maintenance, most tourism and recreation, and hydroelectric power generation where there are no

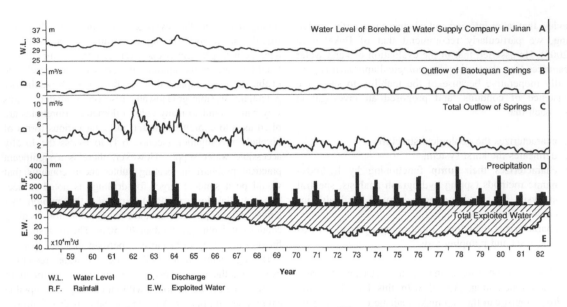

Figure 11.2 Depletion of the discharge of Jinan Spring, Shandong Province, China, following exploitation of groundwater resources in its catchment. Reproduced with permission from Sweeting, M.M., Karst in China: its Geomorphology and Environment © 1995 Springer-Verlag.

interbasin transfers. In agriculture, for example, wet rice (paddy fields) is the primary cereal for human consumption in southern and eastern Asia. In an example from south China, Zhang (1996) showed that ~5 m³ of irrigation water is required to produce 1 kg of rice, twice the common standard in the non-karst areas and four to five times greater than in the best regulated. This is a result of losses into underlying karst aquifers.

The approach to water allocation adopted by different societies varies (see Drew and Hötzl (1999), Richardson (2001) and Hiscock *et al.* (2002) for general discussions of legislation related to water-resource exploitation and protection), but user priorities and needs are usually determined by some combination of legal, political, economic and environmental considerations. For example, Florida State has mandated a '5–3–1 Criterion' for phosphate mines over its main 'Florida Aquifer System' (LaMoreaux 1989): the latter must not be lowered more than 5 ft (1.5 m) at the boundary of the pumped property, the limit for the overlying surficial aquifer is 3 ft (0.9 m), the limit for the nearest pond or river is 1 ft (0.3 m) and the maximum allowance is 1000 gal. per acre per day (10 000 L ha^{-1} day^{-1}). Assessment of real needs involves consideration of the volume and season of peak demand and the consequences of a partial failure of supply in a drought period. In making these assessments, sustainable users may be disadvantaged because of the difficulty of assigning an economic value to their claims

or because of a lack of scientific data permitting identification of optimal and minimal acceptable requirements, particularly in respect to ecosystem maintenance.

The case of water allocation at a large spring complex in an International Union for the Conservation of Nature and Natural Resources (IUCN) Project Aqua site was discussed by Williams (1988b, 2004b). The Waikoropupu Springs yield a combined flow averaging 15 m³ s^{-1} and are fed by a partly artesian system containing a water volume of at least 1.5 km³, but most of the recharge zone lies outside the protected area. Claims on the resource exceed the available spring-flow during dry periods. Although there is abundant phreatic storage in the aquifer, it cannot be 'quarried' without risking the viability of the rare aquatic ecosystem of the spring and the stream it sustains. Thus it is the limitations of the dynamic reserves of the system that determine how much water can be made available for use, rather than the total volume stored in the aquifer. The dilemma for local water resource managers is how to resolve the problem of seasonal shortage in the face of plenty, especially considering their responsibility for a site of international ecological importance. The minimum recorded spring discharge (5.3 m³ s^{-1}) gives some guidance as to what minimum flows the ecosystem can withstand, yet still drier years must have occurred in the past. Thus it can be argued that some proportion of the minimum recorded flow is sufficient for maintenance of the ecology during droughts. What proportion is adopted as appropriate then becomes a political decision. In this case, interim limits were set of $0.5\,m^3\,s^{-1}$ for total abstractions from the recharge zone and $2.9\,m^3\,s^{-1}$ for minimum residual streamflow from the springs, although for most of the time the springs discharge much more than this. Rights to abstract water were allocated for a finite period (but potentially renewable) in order to permit limits to be reassessed in the light of further data and monitored effects.

The problem of defining ecologically acceptable flows has also been faced in the Žrnovnica River in Croatia, which is a karst spring-fed system near Split used for water supply. Its median discharge is about $1\,m^3\,s^{-1}$, but there is considerable seasonal variation. It was concluded that at most no more than 30% of the annual volume of water could be abstracted (Bonacci *et al.* 1998).

In springs, as in rivers, floodflows often discharge water that cannot be easily used. Thus regulation of springs is often practiced in order to impound floodflows and make the volume captured available for future use. The most noteworthy case is that of the Ombla spring near Dubrovnik, Croatia (Bonacci 1995, Milanović 2000), where an underground dam contains underground storage of about $3 \times 10^6\,m^3$ and permits controlled short-term exploitation of up to around twice the mean spring discharge (Figure 11.3). Lu (1986) shows many examples of smaller scale underground impoundments in the caves of China. Valuable subsurface dams are also built beneath the stepped coral terraces of some tropical islands to impound groundwater flow moving at the contact of coral and the underlying impermeable rock. Cases in the Nansei Islands of southern Japan are described by Yoshikawa and Shokohifard (1993). Along the shore of the Alpes Maritime of France springs that were drowned by the post-glacial rise of sea level have been fitted with submarine interception dams to control the ingress of salt water and allow extraction of fresh water (Gilli 2002).

11.3 KARST HYDROGEOLOGICAL MAPPING

The ability of a society to obtain and allocate water resources wisely depends largely on the comprehensiveness of data on the resource and the clarity with which scientists communicate the relevant information. One of the important ways of conveying water resources

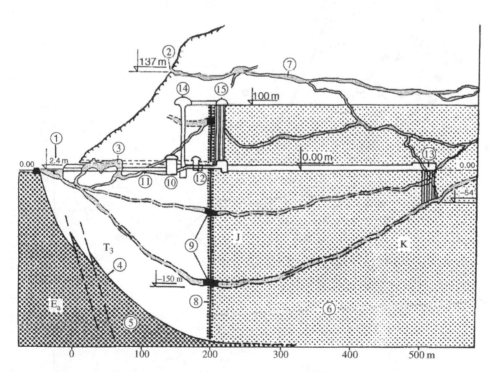

Figure 11.3 The underground dam and reservoir installed at Ombla Spring, near Dubrovnik, Croatia: 1, Ombla Spring at 2.4 m a.s.l.; 2, entrance of an upper cave at 137 m a.s.l.; 3, cave at spring elevation; 4, thrust plane – limestone is thrust over (5) flysch; 6, K is limestone and J is dolomite; 7, cave passages; 8, underground dam, a grout curtain; 9, concrete plugs in lower caves; 10, power plant; 11, outlet tunnel; 12, penstock; 13, water intake tunnel; 14, spillway; 15, upper tunnel. Reproduced from Milanović, P.T. (2000) Geological Engineering in Karst: Dams, Reservoirs, Grouting, Groundwater Protection, Water Tapping, Tunnelling, Zebra, Belgrade, 347 pp.

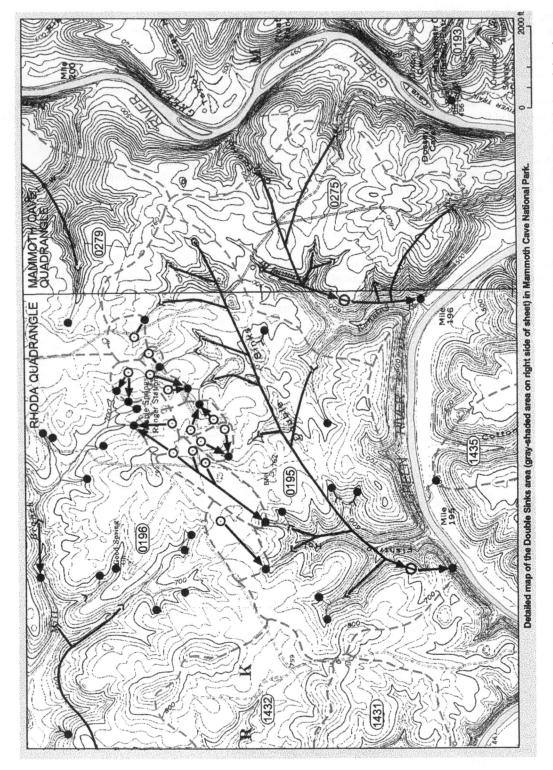

Figure 11.4 An extract from the Beaver Dam quadrangle of mapped karst groundwater basins in Kentucky, USA (Ray and Currens 1998a). The original map is in colour, at a scale of 1 : 100 000, and has a contour interval of 20 m. Depicted are stream-sinks, other tracer-injection points, wells, underflow springs, overflow springs, groundwater flow routes, subsurface and surface overflow routes, and groundwater catchment boundaries. Further information available from http://www.uky.edu/KGS/home.htm

knowledge is through the medium of hydrogeological maps with supporting explanatory notes.

Maps are conceptual representations of the real world. Hence the way in which information is depicted is important for the ideas conveyed. This is not just a matter of the symbols used, but of the mapping objectives and target end-users. Hydrogeological maps are of various types and can be prepared at several scales. They can have a multiplicity of end-users, from politicians and planners to hydraulic engineers, well drillers, hydrogeologists and teachers. The preparation of any hydrogeological map must therefore take into account the likely variation in scientific expertise of its principal users. Small-scale maps are appropriate for atlases, such as the *Hydrogeological Atlas of the People's Republic of China*, which features a map of karst regions in the south of the country. Other excellent examples are the *Hydrogeological Map of France* at 1:1.5M by Margat (1980) and a map of karst water resources in Mediterranean France at about 1:760 000 by Drogue *et al.* (1983). Small-scale maps are necessarily imprecise, but are still useful for reconnaissance planning. A range of larger scale special purpose maps may present data on water quantity, quality and pollution.

Maps at 1:10 000 to 1:200 000 are of the greatest technical interest because at such scales issues of content and its representation also become particularly important. Three types of hydrogeological data are basic to all maps of this kind: information on permeability, aquifer geometry and hydraulic regime. What information can be mapped depends on what is available and the more abundant the data, the larger the scale map needed to convey it clearly. The 1: 100 000 scale maps of karst groundwater basins in the Mammoth Cave area of Kentucky (Ray and Currens 1998a,b) are of particular interest in this context because they show stream sinks, springs, water-traced connections, groundwater divides, overflow routes, wells, etc. (Figure 11.4).

Paloc and Margat (1985) pointed out that two complementary approaches to mapping have been proposed, one more **hydrogeolithological** in emphasis and the other more **hydrogeodynamic**. The first is the more common and led to a recommended international legend (UNESCO 1970). The second is illustrated by Margat's (1980) map of France. Various colour conventions have also been adopted in these schemes.

The hydrogeolithological approach superimposes three kinds of information:

1. lithological types that are assumed to represent permeability classes;
2. piezometric data from which groundwater flow may be inferred;

ORIFICES OF KARSTIC CONDUITS	ENTRANCE		
	impene-trable	penetrable	
		cave	pit
1 – SPRING –permanent	●	◼	▼
–temporary	◐	◖	▽
2 – SWALLET –permanent	○	◻	▽
–temporary	◑	◨	▽
3 – ESTAVELLE –permanent spring–temporary swallet	⊕	◩	▼
–temporary spring–permanent swallet	⊖	◪	▽
4 – KARST WINDOW –on a permanent stream		∩	⩔
–on a temporary stream		∩	⩔
5 – CAVE WITHOUT FLOW		∩	V

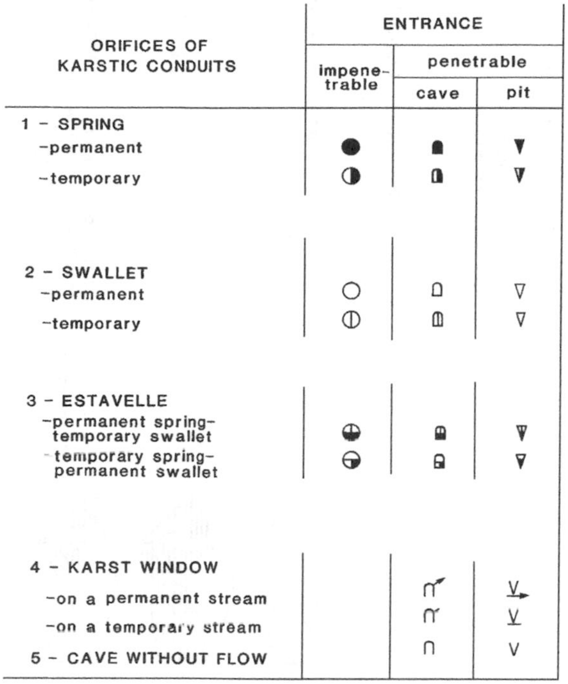

Figure 11.5 Some symbols used in hydrogeological mapping to convey dynamic conditions. Reproduced from Margat, J. (1980) Carte hydrogeologique de la France a 1: 1 500 000. Bureau de Recherches Geologiques et Minieres, Orleans.

3. surface hydrography with data on sites of water exploitation.

For karst areas, Paloc and Margat (1985) emphasized the importance of classifying baseflow discharges of springs and surface streams. Some symbols used are illustrated in Figure 11.5.

The hydrogeodynamic approach presents information concerning:

1. the constitution of aquifer systems, based on the distinction between disposition of and location of the principal rock bodies (taking into account the degree to which they are water-bearing and their possible layering);
2. the boundary conditions of aquifers, distinguishing between
   (a) the direction of water flow (input, output, or static)
   (b) flow conditions as opposed to potential conditions.

Symbols used to convey these ideas are shown in Figure 11.6 and an illustration of a hydrogeological map is provided in Figure 11.7.

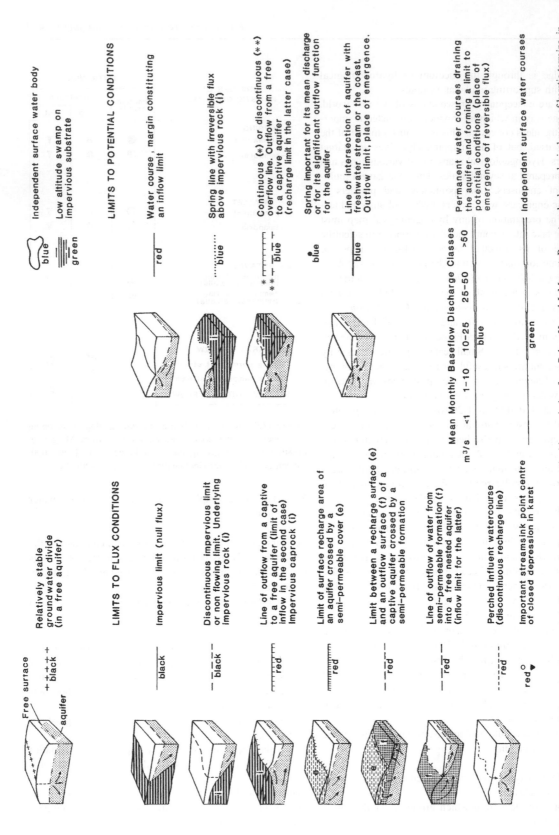

Figure 11.6 Some symbols used on large-scale hydrogeological maps. Reproduced with permission from Paloc, H. and J. Margat, Report on hydrogeological maps of karstic terrains, in Hydrogeological Mapping in Asia and the Pacific Region (eds W. Grimelmann, K.D. Krampe and W. Struckmeier), International Contributions to Hydrogeology, Vol. 7, Heise, Hannover, pp. 301–15. 1985.

Depending on the end-users, it may also be helpful to present other information that complements the hydrogeological data. Relevant material may include topography, rainfall, surface streams, dams, sealed roads and buildings, and waste disposal sites. Some of this can be depicted on small-scale insets, although the use of geographical information systems (GIS) permits such data to be represented in different layers (Longley *et al.* 2005; see Figure 11.14 Lower).

11.4 HUMAN IMPACTS ON KARST WATER

Karst hydrological systems are susceptible to a greater range of environmental impact problems than others, because in addition to surface waters there is another set of difficulties associated with highly developed subterranean conduit networks and their associated fragile ecosystems. In all inhabited karsts, dolines and other sink-points are perceived, unfortunately, as being particularly suited for the dumping of solid or liquid waste (see Figure 11.9), because when it disappears underground 'out of sight is out of mind!' Ekmekçi and Günay (1997) make the point that the general attitude of administrators in Turkey appears to be that karst pollution problems do not exist if they are not seen; and this is sadly true of most other jurisdictions with which we are familiar. Many serious problems of human impacts on karst waters emphatically **do** exist, even if they are not obvious to the casual observer. These are discussed by Yuan (1983), Williams (1993), Drew and Hötzl (1999), Veni and DuChene (2001) and many others. Water contamination may be transmitted into karst by both allogenic and autogenic sources. In the autogenic context, contamination can arise from both dispersed and point sources. Dispersed-source contamination enters the epikarst before being transmitted to the phreatic zone, whereas point-source contamination normally bypasses the epikarst, typically down doline drains, and is rapidly transmitted to the water table.

The most important of the many human impacts on the karst environment is that arising from water pollution, especially where aquifers are unconfined. This is because karst aquifers are notoriously effective in transmitting rather than treating pollutants (Sasowsky and Wicks 2000, COST Action 620 2004), a consequence of the ease of contaminant transport through conduit systems and of the limited capacity for self-treatment found in many karst groundwater systems. A stark reminder of the ease with which unexpected but serious water pollution can occur in karst is given by the 'Walkerton Tragedy' of May 2000. In this small Ontario town of 5000 population, 2300 were affected by illness and seven people died after drinking water from a municipal water supply poisoned

by a strain of *Escherichia coli* and inadequately chlorinated. The water supply was from bores in Silurian and Devonian dolomites. The source of contamination was thought to be stormwater infiltration from a freshly manured field that had penetrated a well casing. Although conventional understanding was that the aquifer in the dolomite bedrock could be modelled using Darcian assumptions, downhole video showed the presence of open bedding planes with turbulent inflows of water. Thus transmission of the bacteria was relatively rapid. Two dye traces yielded velocities about 80 times faster than MODFLOW simulations, clearly demonstrating that pathogenic bacteria could be carried through solutionally enlarged fissures at velocities of hundreds of metres per day and that the possible capture zone for contaminants was much larger than believed (Worthington *et al.* 2001). This also reemphasizes the limitations of much current computer modelling of karst aquifers (section 6.11).

Transport of contaminants through karst aquifers is by a variety of mechanisms depending on the physical and chemical properties of the contaminant. Vesper *et al.* (2000) recognize six categories.

1. Water soluble compounds, both organic and inorganic, such as nitrates and cyanides. These compounds move with the water, forming linear stringers along ribbons of high-velocity flow, but with some detention in eddies as well. Field (2004) discusses means of accurately forecasting or predicting their rates of transport through karst by use of breakthrough curves, storage and diversion models.
2. Slightly soluble organic compounds, less dense than water (light, non-aqueous phase liquids or LNAPLs), such as petroleum hydrocarbons. These will float on the water and tend to pond behind obstructions such as cave siphons. Buried fuel tanks that leak are a worldwide problem. Layers of petroleum several centimetres deep have been discovered at the water table, where cave explorers have been killed by accidentally igniting them.
3. Slightly soluble organic compounds, more dense than water (DNAPLs), such as chlorinated hydrocarbons. The DNAPLs sink to the bottom of the aquifer, or may infiltrate and adhere to clastic sediments. Figure 11.8 is a conceptual model of DNAPL storage and transport in karst aquifers (Loop and White 2001). Common industrial solvents such as trichloroethylene (TCE; $CCl_2{:}CHCl$, density $= 1.47$) are particularly significant because they begin as DNAPLs but degrade over many years to LNAPLs that can rise again from the deep traps; TCE degrades to vinyl chloride (C_2H_3Cl, density $= 0.9$) which is a particularly hazardous substance. Contamination of

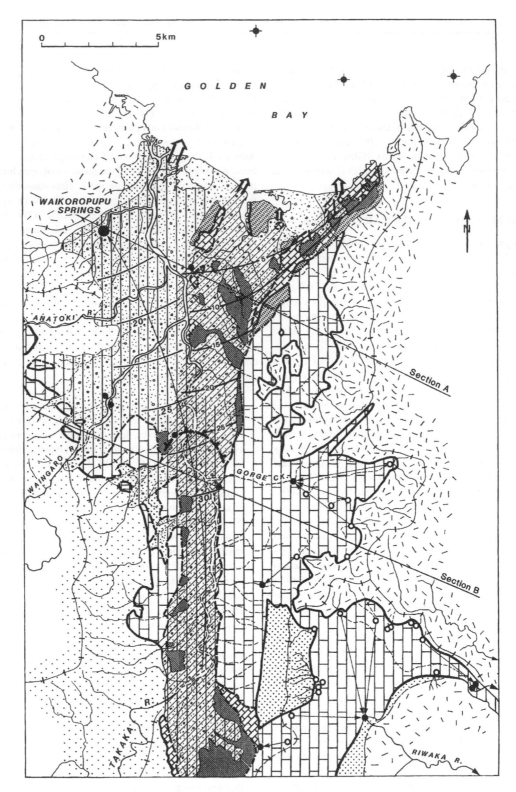

Figure 11.7 Hydrogeological map of the Takaka region, South Island of New Zealand. The main aquifer is in marble with a total water volume of >1.5 km³, partly under confined conditions. A second karst aquifer occurs in the overlying limestones, also partly confined; and a third aquifer is in floodplain gravels. (Data from Grindley 1971, Williams 1977, Mueller 1987.)

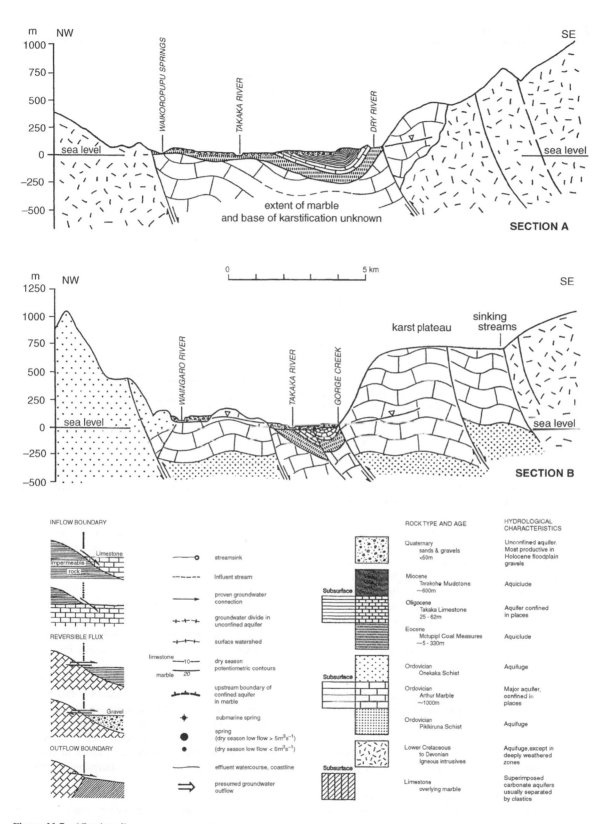

Figure 11.7 (*Continued*)

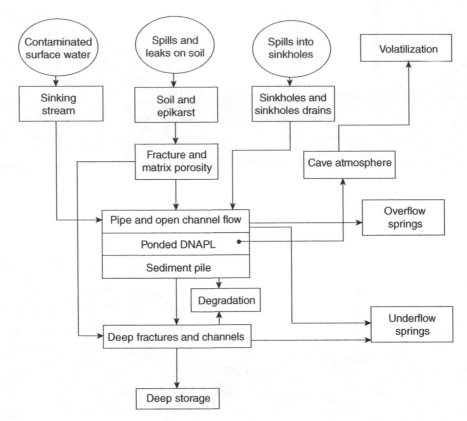

Figure 11.8   A conceptual model for the storage and transport of dense non-aqueous phase liquids (DNAPLs) in a karst aquifer. Reproduced with permission from Loop & White, Ground Water © 2001 Blackwell Publishing

karst by DNAPLs is extremely difficult to clear up (Krešic *et al.* 2005).

4. Pathogens, such as viruses, bacteria and parasites. These are transported readily through conduit systems because most are too small to be filtered.
5. Metals such as chromium, nickel, cadmium and mercury. These tend to precipitate as hydroxides and carbonates or be adsorbed onto clays and organic particles and transported in suspension.
6. Trash such as plastic, animal carcasses, tin cans and bottles. Trash collects at stream sinks and ponds behind underground obstructions, depending on its density. It is readily mobilized during flood events (Figure 11.9).

The natural treatment of waterborne contaminants in karst is relatively ineffective for a number of reasons.

1. The surface area available for colonization of natural micro-organisms as well as for adsorption and ion exchange is much less in dense, fractured karst rocks than in porous clastic sediments.

2. The characteristically rapid infiltration into karst reduces the opportunity for evaporation, a mechanism that is important in the elimination of highly volatile organic compounds, such as the solvents and many pesticides.
3. Physical filtration is relatively ineffective in typically thin karst soils and through rocks with large secondary voids; thus sediment and micro-organisms are readily transported into karst aquifers, as shown in Chapters 5, 6 and 8.
4. Transmission of particulate matter entirely through karst systems is assisted by the turbulent flow regime commonly associated with conduit aquifers.
5. Time-dependent elimination mechanisms (e.g. bacteria and viruses) are curtailed in their effectiveness because of the rapid flow-through times in conduits and reduced retardation by adsorption–desorption processes.

The nature of subsurface processes leading to the purification/attenuation of waterborne contaminants has been investigated by Golwer (1983) and COST Action

Figure 11.9 (Upper) A sink-hole garbage dump; unfortunately, typical of many seen throughout the world. This case is in Newfoundland, Canada. (Lower) Floating trash collected at a swallow hole in Guizhou Province, China. The highly polluted stream water is used further upstream for paddy field irrigation.

620 (2004). Natural purification is brought about by the interaction of numerous physical, chemical and biological reactions and is significantly affected by transport processes and hydrogeological conditions. Various processes act on inorganic, organic and particulate contaminants, but the effectiveness of these processes depends, firstly, upon the properties of the substrate layers through which the contaminants are transmitted and, secondly, on the physical and chemical properties of the contaminants (Figure 11.10). In laboratory experiments to assess the survival of nine bacteria in water at $10 \pm 1°C$, Kaddu-Mulindwa *et al.* (1983) found that *Escherischia coli*, *Salmonella typhimurium*, *Pseudomonas aeruginosa* and other pathogenic or potential pathogenic bacteria survived up to 100 days or even longer in sterilized natural groundwater. Only two bacteria species did not survive 10–30 days (Figure 11.11). Given that the transmission of water through most conduit aquifers is shorter than this (Figure 5.18) and given the neutral buoyancy of bacteria, then the probability of spreading contaminated water through karst is high and potentially serious. Case studies such as by Felton (1996) in Kentucky, USA, and Tranter *et al.* (1997) in Derbyshire, England, provide ample field evidence to confirm these theoretical considerations. Pores in limestones are typically in the range of 10^{-8} to 10^{-5} m and fissures from 10^{-5} to 10^{-3} m; thus viruses and bacteria of 10^{-9} to 10^{-6} m diameter will pass through most fissured limestones, although some protozoa (10^{-6} to 10^{-4} m) such as *Cryptosporidium* and *Giardia* may be filtered. Viruses are especially small ($< 0.25\,\mu m$), so will often be filtered by the small primary pores of the rock matrix, but gain ready access to fissure and conduit networks. Numerous case studies are reported in COST Action 620 (2004).

11.4.1 Agricultural impacts on karst water

In agricultural areas, dispersed-source contamination emanates from spreading manure and fertilizers, spraying agricultural chemicals such as pesticides, and from urine and faeces from grazing animals. Point-source pollution arises from waste dumped in dolines and from discharges from piggery, poultry and dairy sheds into doline ponds and sinking streams. The Asiago Plateau in the Venetian Fore-Alps (home of the celebrated cheese) provides excellent examples (Sauro 1993). Chlorides (conservative) and nitrates (non-conservative) are now widely measured over time as a pair to determine the extent and persistence of such contamination (e.g. Castillo *et al.* 1994, Plagnes and Bakalowicz 2002). Many studies have shown that it is nitrogen fertilizers that are primarily

Key properties \ Key processes	Adsorption	Cation exchange	Filteration	Sedimentation	Biodegradation	Oxidation	Reduction	Complexation	Precipitation	Hydrolysis	Volatilisation	Decay	Die off
Organic matter content	++	+	−	−	++	++	++	++	−	−	−	−	−
Clay content	++	++	−	−	+	−	−	−	−	−	−	−	−
Cation exchange capacity	++	++	−	−	−	−	−	−	−	−	−	−	−
Fe, Mn, Al oxides content	++	−	−	−	−	−	−	−	−	−	−	−	−
Carbonate content	−	−	−	−	−	−	−	−	++	−	−	−	−
Matrix aperture	−	−	++	−	−	−	−	−	−	−	−	−	−
pH value	+	+	−	−	+	−	−	++	++	++	−	−	−
Redox potential	−	−	−	−	++	++	++	+	+	+	−	−	−
Temperature	−	−	−	−	+	−	−	−	+	+	++	−	−

Key properties \ Key processes	Adsorption	Cation exchange	Filteration	Sedimentation	Biodegradation	Oxidation	Reduction	Complexation	Precipitation	Hydrolysis	Volatilisation	Decay	Die off
Solubility	++	++	−	−	+	−	−	+	++	+	−	−	−
Partitioning coefficients	++	++	−	−	−	−	−	+	+	−	−	−	−
Viscosity	−	−	+	−	−	−	−	−	−	−	−	−	−
Degradation half-life	−	−	−	−	++	+	−	−	−	+	−	−	−
Radioactive half-life	−	−	−	−	−	−	−	−	−	−	−	++	−
Biological half-life	−	−	−	−	−	−	−	−	−	−	−	−	++
Stand. Reduction potential	−	−	−	−	+	++	++	+	+	−	−	−	−
Equilibrium constants	+	−	−	−	−	−	−	+	++	+	−	−	−
Vapor pressure	−	−	−	−	−	−	−	−	−	−	++	−	−
Density	−	−	−	++	−	−	−	−	−	−	−	−	−
Particle size	−	−	++	+	−	−	−	−	−	−	−	−	−

Figure 11.10   (Upper) Matrix showing the relationship between the physical and chemical properties of layers through which contaminated water passes and the effectiveness of natural treatment processes (−, indicates little or no correlation; +, significant correlation; ++, strong correlation). (Lower) Matrix showing the relationship between the physical and chemical contaminant properties and the effectiveness of natural treatment processes (−, indicates little or no correlation; +, significant correlation; ++, strong correlation). Reproduced with permission from COST Action 620 (2004) Vulnerability and Risk Mapping for the Protection of Carbonate (Karst) Aquifers, Final report (COST Action 620), Report EUR 20912. Directorate-General Science, Research and Development, European Commission, Office for Official Publications of the European Communities, Luxembourg, 297 pp.

responsible for the major increases in nitrate contamination over the past 50 years or more. Boyer and Pasquarell (1995) found a strong linear relationship between nitrate concentration in karst groundwater and percentage agricultural land in West Virginia. Panno *et al.* (2001) showed anomalously high nitrate concentrations in water from 52% of wells and most springs beneath part of the sink-hole plain of Illinois. Pulido-Bosch *et al.* (1993) found 88% of 230 sampling wells to be above the NO_3 drinking water limit of $50\,mg\,L^{-1}$ during the rainy season in an Andalusian aquifer. Burri *et al.* (1999) present other case studies worldwide. The Green Revolution, which greatly increased food production in southern and eastern Asia in the 1960s and 1970s, relied heavily on herbicides and

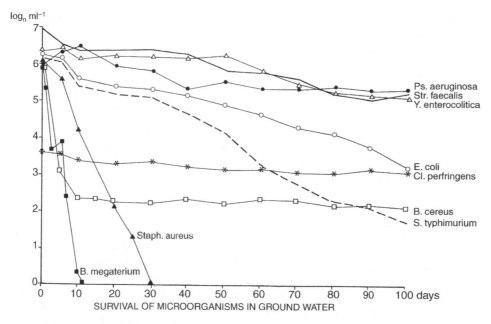

Figure 11.11 Survival of micro-organisms in natural groundwater at 10 1 C. Reproduced with permission from Kaddu-Mulindwa, D., Filip, Z. and Milde, G., Survival of some pathogenic and potential pathogenic bacteria in groundwater, in Ground Water in Water Resources Planning, Publication 42 (UNESCO Koblenz Symposium), © 1983 International Association of Hydrological Sciences, Wallingford, pp. 1137–45.

pesticides; Urich (1993) reported that these were responsible for killing off the fish, crustaceans, wild birds, even water buffalo, in a wet rice region on karst in the Philippines. Sadly, even natural estrogen from the urine of livestock can significantly damage cave ecology, as Wicks *et al.* (2004) show in a study of fish and invertebrates under the Ozark Plateau, Missouri.

Even when there is primary treatment of such wastes in purpose-built settling ponds, the decanted fluid is often still rich in nitrates, and overflows have very high biochemical oxygen demand and faecal coliform counts. Domestic waste water in agricultural communities is often treated by septic tanks, while villages and agricultural industries may have waste-water lagoons for primary treatment. Septic tanks use anaerobic bacteria to treat waste and then effluent is directed through a system of subsoil drains that encourage further treatment by aerobic bacteria, filtration, adsorption and biodegradation before the water percolates to groundwater via the epikarst. However, the processing of waste water is not always effective, especially if the soil is thin and has well-developed macropores, which limit the time available for die-off of pathogenic organisms. Consequently, chemical and biological pollution can often be detected in karst groundwaters from septic tank drainage (Alhajjar *et al.* 1990, Crawford 2001).

Reporting results from a 4-year study of a range of contaminants in a karst area of Kentucky, Felton (1996) concluded that most karst springs in the area are likely to contain sufficient faecal bacteria to be hazardous to human health. Herbicides were also found in groundwaters.

Disposal of waste water after primary treatment in sewage ponds is normally to surface streams, some of which drain underground. The treatment is usually effective in reducing the biochemical oxygen demand, but N and P loads in the discharged effluent can be high. Sometimes waste-water treatment ponds constructed on karst fail. There have been notorious cases of karst collapses beneath them, resulting in their sudden draining and consequential surcharges of highly contaminated waste water into the underlying aquifer. Alexander *et al.* (1993) found that there was a 20% rate of failure over a 20-year span in Minnesota.

Quinlan and Ewers (1985) described how sewage, creamery waste and heavy-metal-containing effluent from the Horse Cave area of the Sinkhole Plain in Kentucky (Figure 6.39) was spread by a karst distributary system to 56 springs in 16 areas along an 8 km reach of the Green River. In spite of that, none of the 23 wells sampled had contaminated water, presumably because they are up gradient from the conduits or have no direct connection to them. This would give a false sense of security concerning the quality of the total groundwater system, if the condition of discharge through the conduits were not also known. This led the authors to stress that reliable monitoring of water quality in karst **must** involve springs, not just wells, because they are places where flow converges at basin outfall points (Quinlan and Ewers 1986). This is a vital message in all well-karstified fissure aquifers, but becomes of lesser significance in more granular limestones even with large conduits **when hydraulic gradients and associated groundwater velocities are low**, as the following example shows.

Waterhouse (1984) mapped nitrate contamination in the porous Tertiary limestone aquifer underlying the small town of Mount Gambier in South Australia (Figure 11.12). Data were obtained from 257 bores over a period of two years. The water table is roughly 14 m above sea level about 25 km inland; so hydraulic gradients are slight and groundwater velocities are low in what is considered a Darcian aquifer (Emmett and Telfer 1994), even though there are phreatic conduits flooded by post-glacial sea-level rise (indicated by flooded collapse dolines; Figure 9.26). Uncontaminated water in the region has a nitrate concentration of about $2 \, \text{mg} \, \text{L}^{-1}$, but a typical polluted groundwater sample had a nitrate concentration of $300 \, \text{mg} \, \text{L}^{-1}$ (maximum $> 450 \, \text{mg} \, \text{L}^{-1}$). In spite of effluent loading for more than 80 years, the lower part of the aquifer still remains substantially free of contaminants, probably mainly attributable to the very slow water movement and diffuse natural recharge of only modest water surplus ($< 200 \, \text{mm} \, \text{a}^{-1}$).

11.4.2 Urban and industrial impacts on karst water

Some cities built on carbonate rocks have limited or no sewer systems, a situation encouraged by the nature of karst with its numerous ready-made sink-holes. As a consequence, considerable municipal and industrial waste has been and in many cases still is being discharged into karst aquifers without treatment. Such was the case for Bowling Green, Kentucky, in 1921 when Charles E. Mace claimed it to be

'The only city in the United States boasting a sewer system in which all the "pipes" were laid by Mother Nature'

and he went on to explain that:

'When a new residence is being built in the Bowling Green region, a "sink finder" is employed who merely goes out in the back yard and digs about in the surface

soil, which is seldom more than 3 ft. deep, until he locates a fissure. A garden hose is then placed in the crevice, and the water is allowed to run until it is free from obstructions. It is then approved by the city inspector, and the house has perfect sewer connection. No city has a more sanitary system. Chemists say the sewage would be purified in a very short distance by passing through the limestone.' (Mace 1921).

Although many engineers, city inspectors and chemists now know better, regrettably this still is not always the case, as is evident in the city of Antalya in Turkey (Kaçaroğlu 1999), to a lesser extent in Lusaka, Zambia (De Waele and Follesa 2003), and in many smaller towns around the world. Impacts on karst that arise from the building of settlements and transport systems are discussed in volumes edited by Williams (1993), Drew and Hötzl (1999), Beck and Herring (2001) and Beck (2005). Deposition, storage and discharge of solid and liquid wastes give rise to many water quality problems; urbanization also exacerbates runoff problems. Thus pollution, flooding and ground collapse are identifiable as the main impacts arising from the urbanization of karst.

It is well known that urbanization considerably increases the size of the mean annual flood in normal surface watersheds, but has progressively less influence on high-magnitude, long recurrence interval events. These accentuated storm hydrograph effects are transmitted to karst when urbanized impervious rock surfaces provide allogenic inputs, and are also seen in autogenic systems following the paving of urban areas. Case studies demonstrating the hydrological impacts of urbanization on karst are provided by Barner (1999), Betson (1977), Crawford (1981, 1984a, 2001) and White *et al.* (1984), and numerous examples of the serious effects of industrial effluents on karst water quality are recorded in Drew and Hötzl (1999).

The city of Bowling Green, Kentucky (population ∼50 000), and subject of the quotation above, is built upon the Sinkhole Plain. Crawford observed that almost all surface storm water generated there escapes either to natural sink-holes or to drainage wells, over 400 of which have been drilled. Urban flooding occurs by three mechanisms:

1. by surface detention when storm flow into sink-holes is greater than their discharge capacity, which may be reduced by clogging during building;
2. when the increased and rapidly concentrated runoff volumes typical of urban storm water so surcharges underground tributary conduits that backflooding is produced upstream of constrictions and extends back to the surface;

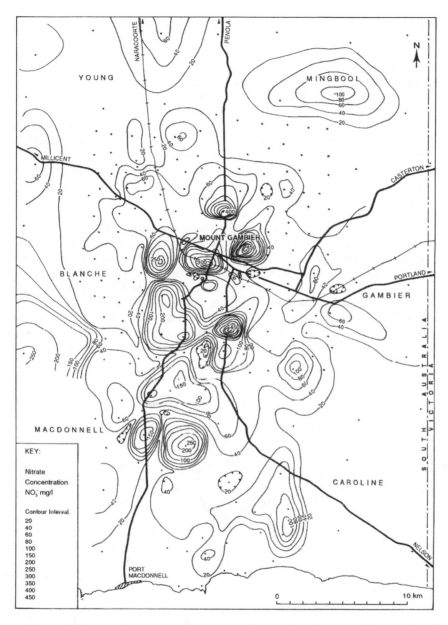

Figure 11.12   Isopleths of nitrate concentration showing dispersion patterns arising from up to 80 years of agricultural practices (mainly animal husbandry) around Mt Gambier, South Australia. Samples were obtained from about 3000 bores. This is a good example of a young (Tertiary) and porous limestone aquifer functioning in a Darcian manner because of low recharge and low hydraulic gradients. Reproduced from Waterhouse, J.D., Investigation of pollution of the karstic aquifer of the Mount Gambier area in South Australia, in Hydrogeology of Karstic Terrains, Vol. 1, pp. 202–5 A. Burger and L. Dubertret (eds) © 1984 International Union of Geological Sciences.

3. when the trunk conduit (the subterranean Barren River) is in flood due to rural runoff, as a consequence of which its backwater effect on vadose feeders causes them to backflood, so exacerbating the other tendencies.

Subsidiary problems are also associated with these processes. Crawford suggested that increased runoff down drainage wells drilled into the bottoms of dolines causes increased subsurface erosion, which in turn led to collapse of residual soils in about 10% of cases, so presenting an additional hazard for buildings. In addition, the quality of urban storm water is poor with consequent deleterious impacts on the subterranean ecology, a problem that can be compounded by the dispersion through karst cavities of toxic and explosive fumes from chemical spills (Crawford 1984b). Flooding damage has been alleviated in Bowling Green by restricting building in flood-prone areas and by requiring developers to construct floodwater detention basins. The concept of a sink-hole 'flood plain' has been adopted for flood insurance purposes, based on a 3-hour, 100-year return period rainfall and assuming no leakage, i.e. complete ponding (Figure 11.13). From the perspective of American law, legal aspects of sink-hole development and flooding in karst terrains have been reviewed by Quinlan (1986b), and legal impediments to utilizing groundwater as a municipal water supply in karst terrains by Richardson (2001).

In central Pennsylvania, White *et al.* (1984) concluded that human activities (i) increase hydraulic gradients through the lowering of the natural water table and (ii) modify stormwater runoff patterns. However, only

the latter has a major effect on the generation of dolines in the area. Urbanization strongly modifies the previously diffuse pattern of rainfall infiltration, focusing stormwater runoff from paved areas into natural sink-holes, swales or into the soil. Prevention of runoff focusing through planning measures is much less expensive than engineering solutions but, where point discharge from paved areas is inevitable, the guiding principle should be to pipe storm water directly to the subsurface bedrock without giving it the opportunity to excavate the soil. Where possible, building lots on karst should be large so as to minimize the volume of rapid runoff from paved surfaces. Barner (1999) evaluated some of the engineering practices associated with managing urban runoff in sink-hole terrain.

Industrial contamination, by accidental spillage or deliberate disposal underground, has been common on karst in the developed world. Mine tailings ponds are especially hazardous. Copper mining contaminated springs in one case in Serbia, killing all flora and fauna downstream as far as the Danube River (Stevanović and Dragišić 1995). Very caustic bauxite waste despoils parts of the principal aquifer in Jamaica. Organic waste from sawmills kills cave fauna. Historically, many early, uncased

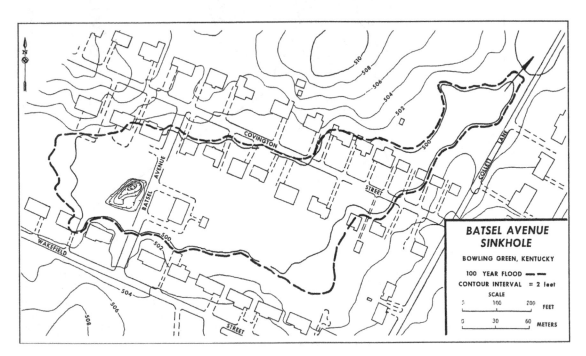

Figure 11.13   An urban floodplain' in an enclosed karst depression in Bowling Green, Kentucky. The dashed line is a contour that defines the boundary of the area that would be inundated by ponded runoff following an intense 3-h, 100 yr rainfall event. The depression is assumed to have an insignificant rate of outflow. Reproduced from Crawford, N.C. (2001) Field trip guide, Part 1: environmental problems associated with urban development upon karst, Bowling Green, Kentucky, in Geotechnical and Environmental Applications of Karst Geology and Hydrology (eds B.F. Beck and J.G. Herring), Balkema, Lisse, pp. 397–424.

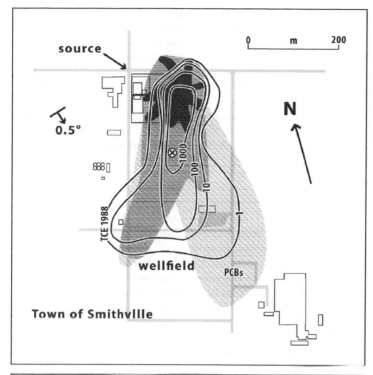

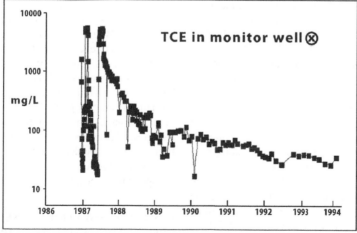

Figure 11.14   Dense non-aqueous phase liquid (DNAPL) contaminant plumes from spills of 1986 and a little earlier spreading in Silurian dolomites at Smithville, Ontario, near Niagara Falls. In dark grey, undissolved poly-chlorinated biphenyls (PCBs) dispersing below the source; lighter shades = dissolved phase PCBs moving down gradient in the two uppermost dolomite units (modeled). Contours = measured plume of dissolved TCE in the dolomites in 1988, in mg L^{-1}. This is a good example of contaminant dispersion where underground karst development is limited (young) and hydraulic gradients are low, but flow is focused along key bedding planes. The bedrocks are covered by 6–7 m of glacial silts and clays exhibiting blocky fracture. The lower frame shows the reduction of the TCE plume achieved at the marked monitoring well as a consequence of eight years of pump-and-treat ground water extraction around the source; cost ~$0.5M/year. (Adapted from Smithville Phase IV Bedrock Remediation Program, Ministry of the Environment, Ontario).

oil wells spilled crude oil or H_2S-rich brine into overlying aquifers (e.g. at Parker Cave on the Kentucky Sinkhole Plain). Cyanide-rich waste oil is widely detected, in trace amounts or greater, in urban spring waters.

Even thermal spas are reporting adverse affects. Dewatering for bauxite mining dried up hot springs in Hungary. Some of the famous travertine terraces at Pamukkale, Turkey, have been seriously discoloured by algal growth due to routing of their hot water first into swimming pools upstream of them, and to faulty septic tanks (Simsek *et al.* 2000).

11.4.3 Landfills and hazardous waste disposal

We have noted that many of the world's karsts are densely peopled. Satisfactory disposal of their wastes can pose severe problems. Dolines are used for casual waste dumping everywhere in rural areas (and many urban areas), so ensuring rapid transmission of contaminants into the conduit aquifers beneath them. Historically, municipal authorities often made the deliberate decision to site their landfills (garbage dumps) where comparatively large dolines were available locally. In many other instances there was dumping onto bare or soil-covered epikarst without any intervening impermeable seal (liner) and/or leachate collecting system. For instance, the notorious Love Canal hazardous chemical waste dump in the city of Niagara Falls, New York, is an infill of an abandoned canal dug in glacial tills just a few metres above karstic dolomites and close to the Niagara River. The US Environmental Protection Agency (EPA) estimates that the cost of cleaning up other individual sites it has investigated on karst in the USA will range from $7 million to $1.2 billion. Figure 11.14 shows a similar situation to Love Canal (though much less extreme), where there was hazardous chemical waste storage on the same dolomites nearby in Ontario, Canada.

Modern practice in developed nations calls for the monitoring of groundwater quality during the active life of any landfill, and in perpetuity after it has been filled and closed. Many jurisdictions followed early US EPA criteria, which required placement of one monitoring well upgradient of a landfill and three downgradient of it spaced across the putative path of any contaminant plume. The late J.F. Quinlan campaigned passionately (and trenchantly – e.g. Quinlan and Ray 1991) against this simplistic monitoring design which, as Figure 6.39 clearly attests, may be grossly misleading where any conduit karst development is known or suspected. He and colleagues were successful in establishing a new standard for groundwater monitoring in karst and fractured-rock aquifers – American Society for Testing and Materials (ASTM), Standard D5717-95 (1996). This calls for thorough site characterization, including location and study of all natural springs that might be draining the potential fill site, measurement of their hydrographs and chemographs and dye tracing to establish their likely contributing areas. Only then can the distribution of monitoring wells (if needed in addition to the springs) and the frequency of sampling for water quality be decided. These points are picked up in US EPA karst wellhead protection guidelines published in 1997.

Disposal of intermediate- and high-level nuclear wastes are particularly problematic. Most nations plan to bury them somewhere. Canada, Sweden and others have investigated disposal in chambers excavated far below the water table in granitic plutons, but found that even there fracturing may permit channelled flow. The USA spent much time and money considering chambers above the water table in volcanic tuffs at Yucca Mountain in the Nevadan desert, but is experiencing great difficulty in establishing the long-term safety of this setting. Attention therefore turned to disposal deep within salt bodies where access conduits might be expected to anneal after closure, sealing in the waste. The Waste Isolation Pilot Project (WIPP) near Carlsbad, New Mexico, utilizes excavated chambers 650 m underground within a salt and anhydrite formation that is ~900 m thick. It has a design capacity of 180 000 m^3 of storage and, since beginning in 1999, has accepted > 2400 shipments of intermediate transuranic waste (protective clothing, tools, etc.) from American nuclear weapons plants. Hill (2003) shows that there has been vigorous evaporite karst activity (solution troughs, breccia pipes) within 10–20 km of the site in the same formations, and that groundwater flow in overlying dolomite and intrastratal dissolution within the formations also pose threats to its long-term stability. There have been too many accidents associated with solution mining (section 12.3) for us to be confident that disposal in salt is a responsible concept.

11.5 GROUNDWATER VULNERABILITY, PROTECTION AND RISK MAPPING

The foregoing discussion has illustrated how the quality of karst groundwaters can be seriously compromised by agricultural, industrial and urban activities. It follows, therefore, that if the considerable values of karst groundwaters for human water supply and aquatic ecosystems are to be preserved, then strenuous efforts are required on the part of local authorities to protect and appropriately manage recharge areas. The urgency of this imperative was increasingly recognized in the closing decades of the 20th century and in many places methods of mapping aquifer vulnerability, hazard and risk were being actively developed and groundwater protection guidelines were being formulated (e.g. Quinlan *et al.* 1991, US EPA 1993, 2000, Lallemand-Barres 1994, COST Action 65 1995, Doerfliger *et al.* 1999, Civita and De Maio 2000, Daly *et al.* 2002, Ettazarini and El Mahmouhi 2004, Perrin *et al.* 2004).

A practical attempt to improve water quality in a karst agricultural catchment was made in Kentucky by implementing a code of best management practices (BMPs). Results after about 3 years of implementation were reported by Currens (2002): the BMPs were judged to be only partly successful when six water quality indicators were compared with pre-BMP values, and it was concluded that future BMP programmes should emphasize buffer strips around sinkholes, excluding stock from streams and karst windows, and withdrawing land from production.

The drive to develop an improved approach to groundwater quality management in Europe was spearheaded by a programme of cooperation in science and technology (COST). The first of these initiatives concerned with karst was COST Action 65 (1995), which dealt with hydrogeological aspects of groundwater protection in carbonate rocks. It was then followed up by COST Action 620 (2004), which was tasked with the development of an improved and consistent approach to the protection of karst groundwaters. This programme drew together karst specialists in hydrogeology, geomorphology, environmental chemistry and microbiology and reviewed the approaches used in 15 participating European countries.

The European approach to vulnerability, hazard and risk mapping when applied to groundwater is based on an origin–pathway–target model. It applies to both groundwater resource and source protection. As explained in COST Action 620, 'origin' is the term used to describe the location of a potential contaminant release; the 'target' is the water that requires protection; and the 'pathway' includes everything between the origin and the target (Figure 11.15). For **resource** protection the target is the surface of the water table, whereas for **source** protection it is the water in the well or spring that is used for potable supply.

A distinction is made in COST Action 620 between **intrinsic** and **specific** vulnerability mapping. Four factors are used to assess intrinsic vulnerability: overlying layers (O), concentration of flow (C), precipitation regime (P) and karst network development (K). The O factor may comprise up to four layers – soil, subsoil, non-karst rock and unsaturated karst rock. The C factor recognizes the complexity of recharge through the unsaturated zone and the potential for runoff to bypass surface protective layers. Thus intrinsic vulnerability of groundwater to contamina-

tion takes into account the inherent geological, hydrological and hydrogeological characteristics of an area, i.e. the characteristics of the system that control its response to an input of contaminant, but is independent of the nature of the contaminants and the contamination scenario.

Specific vulnerability mapping combines two different groups of data:

1. information about the physical and chemical behaviour of the contaminants;
2. information about the physical and chemical properties of the layers through which the contaminants pass.

Using the properties of the contaminants and the layers, the principle of the method is to determine the effectiveness of processes that play a role in the attenuation (retardation and degradation) of the contaminant. A process likely to act on the concentration of a contaminant can be potentially effective if the conditions are met in a given layer, although the extent to which the process will, in practice, be effective also depends on the hydraulic conductivity and thickness of the layer. Specific vulnerability assessment is therefore concerned with the properties of the contaminant (or group of contaminants), its mass and critical concentration in addition to the intrinsic vulnerability of the area. Figure 11.10 shows some of the relationships between physical and chemical properties of **layers** and the effectiveness of various processes. Figure 11.10 also shows the relationship between physical and chemical properties of various **contaminants** and the effectiveness of specific attenuation processes. Figure 11.16 provides hypothetical examples that illustrate the diversity of specific attenuation for several contaminants and geological settings. See COST Action 620 (2004) for more specific evaluation of the behaviour of a wide range of contaminants, including micro-organisms, in carbonate-karst groundwaters.

In the context of groundwater contamination, a **hazard** can be defined as a potential source of contamination resulting from human activities taking place mainly at the land surface (COST Action 620 2004). Consequently, the hazard inventory starts from a differentiation between three main types of land use: urban/infrastructural, industrial and agricultural. Mathematical algorithms were developed under COST Action 620 to calculate the potential degree of harmfulness for each hazard, with five Hazard Index Classes being defined for mapping purposes. It was also proposed that the protection of groundwater resources should be based on a comprehensive **risk analysis**, the term 'risk' being used for the likelihood of a specific adverse consequence that can be expressed as the probability of groundwater contamination to an

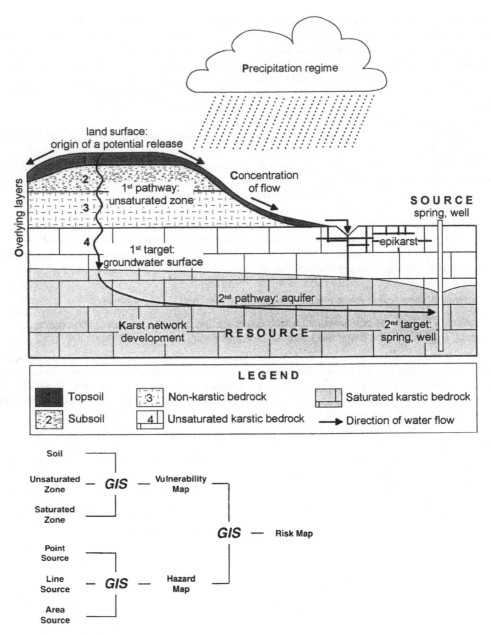

Figure 11.15 (Upper) The European approach to groundwater vulnerability mapping. This is based on an origin-pathway-target conceptual model with the contamination assumed to originate at the land surface. For water resource protectction, the spring or well is the target. Reproduced from COST Action 620 (2004) Vulnerability and Risk Mapping for the Protection of Carbonate (Karst) Aquifers, Final report (COST Action 620), Report EUR 20912. Directorate-General Science, Research and Development, European Commission, Office for Official Publications of the European Communities, Luxembourg, 297 pp. (Lower) Organisation of vulnerability mapping using geographical information systems. Reproduced from Neukum, C. and Hötzl, H. (2005) Standardisation of vulnerability map, in Water Resources and Environmental Problems in Karst (eds Z. Stevanović and P. Milanović), National Committee of the International Association of Hydrogeologists of Serbia and Montenegro, Belgrade, pp. 11–18.

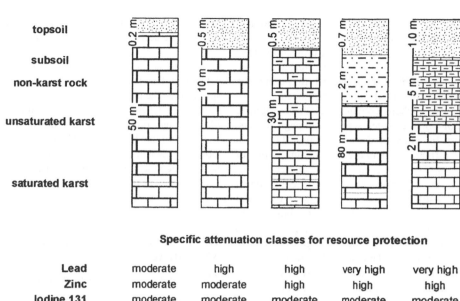

Specific attenuation classes for resource protection

Lead	moderate	high	high	very high	very high
Zinc	moderate	moderate	high	high	high
Iodine 131	moderate	moderate	moderate	moderate	moderate
Nitrate	low	low	low	moderate	moderate
Tetrachloroethene	high	moderate	moderate	high	moderate
Benzene	very high	moderate	high	very high	moderate
Naphtaline	high	high	high	very high	high
Benzo[a]pyrene	moderate	moderate	moderate	high	high
Atrazine	low	low	low	moderate	moderate
Lindane	moderate	moderate	high	high	high
Isoproturon	low	moderate	moderate	moderate	moderate
Cryptosporidium	moderate	moderate	high	moderate	high
E. coli	moderate	moderate	high	very high	very high
Enterovirus	moderate	moderate	high	high	very high

Figure 11.16 An illustration of specific attenuation classes for different water contaminants in a range of possible karst settings. Reproduced from COST Action 620 (2004) Vulnerability and Risk Mapping for the Protection of Carbonate (Karst) Aquifers, Final report (COST Action 620), Report EUR 20912. Directorate-General Science, Research and Development, European Commission, Office for Official Publications of the European Communities, Luxembourg, 297 pp.

unacceptable level. Risk was considered to depend on three elements: (i) the hazards and the probability that a hazardous event could occur; (ii) the vulnerability of the geological sequence; and (iii) the consequences for the groundwater (Figure 11.15). Risk intensity (the potential intensity of the relevant impact reaching groundwater) was separated from risk sensitivity (which considers flow conditions and potential economic and ecological value of the damage).

Epikarst development, Protective cover, Infiltration conditions, aquifer Karstification (EPIK) and 'PI' are similar schemes that are also prominent in Europe (Doerfliger *et al.* 1999; Goldscheider *et al.* 2000). 'VURAAS'

is a variant especially designed for alpine conditions (Chichocki and Zojer, 2005). 'KARSTIC' (Davis and Long 2002) is a variant for karst areas of an older vulnerability mapping scheme, 'DRASTIC' (Aller *et al.* (1987), that was designed for standard Darcian aquifers throughout the USA; it is being tested there currently. Neukum and Hötzl (2005) discuss the international standardization of these various karst vulnerability mapping methods. Many examples of their application can be found in Stevanović and Milanović (2005). A five-level rating system of groundwater sensitivity has also been devised for application in Kentucky (Ray *et al.* 1994), with hydrogeological sensitivity defined as the ease and speed with which a contaminant can move into and within a groundwater system.

Whatever the protocol adopted for the protection of the quality of infiltrating water in a recharge area, a pathway of sufficient length must be achieved to permit the necessary time for die-off of potential pathogenic organisms between sites where contamination may originate and sources of potable water supply. Although at least 30 days must be allowed for this (Figure 11.11), in practice it is difficult to define effective dimensions of protection areas. This is because of the aquifer heterogeneities and the complex processes of contaminant transport in karstified media that we have discussed above, even assuming that the basic facts about catchment areas and flow-through times to springs and wells under a range of recharge conditions in a given area have already been determined. Thus a combination of retirement of land from active production, best management practices (including riparian protection of point-recharge sites such as dolines and stream-sinks), and effective water quality monitoring and treatment at water sources is required. Milanović (2004) discusses zoning criteria that might be used to protect karst water resources and emphasizes the need for flexibility, because of the differing local circumstances that are presented. With that in mind he recommends a hierarchical zoning approach.

1. Protection area of a spring or intake for drinking water. This requires the highest protection and restriction. Fenced boundary located at least 50 m from the intake or spring.
2. Immediate protection area. This requires very severe protection and restriction. It comprises the area with concentrated infiltration that directly connects to the spring (or well) by within 24 h via karst conduits, involving flow velocities that may exceed $5\,cm\,s^{-1}$.
3. Protection area. The area contains all stream-sinks permanently or temporarily active in the catchment, but located outside Zone 2, that have direct underground connections with the tapping structure (spring or well) but need a few to 10 days to reach it. If there are some swallow holes within the general area of Zone 3 that have direct connections to the tapping structure in less than 24 h, they should be set apart in protected enclaves and treated as Zone 2 inliers.
4. External protection area. This zone includes any remaining area between the external boundary line of Zone 3 and the main watershed of the catchment for the spring. From Zone 4 there are no proven direct underground connections to the spring (or water intake structure) demonstrated by tracer tests. Flow through time is much longer and flow velocities are low (e.g.

$< 1\,cm\,s^{-1}$). An isolated stream-sink within this zone could be accorded protection as a Zone 3 enclave.

If there is an area of impervious rock within the catchment that contributes allogenic streamflow to the spring, then the entire non-karst area should be treated as Zone 2 or 3 depending on the flow transmission time. Flow-through times considered should be those experienced under flood conditions when velocities are greatest. Doline fields will be Zone 2 or 3 even where dye tracing has not been used to prove their connections to the spring. Implementation of best management practices for livestock and agriculture in general will be required.

11.6 DAM BUILDING, LEAKAGES, FAILURES AND IMPACTS

11.6.1 Dam construction on carbonate rocks

One of the most important techniques for managing and utilizing karst water resources is through the construction of reservoirs and dams. These permit the water to be stored and then used for flood control, urban and rural water supply, irrigation and generation of hydroelectricity. A large number of dams have been built on karstic limestones and dolomites for these purposes all over the world. Flood control has been particularly important on branches of the Mississippi, where the Tennessee Valley Authority (TVA) was very active in the first half of the 20th Century. Storage to sustain paddy fields, counter prolonged dry seasons or general drought was more important in China, around the Mediterranean and in Iraq, Iran and other semi-arid areas. Hydroelectric power generation was an early priority in alpine sites and is now the principal goal of perhaps the majority of the larger, higher dams. Few nations that constructed them escaped serious problems due to karst leakage, leading to considerable overruns in cost or to outright abandonment in some instances. These are summarized in many engineering design and construction reports. The TVA main report (1949) is still pertinent; Soderberg (1979) gave a more recent review of their work. Therond (1972), Mijatovic (1981), Nicod (1997) and Milanović (2000, 2004) have discussed European experience, which generally has been with geologically more complex mountainous sites.

Therond (1972) identified seven different major factors that may contribute to the general problem. These are: type of lithology, type of geological structure, extent of fracturing, nature and extent of karstification, physiography, hydrogeological situation and the type of dam to be built. For each factor, clearly, there are a number of significantly different conditions. In Therond's estimation, these together yield a combination of 7680 distinct situations that could arise at dam sites on carbonate rocks! It follows that dam design, exploration and construction must be specific to the particular site, and be continually re-evaluated:

> 'There is perhaps no phase of engineering or construction which lends itself less readily to rule-of-thumb or handbook methods than does (dam) foundation' (TVA 1949, p. 93). 'When dealing with karstic foundations **all** geologic features must be evaluated no matter how small or insignificant they appear' and 'design as you construct' (Soderberg 1979, p. 425).

Milanović (2000) suggests that there have been three principal settings for dams on carbonate rocks.

1. In the narrow gorges typically created where large allogenic rivers cross them in steep channels. Here rates of river entrenchment have usually been faster than karst development, with the result that karstification is not a major problem beneath the channels. It may however be hazardous in the gorge walls which will form the dam abutments. Limestone gorges are particularly attractive sites because the rock is mechanically strong enough to support the structure, while the narrow aperture reduces the amount of material required to build the visible dam.
2. Dams and reservoirs in broader valleys where the karst evolution has been as fast as or faster than river entrenchment. The TVA sites are examples. This can cause many problems beneath the dam as well as in the abutments and upstream in valley sides and bottoms. It is particularly hazardous where the valley is hanging at its mouth, as is common in alpine topography, because the natural (pre-dam) groundwater gradient is steepened there. Unfortunately, this will also be an optimum site for hydroelectric power dam location because the reservoir volume, fall height and gradient of the penstock are all maximized there.
3. In poljes to control flooding and store water for dry-season irrigation. This is perhaps the most difficult setting because under natural conditions the dry-season water table will be deep below the polje floor in highly karsted rock. The reservoir floor must be sealed (with clay, shotcrete, PVC, etc.) to retain water but the seals can be blown by air pressure as flood-waters rise in the caves underneath. Ponors must be plugged or walled off by individual dams rising above the reservoir water surface, and estavelles must be fitted with one-way valve systems. Much success has

been achieved in the poljes of former Yugoslavia but the Cernića project there and Taka Polje in Greece are two examples that were abandoned after expensive study.

Many dams are more than 100 m in height and some exceed 200 m. A first, obvious danger of dam construction is that by raising the water table to such extents, an unnaturally steep hydraulic gradient is created with unnatural rapidity across the foundation and abutment rock, and an unnaturally large supply of water is then provided that may follow this gradient. This is a hazardous undertaking because, unless grout curtains penetrate well into unkarstified rock, the increased pressure will drive groundwater movement under the dam and stimulate dissolution. Dreybrodt *et al.* (2002, 2005) have approached this problem with realistic modelling scenarios for limestone and gypsum. In the limestone case solutional conduits are shown to propagate to breakthrough dimensions (turbulent flow – sections 3.10 and 7.2) beneath a 100 m deep grout curtain under a dam in approximately 80 years. Remedial work would then be essential. Table 11.1 provides details of leakage from dams in karst before and after remedial works, and Figure 11.17 shows one example of increasing leakage over time at a dam in Macedonia.

The Hales Bar Dam, Tennessee, is a notorious example of a simple and immediate response to raising the water table, because there was leakage directly under the dam where hydraulic gradient is greatest. Building plans called for dam construction in a period of two years at a cost of $3 M. Because of initial foundation problems in karst, this was extended to eight years and $11.5 M. Two weeks after filling (November 1913) the first serious leaks appeared. Expensive remedial treatments continued for 30 years before leakage was reduced to an acceptable level (TVA 1949). Camarasa Dam, Spain (Table 11.1 and Figure 11.18) provides another example of unexpected leakage via both the foundation and abutments.

However, this is by no means the end of potential problems. Not only may modern fissures and conduits be flushed and enlarged at accelerated rates, but long-dormant relict karst and even sediment-plugged palaeokarst may be reactivated. Therond (1972) cited an instance where raising the water level to +75 m induced only $1.6 \, m^3 \, s^{-1}$ of leakage downstream, but when the level was raised to +85 m leakage increased to an unacceptable $8 \, m^3 \, s^{-1}$. Evidently, sediment-filled conduits (undetected in the exploration) were flushed and reactivated. This emphasizes that conditions of hydraulic conductivity estimated by even the most detailed investigation during the site exploration stage may have to be radically revised

Table 11.1 Leakage from reservoirs reduced after remedial works. Reproduced from Milanović, P.T., Water Resources Engineering in Karst © 2004 CRC Press, Taylor and Francis

Dam/reservoir	After first filling ($m^3 s^{-1}$)	After remedial works ($m^3 s^{-1}$)
Keban (Turkey)	26	< 10
Camarassa (Spain)	11.2	2.6
Mavrovo (FYR Macedonia)	9.5	Considerably reduced
Great Falls (USA)	9.5	0.2
Marun (Iran)	10	Considerably reduced
Canelles (Spain)	8	Negligible
Slano (Yugoslavia) ($34\,m^3 s^{-1}$)*	8	(3.5) Increase till 6
Ataturk (Turkey)	>11	?
Višegrad (Bosnia)	9.4	Remedial work runs
Buško Blato (Bosnia) ($40\,m^3 s^{-1}$)*	5	3
Dokan (Iraq)	6	No leakage
Contreas (Spain)	3–4	?
Hutovo (Herzegovina) ($10\,m^3 s^{-1}$)*	3	1
Gorica (Herzegovina)	2–3	No remedial works
Špilje (FYR Macedonia)	2	No remedial works
El Cajon (Honduras)	1.65	0.1
Krupac (Yugoslavia)	1.4	Negligible
Charmine (France)	0.8	0.02
Kruščica (Sklope) (Croatia)	0.8	0.35
Mornos (Greece)	0.5	Considerably reduced
Piva (Yugoslavia)	0.7–1	No remedial works
Maria Cristina (Spain)	20% of inflow	?
Peruča (Croatia)	1	No remedial works
Sichar (Spain)	20% of inflow	?
La Bolera (Spain)	0.6	?

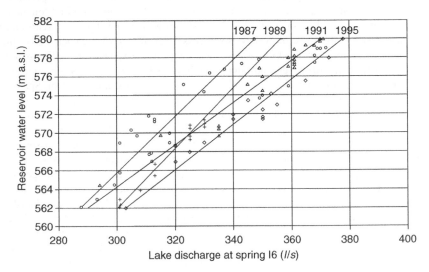

Figure 11.17 Increasing leakage at Spring I6 below the Špilje Dam, Macedonia, related to water level in the reservoir behind the dam. Milanović, P.T. (2000) Geological Engineering in Karst: Dams, Reservoirs, Grouting, Groundwater Protection, Water Tapping, Tunnelling, Zebra, Belgrade, 347 pp.

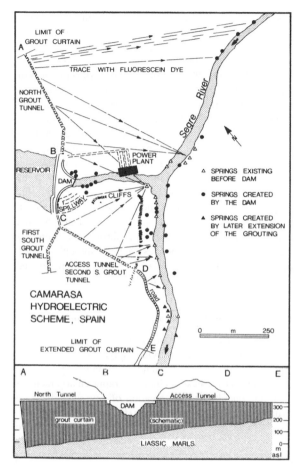

Figure 11.18 Camarasa hydroelectric dam, Noguera, Spain. The dam is built on thick, massive dolomites, dipping at 17° in the upstream direction and with an aquiclude stratum of marls at shallow depth beneath them. The dam is a gravity arc concrete structure 92 m in height and 377 m in length at the crest. It and the abutments were grouted to the marls via north and south grout tunnels. When the reservoir filled 25 new springs with an aggregate flow of $\sim12\,m^3s^{-1}$ were created, which is an exceptional amount. Reproduced from Therond, R., Recherche sur l'etancheite des lacs de barrage en pays karstique, Eyrolles, 1972.

following new evidence that comes to light as the reservoir fills. The Kalecik Dam in Turkey illustrates this. Extra grouting was undertaken after springs broke out downstream following water impoundment, but water losses continue (Turkmen *et al.* 2002).

While leakage through dam foundations and abutments is most feared, it is quite possible that there may be lateral leakage elsewhere in a reservoir. Problems with karst can arise even where the dam itself is built on some other rock, if karst rocks are inundated upstream of it. Montjaques Dam, Spain, was built to inundate a polje. It failed by leaking through tributary passages and the scheme was abandoned (Therond 1972).

In tackling dams on karst the first essential is drilling of exploratory boreholes (with rock core extracted for inspection), and mining of adits (galleries big enough for human entry and inspection) in the abutments. These may later be used for grouting. Surface, downhole and interhole geophysics (Milanović 2000) can amplify the picture but are not in themselves sufficient because they will rarely detect smaller cavities, or even large ones below ~50 m or so. Even intensive drilling and mining may be inadequate. At the Keban Dam site in Turkey, despite 36 000 m of exploratory drilling and 11 km of exploratory adits, a huge cavern of over 600 000 m³ was not detected; 'expect the unexpected!' (Milanović 2000).

Grout curtains are essentially dams built within the rock. 'Due to karst's hydrogeological nature, grout curtains executed in karstified rock mass are more complex and much larger than curtains in other geological formations' (Milanović 2004, p. 81). For example, Ataturk Dam in Turkey has a grout curtain surface area of $1.2 \times 10^6\,m^2$, a length of 5.5 km and extends up to 300 m in depth. The design and density of grout curtains are discussed by Milanović and other authors listed here. A widely adapted principle is that the depth (h) of the curtain under the dam should be $h = H/3 + c$, where H is the height of the dam and c is a constant based on site conditions. It ranges between 8 and 25. Typically, in limestone gorges with little karst development in the floors (setting 1 above) the relation will be $0.3–1.0H$, in simple river valleys (setting 2), $0.5–2.5H$. In hanging valleys it can reach $4H$, and is $6H$ or more in poljes (setting 3). The surest principle is to grout entirely through the limestone into underlying impermeable and insoluble strata where this is possible. Curtains in abutments can also be terminated laterally against such strata (the 'bathtub' solution).

The normal practice is to excavate all epikarst and fill any large caverns discovered by the adits and bores, then place a main curtain beneath the dam, in the abutments and on the flanks (Figure 11.18). A cut-off trench and second, denser curtain may be placed upstream in the foundation if there are grave problems there. In the main curtain a first line of airtrack grout holes will be placed on centres never more than 8–10 m apart and filled until there is back pressure. A second, offset line of holes is then placed and filled between them. Third and fourth lines may be used until the spacing reaches a desirable minimum that is normally not more than 2 m. Adits in the abutments that are used to inject grout should be no more 50 m apart vertically. Standard grouts are cement with clay (particularly bentonite, a clay that expands when wetted), plus sand and gravel for large cavities. Mixtures are made up as slurries with differing proportions of water. Ideally, the goal of all grouting is to reduce leakage of water to one Lugeon unit $(Lu = 1\,L\,min^{-1}\,m^{-1}$ of hole at 10 bars water pressure) under a dam and 2 Lu in the abutments. In practice, in karsted limestones it is often difficult to inject grout where permeability is $< 5.0\,Lu$. Correlation between Lugeon measured during exploration and the amount of grout that will be required can be very poor also; at Grančarevo Dam, Herzegovina, consumption ranged from 1.5 to $1500\,kg\,m^{-1}$ in different holes that had recorded only $\sim 1.0\,Lu$ before grouting began (Milanović 2000).

The Normandy Dam, Tennessee, provides a good example of exploration and grouting at a comparatively simple site. It is an earth-fill dam 34 m in height, footed on horizontal limestones. In preliminary exploration, 4400 m of core holes were drilled and then, in an 80 m wide problem zone, 25 cm diameter holes were drilled on 50-cm centres for inspection by downhole TV camera. The epikarst was then removed along a cut-off trench 6–12 m in depth. Beneath it 100-cm grout holes were emplaced on 120-cm centres (i.e. overlapping), and 12-cm holes were high pressure grouted upstream and downstream of this line. The main grout curtain was then emplaced on 3-m centres and to a depth of 25 m. All holes that accepted grout were reinforced by one to three further holes. The treatment was successful (Soderberg 1979).

All springs and piezometers must be monitored carefully as the reservoir fills behind a completed dam. Operators should be prepared to halt filling and drain the reservoir as soon as serious problems appear. In extreme cases the reservoir floor and sides may be sealed off, e.g. by plastic sheeting. Experience shows that remedial measures after a dam has been completed and tested are much more costly than dense grouting during construction.

Despite such intense effort dams still fail to achieve design levels in karst. A good example is the Lar Dam, Elbruz Mountains, Iran. This is a 105 m high earth-fill dam in a hanging valley at 2440 m a.s.l. The geology is complex. The natural water table was > 200 m beneath the dam, draining to major springs 8 km distant and 350 m lower in elevation. A sequence of international engineering firms tackled it beginning in the 1950s. During the first attempt to fill the reservoir, leakage via the springs rose to 60–80% of the inflow. It was drained and re-grouted, with $1000–40\,000\,kg\,m^{-1}$ of grout being injected in the worst places. A cavern of $> 90\,000\,m^3$ was

also discovered and filled. Water losses remain unacceptably high.

Dams and reservoirs may considerably alter the water regime of rivers and springs in karst, as explained by Milanović (2000, 2004). In the case of a 105 km reach of the Cetina River in Croatia, for example, there are five hydroelectric power plants, five reservoirs and three long tunnels and pipelines. Bonacci and Roje-Bonacci (2003) show that their effect in the upper 65 km has been to redistribute the hydrological regime throughout the year, with low flows increasing and high flows decreasing although the mean annual flow remains the same. In the 40 km reach downstream of Prančevići Reservoir, the original $100\,m^3\,s^{-1}$ flow of the Cetina has been reduced to a 'biological minimum' of only one-tenth of that volume because of its diversion through two tunnels (9.832 km long) and pipelines that cut through a large bend in the river to the Zakučac power station situated near sea level at the mouth of the river.

11.6.2 Dam construction on gypsum and anhydrite

Numerous case histories are provided by A. N. James (1992) that illustrate the wide range of serious difficulties that have been encountered by building dams on evaporite rocks. These include the rapid enlargement of existing conduits and the creation of new ones, because hydraulic gradients are excessive (Figure 11.19); the settling or collapse of foundations or abutments where

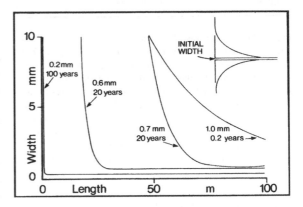

Figure 11.19 Penetration distances or progress of the dissolution front for $\sim$L99 in massive gypsum, calculated for initial fissure widths ranging 0.21–1.0 mm. Time elapsed since initiation is in years. The hydraulic gradient is 0.2 and water temperature is 10°C. (Inset) The form of the dissolution taper into the fissure that is obtained with theoretical calculations such as these. Adapted from James, A.N. and Lupton, A.R.R. (1978) Gypsum and anhydrite in foundations of hydraulic structures. Geotechnique, 28, 249–72.

gypsum is weakened by solution; heave of foundations where anhydrite is hydrated, and attack by sulphate-rich waters upon concrete in the dam itself:

$$CaCO_3 + 2H_2SO_4 = Ca^{2+} + SO_4^{2-} + CO_2 + H_2O$$

In their model analysis, Dreybrodt *et al.* (2002) obtained kinetic breakthrough beneath a 100 m deep grout curtain in gypsum in 20–30 years, the conduits enlarging to give unacceptable rates of leakage within the ensuing five years.

In the USA experience has been gained in simple geological, low-relief terrains in west Texas and New Mexico, where it is possible to avoid truly excessive hydraulic gradients and the problems of complex structure. At one celebrated site, McMillan Dam, gypsum is present only in the abutments, further reducing the difficulties, and no caves were detected when it was built in 1893. Nevertheless, the reservoir drained dry via caves through the left-hand abutment within 12 years. Attempts to seal off the leaking area by a coffer dam failed because new caves developed upstream of it. Between 1893 and 1942 it is estimated that $50 \times 10^6 \, m^3$ of dissolution channels were created (James and Lupton 1978). McMillan Dam is now abandoned, as are nearby Avalon Dam and Hondo Dam.

Dams can be built successfully in gypsum terrains where relief is low and geology simple (or where there are gypsum interbeds in carbonate strata), but comprehensive grouting is necessary and an impermeable covering over all gypsum outcrops is desirable (Pechorkin 1986). Periodic draining and regrouting will probably be needed also.

Dams in mountainous country appear to be especially hazardous. In California, St Francis Dam was built in 1928 on conglomerate containing clay with a high frequency of gypsum veinlets (A. N. James 1992). Shortly after filling an abutment collapsed catastrophically and 400 people were killed downstream. Collapse was attributed to dissolution of the gypsum which weakened the mechanical structure and permitted lubrication of the clay. This is an instance of failure in a rock that would not be classified as karstic at all, merely having a low proportion of gypsum ($< 5\%$?) as fracture filling.

12

Human Impacts and Environmental Rehabilitation

'Nature to be commanded must be obeyed'

Francis Bacon, *Novum Organum* (1620)

12.1 THE INHERENT VULNERABILITY OF KARST SYSTEMS

Natural environmental change and human impacts on the environment have occurred concurrently since the Last Interglacial about 125 000 years ago. *Homo sapiens* is an integral part of the ecosystem, at first minor because our earliest impacts were sustainable hunter–gatherer activities. The earliest cultivators began to change that. Nicod *et al.* (1996) provide a detailed historical review of human–karst relationships, and consider uses of karst and associated impacts from classical Greek and Roman times until present. It was only when there was extensive deforestation and widespread use of fire that our impacts began to leave a permanent imprint. In the past 8000 years, with the rise of settled agriculture and the building of towns, such activities started to have significant effect on natural ecosystems. Many of the impacts on karst are reviewed in special issues of *Catena* (Williams 1993) and *Environmental Geology* (1993, Vol. 21) and in an increasing number of international conference proceedings.

We can distinguish human-induced environmental change from natural environmental change by the rate of change (e.g. of water-table lowering) and the type of change (e.g. landform modification by quarrying), which is often more rapid and different to that encountered in natural systems. Human-induced environmental change is transmitted through karst, often far from the initial point of impact, by hydrological processes that largely operate underground and out of sight; thus their impact often is not obvious until it is well advanced. For example, water extraction via boreholes may cause springs to dry up (Figure 11.2). Activities on adjacent non-karst terrains also often impinge on karst, because allogenic runoff transmits polluting and clogging effects. Figure 12.1 illustrates this point and shows the particularly complex ramifications of such activities as urbanization and quarrying. Sets of human activities whether undertaken on karst or on non-karst terrains upstream can lead to major effects that result in karst ecosystem degradation. Dam building has particularly large impacts as noted in section 11.6, and even when built beyond the outflow boundary of karst can still affect it as a result of backflooding.

Experience shows that karst environments are particularly fragile and vulnerable to damage compared with most other natural systems. The reason for this is the nature of the karst hydrological system. Efficient drainage of the surface down numerous widened joints, dolines and stream-sinks rapidly transmits surface pollution underground and readily evacuates soils stripped from the surface. Filtration of diffuse recharge is minimal because limestone soils are usually thin; transmission of recharge is essentially unfiltered because subterranean conduits have large dimensions and rapid transmission provides minimal opportunity for die-off of pathogenic

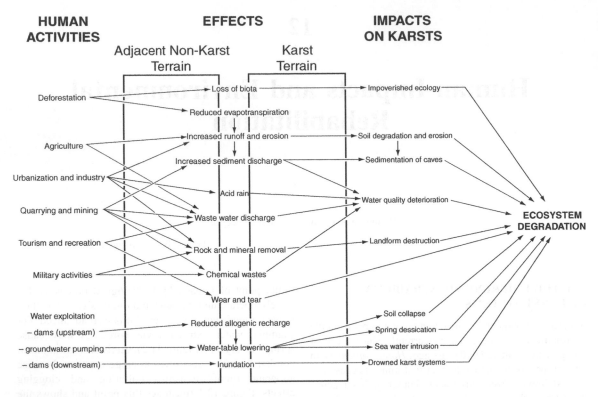

Figure 12.1 Human activities, their effects and impacts on karst terrains. Activities on adjacent, upstream non-karst areas often have an impact on karst as a result of transmission by allogenic drainage into karst. Less frequently an activity on the downstream side also has an impact, such as backflooding from dams. Reproduced from Williams, P.W. (Ed.), Karst terrains; environmental changes and human impacts. Cremlingen-Destedt, Catena Verlag, Catena Supplement 25, 268 pp. © 1993 Elsevier.

organisms. Once thin soils are lost, their replacement time is very long, because there are only small quantities of insoluble residues in karst rocks that might form the inorganic basis of a new soil cover. Whereas inorganic residues might be released at a rate of $50\,t\,km^{-2}\,a^{-1}$, soil erosion in deforested country can easily be one or two orders of magnitude greater. Soil losses have occurred for generations in some temperate and tropical karsts and have led to very severe environmental degradation, in the Gunung Sewu of Java for example. Even in karst lowlands where wet rice paddy field agriculture (supported by irrigation from karst springs) has operated sustainably for centuries, recent changes brought about by population increase, intensification in upland cultivation, deforestation, quarrying, agricultural chemicals and competition for water resources have resulted in severe stress on the resource base and threaten the continued viability of the system, e.g. Urich (1993), Urich and Reeder (1996) provide ample evidence for this from karst areas of the Philippines. Even relatively benign activities such as tourism and recreation can also have severe effects on

karst. Tourist caves in particular are vulnerable because of the focused intensive nature of the visitation. The tourist impact of CO_2 exhalation and lighting was so severe on ancient cave art in Lascaux Cave in France, for example, that this World Heritage site was closed in order to preserve the paintings (section 12.8).

The message is clear: karsts are highly vulnerable to overuse and misuse, have a major subterranean component that requires specialist knowledge to manage properly, and once damaged can be extremely difficult to restore. As a consequence of recognizing this problem, considerable effort has recently been made by countries of the European Union to protect the quality and quantity of groundwater resources, especially karst water resources (COST Action 620 2004), as discussed in Chapter 11. The World Commission on Protected Areas of the International Union for the Conservation of Nature and Natural Resources (IUCN) has also drawn up guidelines for the protection of caves and karst (Watson *et al.* 1997) and for sustainable tourism in protected areas (Eagles *et al.* 2002).

12.2 DEFORESTATION, AGRICULTURAL IMPACTS AND ROCKY DESERTIFICATION

12.2.1 The Kras – stony ground

As described in Chapter 1, the curious name 'karst' itself derives from the impact of early forest clearance and farming on rugged limestone terrains in the northern Dinaric coastal region that created a 'kras' terrain of bare, stony epikarst largely devoid of trees or shrubs. Gams (1991b) describes and illustrates in detail the process followed since classical Greek times in transforming the originally wooded but stony region into farmland. Deciduous forests were cut and burnt by the Illyrians to gain pastures. In Roman times tall pines were logged for shipbuilding. In the succeeding centuries grazing by goats and sheep prevented woodland regeneration on the rocky hillsides and there was severe soil loss by washing into the epikarst and onwards underground. The 'rocky desertification' had become sufficiently intense that in AD 1150 the government in Trieste appointed forest guardians to restrict cutting for firewood, and prohibited the breeding of goats. Strikingly, the Karst is responding well to modern rehabilitation; local regions that showed as little as 5% tree cover in administrative maps of AD 1805 now have more than 40% (Gams 1993).

12.2.2 Modern primary deforestation

The majority of the world's forests today are secondary or later growths upon land that has been cleared of trees at least once for the timber and/or farming. There are few areas where true 'primordial' forest has remained uncut on karst lands until recent times. Two such areas are the rugged coastal rainforests of British Columbia and the adjoining Alaskan 'panhandle', and some similar but smaller areas on the western South Island of New Zealand and in the interior of Tasmania. These are especially significant because some of the primary clear-cutting has been accompanied with impact studies to explore for Best Management Practices; see Kiernan (1984) – Tasmania; Harding (1987) and BC Ministry of Forests (1994) – British Columbia; and Baichtal and Swanston (1996) – Alaska.

Harding and Ford (1993) summarized the results of one detailed impact study. This was on very steep slopes and plateaus in the mountains of northern Vancouver Island, where forests on a thick to massively bedded limestone could be compared with those on adjacent volcanic rocks that are insoluble. The area was glaciated and has veneers of glacial till. Shallow epikarst of postglacial age is well developed on the limestone. The natural forest is coniferous (western hemlock, cedar and fir) with trees that are centuries old and have basal diameters of 1–2 m. Growth is significantly greater on the limestones, where the large trees can root deep into the epikarst. Eight sites in virgin forest (limestone or volcanics) were compared with 16 sites logged between 1970 and 1983, and one on limestone logged as early as 1911. Areas that were deliberately burned after the cutting (to provide some ash to encourage regrowth) were also compared with logged but unburned areas. It was found that there was an average 40% soil loss on the limestone slopes, compared with negligible losses on the control volcanics. Percentage bare rock on the limestone increased from ~2% in the forest to ~25% on the logged terrain (Figure 12.2). There were statistically significant differences between burned and unburned sites, burning always being more deleterious. After 75 years of natural recovery the limestone area cut over in AD 1911 had only 20% of the volume of wood to be found on equivalent uncut areas. The British Columbia forest industry standard cycle period for tree harvest–plant–harvest is 80 years, but it is evident that this cannot be realized on much of the karst because of the soil losses.

In Ontario, Canada, there are very extensive tracts of karstified limestone and dolomite plains, including areas where there is little detrital till cover because of glacial-lake washover. These were densely forested with large white pine until clear-cut between 100 and 160 years ago. All soil and litter was lost over wide areas. In them the forest has been able to re-establish itself naturally (no deliberate planting) by rooting into the epikarst, although the trees are now much smaller or of undesirable species. In contrast to the situation that will prevail on almost all other rocks, therefore, even the complete loss of soil and litter from limestone or dolomite surfaces will not prohibit the re-establishment of forest **if** there is a well-developed epikarst. This is because much soil and litter, plus other nutrients and water, are retained in the epikarst cavities. It is a paradox that surface soil erosion is much more severe on karst rocks than it is on others, yet vegetal recovery on the severely eroded sites can be more complete.

In Ontario and elsewhere the aboriginal hunter–gatherers did not leave all forest on karst untouched. Where drainage of the epikarst was particularly good (e.g. close to escarpment edges) it was common practice to set fires to burn the trees in order to encourage the proliferation of shrubs and grasses in the undergrowth. These attracted deer and other game animals. The result is that there are some extensive tracts of **alvar** vegetation – open grasslands or dwarf juniper shrubs mixed with flowering herbs (Enyedy-Goldner 1994).

Figure 12.2   On micritic limestones in the Benson River valley, northern Vancouver Island, in July 1986. (Upper left) The epikarst, soil, forest floor litter and uncut forest exposed in a small quarry wall. (Lower left) A fresh clear-cut, recently burned. The epikarst is beginning to emerge from the soil cover. Uncut forest in rear. (Right) Dipslopes of 25° that were logged and burned in 1970. The epikarst with rinnen- and rund-karren is exposed.

12.2.3 The impact of historic farming practices

The neolithic and later clearances described by Gams (1991a) in the Classic Karst occurred everywhere else around the Mediterranean, the Fertile Crescent and into northern Europe. The pioneer of groundwater hydrology in the USA, Oscar Meinzer, once remarked that '... in places the Bible reads like a water supply paper', a reference to the significance of karst springs in the spread of small farming and pastoralism. Jericho was a site of very early crop irrigation, for example, built on a spring yielding $10 \times 10^6 \, m^3 \, a^{-1}$. Where the ground sloped or soil was thin there was ready soil loss. In these regions and their equivalent in eastern Asia or middle America, it is commonplace to see epikarst prominences rising 0.5–1.0 m above the remaining soil, with the tips of the

exposed rocks often sharpened by rillenkarren cut into them (Figure 12.3). Vines, olives and fruit trees were planted in holes amongst the rocks. Today, individual banana trees are planted in tiny collapsed caves on some Bahamian islands. In the very rugged (and impoverished) highlands of Mexico each karren hole with a fistful of surviving soil or litter will have a corn plant in it, even where slopes are 35°; the same being true in parts of Guizhou, China.

Farming on karst thus was never easy because of the need to retain soil and accessible water. Technology was limited to human power or that of harnessed animals. Within those limitations there was much modification of the form and function of epikarst and, especially, of dolines. Loose stones were cleared, epikarst pinnacles were broken up and attempts made to level over them

Figure 12.3 'Stone teeth' emerging from soil following deforestation in the highlands of Papua New Guinea.

using terra rossa carried in. Slopes were terraced to retain soil and water, especially around the sides of dolines where there may be flow at the soil–rock interface throughout a summer drought (Figure 5.28). Doline floors received particular attention because they usually contained a deeper natural fill of residual soil, to which the erosion losses during deforestation added a few tens of centimetres or more. Irrigation water could be directed into them by gravity. Thus in many polygonal karsts and upland regions they became the kitchen gardens of the small farmers. Other doline bottoms were puddled with clay to render them impermeable (at least for a few years) in order to convert them into ponds for livestock. Effort was always most intensive in the wet rice regions because of the need to maintain standing water in the paddy fields in face of the threat of soil collapse and abrupt drainage into epikarst or shallow caves. Uvalas and small poljes were much favoured there because of their proportionally greater extent of flat floor. Modification often resulted in the temporary blockage of important sinks, however, causing unwanted flooding.

Dry stone walls and other enclosures are distinctive features of most European farmed karst areas and are seen in a few elsewhere. Walling began with the classical Greek colonizations but the process peaked with the greatly increased population density and intensification of agriculture of the 17th and 18th centuries AD

(Figure 12.4). The order of 100 kg of stones (and sometimes much more) might be removed per metre square of land. Where loose stone was not available for walls, shallow, irregular stone pits might be dug to the bottom of the epikarst and a little below. Now degraded, they have sometimes been mistaken for solution or collapse dolines. There was much cultural variation of form and style. Nicod (1990) and Gams (1991b) give detailed accounts of walling in France and Dalmatia respectively. Krk Island, Croatia, is especially celebrated for its walls and terraces, all built by hand labour.

Many of the poorer karst uplands that were once intensively farmed have now been abandoned because their maintenance is not economic. This is especially true in Europe, some parts of Malaysia and China. In southern France, Corsica and Sardinia they are reverting to **garrigue**, an aromatic but unproductive scrub ecology. In areas of Belize, Guatemala and Mexico, in contrast, modern population pressure is pushing traditional small farmers back onto land that was abandoned during the collapse of the Mayan Empire (AD ~1000–1100) and then reverted to tropical forest for many centuries afterwards. Furley (1987) examined the precise effect over 10 years of soil disturbance by such forest clearance and shifting cultivation on cone karst in Belize. The nutrient and physical properties of the soil were so severely affected, including soil-depth thinning on the upper slopes, that he

Figure 12.4   (Upper) A small field in a doline surrounded by a dry stone wall set in a rocky semi-desertified karst landscape in the Dinaric region. (Photograph by M. Bakalowicz.) (Lower) Small terrace fields in the Dinaric karst with dry stone walls built from stones cleared from the soil.

concluded the magnitude of change resulting from just one cycle of shifting cultivation is so considerable that it throws into question the sense of permitting its long-term continuation in cockpit karst areas.

12.2.4 Mechanized farming and 'rocky desertification'

In the 20th century, especially after 1945, the coming of bulldozers and other heavy equipment has transformed much of the subdued types of karst topography in nations where land prices are high, such as central and southern Europe, Israel, Japan and the Ryukyu Islands. Levelling of ground and filling of shallow dolines has been widespread. Hillslopes are terraced by machine. Bulldozers have torn off epikarst to be ground into gravel, then returned and compacted it with added topsoil; rates of recharge to the underlying karst must be significantly reduced.

The great vineyards of Burgundy and Bordeaux are on weak and porous limestones beneath varying depths of cover gravels, sands and silts. Karstification contributes significantly to the efficient drainage that enhances the quality of the crop. In Bordeaux, for example, the **terroir** where the wine, Graves, is produced is a typical cryptokarst beneath shallow cover; the Medoc terroir (Margaux, Chateau Latour, Lafite, Mouton-Rothschild, etc.) has Eocene carbonate rocks with shallow poljes clogged with sands from the last marine transgression, and Entre-deux-Mers is a mantled polygonal karst. All have been subjected to levelling or other mechanical treatments to aid mechanised tending and harvesting of the grapes. Audra (1999) reported that consequences have included soil erosion, rapid formation of suffosion dolines, blockage and pollution of the aquifers, and backflooding.

War has also contributed in Europe. Areas of the French chalk were densely tunnelled and trenched, 1914–1918, altering the natural karst drainage. Karst plateaus of the Venetian Fore-Alps were also sites of sustained trench warfare. Shell fire there created an 'instant epikarst' of craters 2–10 m in diameter and up to 3.5 m deep. Subsequently, dissolution has exploited the explosion fracturing in many of them, converting them into functioning small dolines (Celi 1991).

Tropical karst lands have seen the introduction of commercial plantations for citrus, bananas, pineapples and other fruits, cacao, coffee, palm oil, etc. for export to the First World. Floodplains of allogenic rivers and floors of border poljes were early targets because they require little levelling work (e.g. Sibun River, Belize; Figure 9.34). There has been expansion into interior poljes and large uvalas such as the 'glades' of Jamaica, with concomitant terracing, flood-control damming and channelization. Soil erosion has increased and many springs have become polluted. In the Chinese, Indonesian and Philippines karsts there are ingenious admixtures of the historic subsistence and modern commercial practices; Figure 12.5 shows an example from Gunung Sewu, Indonesia (Urushibara-Yoshino 1991).

Rocky desertification caused by deforestation is as old as the Classic Karst itself, as we have shown. However, it can be said to have reached a peak during the 'Great Leap Forward' in China in 1958, and in subsequent campaigns there. The Great Leap Forward called for production of iron in small blast furnaces throughout the land. Communal farmers were ordered to strip all trees for conversion into charcoal for the furnaces. Huntoon (1991) studied the results on sample stone forest and peak cluster karsts in the south of the country. One commune reported burning 3000 m³ of

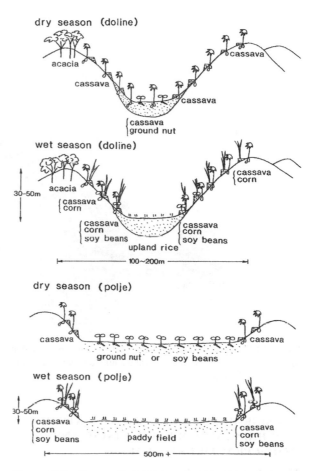

Figure 12.5 The wet and dry season utilization of dolines ('cockpits') and small poljes in the Gunung Sewu karst, Indonesia. Reproduced from Urushibara-Yoshino, K., Miotke, F.-D., Kashima, N., et al. (1999) Solution rate of limestone in Japan. *Physics and Chemistry of the Earth*, Series A, 24(10), 899–903.

wood per day for charcoal there. Even the roots of the trees were grubbed up, destroying the last vestiges of binding strength in the soil (Figure 12.6). Yuan (1996) reported that typical soil losses in the karst ranged from 200 to 2000 t km^{-2} a^{-1}. Conditions deteriorated further during the Cultural Revolution (1966–1976) and immediately following decollectivization in 1979. Loss of the 'green reservoir' on the karst peaks and slopes has increased flood amplitudes and lowered water tables. Reforestation (natural and by planting) is now encouraged everywhere but rural demands for domestic firewood create many conflicts. There is a search for fast-growing trees that will resprout when chopped off just above the roots (see http://www.edu.cn/desert/rockdesert.htm).

Figure 12.6   Rocky desertification in peak cluster karst near Wuxuan, Guangxi Autonomous Region, China. The villagers in the left foreground wisely chose not to strip the hillside above their homes, thus avoiding flooding and soil erosion. (Photograph by Peter Huntoon, with permission.)

12.3 SINK-HOLES INDUCED BY DEWATERING, SURCHARGING, SOLUTION MINING AND OTHER PRACTICES ON KARST

12.3.1 Induced sink-hole formation

It is probably true to write that, after groundwater pollution, induced sink-holes are the most prominent hazardous effect of human activity in karst regions. Agriculture, mining and quarrying, highways and railways, urban and industrial construction all contribute. Induced sink-holes generally develop more rapidly than most natural ones, appearing and enlarging in timespans ranging from seconds to a few weeks. This is faster than most societies can react with preventive or damage-limiting measures, so that such sink-holes are widely described as 'catastrophic'. Although in a few instances the hazardous collapse is of surface bedrock directly into a cave underneath, in 99% of reported cases or more it occurs in an overburden of unconsolidated cover sands, silts or clays, i.e. it is a suffosion or cover-collapse doline; 'subsidence sink-hole' is the widely used alternative term. There are two end-member processes:

1. ravelling, the grain-by-grain (or clump-by-clump) loss of detrital particles into an underlying karst cavity that is transmitted immediately by grain displacement upwards to the surface, where it appears as a funnel that gradually widens and deepens;
2. formation of a soil arch in more cohesive clays and silt–clay mixtures over the karst cavity that then stopes upwards until it breaks through to the surface (section 9.5, Figure 9.25).

The large majority of suffosion sink-holes appear to have formed by a mixture of the two processes, with the soil above an early arch subsiding by downfaulting, suffosion then sapping the fault-weakened mass to create a new arch and repeat the cycle. From the human perspective, pure soil-arch collapses (or 'dropouts') are the most dangerous because they can appear without warning at the surface. There has been much loss of life as a consequence.

The contact between a solution-indented karst surface and overlying cover deposits is the 'rockhead'. Using ground-penetrating radar (GPR) through 2–20 m of cover, Wilson and Beck (1988) estimated that the frequency of karst cavities at the rockhead capable of swallowing ravelled debris varied from $12\,000$ to $730\,000\,\mathrm{km}^{-2}$ in the counties of northern Florida, i.e. the rockhead there is a dense epikarst of solution pits, pipes and shafts. The potential for suffosion if the delicate balance of its natural drainage is disturbed is obvious.

Induced sink-hole formation thus has attracted a good deal of attention in recent decades. Hundreds of thousands of new subsidences have been reported worldwide, ranging from $1 \times 1 \times 1$ m holes to features > 100 m in length or diameter and tens of metres deep. In the USA the abrupt appearance of the Winter Park sink-hole in 1981 (Figure 12.7) led to creation of the Florida Sinkhole Research Institute and to periodic conferences with Proceedings that offer many case studies of collapses, plus groups of papers devoted to their geophysical detection, methods of remediation, appropriate building design, legal regulation, etc. (Beck 2005, and earlier Proceedings edited by Beck). Waltham *et al.* (2005) present a comprehensive and excellent review within one volume.

12.3.2 Groundwater abstraction and dewatering

Dewatering of unconsolidated cover deposits on karst is the most important cause of induced sink-hole formation worldwide. The buoyant support of the water is removed, weakening mechanical stability (section 9.3). Abstraction may be for water supplies, for irrigation, draining a volume of rock for mining or quarrying, or many other purposes. It is most hazardous when the cover is drained entirely so that the water table is depressed below the rockhead into the karst strata. However, sink-holes may also form readily where the lowering is limited to some level within the overburden.

Florida offers many examples of agricultural impacts because there are large plantations of citrus fruits and other tropical crops. To protect these from occasional hard frosts in winter it has been the practice to water heavily on cold nights (ground water temperature is 12°C+). The appearance of small new suffosion and drop-out sink-holes is highly correlated with pumping nights. Irrigation is common every night as well on many golf courses during dry periods. Not a few underlain by limestone have suddenly found themselves with more than the mandatory 18 holes. Collapses are common on corrosion plains and in the bottoms of poljes when these are pumped for irrigation during dry seasons in tropical and mediterranean regions. With or without human intervention, the rapid rate of dissolution of gypsum can yield major problems. For example natural ground collapse over gypsum in northeast England resulted in about $1.5 M worth of damage in the interval 1984–1994 (Cooper 1998).

Mining and quarrying tend to have the greatest impacts because they dewater to the greatest depths, usually far below the rockhead. In many coal mining areas of China the coals are overlain by or interfinger with limestones or gypsum. Tens of thousands of sink-holes, often of large size, have been reported. Table 12.1 gives the example of Enkou Mine, Hunan, where the average rate of sink-hole production was $\sim750\,a^{-1}$ over an eight years span during which the water table was progressively lowered 90 m. Dewatering to -220 m in limestones that are heavily weathered at the top and overlain by 2–30 m of glacial and marine sediments is planned in order to develop a kimberlite pipe for diamonds in northern Ontario. Approximately $300\,km^2$ will be affected, including two large rivers: we shall watch this development with interest.

A notorious case of drop-out doline development was reported by Jennings (1966) and Brink (1984) in the Far West Rand, South Africa. A major pumping programme was initiated to dewater dolomite and dolomitic limestone over gold-bearing conglomerates. Pumping commenced in 1960, and between 1962 and 1966 eight collapse dolines exceeding 50 m wide and 30 m deep formed in the mining area. In the worst incident (December 1962), a three-storey crusher plant with 29 occupants was lost in a few seconds at the West Dreifontein Mine. According to Brink, after 25 years of pumping a total of 38 people had lost their lives in collapsing sinks, and damage to buildings and structures amounted to tens of millions of Rand. The depressions are the result of soil-arch collapse in 20–40 m of regolith overlying dissolution-widened joints which lead downwards into large caverns. Geophysical exploration for new cavities proved to be nearly useless. Arrays of simple telescopic benchmarks were set out in the hope of detecting the stoping arches before they breached the surface. Wagener and Day (1986) discussed construction techniques on these rocks.

12.3.3 Surcharging with water

In the case of surcharging it is the addition of water at particular points that causes ravelling of overburden into karst cavities. It will therefore be especially potent where the unsurcharged water table is below the rockhead, but can also be effective with water levels in the overburden. In modern cities such point-located surcharging will be widespread unless precautions are taken, caused by drainage from individual downspouts on buildings, leaks from water supply and sewer pipes, leaking stormwater management ponds, parking lots, etc. Sink-hole formation is fastest where the overburden is thin, and is chiefly by ravelling. Sink-holes tend to be smaller than those associated with substantial dewatering, being mostly less than 10 m in diameter. Nevertheless, there are many

Figure 12.7 The Winter Park suffosion doline, Florida. This feature developed during a 72-h period in May 1981, consuming a dwelling, part of a road, automobiles, etc. It developed in 30 m of sands and clayey sands burying densely karstified limestone and is believed to have been induced when the water table was lowered 6 m within the sands. The feature was 106 m in diameter and 30 m deep. The lower photograph (August 1984) shows the doline after remedial action, including restoration of the water table to its original elevation.

Table 12.1   Relationships between water extraction, water-table drawdown and accumulated induced sink-holes at Enkou Coal Mine, Hunan, China (Data from Lei *et al.* 2001)

Period	Pumping rate ($m^3 h^{-1}$)	Lowering of water table (m)	Accumulating total of sink-holes
1974	1270	14.24	317
1976	3388	48.67	1329
1979	3868	62.25	4924
1982	4130	90.42	5811

reports of building foundations being undermined and collapsed (Figure 12.8), and damage to roads and railway tracks due to neglect of soakaways or other means of dispersing stormwaters. Natural surcharging occurs on river floodplains, corrosion plains and poljes when they are inundated and is often accompanied by collapse and suffosion as the waters recede (section 9.3)

General raising of the water table can also create collapse or subsidence by destroying the cohesion of susceptible clay soils. However, this is comparatively rare in karst areas. The load or vibration from heavy equipment can induce small collapses locally, especially beneath it. In historic times plough-horse teams have dropped; in modern times many tractors, haulage trucks,

drilling rigs and military tanks have fallen. Rock blasting from quarries or foundation cutting, etc. often causes collapses of small to intermediate scale.

12.3.4 Solution mining

Salt mining has induced many collapses and subsidences over the centuries. Normally this will involve significant thicknesses of overlying consolidated rocks. Suffosion in superficial unconsolidated deposits thus is not usually predominant, as it is in the dewatering and surcharging situations considered above.

Traditionally, extraction has been by one of two methods:

Figure 12.8   Subsidence of a timber-frame house into a collapse sink-hole, Phillipsburg, New Jersey. The sink-hole was created when a water main leaked, surcharging soil-mantled epikarst. (Photograph by Rick Rader, with permission.)

1. conventional mining via shafts and adits, removing the product by hand or machine at the workface, as in a coal mine, etc.;
2. by pumping the water from natural salt springs ('wild brine').

Recently, where feasible these have been replaced by solution mining, in which water is injected via one set of boreholes and extracted as brine via another, i.e. no workers or equipment are committed underground. The planimetric extent and volumes of cavities that will be created in the salts by both the wild brine and the injection methods are always uncertain and can be hazardous.

The most celebrated historic examples of induced subsidence in the English-speaking world are in the county of Cheshire, England, where wild brine mining began in Roman times and adit mining of salt became important with the beginning of the Industrial Revolution. Large areas of the surface have now subsided, over both open mines and solution mines, with much property damage, although catastrophically rapid collapse of ground is comparatively rare due to local geological conditions (Cooper 2001).

There was a spectacular collapse at the Bereznikovsky No. 3 Mine (Urals, Russia) in 1986. At the time this was the world's largest potash mine, a pit-and-stall mine in mixed potash–salt–anhydrite–marl strata at ~425 m depth, overlain by limestones containing much groundwater, and upper marls and shales. Leakage from the limestone began in January 1986 at rates of $10–30\,m^3\,h^{-1}$, increasing to $\sim100\,m^3\,h^{-1}$ by March when the mine was abandoned. It then filled progressively with water, dissolving the stalls (salt pillars) left to support the roof. Evidently upward stoping soon began because at midnight 24–25 July there was an abrupt collapse of the surface accompanied by explosive release of compressed gases that hurled rocks of decimetric sizes several hundred metres from the crater. During the next few weeks the cavity stabilized as an arch collapse measuring $100 \times 50\,m$ in the bedrocks and 50–60 m deep to the waterline. Andrejchuk (2002) gives an outstanding analysis.

This was an inadvertent, but perhaps expectable, collapse over a deliberately mined void. Corporations drilling exploratory and extractive oil wells do not expect to be involved in salt solution mining as well but this has happened in many instances in recent decades. Figure 12.9 is an interpretation by Johnson (1989) of the events leading to a catastrophic collapse, Wink Sink, in Texas in June 1980. It was 110 m in diameter and 24 m deep to a debris floor below the waterline. It is attributed to leakage down a borehole drilled as long ago as

1928 to explore for oil in Tansill and Yates dolomites beneath the Salado Salt Formation. The Salado hosts the WIPP hazardous waste disposal facility nearby in New Mexico (section 11.4).

12.3.5 *Hazard prediction, detection and regulation*

Understandably, individuals and organizations of all kinds will wish to assess the risk of subsidence and collapse in karst terrains. It is recognized that accurate prediction of such hazards in advance of, say, a major de-watering, is very difficult because of the complex multivariate nature of the problem. Accordingly it is becoming standard practice to take highly generalized, semi-quantitative and probabilistic approaches rather than attempting precise site-specific predictions. Risk then can be defined as:

$$\text{risk} = \text{magnitude of hazard} \\ \times \text{probability in a given time} \times \text{given area}$$

Tolmachev *et al.* (1986, 2005) and Ragozin *et al.* (2005) have utilized such approaches in evaluating risk of natural suffosion sink-hole formation under structures such as railbeds or new chemical plants in the extensive covered karsts of the Russian Plain. The occurrence of such sink-holes is assumed to be random (Poisson) in time. Historical records are used to obtain mean frequencies, and the risk assessment then becomes a matter of specifying the extent of the area to be considered and the amount of time (e.g. the planned lifetime of a factory). A refinement is to specify a scale of damage. Tolmachev *et al.* (1986) adopted four levels, ranging from trivial damage to total destruction; the British National Coal Board recognizes seven. There are many similar schemes in use in different parts of the world.

However, such methods require good historical records of collapses but these are not often available and, in any case, will be of little use where, for example, a major new dewatering is being proposed. Song (1987) tackled the prediction problem here by compiling records of >18 000 dewatering suffosion collapses in southern China between 1974 and 1986, mostly around coal mines, and proposed the following experimental approach:

$$R = \frac{\alpha R_1}{S_1} S \tag{12.1}$$

where R is the uncorrected radius of the area that will be at hazard, R_1 is the radius of effects noted in a first

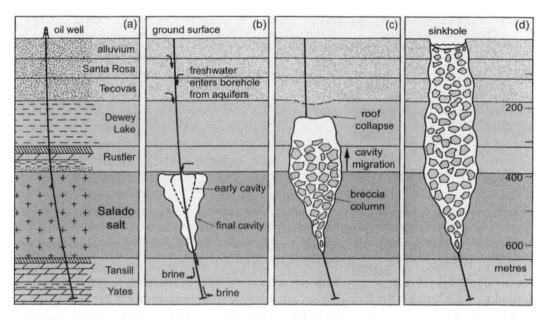

Figure 12.9 Schematic cross-sections to show the interpreted relationship between an oil well drilled in 1928 in west Texas: (a) borehole leakage creating a dissolution cavity in intercepted salt; (b) stoping in overlying rocks; and (c) the abrupt appearance of 'Wink Sink' in 1980.

experimental withdrawal pumping, S_1 is the depth of drawdown during that experiment, and S is the intended final depth of drawdown. Variables R_2 and S_2 are radius of effects and depth of drawdown in a second experiment:

$$\alpha = \frac{S_1 R_2}{S_2 R_1} \qquad (12.2)$$

The corrected radius of the area at hazard, R', is given by

$$R' = X + (R - X)K_1/K_2 \qquad (12.3)$$

where X is the distance between the pumping point and the boundary of karst rock, K_1 and K_2 are the numbers of subsidences produced in the two experiments. Song (1987) was unable to undertake these experiments at an appropriate scale but did publish interesting tables illustrating the natural variability that is found (Table 12.2).

Others have suggested more precise geotechnical solutions, with their focus primarily on cases where there is rapid drawdown by pumping (i.e. much faster than would occur naturally). Tharp (2001) proposed that such drawdown induces hydraulic fracturing, which then becomes the chief process in development of the compound ravelling and arch collapse form that is the most common type of cover subsidence sink-hole. In contrast, He *et al.* (2001) suggested that the air pressure differential

between the open surface and a suddenly drained incipient cavity over the karst will trigger some collapse by a vacuum suction mechanism. Anikeev (1999) considered the case where there is a comparatively impervious clay between the rockhead and the main overburden. This is common on many corrosion plains or in areas of recent marine transgression such as Entre-deux-Mers (section 12.2). He found that rapid drawdown in the overburden will induce spalling from arches in the clay. These three examples are representative of the range of ongoing investigations today.

A wide range of surface and airborne mapping techniques, ground-based and downhole geophysics have been brought in to play to attempt to detect karst hazards beneath cover deposits; see Waltham *et al.* (2005) for comprehensive discussion. Black-and-white and infrared aerial photographs are basic tools; the infrared may detect vegetation stress (drought) above an unseen stoping cavity. Satellite radar can record slow subsidences, year by year.

There is much focus on ground-based geophysical detection when dangerous rockhead conditions are suspected within small areas such as construction sites. Section 6.2 introduced some methods; Table 12.3 gives a summary. Seismic methods are standard. They use hammers, explosives, etc. to generate shock waves that are recorded at arrays of geophones. Anomalies in wave transmission rates may signal cavities, etc. Depth

Table 12.2   Dewatering, drawdown and subsidence in covered karst around some Chinese coal mines (Modified from Song, L. (1987) Pumping subsidence of surface in some karst areas of China, Proceedings of the International Symposium on Karst and Man, Lubljana, pp. 49–64)

Site	Depth of drawdown (m)	R radius of drawdown (m)	L – radius of area with collapses (m)	L/R
Dalinjing	80	950	800	0.8
Yehuaxiang	109	3000	2300–2600	0.9
Tiantanjing	204	830	630	0.7
Qiaotonha	280	10 000	10 000	1.0
Jinjiang	1200	600	175	0.3

penetration is only about one-third of the geophone spread but can be partly overcome by downhole (cross-hole) applications. Cross-hole electrical tomography is a new development that appears more promising but is expensive. Electrical resistivity/conductivity traverses are similar in concept to surface seismic but instead utilize induced electrical pulses to detect the anomalies, e.g. air-filled cavities have very high resistance, saturated clays are very low. Depth of penetration is better and some large vadose caves have been found as deep as 50 m. Small features are poorly depicted. High-resolution gravimetry (microgravity) is quick and thus cheap; it can detect shallow cavities beneath comparatively flat surfaces, but requires many corrections there and is unsatisfactory in rugged terrain. Ground-penetrating radar (GPR) fires high-frequency electromagnetic pulses from a device little larger or more cumbersome than a domestic vacuum cleaner, and records anomalies in the return reflections. It was developed for the detection of buried pipes, cables, etc. Wilson and Beck (1988) had great success in detecting rockhead hazards as deep as −30 m below cover sands in northern Florida, but GPR cannot penetrate clays in

most instances. It is now widely used in combination with one of the other methods. All of them must be treated as **indicative** but not **definitive**. If it is believed that they have detected a threatening feature, it must be explored by boreholes. Exploration by boreholes alone (no prior geophysics) is economically unfeasible in most instances; Zisman (2001) showed that > 2000 borings per hectare would be needed for a 90% certainty of detecting a cavity 2–3 m in diameter. The rapid increase in induced sinkhole formation that marked the 20th Century is causing many governments to introduce regulations. Dewatering in particular may be closely controlled; in Florida dewatering operations are required to establish a 'Zone of Influence' and compensate other stakeholders within it. All Florida insurance companies must include subsidence in their standard policies now. Elsewhere in the USA these are optional extras, or are ignored with the legal implication that there is no cover. The UK mandates that subsidence must be included in property insurance, but there is often a large deductible charge. Insurance companies there make use of national geohazards maps at 1:50 000 to assess risk and set premiums.

Table 12.3   Recommended methods for the geophysical location of specific dissolution features in karst (Reproduced from Waltham, A.C., Bell, F. and Culshaw, M. (2005) Sinkholes and Subsidence: Karst and Cavernous Rocks in Engineering and Construction. Praxis, 382 pp © 2005 Praxis)

Karst feature	Dimensions	Recommended methods	Factors to consider
Pipes and hollows, with clay fill	Depth:diameter < 2:1 Depth < 30 m	Conductivity traversing Magnetic	Coil separation, cf. depth Local magnetic gradient
Pipes and hollows, with sand fill	Depth < 5 m	Ground-penetrating radar	Conductivity of cover and fill, and cover thickness
Small open caves	Depth:diameter < 2:1 Depth < 30 m Depth > 30 m	Conductivity traversing Microgravity Cross-hole seismic	Coil separation, cf. depth Density and nature of fill Borehole spacing
Large open caves	Depth < 10 m Depth > 10 m	Ground-penetrating radar Conductivity traversing Gravity and microgravity Cross-hole seismic	Ground conductivity Cavity infill Cavity infill, terrain relief Borehole spacing

12.4 PROBLEMS OF CONSTRUCTION ON AND IN THE KARST ROCKS – EXPECT THE UNEXPECTED!

Karst processes and landforms pose many different problems for construction and other economic development. Every nation with karst rocks has its share of embarrassing failures such as collapse of buildings or construction of reservoirs that never held water. It is probably true to write that the global cost of extra preventive measures or of unanticipated remedial measures in karst terrains now amounts to some billions of dollars (US) each year. The problems encountered can be classified by the extent of the impact of construction or other development upon karst features already existing at a site. There may be no impact, as in the case of construction in the path of a potential landslide. There can be small- to large-scale impact where foundations for bridges, buildings, roads and railways are placed upon karst without much effect upon the local water table. Large-scale impact is more common if the water table is raised or lowered. This may become extreme in the construction of tunnels, mines and dams, as we have emphasised above and in Chapter 11. Careless road construction can locally overwhelm sink-holes and caves with detritus and alter karst drainage patterns; James (1993) described the dramatic example of an access road to a gold mine in Papua New Guinea.

12.4.1 Rock slide-avalanche hazards in karst

A **landslide** or **rock slide-avalanche** is the catastrophically rapid fall or slide of large masses of fragmented bedrock such as limestone (Cruden 1985). 'Landslide' is more widely used but is also applied to slides of unconsolidated rocks. Rock slides take place at **penetrative discontinuities**, the mechanical engineering term for any kind of surface of failure within a mass. Once initiated, there is powerful momentum transfer within the falling mass and it may partially ride on a cushion of compressed air that can permit it to run for some hundreds of metres upslope on the other side of a valley (van Gassen and Cruden 1989).

Carbonate rocks and gypsum are especially prone for two reasons.

1. While faults and joints are the only important penetrative discontinuities in most other rocks, in karst strata there is also major penetration via bedding planes. In fact they are particularly favoured as surfaces of failure because of their great extent.

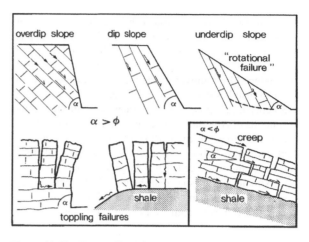

Figure 12.10 Types of landslides (or rock slide-avalanches) in carbonate rocks. $\emptyset$ is the internal angle of friction of the rock. Failures on dip and overdip slopes are termed 'slab slides'.

2. Large quantities of water may pass rapidly through the rock via its karst cavities to saturate or lubricate interlaminated or underlying weak or impermeable strata such as clays. The forces that resist catastrophic failure within a particular rock are defined by an **internal angle of friction**. Minimum angles for relatively hard carbonates without shale interbeds range from 14° to 32°.

The principal settings of landslides in karst rocks are shown in Figure 12.10. Slab slides are particularly common because they are bedding-plane failures. They are especially frequent and dangerous in the overdip situation. Rotational failures within massive carbonates are comparatively rare but there are large ones in dolomites in the Mackenzie Mountains, Canada. Toppling cliffs are common in all rocks; see Cruden (1989) for formal analysis. Toppling or rotational failures are quite common along escarpment fronts where the permeable karst rock rests on a weak but impermeable base such as a shale; Ali (2005) describes 12 limestone failures of up to 800×10^6 t each along a 20 km frontage near the city of Sulaimaniya in northern Iraq, caused by spring sapping at the contact with underlying shales.

Downslope detachment and creep of karst rock formations resting on slick but impermeable strata beneath them (Figure 12.10) may proceed slowly for long periods and then suddenly accelerate into a landslide, usually as a consequence of heavy rains or an earthquake. At the Vajont Dam disaster of 1963 in the Italian Alps, in which 2000 lives were lost, rise of water level in a reservoir may have contributed by increasing porewater pressures on the slide plane. The Ok Ma Landslide (Papua New Guinea) was a slide of $\sim 36 \times 10^6$ m^3 of fractured massive limestone on clay dipping into a river valley that was induced by removal of the toe of a previous slide in order to install a dam for a gold mine.

Steeply dipping carbonates predominate in the Rocky Mountains of Alberta. On average, there have been one or two rock slides in each 100 km^2 there since regional deglaciation $\sim 10\,000$ year ago. The volumes of rock detached and the areas that they bury are found to approximate a Poisson distribution. The largest slide contained 30×10^6 m^3 of limestone; it buried an area of ~ 3 km^2 to a mean depth of 14 m in a timespan of only

100 s. This was the Frank Slide, which occurred one night in AD 1904, crushing the small town of Frank and taking 70 lives. On the Yangtse River, China, there are a total of 283 old landslides (in bedrocks and/or alluvial and colluvial deposits) along that part of its course that will be inundated by the Three Gorges Dam, approximately 15% of them being in limestone (Lu 1993).

12.4.2 *Setting foundations for buildings, bridges, etc.*

Setting foundations where there are soils, etc., covering maturely dissected epikarst can encounter many problems. Figure 12.11 illustrates the range of different methods that are used to overcome them by compacting

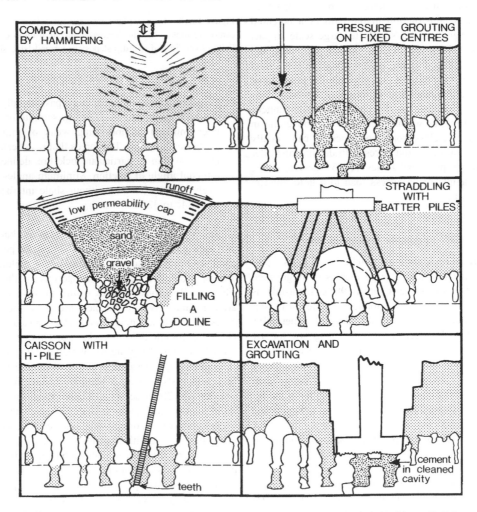

Figure 12.11 Illustrations of some of the principal types of foundation treatments in a soil-mantled karst. Based upon figures in Sowers, G.F. (1984) Correction and protection in limestone terrace. in Sinkholes: their Geology, Engineering and Environmental Impact (ed. B.F. Beck), Balkema, Boston, pp. 373–8.

the soil or pinning the footings to (comparatively) firm bedrock. Under large or heavy structures the majority of these methods can be very expensive. Reinforced concrete slabs ('rafts' floating on the soil following its mechanical compaction) are now much used as alternatives under buildings. For roads on mantled epikarst or spanning infilled solution and suffosion dolines strong synthetic plastic sheeting, strips or meshes ('geofabrics') are being substituted because they are cheaper: their long-term reliability is not yet established, however. Much has been written on these subjects; see Beck (2005) and Waltham *et al.* (2005) for recent surveys.

Building calamities remain frequent worldwide. A celebrated example was the collapse of 'Corporate Plaza', a major downtown redevelopment with an adjoining multi-storey parking garage, in Allentown, Pennsylvania, one night in February 1994. It was destroyed by arch collapse sink-holes in glacial deposits resting on karstic dolomites. The buildings were footed on columns with 2.5 m bases placed 1.5–1.8 m into the glacial drift; no attempts had been made to anchor them to solid bedrock or to spread the loads (Dougherty 2005).

Cavities entirely within bedrock can also pose dangers if they are at very shallow depth or if the planned structural load is considerable. For typical strong limestones with caves, Waltham *et al.* (2005) recommend a minimum of 3 m bedrock above a cavity 5 m wide, 7 m for widths >10 m (and see Figure 7.50); for chalk and gypsum, at least 5 m of rock above a cave 5 m wide. Severe difficulties were encountered in footings for four piers in limestone for a motorway viaduct in Belgium, increasing its overall cost by 15%. One pier had to be shifted 15 m onto stronger rock. A standard programme of exploratory drilling had missed cavities 3 m wide (Waltham *et al.* 1986).

Construction on gypsum requires particular care. Gutierrez (1996) and Gutierrez and Cooper (2002) discuss the rich example of Calatayud, a town of 17 000 persons in the Ebro Valley, Spain. It is built on a fan of gypsiferous silts interfingering with floodplain alluvium, and underlain by a main gypsum formation ~500 m thick. The existing buildings are 12th century to modern in age. Many (of all ages including recent) display subsidence damage that ranges from minor to very severe (Figure 12.12). The primary cause is believed to be dissolution of the gypsum bedrock, which is abetted by local compaction of overburden accumulated since the town was founded in AD 716, collapse of some abandoned cellars and dissolution of the silts. Al-Kaisy (2005) describes similar problems at Tikrit, Iraq. Numerous case studies concerning evaporite-related engineering and environmental problems in the USA are presented by Johnson and Neal (2003).

12.4.3 Tunnels and mines in karst rocks

Tunnels and mine galleries (adits or levels) will be cut through rocks in one of three hydrogeological conditions: (i) vadose; (ii) phreatic but at shallow depth or where discharge is limited, so that the tunnel serves as a transient drain that permanently draws down the water table along its course, as in Figure 5.7b; (c) phreatic, as a steady-state drain, i.e. permanently water-filled unless steps are taken drain it (Figure 5.7a). Long tunnels in mountainous country may start in the vadose zone at each end but pass into a transient zone, or even a steady-state phreatic zone, in their central parts.

Vadose and transient zone tunnels are cut on gentle inclines to permit them to drain gravitationally. It was by this means that lead–zinc ore bodies within interfluves in the hilly country of Bohemia and Derbyshire were drained below the natural water tables in the 15th and 16th centuries in order to mine them. A modern example is the water-supply tunnel of the city of Yalta in the Crimea. It passes through faulted limestone as a transient phreatic drain, and is 7 km in length with a fall of 50 m. Groundwater discharges were $\sim$1000 m^3 h^{-1} in the first year, declining to $\sim$350 m^3 h^{-1} over the next several years.

Where the tunnel or mine is a deep transient drain or is in the steady-state phreatic zone, gravitational drainage will not suffice, e.g. if the tunnel is below sea level. Three alternative strategies can then be adopted. The first is to pump from the tunnel itself, when necessary. The first steam engines were developed to apply this simple strategy below sea level in the coastal tin mines of Cornwall, UK. It continues to be the most popular method in smaller mines and some short, shallow tunnels. It is prone to failure if the pumps fail and to disaster (for the miners) if large water-filled cavities are intercepted, causing catastrophic inrushes of water.

The second means is to grout the tunnel and then to pump any residual leakage as necessary. It is the essential method for transportation tunnels. Traditionally, tunnel surfaces were rendered impermeable by applying a sealant (e.g. concrete) as they became exposed. This does not deal with the catastrophic inrush problem. The first undersea tunnel was the Severn Railway Tunnel, cut in the 1860s in thick to massively bedded limestones beneath the Severn River estuary of Britain. It used the cut-and-seal method. At the halfway point (~5 km from shore beneath a saltwater estuary) a large freshwater spring was encountered. It flooded the tunnel, delaying completion for one year. It has been necessary to pump the site continuously ever since.

Modern practice is to drill a 360° array of grouting holes forward horizontally, then blast out and seal a

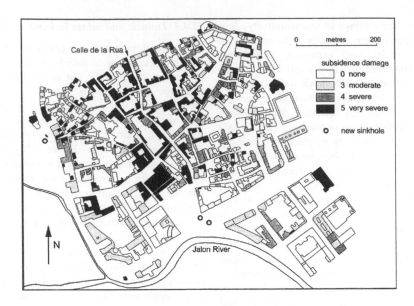

Figure 12.12   (Upper) Subsidence of buildings in Calatayud, Spain. Damage is mapped on a four-part scale, from None to Very Severe. Reproduced from Gutierrez, F. and Cooper, A.H. (2002) Evaporite dissolution subsidence in the historical city of Calatayud, Spain: Damage Appraisal and Prevention. Natural Hazards, 25, 259–88. (Lower) Moderate damage in the main square of the town.

section of tunnel inside this completed grout curtain. This largely deals with the hazard of catastrophic inrush, i.e. a flooded cavity should be first encountered by a narrow bore drill hole that can be sealed off quickly. For example, the cooling water intake tunnel for Bruce B atomic power station, Ontario, was an 8 m diameter tunnel extending 600 m from shore beneath Lake Huron. It followed a corallian limestone formation just below the lake bed. Grouting forward proceeded in 20 m sections and the tunnel was cut in 8 m sections, i.e. there was 60% overlap of successive grout curtains. A cavity

was encountered that could not be grouted because it was too large. It was sealed off and the tunnel was then deflected around it without serious difficulty, but at substantial extra cost. Milanović (2000) and Marinos (2005) discuss tunnel protection thoroughly, with many examples.

Grouting is not feasible in the extracting galleries of a mine. Here, a third and most elaborate strategy is to dewater the mine zone entirely, i.e. maintain a cone of depression about it for as long as the mine is worked. A good modern example of the method is provided by the development of lead–zinc mines at Olkusz, Poland (Wilk 1989). The ores are contained in filled dolines and cavities in a dolomite palaeokarst at a depth of 200 to 300 m below Quaternary sands that were in hydrological contact with the bedrock. Potentially, this was a very hazardous situation. An area of 500 km^2 was surveyed around the potential mine. It contained 70 natural springs and 600 wells. A further 1700 exploration boreholes were drilled. Piezometers were installed in 300 wells and bores in order to conduct pumping tests. From the latter it was estimated that 300×10^6 m^3 s^{-1} of groundwater would have to be pumped to establish the cone of depression for the mine. The cone was pumped out via vertical wells plus drainage adits with high capacity pumps that were cut beneath each extraction level before ore extraction began. By these means, maximum local inrushes of water were held to 1.5 m^3 s^{-1}, within the capacity of the pumps.

As expected, dewatering the Olkusz mine dried up springs and wells. It induced many suffosion dolines in the overlying sands, one of them in the mill-tailings pond which, as a result, discharged 30×10^3 m^3 of sludge into a sector of the mine. In addition, a paper mill 6 km north of the mine had disposed of highly contaminated waste water into the local sands aquifer. With the drawdown these contaminants were drained into part of the mine water, ruining it as a source of replacement water for the springs.

12.5 INDUSTRIAL EXPLOITATION OF KARST ROCKS AND MINERALS

12.5.1 Limestone and dolomite

Limestone and dolomite are the world's principal mined or quarried rocks and are used for a wider range of purposes than any other rocks. To begin with, most stone sculpture and interior decorative stonework such as stairs uses the highest quality limestone, i.e. with fewest imperfections. The preferred stone is true marble (section 2.3) because of its homogeneous crystalline composition and colour. However, any well-textured limestone or dolomite that will take a good polish is marketed as 'marble' in many nations.

Limestone blocks are used for the construction of entire buildings, normally with cheaper grades of stone (more porous and friable) plus marble facings where desired. An early example was the use of a gleaming white limestone for facing the sandstone masses of the pyramids at Giza, Egypt. The Sphinx is of a local limestone. More significant, perhaps, was the fact that limestone is the predominant local rock in Crete and Greece. Cretan palaces of the Minoan age and the Hellenic

temples and public buildings of Greece used it almost exclusively (Figure 12.13). The Romans followed with their buildings and statuary; at Rome itself a locally available lake travertine was especially favoured because it was soft and easy to cut. It is comparatively porous and friable and thus the classical buildings of Rome have not worn so well as those of Athens.

Classical use set the taste for building throughout much of Western history. In a majority of nations the principal palaces, churches, parliament buildings, etc. of the 12th to 19th centuries use limestone, e.g. St Peter's, Rome and St Paul's, London. Entire cities such as Bath and Venice are built of it. Although much was taken from open quarries, underground stone mines that selectively removed the best quality limestone beds (cut to standard sizes by saws) were also common; there are many tens of kilometres of galleries in the Jurassic oolites around Bath and >350 km in Champagne and Alsace. Often, the stone was transported considerable distances. Eighteenth and 19th century buildings in Budapest and Vienna used 'Aurisina marble' from Trieste; see Cucchi and Gerdol (1985) for a thorough review of limestone masonry, with the Aurisina as an example. Limestone and dolomite were the principal choices for the great public buildings in the USA and Canada. Nowadays, however, the use of whole stone for building is uncommon in western countries. Concrete blocks with thin facings of the rock have replaced it.

Marble is equally prized in the Oriental cultures, e.g. the Taj Mahal, India. The Chinese and Japanese built with wood, but limestone and marble were used for courtyards and stairs in the imperial palaces. Temple builders of the Western Hemisphere (Aztec, Inca, Maya, Toltec) used any local stone. In many instances this was limestone. The Mayan structures of the Yucatan Peninsula, for example, are of soft, very permeable Tertiary and Quaternary limestones that have stood up surprisingly well to 800–1200 year of burial beneath secondary jungle (Figure 12.14).

However, many fine historic buildings worldwide, some of them now cultural World Heritage sites, are suffering the ravages of human impact through acid rain. An example from Beijing is shown in Figure 12.14. The processes of weathering of building stone, and repair and conservation measures are now the subject of much study. See Trudgill and Inkpen (1993) for a discussion of the processes and effects, and Viles (2000) for techniques used in its detailed investigation.

At a more humble level, limestones are widely used to build dry stone walls, as noted in section 12.2. As field or garden walls these are features of rural landscapes in places as far apart as Galway Bay and Okinawa.

Figure 12.13   The Parthenon at Athens, the most celebrated building in Western architecture, is built of massive limestone with marble facings and sculpture, and stands on a denuded epikarst surface, now rounded down by the tread of countless visitors.

In bulk terms, the overwhelming modern use of limestone and dolomite in most of the developed world is as aggregate, coarsely ground to rubble or gravel sizes to serve as the foundations for road and rail beds, buildings, parking lots, etc. A quarry that is cutting good stone for facings or very pure limestone for cement may condemn less desirable intervening beds to aggregate. As an example, about 100×10^6 t of limestone and dolomite

Figure 12.14 (Upper) Façade of a Mayan temple at Chichen Itza, Mexico. It is built of soft Tertiary limestone facings with trash stone filling behind, a common method worldwide. Note that weathering discolouration is distinctly less beneath projecting lintels that offer some protection from rain. The building was buried in second growth forest for ∼800 year. (Lower) Detail of sulphation breaking up ornamental carving of the Ming dynasty on an outside staircase in the Winter Palace, Beijing, China.

are quarried each year in the UK. Roughly 40% of this is used for roads, 40% for other aggregate needs and 10% for cement. Agriculture (for lime), iron and steel production (for flux), paints and plastics industries (for fillers) are the other chief users. Only the tiny proportion of 0.2% is now used for building stones or facings (Gunn 2004b).

High purity, low-Mg, limestone is the principal ingredient of Portland cement. It is pulverized and fired at 1400–1650°C with proportions of iron compounds and silica for additional strength. Gypsum is then added for quick setting. Mixed with fine-grained aggregate such as sand it forms concrete, which is now the world's principal building material. In western countries, modern concrete consumption averages between 0.1 and 0.5 t per person per year. Concrete is a karstifiable rock, as growth of stalactites from cracks in bridges everywhere attests; see Reardon (1992) for a comprehensive review of cement–water interaction.

Historically, in the European and Oriental cultures limestone was burned in kilns to produce 'quick lime' (CaO – section 2.3) as a calcium fertilizer. In long-settled rural areas the landscapes are dotted with small extraction pits as a consequence; these are sometimes mapped as dolines! With the availability of powerful grinding machinery, pulverized limestone is now replacing burned lime.

12.5.2 Gypsum and anhydrite

Gypsum is used as a whole rock for statuary and interior decorative facings. It is too soft and soluble to be suitable for exterior use in most climates. It is most prized when it is lustrous and macrocrystalline, pure white ('alabaster') or pink ('selenite'). As such, it was used in the baths of Minoan palaces, and can be found as small sculptures in tourist shops everywhere in the world today.

The principal use of gypsum is as Plaster of Paris. With a proportion of silt or sand added, it is used for interior wall and ceiling plaster work, mouldings, etc. In North America it is most often made up as plaster board, to be cut to size during construction. As noted above, pulverized gypsum is also an important ingredient of cement.

12.5.3 Oil and natural gas

Approximately 50% of the current production and known reserves of pool oil and natural gas are contained in carbonate rocks. Roehl and Choquette (1985) and Moore (2001) have provided detailed reviews.

Carbonate reservoirs display a wider range of conditions than do reservoirs in the clastic rocks. The most productive oil wells on record ($10^5 m^3$ daily from 36 wells at Agha Jari in Iran) are from limestone; at the other extreme are tight, normally siliceous, limestones that will not yield at all unless artificially fractured and treated with acid.

There are three principal types of oil or gas traps. Structural geological traps may be in limestone, dolomite, chalk or even marl. They are especially productive if naturally fractured. Carbonate caprocks above salt diapirs or anticlines belong to this category. Lithological traps are intraformational, being individual beds or sequences of beds with good storage porosity that is attributed to regional eogenetic or mesogenetic processes of dissolution and recrystallization.

Geological unconformities are the third category of traps. They may merge laterally or vertically with either of the other types. It is recognized that 40% or more of recoverable hydrocarbons in carbonates are trapped at unconformities which, in most cases, will be of karst origin (i.e. they are now palaeokarst). The most simple examples, which may involve the least amount of karstification, are buried coral reefs. They are always targeted during exploration for oil. Many are dolomitized, with large amounts of vuggy porosity plus occasional dedolomite patches containing large open voids. Maximovich and Bykov (1976, p. 47) recorded 36 buried pinnacle reefs along a 100 km transect at the western edge of the Ural Mountains, of which 19 were productive. Craig (1987) gave an excellent account of a one billion barrel oilfield in west Texas that was a coral island karst subsequently buried and dolomitized. Breccias are a second type of unconformable trap. Some are created by the collapse, and (perhaps) wave erosion, of epikarst and shallow caves during marine transgressions onto past karst terrains but they are more commonly due to deep inter- or intraformational dissolution that generated stoping upwards. Edgell (1991), for example, argues that deep dissolution of Proterozoic basinal salt deposits around the Persian Gulf created major traps in the overlying, younger rocks, chiefly by brecciation.

The third and most complex type of unconformity is the buried but preserved karst terrain. This may be one single epikarst zone created during a brief marine emergence that attained high karstic porosity under humid climatic conditions (section 2.10). In contrast, subaerial exposure under tropical semi-arid conditions may lead to the surficial porosity being much reduced by calcrete infilling, which creates an impermeable seal that traps oil beneath it (Moore 2001). Many productive oilfields consist of a number of small karst weathering unconformities of either type stacked up in longer depositional sequences (Fritz *et al.* 1993). The greatest unconformities, however,

are more rugged karst terrains buried in their entirety, with or without some trimming by marine erosion during the submergence. The Liuhua Field in the South China Sea appears to be a polygonal karst (Yuan *et al.* 1991) but with chaotic (breccia?) facies as well. The El Paso Field, Texas, features a 30 M year depositional unconformity that includes many collapsed caves (Wright *et al.* 1991). The Rospo Mare Field on the Adriatic coast of Italy (Figure 2.18) was a one billion barrel strike in a palaeokarst with well-developed epikarst and closed depressions (Soudet *et al.* 1994). The Renqiu Field in the Bohai geological basin south of Beijing is interpreted as a buried tower karst; local relief in it is as great as 800 m, however, so there will also be fault displacement.

The world's first **drilled** oil well (as opposed to a dug well) was in southern Ontario in AD 1857. It explored Silurian dolomites that had subsided and fractured where salt was dissolved beneath them, i.e. a breccia trap. This was sealed tight by a cover of glacial clay. It can be described as an ancient palaeokarst feature enhanced by Quaternary glacial action. It yielded a prolific flow of oil when drilled to a depth of a mere 60 m.

12.5.4 Carbonate-hosted ores and other economic deposits

Almost every metallic ore and most kinds of non-metallic economic deposits have been extracted from karst in the past. Historically, karstic traps were perhaps the principal sources for most of them until the dawn of the Industrial Revolution. Nicod (1996) has reviewed the history in Europe and the Middle East. Lead was worked cold from palaeokarst exposures in Turkey as early as ~8000 year BP. The first copper for bronze came from small karstic traps around the Mediterranean and the first iron from the Caucasus ~3500 year BP. Natural caves were explored for hydrothermal precipitates on the walls; in England the Romans mined Treak Cliff Caverns for fluorite and Speedwell Caverns for lead.

The deposits are found in the three different settings illustrated in Figure 12.15. Surficial **placer** deposits (or **supergene**) are of clastic detritus, precipitates or both that have accumulated in karst depressions such as dolines, uvalas and poljes on the modern surface. A subtype is the subrosion trough on evaporites that may enlarge into a substantial sedimentary basin. The deposits may be allogenic and transported to these traps by fluvial, sheetwash, colluvial, aeolian or marine processes, or weathering residues of local origin. There may be greater or lesser diagenetic alteration once the deposit is in the trap; e.g. finely divided iron tends to aggregate into

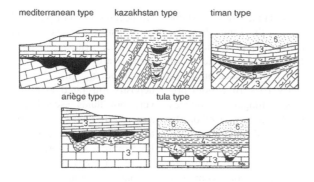

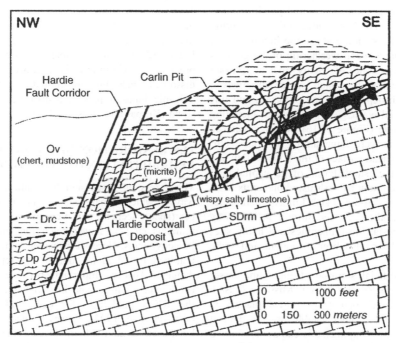

Figure 12.15 The contrasted settings of some economic ore deposits in karst. (Upper) Types of placer and buried placer bauxites (Bárdossy 1989). (Lower) The Carlin and Hardie Footwall gold deposits, Nevada. Hot acidic fluids from nearby volcanic intrusions ascended the Hardie Fault Corridor and discharged along the contact between more permeable silty limestones and overlying dense micrites. Teal, L. and Jackson, M. (1997) Geologic overview of the Carlin Trend gold deposits and descriptions of recent discoveries. Society of Economic Geologists Newsletter, 31, 13–25.

pisolites or nodules. The second setting is the former surficial placer (all origins) that has subsided and been buried by later strata, as is the case with most bauxites in Hungary for example. The third setting is where **hypogene** precipitates are emplaced deep underground in the karst rocks as consequences of invasion by hydrothermal waters, intrusion of magmatic fluids, or discharge of H_2S gas and mixing processes close to the water table as explained in section 7.8.

Bauxite is the ore of alumina, concentrated as a weathering residuum. The deposits are red, porous and earthy, interlayered with other continental sediments. Average composition is 35–50% Al_2O_3 plus lesser Fe_2O_3, trace minerals and earth. Individual deposits are rarely more than 20–30 m in thickness. Deposition may have taken place above the water table or below it, the former being more favourable. The oldest known bauxites are from basal Cambrian strata and the youngest are Quaternary. There is an obvious geographical association with warm climates (D'Argenio and Mindszenty 1992). Approximately 10% of global production is obtained from karst sites, the balance from laterites.

The mineral is named after Les Baux, France, where mining began in karst depressions in the 1860s. It is the prime modern example of economic karst placer deposits, both in surficial and buried settings. World production of karst bauxite is now $\sim 30 \times 10^6 \, t \, a^{-1}$; Jamaica produces $10–13 \times 10^6 \, t$ of this and China $9–10 \times 10^6 \, t$, both primarily from surface placers. Hungary produces $\sim 2 \times 10^6 \, t$ from buried placers. It requires 6 t of ore to produce 1 t of aluminium; the chief residue is a hazardous alkaline red mud.

Most of the world's major economic tin deposits are found in very large, deep closed depressions formed at the contact between carbonate rocks and granitic stocks and bosses that intruded them. Differential weathering leaves the bosses, etc. standing above the limestones. In tropical humid climates, deep weathering of the granites and adjoining skarn zone of highly metamorphosed carbonates generates acidic runoff which dissolves a rugged epikarst with depressions and pinnacles and fills hollows with colluvial wastes and precipitates, including the tin and wolframite (Figure 12.16). Major cassiterite deposits in Malaysia were found amongst 'stone teeth' on the corroded rockhead surface of the Kinta valley and were significant in the establishment of the capital city, Kuala Lumpur, which now has exhausted open pits in its suburbs (Fig. 12.16). About 14 000 km² of peninsular Malaysia has been alienated for mining in this way (Yeap 1987). Indonesia, the Philippines and many sites in southern China have also been important and some remain so.

Placer coal in surficial subrosion troughs and lesser basins is important in Russia, Poland, Missouri (USA), and in association with bauxite as buried placers in Hungary and China. The other notable organic deposit is phosphate from bird droppings; this can accumulate to substantial depths in epikarst surfaces on tropical islands such as Nauru (Bourrouilh-Le Jan 1989). Major placer iron ores (siderite, limonite, haematite and goethite) are mined in Russia and Vietnam. Moulding sands, pottery clays and kaolin for porcelain are also being mined from infilled karst depressions today (Bosák *et al.* 1989). More exotic is the extraction of placer diamonds from dolines or deep epikarst in Siberia and South Africa (Filippov 2004a), and rubies and sapphires from dolines in Myanmar and Sri Lanka.

The world's principal sources of lead and zinc are in hypogene karst deposits in carbonate rocks (Sangster 1988, Dżulynski and Sass-Gutkiewicz 1989). Mississippi Valley Type (MVT) deposits are associated with interstratal dissolution of evaporite beds that create breccia zones or chambers in the carbonates (Figure 7.29), or mazes of open galleries in faults, joints or bedding planes.

All become filled or partly filled with precipitated pyrite, galena, sphalerite and secondary (**gangue**) minerals such as fluorite, barite and coarse crystalline dolomite. Hydrocarbons are often associated with them, indicated by a strong odour. The Nanisivik (Baffin Island, Canada) deposits illustrated in Figure 10.21 are an example of very hot fluids ascending boundary faults in a horst structure in dolomite and migrating out into the body of the rock where there were palaeokarst target voids of Precambrian age (Ford 1995).

Breccia pipes may host many different precipitates. Amongst the most celebrated are the uranium ores of Tyuya Muyun, Khirgizstan, and the Hualapai Reserve on the north side of Grand Canyon, Arizona. 'Carlin' gold (named for the Carlin Trend, Nevada, and shown in Figure 12.15) is an example of more widely dispersed precipitation within fractures and bedding planes that are dissolved open only to limited extents by ascending volcanic fluids trapped and dispersed beneath aquitard strata (Korpas and Hofstra 1999).

12.6 RESTORATION OF KARSTLANDS AND REHABILITATION OF LIMESTONE QUARRIES

The foregoing discussion has indicated the numerous ways in which human occupation and use of karst resources (space, scenery, soil, vegetation, water, rock, caves, minerals, etc.) can have a deleterious impact on karst and its ecosystems. Here we consider some of the measures that can be taken to contain and repair unacceptable damage. This is important, because karst has many economic, scientific and human values as discussed in the IUCN guidelines for cave and karst protection (Watson *et al.* 1997). However, where karst areas have been used for atomic bomb tests, as the British did along the northeast margin of the Nullarbor Plain (Gillieson 1993), or the Americans on Bikini and Eniwetok atolls and the French on Mururoa and Fangataufa atolls, the physical damage and dispersal of plutonium isotopes present effectively insuperable restoration problems.

12.6.1 Restoration of karst watersheds

Basic principles for karst rehabilitation are the same as for catchment restoration in general, but with the added consideration that the measures undertaken must also assist subterranean restoration. It is important, first, to identify the objective of the rehabilitation, because full ecosystem restoration requires much more comprehensive measures than, say, water-quality restoration. Partial rehabilitation and water-quality management is all that is usually needed in an agricultural landscape, but we must

Figure 12.16   (Upper) Placer tin mining in a depression at the contact of granite and marble. New Lahat Mine near Ipoh, Kinta valley, Malaysia. Note bedrock bosses revealed by the excavation. (Lower) Abandoned placer mining quarry near Kuala Lumpur, Malaysia, being reclaimed for urban development and recreation.

recognize that this may be impossible to achieve in areas of extreme population pressure, where rocky desertification may already have occurred. Some rehabilitation gains may also be at the cost of some further environmental damage.

The Burren in western Ireland is an example of a karst that was semi-desertified by deforestation and severe soil erosion following settlement that commenced 4500–6500 year ago (Drew 1983). But despite the millennia of impacts in a small area (367 km^2), the Burren still contains more than half Ireland's species of native vascular plants and a rich archaeological heritage. Drew and Magee (1994) explained that between 1981 and 1991 about 4% of the area was reclaimed, i.e. land was converted from scrub or rocky pasture into uniform, seeded, manageable fields, most of the land being used to provide silage, but that the resulting gains in agricultural productivity were at some expense to the wider natural and cultural environment (Figure 12.17). For example, there has been a loss of environmental variety as scrub and semi-natural grassland, limestone pavement and ancient field boundaries have been replaced by uniform grassland fields. The great increase in fertilizer and silage use is also likely to pose a threat to groundwater quality in the area. The message, therefore, is that we must consider the wider costs of reclamation/rehabilitation as well as the potential benefits before undertaking reclamation/rehabilitation, and be prepared to make adjustments to the style and location of proposed changes in order to reduce environmental costs. We should also monitor the effects, to see if the desired outcome is being achieved. When hydrological restoration is involved, attention must be given to both quantity and quality characteristics of the waters, and monitoring must be undertaken at karst springs.

12.6.2 Water quantity and drainage networks

As we saw in section 12.2, deforestation increases the total volume of runoff because natural evapotranspiration losses are reduced. This can amount to as much as another 700 mm runoff in a temperate region with 2000 mm annual rainfall. Human activities usually also reduce infiltration because of compaction of soils with heavy machinery and sealing of surfaces by roads and buildings. The extra water surplus and more rapid runoff results in hydrograph peaks that arrive earlier and rise much higher than before; thus increasing flooding both on the surface and underground, as discussed in section 12.4. Consequently, management of water quantity requires major replanting, in extreme cases may require construction of floodwater detention dams, and surfaces may have to be deeply ploughed to restore permeability. In some cases choked dolines and stream-sinks may require unblocking. Appropriate management responses depend on the final objectives and extent of the environmental damage.

12.6.3 Water quality

Removal of natural vegetation cover, agricultural activities and human habitation always reduce water quality. In any environment this presents a significant problem,

but in karst it becomes critical because most drainage disappears underground; thus the insidious effects of chemical and particulate pollution can go unnoticed in the place where the problem originates, but can emerge at distant sites downstream. Spatially concentrated point-source pollution usually stems from settlements, e.g. septic tank effluent, industrial and sewage discharges, pig- and dairy-shed waste waters, and rubbish tip leachate. In a karst context garbage dumps in dolines are another frequent culprit, as noted. Spatially diffuse or non-point-source pollution arises from widespread soil erosion, fertilizer and pesticide runoff and waste from grazing livestock.

Creeping water-quality deterioration often occurs progressively over several generations as population increases, land use intensifies and new or more agricultural chemicals are used; thus the original reference for natural water quality is lost and forgotten and communities tolerate poor water because they have known no better. Many agricultural communities are oblivious to the damage their practices inflict on water quality, and so the first step in making improvements is to raise their awareness and harness their cooperation.

Some of the measures that can be used to protect water quality were discussed in section 11.5. Improvement of water quality once deterioration has set in is more difficult, but the measures needed are essentially the same as those practiced in surface catchments. The first requirement is to prevent contaminants getting into waterways, for instance by ensuring that waste-water discharges are directed into community water treatment plants. However, even when satisfactorily treated for bacteria, etc., the treated water may still require discharge into wetlands if nutrient stripping is required to prevent eutrophication of receiving water bodies. A different approach is required for diffuse-source pollutants, the most effective preventive measure being to ensure natural filtration of runoff by thick vegetation along water courses and in dolines. Many jurisdictions now require 10–50 m buffer zones of undisturbed ground around doline perimeters. This will reduce the suspended load reaching streams and will biologically strip some nutrients, but will have little effect on pathogenic organisms. The more comprehensive the implementation of surface revegetation, the greater the improvement of water quality. Land abandoned following rural depopulation can show remarkable spontaneous secondary regrowth of woodland species, as seen in the Slovenian karst for example; so passive measures can sometimes be effective in both ecological and water-quality rehabilitation.

Proactive water-quality management requires objective assessment of changes accomplished. The best place

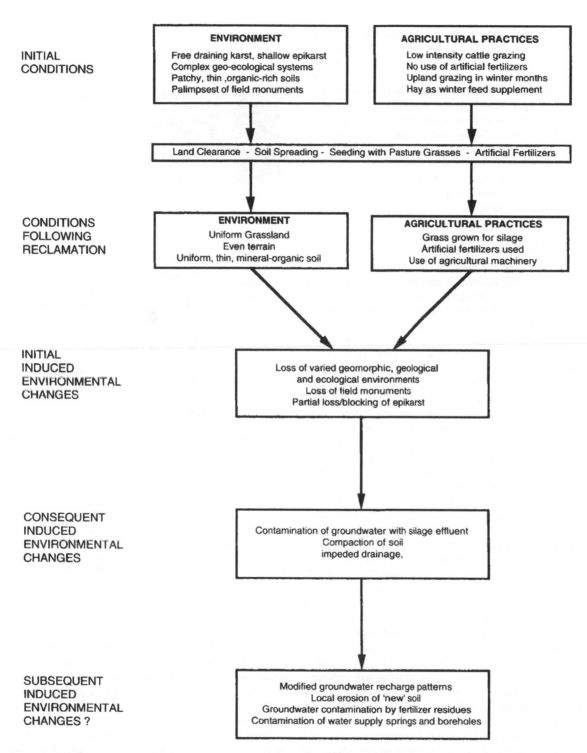

Figure 12.17   The sequence of direct, indirect and possible changes to the Burren karst consequent upon intensive land reclamation. Reproduced with permission from Drew, D. and Magee, E., Environmental implications of land reclamation in the Burren, Co. Clare: a preliminary analysis. Irish Geography 27(2), 81–96 © 1994 Geographical Society of Ireland.

to monitor the effectiveness of karstland rehabilitation practices is in streams and especially at karst springs. The condition of natural water as assessed by both biological and chemical indicators is the best measure of the ecological health of a catchment. However, Rice and Hartowicz (2003) point out that springs have a naturally low biotic diversity and that research on the use of these biota for assessing groundwater quality at springs is not as advanced as it is for using biota to assess water quality of surface streams. Nevertheless, they conclude that using biota is a more reliable means of assessing emerging groundwater quality than chemical analysis of periodic grab samples.

12.6.4 Quarrying

Open-pit quarrying presents most challenging restoration problems, because it represents an extreme case of human impact on karst – large volumes of it are removed entirely. No one doubts the importance of concrete and cement to modern society and therefore the importance of having access to the resources afforded by limestone quarries. Nevertheless, they are often associated with total destruction of ecosystems, smothering of nearby places by dumping of overburden and distribution of tailings and dust, and severe groundwater pollution by silt and oil; so their impact can spread well beyond the immediate confines of the quarry site. For these reasons a guiding principle when siting new quarries is that they should be located only on the outflow margins of karsts, because that will at least protect upstream areas from significant water pollution and water-transmitted ecosystem damage, although if quarries are deep so that water tables are lowered, the cones of depression will extend upstream as well as downstream. When limestone extraction ceases large holes are left, often with deep lakes with steep sides that present significant safety hazards.

Although natural processes may partially and slowly rehabilitate quarries by softening and revegetating slopes, appearances can be enhanced and processes accelerated by human intervention aimed at replicating natural landforms and improving the quality of runoff and infiltrating waters. The key elements found in the natural landforms of the region are first identified and then replicated as far as possible using restoration blasting and replanting with native species. Natural looking slope profiles with spurs and swales, mantled with appropriate grain-sized rock fragments, for example, can be relatively easily created (Figure 12.18). The aim is to mimic natural slopes and to provide a substrate that will permit successful replanting by native species. The resulting well-vegetated slopes yield clean runoff and have habitat value. Major

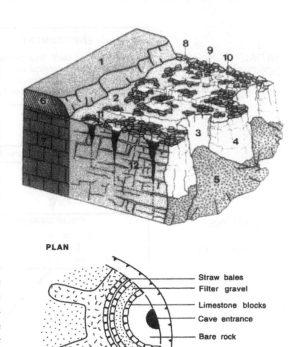

Figure 12.18   (Upper) The typical topography of an abandoned limestone or dolomite quarry, emphasizing development of artificial epikarst and dolines. Reproduced from Gunn, J. (1993) The geomorphological impacts of limestone quarrying, in Karst Terrains: Environmental Changes and Human Impact (ed. P.W. Williams). Catena Supplement, 25, 187–97. Designed blasting to create gullies and talus slopes will improve the restoration. (Lower) Layout and construction details of treatment of stream-sinks and infiltration areas, Lune River Quarry rehabilitation, Tasmania. Reproduced from Gillieson, D. (1996) Caves: Processes, Development and Management, Blackwell, Oxford, 324 pp.

contributions in this field have been made by Gunn and his co-workers (Gunn 1993, Gunn and Bailey 1993, Gunn *et al.* 1997, Hobbs and Gunn 1998) and have been reviewed by Hess and Slattery (1999) and Bradshaw (2002). Filter systems using natural materials also can be designed around sites of infiltration (Gillieson and Household 1999) in order to improve the quality of point recharge. This is a karst-specific form of riparian zone management (Figure 12.18).

There are other adaptations. Where a lake is created by quarrying or mining it will often be redeveloped for recreational purposes. A spectacular example is the adaptation of some old tin mines (deep open pits) around

Kuala Lumpur into ornamental lakes with theme parks and adjoining residential complexes; the tallest surviving karst pinnacle overlooking one lake is now topped with a fairy castle (Figure 12.16). Unfortunately, amongst the most frequent adaptations will be the use of the abandoned holes for 'sanitary' landfills. In many nations today there is more profit to be made in putting 1 t of garbage back into the hole than there was in taking the rock out in the first place! Landfills threaten permanent pollution of aquifers down gradient. The long-term stability of engineered seals (clay liners, etc.) and/or geofabrics to contain them is nowhere proven. If the pit extends below the water table it warrants pumping in perpetuity – which will not occur.

12.7 SUSTAINABLE MANAGEMENT OF KARST

12.7.1 Agriculture

Sustainable management of rural systems depends more on the management of human behaviour than on the management of the physical environment. Urich (1989, 1993) has reminded us of this in the context of wet rice cultivation in the Philippines. The social and cultural values and survival imperatives of the people occupying the land will more than anything else determine whether a given karst area can be managed sustainably. The more subsistence the economy, the more obvious this becomes. The more severe the population pressure, the more survival rather than sustainability becomes the focus of daily life. People in need apply conservation measures if they perceive a short-term material advantage, otherwise they cannot risk changing well-tested traditional practices and threatening their food supply and very existence. Most regions that have succumbed to or are currently succumbing to rocky desertification were or are regions of peasant economy where population pressure has driven devegetation, cultivation of steep slopes, major soil erosion and serious water-quality deterioration. Only alternative forms of livelihood or external aid will permit the pressure on the land to be reduced long enough to enable steep hills to be revegetated and to give karst rehabilitation a chance. Rural to urban population drift around the Mediterranean has seen spontaneous recovery of the woodland vegetation in many karsts. Having achieved a viable ecosystem, even if somewhat species-impoverished compared with its natural state, then sustainable management becomes a possibility. But whether it can succeed depends again upon the willingness and ability of the local people to try new techniques and make it work.

The difficulty of achieving a sustainable agricultural system in some karst regions can be illustrated by the case of south China, where karst covers 500 000 km^2 and is occupied by > 100 million people of whom about one-third are considered very poor. In this karst, the population is increasing at an annual rate of 1.3 to 2% and maximum population density reaches 280 people km^{-2}. Considerable experience of the area led Song (1999) to conclude that the following measures must be adopted if sustainable agriculture is to be achieved:

1. serious control of population increase;
2. environmental education of the inhabitants;
3. terracing of slopes < 25° to improve soil conservation and provide more cultivatable land;
4. improvement of energy supplies by conservation and development of new sources (solar, biological, etc.);
5. reforestation of karst, especially slopes > 25°, with species of ecological and economic value;
6. improved utilisation of water resources including rainwater.

Even the last measure is difficult, because in Guangxi Province alone over 60% of 1252 reservoirs that were built are dry because of karst leakage. Energy is also a major problem because the traditional rural source is wood, but domestic fuel requirements exceed natural regrowth by about 80%. Unfortunately, similar problems to those faced in south China are rife in Third World agricultural economies based on karst, with many serious and looming problems throughout southeast Asia and the Caribbean in particular.

12.7.2 Forests

Nevertheless, progress is being made in raising the awareness of authorities to the particular problems associated with management of karst in many situations. Forest management is now quite well understood and appreciated in reforestation areas of Europe, North America and Australasia, as noted in section 12.2. Gillieson (1996) reviewed some of the measures being taken. Following two decades of research and campaigning by P. and K. Griffiths and associates, the Province of British Columbia in Canada released *Karst Inventory Standards and Vulnerability Assessment Procedures* (2000) and a *Karst Management Handbook for British Columbia* (2003) (www.for.gov.bc.ca/hfp/fordev/karst/karst-final-Aug1-web.pdf).

The recommended best management practices in the British Columbia handbook are designed to promote sustainable forest practices on karst, while minimizing impacts to timber supply and operational costs, with the aim of achieving the following objectives:

1. maintain the capability of karst landscapes to regenerate healthy and productive forests after harvesting;
2. maintain the high level of biodiversity associated with karst ecosystems, including surface and subsurface habitats;
3. maintain the natural flows and water quality of karst hydrological systems;
4. maintain the natural rates of air exchange between the surface and subsurface;
5. manage and protect significant surface karst features and subsurface karst resources;
6. provide recreational opportunities where appropriate.

There are, of course, many forests on karst that should never be logged because of their outstanding universal conservation value, but where sustainable forestry is deemed appropriate, the above objectives are endorsed by us as eminently achievable and appropriate. The handbook also provides guidance on how the significance of karst features can be assessed and how best management practices might be applied. Effective implementation, however, always depends on the human factor.

12.7.3 The protection, management and restoration of caves

Caves are fragile landforms. No matter how careful they may be, cave explorers trample silt and clay floors into broken mud, dirty the walls and speleothems, perhaps breaking the latter by accident while passing constrictions. Historical use of torches and lanterns (burning animal fat or coal oil) blackened the ceilings. Speleothems were commonly taken as souvenirs, while there are plenty of instances of their being deliberately smashed in acts of vandalism. Many nations now attempt to protect caves by various means. **Wild caves** (the large majority that are not developed for show to tourists, etc.) may be gated or entry to them otherwise regulated; and there is often rationing of numbers of caving visitors. Organizations or individuals sometimes buy caves outright. Explorers are becoming increasingly careful. Oil lanterns and carbide lamps have been replaced with electric bulb and LED lights. It is becoming mandatory to take out all waste, including excrement. Restoration has been attempted for many decades; the authors wielded scrubbing brushes in caves of the Mendip Hills, England, in the 1950s and 1960s, for example, and produced scientifically based guidelines for cave management in the 1970s. Broken speleothems can be repaired with careful use of epoxy glues. The IUCN has issued sets of guidelines for protection and restoration (Watson *et al.* 1997).

Understandably, management interest has focused upon **show caves** for tourists. There are now approximately 650 of them worldwide, with an estimated gross annual income of ~US$2.5 billion. The number of visitors ranges from a few thousand per year in remote places to 0.4–1.0 million per year in high-profile resort areas such as Mammoth Cave National Park, USA, or Guilin, China. The quality of the development for visitors (lay out and construction of paths, type and placement of lights, etc.) ranges from the very good (e.g. Carlsbad Caverns) to the frankly awful. Natural air circulation in many has been seriously disturbed by opening artificial entrances, which may cause desiccation. Some caves have even been reamed out with tunnel boring equipment. The need for better design of installations and for management of the numbers of visitors became apparent when it proved necessary to close the famous painted caves of Lascaux, France, in 1963 because the lighting and the CO_2 from visitors was seriously damaging the paintings by growth of moulds. An International Show Caves Association was formed in 1989 and there are now effective national organizations in many countries; e.g. the Australasian Cave and Karst Management Association (http://ackma.org) publishes a quarterly journal, while in the USA the National Caves Association (www.cavern.com) hosts biennial meetings with Proceedings that report the latest research and development findings.

The important questions, for both wild caves and show caves, are: How fragile are they? What is their carrying capacity (i.e. how many visitors per hour, per day, etc)? Heaton (1986) suggested that caves fall into three simple groups:

1. high energy, where the natural energy flows will far exceed those of visitors, such as large river caves that flood frequently and are thus flushed clean;
2. intermediate energy, where smaller streams, seepage, airflow, etc. will supply roughly the same magnitude as the energy output from quite large numbers of visitors;
3. low energy, where the natural energy flow is extremely low – these are cul-de-sac caves with no streams, near-constant temperatures and weak air circulation, such as the ice crystal chamber of Figure 8.18 where the crystals are melted by body heat if one or two visitors stay for more than a few minutes.

Cigna has made important contributions by addressing these questions from a scientific management perspective (see e.g. Cigna 1993, Cigna and Burri 2000, Kranjc 2002). Ideally, before developing a show cave, its natural dynamic parameters will be established by physical measurement and its ecosystem and important species will be established by biological survey. Fossil, archaeological and indigenous cultural values may also require assessment. Physical dynamic parameters are, principally, air temperature, relative humidity, atmospheric CO_2, water flow and water quality. The natural range of each parameter over a hydrological year should be determined. The goal then will be to develop the cave so that the upper limit of the variable with the least range (e.g. air temperature) will not be exceeded during visits. This may be critical for the maintenance of cave organisms. This procedure was more or less followed at Grotta Grande del Vento (or Grotta Frasassi) in Italy under Cigna's supervision; it is a mixed meteoric-H_2S cave by genesis, possessing large chambers with superlative calcite speleothem formations that require careful protection. Similar procedures have been followed in New Zealand where biota (especially cave glow worms) are fragile and of special significance.

Every visitor to a cave radiates heat and moves inside his or her cloud of exhaust gases, vapor and shed particulates such as dust and lint (Michie 1997). The heat energy introduced by one visitor is $\sim$0.1–0.2 kWh (100–200 joules s^{-1}) when standing or walking. The annual visitor heat input to a cave, E (joules s^{-1}), thus can be approximated by

$$E = 170 \times t \times 3600 \times N \tag{12.4}$$

where t is the average duration of a visit and N is the total number of visitors (Villar *et al.* 1984). At Grotta Grande del Vento, with $\sim$500 000 persons staying $\sim$1.5 h each, it amounts to 4.6×10^{11} joules s^{-1} (128 MWh). This is considerable, indicating that the effects in an intermediate energy cave may be large. There can also be longer term warming, as Figure 12.19 from Grotta di Castellana, Italy, clearly shows. This may reduce the relative humidity, drying and dulling the appearance of many speleothems, disturbing the biota, etc.

Since the closure of Lascaux much attention has been paid to visitor-generated CO_2 concentrations in cave atmospheres. These are estimated by:

$$C(T) = (1.7 \times 10^4 \times N \times t)V \tag{12.5}$$

$C(T)$ is the change of CO_2 concentration (ppm, vol.) at time T, and V is the volume of the cave or chamber being considered (Villar *et al.* 1986). Measurement during visiting hours in many show caves has established that there are usually increases of CO_2 significantly above the standard atmospheric value of $\sim$380 ppm. Figure 12.19

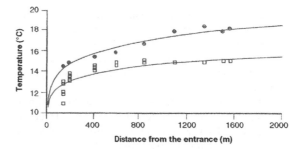

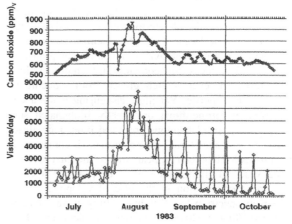

Figure 12.19 (Upper) Mean air temperature profile measured along the tourist route in Grotta Castellana, in 1958–1960 (open squares) compared with 1982 (dots). (Lower) Daily numbers of visitors compared with daily mean CO_2 concentrations in Ancona Hall, Grotta Grande del Vento. Reproduced from Cigna, A. A., Environmental management of tourist caves: the examples of Grotta di Castellana and Grotta Grande del Vento. Environmental Geology 21, 173–180. © 1993 Springer-Verlag.

shows the Grotta Grande del Vento example for part of 1983; CO_2 is generally about double the standard, the impact of weekend visitors is quite clear, as is the effect of opening an airlock to permit gravitational drainage of the dense gas one day in August that year. Concentrations $\leq$ 1000–2000 ppm are commonly found in some other show caves and as much as 5000 ppm in heavily visited caves in China. There may be condensation corrosion derived from this CO_2 excess as a cave cools after the visitors have left. Condensation processes in a tourist cave in New Zealand have been measured by de Freitas and Schmekal (2003).

Conversely, there is concern that radon (^{222}Rn) released from the decay of ^{234}U and ^{238}U in the enclosing rocks and mineral deposits such as galena may harm cave guides and others who must spend many hours each year in caves, especially where there is low energy with little air flow to clear accumulations. Radon-222 is a noble gas (i.e. inert) but it decays to particulate ^{218}Po, ^{214}Po and ^{214}Bi, which may adhere to aerosols and so build up in the lungs. This has been a major cause of lung cancer in uranium miners and workers in nuclear facilities. Most nations now define Working Limits for annual uptake of ^{222}Rn and its derivatives by individuals, and show caves are tested for concentrations from time to time. In most instances the latter will be well below danger levels but in caves with lead, zinc or uranium ores in veins there may be danger. Gunn (2004c) gives details.

The most obvious impacts of development in many show caves are the deleterious effects of the lights. Standard electric lamps with a broad emission spectrum (orange or 'warm' light) will warm surrounding air and nearby rock surfaces. Speleothems can then dry and lose their lustre, or even switch growth from stalactite to helictite or evaporite. Yet more disfiguring is the accumulation of **lampenflora** on brightly illuminated surfaces. These are filamentous green and blue-green algae, plus mosses and ferns where growth becomes well established. They can appear within the first two or three years of illumination, even deep inside a cave, and are very common in the older developments. Use of narrow spectrum 'cool' lights emitting at wavelengths between 500 and 620 nm reduces the problem. The algae can be destroyed by steam or scrubbing with water. Solutions of bleach or calcium hypochlorite ($CaClO^-$) are also effective, but it is essential to use dilute solutions ($< 2\%$) and keep them away from any cave fauna.

12.8 SCIENTIFIC, CULTURAL AND RECREATIONAL VALUES OF KARSTLANDS

By focusing on only two of its facets, this book provides no more than a glimpse of the wider environmental significance of karst. Karst rocks are important in almost every aspect of our lives. The calcium carbonate cycle is fundamental to the CO_2 concentration of our atmosphere; carbonate rocks host water supplies and building materials of immense value to many millions of people; limestones hold the key to the sea-level history of the Earth for at least the past 1 Ma; speleothems contain the palaeoenvironmental record of the continents over a similar period; cave systems and associated aquifers provide the habitats of troglobitic ecosystems that are barely investigated; and karst supports a morphology that is a microcosm of geomorphology as a whole. Carbonate karst rocks that cover about 10–15% of the ice-free continental land thus have a value very much greater than their limited area might suggest.

The ravages of 20th century industrialization consumed and polluted much of the world's karst, including the 'classic' terrain between Italy and Slovenia, although there have been notable recent attempts at its restoration. The trend of despoliation seems likely to continue in the present century, but accompanied by stronger national and international attempts at conservation. Goldie (1993) provides an example from the UK, where national legislation is helping to protect limestone pavements. However, it seems inevitable that limestone quarries will expand and karst groundwater resources will be still more heavily exploited in the years ahead with dire consequences for water supplies, spring flows and subterranean ecosystems.

Caves are immensely important as nature's vaults. Whether in the 'Old World' or the 'New', they contain irreplaceable and datable records of biological, climatic and landscape history. Palaeontological discoveries in European caves since the eighteenth and nineteenth centuries (e.g. see *Reliquiae Diluvianae* by Buckland, 1823) revealed some of our first evidence for environmental change, because they found in the modern temperate zone remains of rhinoceros, hyaena, hippopotamus, etc; animals that are more associated with Africa. Ancient rock drawings of animals and hunting scenes (Figure 12.20) provided other clues to the ecology of past times, as well as being superb examples of the artistic ability of our forebears. Most famous amongst locations where these can be seen are Altamira Cave near Santander in northern Spain, and many caves in southern France such as Lascaux and Cussac in the valley of the Dordogne, Niaux Cave in the Pyrénées and Grotte Chauvet in the Ardèche Gorge. Some of their paintings may date back to 32 000 years BP. The entrance to Grotte Cosquer on the Mediterranean coast is now 37 m beneath the sea; its paintings are dated to two periods between 18 500 and 27 000 years BP. Measured on any scale, these are internationally important sites of cultural heritage, but sadly they face conservation problems of great complexity, being prone to inadvertent damage by the very tourists who go to admire them.

It was in these caves that the intricate and conflicting demands of tourist cave development and conservation was first appreciated. As we have noted, Lascaux Cave had to be closed in 1963 to protect the prehistoric art from lampenflora and crystal growth, because the combination of lighting and respiration of visitors caused deterioration of the famous cave paintings that they had come to enjoy. A meticulously constructed replica cave (Lascaux II) was built alongside to absorb the visitor demand. Subterranean ecological attractions have also been found to be vulnerable. Thus the Glowworm Cave in New Zealand was temporarily closed in 1973 because the heat energy input of tourism reduced the humidity of the air, stressing the glow worms and drastically

Figure 12.20   Painting of bison, Grotte de Niaux, Ariège, France.

reducing their numbers. Scientific investigation has since established principles for management and the situation is restored. The importance of conserving representative karst areas for science and recreation has been recognized in many countries by the designation of national parks and reserves. Of great value amongst those recently preserved is the Gunung Mulu National Park of Sarawak in northern Borneo. It contains magnificent humid tropical karst with its ecosystem intact – both below and above ground – a feature of immense significance at a time when tropical forests around the world are being destroyed. Rising precipitously above the forest are the spire-like pinnacle karren of Mount Api (Figure 9.12), one of the best known examples of this landform in the world. And piercing the mountain are huge allogenic river caves, including Good Luck Cave with its enormous void of Sarawak Chamber (Figure 7.51). But there is still much to do in conserving karsts of universal value, especially in the Third World. The internationally important type-site of 'cockpit karst' in Jamaica, for example, remains under threat because of unresolved issues of land tenure and inadequate local protective legislation (Chenoweth *et al.* 2001) and rapid tourist developments threaten the wonderful drowned caves of the Yucatan coast. However, progress is being made in systematically assessing the World Heritage potential of different karst regions, as in the Asia–Pacific region for example (Wong *et al.* 2001).

If any geomorphologist were asked to nominate a short list of the great landscapes of the world, the celebrated tower karst of southern China would certainly rank high upon it. This landscape has been an enduring source of inspiration, entrancement and curiosity for travellers, artists and scientists of many dynasties from before Xu Xiake's time in the 17th century to the present. We remain equally captivated by its beauty and mystery. It would be our first choice on a World Heritage list of karst. The great karsts of China also contain some of our most important sites for the study of human evolution. The discovery of 'Peking Man' or *Homo erectus pekinensis* in the late 1920s drew the attention of world science to excavations in cave deposits at Zhoukoudian, about 40 km southwest of Beijing. From New Cave at Zhoukoudian the regional transition from *H. erectus* to archaic *H. sapiens* appears to be around 400 ka and archaic *H. sapiens* is dated to >269 ka (Shen *et al.* 2004). Elsewhere in China, hominid remains found in caves are known back to about 1.9 M year (Zhu *et al.* 2003). Yet still older deposits of ape-like hominoids *Australopithecus* that preceded the genus *Homo* have been found in breccias in Sterkfontein Cave in South Africa dating back to 3.3 to 3.6 Ma, in addition to remains of *H. habilis* from about 2–1.7 Ma (Martini *et al.* 2003). *Australopithecus* and *H. erectus* have been excavated in similar circumstances in nearby Swartzkrans Cave. In northern Namibia, karst breccias have yielded a Miocene hominid dating from about 13 ± 1 Ma (Conroy *et al.* 1992). The oldest hominin fossil in Europe is *H. antecessor*, many remains of which were found in sediments of a doline and cave complex at Atapuerca in Spain and palaeomagnetically dated at >780 ka (Parés and Pérez-González 1995). See Berger (2001) for a discussion of hominid/hominin terminology.

Teeth and mandibles of the largest known primate, *Gigantopithecus blacki*, have also been found in caves in China and Vietnam (Schwartz *et al.* 1995). When standing erect, this Orangutang-like creature may have been about 4 m tall. The world's largest bird, *Dinornis maximus*, the now extinct giant Moa, was of up to 240 kg live mass and is represented in many cave deposits in New Zealand, sometimes alongside the great extinct eagle *Harpagornis*, with a wingspan at least as large as the Condor (Worthy and Holdaway 2002).

Fortunately the establishment of a series of karst World Heritage sites is underway. The establishment of an international network of Geoparks has also been proposed for important sites that do not meet the broader World Heritage requirements. In a recent review, Hamilton-Smith (2004) noted that worldwide there are about 50 World Heritage sites that are located upon or feature karst phenomena. Nine of these were inscribed specifically for their cave and karst features (Puerto-Princesa, Philippines; Gunung Mulu, Malaysia; Desembarco del Gramma-Cabo Cruz, Cuba; Carlsbad Caverns and Mammoth Cave, USA; Plitvice Lakes, Croatia; Aggtelek and Slovak Karst, Hungary–Slovakia; Skocjanske Jame, Slovenia; Ha Long Bay and Phong Nha Ke Bang, Vietnam). Twenty-six sites were inscribed for other natural reasons, but contain significant karst features (including the Canadian Rockies and Nahanni, Canada; Pyrenees-Mont Perdu, France–Spain; Tsingy de Bemaraha, Madagascar; Western Caucasus, Russian Federation; Pamukkale, Turkey); and nine sites were inscribed for cultural reasons, but contain important karst features (including Zhoukoudian, China; Altamira, Spain; Caves of the Vézère, France). Work is continuing through the efforts of the IUCN to ensure that there will be a representative coverage of the great karsts of the world within the World Heritage system. A new cluster of sites embracing the great karsts of south China has been proposed and we eagerly await its inscription.

In addition to conservation at a global level, there are many karst features worthy of protection at a national level, and efforts should be made to ensure their future using the local national park system and possibly the new Geopark concept. Groups associated with the International Association of Geomorphologists, International Association of Hydrogeologists and the International Speleological Union are amongst the best placed organizations to ensure that this is achieved. It is essential that this be done, because we are constantly reminded of the immense importance of caves as protected repositories of natural and cultural history. The recent discovery of a new human species, *Homo floresiensis*, in Liang Bua Cave in Flores Island, Indonesia, is a case in point (Morwood *et al.* 2004). Who knows what other discoveries await us in the intriguing world of karst?

References

Abel, O. & G. Kyrle (eds.) (1931) *Die Drachenhohle bei Mixnitz*, Speleologisches Monographes 7, 8 and 9, Wien.

Adams, A.E. and MacKenzie, W.S. (1998) *A Colour Atlas of Carbonate Sediments and Rocks under the Microscope*, London, Manson.

Adams, A.E., MacKenzie, W.S. and Guilford, C. (1984) *Atlas of Sedimentary Rocks under the Microscope*, John Wiley & Sons, New York, 104 pp.

Adams, R. and Parkin, G. (2002) Development of a coupled surface-groundwater-pipe network model for the sustainable management of karstic groundwater. *Environmental Geology*, **42**, 513–17.

Agostini, S., Forti, P. and Postpischl, D. (1994) Gli studi sismotettonici e palcosismici effetuati nella Grotta del Cervo di Petrasecca nel periodo 1987–1991. *Memoires Instituto Italiano di Speleologia*, **2**(5), 97–104.

Ahnert, F. (1996) The point of modelling geomorphological sysyems, in *Geomorphology Sans Frontières* (eds B. McCann and D.C. Ford), J. Wiley & Sons, Chichester, pp. 91–113.

Ahnert, F. and Williams, P.W. (1997) Karst landform development in a three-dimensional theoretical model. *Zeitschrift für Geomorphologie, Supplementband*, **108**, 63–80.

Aitken, J.D. and Cook, D.G. (1969) *Geology, Lac Belot, District of Mackenzie*. Geological Survey of Canada: Map 6.

Akerman, J.H. (1983) Notes on chemical weathering, Kapp Linne, Spitzbergen, *Proceedings 4th International Conference on Permafrost*, National Academy Press, Washington, DC, pp. 10–15.

Alexander, E.C., Broberg, J.S., Kehren, A.R., *et al.* (1993) Bellechester Minnesota lagoon collapse, in *Applied Karst Geology* (ed. B. Beck), Balkema, Rotterdam, pp. 63–72.

Alexeev, S.V. and Alexeeva, L.P (2002) Ground ice in the sedimentary rocks and kimberlites of Yakutia, Russia. *Permafrost and Periglacial Processes*, **13**(1), 53–9.

Al-fares W., Bakalowicz M., Guerin R. and Dukhan M. (2002) Analysis of the karst aquifer structure of the Lamalou area (Herault, France) with ground penetrating radar. *Journal of Applied Geophysics* **51**(2–4), 97–106.

Alhajjar, B.J., Chesters, G. and Harkin, J.M. (1990) Indicators of chemical pollution from septic systems. *Ground Water*, **28**(4), 559–68.

Ali, S.S. (2005) Effect of slides masses on ground water occurrence in some areas of Sharazoor Plain, N.E. Iraq, in *Water Resources and Environmental Problems in Karst* (eds Z. Stevanović and P. Milanović), International Association of Hydrogeologists, Belgrade, pp. 215–22.

Al-Kaisy, S.A.S. (2005) Using of groundwater to reduce the problem of cavitation under the foundations in gypsiferous soil in the site of Tikrit University, North Iraq, In in *Water Resources and Environmental Problems in Karst* (eds Z. Stevanović and P. Milanović), International Association of Hydrogeologists, Belgrade, pp. 697–704.

Alkattan, M., Oelkers, E.H., Dandurand, J.-L. and Schott (1997) Experimental studies of halite dissolution kinetics, 1. The effect of saturation state and the presence of trace metals. *Chemical Geology*, **137**, 201–19.

Allen, J.R.L. (1972) On the origin of cave flutes and scallops by the enlargement of inhomogeneities. *Rassegna Speleologica Italiana*, **24**, 3–23.

Allen, J.R.L. (1977) *Physical Processes of Sedimentation*, 4th impression, George Allen & Unwin, London, 248 pp.

Aller, L., Bennet, T., Lehr, J.H., *et al.* (1987) *DRASTIC: a Standardized System for Evaluating Groundwater Pollution Potential using Hydrogeologic Settings*. Office of Research and Development, US Environmental Protection Agency (EPA-600/2-87/035).

Alley, N.F., Clarke, J.D.A., Macphail, M. and Truswell, E.M. (1999) Sedimentary infillings and development of major Tertiary palaeodrainage systems of south-central Australia. *Special Publications of the International Association of Sedimentologists*, **27**, 337–366.

Allison, R.J. and Goudie, A.S. (1990) Rock control and slope profiles in a tropical limestone environment: the Napier Range of Western Australia. *The Geographical Journal*, **156**(2), 200–11.

Allred, K. (2004) Some carbonate erosion rates of southeast Alaska. *Journal of Cave and Karst Studies*, **66**(3), 89–97.

Alsharhan, A.S. and Kendall, C.G. St. C. (2003) Holocene coastal carbonates and evaporites of the southern Arabian Gulf and their ancient analogues. *Earth-Science Reviews*, **61**, 191–243.

Alsop, G.I. Blundell, D.J. and Davison, I. (eds) (1996) *Salt Tectonics*. Special Publication 100, Geological Society Publishing House, Bath.

American Society for Testing and Materials (1995) *Standard Guide for Design of Ground-Water Monitoring Systems in Karst and Fractured-Rock Aquifers*, D 5717 – 95, Annual Book of ASDTM Standards, pp. 435–51.

Anderson, G.M. (1991) Organic maturation and ore precipitation in southeast Missouri. *Economic Geology*, **86**(5), 909–26.

Anderson, N.L. and Hinds, R.C. (1997) Glacial loading and unloading: a possible cause of rock-salt dissolution in the Western Canadian Basin. *Carbonates and Evaporites*, **12**(1), 43–52.

Andre, B. and Rajaram, H. (2005) Dissolution of limestone fractures by cooling waters: early development of hypogene karst systems. *Water Resources Research*, **41**(1), W01015 10.1029/2004WR003331.

André, M.-F. (1996a) Vitesses de dissolution aréolaire postglaciaire dans les karsts polaires et haut-alpins – de l'Arctique scandinave aux alpes de Nouvelle-Guinée. *Revue d'Analyse Spatiale Quantitative et Appliquée*, **38–39**, 99–107.

André, M.-F. (1996b) Rock weathering rates in arctic and subarctic environments (Abisko Mts., Swedish Lappland). *Zeitschrift für Geomorphologie*, **40**(4), 499–517.

André, M-F. (2001) Tors et roches moutonnées en laponie Suédoise: antagonisme ou filiation ? *Géographie physique et Quaternaire*, **55**(3), 229–42.

Andrejchouk, V. (2002) Collapse above the world's largest potash mine (Ural, Russia). *International Journal of Speleology*, **31**(1/4), 137–58.

Andrejchouk, V.N. and Klimchouk, A.B. (2001) Geomicrobiology and redox geochemistry of the karstified Miocene gypsum aquifer, Western Ukraine: The Study of Zoloushka Cave. *Geomicrobiology Journal*, **18**, 275–95.

Andrejchouk, V. (1999) *Collapses above Gypsum Labyrinth Caves and Stability Assessment of Karstified Terrains*, Prut, Chernovtsy, 51 pp. [In Russian]

Andreo, B., Carrasco, F., Bakalowicz, M., *et al.* (2002) Use of hydrodynamic and hydrochemistry to characterise carbonate aquifers. Case study of the Blanca-Mijas unit (Málaga, southern Spain). *Environmental Geology*, **43**, 108–19.

Andrews, L.M. and Railsback, L.B. (1997) Controls on stylolite development: morphologic, lithologic temporal evidence from bedding-parallel and transverse stylolites from the U.S. Appalachians. *Journal of Geology*, **105**, 59–73.

Andrieux, C. (1963) Etude crystallographique des pavements polygonaux des croutes polycristallines de calcite des grottes. *Bulletin de la Societe Francaise de Mineralogie et de Cristallographie*, **86**, 135–8.

Anikeev, A. (1999) Casual hydrofracturing theory and its application for sinkhole development prediction in the area of Novovoronezh Nuclear Power House 2 (NV NPH–2), Russia,

in *Hydrogeology and Engineering Geology of Sinkholes and Karst* (eds B.F. Beck, A.J. Pettit, and J.G. Herring), A.A. Balkema, Rotterdam, pp. 77–83.

Annable, W.K. (2003) *Numerical analysis of conduit evolution in karstic aquifers*. Univ. of Waterloo PhD thesis, 139 pp.

Anthony, D.M. and Granger, D.E. (2004) A Tertiary origin for multilevel caves along the western escarpment of the Cumberland Plateau, Tennessee and Kentucky, established bt cosmogenic [26]Al and [10]Be. *Journal of Cave and Karst Studies*, **66**(2), 46–55.

Antonioli, F., Bard, E., Potter, E.-K., *et al.* (2004) 215-ka history of sea level oscillations from marine and continental layers in Argenterola Cave speleothems, Italy. *Global and Planetary Change*, **43**(1–2), 57–78.

Appelo, C.A.J. and Postma, D. (1994) *Geochemistry, Groundwater and Pollution*, Balkema, Rotterdam, 536 pp.

Arandjelovic, D. (1984) Application of geophysical methods to hydrogeological problems in Dinaric karst of Yugoslavia, in *Hydrogeology of the Dinaric Karst* (ed. B.F. Mijatovic), International Contributions to Hydrogeology 4, Heise, Hannover, pp. 143–59.

Arandjelovic, D. (1966) Geophysical methods used in solving some geological problems encountered in construction of the Treblinisca water power plant (Yugoslavia). *Geophysical Prospecting*, **14**(1), 80–97.

Arfib, B., de Marsily, G and Ganoulis, J. (2002) Les sources karstiques côtières en Méditerranée: étude des mécanismes de pollution saline de l'Almyros d'Héraklion (Crète), observations et modélisation. *Bulletin de la Societe Geologique de France*, **173**(3), 245–53.

Ashton, K. (1966) The analyses of flow data from karst drainage systems. *Transactions of the Cave Research Group of Great Britain*, **7**(2), 161–203.

Association Francaise de Karstologie (1981) Formations Carbonates Externes, Tufs et Travertins. *Bulletin de l' Association Geographique France, Memoire* **3**.

Astier, J.L. (1984) Geophysical prospecting, in *Guide to the Hydrology of Carbonate Rocks* (eds P.E. LaMoreaux, B.M. Wilson and B.A. Memon), Studies and Reports in Hydrology No. 41, UNESCO, Paris, pp. 171–96.

Atkinson, T.C. (1977a) Carbon dioxide in the atmosphere of the unsaturated zone: an important control of groundwater hardness in limestones. *Journal of Hydrology*, **35**, 111–23.

Atkinson, T.C. (1977b) Diffuse flow and conduit flow in limestone terrain in the Mendip Hills, Somerset (Great Britain). *Journal of Hydrology*, **35**, 93–110.

Atkinson, T.C. (1983) Growth mechanisms of speleothems in Castleguard Cave, Columbia Icefields, Alberta, Canada. *Arctic and Alpine Research*, **15**(4), 523–36.

Atkinson, T.C. (1985) Present and future directions in karst hydrogeology. *Annales de la Societe Géologique de Belgique*, **108**, 293–96.

Atkinson, T.C. and Rowe, P.J. (1992) Applications of dating to denudation chronology and landscape evolution, in *Uranium Series Disequilibrium. Applications to Marine, Earth and Environmental Sciences* (eds M. Ivanovich and R.S. Harmon), Clarendon Press, Oxford, pp. 669–703.

Atkinson, T.C. and Smith, D.I. (1976) The erosion of limestones, in *The Science of Speleology* (eds T.D. Ford and C.H.D. Cullingford), Academic Press, London, pp. 151–177.

Atkinson, T.C., Ward, R.S. and O'Hannelly, E. (2000) A radial-flow tracer test in Chalk: comparison of models and fitted parameters, in *Tracers and Modelling in Hydrogeology* (ed. A. Dassargues), Publication 262, International Association of Hydrological Sciences, Wallingford, pp. 7–15.

Aubert, D. (1967) Estimation de la dissolution superficielle dans le Jura. *Bulletin de la Societe Vaudoise des Sciences Naturelles*, No. 324, **69**(8), 365–76.

Aubert, D. (1969) Phenomenes et formes du karst jurassien. *Eclogae Geologicae Helvetiae*, **62**(2), 325–99.

Audra, P. (1994) Karsts alpins. Gènese des grands rèseaux souterrains. *Karstologia, Mémoires* **5**.

Audra, P. (1999) Soil erosion and water pollution in an intensive vine cultivation area: The Entre-deux-Mers example (Gironde, France), in *Karst Hydrogeology and Human Activities: Impacts, Consequences and Implications* (eds D. Drew and H. Hötzl), Balkema, Rotterdam, pp. 70–2.

Audra, P. (2000) Le karst haut alpin du Kanin (Alpes juliennes, Slovénie-Italie). *Karstologia*, **35**(1), 27–38.

Audra, P., Bigot, J.-Y. and Mocochain, L. (1993) Hypogenic caves in Provence (France): specific features and sediments. *Speleogenesis and Evolution of Karst Aquifers*, **1**(1), 10 pp.

Audra, P., Camus, H. and Rochette, P. (2001) Le karst des plateaux jurassique de la moyenne vallée de l'Ardèche: datation par paléomagnétisme des phases d'évolution plio quaternaires (aven de la Combe Rajeau). *Bullétin Sociète Géologique de France*, **172**(1), 121–9.

Auler, A.S. and Smart, P.L. (2003) The influence of bedrock-derived acidity in the development of surface and underground karst: evidence from the Precambrian carbonates of semi-arid northeastern Brazil. *Earth Surface Processes and Landforms*, **28**, 157–68.

Auler, A.S., Smart, P.L., Tarling, D.H. and Farrant, A.R. (2002) Fluvial incision rates derived from magnetostratigraphy of cave sediments in the cratonic area of eastern Brazil. *Zeitschrift für Geomorphologie*, **46**(3), 391–403.

Ayalon, A., Bar-Matthews, M. and Kaufman, A. (2002) Climatic conditions during marine oxygen isotope stage 6 in the eastern Mediterranean region from the isotopic composition of speleothems of Soreq Cave, Israel. *Geology*, **30**(4), 303–6.

Ayers, J.F. and Vacher, H.L. (1986) Hydrogeology of an atoll island: a conceptual model from detailed study of a Micronesian example. *Ground Water*, **24**(2), 185–98.

Ayliffe, L.K., Marianelli, P.C., Moriarty, K., *et al.* (1998) 500 ka precipitation record from southeastern Australia: evidence for interglacial relative aridity. *Geology*, **26**, 147–50.

Ayora, C., Taberner, C., Saaltink, A.B. and Carrera, J. (1998) The genesis of dedolomites: a discussion based on reactive transport modelling. *Journal of Hydrology*, **209**, 346–65.

Bachman, G.O. (1976) Cenozoic deposits of southeastern New Mexico and an outline of the history of evaporite dissolution. *Journal of Research, US Geological Survey*, **4**(2), 135–49.

Bachman, G.O. (1984) Regional geology of the Ochoan evaporites, northern part of Delaware Basin. *New Mexico Bureau of Mines and Mineral Resources, Circular*, **184**. 24 pp.

Bachman, G.O. (1987) *Karst in Evaporites in Southeastern New Mexico*, Report SAND 86-7078, Sandia National Laboratories.

Back, W., Hanshaw, B.B. and van Driel, J.N. (1984) Role of groundwater in shaping the Eastern Coastline of the Yucatan Peninsula, Mexico, in *Groundwater as a Geomorphic Agent* (ed. R. G. LaFleur), Allen & Unwin, London, pp. 280–93.

Backshall, D.G., J. Barnett, P.J. Davies, D.C. *et al.* (1979) Drowned dolines – the blue holes of the Pompey Reefs, Great Barrier Reef. *BMR Journal of Australian Geology and Geophysics*, **4**, 99–109.

Badino, G. and Romeo, A. (2005) Crio-Karst in the Hielo Continental Sur, in *Glacier Caves and Glacial Karst in High Mountains and Polar Regions* (ed. B.R. Mavlyudov), Institute of Geography, Russian Academy of Sciences, Moscow, pp. 13–18.

Baedke, S.J. and Krothe, N.C. (2001) Derivation of effective hydraulic parameters of a karst aquifer from discharge hydrograph analysis. *Water Resources Research*, **37**(1), 13–19.

Bagnold, R.A. (1966) An approach to the sediment transport problem from general physics. *US Geological Survey Professional Paper*, **422-I**, 37 pp.

Baichtal, J.F. and Swanston, D.N. (1996) *Karst Landscapes and Associated Resources: a Resource Assessment*. Report PNW-GTR–383, US Department of Agriculture Forestry Service.

Bakalowicz, M. (1973) Les grandes manifestations hydrologiques des karsts dans le monde. *Spelunca 2*, 38–40.

Bakalowicz, M., Blavoux, B and Mangin, A. (1974) Apports du tracage isotopique naturel à la connaissance du fonctionnement d'un systeme karstique-teneurs en oxygène–18 de trois systemes des Pyrenees, France. *Journal of Hydrology*, **23**, 141–58.

Bakalowicz, M. (1976) Géochimie des eaux karstiques. Une methode d'etude de l'organisation des ecoulements souterrains. *Annales Scientifiques de l' Universite de Besancon*, **25**, 49–58.

Bakalowicz, M. (1977) Etude du degre d'organisation des coulements souterrains dans les aquiferes carbonates par une methode hydrogochimique nouvelle. *Compte Rendus Academie des Sciences Paris, Series D*, **284**, 2463–6.

Bakalowicz, M. (1979) *Contribution de la geochimie des eaux a la connaissance de l'aquifere karstique et de la karstification*. Univ. Paris-6 These Doctorate Sciences, 269 pp.

Bakalowicz, M. and Mangin, A. (1980) L'aquifere karstique. Sa definition, ses characteristiques et son identification. *Memoires de la Societe Geologique de France*, **11**, 71–9.

Bakalowicz, M. (1984) Water chemistry of some karst environments in Norway. *Norsk Geografisk Tidsskrift*, **38**, 209–14.

Bakalowicz, M., Mangin, A., Rouch, R., *et al.* (1985) *Caractere de l'environnement souterrain de la galerie d'entree de la Grotte de Bedeilhac, Ariege*. Laboratoire Souterrain du C.N.R.S., Moulis, 67 pp.

Bakalowicz, M.J., Ford, D.C., Miller, T.E., *et al.* (1987) Thermal genesis of solution caves in the Black Hills, South Dakota. *Geological Society of America Bulletin*, **99**, 729–38.

Bakalowicz, M.J. and Jusserand, C. (1987) Etude de l'infiltration en milieu karstique par les methodes geochemiques et isotopiques. Cas de la Grotte de Niaux (Ariege, France). *Bulletin du Centre d'Hydrogeologie, Universite Neuchatel*, **7**, 265–83.

Bakalowicz, M. (1992) Géochemie des eaux et flux de matières dissoutes: l'approche objective du rôle du climat dans le karstogénèse, in Salomon, J.-N. and Maire, R. (eds). *Karst et évolutions climatiques*. Presses Universitaires, Bordeaux, 61–74.

Bakalowicz, M., Drew, D., Orvan, J., *et al.* (1995) The characteristics of karst groundwater systems, in *COST Action 65 – Hydrogeological Aspects of Groundwater Protection in Karstic Areas*. Final Report, European Commission Report EUR 16547, Luxembourg (ISBN 92-827-4682-8), pp. 349–69.

Bakalowicz, M. (1995) La zone d'infiltration des aquifères karstiques. Méthodes d'étude. Structure et fonctionnement. *Hydrogéologie* **4**, 3–21.

Bakalowicz, M. (2001) Exploration techniques for karst groundwater resources, in *Present State and Future Trends of Karst Studies* (eds G. Günay, K.S. Johnson, D. Ford and A.I. Johnson), IHP-V, Technical Documents in Hydrology, 49(II), UNESCO, Paris, pp. 31–44.

Bakalowicz, M. (2005) Karst groundwater: a challenge for new resources. *Hydrogeology Journal*, **13**, 148–60.

Bakalowicz, M., Fleyfel, M. and Hachache, A. (2002) Une histoire ancienne: le captage de la source de Ras el Aïn et l'alimentation en eau de la ville de Tyr (Liban). *La Houille Blanche*, **4**(5), 157–60.

Baker, A. and Brunsdon, C. (2003). Non-linearities in drip water hydrology: an example from Stump Cross Caverns, Yorkshire. *Journal of Hydrology*, **277**, 151–63.

Baker, A. and Smart, P.L. (1995) Recent flowstone growth rates: field measurements and comparison to theoretical results. *Chemical Geology*, **122**, 121–8.

Baker, A., Smart, P.L. and Ford, D.C. (1993a) Northwest European paleoclimate as indicated by growth frequency variations of secondary calcite deposits. *Paleogeography, Paleoclimatology, Paleoecology*, **100**, 291–301.

Baker, A., Smart, P.L., Edwards, R.L. and Richards, D.A. (1993b) Annual growth banding in a cave stalagmite. *Nature*, **364**, 518–20.

Baker, A., Ito, E., Smart, P.L. and McEwan, R.F. (1997) Elevated and variable values of ^{13}C in speleothems in a British cave system. *Chemical Geology*, **136**, 263–70.

Baker, A., Genty, D., Dreybrodt, W., *et al.* (1998) Testing theoretically predicted stalagmite growth rates with Recent annually laminated samples: Implications for past stalagmite deposition. *Geochimica et Cosmochimica Acta*, **62**(3), 393–404.

Baker, A., Genty, D. and Fairchild, I.J. (2000) Hydrological characterisation of stalagmite dripwaters at Grotte de Villars, Dordogne, by the analysis of inorganic species and luminescent organic matter. *Hydrology and Earth System Sciences*, **4**(3), 439–49.

Balazs, D. (1973) Relief types of tropical karst areas, in *IGU Symposium on Karst Morphogenesis* (ed. L. Jakucs), Attila Jozsef University, Szeged, pp. 16–32.

Balch, E.S. (1900) *Glacières or Freezing Caverns*. Lane & Scott, Philadelphia, 337 pp.

Baldini, J.U.L., McDermott, F. Fairchild, I.J. (2002) Structure of the 8200-year cold event revealed by a speleothem trace element record. *Science, 296*, 2203–2206.

Banks, D, Davies, C. and Davies, W (1995) The Chalk as a karstic aquifer: evidence from a tracer test at Stanford Dingley, Berkshire, UK. *Quarterly Journal of Engineering Geology*, **28**, S31–S38.

Bárdossy, G. (1982) *Karst Bauxites. Bauxite Deposits on Carbonate Rocks*, Akademia Kiado-Elsevier, Budapest-Amsterdam, 441 pp.

Bárdossy, G. (1989) Bauxites, in *Paleokarst – a Systematic and Regional Review* (eds P. Bosák, D.C. Ford, J. Głazek and I. Horáček), Academia Praha/Elsevier, Prague/Amsterdam, pp. 399–418.

Barker, J.A. and Black, J.H. (1983) Slug tests in fissured aquifers. *Water Resources Research*, **19**, 1558–1564.

Bar-Matthews, M. and Ayalon, A. (1997) Late Quaternary paleoclimate in the eastern Mediterranean region from stable isotope analysis of speleothems at Soreq Cave, Israel. *Quaternary Research*, **47**, 155–68.

Bar-Matthews, M., Ayalon, A., Matthews, A., *et al.* (1996) Carbon and oxygen isotope study of the active water-carbonate system in a karstic Mediterranean cave: implications for paleoclimate research in semiarid regions. *Geochimica et Cosmochimica Acta*, **60**(2), 337–47.

Barner, W.L. (1999) Comparison of stormwater management in a karst terrane in Springfield, Missouri – case histories. *Engineering Geology*, **52**, 105–12.

Barrows, T.T. and Juggins, S. (2005) Sea-surface temperatures around the Australian margin and Indian Ocean during the Last Glacial Maximum. *Quaternary Science Reviews*, **24**, 1017–47.

Barton, N. and Stephansson, O. (eds) (1990) *Rock Joints*, A.A. Balkema, Rotterdam, 994 pp.

Baskaran, M. and Krishnamurthy, R.V. (1993) Speleothems as proxy for the carbon isotope composition of atmospheric CO_2. *Geophysical Research Letters*, **20**(24), 2905–8.

Bastin, B. (1979) L'analyse pollinique des stalagmites. *Annales de la Societe Geologique de Belgique*, **101**, 13–19.

Bastin, B. (1990) L'analyse pollinique des concretions stalagmitiques: méthodologier et resultants en provenance des grottes belges. *Karstologia, Mémoires*, **2**, 3–10.

Batsche, H., F. Bauer, H. Behrens, K., *et al.* (1970) Kombinierte Karstwasser-untersuchungen in Gebiet der Donauversick erung (Baden-Wuttemberg) in den Jahren, 1966–1969, 1970. *Steirische Beitrage zur Hydrogeologie*, **22**(B90), 5–165.

Bauer, F. (1962) NacheiszeitlicheKarstformen in der osterreichischen Kalkalpen, *Proceedings of the 2nd International Congress of Speleology, Bari*, pp. 299–328.

Bauer, S. (2002) Simulation of the genesis of karst aquifers in carbonate rocks. *Tübingen Geowissenschaftliche Arbeiten*, **C62**, 143 pp.

Bear, J. (1972) *Dynamics of Fluids in Porous Media*, Elsevier, New York.

Beck, B.F. (ed.) (2003) *Sinkholes and the Engineering and Environmental Impacts of Karst, Proceedings* 9th Multidisciplinary Conference, Huntsville, Alabama. Geotechnical

Special Publication 122, American Society of Civil Engineers, 737 pp.

Beck, B.F. (ed.) (2005) *Sinkholes and the Engineering and Environmental Impacts of Karst*, Geotechnical Special Publication No. 144, American Society of Civil Engineers, 677 pp.

Beck, B.F. and Herring, J.G. (eds) (2001) *Geotechnical and Environmental Applications of Karst Geology and Hydrology*, Balkema, Lisse, 437 pp.

Beddows, P.A. (2004) Yucatán phreas, Mexico, in *Encyclopedia of Caves and Karst Science* (ed. J. Gunn), Fitzroy Dearborn, New York, pp. 786–8.

Beddows, P.A., Smart, P.L., Whitaker, F.F. and Smith, S.L. (2002) Density stratified groundwater circulation on the Caribbean coast of the Yucatan Peninsula, Mexico, in *Hydrogeology and Biology of Post-Paleozoic Carbonate Aquifers* (eds J.B. Martin, C.M. Wicks, and I.D. Sasowsky), Special Publication 7, Karst Waters Institute, Charles Town, WV, pp. 129–34.

Beede, J.W. (1911) The cycle of subterranean drainage as illustrated in the Bloomington Quadrangle (Indiana), *Proceedings Indiana Academy of Science*, **20**, 81–111.

Behrens, H. (1998). Radioactive and activable isotopes, in *Tracing Technique in Geohydrology* (ed. W. Käss), Rotterdam, Balkema, pp. 167–87.

Bell, K. (ed.) (1989) *Carbonatites: Genesis and Evolution*, Unwin Hyman, London, 618 pp.

Belloni, S., Martins, B. and Orombelli, G. (1972) Karst of Italy, in *Karst: Important Karst Regions of the Northern Hemisphere* (eds M. Herak and V.T. Stringfield), Elsevier, Amsterdam, pp. 85–128.

Benderitter, Y., Roy, B. and Tabbagh, A. (1993) Flow characterization through heat transfer evidence in a carbonate fractured medium: first approach. *Water resources Research*, **29**(11), 3741–7.

Beniawski, Z.T. (1976) Rock mass classification in rock engineering, in Z.T. Beniawski (ed.). *Exploration for rock engineering*. A.A. Balkema, Cape Town, pp. 97–106.

Benito, G., Perez del Campo, P., Gutierrez-Elzora, M. and Sancho, C. (1995) Natural and human-induced sinkholes in gypsum terrain and associated environmental problems in NE Spain. *Environmental Geology*, **25**, 156–64.

Benson, R.C., Yuhr, L. and Kaufmann, R.D. (2003) Assessing the risk of karst subsidence and collapse, in *Sinkholes and the Engineering and Environmental Impacts of Karst* (ed. B.E. Beck), *Proceedings* of 9th Multidisciplinary Conference, Huntsville, Alabama, Geotechnical Special Publication 122, American Society of Civil Engineers, pp. 31–9.

Berger, L.R. (2001) Viewpoint: is it time to revise the system of scientific naming? *National Geographic News*, December 4. [http:// news.nationalgeographic.com/news/2001/12/1204_hominin_id.html]

Berner, E.K. and Berner, R.A. (1996) *Global Environment. Water, Air and Geochemical Cycles*, Prentice Hall, Upper Saddle River, NJ.

Berner, R.A. and Morse, J.W. (1974) Dissolution kinetics of calcium carbonate in seawater. IV. Theory of calcite solution. *American Journal of Science*, **274**, 108–34.

Berstad, I.M., Lundberg, J., Lauritzen, S.-E. and Linge, H.C. (2002) Comparison of the climate during marine isotope stage 9 and 11 inferred from a speleothem isotope record from northern Norway. *Quaternary Research*, **58**, 361–71.

Betson, R.P. (1977) The hydrology of karst urban areas, in *Hydrologic Problems in Karst Regions* (eds R.R. Dilamarter and S.C. Csallany), Western Kentucky University, Bowling Green, pp. 162–75.

Bini, A. (1978) Appunti di geomorfologia ipogea: le forme parietali, *5th Convention on Regional Speleologia of Trentino–Alto Adige*, pp. 19–46.

Bini, A., Meneghel, M. and Sauro, U. (1998a) Karst geomorphology of the Altopiani Ampezzani. *Zeitschrift für Geomorphologie NF, Supplement-Band*, **109**, 1–21.

Bini, A., Tognini, P. and Zuccoli, L. (1998b) Rapport entre karst et glaciers durant les glaciations dans les vallées préalpines du Sud des Alpes. *Karstologia*, **32**(2), 7–26.

Biondic, B., Dukaric, F. and Biondic, R., 1999) Impact of the sea on the Perilo abstraction site in Bakar Bay – Croatia, in *Karst hydrogeology and human activities: impacts, consequences and implications* (eds D. Drew and H. Hötzl), Balkema, Rotterdam, pp. 244–51.

Bird, J.B. (1967) *The Physiography of Arctic Canada with Special Reference to the Area South of Parry Channel*, Johns Hopkins University Press, Baltimore, 336 pp.

Birk, S., Liedl, R. and Sauter, M. (2000) Characterisation of gypsum aquifers using a coupled continuum-pipe flow model, in *Calibration and Reliability in Groundwater Modelling* (eds Stauffer, F., Kinzelbach, W., Kovar, K. and Hoehn, E.), Publication 265, International association of Hydrological Sciences, Wallingford, pp. 16–21.

Bischof G. (1854) *Chemical and Physical Geology* (translation), Paul & Drummond, London.

Blanc, J.-J. (1992) Importance geodynamique des surfaces d'aplanissement en Provence, in *Karst et Evolutions Climatiques* (ed. J.-N. Salomon), Presses Universitaires de Bordeaux, pp. 191–207.

Bland, W. and Rolls, D. (1998) *Weathering: an Introduction to the Scientific Principles*, Arnold, London, 271 pp.

Blavoux, B., Mudry, J. and Puig, J.M. (1992). Water-budget, functioning and protection of the Fontaine-de-Vaucluse karst system (southeastern France). *Geodinamica Acta*, **5**(3), 153–72.

Bloom, A.L. (1974) Geomorphology of reef complexes, in *Reefs in Time and Space* (ed. L.F. Laporte), Special Publication 18, Society of Economic Palaeontologists and Mineralogists, Tulsa, OK, pp. 1–8.

Bocker, T. (1977) Economic significance of karst water research in Humgary. *Karszt es Barlang, Special Issue*, 27–30.

Bögli, A. (1960) Kalklosung und Karrenbildung. *Zeitschrift für Geomorphologie, Supplement-band*, **2**, 4–21.

Bögli, A. (1961) Karrentische, ein Beitrag sur Karstmorphologie. *Zeitschrift für Geomorphologie*, **5**, 185–93.

Bögli, A. (1964) Mischungskorrosion; ein Beitrag zum Verkarstungsproblem. *Erdkunde*, **18**(2), 83–92.

Bögli, A. (1970) *Le Holloch et son Karst*, Editions la Baconnière, Neuchatel.

Bögli, A. (1980) *Karst Hydrology and Physical Speleology,* Springer-Verlag, Berlin, 284 pp.

Bolner-Takacs, K. (1999) Paleokarst features and other climatic relics in Hungarian caves. *Acta Carsologica,* **28**(1), 27–37.

Bonacci, O. (1987) *Karst Hydrology: with Special Reference to the Dinaric Karst.* Springer-Verlag, Berlin, 184 pp.

Bonacci, O. (1993) Karst springs hydrographs as indicators of karst aquifers. *Hydrological Sciences Journal–Journal Des Sciences Hydrologiques,* **38**(1), 51–62.

Bonacci, O. (1995) Ground water behaviour in karst: example of the Ombla Spring (Croatia). *Journal of Hydrology,* **165**, 113–34.

Bonacci, O. (2001a) Analysis of the maximum discharge of karst springs. *Hydrogeology Journal,* **9**, 328–338.

Bonacci, O. (2001b) Monthly and annual effective infiltration coefficients in Dinaric karst: example of the Gradole karst spring catchment. *Hydrological Sciences Journal,* **46**(2), 287–99.

Bonacci, O. (2004) Poljes, in *Encyclopedia of Caves and Karst Science* (ed. J. Gunn), Fitzroy Dearborn, New York, pp. 599–600.

Bonacci, O. and Bojanic, D. (1991) Rhythmic karst springs. *Hydrological Sciences Journal,* **36**(1), 35–47.

Bonacci, O. and Roje-Bonacci, T. (2000) Interpretation of groundwater level monitoring results in karst aquifers: examples from the Dinaric karst. *Hydrological Processes,* **14**, 2423–38.

Bonacci, O. and Roje-Bonacci, T. (2003) The influence of hydroelectrical development on the flow of the karstic river Cetina. *Hydrological Processes,* **17**, 1–15.

Bonacci, O. and Roje-Bonacci, T. (2004) Plitvice Lakes, Croatia, in *Encyclopedia of Caves and Karst Science* (ed. J. Gunn), Fitzroy Dearborn, New York, pp. 597–598.

Bonacci, O., Kerovec, M., Roje-Bonacci, T., *et al.* (1998) Ecologically acceptable flows definition for the Žrnovnica River (Croatia). *Regulated Rivers: Research and Management,* **14**, 245–56.

Bosák, P. (1989) Problems of the origin and fossilization of karst forms, in *Paleokarst – a Systematic and Regional Review* (eds P. Bosák, D.C. Ford, J. Głazek and I. Horáček), Academia Praha/Elsevier, Prague/Amsterdam, pp. 577–98.

Bosák, P., Ford, D.C., Glazek, J. and Horacek, I. (eds) (1989) *Paleokarst – a Systematic and Regional Review,* Academia Praha/Elsevier, Prague/Amsterdam.

Bosák, P., Pruner, P. and Zupan Hajna, N. (1998) Paleomagnetic research of cave sediments in S.W.Slovenia. *Acta Carsologica,* **28**(2), 151–79.

Bosák, P., Bruthans, J., Filippi, M., *et al.* (1999) Karst and Caves in the Salt Diapirs, SE Zagros Mts., Iran. *Acta Carsologica,* **2820**, 41–75.

Bosák, P. (2002) Karst processes from the beginning to the end: how can they be dated?, in *Evolution of Karst: from Prekarst to Cessation* (ed. F. Gabrovsek), Institut za raziskovanje krasa, ZRC SAZU, Postojna-Ljubljana, pp. 191–223.

Bosák, P., Bella, P., Cilek, V., *et al.* (2003) Ochtina Aragonite Cave, Slovakia: morphology, mineralogy and genesis. *Geologica Carpathica,* **53**(6), 399–441.

Bosák, P., Mihevc, A. and Pruner, P. (2004) Geomorphological evolution of the Podgorski karst, S.W. Slovenia. *Acta Carsologica,* **33**(1), 175–204.

Bosch, R.F. and White, W.B. (2004) Lithofacies and transport of clastic sediments in karstic aquifers, in *Studies of Cave Sediments: Physical and Chemical Records of Paleoclimate* (eds I.D. Sasowsky and J. Mylroie), Kluwer Academic, New York, pp. 1–22.

Bottrell, S.H. and Atkinson, T.C. (1992) Tracer study and storage in the unsaturated zone of a karstic limestone aquifer, in *Tracer Hydrology* (eds H. Hotzl and A. Werner), Balkema, Rotterdam, pp. 207–11.

Bourdon, B., Henderson, G.M., Lundstrom, C.C. and Turner, S.P. (eds) (2003) *Uranium-series Geochemistry. Reviews in Mineralogy and Geochemistry,* **52**, 656 pp.

Bourrouilh-Le Jan, F.G. (1989) The oceanic karst: modern phosphate and bauxite ore deposits on the high carbonate islands (so-called 'uplifted atolls') of the Pacific Ocean, in *Paleokarst – a Systematic and Regional Review* (eds P. Bosák, D.C. Ford, J. Głazek and I. Horáček), Academia Praha/Elsevier, Prague/Amsterdam, pp. 443–71.

Boussinesq, J. (1903) Sur un mode simple d'écoulement des nappes d'eau d'infiltration à lit horizontal, avec rebord vertical tout autour lorsqu'une partie de ce rebord est enlevée depuis la surface jusqu'au fond. *Comptes Rendus, Académie des Sciences (Paris),* **137**, 5–11.

Boussinesq, J. (1904) Recherches théoretiques sur l'écoulement des nappes d'eau infiltrées dans le sol et sur le débit des sources. *Journal de Mathematiques Pures et Appliquees,* **10**, 5–78.

Bowen, G.J. and Wilkinson, B. (2002) Spatial distribution of $\delta^{18}O$ in meteoric precipitation. *Geology,* **30**(4), 315–18.

Boyer, B.W. (1997) Sinkholes, soils, fractures drainage: Interstate 70 near Frederick, Maryland. *Environmental and Engineering Geoscience,* **III**(4), 469–85.

Boyer, D.G. and Pasquarell, G.C. (1995) Nitrate concentrations in karst springs in an extensively grazed area. *Water Resources Bulletin,* **31**(4), 729–36.

Bozicevic, S. and Biondic, B. (1999) The Plitvice Lakes, in *Karst Hydrogeology and Human Activities* (eds D. Drew and H. Hötzl), Balkema, Rotterdam, pp. 174–78.

Bradley, R.S. (1999) *Paleoclimatology: Reconstructing the Climates of the Quaternary,* San Diego, Harcourt Academic Press. 610 pp.

Bradshaw, P. (ed.) (2002) *Reclamation of Limestone Quarries by Landform Simulation: Summary of Lessons Learnt from Trial Sites,* Department for Transport, Local Government and the Regions, London.

Brady, B.M.G. and Brown, E.T. (1985) *Rock Mechanics for Underground Mining,* George Allen & Unwin, London, 527 pp.

Breisch, R.L. and Wefer, F.L. (1981) The shape of 'gypsum bubbles', *Proceedings of the 8th International Congress of Speleology, Bowling Green, Kentucky,* Vol. 2, pp. 757–9.

Bretz, J.H. (1942) Vadose and phreatic features of limestone caves. *Journal of Geology,* **50**(6), 675–811.

Briceño, H.O. and Schubert, C. (1990) Geomorphology of the Gran Sabana, Guyana Shield, southeastern Venezuela. *Geomorphology,* **3**, 125–41.

Bricelj, M. (1997) Results with phages, in Kranjc, A. (ed.) *Karst Hydrogeological Investigations in South-western Slovenia.*

Acta Carsologica Krasoslovni Zbornik XXVI/1, Ljubljana, pp. 307–14.

Brink, A.B.A. (1984) A brief review of the South African sinkhole problem, in *Sinkholes: their Geology, Engineering and Environmental Impact* (ed. B.F. Beck), Balkema, Rotterdam, pp. 123–7.

British Columbia Ministry of Forest (1994) *Cave/Karst Management Handbook for the Vancouver Forest Region*, Province of British Columbia, Victoria, BC.

British Columbia Ministry of Forest (2003) *Karst Management Handbook for British Columbia*, Province of British Columbia, Victoria, BC., 69 pp.

Brod, L.G. (1964) Artesian origin of fissure caves in Missouri. *Bulletin of the National Speleological Society*, **26**(3), 83–112.

Brook, G.A. (1981) An approach to modelling karst landscapes. *South African Geography Journal*, **63**(1), 60–76.

Brook, G.A. and Ford, D.C. (1978) The origin of labyrinth and tower karst and the climatic conditions necessary for their development. *Nature*, **275**, 493–96.

Brook, G.A. and Ford, D.C. (1980) Hydrology of the Nahanni karst, northern Canada the importance of extreme summer storms. *Journal of Hydrology*, **46**, 103–21.

Brook, G.A. and Hanson, M. (1991. Double Fourier series analysis of cockpit and doline karst near Browns Town, Jamaica. *Physical Geography*, **12**(1), 37–54.

Brook, G.A., Folkoff, M.E. and Box, E.O. (1983) A world model of soil carbon dioxide. *Earth Surface Processes and Land forms*, **8**, 79–88.

Brook, G.A., Burney, D.A. and Cowart, J.B. (1990) Desert paleoenvironmental data from cave speleothems with examples from the Chihuahuan, Somali-Chalbi Kalahari deserts. *Paleogeography, Paleoclimatology, Paleoecology*, **76**, 311–29.

Brook, G.A., Embabi, N.S., Ashour, M.M., et al. (2003) Quaternary environmental change in the western desert of Egypt: evidence from cave speleothems, spring tufas and playa sediments. *Zeitschrift für Geomorphologie, Supplment-Band*, **131**, 59–87.

Brooks, S. and Gebauer, D. (2004) Indian subcontinent, in *Encyclopedia of Caves and Karst Science* (ed. J. Gunn), Fitzroy Dearborn, New York, pp. 442–45.

Brown, M.C. (1972) *Karst Hydrology of the Lower Maligne Basin, Jasper, Alberta*, Cave Studies 13, Cave Research Association, California, 97 pp.

Brown, M.C. (1973) Mass balance and spectral analysis applied to karst hydrologic networks. *Water Resources Research*, **9**(3), 749–52.

Brown, R.H., Konoplyantsev, A., Ineson, J. and Kovalevsky, V.S. (eds) (1972) *Groundwater Studies*, Studies and Reports in Hydrology 7, Unesco, Paris.

Brush, D.J. and Thomson, N.R. (2003) Fluid flow in synthetic rough-walled fractures: Navier–Stokes, Stokes, and local cubic law simulations. *Water Resources Research*, **39**, 1085, doi:10.1029/2002WR001346.

Bruthans, J. and Zeman, O. (2003) Factors controlling exokarst morphology and transport through caves; comparison of carbonate and salt karst. *Acta Carsologica*, **32**(1), 83–99.

Buchtela, K. von (1970) Aktivierungsanalyse in der Hydrogeologie. *Steirische Beitrage zur Hydrogeologie*, **22**, 189–98.

Buckland, W. (1823) *Reliquae Diluvianae*, Oxford.

Budel, J. (1982) *Climatic Geomorphology*, Princeton University Press, Princeton, NY.

Bull, P.A. (1977a) Boulder chokes and doline relationships, *Proceedings of the 7th International Congress of Speleology, Sheffield*, pp. 93–96.

Bull, P.A. (1977b) Laminations or varves? Processes of fine grained sediment deposition in caves, *Proceedings of the 7th International Congress of Speleology, Sheffield*, pp. 86–87.

Bull, P.A. (1978) Surge mark formation and morphology. *Sedimentology*, **25**, 877–86.

Bull, P.A. (1982) Some fine-grained sedimentation phenomena in caves. *Earth Surface Processes and Landforms*, **6**, 11–22.

Bull, P.A. and Laverty, M. (1982) Observations on phytokarst. *Zeitschrift für Geomorphologie*, **26**, 437–457.

Burdon, D.J. and Papakis, N. (1963) *Handbook of Karst Hydrogeology*, Institute for Geology and Subsurface Research/FAC, Athens.

Burdon, D.J. and Safadi, C. (1963) Ras-el-Ain: the great spring of Mesopotamia. *Journal of Hydrology*, **1**, 58–95.

Burin, K., Spassov, K., Kolev, D., et al. (1976) Two experiments in tracing karst underground waters with bromine, using neutron activation analysis in Bulgari,, *Proceedings of the 3rd International Symposium on Underground Water Tracing*, Institute of Karst Research, Postojna, Ljubljana, B159, pp. 35–45.

Burney, D.A. and Burney, L.P. (1993) Modern pollen deposition in cave sites: experimental results from New York State. *New Phytology*, **124**, 523–35.

Burri, E., Castiglioni, B. and Sauro, U. (1999) Karst and agriculture in the world. *International Journal of Speleology*, **28B**(1/4), theme issue; 5–185.

Busche, D. and Erbe, W. (1987) Silicate karst landforms of the southern Sahara. *Zeitschrift für Geomorphologie, N.F. Supplement-Band*, **64**, 55–72.

Busche, D. and Sponholz, B. (1992) Morphological and micromorphological aspects of the sandstone karst of eastern Niger. *Zeitschrift für Geomorphologie, N.F. Supplement-Band*, **85**, 1–18.

Busenberg, E. and Plummer, L.N. (1982) The kinetics of dissolution of dolomite in CO_2–H_2O systems at 1.5° to 65° C and 0 to 1 atm PCO_2. *American Journal of Science*, **282**, 45–78.

Busenberg, E. and Plummer, L.N. (1986) A comparative study of the dissolution kinetics of calcite and aragonite, in *Studies in Diagenesis* (ed. F.A. Mumton). *US Geological Survey Bulletin*, **1578**, 139–168.

Butler, J.J. (1998) *The Design, Performance Analysis of Slug Tests*, Lewis, Boca Raton, 252 pp.

Cabrol, P. (1978) *Contribution à L'étude du concretionnement carbonate des grottes du sud de la France morphologie, genese, diagenese*. C.E.R.G.H., Univ. Montpellier, tome XII, 275 pp.

Cabrol, P. and Mangin, A. (2000) *Fleurs de pierre: Les plus belles concretions des grottes de France*, Delachaux et Niestlé, Lausanne, 191 pp.

Cai, Z., Tang, W. and Maire, R. (1993) Le géographe chinoise Xu Xiake. *Karstologia*, **21**(1), 43–50.

Cailleux, A. and J. Tricart (1963) *Initiation a l'etude des sables et des galets*, 3 vols, Centre Doc. Universitaire, Paris, 765 pp.

Calaforra, J.M. and Pulido-Bosch, A. (1999) Gypsum karst features as evidence of diapiric processes in the Betic Cordillera, Southern Spain. *Geomorphology*, **19**, 251–64.

Campy, M. (1990) L'enregistrement du temps et climat dans les remplissages karstiques: l'apport de la sédimentologie. *Karstologia Mémoires*, **2**, 11–22.

Cardona, F. and Viver, J. (2002) *Sota la sal de Cardona*, Espeleo Club de Gracia, Barcelona, 128 pp.

Carew, J.L., Mylroie, J., Wehmiller, J.F. and Lively, R.S. (1984) Estimates of late Pleistocene sea level high stands from San Salvador, Bahamas, *Proceedings, 2nd Symposium on the Geology of the Bahamas* (ed. J.W. Teeter), pp. 153–75.

Carozzi, A.V., Scholle, P.A. and James, N.P. (eds) (1996) *Carbonate Petrography: Grains, Textures Case Studies*, Society of Economic Paleontologists and Mineralogists Publications, Denver, 700 pp.

Carter, W.L and Dwerryhouse, A.R. (2004) The underground waters of northwest Yorkshire (Part 2). The underground waters of Ingleborough. *Proceedings of the Yorkshire Geological Society* **15**(2), 248–92. [Part 1, 1900.]

Carthew, K.D., Taylor, M.P. and Drysdale, R.N. (2006) An environmental model of fluvial tufas in the monsoonal tropics, Barkly karst, northern Australia. *Geomorphology*, **73**, 78–100.

Castellani, V. and Dragoni, W. (1986) Evidence for karstic mechanisms involved in the evolution of Moroccan Hamadas. *International Journal of Speleology*, **15**(1–4), 57–71.

Castany, G. (1982) *Principes et Methodes de l'Hydrogologie*, Dunod, Paris.

Castany, G. (1984a) Hydrogeological features of carbonate rocks, in *Guide to the Hydrology of Carbonate Rocks* (eds P.E. LaMoreaux, B.M. Wilson and B.A. Memon), Studies and Reports in Hydrology 41, UNESCO, Paris, pp. 47–67.

Castany, G. (1984b) Determination of aquifer characteristics, in *Guide to the Hydrology of Carbonate Rocks* (eds P.E. LaMoreaux, B.M. Wilson and B.A. Memon), Studies and Reports in Hydrology 41, UNESCO, Paris, pp. 210–37.

Castillo, A., Lopez Chicano, M. and Pulido-Bosch, A. (1993) Temporal evolution of Riofrio nitrate content (Sierra Gorda, Granada), in *Some Spanish Karstic Aquifers* (ed. A. Pulido-Bosch), University of Granada Press, Granada, pp. 117–26.

Caumartin, V. (1963) Review of the microbiology of underground environments. *Bulletin of the National Speleological Society*, **25**, 1–14.

Celi, M. (1991) The impact of bombs of World War I on limestone slopes of Monte Grappa, *Proceedings, International Conference on Environmental Changes in Karst Areas*, International Geographical Union/International Speleological Union, pp. 279–87.

Celico, P., Gonfiantini, R., Koizumi, M. and Mangano, F. (1984) Environmental isotope studies of limestone aquifers in central Italy, in *Isotope Hydrology 1983*, International Atomic Energy Agency, Vienna, pp. 173–92.

Cerling, T.E. (1984) The stable isotopic composition of modern soil carbonate and its relationship to climate. *Earth and Planetary Science Letters*, **71**, 229–40.

Chafetz, H.S. and R.L. Folk (1984) Travertines: depositional morphology and the bacterially constructed constituents. *Journal of Sedimentary Petrology*, **54**(1), 189–316.

Chafetz, H.S., Srdoč, D. and Horvatinčič, N. (1994) Early diagenesis of Plitviče Lakes waterfall and barrier travertine deposits. *Gèographie physique et Quaternaire*, **43**(3); 247–55.

Chaix, E. (1895) Contribution a l'etude des lapies: la topographie du desert de Plate. *Le Globe*, **XXXIV**, 67–108.

Chappell, J. (1983) A revised sea-level record for the last 300,000 years from Papua, New Guinea. *Search*, **14**, 99–101.

Chappell, J. and Shackleton, N.J. (1986) Oxygen isotopes and sea level. *Nature*, **324**, 137–40.

Chapellier, D. (1992) *Well Logging in Hydrogeology*, Balkema, Rotterdam, 175 pp.

Chapman, J.B., Ingraham, N.L. and Hess, J.W. (1992) Isotopic investigation of infiltration and unsaturated zone processes at Carlsbad Cavern, New Mexico. *Journal of Hydrology*, **133**, 343–63.

Chardon, M. (1992) Evolution actuelle et recent des karsts de la Vanoise orientale. in *Karst et Evolutions Climatiques* (eds J-N. Salomon, and R. Maire), Presses Universitaires de Bordeaux, pp. 293–308.

Chen, W., Zhu D. and Zhu X. (2003) *Study on Karst Landscape and World Natural Heritage Value of Tiankeng-Difeng Tourist Site, Fengjie, Chongqing*. Geological Printing House, Beijing, 133 pp. [In Chinese.]

Chen, X. (ed.) (1998) *South China Karst 1*. ZRC SAZU, Lubljana, 247 pp.

Chen, Z., Song, L. and Sweeting, M.M. (1986) The pinnacle karst of the Stone Forest on Lunan, Yunnan, China: an example of a subjacent karst, in *New Directions in Karst* (eds M.M. Sweeting and K. Paterson), Geo Books, Norwich, pp. 587–607.

Chenoweth, S., Day, M., Koenig, S., *et al.* (2001) Conservation issues in the Cockpit Country, Jamaica, *Proceedings of the 13th International Congress of Speleology, Brazilia*, Vol. 2, pp. 237–41.

Chernyshev, S.N. (1983) *Fissures in Rocks*. Nauka, Moscow, 240 pp. [In Russian.]

Chicocki, G. and Zojer, H. (2005) VURAAS – vulnerability and risk analysis in Alpine aquifer systems, in *Water Resources and Environmental Problems in Karst. Belgrade* (eds Z. Stevanović, and P. Milanović), National Committee of the International Association of Hydrogeologists of Serbia and Montenegro, pp. 91–6.

Choquette, P.W. and Pray, L.C. (1970) Geological nomenclature and classification of porosity in sedimentary carbonates. *American Association of Petroleum Geologists Bulletin*, **54**, 207–50.

Chorley, R.J. and B.A. Kennedy (1971) *Physical geography, a Systems Approach*. Prentice-Hall, London, 370 pp.

Chou, L., Garrels, R.M. and Wollast, R. (1989) Comparative study of the kinetics and mechanisms of dissolution of carbonate minerals. *Chemical Geology*, **78**, 269–82.

Christiansen, E.A. (1967) Collapse structures near Saskatchewan, Canada. *Canadian Journal of Earth Sciences*, **4**, 757–67.

Christiansen, E.A. and Sauer, E.K. (2001) Stratigraphy and structure of a Late Wisconsin salt collapse in the Saskatoon Low south of Saskatoon, Saskatchewan, Canada: an update. *Canadian Journal of Earth Sciences*, **38**, 1601–13.

Christopher, N.S.J. (1980) A preliminary flood pulse study of Russett Well, Derbyshire. *Transactions of the British Cave Research Association*, **7**(1), 1–12.

Cigna, A.A. (1986) Some remarks on phase equilibria of evaporites and other karstifiable rocks. *Le Grotte d'Italia*, **4**(XII), 201–8.

Cigna, A.A. (1993) Environmental management of tourist caves: the examples of Grotta di Castellana and Grotta Grande del Vento. *Environmental Geology*, **21**, 173–80.

Cigna, A.A. and Burri, E. (2000) Development, Management and Economy of Show Caves. *International Journal of Speleology*, **29**B9(1/4), 1–27.

Cinq-Mars, J. and Lauriol, B. (1985) Le karst de Tsi-It-Toh-Choh: notes preliminaire sur quelques phenomenes karstiques du Yukon septentrional, Canada. *Comptes Rendus du Colloque International de Karstologie Applique, Universite de Liege*, 185–96.

Ciry, R. (1962) Le role du froid dans la speleogenese. *Spelunca Memoires*, **2**(4), 29–34.

Civita, M. and De Maio, M. (2000) *Valutazione e cartografia automatica della vulnerabilita degli acquiferi all'inquinamento con il sistema parametrico SINTACS R5*, Pitagora Editore, Bologna.

Clark, I.D. and Fritz, P. (1997) *Environmental Isotopes in Hydrogeology*. Lewis, New York, 328 pp.

Clark, I.D. and Lauriol, B. (1999) Aufeis of the Firth River Basin, Northern Yukon, Canada: Insights into Permafrost Hydrogeology and Karst. *Arctic and Alpine Research*, **29**(2), 240–52.

Clark, P.J. and Evans, F.C. (1954) Distance to nearest neighbour as a measure of spatial relationships in populations. *Ecology*, **35**, 445–53.

Clark, P.U. and Mix, A.C. (2002) Ice sheets and sea level of the Last Glacial Maximum. *Quaternary Science Reviews*, **21**, 1–7.

Clarke, F.W. (1924) The data of geochemistry. *US Geological Survey Bulletin*, **770**, 84 pp.

Clarke, J.D.A. (1994) Evolution of the Lefroy and Cowan palaeodrainage channels, Western Australia. *Australian Journal of Earth Sciences*, **41**, 55–68.

Clemens, T., Hückinghaus, M., Sauter, M., Liedl, R. and Teutsch, G. (1997) Simulation of the evolution of maze caves. *Bulletin d'Hydrogéologie*, **16**, 201–9.

Clemens, T., Huckinghaus, D., Liedl, R. and Sauter, M. (1999) Simulation of the development of karst aquifers: role of the epikarst. *International Journal of Earth Sciences*, **88**(1), 157–62.

Cockburn, H.A.P. and Summerfield, M.A. (2004) Geomorphological applications of cosmogenic isotope analysis. *Progress in Physical Geography*, **28**(1), 1–42.

Collins, M. E., Cum, M. and Hanninen, P. (1994). Using ground-penetrating radar to investigate a subsurface karst landscape in north-central Florida. *Geoderma*, **61**(1–2): 1–15.

Committee on Fracture Characterization and Fluid Flow (1996) *Rock Fractures and Fluid Flow: Contemporary Understanding and Applications*, National Academy Press, Washington, DC, 551 pp.

Condomines, M., Brouzes, C. and Rihs, S. (1999) Le radium et ses descendants dans quelques carbonates hydrothermaux d'Auvergne: origin et utilization pour la datation. *Comptes Rendus, Académie des Sciences (Paris)*, **328**, 23–8.

Conroy, G.C, Pickford, M., Senut, B., van Couvering, J. and Mein, P. (1992) *Otavipithecus namibiensis*, first Miocene hominoid from Southern Africa. *Nature*, **356**, 144–8.

Contos, A. and James, J.M. (2001) Nullarbor rafts and their paleo-environmental significance, *Proceedings of the 13th International Congress of Speleology, Brasilia*, Vol. 1, p. 53.

Contos, A., James, J., Holmes, A., *et al.* (2001) Calcite biomineralisation in the caves of Nullarbor Plains, Australia, *Proceedings of the 13th International Congress of Speleology, Brazilia*, Vol. 1, pp. 33–8.

Cooper, A.H. (1995) Subsidence hazards due to the dissolution of Permian gypsum in England: Investigation and remediation, in *Karst Geohazards* (ed. B.F. Beck), Balkema, Rotterdam, pp. 23–9.

Cooper, A.H. (1998) Subsidence hazards caused by the dissolution of Permian gypsum in England: geology, investigation and remediation, in *Geohazards in Engineering Geology* (eds J.G. Maund, and M. Eddleston), Engineering Geology Special Publications 15, Geological Society Publishing House, Bath, pp. 265–75.

Cooper, A.H. (2001) Natural and induced halite karst geohazards in Great Britain, in *Geotechnical and Environmental Applications of Karst Geology and Hydrology* (eds B.F. Beck and J.G. Herring), Balkema, Lisse, pp. 119–124.

Coplen, T.B. (1994) Reporting of stable hydrogen, carbon and oxygen abundances. *Pure and Applied Chemistry*, **66**, 2423–44.

Coplen, T.B., Winograd, I.J., Landwehr, J.M. and Riggs, A.C. (1994) 500,000 year stable carbon isotopic record from Devils Hole, Nevada. *Science*, **263**, 361–5.

Corbel, J. (1956) A new method for the study of limestone regions. *Revue Canadienne de Geographie*, **10**, 240–2.

Corbel, J. (1957) *Les Karsts du Nord-Ouest de l'Europe*, Institut Etudes Rhodaniennes Univ. Lyon 12 Mms. Docs, 531 pp.

Corbel, J. (1959) Erosion en terrain calcaire. *Annales de Geographie*, **68**, 97–120.

Correa Neto, A.V. (2000) Speleogenesis in quartzite in south-eastern Minas Gerais, Brazil, in *Speleogenesis; Evolution of Karst Aquifers* (eds A.V. Klimchouk, D.C. Ford, A.N. Palmer and W. Dreybrodt), National Speleological Society of America, Huntsville, AL, pp. 452–7.

COST Action 65 (1995a) *Hydrogeological Aspects of Groundwater Protection in Karstic Areas*, Final report (COST Action 65), Report EUR 16547, Directorate-General Science, Research and Development, European Commission, Office for Official Publications of the European Communities, Luxembourg, 446 pp.

COST Action 65 (1995b) *Hydrogeological Aspects of Groundwater Protection in Karstic Areas – Guidelines*, EUR 16526,

Directorate-General Science, Research and Development, European Commission, Office for Official Publications of the European Communities, Luxembourg, 15 pp.

COST Action 620 (2004) *Vulnerability and Risk Mapping for the Protection of Carbonate (Karst) Aquifers*, Final report (COST Action 620), Report EUR 20912. Directorate-General Science, Research and Development, European Commission, Office for Official Publications of the European Communities, Luxembourg, 297 pp.

Courbon, P., Chabert, C., Bosted, P. and Lindsley, K. (1989) *Atlas of the Great Caves of the World*, Cave Books, St Louis, 369 pp.

Courrèges, M. and Maire, R. (1996) Karsts et vignobles en Bordelais. *Rapports de recherches, Institut de Géographie, Université de Fribourg*, **8**, 9–22.

Coward, J.M.H. (1975) *Paleohydrology and streamflow simulation of three karst basins in southeastern West Virginia.* McMaster Univ. PhD thesis, Canada.

Cowell, D.W. and Ford, D.C. (1980. Hydrochemistry of a dolomite terrain; the Bruce Peninsula, Ontario. *Canadian Journal of Earth Sciences*, **17**(4), 520–6.

Coxon, C.E. (1987) The spatial distribution of turloughs. *Irish Geography*, **20**, 11–23.

Craig, D.H. (1987) Caves and other features of Permian karst in San Andres dolomites, Yates field reservoir, West Texas, in *Paleokarst* (eds N.P. James and P.W. Choquette), Springer-Verlag, New York, pp. 342–63.

Craig, H. (1961) Isotopic variations in meteoric waters. *Science*, **133**, 1702–3.

Cramer, H. (1941) Die Systematik der Karstdolinen. *Neues Jahrbuch für Mineralogie, Geologie und Palaontogie*, **85**, 293–382.

Crampon, N., Roux, J.C. and Bracq, P. (1993) France, in *The Hydrogeology of the Chalk of North-west Europe* (eds R.A. Downing, M. Price and G.P. Jones), Clarendon, Oxford, pp. 113–52.

Crawford, N.C. (1981) Karst flooding in urban areas: Bowling Green, Kentucky, *Proceedings of the 8th International Congress of Speleology, Bowling Green, Kentucky*, Vol. 2, pp. 763–5.

Crawford, N.C. (1984a) Sinkhole flooding associated with urban development upon karst terrain: Bowling Green, Kentucky, in *Sinkholes: their Geology, Engineering and Environmental Impact* (ed. B.F. Beck), Balkema, Rotterdam, pp. 283–92.

Crawford, N.C. (1984b) Toxic and explosive fumes rising from carbonate aquifers: a hazard for residents of sinkhole plains, in *Sinkholes: their Geology, Engineering and Environmental Impact* (ed. B.F. Beck), Balkema, Rotterdam, pp. 297–304.

Crawford, N.C. (2001) Field trip guide, Part 1: environmental problems associated with urban development upon karst, Bowling Green, Kentucky, in *Geotechnical and Environmental Applications of Karst Geology and Hydrology* (eds B.F. Beck and J.G. Herring), Balkema, Lisse, pp. 397–424.

Crawford, N.C., Lewis, M.A., Winter, S.A. and Webster, J.A. (1999) Microgravity techniques for subsurface investigations of sinkhole collapses and for detection of groundwater flow paths through karst aquifers, in *Hydrology and Engineering*

Geology of Sinkholes and Karst (eds B.F. Beck, A.J. Pettit and J.G. Herring), Balkema, Rotterdam, pp. 203–18.

Crawford, S.J. (1994) *Hydrology and geomorphology of the Paparoa Karst, north Westland, New Zealand*. Auckland Univ. Unpublished PhD thesis, New Zealand, 240 pp.

Crowther, J. (1983) A comparison of the rock tablet and water hardness methods for determining chemical erosion rates on karst surfaces. *Zeitschrift für Geomorphologie NF.*, **27**(1), 55–64.

Crowther, J. (1989) Groundwater chemistry and cation budgets of tropical karst outcrops, peninsular Malaysia, 1 calcium and magnesium. *Journal of Hydrology*, **107**, 169–92.

Crowther, J. (1998) New methodologies for investigating rillen-karren cross-sections: a case study at Lluc, Mallorca. *Earth Surface Processes and Landforms*, **23**, 333–44.

Cruden, D.M. (1985) Rock slope movements in the Canadian Cordillera. *Canadian Geotechnical Journal*, **22**, 528–40.

Cruden, D.M. (1989) Limits to common toppling. *Canadian Geotechnical Journal*, **26**, 737–42.

Cruz, F.W., Burns, S.J., Karmann, I., *et al.* (2005a) Insolation-driven changes in atmospheric circulation over the past 116,000 years in subtropical Brazil. *Nature*, **434**, 63–66.

Cruz, F.W., Karmann, I., Viana, O., Burns, S.J., Ferrari, J.A., Vuille, M., Sial, A.N. and Moreira, M.Z. (2005b) Stable isotope study of cave percolation waters in subtropical Brazil: Implications for palaeoclimate inferences from speleothems. *Chemical Geology*, **220**, 245–62.

Cser, F. and Maucha, L. (1966) Contribution to the origin of 'excentric' concretions. *Karszt- es Barlangkutatas*, **6**, 83–100.

Cucchi, F. and Forti, F. (1994) Degradation by dissolution of carbonate rocks. *Acta Carsologica*, **XXIII**, 55–62.

Cucchi, F. and Gerdol, S. (1985) *I marmi del Carso triestino*, Camera di Commercio, Trieste, 195 pp.

Curl, R.L. (1966) Scallops and flutes. *Transactions of the Cave Research Group of Great Britain*, **7**(2), 121–60.

Curl, R.L. (1973) Minimum diameter stalagmites. *Bulletin of the National Speleological Society*, **35**(1), 1–9.

Curl, R.L. (1974) Deducing flow velocity in cave conduits from scallops. *Bulletin of the National Speleological Society*, **36**(2), 1–5.

Currens, J.C. (2002) Changes in groundwater quality in a conduit-flow-dominated karst aquifer, following BMP implementation. *Environmental Geology*, **42**, 525–31.

Cvijic, J. (1893) *Das Karstphanomen*. Versuch einer morpholo-gischen Monographie, Geographische Abhandlungen heraus-gegeben von A Pench, Bd., V.H, 3. Wien, pp. 218–329. [The section on dolines, pp. 225–76, is translated into English in Sweeting (1981).]

Cvijic, J. (1901) Morphologische und glaciale Studien aus Bosnien, der Hercegovina und Montenegro: die Karst-Poljen. *Abhandlungen der Geographie Gesellschaft Wien*, **3**(2), 1–85.

Cvijic, J. (1918) Hydrographie souterraine et evolution morpho-logique du karst. *Hydrographie souterraine et evolution morphologique du karst*, **6**(4), 375–426.

Cvijic, J. (1925) Types morphologiques des terrains calcaires. *Comptes Rendus, Académie des Sciences (Paris)*, **180**, 592, 757, 1038.

D'Argenio, B. and Mindszenty, A. (1992) Tectonic and climatic control on paleokarst and bauxites. *Giornale di Geologia*, **54**(1), 207–18.

Daly, D. Dassargues, A., Drew, D., *et al.* (2002) Main concepts of the 'European approach' to karst-groundwater-vulnerability assessment and mapping. *Hydrogeology Journal*, **10**, 340–5.

Damblon, F. (1974) Observations palynologiques dans la Grotte de Remouchamps. *Bulletin de la Societe Royale Belge de Anthropologie et Prehistorie*, **85**, 131–55.

Danes, J.V. (1908) Geomorphologische Studien in Karstgebiete Jamaikas, *Proceedings of the 9th International Geogological Congress*, Vol. 2, pp. 178–82.

Danes, J.V. (1910) Die Karstphanomene im Goenoeng Sewoe auf Java. *Tijdschrift der Koninklijke Nederlandsche Aardrijkskundig Genootschap*, **27**, 247–60.

Danin, A. (1983) Weathering of limestone in Jerusalem by cyanobacteria. *Zeitschrift für Geomorphologie*, **27**(4), 413–21.

Danin, A. (1993) Biogenic weathering of marble monuments in Didim, Turkey in Trajan Column, Rome. *Water Science Technology*, **27**, 557–63.

Danin, A. and Garty, J. (1983) Distribution of cyanobacteria and lichens on hillsides in the Neger Highlands and their impact on biogenic weathering. *Zeitschrift für Geomorphologie*, **27**(4), 423–44.

Darabos, G. (2003) Observation of microbial weathering resulting in peculiar 'exfoliation-like' features in limestone from Hirao-dai karst, Japan. *Zeitschrift für Geomorphologie, Supplement-Band*, **131**, 33–42.

Darcy, H. (1856) *Les fontaines publiques de le ville de Dijon*, Dalmont, Paris.

Dassargues, A. (ed.) (2000) *Tracers and Modelling in Hydrology*, Publication 262, International Association of Hydrological Sciences, Wallingford, 571 pp.

Dassargues, A. and Derouane, J. (1997) A modelling approach as an intermediate step for the study of protection zones in karstified limestones, in *Karst Hydrology* (eds C. Leibundgut, J. Gunn and A. Dassargues), Publication 247, International Association of Hydrological Sciences, Wallingford, pp. 71–80.

Davies, W.E. and LeGrand, H. (1972) Karst of the United States, in *Karst: Important Karst Regions of the Northern Hemisphere* (eds M. Herak and V.T. Stringfield), Elsevier, Amsterdam, pp. 467–505.

Davis, A.D. and Long, A.J. (2002) KARSTIC: a sensitivity method for carbonate aquifers in karst terrains. *Environmental Geology*, **42**, 65–72.

Davis, S.N. (1966) Initiation of groundwater flow in jointed limestone. *Bulletin of the National Speleological Society*, **28**, 111.

Davis, W.M. (1899) The geographical cycle. *Geographical Journal*, **14**, 481–504.

Davis, W.M. (1930) Origin of limestone caves. *Geological Society of America Bulletin*, **41**, 475–628.

Davison, I., Bosence, D., Alsop, G.I. and Al-Aawah, M.H. (1996) Deformation and sedimentation around active Miocene salt diapers on the Tihama Plain, northwest Yemen, in *Salt Tectonics* (eds G.I. Alsop, D.J. Blundell and I. Davison), Special Publication 100, Geological Society Publishing House, Bath, pp. 23–39.

Day, M. (1977) Surface roughness in tropical karst terrain, *Proceedings of the 7th International Congress of Speleology, Sheffield*, pp. 139–43.

Day, M.J. (1978) Morphology and distribution of residual limestone hills (mogotes) in the karst of northern Puerto Rico. *Geological Society of America Bulletin*, **89**, 426–32.

Day, M.J. (1979) Surface roughness as a discriminator of tropical karst styles. *Zeitschrift für Geomorphologie, Supplement-band*, **32**, 1–8.

Day, M.J. (1983) Doline morphology and development in Barbados. *Annals of the Association of American Geographers*, **73**(2), 206–19.

Day, M.J. (1984) Carbonate erosion rates in southwestern Wisconsin. *Physical Geography*, **5**(2), 142–9.

Day, M.J. (2001) Sandstone caves in Wisconsin. Brasilia, *Proceedings of the 13th International Congress of Speleology*, Vol. 1, pp. 88–92.

De Bellard-Pietri, E. (1981) Stalactite growth in the tropics under artificial conditions, *Proceedings of the 8th International Congress of Speleology*, pp. 221–2.

De Freitas, C.R. and Schmekal, A. (2003) Condensation as a microclimate process: measurement, numerical simulation and prediction in the Glowworm Cave, New Zealand. *International Journal of Climatology*, **23**, 557–75.

De Waele, J. and Follesa, R. (2003) Human impact on karst: the example of Lusaka (Zambia). *International Journal of Speleology*, **32**(1/4), 71–83.

Debenham, N.C. and Aitken, M.J. (1984) Thermoluminescence dating of stalagmitic calcite. *Archaeometry*, **26**(2), 155–70.

Deike, G.H. and White, W.B. (1969) Sinuosity in limestone solution conduits. *American Journal of Science*, **267**, 230–41.

Delannoy, J-J. and Caillault, S. (1998) Les apports de l'endokarst dans la reconstitution morphogénique d'un karst: exemple de l'Antre de Vénus. *Karstologia*, **31**, 27–41.

Delannoy, J.-J., Guendon, J.-L., Quinif, Y and Roiron, P. (1997) Les formations travertineuses: Des temoins paleoenvironmentaux et morphogeniques. Exemple du Piemont Mediterraneen de La Serrania de Ronda (Province de Malaga, Espagne). *Bulletin de la Societe belge de Geologie*, **106**, 79–96.

DeMille, G., Shouldice, J.R. and Nelson, H.W. (1964) Collapse structures related to evaporites of the Prairie Formation, Saskatchewan. *Geological Society of America Bulletin*, **75**, 307–16.

Denizman, C. (2003) Morphometric and spatial distribution parameters of karstic depressions, lower Suwannee River basin, Florida. *Journal of Cave and Karst Studies*, **65**(1), 29–35.

Dennis, P.R., Rowe, P.J. and Atkinson, T.C. (2001) The recovery and isotopic measurement of water from fluid inclusions in speleothems. *Geochimica et Cosmochimica Acta*, **65**, 871–84.

Denniston, R.F., Gonzalez, L.A., Baker, R., *et al.* (1999) Speleothem evidence for Holocene fluctuations of the prairie-forest ecotone, north-central USA. *The Holocene*, **9**(6), 671–6.

Denniston, R.F., Gonzalez, L.A., Asmerom, Y., *et al.* (2000) Speleothem carbon isotopic records of Holocene environments in the Ozark Highlands, USA. *Quaternary International*, **67**, 21–7.

Dewandel, B., Lachassagne, P., Bakalowicz, M., *et al.* (2003) Evaluation of aquifer thickness by analysing recession hydrographs. Application to the Oman ophiolite hard-rock aquifer. *Journal of Hydrology*, **274**, 248–69.

Dilsiz, C. and Günay, G. (2004) Pamukkale, Turkey, in *Encyclopedia of Caves and Karst Science* (ed. J. Gunn), Fitzroy Dearborn, New York, pp. 568–9.

Dincer, T., Payne, B.R., Yen, C.K. and Zotl, J. (1972) Das Tote Gebirge als – Entwasserungstypus der Karstmassive der nordostlichen Kalkhochalpen (Ergebnisse von Isotopenmessungen). *Steirische Beitrage zur Hydrogeologie*, **24**, 71–109.

Doerfliger, N., Jeannin, P.-Y., and Zwahlen, F. (1999) Water vulnerability assessment in karst environments: a new method of defining protection areas using a multi-attribute approach and GIS tools (EPIK method). *Environmental Geology*, **39**(2), 165–76.

Dogan, U. and Yesilyurt, S. (2004) Gypsum karst south of Imranli, Sivas, Turkey. *Cave and Karst Science*, **31**(1), 7–14.

Dogwiler, T. and Wicks, C.M. (2004) Sediment entrainment and transport in fluviokarst systems. *Journal of Hydrology*, **295**(1–4), 163–72.

Domenico, P.A. and Schwartz, F.W. (1998) *Physical and Chemical Hydrogeology*, 2nd edn, John Wiley & Sons, New York, 506 pp.

Dorale, J.A., Gonzalez, L.A., Reagan, M.K., *et al.* (1992) A high resolution record of Holocene climate change in speleothem calcite from Cold Water Cave, northeast Iowa. *Science*, **258**, 1626–30.

Dorale, J.A., Edwards, R.L., Ito, E. and Gonzalez, L.A. (1998) Climate and vegetation history of the mid-continent from 75 to 25 ka: a speleothem record from Crevice Cave, Missouri, USA. *Science*, **282**, 1871–4.

Dorale, J.A., Edwards, R.L., Alexander, E.C., *et al.* (2004) Uranium-series dating of speleothems: current techniques, limits applications, in *Studies of Cave Sediments: Physical and Chemical Records of Paleoclimate* (eds I.D. Sasowsky and J. Mylroie), Kluwer Academic, New York, pp. 177–97.

Dougherty, P. (2005) Case Study #7. Sinkhole destruction of Corporate Plaza, Pennsylvania, in *Sinkholes and Subsidence: Karst and Cavernous Rocks in Engineering and Construction* (eds A.C. Waltham, F. Bell, and M. Culshaw), Praxis Publishing, Chichester, pp. 304–8.

Douglas, I. (1968) Some hydrologic factors in the denudation of limestone terrains. *Zeitschrift für Geomorphologie*, **12**(3), 241–55.

Dove, P.M. and Rimstidt, J.D. (1994) Silica–water interactions, in *Silica* (eds P.J. Heaney, C.T. Prewitt and G.V. Gibbs). *Reviews in Mineralogy*, **29**, 259–308.

Drake, J.J. (1984) Theory and model for global carbonate solution by groundwater, in *Groundwater as a Geomorphic Agent* (ed. R. G. LaFleur), Allen & Unwin, London, pp. 210–26.

Drake, J.J. and D.C. Ford (1972) The analysis of growth patterns of two generations; the example of karst sinkholes. *Canadian Geographer*, **16**, 381–4.

Drake, J.J. and D.C. Ford (1973) The dissolved solids regime and hydrology of two mountain rivers, *Proceedings of the 6th International Congress of Speleology, Olomouc, CSSR*, Vol. 4, pp. 53–6.

Drake, J.J. and Harmon, R.S. (1973) Hydrochemical environments of carbonate terrains. *Water Resources Research*, **9**(4), 949–957.

Drake, J.J. and Wigley, T.M.L. (1975) The effect of climate on the chemistry of carbonate groundwater. *Water Resources Research*, **11**, 958–62.

Dreiss, S.J. (1982) Linear Kernels for karst aquifers. *Water Resources Research*, **18**(4), 865–76.

Dreiss, S.J. (1989) Regional scale transport in a karst aquifer, 1. Component separation of spring flow hydrographs. *Water Resources Research*, **25**, 117–25.

Drew, D.P. (1983) Accelerated soil erosion in a karst area: the Burren, western Ireland. *Journal of Hydrology*, **61**, 113–24.

Drew, D.P. and Daly, D. (1993) *Groundwater and Karstification in mid-Galway, South Mayo and North Clare*, Report Series 93/3 (Groundwater), Geological Survey of Ireland, 86 pp.

Drew, D. and Hötzl, H. (eds). (1999) *Karst Hydrogeology and Human Activities: Impacts, Consequences and Implications*, Rotterdam, Balkema, 322 pp.

Drew, D.P. and Jones, G.L. (2000) Post-Carboniferous pre-Quaternary karstification in Ireland, *Proceedings of the Geologists' Association*, **111**, 345–53.

Drew, D.P. and Magee, E. (1994) Environmental implications of land reclamation in the Burren, Co. Clare: a preliminary analysis. *Irish Geography*, **27**(2), 81–96.

Drew, D.P. and Smith, D.I. (1969) Techniques for the tracing of subterranean water. *British Geomorphological Research Group, Technical Bulletin*, **2**, 36 pp.

Drew, D.P., Doerfliger, N. and Formentin, K. (1997) The use of bacteriophages for multi-tracing in a lowland karst aquifer in western Ireland, in *Tracer Hydrology 97* (ed. A. Kranjc), Balkema, Rotterdam, pp. 33–7.

Drewry, D. (1986) *Glacial Geologic Processes*, Edward Arnold, London.

Dreybrodt, W. (1988) *Processes in Karst Systems*, Physics, Chemistry and Geology Series, Springer-Verlag, Berlin, 288 pp.

Dreybrodt, W. (1990) The role of dissolution kinetics in the development of karst aquifers in limestone: a model simulation of karst evolution. *Journal of Geology*, **98**, 639–55.

Dreybrodt, W. (1996a) Principles of early development of karst conduits under natural and man-made conditions revealed by mathematical analysis of numerical models. *Water Resources Research*, **32**, 2923–35.

Dreybrodt, W. (1996b) Chemical kinetics, speleothem growth and climate, in *Climate Change: the Karst Record* (ed. S.-E. Lauritzen), Special Publication 2, Karst Waters Institute, Charles Town, WV, pp. 23–5.

Dreybrodt, W. (2003) Viewpoints and comments on feasibility of condensation processes in hypogenic caves. *www. Speleogenesis*, **1**(2), 2 pp.

Dreybrodt, W. (2004) Speleogenesis: computer models, in *Encyclopedia of Caves and Karst Science* (ed. J. Gunn), Fitzroy Dearborn, New York, pp. 677–81.

Dreybrodt, W. and Buhmann, D. (1991) A mass transfer model for dissolution and precipitation of calcite from solutions in turbulent motion. *Chemical Geology*, **90**, 107–22.

Dreybrodt, W. and Gabrovšek, F. (2000a) Influence of fracture roughness on karstification times, in *Speleogenesis; Evolution of Karst Aquifers* (eds A.V. Klimchouk, D.C. Ford, A.N. Palmer and W. Dreybrodt), National Speleological Society of America, Huntsville, AL, pp. 220–3.

Dreybrodt, W. and Gabrovsek, F. (2000b) Dynamics of the evolution of single karst conduits, in *Speleogenesis; Evolution of Karst Aquifers* (eds A.V. Klimchouk, D.C. Ford, A.N. Palmer and W. Dreybrodt), National Speleological Society of America, Huntsville, AL, pp. 184–93.

Dreybrodt, W. and Gabrovšek, F. (2002) Basic processes and mechanisms governing the evolution of karst, in *Evolution of Karst: from Prekarst to Cessation* (ed. F. Gabrovsek), Institut za raziskovanje krasa, ZRC SAZU, Postojna-Ljubljana, pp. 115–54.

Dreybrodt, W., Gabrovšek, F. and Siemers, J. (1999) Dynamics of the early evolution of karst, in *Karst Modelling* (eds A.N. Palmer, M.V. Palmer and I.D. Sasowsky), Special Publication 5, Karst Waters Institute, Charles Town, WV, pp. 106–19.

Dreybrodt, W., Romanov, D. and Gabrovsek, F. (2002) Karstification below dam sites: a model of increasing leakage from reservoirs. *Environmental Geology*, **42**, 518–24.

Dreybrodt, W., Gabrovšek, F. and Perne, M. (2005a) Condensation corrosion: a theoretical approach. *Acta Carsologica*, **34**(2), 317–48.

Dreybrodt, W., Gabrovšek, F. and Romanov, D. (2005b) *Processes of Speleogenesis: a Modeling Approach*, ZRC Publishing, Lubljana.

Drogue, C. (1980) Essai d'identification d'un type de structure de magasins carbonatés fissures. Application à l'interprétation de certains aspects du fonctionnement hydrogéologique. *Memoire hors série Société géologique de la France*, **11**, 101–8.

Drogue, C. (1989) Continuous inflow of seawater and outflow of brackish water in the substratum of the karstic island of Cephalonia, Greece. *Journal of Hydrology*, **106**, 147–53.

Drogue, C. (1993) Absorption massive d'eau de mer par des aquiferes karstiques côtiers, in *Hydrogeological Processes in Karst Terranes* (eds G. Günay, A.I. Johnson and W. Back), Publication 207, International Association of Hydrological Sciences, Wallingford, pp. 119–28.

Drogue, C., Laty, A.-M. and Paloc, H. (1983) Les eaux souterraines des karsts mediterraneens. Exemple de la region pyreneo-provencale (France meridionale). *Bulletin Bureau de Recherches Geologique et Minieres, Hydrogeologie-geologie de l'ingenieur*, **4**, 293–311.

Droppa, A. (1966) The correlation of some horizontal caves with river terraces. *Studies in Speleology*, **1**(4), 186–92.

Drost, W. and Klotz, D. (1983) Aquifer characteristics, in *Guidebook on Nuclear Techniques in Hydrology*, Technical Reports Series No. 91, International Atomic Energy Agency, Vienna, pp. 223–56.

Drysdale, R.N., Taylor, M.P. and Ihlenfeld, C. (2002) Factors controlling the chemical evolution of travertine-depositing rivers of the Barkly karst, northern Australia. *Hydrological Processes*, **16**, 2941–62.

Du Chene, H. and Hill, C.A. (eds) (2000) The caves of the Guadalupe Mountains. *Journal of Cave and Karst Studies*, **62**(2), 52–158.

Dubljansky, V.N. (2000a) A giant hydrothermal cavity in the Rhodope Mountains, in *Speleogenesis; Evolution of Karst Aquifers* (eds A.V. Klimchouk, D.C. Ford, A.N. Palmer and W. Dreybrodt), National Speleological Society of America, Huntsville, AL, pp. 317–8.

Dubljansky, V.N. (2000b) *Fascinating Speleology*, Ural Ltd, Perm, 527 pp. [In Russian.]

Dubljansky, V.N. and Dubljansky, Y.V. (2000) The role of condensation in karst hydrogeology and speleogenesis, in *Speleogenesis; Evolution of Karst Aquifers* (eds A.V. Klimchouk, D.C. Ford, A.N. Palmer and W. Dreybrodt), National Speleological Society of America, Huntsville, AL, pp. 100–112.

Dubljansky, Y.V. and Dubljansky, V.N. (1997) Hydrothermal cave minerals, in *Cave Minerals of the World*, 2nd edn (eds C.A. Hill and P. Forti), National Speleological Society of America, Huntsville, AL, pp. 252–5.

Dublyansky,V.N. and Kiknadze,T.Z. (1983) *Hydrogeology of Karst of the Alpine Folded Region of the South of the U.S.S.R.*, Nauka, Moscow, 125 pp. [in Russian.]

Dulinski, M. and Rosanski, K. (1990) Formation of $^{13}C/^{12}C$ isotope ratios in speleothems: a semi-dynamic model. *Radiocarbon*, **32**, 7–16.

Dunham, R.J. (1962) Classification of carbonate rocks. *Memoirs, American Association of Petroleum Geologists*, **1**, 108–21.

Dunkerley, D.L. (1983) Lithology and microtopography in the Chillagoe karst, Queensland, Australia. *Zeitschrift für Geomorphologie*, **27**(2), 191–204.

Dunne, J.R. (1957) Stream tracing: mid-Appalachian region. *Bulletin of the National Speleological Society*, **2**, 7 pp.

Dunne, T.R. and Leopold, L.B. (1978) *Water in Environmental Planning*, Freeman, San Francisco.

Durán Valsero, J.J. and López Martinez, J. (1998) *Karst en Andalucia*, Instituto Tecnologico GeoMinero de España, Madrid, 192 pp.

Durozoy, G. and Paloc, H. (1973) *Le regime des eaux de la fontaine de Vaucluse*, Bureau de recherches gol. minières, Min. du Dev. Industriel et Scientifique, 31 pp.

Dykoski, C.A., Edwards, R.L. Cheng H., *et al.* (2005) A high-resolution, absolute-dated Holocene and deglacial Asian monsoon record from Dongge Cave, China. *Earth and Planetary Science Letters*, **233**, 71–86.

Dżulynski, S. and Sass-Gutkiewicz, M. (1989) Pb-Zn ores, in *Paleokarst – a Systematic and Regional Review* (eds P. Bosák, D.C. Ford, J. Głazek and I. Horáček), Academia Praha/Elsevier, Prague/Amsterdam, pp. 377–97.

Eagles, P.F.J., McCool, S.F. and Haynes, C.D. (2002) *Sustainable Tourism in Protected Areas: Guidelines for Planning and Management*, International Union for the Conservation of Nature and Natural Resources (IUCN), Gland and Cambridge, xv + 183 pp.

Eckert, M. (1895) Das Karrenproblem. Die Geschichte seiner Loesung. *Zeitschrift für Naturwissenschaften* (Leipzig), **68**, 321–432.

Edgell, H.S. (1991) Proterozoic salt basins of the Persian Gulf area and their role in hydrocarbon generation. *Precambrian Research*, **54**, 1–14.

Edgell, H.S. (1993) Karst and Water Resources in the Hyper-Arid Areas of Northeastern Saudi Arabia, *Proceedings, International Symposium on Water Resources in Karst, with Special Emphasis on Arid and Semi-Arid Zones, Shiraz*, pp. 309–26.

Edmond, J.M. and Huh, Y. (2003) Non-steady-state carbonate recycling and implications for the evolution of atmospheric PCO_2. *Earth and Planetary Science Letters*, **216**, 125–39.

Edwards, R.L., Chen, J.H. and Wasserburg, G.J. (1986/7) ^{238}U–^{234}U–^{230}Th–^{232}Th systematics and the precise measurement of time over the past 500,000 years. *Earth and Planetary Science Letters*, **81**, 175–92.

Edwards, R.L., Cheng, H., Murrell, M.T. and Goldstein, S.J. (1997) Protactinium-231 dating of carbonates by thermal ionization mass spectrometry: Implications for Quaternary climate change. *Science*, **276**, 782–6.

Egemeier, S.J. (1981) Cavern development by thermal waters. *Bulletin of the National Speleological Society*, **43**, 31–51.

Eggins, S.M., Grün, R., McCulloch, M.T., *et al.* (2005) *In situ* U-series dating by laser-ablation multi-collector ICPMS: new prospects for Quaternary geochronology. *Quaternary Science Reviews*, **24**, 2523–38.

Ehrlich, H.L. (1981) *Geomicrobiology*, Marcel Dekker, New York.

Eisenlohr, L., Madry, B. and Dreybrodt, W. (1997a) Changes in the dissolution kinetics of limestone by intrinsic inhibitors adsorbing to the surface, *Proceedings of the 12th International Congress of Speleology*, Vol. 2, pp. 81–4.

Eisenlohr, L., Bouzelboudjen, M., Kiraly, L. and Rossier, Y. (1997b) Numerical versus statistical modelling of natural response of a karst hydrogeological system. *Journal of Hydrology*, **202**(1–4), 244–62.

Eisenlohr, L., Kiraly, L., Bouzelboudjen, M. and Rossier, Y. (1997c) Numerical simulation as a tool for checking the interpretation of karst spring hydrographs. *Journal of Hydrology*, **193**(1–4), 306–15.

Ek, C. (1961) Conduits souterrains en relation avec les terrasses fluviales. *Annales de la Societe Géologique de Belgique*, **84**, 313–40.

Ek, C. (1973) *Analyses d'Eaux des Calcaires Paleozoiques de la Belgique*, Professional Paper 13, Service Géologique de Belgique.

Ek, C. and Gewelt, M. (1985) Carbon dioxide in cave atmospheres. New results in Belgium and comparison with some other countries. *Earth Surface Processes and Landforms*, **10**, 173–87.

Ekmekçi, M. and Günay, G. (1997) Role of public awareness in groundwater protection. *Environmental Geology*, **30**(1/2), 81–7.

El Aref, M.M., Abou Khadrah, A.M. and Lotfy, Z.H. (1987) Karst topography and karstification processes in the Eocene limestone plateau of El Bahariya Oasis, Western Desert, Egypt. *Zeitschrift für Geomorphologie*, **31**, 45–64.

Elkhatib, H. and Günay, G. (1993) Analysis of sea water intrusion associated with karstic channels beneath Ovacik Plain, southern Turkey, in *Hydrogeological Processes in Karst Terranes* (eds G. Günay, A.I. Johnson and W. Back), Publication 207, International Association of Hydrological Sciences, Wallingford, pp. 129–32.

Ellaway, M., Smith, D.I., Gillieson, D.S. and Greenaway, M.A. (1990) Karst water chemistry – Limestones Ranges, Western Australia. *Helictite*, **28**(2), 25–36.

Embry, A.F. and Klovan, J.E. (1971) A Late Devonian reef tract in northeastern Banks Island, Northwest Territories. *Canadian Petroleum Geologists Bulletin*, **19**, 730–81.

Emmett, A.J. and Telfer, A.L. (1994) Influence of karst hydrology on water quality management in southeast South Australia. *Environmental Geology*, **23**, 149–55.

Enyedy-Goldner, S.R. (1994) *The karst geomorphology of Manitoulin Island, Ontario*. McMaster Univ. MSc thesis, 312 pp.

Ettazarini, S. and El Mahmouhi, N. (2004) Vulnerability mapping of the Turonian limestone aquifer in the Phosphates Plateau (Morocco). *Environmental Geology*, **46**, 113–17.

Eugster, H.P. and Hardie, L.A. (1978) Saline Lakes. in *Lakes – Chemistry, Geology and Physics* (ed. A. Lerman), Springer-Verlag, New York, 237–93.

Even, H., Carmi, I., Magaritz, M. and Gerson, R. (1986) Timing the transport of water through the upper vadose zone in a karstic system above a cave in Isreal. *Earth Surface Processes and Landforms*, **11**, 181–91.

Ewers, R.O (1973) A model for the development of subsurface drainage routes along bedding planes, *Proceedings of the 6th International Congress of Speleology, Olomouc*, Vol. III, pp. 79–82.

Ewers, R.O (1978) A model for the development of broadscale networks of groundwater flow insteeply dipping carbonate aquifers. *Transactions of the British Cave Research Association*, **5**, 121–5.

Ewers, R.O. (1982) *Cavern development in the dimensions of length and breadth*. McMaster Univ. PhD thesis, 398 pp.

Ewing, A. (1885) Attempt to determine the amount and rate of chemical erosion taking place in the limestone valley of Center Co., Pennsylvania. *American Journal of Science*, **3**(29), 29–31.

Eyles, N. (ed.) (1983) *Glacial Geology*, Pergamon Press, Oxford, 409 pp.

Fabel, D., Henricksen, D. and Finlayson, B.L. (1996) Nickpoint recession in karst terrains: an example from the Buchan karst, southeastern Australia. *Earth Surface Processes and Landforms*, **21**, 453–66.

Fang, L. (1984) Application of distances between nearest neighbours to the study of karst. *Carsologica Sinica* **3**(1), pp. 97–101.

Fairbanks, R.G. and Matthews, R.K. (1978) The marine oxygen isotope record in Pleistocene coral, Barbados, West Indies. *Quaternary Research*, **10**, 181–96.

Fairchild, I.J., Borsato, A., Tooth, A.F., *et al.* (2000) Controls on trace element (Sr–Mg) compositions of carbonate cave waters:

implications for speleothem climate records. *Chemical Geology*, **166**, 255–69.

Fairchild, I.J., Smith, C.L., Baker, A., *et al.* (2006) Modification and preservation of environmental signals in speleothems. *Earth-Science Reviews*, **75**, 105–53.

Farnsworth, R.K., Barrett, E.C. and Dhanju, M.S. (1984) *Application of Remote Sensing to Hydrology including Ground Water*. Technical Documents in Hydrology, UNESCO, Paris, 122 pp.

Farrant, A.R. and Ford, D.C. (2004) Mendip Hills, England, in *Encyclopedia of Caves and Karst Science* (ed. J. Gunn), Fitzroy Dearborn, New York, pp. 503–5.

Farrant, A.R., Smart, P.L., Whitaker, F.F. and Tarling, D.H. (1995) Long-term Quaternary uplift rates inferred from limestone caves in Sarawak, Malaysia. *Geology*, **23**(4), 357–60.

Faulkner, G.L. (1976) Flow analysis of karst systems with well developed underground circulation, *Proceedings of the Yugoslavian Symposium on Karst Hydrology, June 1975*, Vol. 1, p. 137–64.

Fay, R.O. and Hart, D.L. (1978) Geology and mineral resources (exclusive of petroleum) of Custer County, Oklahoma. *Oklahoma Geological Survey Bulletin*, **114**, 88 pp.

Felton, G.K. (1996) Agricultural chemicals at the outlet of a shallow carbonate aquifer. *Transactions of the American Society of Agricultural Engineers*, **39**(3), 873–82.

Feng, X. and Epstein, S. (1995: Carbon isotopes of trees from arid environments and implications for reconstructing atmospheric CO_2 concentration. *Geochimica et Cosmochimica Acta*, **59**, 2599–608.

Ferrarese, F., Sauro, U. and Tonello, C. (1997) The Montello Plateau: evolution of an alpine neotectonic morphostructure. *Zeitschrift für Geomorphologie Supplement-band*, **109**, 41–62.

Field, M.S. (ed.) (1999) *A Lexicon of Cave and Karst Terminology with Special Reference to Environmental Karst Hydrology*. Report EPA/600/R–99/006, US Environmental Protection Agency, Washington, DC, 195 pp.

Field, M.S. (2004) Forecasting versus predicting solute transport in solution conduits for estimating drinking-water risks. *Acta Carsologica*, **33**(2) 115–49.

Field, M.S., Wilhelm, R.G., Quinlan, J.F. and Aley, T.J. (1995) An assessment of the potential adverse properties of fluorescent tracer dyes used for groundwater tracing. *Environmental Monitoring and Assessment*, **38**, 75–96.

Filippov, A.G. (2000) Speleogenesis of Botovskaya Cave, eastern Siberia, Russia, in *Speleogenesis; Evolution of Karst Aquifers* (eds A.V. Klimchouk, D.C. Ford, A. N. Palmer and W. Dreybrodt), National Speleological Society of America, Huntsville, AL, pp. 282–6.

Filippov, A.G. (2004a) Siberia, Russia, in *Encyclopedia of Caves and Karst Science* (ed. J. Gunn), Fitzroy Dearborn, New York, pp. 645–7.

Filippov, A.G. (2004b) Mineral deposits in karst, in *Encyclopedia of Caves and Karst Science* (ed. J. Gunn), Fitzroy Dearborn, New York, pp. 514–15.

Finlayson, B. and Hamilton-Smith, E. (eds) (2003) *Beneath the Surface: a Natural History of Australian Caves*, UNSW Press, Sydney, 182 pp.

Fiol, L., Fornos, J.J. and Gines, A. (1996) Effects of biokarstic processes on the development of solutional rillenkarren in limestone rocks. *Earth Surface Processes and Landforms*, **21**, 447–52.

Fish, J.E. (1977) *Karst hydrogeology and geomorphology of the Sierra de El Abra and the Valles-San Luis Potosi Region, Mexico*. McMaster Univ. PhD thesis, 469 pp.

Fleitmann, D., Burns, S.J., Neff, U., *et al.* (2003) Changing moisture sources over the last 330,000 years in Northern Oman from fluid-inclusion evidence in speleothems. *Quaternary Research*, **60**, 223–232.

Fleitmann, D., Burns, S.J., Neff, U., *et al.* (2004) Palaeoclimate interpretation of high-resolution oxygen isotope profiles derived from annually laminated speleothems from Southern Oman. *Quaternary Science Reviews*, **23**(7–8), 935–45.

Florea, L.J., Paylor, R.L., Simpson, L. and Gulley, J. (2002) Karst GIS advances in Kentucky. *Journal of Cave and Karst Studies*, **64**(1), 58–62.

Folk, R.L. (1962) Spectral subdivision of limestone types. *Memoirs, American Association of Petroleum Geologists*, **1**, 62–84.

Folk, R.L. and Assereto, R. (1976) Comparative fabrics of length-slow and length-fast calcite and calcitized aragonite in a Holocene speleothem, Carlsbad Caverns, New Mexico. *Journal of Sedimentary Petrology*, **46**(3), 486–96.

Folk, R.L., Roberts, H.H. and Moore, C.M. (1973) Black phytokarst from Hell, Cayman Islands, West Indies. *Geological Society of America Bulletin*, **84**, 2351–60.

Fontes, J. (1980) Environmental isotopes in groundwater hydrology, in *Handbook of Environmental Isotope Geochemistry* (eds P. Fritz and J. Fontes), Elsevier, Amsterdam, Vol. 1, pp. 75–140.

Fontes, J.-Ch. (1983) Dating of groundwater, in *Guidebook on Nuclear Techniques in Hydrology*, Technical Reports Series No. 91, International Atomic Energy Agency, Vienna, pp. 285–317.

Ford, D.C. (1965a) The Origin of Limestone Caves: a model from the central Mendip Hills, England. *Bulletin of the National Speleogical Society*, **27**(4), 107–32.

Ford, D.C. (1965b) Stream potholes as indicators of erosion phases in caves. *Bulletin of the National Speleogical Society*, **27**(1), 27–32.

Ford, D.C. (1968) Features of cavern development in central Mendip. *Transactions of the Cave Research Group of Great Britain*, **10**, 11–25.

Ford, D.C. (1971a) Characteristics of limestone solution in the southern Rocky Mountains and Selkirk Mountains, Alberta and British Columbia. *Canadian Journal of Earth Science*, **8**(6), 585–609.

Ford, D.C. (1971b) Geologic structure and a new explanation of limestone cavern genesis. *Transactions of the Cave Research Group of Great Britain*, **13**(2): 81–94.

Ford, D.C. (1973) Development of the canyons of the South Nahanni River, N.W.T. *Canadian Journal of Earth Sciences*, **10**(3), 366–78.

Ford, D.C. (1979) A review of alpine karst in the southern Rocky Mountains of Canada. *Bulletin of the National Speleological Society*, **41**, 53–65.

Ford, D.C. (1980) Threshold and limit effects in karst geomorphology, in *Thresholds in Geomorphology* (eds D. L. Coates and J. D. Vitek), George Allen & Unwin, London, pp. 345–62.

Ford, D.C. (1983a) Effects of glaciations upon karst aquifers in Canada. *Journal of Hydrology*, 61, 149–58.

Ford, D.C. (1983b) The Winnipeg Aquifer, *Journal of Hydrology*, 61(1/3), 177–80.

Ford, D.C. (1983c) Alpine karst systems at Crowsnest Pass, Alberta-British Columbia, *Journal of Hydrology*, 61(1/3), 187–92.

Ford, D.C. (1984) Karst groundwater activity and landform genesis in modern permafrost regions of Canada, in *Groundwater as a Geomorphic Agent* (ed. R.G. LaFleur), Allen & Unwin, London, pp. 340–50.

Ford, D.C. (1987) Effects of glaciations and permafrost upon the development of karst in Canada. *Earth Surface Processes and Landforms*, 12(5), 507–21.

Ford, D.C. (1991) Features of the genesis of Jewel Cave and Wind Cave, Black Hills, South Dakota. *Bulletin of the National Speleological Society*, 51, 100–10.

Ford, D.C. (1996a) Paleokarst phenomena as 'targets' for modern karst groundwaters: the contrasts between thermal water and meteoric water behaviour. *Carbonate and Evaporites*, 10(2), 138–47.

Ford, D.C. (1996b) Karst in a cold climate, in *Geomorphology sans Frontières* (eds S.B. McCann and D.C. Ford), John Wiley & Sons, Chichester, pp 153–79.

Ford, D.C. (1997a) Dating and paleo-environmental studies of speleothems, in *Cave Minerals of the World*, 2nd edn (eds C.A. Hill and P. Forti), National Speleological Society of America Press, Huntsville, AL, pp. 271–84.

Ford, D.C. (1997b) Principal features of evaporite karst in Canada. *Carbonates and Evaporites*, 12(1); 15–23.

Ford, D.C. (1998) Perspectives in karst hydrogeology and cavern genesis. *Bulletin d'Hydrogeologie*, 16, 9–29.

Ford D.C. (2000a) Speleogenesis Under Unconfined Settings, in *Speleogenesis; Evolution of Karst Aquifers* (eds A.V. Klimchouk, D.C. Ford, A.N. Palmer and W. Dreybrodt), National Speleological Society of America, Huntsville, AL, pp. 319–24.

Ford D.C. (2000b) Deep phreatic caves and groundwater systems of the Sierra del Abra, Mexico, in *Speleogenesis; Evolution of Karst Aquifers* (eds A.V. Klimchouk, D.C. Ford, A.N. Palmer and W. Dreybrodt), National Speleological Society of America, Huntsville, AL, pp. 325–31.

Ford, D.C. (2000c) Caves Branch, Belize the Baradla–Domica System, Hungary and Slovakia, in *Speleogenesis; Evolution of Karst Aquifers* (eds A.V. Klimchouk, D.C. Ford, A.N. Palmer and W. Dreybrodt), National Speleological Society of America, Huntsville, AL, pp. 391–6.

Ford, D.C. (2002) Depth of conduit flow in unconfined carbonate aquifers: comment. *Geology*, 30(1), 93.

Ford, D.C. and Ewers, R.O. (1978) The development of limestone cave systems in the dimensions of length and breadth. *Canadian Journal of Earth Science*, 15, 1783–98.

Ford, D.C. and Lundberg, J.A. (1987) A review of dissolutional rills in limestone and other soluble rocks. *Catena Supplement*, 8, 119–40.

Ford, D.C. and Stanton, W.I. (1968) Geomorphology of the south-central Mendip Hills, *Proceedings of the Geologists' Association*, 79(4), 401–27.

Ford, D.C. and Williams, P.W. (1989) *Karst Geomorphology and Hydrology*, Unwin Hyman, London, 601 pp.

Ford, D.C., Harmon, R.S., Schwarcz, H.P., et al. (1976) Geohydrologic and thermometric observations in the vicinity of the Columbia Icefields, Alberta and British Columbia. *Journal of Glaciology*, 16(74); 219–30.

Ford, D.C., H.P. Schwarcz, J.J. Drake, et al. (1981) Estimates of the age of the existing relief within the Southern Rocky Mountains of Canada. *Arctic and Alpine Research*, 13(1), 1–10.

Ford, D.C., Lundberg, J., Palmer, A.N., et al. (1993) Uranium-series dating of the draining of an aquifer: the example of Wind Cave, Black Hills, South Dakota. *Geological Society of America Bulletin*, 105, 241–250.

Ford, D.C., Salomon, J.-N., and Williams, P.W. (1996) Les 'Forêts de Pierre' ou 'Stone forests' de Lunan (Yunnan, Chine). *Karstologia*, 28(2), 25–40.

Ford, T.D. (1989) Tufa – the Whole Dam Story. *Cave Science*, 16(2); 39–49.

Ford, T.D. (1995) Some thoughts on hydrothermal caves. *Cave and Karst Science*, 22, 107–18.

Ford, T.D. and Pedley, H.M. (1996) A review of tufa and travertine deposits of the world. *Earth Science Reviews*, 41, 117–75.

Formentin, K., Rossi, P., Aragno, M. and Müller, I. (1997) Determination of bacteriophage migration and survival potential in karstic groundwaters using batch agitated experiments and mineral colloidal particles, in *Tracer Hydrology 97* (ed. A. Kranjc), Rotterdam, Balkema, pp. 39–46.

Fornós, J.-J. and Ginés, A. (eds.) (1996) *Karren Landforms*, Universitat de les Illes Balears Press, Palma, 450 p.

Fornós, J.J., Gelabert, B., Ginés, A., et al. (2002) Phreatic overgrowths on speleothems: a useful tool in structural geology in littoral karst landscapes. The example of eastern Mallorca (Balearic Islands). *Geodinamica Acta*, 15, 113–25.

Forti, P. (1997) Speleothems and earthquakes, in *Cave Minerals of the World*, 2nd edn (eds C.A. Hill and P. Forti), National Speleological Society of America, Huntsville, AL, pp. 284–5.

Forti, P. and Grimandi, P. (eds) (1986) *Atti del symposio internazionale sul carsismo delle evaporiti. Bologna, 1985. Le Grotte d'Italia*, 4(XII), 420 pp.

Forti, P. and Postpischl, D. (1984) Seismotectonic and paleoseismic analyses using karst sediments. *Marine Geology*, 55, 145–61.

Forti, P. and Postpischl, D. (1985) Relazioni tra terremoti e deviazioni degli assi di accrescimento delle stalagmiti. *Le Grotte d'Italia*, 4(XII), 287–303.

Fowler, A. (2002) Assessment of the validity of using mean potential evaporation in computations of the long-term soil water balance. *Journal of Hydrology*, 256, 248–63.

Franke, M.W. (1965) The theory behind stalagmite shapes. *Studies in Speleology*, 1, 89–95.

Frappier, A., Sahagian, D., Gonzalez, L.A. Carpenter, S.J. (2002) El Niño events recorded by stalagmite carbon isotopes. *Science*, **298**, 565.

Freeze, R.A. and Cherry, J.A. (1979) *Groundwater*, Prentice-Hall, New Jersey, 604 pp.

Freiderich, H. and Smart, P.L. (1981) Dye tracer studies of the unsaturated-zone recharge of the Carbonifereous Limestone aquifer of the Mendip Hills, England, *Proceedings of the 8th International Congress of Speleology*, Vol. 1, pp. 283–6.

French, H.M. (1996) *The Periglacial Environment*, 2nd edn, Harlow, Addison Wesley Longman; 341 pp.

Friederich, H. and Smart, P.L. (1982) The classification of autogenic percolation waters in karst aquifers: a study in G.B. Cave, Mendip Hills, England, *Proceedings of the University of Bristol Speleological Society*, **16**(2), 143–59.

Frisia, S., Borsato, A., Fairchild, I.J. and McDermott, F. (2000) Calcite fabrics, growth mechanisms environments of formation in speleothems from the Italian Alps and southwestern Ireland. *Journal of Sedimentary Research*, **70**(5), 1183–96.

Fritz, P. and Fontes, J.C. (eds) (1980) *Handbook of Environmental Isotope Geochemistry, Vol. 1, The Terrestrial Environment*, Elsevier, Amsterdam.

Fritz, R.D., Wilson, J.L and Yurewicz, D.A. (1993) *Paleokarst Related Hydrocarbon Reservoirs*. Core Workshop 18, Society of Ecomic Paleontologists and Mineralogists, New Orleans, 275 pp.

Frumkin, A. (1992) *The karst system of the Mt Sedom salt diapir*. Hebrew University PhD thesis, Jerusalem, 135 pp.

Frumkin, A. (1994) Hydrology and denudation rates of halite karst. *Journal of Hydrology*, **162**, 171–89.

Frumkin, A. (1995) Morphology and development of salt caves. *Bulletin of the National Speleological Society*, **56**, 82–95.

Frumkin, A. (1996) Uplift rate relative to base-levels of a salt diapir (Dead Sea Basin, Israel) as indicated by cave levels, in *Salt Tectonics* (eds G.I. Alsop, D.J. Blundell and I. Davison), Special Publication 100, Geological Society Publishing House, Bath, pp. 41–7.

Frumkin, A. (2000a) Dissolution of salt, in *Speleogenesis; Evolution of Karst Aquifers* (eds A.V. Klimchouk, D.C. Ford, A.N. Palmer and W. Dreybrodt), National Speleological Society of America, Huntsville, AL, pp. 169–70.

Frumkin, A. (2000b) Speleogenesis in salt – the mount Sedom area, Israel, in *Speleogenesis; Evolution of Karst Aquifers* (eds A.V. Klimchouk, D.C. Ford, A.N. Palmer and W. Dreybrodt), National Speleological Society of America, Huntsville, AL, pp. 443–51.

Frumkin, A. and Ford, D.C. (1995) Rapid entrenchment of stream profiles in the salt caves of Mount Sedom, Israel. *Earth Surface Processes and Landforms*, **20**, 139–52.

Frumkin, A., Ford, D.C. and Schwarcz, H.P. (1999) Continental oxygen isotope record of the last 170,000 years in Jerusalem. *Quaternary Research*, **51**, 317–27.

Furley, P.A. (1987) Impact of forest clearance on the soils of tropical cone karst. *Earth Surface Processes and Landforms*, **12**, 523–9.

Gabrovšek, F. (ed.) (2002) *Evolution of Karst: from Prekarst to Cessation*, Institut za raziskovanje krasa, ZRC SAZU, Postojna-Ljubljana, 448 pp.

Gabrovsek, F. and Dreybrodt, W. (2000) The role of mixing corrosion in calcite-aggressive H_2O-CO_2-$CaCO_3$ solutions in the early evolution of karst aquifers. *Water Resources Research*, **36**, 1179–88.

Galan, C. (ed.) (1995) Exploracion y estudio de cavidades en rocas siliceas Precambricas del Grupo Roraima, Guayana Venezolana: una sintesis actual. *Karaitza*, **4**, 1–52.

Galan, C. and Lagarde, J. (1988) Morphologie et evolution des caverns et formes superficielles dans les quartzites du Roraima (Venezuela). *Karstologia*, **11–12**, 49–60.

Galdenzi, S. and Menichetti, M. (1990) Un modello genetico per La Grotta Grande del Vento. *Instituto Italiana Speleologica*, **4**(Series II), 123–42.

Gale, S.J. (1984) The hydraulics of conduit flow in carbonate aquifers. *Journal of Hydrology*, **70**, 309–327.

Gams, I. (1962) Measurements of corrosion intensity in Slovenia and their geomorphological significance. *Geografski vestnik*, **34**, 3–20.

Gams, I. (1972) Effect of runoff on corrosion intensity in the northwest Dinaric karst. *Transactions of the Cave Research Group of Great Britain*, **14**(2), 78–83.

Gams, I. (1973a) *Slovenska kraska terminologija (Slovene karst terminology)*, Zveza Geografskih Institucij Jugoslavije, Ljubljana.

Gams, I. (1973b) *The Terminology of the Types of Polje*. Slovenska Kraska Terminologija, Zveza Geografskih Institucij Jugoslavije, Ljubljana, pp. 60–7.

Gams, I. (1976) Hydrogeographic review of the Dinaric and alpine karst in Slovenia with special regard to corrosion, in *Problems of Karst Hydrology in Yugoslavia*, Memoir 13, Serbia Geographical Society, pp. 41–52.

Gams, I. (1978) The polje: the problem of its definition. *Zeitschrift für Geomorphologie*, **22**, 170–181.

Gams, I. (1980) Poplave na Planinskem polju (Inundations in Planina polje). *Geografski Zbornik*, **XX**, 4–30.

Gams, I. (1981) Comparative research of limestone solution by means of standard tablets, *Proceedings of the 8th International Congress of Speleology, Bowling Green, Kentucky*, Vol. 1, pp. 273–5.

Gams, I. (1985) International comparative measurement of surface solution by means of standard limestone tablets. *Zbornik Ivana Rakovica, Razprave 4*, Razreda Sazu 26, pp. 361–86.

Gams, I. (1991a) The origin of the term karst in the time of transition of karst (kras) from deforestation to forestation, *Proceedings of the International Conference on Environmental Changes in Karst Areas* (IGU/UIS), Quaderni del Dipartimento di Geografia 13, Universita di Padova, pp. 1–8.

Gams, I. (1991b) Systems of adapting the littoral Dinaric karst to agrarian land use. *Geografski Zbornik*, **XXXI**, 5–106.

Gams, I. (1993) Origin of the term 'karst' and the transformation of the Classical karst (kras). *Environmental Geology*, **21**, 110–4.

Gams, I. (1994) Types of contact karst. *Geografia Fisica e Dinamica Quateraria*, **17**, 37–46.

Gams, I. (2003) *Kras v Sloveniji v prostoru in času*, Založba ZRC, ZRC SAZU, Ljubljana, 516 pp.

Gams, I. and Kogovšek, J. (1998) The dynamics of flowstone deposition in the caves Postojnska, Planinska, Taborska and Škocjanske, Slovenia. *Acta Carsologica*, **38**(1), 299–324.

Gandino, A. and Tonelli, A.M. (1983) Recent remote sensing technique in freshwater marine springs monitoring: qualitative and quantitative approach, in *Methods and Instrumentation for the Investigation of Groundwater Systems, Proceedings* International Symposium, Noordwijkerhout, Netherlands, UNESCO/IAHS, pp. 301–10.

Gao, Y., Alexander, E.C. and Tipping, R.G. (2005a) Karst database development in Minnesota: design and data assembly. *Environmental Geology*, **47**, 1072–82.

Gao, Y., Alexander, E.C. and Tipping, R.G. (2005b) Karst database development in Minnesota: analysis of sinkhole distribution. *Environmental Geology*, **47**, 1083–98.

Gao, Y., Alexander, E.C. and Tipping, R.G. (2002) The development of a karst feature database for southeastern Minnesota. *Journal of Cave and Karst Studies*, **64**(1), 51–7.

Gargani, J. (2004) Modelling of the erosion in the Rhone valley during the Messinian crisis (France). *Quaternary International*, **131**, 13–22.

Garrels, R.M. and Christ, C.L. (1965) *Solutions, Minerals and Equilibria*, Harper & Row, New York, 450 pp.

Gary, M.O., Sharp, J.M., Havens, R.S. and Stone, W.C. (2002) Sistema Zacatón: identifying the connection between volcanic activity and hypogenic karst in a hydrothermal phreatic cave system. *Geo²*, **29**(3–4), 1–14.

Gascoyne, M. (1977) Trace element geochemistry of speleothems, *Proceedings of the 7th International Congress of Speleology, Sheffield*, pp. 205–8.

Gascoyne, M. (1992) Palaeoclimate determination from cave calcite deposits. *Quaternary Science Reviews*, **11**, 609–32.

Gascoyne, M., Benjamin, G.J., Schwarcz, H.P. and Ford, D.C. (1979) Sea-level lowering during the Illinoian glaciation: evidence from a Bahama 'blue hole'. *Science*, **205**, 806–808.

Gascoyne, M., Ford, D.C. and Schwarcz, H.P. (1983) Rate of cave and landform development in the Yorkshire Dales from speleothem age data. *Earth Surface Processes and Landforms*, **8**, 557–68.

Gaspar, E. (1987) *Modern Trends in Tracer Hydrology*, CRC Press, Boca Raton, 145 pp.

Gat, J.R. (1980) The isotopes of hydrogen and oxygen in precipitation, in *Handbook of Environmental Isotope Geochemistry* (eds P. Fritz and J. Fontes), Elsevier, Amsterdam, Vol. 1, pp. 21–47.

Gat, J.R. and Carmi, I. (1987) Effect of climate changes on the precipitation patterns and isotopic composition of water in a climate transition zone: case of the eastern Mediterranean Sea area, in *The Influence of Climate Change and Climate Variability on the Hydrologic Regime and Water Resources*, Publication 168, International Association of Hydrological Sciences, Wallingford, pp. 513–23.

Gavrilovic, D. (1970) Intermittierende Quellen in Jugoslawien. *Die Erde*, **101**(4),B381 284–298.

Genty, D. (1992) Les spéléothemes du tunnel de Godarville (Belgique) – un exemple exceptionnel de concrétionnement moderne. *Spéléochronos*, **4**(6), 3–29.

Genty, D. and Deflandre, G. (1998) Drip flow variations under a stalactite of the Pere Noel cave (Belgium). Evidence of seasonal variations and air pressure constraints. *Journal of Hydrology*, **211**(1–4), 208–32.

Genty, D. and Quinif, Y. (1996) Annually laminated sequences in the internal structure of some Belgian stalagmites – importance for paleoclimatology. *Journal of Sedimentary Research*, **66**(1), 275–288.

Genty, D., Bastin, B. and Ek, C. (1995) Nouvel exemple d'alternance delamines annuelles dans une stalagmite (Grotte de Dinant 'La Merveilleuse,' Belgique). *Spéléochronos*, **6**, 9–22.

Genty, D., Baker, A., R., Massault, M., Proctor, C.,Gilmour, M., Pons-Branchu, E. Hamelin, B. (2001a) Dead carbon in stalagmites: Carbon bedrock paleodissolution vs. ageing of soil organic matter. Implications for ^{13}C variations in speleothems. *Geochimica et Cosmochimica Acta*, **65**(20), 3443–57.

Genty, D., Baker, A. Vokal, B. (2001b) Intra- and inter-annual growth rate of modern stalagmites. *Chemical Geology*, **176**, 191–212.

Genty, D., Plagnes, V., Causse, C., *et al.* (2002) Fossil water in large stalagmite voids as a tool for paleoprecipitation stable isotope composition reconstitution and paleotemperature calculation. *Chemical Geology*, **184**, 83–95.

Genty, D., Blamart, D., Ouahdi, R. *et al.* (2003) Precise dating of Dansgaard–Oeschger climate oscillations in western Europe from stalagmite data. *Nature*, **421**, 833–7.

Gerba, C.P., Wallis, C. and Melnick, J.L. (1975) Fate of wastewater bacteria and viruses in soil. *ASCE Journal of the Irrigation and Drainage Division*, **101**, 157–74.

Gerson, R. (1976) Karst and fluvial denudation of carbonate terrains under sub-humid Mediterranean and arid climates - principles, evaluation and rates (examples from Israel), in *Karst Processes and Relevant Landforms* (ed. I. Gams), *Proceedings* of the Karst Denudation Symposium, International Speleology Union, University of Ljubljana, pp. 71–9.

Geurts, M.A. (1976) Genese et stratigraphie des travertins au fond de vallee en Belgique. *Acta Geographica Lovaniensa*, **16**, 87 pp.

Ghannam, J., Ayoub, G.M. and Acra, A. (1998) A profile of the submarine springs in Lebanon as a potential water resource. *Water International*, **23**, 278–86.

Ghazban, F., Schwarcz, H.P. and Ford, D.C. (1992a) Multistage dolomitization in the Society Cliffs Formation, northern Baffin Island, Northwest Territories, Canada. *Canadian Journal of Earth Sciences*, **29**, 1459–73.

Ghazban, F., Schwarcz, H.P and Ford, D.C. (1992b) Correlated strontium, carbon and oxygen isotopes in carbonate gangue at the Nanisivik zinc-lead deposits, northern Baffin Island, Canada. *Chemical Geology (Isotope Geoscience Section)*, **87**, 137–46.

Ghergari, L., Onac, B.P., Vremir, M. and Strusievicz, R. (1998) La cristallogenèse des spéléothèmes, Monts Pădurea Crailui, Roumanie. *Karstologia*, **31**, 19–26.

Ghyben, W.B. (1889) *Nota in verband met de voorgenomen put boring nabij Amsterdam* (Notes on the probable results of the proposed well drilling near Amsterdam). Koninski. Instituut Ingenieur Tijdschrift, The Hague, 21 pp.

Gilli, E. (2002) Les karsts littoraux des Alpes-Maritimes: inventaire des emergences sous-marines et captage expértal de Cabbé. *Karstologia*, **40**(2), 1–12.

Gillieson, D. (1986) Cave sedimentation in the New Guinea highlands. *Earth Surface Processes and Landforms*, **11**, 533–543.

Gillieson, D. (1993) Environmental change and human impact on karst in arid and semi-arid Australia, in *Karst terrains; Environmental Changes and Human Impacts* (ed. P.W. Williams). *Catena Supplement*, **25**, 127–46.

Gillieson, D. (1996) *Caves: Processes, Development and Management*, Blackwell, Oxford, 324 pp.

Gillieson, D. (2004) Nullarbor Plain, Australia, in *Encyclopedia of Caves and Karst Science* (ed. J. Gunn), Fitzroy Dearborn, New York, pp. 544–6.

Gillieson, D. and Household, I. (1999) Rehabilitation of the Lune River Quarry, Tasmanian Wilderness World Heritage area, Australia, in *Karst Hydrogeology and Human Activities: Impacts, Consequences and Implications* (eds D. Drew and H. Hotzl), Rotterdam, Balkema, pp. 201–5.

Gillieson, D. Spate, A. (1992) The Nullarbor karst, in *Geology, Climate, Hydrology and Karst Formation: Field Symposium in Australia; Guidebook* (ed. D.S. Gillieson), Special Publication 4, Department of Geography and Oceanography, University College, Australian Defense Force Academy, Canberra, pp. 65–99.

Gillieson, D., Smith, D.I., Greenaway, M. and Ellaway, M. (1991) Flood history of the limestone ranges in the Kimberley region, Western Australia. *Applied Geography*, **11**, 105–23.

Gladysz, K. (1987) *Karren on the Quatsino Limestone, Vancouver Island*. McMaster Univ. BSc thesis.

Glazek, J., Rudnicki, J. and Szynkiewicz, A. (1977) Proglacial caves – a special genetic type of cave in glaciated areas, *Proceedings of the 7th International Congress of Speleology, Sheffield*, pp. 215–7.

Glennie, E.A. (1954) Artesian flow and cave formation. *Transactions of the Cave Research Group of Great Britain*, **3**, 55–71.

Glew, J.R. and Ford, D.C. (1980) A simulation study of the development of rillenkarren. *Earth Surface Processes*, **5**(B404), 25–36.

Glover, R.R. (1974) Cave development in the Gaping Ghyll System, in *Limestones and Caves of Northwest England* (ed. A.C. Waltham), David and Charles, Newton Abbot, pp. 343–84.

Ginés, A. (2004) Karren, in *Encyclopedia of Caves and Karst Science* (ed. J. Gunn), Fitzroy Dearborn, New York, pp. 470 –73.

Goede, A. Green, D.C. and Harmon, R.S. (1982) Isotopic composition of precipitation, cave drips and actively forming speleothems at three Tasmanian cave sites. *Helictite*, **20**, 17–29.

Goede, A., Green, D.C. and Harmon, R.S. (1986) Late Pleistocene paleotemperature record from a Tasmanian speleothem. *Australian Journal of Earth Science*, **33**, 333–42.

Goldie, H.S. (1993) The legal protection of limestone pavements in Great Britain. *Environmental Geology*, **21**, 160–66.

Goldie, H.S. (2005) Erratic judgements: re-evaluating solutional erosion rates of limestones using erratic-pedestal sites, including Norber, Yorkshire. *Area*, **37**(4), 433–42.

Goldie, H.S. and Cox, N.J. (2000) Comparative morphometry of limestone pavements in Switzerland, Britain and Ireland. *Zeitschrift für Geomorphologie, N.F. Supplement-Band*, **122**, 85–112.

Goldscheider, N., Klute, M., Sturm, S. and Hötzl, H. (2000) The PI method: a GIS based approach to mapping groundwater vulnerability with special consideration of karst aquifers. *Zeitschrift für Angewendte Geologie*, **463**, 157–66.

Golwer, A. (1983) Underground purification capacity, in *Ground Water in Water Resources Planning*, Publication 42 (UNESCO Koblenz Symposium), International Association of Hydrological Sciences, Wallingford, pp. 1063–72.

Goodchild J.G. (1875) Glacial erosion. *Geological Magazine*, **II**, 323 8 and 356–62.

Goodchild, J.G. (1890) Notes on some observed rates of weathering of limestone. *Geological Magazine*, **27**, 463–6.

Gorbunova, K.A. (1977) Exogenetic gypsum tectonics, *Proceedings of the 7th International Congress of Speleology, Sheffield*, pp. 222–3.

Gorbunova, K.A. (1979) *Morphology and Hydrogeology of Gypsum Karst*. All-Union Karst and Speleology Institute, Perm, 93 pp. [In Russian.]

Gospodarič, R. (1976) The Quaternary caves development between the Pivka basin and Polje of Planina. *Acta Carsologica*, **7**, 5 135.

Gospodaric, R. and Habic, P. (1976) *Underground Water Tracing*, Institute Karst Research, Postojna, Ljubljana.

Gospodarič, R. and Habic, P. (1978) *Kraski pojavi Cerkniskega polja* (Karst phenomena of Cerknisko polje). *Acta Carsologica*, **8**(1), 6–162.

Goudie, A.S. (1983) Calcrete, in *Chemical Sediments and Geomorphology*, (eds A. S. Goudie and K. Pye), Academic Press, London, pp. 93–131.

Goudie, A.S. (1999) A comparison of the relative resistance of limestones to frost and salt weathering. *Permafrost and Periglacial Processes*, **10**, 309–16.

Goudie, A.S., Bull, P.A. and Magee, A.W. (1989) Lithological control of rillenkarren development in the Napier Range, Western Australia. *Zeitschrift für Geomorphologie*, **75**, 95–114.

Goudie, A.S., Viles, H., Allison, R., Day, M., *et al.* (1990) The geomorphology of the Napier Range, Western Australia. *Transactions of the Institute of British Geographers, New Series*, **15**(3), 308–22.

Grabau, A.W. (1913) *Principles of Stratigraphy*. Seiler, New York.

Gracia, F.J., Gutiérrez, F. and Gutiérrez, M. (2003) The Jiloca karst polje-tectonic graben (Iberian Range, NE Spain). *Geomorphology*, **52**, 215–31.

Gradzinski, M. and Kicinska, D. (2002) Morphology of Czarna Cave and its significance for the geomorphic evolution of the Koscielska Valley (Western Tatra Mountains). *Annales Societatis Geologorum Poloniae*, **72**, 255–62.

Gradziński, M., Rospondek, M. and Szulc, J. (1997) Paleoenvironmental controls and microfacies variability of the flowstone cover from the Zvonivá Cave in the Slovakian Karst. *Slovak Geological Magazine*, 3(4), 299–313.

Graf, C.G. (1999) Hydrogeology of Kartchner Caverns State Park, Arizona. *Journal of Cave and Karst Studies*, 61(2), 59–67.

Granger, D.E. and Muzikar, P.F. (2001) Dating sediment burial with *in situ* produced cosmogenic nuclides: theory, techniques, limitations. *Earth and Planetary Science Letters*, 188(1–2), 269–281.

Granger, D. E. and Smith, A.L. (2000) Dating buried sediments using radioactive decay and muogenic production of ^{26}Al and ^{10}Be. *Nuclear Instruments and Methods in Physics Research, Series B*, 172, 822–6.

Granger, D.E., Kirchner, J. and Finkel, R. (1997) Quaternary downcutting rate of the New River, Virginia, measured from differential decay of cosmogenic ^{26}Al and ^{10}Be in cave-deposited alluvium. *Geology*, 25(2), 107–10.

Granger, D.E., Fabel, D. and Palmer, A.N. (2001) Plio-Pleistocene incision of the Green River, KY, from radioactive decay of cosmogenic ^{26}Al and ^{10}Be in Mammoth Cave sediments. *Bulletin, Geological Society of America*, 113(7), 825–36.

Grasby, S.E., Chen, Z. and Betcher, R. (2002) Impact of Pleistocene glaciation on the hydrodynamics of the Western Canada Sedimentary Basin. *Annual Meeting of the Geological Association of Canada, abstract*.

Grigg, R.W., Grossman, E.E., Earle, S.A., *et al.* (2002) Drowned reefs and antecedent karst topography, Au'au Channel, S.E. Hawaiian Islands. *Coral Reefs*, 21(1), 73–82.

Grimes, K.G. (1994) The south-east karst province of South Australia. *Environmental Geology*, 23, 134–48.

Grimes, K.G. (2002) Syngenetic and eogenetic karst: an Australian viewpoint, in *Evolution of Karst from Prekarst to Cessation* (ed. F. Gabrovšek), ZRC SAZU, Postojna, pp. 407–14.

Grimes, K.G., Mott, K. and White, S. (1999) The Gambier Karst Province, *Proceedings of the 13th Australian Conference on Cave and Karst Management*, pp. 1–7.

Grindley, G.W. (1971) *Sheet S8 Takaka. Geological Map of New Zealand*, 1:63360, Department of Scientific and Industrial Research, Wellington.

Groves, C. and Meiman, J. (2005) Weathering, geomorphic work, and karst landscape evolution in the Cave City groundwater basin, Mammoth Cave, Kentucky. *Geomorphology*, 67, pp. 115–126.

Grün, R. Moriarty, K. and Wells, R. (2001) Electron spin resonance dating of the fossil deposits in the Naracoorte Caves, South Australia. *Journal of Quaternary Science*, 16(1), 49–59.

Grund, A. (1903) *Die Karsthydrographie: Studien aus Westbosnien*. Geographischen Abhandlungen, Band VII, Heft 3, von A. Penck, 7, pp. 103–200.

Grund, A. (1914) Der geographische Zyklus im Karst. *Gesellschaft für Erdkunde*, 52, 621–40. [Translated into English in Sweeting (1981).]

Günay, G. (2002) Gypsum karst, Sivas, Turkey. *Environmental Geology*, 42, 387–98.

Günay, G. and Şimşek, Ş. (2000) Karst hydrogeology in hydrothermal systems, in *Present State and Future Trends of Karst Studies* (eds G. Günay, K.S. Johnson, D. Ford and A.I. Johnson), IHP-V, Technical Documents in Hydrology, 49(II), UNESCO, Paris, pp. 501–13.

Gunn, J. (1978) *Karst hydrology and solution in the Waitomo District, New Zealand*. Auckland Univ PhD thesis.

Gunn, J. (1981a) Limestone solution rates and processes in the Waitomo district, New Zealand. *Earth Surface Processes and Landforms*, 6, 427–45.

Gunn, J. (1981b) Hydrological processes in karst depressions. *Zeitschrift für Geomorphologie*, NF, 25(3), 313–31.

Gunn, J. (1982) Magnitude and frequency properties of dissolved solids transport. *Zeitschrift für Geomorphologie*, 26(4), 505–11.

Gunn, J. (1983) Point recharge of limestone aquifers – a model from New Zealand karst. *Journal of Hydrology*, 61, 19–29.

Gunn, J. (1993) The geomorphological impacts of limestone quarrying, in *Karst Terrains: Environmental Changes and Human Impact* (ed. P.W. Williams). *Catena Supplement*, 25, 187–97.

Gunn, J. (ed.) (2004a) *Encyclopedia of Caves and Karst Science*, Fitzroy Dearborn, New York, 902 pp.

Gunn, J. (2004b) Limestone as a mineral resource, in *Encyclopedia of Caves and Karst Science* (ed. J. Gunn), Fitzroy Dearborn, New York, pp. 489–90.

Gunn, J. (2004c) Radon in caves, in *Encyclopedia of Caves and Karst Science* (ed. J. Gunn), Fitzroy Dearborn, New York, pp. 617–8.

Gunn, J. and Bailey, D. (1993) Limestone quarrying and quarry reclamation in Britain. *Environmental Geology*, 21(3), 167–72.

Gunn, J. and Lowe, D.J. (2000) Speleogenesis on tectonically active carbonate islands, in *Speleogenesis; Evolution of Karst Aquifers* (eds A.V. Klimchouk, D.C. Ford, A. N. Palmer and W. Dreybrodt), National Speleological Society of America, Huntsville, AL, pp. 238–43.

Gunn, J., Bailey, D. and Handley, J. (1997) *The Reclamation of Limestone Quarries using Landform Replication*. Department of the Environment, Transport and the Regions, London.

Gutierrez, F. (1996) Gypsum karstification-induced subsidence: effects on alluvial systems and derived geohazards (Calatayud Graben, Iberian Range, Spain). *Geomorphology*, 16, 277–93.

Gutierrez, F. and Cooper, A.H. (2002) Evaporite dissolution subsidence in the historical city of Calatayud, Spain: Damage Appraisal and Prevention. *Natural Hazards*, 25, 259–88.

Gutièrrez, M. and Gutièrrez, F. (1998) Geomorphology of the Tertiary gypsum formations in the Ebro Depression (Spain). *Geoderma*, 87, 1–29.

Hagen, G. (1839) Uber die Bewegung des Wassers in engen cylindrischen Rohren. *Poggendorff Annalen*, 46(B444), 423–42.

Haid, A. (1996) Yonne. *Spelunca*, 62, 14.

Halihan, T. and Wicks, C.M. (1998) Modeling of storm responses in conduit flow aquifers with reservoirs. *Journal of Hydrology*, 208, 82–91.

Halihan, T., Wicks, C.M. and Engeln, J.F. (1998) Physical response of a karst drainage basin to flood pulses: example of the Devil's Icebox cave system (Missouri, USA). *Journal of Hydrology*, **204**, 24–36.

Halihan, T., Sharp, J.M. and Mace, R.E. (1999) Interpreting flow using permeability at multiple scales, in *Karst Modeling* (eds A.N. Palmer, M.V. Palmer and I.D. Sasowsky) (1999), Special Publication 5, Karst Waters Institute, Charles Town, WV, pp. 82–96.

Hallet, B. (1976) Deposits formed by subglacial precipitation of $CaCO_3$. *Bulletin, Geological Society of America*, **87**, 1003–15.

Hallet, B. (1979) Subglacial regelation water film. *Journal of Glaciology*, **23**(89), 321–34.

Halliday, W.R. and Anderson, C.H. (1970) Glacier caves: a new field of speleology. *Studies in Speleology*, **220**, 53–9.

Hamilton, J. and Ford, D.C. (2002) Karst geomorphology and hydrogeology of the Bear Rock formation – a remarkable dolostone and gypsum megabreccia in the continuous permafrost zone of Northwest Territories, Canada. *Carbonates and Evaporites*, **17**(2), 54–56.

Hamilton-Smith, E. (2004) *The World Heritage Context*. IUCN International Workshop on China World Heritage Biodiversity Programme, Kunming, China, 12 pp.

Han, B.P. (1998) *Study on Microkarstification Process*, Geological Publishing House, Beijing, 156 pp. [In Chinese; summary translation courtesy of the author.]

Hanna, R.B. and Rajaram, H. (1998) Influence of aperture variability on dissolutional growth of fissures in karst formations. *Water Resources Research*, **34**(1), 2843–53.

Hanshaw, B.B. and Back, W. (1979) Major geochemical processes in the evolution of carbonate–aquifer systems. *Journal of Hydrology*, **43**, 287–312.

Harbaugh, A.W. and McDonald, M.G. (1996) User's documentation for MODFLOW-96, an update to the US Geological Survey modular finite-difference ground-water flow model. *US Geological Survey, Open-File Report*, **96-485**, 56.

Harbor, J. (ed.) (1999) Cosmogenic isotopes in geomorphology. *Geomorphology* (Special Issue), **27**, 1–172.

Harding, K. (1987) *Deforestation of limestone slopes on northern Vancouver Island*. McMaster Univ. MSc thesis, 188 pp.

Harding, K. and Ford, D.C. (1993) Impacts of primary deforestation on limestone slopes in northern Vancouver Island, British Columbia. *Environmental Geology*, **21**, 137–43.

Harmon, R.S., Ford, D.C. and Schwarcz, H.P. (1977) Interglacial chronology of the Rocky and Mackenzie Mountains based on ^{230}Th/^{234}U dating of calcite speleothems. *Canadian Journal of Earth Sciences*, **14**, 2543–52.

Harmon, R.S., Thompson, P., Schwarcz, H.P. and Ford, D.C. (1978) Late Pleistocene paleoclimates of North America as inferred from stable isotope studies of speleothems. *Quaternary Research*, **9**, 54–70.

Harmon, R.S., Schwarcz, H.P. Gascoyne, M. Hess, J.W. and Ford, D.C. (2004) Paleoclimate information from speleothems: the present as a guide to the past, in *Studies of Cave Sediments: Physical and Chemical Records of Paleoclimate* (eds I.D. Sasowsky and J. Mylroie), Kluwer Academic, New York, pp. 199–226.

Harmon, R.S., Atkinson, T.C. and Atkinson, J.J. (1983) The mineralogy of Castleguard Cave, Columbia Icefields, Alberta, Canada. *Arctic and Alpine Research*, **15**(4), 503–16.

Haryono, E. and Day, M. (2004) Landform differentiation within the Gunung Kidul kegelkarst, Java, Indonesia. *Journal of Cave and Karst Studies*, **66**(2), 62–9.

Hauns M., Jeannin, P.Y. and Atteia, O. (2001) Dispersion, retardation and scale effect in tracer breakthrough curves in karst conduits. *Journal of Hydrology*, **241**(3–4), 177–193.

Häuselmann, P. and Granger, D.E. (2004) Dating caves with cosmogenic nuclides: methods, possibilities and the Siebenhengste example, in *Dating of Cave Sediments* (eds A. Mihevc and N. Zupan Hajna), SAZU, Postojna, p. 50.

Häuselmann, P., Jeannin, P-Y. and Monbaron, M. (2003) Role of epiphreatic flow and soutirages in conduit morphogenesis: the Bärenschacht example (BE, Switzerland). *Zeitschrift für Geomorphologie*, **47**(2), 171–90.

Hays, P.D. and Grossman, E.L. (1991) Oxygen isotopes in meteoric calcite cement as indicators of continental paleoclimate. *Geology*, **19**, 441–4.

He, K., Liu, C. and Wang, S. (2001) Karst collapse mechanism and criterion for its stability. *Acta Geologica Sinica*, **75**(3), 330–5.

Heaton, T. (1986) Caves: a tremendous range of energy environments on Earth. *National Speleological Society News*, **August**, 301–4.

Heim, A. (1877) Uber die Karrenfelder. *Jahrbuch des Schweizer Alpenclub*, **XIII**, 421–33.

Hellden, U. (1973) Limestone solution intensity in a karst area in Lapland, northern Sweden. *Geografiska Annaler*, **54A**(3/4), 185–96.

Hendy, C.H. (1971) The isotopic geochemistry of speleothems – I. The calculation of the effects of different modes of formation on the isotopic composition of speleothems and their applicability as palaeoclimatic indicators. *Geochimica et Cosmochimica Acta*, **35**, 801–24.

Hennig, G.J. and Grün, R. (1983) ESR dating in Quaternary geology. *Quaternary Science Reviews*, **2**, 157–238.

Herak, M. (1972) Karst of Yugoslavia, in *Karst: Important Karst Regions of the Northern Hemisphere* (eds M. Herak and V.T. Stringfield), Elsevier, Amsterdam, pp. 25–83.

Herak, M. and Stringfield, V.T. (1972) Historical review of hydrogeologic concepts, in *Karst: Important Karst Regions of the Northern Hemisphere* (eds M. Herak and V.T. Stringfield), Elsevier, Amsterdam, pp. 19–24.

Hercman, H. (2000) Reconstruction of paleoclimatic changes in central Europe between 10 and 200 thousand years BP, based on analysis of growth frequency of speleothems. *Studia Quaternaria*, **17**, 35–70.

Herman, J.S. and White, W.B. (1985) Dissolution kinetics of dolomite: effects of lithology and fluid flow velocity. *Geochimica et Cosmochimica Acta*, **49**, 2017–26.

Herold, T., Jordan, P. and Zwahlen, F. (2000) The influence of tectonic structures on karst flow patterns in karstified limestones and aquitards in the Jura Mountains, Switzerland. *Eclogae Geologicae Helvetiae*, **93**(3), 349–62.

Hertbert, T.D., Schuffert, J.D., Andreasen, D., *et al.* (2001) Collapse of the California Current during glacial maxima linked to climate change on land. *Science*, **293**, 71–6.

Herzberg, A. (1901) Die Wasserversorgung einiger Nordsee Bader. *Journal für Gasbeleuchtung und Verwandte Beleuchtungsarten sowie für Wasserversorgung*, **44**, 815–9, 842–4.

Hess, J.W. and Slattery, L.D. (1999) Extractive industries impact, in *Karst Hydrogeology and Human Activities: Impacts, Consequences and Implications* (eds D. Drew and H. Hotzl), Rotterdam, Balkema, pp. 187–201.

Hewlett, J.D. and Hibbert, A.R. (1967) Factors affecting the response of small watersheds to precipitation in humid areas, in *Forest Hydrology* (eds W.E. Sopper and H.W. Lull), Pergamon, Oxford, pp. 275–90.

High, C. and Hanna, G.K. (1970) A method for the direct measurement of erosion of rock surfaces. *Brititish Geomorphological Research Group Technical Bulletin*, **5**, 24 pp.

Hill, C.A. (ed.) (1981) Saltpeter: a symposium. *Bulletin of the National Speleological Society*, **43**(4), 83–131.

Hill, C.A. (1982) Origin of black deposits in caves. *Bulletin of the National Speleological Society*, 44, pp. 15–19.

Hill, C.A. (1987) Geology of Carlsbad Caverns and other caves of the Guadalupe Mountains. *New Mexico Bureau of Mines and Minerals Bulletin*, **117**, 150 pp.

Hill, C.A. (1995) Sulfur redox reactions: hydrocarbons, native sulfur, Mississippi Valley-type deposits sulfuric acid karst in the Delaware Basin, New Mexico and Texas. *Environmental Geology*, **25**, 16–23.

Hill, C.A. (2003) Intrastratal Karst at the Waste Isolation Pilot Plant Site, Southeastern New Mexico. *Oklahoma Geological Survey Circular*, **109**, 197–209.

Hill, C.A. and Forti, P. (1997) *Cave Minerals of the World*, 2nd edn, National Speleological Society of America, Huntsville, AL, 463 pp.

Hillaire-Marcel, C., Soucy, J.M. and Cailleux, A. (1979) Analyse isotopique de concretions sous-glaciaires de l'inlandsis laurentidien et teneur en oxygene 18 de la glace. *Canadian Journal of Earth Sciences*, **16**, 1494–8.

Hillel, D. (1982) *Introduction to Soil Physics*, Academic Press, New York.

Hillner, P.E., Gratz, A.J., Manne, S. and Hansma, P.K. (1992) Atomic-scale imaging of calcite growth and dissolution in real time. *Geology*, **20**, 359–62.

Hiscock, K.M., Rivett, M.O. and Davison, R.M. (eds) (2002) *Sustainable Groundwater Development*, Special Publication 193, Geological Society Publishing House, Bath, 352 pp.

Hobbs, S.L. and Gunn, J. (1998) The hydrogeological impacts of quarrying karstified limestone, options for prediction and mitigation. *Quarterly Journal of Engineering Geology*, **31**, 47–157.

Hobbs, S.L. and Smart, P.L. (1986) Characterization of carbonate aquifers: a conceptual base, *Proceedings of the 9th International Congress of Speleology, Barcelona*, Vol. 1, pp. 43–6.

Hoblea, F., Jaillet, S. and Maire, R. (2001) Érosion et ruissellem,ent sur karst nu en contexte subpolaire océanique: les îles de Patagonie (Magallanes, Chili). *Karstologia*, **38**(2); 13–18.

Holmgren, K., Lee-Thorp, J. A., Cooper, G. R. J *et al.* (2003) Persistent millennial-scale climatic variability over the past 25,000 years in Southern Africa. *Quaternary Science Reviews*, **22**, 2311–26.

Homann, W. (1969) Experimentelle Ergebnisse zum Wachstum rezenter Hohlenperlen, *Proceedings of the 5th International Congress of Speleology, Stuttgart*, Vol. 2, pp. 5/1–5/19.

Hopley, D. (1982) *The Geomorphology of the Great Barrier Reef*, John Wiley & Sons, New York.

Hose, L. (2004) Cueva de Villa Luz, Mexico, in *Encyclopedia of Caves and Karst Science* (ed. J. Gunn), Fitzroy Dearborn, New York, pp. 758–9.

Hose, L.D., Palmer, A.N., Palmer, M.V., *et al.* (2000) Microbiology and geochemistry in a hydrogen-sulphide-rich karst environment. *Chemical Geology*, **169**, 399–423.

Hotzl, H. and Werner, A. (eds) (1992) *Tracer Hydrology*. Balkema, Rotterdam.

Howard, A.D. (1963) The development of karst features. *Bulletin of the National Speleological Society*, **25**, 45–65.

Howard, A.D., Kochel, R.C. and Holt, H.E. (1988) *Sapping Features of the Colorado Plateau: a Comparative Planetary Geology Field Guide*, NASA, Washington, DC, 108 pp.

Hsü, K.J., Ryan, W.B.F. and Cita, M.B. (1973) Late Miocene desiccation of the Mediterranean. *Nature*, **242**, 240–4.

Huang, Y., Fairchild, I.J., Borsato, A., *et al.* (2001) Seasonal variations in Sr, Mg and P in modern speleothems (Grotta di Ernesto, Italy). *Chemical Geology*, **175**, 429–48.

Hubbard, B. and Hubbard, A. (1998) Bedrock surface roughness and the distribution of subglacially precipitated carbonate deposits: implications for formation at Glacier de Tsanfleuron, Switzerland. *Earth Surface Processes and Landforms*, **23**, 261–70.

Hubbert, M.K. (1940) The theory of groundwater motion. *Journal of Geology*, **48**, 785–944.

Huizing, T., Jarnot, M., Neumeier, G., *et al.* (eds) (2003) *Calcite: the Mineral with the Most Forms*, Christian Weise Verlag, Munich, 114 pp.

Hunt, R.E. (1984) *Geotechnical Engineering Investigation Manual*, McGraw-Hill, New York.

Huntoon, P.W. (1991) Chairman Mao's Great Leap Forward and the deforestation ecological disaster in the South China Karst Belt, *Proceedings, Third Conference on Hydrogeology, Ecology, Monitoring Management of Ground Water in Karst Terranes, Nashville, TN*, pp. 149–60.

Huppert, G., Burri, E., Forti, P. and Cigna, A. (1993) Effects of tourist development on caves and karst, in *Karst Terrains: Environmental Changes and Human Impact* (ed. P.W. Williams). *Catena Supplement*, **25**, 251–68.

Hutton, J. (1795) *Theory of the Earth, with Proofs and Illustrations*, Vol. II, Edinburgh.

Hyatt, J.A. and Jacobs, P.M. (1996) Distribution and morphology of sinkholes triggered by flooding following Tropical Storm Alberto at Albany, Georgia, USA. *Geomorphology*, **17**, 305–16.

Hyatt, J.A., Wilson, R., Givens, J.S. and Jacobs, P.M. (2001) Topographic, geologic hydrogeologic controls on dimensions and locations of sinkholes in thick covered karst, Lowndes

County, Georgia, in *Geotechnical and Environmental Applications of Karst Geology and Hydrology* (eds B.F. Beck, and J.G. Herring), Balkema, Lisse, pp. 37–45.

IAEA (1983) *Guidebook on Nuclear Techniques in Hydrology,* Technical Reports Series No. 91, International Atomic Energy Agency, Vienna.

IAEA (1984) *Isotope Hydrology 1983,* International Atomic Energy Agency, Vienna.

Indermühle, A., Stocker, T. F., Joos, F., *et al.* (1999) Holocene carbon-cycle dynamics based on CO_2 trapped in ice at Taylor Dome, Antarctica. *Nature,* **398,** 121–6.

Ingraham, N.L. (1998). Isotopic variations in precipitation, in *Isotope Tracers in Catchment Hydrology* (eds C. Kendall, and J.J. McDonnell), Elsevier, Amsterdam, pp. 87–118.

Ireland, P. (1979) Geomorphological variations of 'case hardening' in Puerto Rico. *Zeitschrift für Geomorphologie, Supplement-Band,* **32,** 9–20.

Issar, A. (1983) Emerging groundwater, a triggering factor in the formation of the makhteshim (erosion cirques) in the Negev and Sinai. *Israel Journal of Earth Sciences,* **32,** 53–61.

Iurkiewicz, A. and Mangin, A. (1993) Utilisation de l'analyse systemique dans l'étude des aquifères karstiques des Monts Vâlcan (Roumanie). *Theoretical and Applied Karstology,* **7,** 9–96.

Ivanovich, M. and Harmon, R.S. (1992) *Uranium-series Disequilibrium: Applications to Earth, Marine Environmental Sciences,* Oxford Geoscience Publications, 910 pp.

Ivanovich, M. and Ireland, P. (1984) Measurements of uranium series disequilibrium in the case-hardened Aymamon limestone in Puerto Rico. *Zeitschrift für Geomorphologie,* **28,** 305–19.

Jackson, M.P.A., Roberts, D.G. and Snelson, S. (eds) (1995) *Salt Tectonics: a Global Perspective,* Memoir 65, American Association of Petroleum Geologists, Tulsa, OK.

Jacobson, G. and Hill, P.J. (1980) Hydrogeology of a raised coral atoll, Niue Island, South Pacific Ocean. *Journal of Australian Geology and Geophyics,* **5**(4), 271–8.

Jacobson, G., Hill, P.L. and Ghassemi, F. (1997) Geology and hydrogeology of Nauru Island, in *Geology and Hydrogeology of Carbonate Islands* (eds H.L. Vacher and T. Quinn), Developments in Sedimentology 54, Elsevier, pp. 707–42.

Jahn, B. and Cuvellier, H. (1994) Pb–Pb and U–Pb geochronology of carbonate rocks: an assessment. *Chemical Geology (Isotope Geoscience Section),* **115,** 125–51.

Jakucs, L. (1959) Neue Methoden der Hohlenforschung in Ungarn und ihre Ergebnisse. *Hohle,* **10**(4), 88–98.

Jakucs, L. (1977) *Morphogenetics of Karst Regions: Variants of Karst Evolution,* Akademiai Kiado, Budapest, 284 pp.

James, A.N. (1992) *Soluble Materials in Civil Engineering,* Ellis Horwood, Chichester, 434 pp.

James, A.N. and Lupton, A.R.R. (1978) Gypsum and anhydrite in foundations of hydraulic structures. *Geotechnique,* **28,** 249–72.

James, J.M. (1992) Corrosion par melange des eaux dans les grottes de la Plaine de Nullarbor, Australie, in *Karst et évolutions climatiques* (eds J.-N. Salomon and R. Maire), Presses Universitaires, Bordeaux, pp. 333–48.

James, J.M. (1993) Burial and infilling of a karst in Papua New Guinea by road erosion sediments. *Environmental Geology,* **21,** 144–51.

James, J.M. (1994) Microorganisms in Australian caves and their influence on speleogenesis, in *Breakthroughs in Karst Geomicrobiology and Redox Geochemistry* (eds I.D. Sasowsky and M.V. Palmer), Special Publication 1, Karst Waters Institute, pp. 31–4.

James, J.M. (1997) Minor, trace and ultra-trace constituents of speleothems, in *Cave Minerals of the World,* 2nd edn (eds C.A. Hill and P. Forti), National Speleological Society of America, Huntsville, AL, pp. 236–37.

James, N.P. and Choquette, P.W. (1984) Diagenesis (9). Limestones – the meteoric diagenetic environment. *Geoscience Canada,* **11,** 161–94.

James, N.P. and Choquette, P.W. (1988) *Paleokarst,* Springer-Verlag, New York, 416 pp.

Jarvis, R.S. (1981) Specific geomorphometry, in *Geomorphological Techniques* (ed. A. Goudie), Allen & Unwin, London, pp. 42–6.

Jeanin, P-Y, Bitterli, T. and Häuselmann, P. (2000) Genesis of a large cave system: case study of the North of Lake Thun System (Canton Bern, Switzerland), in *Speleogenesis; Evolution of Karst Aquifers* (eds A.V. Klimchouk, D.C. Ford, A.N. Palmer and W. Dreybrodt), National Speleological Society of America, Huntsville, AL, pp. 338–47.

Jeannin, P.-Y. and Sauter, M. (1998) Analysis of karst hydrodynamic behaviour using global approaches: a review. *Bulletin d'Hydrogéologie* (Neuchâtel), **16,** 31–48.

Jeannin, P-Y., Liedl, R. and Sauter, M. (1997 Some concepts about heat transfer in karstic systems, *Proceedings of the 12th International Congress of Speleology,* Vol. 2, pp. 195–9.

Jennings, J.E. (1966) Building on dolomites in the Transvaal. *Transactions of the South African Institute of Civil Engineers,* **8**(2), 41–62.

Jennings, J.N. (1968) Syngenetic karst in Australia, in *Contributions to the Study of Karst* (eds P.W. Williams and J.N. Jennings), Publication G5, Research School for Pacific Studies Australian National University, Canberra, pp. 41–110.

Jennings, J.N. (1972a) The Blue Waterholes, Cooleman Plain, N.S.W. the problem of karst denudation rate determination. *Transactions of the Cave Research Group of Great Britain,* **14,** 109–17.

Jennings, J.N. (1972b) Observations at the Blue Waterholes, March 1965 to April 1969 limestone solution on Cooleman Plain, N.S.W. *Helictite,* **10**(1–2), 1–46.

Jennings, J.N. (1975) Doline morphometry as a morphogenetic tool: New Zealand examples. *New Zealand Geographer,* **31,** 6–28.

Jennings, J.N. (1976) A test of the importance of cliff-foot caves in tower karst development. *Zeitschrift für Geomorphologie, Supplement-Band,* **26,** 92–7.

Jennings, J.N. (1982) *Karst of Northeastern Queensland Reconsidered.* Tower Karst Occasional Paper No. 4, Chillagoe Caving Club, pp. 13–52.

Jennings, J.N. (1983a) The disregarded karst of the arid and semiarid domain. *Karstologia,* **1,** 61–73.

Jennings, J.N. (1983b) Sandstone pseudokarst or karst? in *Aspects of Australian Sandstone Landscapes* (eds R.W. Young, and G.C. Nanson), Special Publication 1, Australian and New Zealand Geomorphology Group, pp. 21–30.

Jennings, J.N. (1985) *Karst Geomorphology*, Basil Blackwell, Oxford, 293 pp.

Jennings, J.N. and Bik, M.J. (1962) Karst morphology in Australian New Guinea. *Nature*, **194**, 1036–8.

Jennings, J.N. and Sweeting, M.M. (1963) The limestone ranges of the Fitzroy Basin, Western Australia. *Bonner Geographische Abhandlungen*, **32**, 60 pp.

Jeschke, A.A., Vosbeck, K. and Dreybrodt, W. (2001) Surface controlled dissolution rates of gypsum in aqueous solutions exhibit nonlinear dissolution kinetics. *Geochimica et Cosmochimica Acta*, **65**, 13–20.

Johnson, K.S. (1989) Development of the Wink Sink in west Texas, USA, due to salt dissolution and collapse. *Environmental Geology Water Science*, **14**, 81–92.

Johnson, K, S. (1996) Gypsum karst in the United States. *International Journal of Speleology*, **25**(3–4), pp. 183–193.

Johnson, K.S. and Neal, J.T. (2003) *Evaporite Karst and Engineering/Environmental Problems in the United States*, Oklahoma Geological Survey Circular 109, US Geological Survey and National Cave and Karst Research Institute, National Park Service, 353 pp.

Jones, B. (2001) Microbial activity in caves – a geological perspective. *Geomicrobiology Journal*, **18**, 345–57.

Jones, I.C. and Banner, J.L. (2000) Estimating recharge in a tropical karst aquifer. *Water Resources Research*, **36**(5), 1289–99.

Jones, W.K., Culver, D.C. and Herman, J.S. (eds) (2004) *Epikarst*, Special Publication 9, Karst Waters Institute, Charles Town, WV, 160 pp.

Julian, M., Martin, J. and Nicod, J. (1978) Les karsts Mediterraneens. *Mediterranee*, **1–2**, 115–31.

Kaçaroğlu, F. (1999) Review of groundwater pollution and protection in karst areas. *Water, Air, and Soil Pollution*, **113**, 337–56.

Kaddu-Mulindwa, D., Filip, Z. and Milde, G. (1983) Survival of some pathogenic and potential pathogenic bacteria in groundwater, in *Ground Water in Water Resources Planning*, Publication 42 (UNESCO Koblenz Symposium), International Association of Hydrological Sciences, Wallingford, pp. 1137–45.

Karanjac, J. and Gunay, G. (1980) Dumanli Spring, Turkey – the largest karstic spring in the world? *Journal of Hydrology*, **45**, 219–31.

Karolyi, M.S. and Ford, D.C. (1983) The Goose Arm Karst, Newfoundland. *Journal of. Hydrology*, **61**(1/3), 181–6.

Karp, D. (2002) *Land Degradation Associated with Sinkhole Development in the Katherine Region*, Technical Report No.11/2002, Department of Infrastructure, Planning and Development, Northern Territories Government, 80 pp.

Käss, W. (1967) Erfahrungen mit Uranin bei Farbversuchen. *Steirische Beitrage zur Hydrogeologie*, **18/19**, 123–34.

Käss, W. (1998) Tracing technique in geohydrology. Rotterdam, Balkema, 581p.

Kastning, E.H. (1983) Relict caves as evidence of landscape and aquifer evolution in a deeply dissected carbonate terrain: south-

west Edwards Plateau, Texas, U.S.A. *Journal of Hydrology*, **61**, 89–112.

Katz, B.G. (2004) Sources of nitrate contamination and age of water in large karstic springs of Florida. *Environmental Geology*, **46**, 689–706.

Katz, B.G., Bohlke, J.K. and Hornsby, H.D. (2001) Timescales for nitrate contamination of spring waters, northern Florida, USA. *Chemical Geology*, **179**, 167–86.

Katzer, E. (1909) Karst und Karsthydrograph. *Zur Kunde der Balkan halbinsel* (Sarajevo), **8**.

Kaufman, G. (2002) Ghar Alisadr, Hamadan, Iran: first results on dating calcite shelfstones. *Cave and Karst Science*, **29**(3), 129–33.

Kaufmann, G. (2003) Stalagmite growth and paleoclimate: the numerical perspective. *Earth and Planetary Letters*, **214**(1–2), 251–66.

Kaufmann, G. and Braun, J. (2000) Karst aquifer evolution in fractured, porous rocks. *Water Resources Research*, **36**, 1381–91.

Kaufmann, O. and Quinif, Y. (1999) Cover-collapse sinkholes in the 'Tournaisis' area, southern Belgium. *Engineering Geology*, **52**, 15–22.

Kemmerly, P.R. (1982) Spatial analysis of a karst depression population: clues to genesis. *Geological Society of America Bulletin*, **93**, 1078–86.

Kemmerly, P.R. and Towe, S.K. (1978) Karst depressions in a time context. *Earth Surface Processes and Landforms*, **35**, 355–62.

Kempe, S. and Spaeth, C. (1977) Excentrics: their capillaries and growth rates, *Proceedings of the 7th International Congress of Speleology, Sheffield*, pp. 259–62.

Kempe, S., Brandt, A., Seeger, M. and Vladi, F. (1975) 'Facetten' and 'Lungdecken', the typical morphological elements of caves developed in standing water. *Annales de Speleologie*, **30**(4), 705–8.

Kendall, C. and Caldwell, E.A. (1998) Fundamentals of isotope geochemistry, in *Isotope Tracers in Catchment Hydrology* (eds C. Kendall, and J.J. McDonnell), Elsevier, Amsterdam, pp. 51–86.

Kendall, C. and McDonnell, J.J. (eds) (1998) *Isotope Tracers in Catchment Hydrology*, Elsevier, Amsterdam, 839 pp.

Khang, P. (1991.) Présentation des régions karstiques du Vietnam. *Karstologia*, **18**, 1–12.

Kiernan, K. (1984) *Towards a Forestry Commission Karst and Karst Catchment Management Policy*. Forestry Commission, Tasmania.

Kindinger, J.L., Davis, J.B. and Flocks, J.G. (1999) Geology and evolution of lakes in north-central Florida. *Environmental Geology*, **38**(4), 301–21.

Kiraly, L. (1975) Rapport sur l'etat actuel des connaissances dans le domaine des caracteres physiques des roches karstiques, in *Hydrogeology of Karstic Terrains* (eds A. Burger and L. Dubertret), International Union of Geological Sciences, Series B, Vol. 3, pp. 53–67.

Kiraly, L. (1998) Modelling karst aquifers by the combined discrete channel and continuum approach. *Bulletin d'Hydrogeologie* (Neuchatel), **16**.

Kiraly, L. (2002) Karstification and groundwater flow, in *Evolution of Karst: from Prekarst to Cessation* (ed. F. Gabrovsek), Institut za raziskovanje krasa, ZRC SAZU, Postojna-Ljubljana, pp. 155–90.

Kirkland, D.W. and Evans, R. (1980) Origin of castiles on the Gypsum Plain of Texas and New Mexico. *New Mexico Geological Society, Guidebook*, **31**, 173–8.

Klimchouk, A.B. (1995) Karst morphogenesis in the epikarstic zone. *Cave and Karst Science*, **21**(2), 45–50.

Klimchouk, A.B. (1996) The dissolution and conversion of gypsum and anhydrite. *International Journal of Speleology*, **25**(3–4), 21–36.

Klimchouk, A.B. (2000a) The formation of epikarst and its role in vadose speleogenesis, in *Speleogenesis; Evolution of Karst Aquifers* (eds A.V. Klimchouk, D.C. Ford, A.N. Palmer and W. Dreybrodt), National Speleological Society of America, Huntsville, AL, pp. 91–9.

Klimchouk, A.B. (2000b) Speleogenesis of the Great Gypsum Mazes in the Western Ukraine, in *Speleogenesis; Evolution of Karst Aquifers* (eds A.V. Klimchouk, D.C. Ford, A.N. Palmer and W. Dreybrodt), National Speleological Society of America, Huntsville, AL, pp. 261–73.

Klimchouk, A.B. (2003) Conceptualisation of speleogenesis in multi-storey artesian systems: a transverse speleogenesis. *www. Speleogenesis*, **1**(2), 21 pp.

Klimchouk, A. (2004) Evaporite karst, in *Encyclopedia of Caves and Karst Science* (ed. J. Gunn), Fitzroy Dearborn, New York, pp. 343–7.

Klimchouk, A.B. and Aksem, S.D. (2000) Gypsum karst in the western Ukraine, in *Present State and Future Trends of Karst Studies* (eds G. Günay, K.S. Johnson, D. Ford and A.I. Johnson), IHP-V, Technical Documents in Hydrology, 49(II), UNESCO, Paris, pp. 67–80.

Klimchouk, A.B. and Andrejchouk, V.N. (1986) Geological and hydrogeological conditions of gypsum karst development in the western Ukraine. *Le Grotte d'Italia*, **4**(XII), 349–58.

Klimchouk, A.B. and Andrejchuk, V. (1996) Sulphate rocks as an arena for karst development. *International Journal of Speleology*, **25**(3–4), 9–20.

Klimchouk, A.B. and Andrejchuk, V.N. (2003) Karst breakdown mechanisms from observations in the gypsum caves of the Western Ukraine: implications for subsidence hazard assessment. *www. Speleogenesis*, **1**(4), 20 pp.

Klimchouk, A.B. and Ford, D.C. (2000) Types of Karst and Evolution of Hydrogeologic Settings. in Klimchouk, A.B., Ford, D.C., Palmer, A.N. and Dreybrodt, W. (Editors). *Speleogenesis; Evolution of Karst Aquifers*. Huntsville, Al. National Speleological Society of America, pp. 45–53.

Klimchouk, A.B., Cucchi, F., Calaforra, J.M., *et al.* (1996a) Dissolution of gypsum from field observations. *International Journal of Speleology*, **25**(3–4), 37–48.

Klimchouk, A.B., Sauro, U. and Lazzarotto, M. (1996b) 'Hidden' shafts at the base of the epikarstic zone: a case study from the Sette Communi plateau, Venetian Pre-Alps, Italy. *Cave and Karst Science*, **23**(3), 101–7.

Klimchouk, A.B., Ford, D.C., Palmer, A.N. and Dreybrodt, W. (eds) (1996c) *Speleogenesis; Evolution of Karst Aquifers*, National Speleological Society Press, Huntsville, AL, 527 pp.

Knez, M. (1996) *Bedding-plane impact on the development of karst caves*. Univ. of Lubljana PhD thesis, 186 pp.

Knez, M., Otoničar, B., and Slabe, T. (2003) Subcutaneous stone forest (Trebnje, central Slovenia). *Acta Carsological*, **32**(1), 29–38.

Knisel, W.G. (1972) *Response of Karst Aquifers to Recharge*, Hydrological Paper 60, Colorado State University, Fort Collins, 48 pp.

Kogovsek, J. (1997) Water tracing tests in the vadose zone, in *Tracer Hydrology 97* (ed. A. Kranjc), Rotterdam, Balkema, pp. 167–72.

Kohn, M.J. and Welker, J.M. (2005) On the temperature correlation of $\delta^{18}O$ in modern precipitation. *Earth and Planetary Science Letters*, **231**, 87–96.

Kolodny, Y., Bar-Matthews, M., Ayalon, A. and McKeegan, K.D. (2003) High spatial resolution $\delta^{18}O$ profile of a speleothem using an ion-microprobe. *Chemical Geology*, **197**, 21–8.

Konzuk, J.S. and Kueper, B.H. (2004) Evaluation of cubic law based models describing single-phase flow through a rough-walled fracture. *Water Resources Research* **40**, W02402, doi:10.1029/2003WR002356.

Korotkov, A.N. (ed.) (1974) *Caves of the Pinego-Severodvinskaja Karst*. Geographical Society of the USSR, Leningrad, 191 pp. [In Russian.]

Korpás, L. and Hofstra, A.H. (eds) (1999) *Carlin Gold in Hungary*, T. 24, Geologica Hungarica, Budapest, 331 pp.

Korshunov, V. and Semikolennyh, A. (1994) A model of speleogenic processes connected with bacterial redox in sulfur cycles in the caves of Kugitangtau Ridge, Turkmenia, in *Breakthroughs in Karst Geomicrobiology and Redox Geochemistry* (eds I.D. Sasowsky and M.V. Palmer), Special Publication 1, Karst Waters Institute, pp. 43–4.

Kósa, A. (1995/6) *The Cavers' Living Dictionary*, ER-PETRO, Budapest, 157 pp. [Terms in English, French, German and Hungarian. Updates available online at www.uis-speleo.org]

Kósa, A. (ed.) (1981) *Bir Al Ghanam Karst Study Project*, Jamahiriya, Tripoli, 79 pp.

Kovács, A. (2003). *Geometry and hydraulic parameters of karst aquifers: a hydrodynamic modeling approach*. Faculté des Sciences, Université de Neuchâtel Thesis, 131 pp. [Available in pdf at www.unine.ch/biblio/]

Kowalski, K. (1965) Cave studies in China today. *Studies in Speleology*, **1**(2–3), 75–81.

Kozary, M.T., Dunlap, J.C. Humphrey, W.E. (1968) Incidence of saline deposits in geologic time. *Geological Society of America Special Paper*, **88**, 43–57.

Kranjc, A. (1981) Sediments from Babja Jama near Most na Soci. *Acta Carsologica*, **X**(9), 201–11.

Kranjc, A. (1982) Prod iz Kacne jame. *Nase jame*, **23**(4), 17–23.

Kranjc, A. (1985) The lake of Cerknisko and its floods. *Geografski Zbornik*, **25**(2), 71–123.

Kranjc, A. (ed.) (1989a) *Cave Tourism*, Institute of Karst Research, Postojna, 204 pp.

Kranjc, A. (1989b) *Recent fluvial cave sediments, their origin and role in speleogenesis*. Slovenian Academy of Sciences, Lubljana, 167 pp.

Kranjc, A. (ed.) (1997a) *Tracer Hydrology 97*, Balkema, Rotterdam, 450 pp.

Kranjc, A. (ed.) (1997b) *Karst Hydrogeological Investigations in South-western Slovenia*. *Acta Carsologica Krasoslovni Zbornik* (Ljubljana), **XXVI**(1), 388 pp.

Kranjc, A. (2001a) About the name kras (karst) in Slovenia, *Proceedings of the 13th International Congress of Speleology, Brazilia*, Vol. 2, 140–2.

Kranjc, A. (ed.) (2001b) Classical Karst – Contact Karst; a Symposium. *Acta Carsologica*, **30**(2), 13–164.

Kranjc, A. (ed.) (2002) Monitoring of Karst Caves. *Acta Carsologica, 31*. 177 pp.

Kranjc, A. and Lovrencak, F. (1981) Poplavni svet na Kocevskem polju (Floods in Kocevsko polje). *Geografski zbornik*, **21**, 1–39.

Krawczyk, W.E. (1996) *Manual for Karst Water Analysis*, Handbook 1 – Physical Speleology, International Journal of Speleology, 51 pp.

Krawczyk, W.E. and Ford, D.C. (2006) Correlating specific conductivity with total hardness in limestone and dolomite karst waters. *Earth Surface Processes and Landforms*, **31**, 221–34.

Krawczyk, W.E., Glowacki, P. and Niedzwiedz, T. (2002) Charakterystyka chemiczna opadów atmosferycznych w rejonie Hornsundu (SW Spitsbergen) latem 2000 r. na tle cyrkulacji atmosferycznych, in *Funkcjonowanie i monitoring geoekosystemów obszarów polarnych* (eds A. Kostrzewski and G. Rachlewicz), Polish Polar Studies, Poznań, pp. 187–202. [In Polish.]

Kresic, N. (1995). Remote sensing of tectonic fabric controlling groundwater flow in Dinaric karst. *Remote Sensing of the Environment*, **53**, 85–90.

Kresic, N., O'Laskey, R., Deeb, R., *et al.* (2005) Technical impracticability of DNAPL remediation in karst, in *Water Resources and Environmental Problems in Karst* (eds Z. Stevanovic and P. Milanovic), pp. 63–6.

Kruse, P.B. (1980) *Karst investigations of Maligne Basin, Jasper National Park, Alberta*. Univ. of Alberta MSc thesis, 120 pp.

Kunaver, J. (1982) Geomorphology of the Kanin Mountains with special regard to the glaciokarst. *Geografsky Zbornik*, **XXII**, 200–343.

Kunsky, J. (1950) *Kras a Jeskyne*, Priro, Naklad. Praz, Prague, 263 pp.

Labat, D. Ababou, R. and Mangin, A. (2001) Multifractal and wavelet analyses in karstic hydrology: concepts and applications, in *Present State and Future Trends of Karst Studies* (eds G. Günay, K.S. Johnson, D. Ford and A.I. Johnson), IHP-V, Technical Documents in Hydrology, 49(II), UNESCO, Paris, pp. 441–50.

Lageat, Y., Sellier, D. and Twidale, C.R. (1994) Mégalithes et météorisation des granites en Bretagne littorale, France du nord-ouest. *Géographie physique et Quaternaire*, **48**(1), 107–13.

Lakey, B. and Krothe N. C. (1996) Stable isotopic variation of storm discharge from a perennial karst spring, Indiana. *Water Resources Research*, **32**(3), 721–731.

Lalkovič, M. (1995) On the problem of the ice filling in Dobšina Ice Cave. *Acta Carsologica*, **24**, 314–22.

Lallemand-Barres, A. (1994) *Standardization of Criteria of Establishment of Vulnerability to Pollution Maps. Preliminary Documentary Study*, Report R37928, Bureau de Recherche Geologiques et Minieres, 21 pp.

Lambeck, K. and Chappell, J. (2001) Sea level change through the last glacial cycle. *Science*, **292**, 679–86.

LaMoreaux, P.E. (1989) Water development for phosphate mining in a karst setting in Florida – a complex environmental problem. *Environmental Geology and Water Science*, **14**(2), 117–53.

LaMoreaux, P.E. and Wilson, B.M. (1984) Remote sensing, in *Guide to the Hydrology of Carbonate Rocks* (eds P.E. LaMoreaux, B.M. Wilson and B.A. Memon), Studies and Reports in Hydrology No. 41, UNESCO, Paris, pp. 166–71.

Land, L.A. and Paull, C.K. (2000) Submarine karst belt rimming the continental slope in the Straits of Florida. *Geo-Marine Letters*, **20**, 123–32.

Lange, A.L. (1968) The changing geometry of cave structures. *Cave Notes*, **10**(1–3), 1–10, 13–19, 26–7, 29–32.

Lange, A.L. and Barner, W.L.(1995). Application of the natural electric field for detection of karst conduits on Guam, in *Karst Geohazards* (ed. B. F. Beck), Balkema, Rotterdam, pp. 425–41.

Langmuir, D. (1971) The geochemistry of some carbonate groundwaters in central Pennsylvania. *Geochim et Cosmochim Acta*, **35**, 1023–45.

Langmuir, D. (1996) *Aqueous Environmental Geochemistry*, Prentice Hall, New Jersey, 600 pp.

Lascu, C. (2004) Movile Cave, Romania, in *Encyclopedia of Caves and Karst Science* (ed. J. Gunn), Fitzroy Dearborn, New York, pp. 528–30.

Lasserre, G. (1954) Notes sur le karst de la Guadeloupe. *Erdkunde*, **VIII**(1/4), 115–17.

Last, W.M. (1990) Lacustrine dolomite – an overview of modern, Holocene and Pleistocene occurrences. *Earth Science Reviews*, **27**, 221–63.

Latham A.G. and Ford, D.C. (1993) The paleomagnetism and rock magnetism of cave and karst deposits, in *Applications of Paleomagnetism to Sedimentary Geology*, Special Publication 49, Society of Economic Paleontologists and Mineralogists, Tulsa, OK, pp. 149–155.

Latham, A.G., H.P. Schwarcz, D.C. Ford and W.G. Pearce (1979) Palaeo-magnetism of stalagmite deposits. *Nature*, **280**(5721), 383–5.

Latham, A.G. Schwarcz, H.P. and Ford, D.C. (1986) The paleomagnetism and U–Th dating of Mexican stalagmite, DAS 2. *Earth and Planetary Science Letters*, **79**, 195–207.

Lattman, L.H. and Parizek, R.P. (1964) Relationship between fracture traces and the occurrence of groundwater in carbonate rocks. *Journal of Hydrology*, **2**, 73–91.

Laudermilk, J.D. and Woodford, A.O. (1932) Concerning Rillensteine. *American Journal of Science*, **223**, 135–54.

Lauriol, B. and Clark, I.D. (1993) An approach to determine the origin and age of massive ice blockages in two arctic caves. *Permafrost and Periglacial Processes*, **4**, 77–85.

Lauriol, B. and Clark, I.D. (1999) Fissure calcretes in the arctic: a paleohydrologic indicator. *Applied Geochemistry*, **14**, 775–85.

Lauriol, B. and Gray, J.T. (1990) Drainage karstique en Mileu de Pergélisol: le cas de l'ile d'Akpatok, T.N.O. Canada. *Permafrost and Perigalcial Processes*, **1**, 129–44.

Lauriol, B., Ford, D.C., Cinq-Mars, J. and Morris, W.A. (1997) The chronology of speleothem deposition in Northern Yukon and its relationship to permafrost. *Canadian Journal of Earth Sciences*, **34**(7), 902–11.

Lauritzen, S.-E. (1981) Simulation of rock pendants – small scale experiments on plaster models, *Proceedingsof the 8th International Congress of Speleology, Kentucky*, pp. 407–9.

Lauritzen, S.-E. (1982) The paleocurrents and morphology of Pikhaggrottene, Svartisen, North Norway. *Norsk Geografisk Tidsskrift*, **4**, 184–209.

Lauritzen, S.-E. (1984a) A symposium: arctic and alpine karst. *Norsk Geografisk Tidesskrift*, **38**, 139–214.

Lauritzen, S.-E. (1984b) Evidence of subglacial karstification in Glomdal, Svartisen. *Norsk Geografisk Tidesskrift*, **38**(3–4), 169–70.

Lauritzen, S.-E. (1986) Kvithola at Fauske, northern Norway: an example of ice-contact speleogenesis. *Norsk Geografisk Tidesskrift*, **66**, 153–61.

Lauritzen, S.-E. (1990) Autogenic and allogenic denudation in carbonate karst by the multiple basin method: an example from Svartisen, north Norway. *Earth Surface Processes and Landforms*, **15**, 157–67.

Lauritzen, S.-E. (1995) High-resolution paleotemperature proxy record for the Last Interglaciation based on Norwegian speleothems. *Quaternary Research*, **43**, 133–46.

Lauritzen, S.-E. (1996a) Interaction between glacier and karst aquifers: Preliminary results from Hilmarfjellet, South Spitsbergen. *Kras I Speleologica*, **XVII**, 17–28.

Lauritzen, S.-E. (1996b) Karst landforms and caves of Nordland, North Norway, in *Climate Change: the Karst Record*, Guide to Excursion 2, University of Bergen, 160 pp.

Lauritzen, S.-E. and Karp, D. (1993) *Speleological assessment of Karst Aquifers developed within the Tindall Limestone, Katherine, N.T.* Report 63/1993, Power and Water Authority, Darwin, NT, 60 pp.

Lauritzen, S.-E. and Lundberg, J. (1999a) Speleothems and climate: a special issue of The Holocene. *The Holocene*, **9**(6), 643–7.

Lauritzen, S.-E. and Lundberg, J. (1999b) Calibration of the speleothem delta function: an absolute temperature record for the Holocene in northern Norway. *The Holocene*, **9**(6), 659–69.

Lauritzen, S.-E. and Lundberg, J. (2000) Meso- and micromorphology of caves, in *Speleogenesis; Evolution of Karst Aquifers* (eds A.V. Klimchouk, D.C. Ford, A.N. Palmer and W. Dreybrodt), National Speleological Society of America, Huntsville, AL, pp. 407–26.

Lauritzen, S.-E., Abbott, J., Arnesen, R. *et al.* (1985) Morphology and hydraulics of an active phreatic conduit. *Cave Science*, **12**(4), 139–46.

Lauritzen, S-E., Haugen, J.E., Løvlie, R. and Gilje-Nielsen, H. (1994) Geochronological potential of isoleucine epimerization in calcite speleothems. *Quaternary Research*, **41**, 52–8.

Lea, D.W., Pak, D.K. and Spero, H.J. (2000) Climate impact of late Quaternary equatorial Pacific sea surface temperature variations. *Science*, **289**, 1719–24.

Lee, E.S. and Krothe, N.C. (2001) A four-component mixing model for water in a karst terrain in south-central Indiana, USA. Using solute concentration and stable isotopes as tracers. *Chemical Geology*, **179**(1–4), 129–43.

Lee, E.S. and Krothe N.C. (2003) Delineating the karstic flow system in the upper Lost River drainage basin, south central Indiana: using sulphate and delta S–34(SO4) as tracers. *Applied Geochemistry*, **18**(1), 145–53.

Lee, M.R. and Bland, P.A. (2002) Dating climatic change in hot deserts using desert varnish on meteorite finds. *Earth and Planetary Science Letters*, **6487**, 1–12.

Lehmann, H. (1936) *Morphologische Studien auf Java*. Series 3, No. 9, Geographische. Abhandlungen, Stuttgart, 114 pp.

Lehmann, H. (ed.) (1954) Das Karstphanomen in den verschiedenen Klimazonen. *Erdkunde*, **8**, 112–39.

Lehmann, H.W., Krommelbein, H.K. and Lotschert, W. (1956) Karstmorphologische, geologische und botanische Studien in der Sierra de los Organos auf Cuba. *Erdkunde*, **10**, 185–204.

Lei, M., Jiang, X. and Li, Y. (2001) New advances of karst collapse research in China, in *Geotechnical and Environmental Applications of Karst Geology and Hydrology* (eds B.F. Beck, and J.G. Herring), Balkema, Lisse, pp. 145–51.

Leibundgut, Ch. and Hadi, S. (1997) A contribution to toxicity of fluorescent tracers, in *Tracer Hydrology 97* (ed. A. Kranjc), Rotterdam, Balkema, pp. 69–75.

Leighton, M.W. and Pendexter, C. (1962) Carbonate rock types. *Memoirs, American Association of Petroleum Geologists*, **1**, 33–61.

Leutscher, M. and Jeannin, P.-Y. (2004) Temperature distribution in karst systems: the role of air and water fluxes. *Terra Nova*, **16**, 344–50.

Lewis, D.C., Kriz, G.J. and Burgy, R.H. (1966) Tracer dilution sampling technique to determine hydraulic conductivity of fractured rock. *Water Resources Research*, **2**(3), 533–42.

Lewis, I. and Stace, P. (1980) *Cave Diving in Australia*, I. Lewis. Adelaide.

Ley, R.G. (1979) The development of marine karren along the Bristol Channel coastline. *Zeitschrift für Geomorphologie, Supplement-Band*, **32**, 75–89.

Li, W-X., Lundberg, J., Dickin, A.P., Ford, D.C., *et al.* (1989) High precision mass spectrometric dating of speleothem and implications for paleoclimate studies. *Nature*, **339**(6225), 534–6.

Liedl, R. and Sauter, M. (1998) Modelling of aqifer genesis and heat transport in karst systems. *Bulletin d'Hydrogéologie* (Neuchâtel), **16**, 185–200.

Lignier, V. and Desmets, M. (2002) Les archives sedimentaires quaternaries de la grotte sous les Sangles, Bas-Bugey, Jura méridional, France. *Karstologia*, **39**(1), 27–46.

Linge, H., Lauritzen, S.-E., Lundberg, J. and Berstad, I.M. (2001) Stable isotope stratigraphy of Holocene speleothems:

examples from a cave system in Rana, northern Norway. *Palaeogeography, Palaeoclimatology, Palaeoecology,* **167,** 209–24.

Lismonde, B. (2003) Limestone wall retreat in a ceiling cupola controlled by hydrothermal dissolution with wall condensation (Szunyogh model). *www. Speleogenesis,* **1**(4), 3 pp.

Longley, P.A., Goodchild, M.F., Maguire, D.J. and Rhind, D.W. (2005) *Geographic Information Systems and Science,* 2nd edn, J. Wiley & Sons, New York.

Loop, C.M. and White, W.B. (2001) A conceptual model for DNALP transport in karst ground water basins. *Ground Water,* **39**(1), 119–27.

López-Chicano, A., Calvache, M. L., Martīn-Rosales, W. and Gisbert, J. (2002) Conditioning factors in flooding of karstic poljes – the case of the Zafarraya polje (South Spain). *Catena,* **49**(4), 331–52.

López-Chicano, M. and Pulido-Bosch, A. (1993) The fracturing in the Sierra Gorda karstic system (Granada), in *Some Spanish Karstic Aquifers* (ed. A. Pulido-Bosch), University of Granada, pp. 95–116.

Lowe, D.J. (2000) The speleo-inception concept, in *Speleogenesis; Evolution of Karst Aquifers* (eds A.V. Klimchouk, D.C. Ford, A.N. Palmer and W. Dreybrodt), National Speleological Society of America, Huntsville, AL, pp. 65–75.

Lowe, D.J. and Waltham, A. (2002) *A Dictionary of Karst and Caves,* Cave Studies 10, British Cave Research Association, London.

Lowry, D.C. and Jennings, J.N. (1974) The Nullarbor karst, Australia. *Zeitschrift für Geomorphologie,* **18,** 35–81.

Lu, G., Zheng, C., Donahoe, R.J. and Berry Lyons, W. (2000) Controlling processes in a $CaCO_3$ precipitating stream in Huanglong natural Scenic District, Sichuan, China. *Journal of Hydrology,* **230,** 34–54.

Lu, Y. (1985) *Karst in China: Landscapes, Types, Rules.* Geological Publishing House, Beijing, 288 pp.

Lu, Y. (1993) *Comparative Research on Evolution of Karst Environments in the Main Construction Regions of China,* Institute of Hydrogeology and Engineering Geology, Chinese Academy of Geological Sciences, Beijing, 13 pp.

Ludwig, K.R., Simmons, K.R., Szabo, B.J., *et al.* (1992) Mass-spectrometric ^{230}Th–^{234}U–^{238}U dating of the Devils Hole Calcite Vein. *Science,* **258,** 284–7.

Lundberg, J., Ford, D.C and Hill, C.A. (2000) A preliminary U–Pb date on cave spar, Big Canyon, Guadalupe Mountains, New Mexico, U.S.A. *Journal of Cave and Karst Studies,* **62**(2), 144–8.

Lynch, F.L., Mahler, B.J. and Hauwert, N.N. (2004) Provenance of suspended sediment discharged from a karst aquifer determined by clay mineralogy, in *Studies of Cave Sediments: Physical and Chemical Records of Paleoclimate* (eds I.D. Sasowsky and J. Mylroie), Kluwer Academic, New York, pp. 83–94.

Macaluso, T. and Sauro, U. (1996) Weathering crust and karren on exposed gypsum surfaces. International *Journal of Speleology,* **25**(3–4), 115–26.

Mace, C.E. (1921) Sewer system more than a million years old. *Popular Mechanics Magazine* **35**(5), 687.

Machel, H.G. (2001) Bacterial and thermochemical sulphate reduction in diagenetic settings – old and new insights. *Sedimentary Geology,* **140,** 143–75.

Maclay, R.W. and Small, T.A. (1983) Hydrostratigraphic subdivisions and fault barriers of the Edwards aquifer, south-central Texas, U.S.A. *Journal of Hydrology,* **61,** 127–46.

Magdalene, S. and Alexander, E.C. (1995) Sinkhole distribution in Winona County, Minnesota revisited, in *Karst Geohazards* (ed. B.F. Beck), Balkema, Rotterdam, pp. 43–51.

Maillet, E. (1905) *Essais d'Hydraulique souterraine et fluviale,* Hermann, Paris.

Mainguet, M. (1972) *Le Modele des Gres: Problemes Generaux,* Institut Geographique National, Paris.

Maire, R. (1981a) Giant shafts and underground rivers of the Nakanai Mountains (New Britain). *Spelunca,* **3**(Supplement), 8–9.

Maire, R. (1981b) Karst and hydrogeology synthesis. *Spelunca,* **3**(Supplement), 23–30.

Maire, R. (1981c) Inventory and general features of PNG karsts. *Spelunca,* **3**(Supplement), 7–8.

Maire, R. (1986) A propos des karsts sous-marins. *Karstologia,* **7,** 55.

Maire, R. (1990) La Haute Montagne Calcaire. *Karstologia Memoire,* **3,** 731 pp.

Maire, R. (1999) Les glaciers de marbre de Patagonie, Chili. Un karst subpolaire oceanique de la zone australe. *Karstologia,* **33,** 25–40.

Maire, R. et l'equipe Ultima Esperanza (1999) Les 'glaciers de marbre' de Patagonie, Chili. *Karstologia,* **33,** 25–40.

Malis, C.P. (1997) Littoral Karren along the Western Shore of Newfoundland,. MSc thesis, McMaster University. 310 p.

Mangin, A. (1969a) Nouvelle interprtation du mcanisme des sources intermittentes. *Comptes Rendus, Académie des Sciences (Paris),* **269,** 2184–6.

Mangin, A. (1969b) Etude hydraulique du mecanisme d'intermittence de Fontestorbes (Blesta-Ariege). *Annales et Speleologie,* **24**(2), 253–98.

Mangin, A. (1973) Sur la dynamiques des transferts en aquifere karstique, *Proceedings of the 6th International Congress of Speleology, Olomouc, CSSR,* Vol. 6, 157–62.

Mangin, A. (1975) *Contribution a l'etude hydrodynamique des aquiferes karstiques.* Univ. Dijon These Doct. es. Sci. [*Annales et Speleologie,* **29**(3), 283–332; **29**(4), 495–601, 1974; **30**(1), 21–124, 1975.]

Mangin, A. (1981a) Utilisation des analysis correlatoire et spectrale dans l'approche des systemes hydrologiques. *Comptes Rendus, Académie des Sciences (Paris),* **293,** 401–4.

Mangin, A. (1981b) Apports des analyses correlatoire et spectrale croisees dans la connaissance des sytsemes hydrologiques. *Comptes Rendus, Académie des Sciences (Paris),* **293**(II), 1011–14.

Mangin, A. (1984) Pour une meilleure connaissance des systemes hydrologiques à partir des analyses correlatoire et spectrale. *Journal of Hydrology,* **67,** 25–43.

Mangin, A. (1998) L'approche hydrogéologique des karsts. *Spéléochronos,* **9,** 3–26.

Mangini, A., Spotl, C. and Verdes, P. (2005) Reconstruction of temperature in the Central Alps during the past 2000 yr from a $\delta^{18}O$ stalagmite record. *Earth and Planetary Science Letters*, **235**, 741–51.

Margat, J. (1980) *Carte hydrogeologique de la France a 1: 1 500 000*. Bureau de Recherches Geologiques et Minieres, Orleans.

Margrita, R., Guizerix, J., Corompt, P., *et al*. (1984) Reflexions sur la theorie des traceurs: applications en hydrologie isotopique, in *Isotope Hydrology 1983*, International Atomic Energy Agency, Vienna, pp. 653–78.

Marinos, P.G. (2005) Experiences in tunnelling through karstic rocks, in *Water Resources and Environmental Problems in Karst* (eds Z. Stevanović and P. Milanović), International Association of Hydrogeologists, Belgrade, pp. 617–44.

Marker, M.E. (1976) Cenotes: a class of enclosed karst hollows. *Zeitschrift für Geomorphologie Supplement-Band*, **26**, 104–23.

Marquette, G.C., Gray, J.T., Gosse, J.C., *et al*. (2004) Felsenmeer persistence under non-erosive ice in the Torngat and Kaumajet mountains, Quebec and Labrador, as determined by soil weathering and cosmogenic nuclide exposure dating. *Canadian Journal of Earth Sciences*, **41**(1), 19–38.

Martel, E.A. (1894) *Les Abimes*, Delagrave, Paris.

Martel, E.A. (1921) *Nouveau trait des eaux sonterraines*, Editions Doin, Paris, 840 pp.

Martin, J.B., Wicks, C.M. and Sasowsky, I.D. (eds) (2002) *Hydrogeology and Biology of Post-Paleozoic Carbonate Aquifers*, Special Publication 7, Karst Waters Institute, Charles Town, WV, 212 pp.

Martin-Algarra, A., Martin, M., Andreo, B., *et al*. (2003) Sedimentary patterns in perched spring travertines near Granada (Spain) as indicators of the paleohydrological and paleoclimatological evolution of a karst massif. *Sedimentary Geology*, **161**, 217–28.

Martini, J.E.J. (2000) Dissolution of quartz and silicate minerals, *Speleogenesis; Evolution of Karst Aquifers* (eds A.V. Klimchouk, D.C. Ford, A.N. Palmer and W. Dreybrodt), National Speleological Society of America, Huntsville, AL, pp. 171–4.

Martini, J.E.J. (2004) Silicate karst, in *Encyclopedia of Caves and Karst Science* (ed. J. Gunn), Fitzroy Dearborn, New York, pp. 649–53.

Martini, J.E.J., Wipplinger, P.E., Moen, H.F.G. and Keyser, A. (2003) Contribution to the speleology of Sterkfontein Cave, Gauteng Province, South Africa. *International Journal of Speleology*, **32**(1/4), 43–69.

Martini, P., Brookfield, M.E. and Sadura, S. (2001) *Principles of Glacial Geomorphology and Geology*, New Jersey, Prentice Hall, 381 pp.

Matthews, M.C., Clayton, C.R.I. and Rigby-Jones, J. (2000) Locating dissolution features in the Chalk. *Quarterly Journal of Engineering Geology and Hydrogeology*, **33**, 125–40.

Maurin, V. and Zotl, J. (1959) Die Untersuchung der Zusammenhange unterirdischer Wasser mit besonderer Berucksichtigung der Karstvarheltnisse. *Steirische Beitrage zur Hydro-geologie* (Graz), **11**, 5–184.

Maurin, V. and Zotl, J. (1963) Karsthydrologische Untersuchungen auf Kephallenia. *Osterreische Hochschulz*, **15**, 6.

Mavlyudov, B.R. (ed.) (2005) *Glacier Caves and Glacial Karst in High Mountains and Polar Regions*, Institute of Geography, Russian Academy of Sciences, Moscow, 177 pp.

Maximovich, G.A. and Bykov, V.N. (1976) *Karst of Carbonate Oil- and Gas-bearing Series*, Ministry of Higher Education, Perm, 96 pp. [In Russian.]

Mayr, A. (1953) Blutenpollen und pflanzliche Sporen als Mittel zur Untersuchung von Quellen und Karstwassern. *Anzeiger der Osterreichischen Akademie der Wissenschaften, Mathimatisch-Naturwissenschaftliche Klasse* (Wien).

McConnell, H. and Horn, J.M. (1972) Probabilities of surface karst, in *Spatial Analysis in Geomorphology* (ed. R.J. Chorley), London, Methuen, pp. 111–33.

McDermott, F. (2004) Palaeo-climate reconstruction from stable isotope variations in speleothems: a review. *Quaternary Science Reviews*, **23**, 901–918.

McDermott, F., Frisia, S, Huang, Y, *et al*. (1999) Holocene climate variability in Europe: evidence from $\delta^{18}O$, textural and extension-rate variations in three speleothems. *Quaternary Science Reviews*, **18**, 1021–38.

McDermott, F., Mattey, D.P. and Hawkesworth, C. (2001) Centennial-scale Holocene climate variability revealed by a high-resolution speleothem $\delta^{18}O$ record from SW Ireland. *Science*, **294**, 1328–31.

McDonald, B.S. and Vincent, J.S. (1972) Fluvial sedimentary structures formed experimentally in a pipe: their implications for interpretation of subglacial sedimentary environments. *Geological Survey of Canada Paper*, **72-77**, 30 pp.

McDonald, J., Drysdale, R. and Hill, D. (2004) The 2002–2003 El Niño recorded in Australian cave drip waters: Implications for reconstructing rainfall histories using stalagmites. *Geophysical Research Letters*, **31**, L22202.

McDonald, R.C. (1976) Limestone morphology in south Sulawesi, Indonesia. *Zeitschrift für Geomorphologie, Supplement-Band*, **26**, 79–91.

McDonald, R.C. (1979) Tower karst geomorphology in Belize. *Zeitschrift für Geomorphologie, Supplement-Band*, **32**, 35–45.

McGarry, S.F. and Caseldine, C. (2004) Speleothem palynology: an undervalued tool in Quaternary Studies. *Quaternary Science Reviews*, **23**, 2389–2404.

McGarry, S., Bar-Matthews, B., Matthews, A., *et al*. (2004) Constraints on hydrological and paleotemperature variations in the Eastern Mediterranean region in the last 140 ka given by the δD values of speleothem fluid inclusions. *Quaternary Science Reviews*, **23**, 919–34.

McGrath R. J., Styles P., Thomas E. Neale S. (2002) Integrated high-resolution geophysical investigations as potential tools for water resource investigations in karst terrain. *Environmental Geology*, **42**(5), 552–557.

Mecchia, M. and Piccini, L. (1999) Hydrogeology and SiO_2 geochemistry of the Aonda Cave System, Auyan-Tepui, Bolivar, Venezuela. *Bollettino Sociedad Venezolana Espeleologia*, **33**, 1–18.

Meijerink, A.M.J. (2000). Groundwater, in *Remote Sensing in Hydrology and Water Management* (eds G.A. Schultz and E.T. Engman), Springer-Verlag, Berlin, pp. 305–25.

Meiman, J. and Ryan, M.T. (1999) The development of basin-scale conceptual models of the active-flow conduit system, in *Karst Modeling* (eds A.N. Palmer, M.V. Palmer and I.D. Sasowsky) (1999), Special Publication 5, Karst Waters Institute, Charles Town, WV, pp. 203–12.

Mellett, J.S. and Maccarillo, B. J. (1995). A model for sinkhole formation on interstate and limited access highways, with suggestions on remediation, in *Karst Geohazards* (ed. B.F. Beck), Balkema, Rotterdam, pp. 335–9.

Merrill, G.K. (1960) Additional notes on vertical shafts in limestone caves. *Bulletin of the National Speleological Society*, **22**(2), 101–5.

Meyer-Peter, E. & Muller, R. (1948) Formulas for bedload transport, *Proceedings of the 3rd Conference of the International Association for Hydraulic Research, Stockholm*, pp. 39–64.

Miall, L.C. (1870) On the formation of swallow-holes or pits with vertical sides in mountain limestone. *Nature*, **ii**, *526* (and *Geological Magazine*, **ii**, 513).

Michie, N. (1997) The threat to caves of the human dust sources, *Proceedings of the 12th International Congress of Speleology*, pp. 43–6.

Mickler, P.J., Banner, J.L., Stern, L., *et al.* (2004a) Stable isotope variations in modern tropical speleothems: evaluating equilibrium vs. kinetic isotope effects. *Geochimica et Cosmochimica Acta*, **68**(21), 4381–93.

Mickler, P.J., Ketcham, R.A., Colbert, M.W. and Banner, J.L. (2004b) Application of high-resolution X-ray computed tomography in deteremining the suitability of speleothems for use in paleoclimatic, paleohydrologic reconstructions. *Journal of Cave and Karst Studies*, **66**(1), 4–8.

Middleton, G.V. (ed.) (2003) *Encyclopedia of Sediments and Sedimentary Rocks*. Kluwer Academic Publishers, Dordrecht, 821 pp.

Middleton, G. (2004) Madagascar, in *Encyclopedia of Caves and Karst Science* (ed. J. Gunn), Fitzroy Dearborn, New York, pp. 493–5.

Mijatovic, B.F. (1984a) Problems of sea water intrusion into aquifers of the coastal Dinaric karst, in *Hydrogeology of the Dinaric Karst* (ed. B.F. Mijatovic), International Contributions to Hydrogeology 4, Heise, Hannover, pp. 115–42.

Mijatovic, B.F. (1984b) Karst poljes in Dinarides, in *Hydrogeology of the Dinaric Karst* (ed. B.F. Mijatovic), International Contributions to Hydrogeology 4, Heise, Hannover, pp. 87–109.

Milanović, P. (1976) Water regime in deep karst. Case study of the Ombla spring drainage area, in *Karst Hydrology and Water Resources: Vol. 1 Karst Hydrology* (ed. V. Yevjevich), Water Resources Publications, Colorado, pp. 165–91.

Milanović, P.T. (1981) *Karst Hydrogeology*, Water Resources Publication, Colorado, 434 pp.

Milanović, P.T. (1993) The karst environment and some consequences of reclamation projects, in *Proceedings of the International Symposium on Water Resources in Karst with Special Emphasis on Arid and Semi Arid Zones* (ed. A. Afrasiabian), Shiraz, Iran, pp. 409–24.

Milanović, P.T. (2000) *Geological Engineering in Karst: Dams, Reservoirs, Grouting, Groundwater Protection, Water Tapping, Tunnelling*, Zebra, Belgrade, 347 pp.

Milanović, P.T. (2002) The environmental impacts of human activities and engineering constructions in karst regions. *Episodes*, **25**(1), 13–21.

Milanović, P.T. (2003) Prevention and remediation in karst engineering, in *Sinkholes and the Engineering and Environmental Impacts of Karst* (ed. B.F. Beck), Proceedings of the 9th Multidisciplinary Conference, Huntsville, Alabama, Geotechnical Special Publication 122, American Society of Civil Engineers, pp. 3–28.

Milanović, P.T. (2004) *Water Resources Engineering in Karst*, CRC Press, Boca Raton, 312 pp.

Miller, G.H. (1980) Amino acid geochronology: integrity of the carbonate matrix and potential of molluscan fossils, in *Biogeochemistry of Amino Acids* (eds P.E. Hare, T.C. Hoering and J. King), John Wiley & Sons, New York, pp. 415–43.

Miller, T.E. (1982) *Hydrochemistry, hydrology and morphology of the Caves Branch karst, Belize*. Mcmaster Univ. PhD thesis, 280 pp.

Mills, W.R.P. (1981) Karst development and groundwater flow in the Quatsino Formation, northern Vancouver Island. MSc. thesis, McMaster Univ. 170 pp.

Miotke, F.D. (1968) Karstmorphologische Studien in der glazial-uberformten Hohenstufe der 'Picos de Europa', Nordspanien. *Jahrbuch der geographischen Gesellschaft zu Hannover/Sonderhef*, **4**, 161 pp.

Miotke, F.-D. (1974) Carbon dioxide and the soil atmosphere. *Abhandlungen zur Karst-und Höhlenkunde, Reihe A, Speläologie*, **9**, 52 pp.

Monroe, W.H. (1964) The origin and interior structure of mogotes, *Proceedings of the 20th International Geographical Congress, Abstracts*, p. 108.

Monroe, W.H. (1966) Formation of tropical karst topography by limestone solution and reprecipitation. *Caribbean Journal of Science*, **6**, 1–7.

Monroe, W.H. (1968) The karst features of northern Puerto Rico. *Bulletin of the National Speleological Society*, **30**, 75–86.

Monroe, W.H. (1974) Dendritic dry valleys in the cone karst of Puerto Rico. *Journal of Research, U.S. Geological Survey*, **2**(2), 159–63.

Moon, B.P. (1985) Controls on the form and development of rock slopes in fold terrace, in *Hillslope Processes* (ed. A.D. Abrahams), *16th Annual Binghamton Symposium*, Program and Abstracts, 22 pp.

Moore, C.H. (2001) *Carbonate Reservoirs: Porosity, Evolution and Diagenesis in a Sequence Stratigraphic Framework*, Developments in Sedimentology 55, Elsevier, Amsterdam, 444 pp.

Moriarty, K.C., McCulloch, M.T., Wells, R.T. and McDowell, M.C. (2000) Mid-Pleistocene cave fills, megafaunal remains and climate change at Naracoorte, South Australia: towards a predictive model using U–Th dating of speleothems. *Palaeogeography, Palaeoclimatology, Palaeoecology*, **159**, 113–43.

Mörner, N.-A. (2005) Sea level changes and crustal movements with special aspects on the eastern Mediterranean. *Zeitschrift für Geomorphologie*, **137**, 91–102.

Morrow, D. (1998) Regional subsurface dolomitization: models and constraints. *Geoscience Canada*, **25**(2), 57–70.

Morse, J.W. and Arvidson, R.S. (2002) The dissolution kinetics of major sedimentary carbonate minerals. *Earth Science Reviews*, **58**, 51–84.

Morwood, M.J., Soejono, R.P., Roberts, R.G., *et al.* (2004) Archaeology and age of a new hominin from Flores in eastern Indonesia. *Nature*, **431**, 1087–91.

Moser, H. (1998a) Environmental Isotopes, in *Tracing Technique in Geohydrology* (ed. W. Käss), Balkema, Rotterdam, pp. 279–303.

Moser, H. (1998b) Radiohydrometrical single-well-methods, in *Tracing Technique in Geohydrology* (ed. W. Käss), Balkema, Rotterdam, pp. 382–96.

Moser, H., Rajner, V., Rank, D. and Stichler, W. (1976) Results of measurements of the content of deuterium, oxygen-18 and tritium in water samples from test area taken during 1972–1975, in *Underground Water Tracing* (eds R. Gospodaric and P. Habic), Institute of Karst Research, Postojna, Ljubljana, pp. 93–117.

Moser, M. and Geyer, M. (1979) Seismospelaologic-Erdbebenzerstorungen in Hoblen am Beispiel des Gaislochs bei Oberfellendorf (Oberfranken, Bayern). *Die Hohle*, **4**, 89–102.

Mottershead, D.N., Moses, C.A. and Lucas, G.R. (2000) Lithological control of solution flute form: a comparative study. *Zeitschrift fur Geomorphologie*, **44**(4), 491–512.

Mueller, M. (1987) *Takaka Valley Hydrogeology (Preliminary Assessment)*, Nelson Catchment Board and Regional Water Board.

Muhs, D.R. (2002) Evidence for the timing and duration of the last interglacial period from high-precision uranium-series ages of corals on tectonically stable coastlines. *Quaternary Research*, **58**, 36–40.

Muldoon, M.A., Simo, J.A. and Bradbury, K.R. (2001) Correlation of hydraulic conductivity with stratigraphy in a fractured-dolomite aquifer, northeastern Wisconsin, USA. *Hydrogeology Journal*, **9**, 570–583.

Muller, P. and I. Sarvary (1977) Some aspects of developments in Hungarian speleology theories during the last ten years. *Karsztés Barlang*, Special Issue, pp. 53–59.

Myers, A.J. (1962) A fossil sinkhole. *Oklahoma Geolological Notes*, **22**(B677), 13–15.

Mylroie, J.E. (1984) Hydrologic classification of caves and karst, in *Groundwater as a Geomorphic Agent* (ed. R. G. LaFleur), Allen & Unwin, London, pp. 157–72.

Mylroie, J.E. and Carew, J.L. (1990) The flank margin model for dissolution cave development in carbonate platforms. *Earth Surface Processes and Landforms*, **15**, 413–24.

Mylroie, J.E. and Carew, J.L. (1995) Karst development on carbonate islands, in *Unconformities and Porosity in Carbonate Strata* (eds D.A. Budd, P.M. Harris and A. Saller), Memoir

63, American Association of Petroleum Geologists, Tulsa, OK, pp. 55–76.

Mylroie, J.E. and Carew, J.L. (2000) Speleogenesis in Coastal and Oceanic Settings, in *Speleogenesis; Evolution of Karst Aquifers* (eds A.V. Klimchouk, D.C. Ford, A.N. Palmer and W. Dreybrodt), National Speleological Society of America, Huntsville, AL, pp. 226–33.

Mylroie, J.E. and Jenson, J.W. (2002) Karst flow systems in young carbonate islands, in *Hydrogeology and Biology of Post-Paleozoic Carbonate Aquifers* (eds J.B. Martin, C.M. Wicks and I.D. Sasowsky), Special Publication 7, Karst Waters Institute, Charles Town, WV, pp. 107–10.

Mylroie, J.E. and Vacher, H.L. (1999) A conceptual view of carbonate island karst, in *Karst Modelling* (eds A.N. Palmer, M.V. Palmer and I.D. Sasowsky), Special Publication 5, Karst Waters Institute, Charles Town, WV, pp. 48–57.

Mylroie, J.E., Carew, J.L. and Vacher, H.L. (1995) Karst development in the Bahamas and Bermuda, in *Terrestrial and Shallow Marine Geology of the Bahamas and Bermuda* (eds H.A. Curran and B. White). *Geological Society of America Special Paper*, **300**, 251–67.

Nash, D.J. (2004) Calcrete, in *Encyclopedia of Geomorphology* (ed. A.S. Goudie), Routledge, London and New York, pp. 108–11.

Naylor, L.A. and Viles, H.A. (2002) A new technique for evaluating short-term rates of coastal bioerosion and bioprotection. *Geomorphology*, **47**, 31–44.

Neumann, A.C. (1968) Biological erosion of limestone coasts, in *Encyclopedia of Geomorphology* (ed. R.W. Fairbridge), Reinhold, New York, pp. 75–81.

Neumeier, U. (1999) Experimental modelling of beach rock cementation under microbial influence. *Sedimentary Geology*, **126**, 35–46.

Neukum, C. and Hötzl, H. (2005) Standardisation of vulnerability map, in *Water Resources and Environmental Problems in Karst* (eds Z. Stevanović and P. Milanović), National Committee of the International Association of Hydrogeologists of Serbia and Montenegro, Belgrade, pp. 11–18.

Newitt, D.M., Richardson, J.F., Abbott, M. and Turtle, R.B. (1955) Hydraulic conveying of solids in horizontal pipes. *Transactions Institute of Chemical Engineers*, **33**, 93–110.

Nicod, J. (1976) Karst des gypses et des evaporites associees. *Annales de Geographie*, **471**, 513–54.

Nicod, J. (1990) Murettes et terrasses de culture dans les regions karstiques méditerranéenes. *Méditerranée*, **3–4**, 43–50.

Nicod, J. (1992) Barrages en terrains karstiques; problèmes géomorphologiques et géotechniques dans le domaine Méditerranéen, in *Géo-Méditer, Géographie physique et Méditerranée* (eds M. Tabeau, P. Pech, and L. Simon) (Hommage à G. Beaudet et E. Miossenet), Sorbonne, Paris, pp. 185–200.

Nicod, J. (1996) Karst et mines en France et en Europe. *Karstologia*, **27**(1), 1–20.

Nicod, J. (1997) Les canyons karstique "nouvelles approches de problèmes géomorphologiques classigues" (spécialement dans les domaines méditerranéens et tropicau). *Quateraire*, **8**(2–3), pp. 71–89.

Nicod, J. (2002) Pamukkale (Hiéropolis): un site de travertins hydrothermaux exceptionnel de Turquie. *Karstologia*, **39**, 51–4.

Nicod, J., Julian, M. and Anthony, E. (1996) A historical review of man–karst relationships: miscellaneous uses of karst and their impact. *Rivista di Geografia Italiana*, **103**, 289–338.

Nishiikumi, K. (1993) Role of in situ cosmogenic nuclides [10]Be and [26]Al in the study of diverse geomorphic processes. *Earth Surface Processes and Landforms*, **18**, 407–25.

Noel, M. and Bull, P.A. (1982) The palaeomagnetism of sediments from Clearwater Cave, Mulu, Sarawak. *Cave Science*, **9**(2),134–41.

Noller, J.S., Sowers, J.M. and Lettis, W.R. (2000) *Quaternary Geochronology: Methods and Applications*, American Geophysical Union, Washington, DC, 582 pp.

Nordstrom D.K. (2004) Modeling low-temperature geochemical processes, in *Treatise in Geochemistry, 5, Surface and Ground Water, Weathering Soils* (ed. I. Drever), Elsevier-Pergamon, Oxford, 626 pp.

Northup, D.E. and Lavoie, K.H. (2001) Geomicrobiology of caves: a review. *Geomicrobiology Journal*, **18**, 199–222.

Nunn, P. (1986) Implications of migrating geoid anomalies for the interpretation of high-level fossil coral reefs. *Bulletin, Geological Society of America*, **97**, 946–52.

Olive, W.W. (1957) Solution subsidence troughs, Castile Formation of gypsum plain, Texas and New Mexico. *Geological Society of America Bulletin*, **68**(B693), 351–8.

Ollier, C.D. (1975) Coral island geomorphology:the Trobriand Islands. *Zeitschrift fur Geomorphologie*, **19**(2), 164–90.

Onac, B.P. (1997) Crystallography of speleothems, in *Cave Minerals of the World*, 2nd edn (eds C.A. Hill and P. Forti), National Speleological Society of America, Huntsville, AL, pp. 230–36.

Onac, B.P. and Vereş, D.Ş. (2003) Sequence of secondary phosphate deposition in a karst environment: evidence from Măgurici Cave, Romania. *European Journal of Mineralogy*, **15**, 741–745.

Onac, B.P., Viehmann, I., Lundberg, J., *et al.* (2005) U–Th ages constraining the Neanderthal footprint at Vârtop Cave, Romania. *Quaternary Science Reviews*, **24**, 1151–7.

Onac, B.P., Tudor, T., Constantin, S. and Persoiu, A. (eds) (2006) *Archives of Climate Change in Karst*, Special Publication 10, Karst Waters Institute, Charles Town, WV, 246 pp.

O'Neill, J.R., Clayton, R.N. and Mayeda, T. (1969) Oxygen isotope fractionation in divalent metal carbonates. *Journal of Chemistry and Physics*, **51**, 5547–8.

Ordonez, S., Gonzalez Martin, J.A., Garcia del Cura, M.A. and Pedley, H.H. (2005) Temperate and semi-arid tufas in the Pleistocene to Recent fluvial barrage system in the Mediterranean area: The Ruidera Lakes Natural Park (Central Spain). *Geomorphology*, **69**, 332–50.

Osborne, R.A.L (1999) The origin of the Jenolan Caves: elements of a new synthesis and framework chronology, *Proceedings of the Linnean Society of New South Wales*, **121**, 1–27.

Osborne, R.A.L. (2001) Petrography of lithified cave sediments, *Proceedings of the 13th International Congress of Speleology, Brasilia*, Vol. 1, pp. 101–4.

Osborne, R.A.L. (2004) The trouble with cupolas. *Acta Carsologica*, **33**(2), 9–36.

Osmaston, H. and Sweeting, M.M. (1982) Geomorphology (of the Gunung Mulu National Park). *Sarawak Museum Journal*, **30**(51, new series), 75–93.

Ota, Y. and Chappell, J. (1999) Holocene sea-level rise and coral reef growth on a tectonically rising coast, Huon Peninsula, Papua New Guinea. *Quaternary International*, **55**, 51–9.

Padilla A., Pulidobosch A. Mangin A. (1994) Relative importance of baseflow and quickflow from hydrographs of karst spring. *Ground Water*, **32**(2), 267–77.

Palmer, A.N. (1975) The origin of maze caves. *Bulletin of the National Speleological Society*, **37**(3), 56–76.

Palmer, A.N. (1984) Geomorphic interpretation of karst features, in *Groundwater as a Geomorphic Agent* (ed. R.G. LaFleur), Allen & Unwin, London, pp. 173–209.

Palmer, A.N. (1991) Origin and morphology of limestone caves. *Geological Society of America Bulletin*, **103**, 1–21.

Palmer, A.N. (1995) Geochemical models for the origin of macroscopic solution porosity in carbonate rocks. in *Unconformities and Porosity in Carbonate Strata* (eds D.A. Budd, P.M. Harris and A. Saller), Memoir 63, American Association of Petroleum Geologists, Tulsa, OK, pp. 77–101.

Palmer, A.N. (1999) Anisotropy in carbonate aquifers, in *Karst Modeling* (eds A.N. Palmer, M.V. Palmer and I.D. Sasowsky) (1999), Special Publication 5, Karst Waters Institute, Charles Town, WV, pp. 223–7.

Palmer, A.N. (2000) Hydrogeologic control of cave patterns, in *Speleogenesis; Evolution of Karst Aquifers* (eds A.V. Klimchouk, D.C. Ford, A.N. Palmer and W. Dreybrodt), National Speleological Society of America, Huntsville, AL, pp. 77–90.

Palmer, A.N. (2002) Speleogenesis in carbonate rocks, in *Evolution of Karst: from Prekarst to Cessation* (ed. F. Gabrovsek), Institut za raziskovanje krasa, ZRC SAZU, Postojna-Ljubljana, pp. 43–59.

Palmer, A.N. and Audra, P. (2003) Patterns of caves, in *Encyclopedia of Caves and Karst Science* (ed. J. Gunn), Fitzroy Dearborn, New York, pp. 573–5.

Palmer, A.N. and Palmer, M.V. (1989) Geologic history of the Black Hills Caves, South Dakota. *National Speleological Society, Bulletin*, **51**, 72–99.

Palmer, A.N. and Palmer, M.V. (1995) The Kaskaskia paleokarst of the Northern Rocky Mountains and the Black Hills, northwestern U.S.A. *Carbonates and Evaporites*, **10**, 148–60.

Palmer, A.N. and Palmer, M.V. (2000) Speleogenesis of the Black Hills Maze Caves, South Dakota, U.S.A, in *Speleogenesis; Evolution of Karst Aquifers* (eds A.V. Klimchouk, D.C. Ford, A.N. Palmer and W. Dreybrodt), National Speleological Society of America, Huntsville, AL, pp. 274–81.

Palmer, A.N. and Palmer, M.V. (2003) Geochemistry of capillary seepage in Mammoth Cave. *www.Speleogenesis*, **1**(4), 10 pp.

Palmer, A.N., Palmer, M.V. and Sasowsky, I.D. (eds) (1999) *Karst Modeling*, Special Publication 5, Karst Waters Institute, Charles Town, WV, 265 pp.

Palmer, M.V. and Palmer, A.N. (1975) Landform development in the Mitchell Plain of southern Indiana: origin of a partially karsted plain. *Zeitschrift für Geomorphologie*, N.F., **19**(1), 1–39.

Palmer, M.V. and Palmer, A.N. (1989) Paleokarst of the United States, in *Paleokarst; a Systematic and Regional Review* (eds P. Bosak, D.C. Ford, J. Glazek and I. Horacek), Academia Praha/Elsevier, Prague, pp. 337–63.

Palmer, R.J. and Heath, L.M. (1985) The effect of anchialine factors and fracture control on cave development below eastern Grand Bahama. *Cave Science*, **12**(3), 93–7.

Paloc, H. and J. Margat (1985) Report on hydrogeological maps of karstic terrains, in *Hydrogeological Mapping in Asia and the Pacific Region* (eds W. Grimelmann, K.D. Krampe and W. Struckmeier), International Contributions to Hydrogeology, Vol. 7, Heise, Hannover, pp. 301–15.

Panno, S.V., Hackley, K.C., Hwang, H.H. and Kelly, W.R. (2001) Determination of the sources of nitrate contamination in karst springs using isotopic and chemical indicators. *Chemical Geology*, **179**, 113–28.

Panoš, V. (2001) *Karstological and Speleological Terminology*. Slovak Caves Administration and Geological Institute of the Czech Academy of Sciences, Zilina, Slovakia, 352 pp. [Definitions in the Slovak language, with the equivalent terms in English, French, German, Italian, Russian and Spanish.]

Panos, V. and Stelcl O. (1968) Physiographic and geologic control in development of Cuban mogotes. *Zeitschrift für Geomorphologie*, **12**(2), 117–73.

Parés, J.M. and Pérez-González, A. (1995) Paleomagnetic age for hominid fossils at Atapuerca archaeological site, Spain. *Science*, **269**, 830–2.

Parizek, R.P. (1976) On the nature and significance of fracture traces and lineaments in carbonate and other terranes, in *Karst Hydrology and Water Resources: Vol. 1 Karst Hydrology* (ed. V. Yevjevich), Water Resources Publications, Colorado, pp. 47–108.

Parkhurst, D.L. and Appelo, C.A.J. (1999) *User's guide to PHREEQC (Version 2) – a computer program for Speciation, Batch Reaction, One-Dimensional Transport and Inverse Geochemical Calculations*, Report 99–4259, Water Resources Investigations, US Geological Survey.

Pasini, G. (1973) Sull'importanza speleogenetica dell' 'erosione antigravitativa'. *Le Grotte d'Italia*, **4**, 297–326.

Paterson, K. (1979) Limestone springs in the Oxfordshire Scarplands: the significance of spatial and temporal variations in their chemistry. *Zeitschrift für Geomorphologie, N.F. Supplement-Band*, **32**, 46–66.

Patterson, D.A., Davey, J.C. Cooper, A.H. and Ferris, J.K. (1995) The investigation of dissolution subsidence incorporating microgravity geophysics at Ripon, Yorkshire. *Quaterly Journal of Engineering Geology*, **28**, 83–94.

Paulsen, D.E., Li, H.-C. and Ku, T.-L (2003) Climate variability in central China over the last 1270 years revealed by high-resolution stalagmite records. *Quaternary Science Reviews*, **22**, 691–701.

Pease, P.P. and Gomez, B. (1997) Landscape development as indicated by basin morphology and the magnetic polarity of cave sediments, Crawford Upland, south-central Indiana. *American Journal of Science*, **297**, 842–58.

Pechorkin, A.N. (1986) On gypsum and anhydrite distribution in zones near to the surface of sulphate massifs. *Le Grotte d'Italia*, **4**(XII), 397–406.

Pechorkin, A.N. and Bolotov, G.V. (1983) *Geodynamics of Relief in Karstified Massifs*, University of Perm, 83 pp. [In Russian.]

Pechorkin, I.A. (1986) Engineering geological investigations of gypsum karst. *Le Grotte d'Italia*, **4**(XII), 383–8.

Penck, A. (1900) Geomorphologische Studien aus der Hercegovina. *Zeitschrift der Deutschen Osterreichischer Alpenver*, **31**, 25–41.

Pentecost, A. (1992) Carbonate chemistry of surface waters in a temperate karst region: the southern Yorkshire Dales, UK. *Journal of Hydrology*, **139**, 211–32.

Pentecost, A. (1994) Travertine-forming cyanobacteria, in *Breakthroughs in Karst Geomicrobiology and Redox Geochemistry* (eds I.D. Sasowsky and M.V. Palmer), Special Publication 1, Karst Waters Institute, Charles Town, WV, p. 60.

Pentecost, A. (1995) The Quaternary travertine deposits of Europe and Asia Minor. *Quaternary Science Reviews*, **14**, 1005–28.

Perna, G. (1990) Forme di corrosione carsica superficiale al Lago di Loppio (Trentino). *Natura Alpina*, **XLI**(4), 17–27.

Perna, G. and Sauro, U. (1978) *Atlante delle microforme di dissoluzione carsica superficiale del Trentino e del Veneto*, Museo Tridentino, Trento, 176 pp.

Perrin, J., Pochon, A., Jeannin, P.-Y., and Zwahlen, F. (2004) Vulnerability mapping in karstic areas: validation by field assessments. *Environmental Geology*, **46**, 237–45.

Perritaz, L. (1996) Le 'karst en vagues' des Ait Abdi (Haut-Atlas central, Maroc). *Karstologia*, **28**(1), 1–12.

Perry, E., Marin, L., McClain, J. and Guadalupe, V. (1996) Ring of Cenotes (sinkholes), northwest Yucatan, Mexico: its hydrogeologic characteristics and possible association with the Chicxulub impact crater. *Geology*, **23**, 17–20.

Peterson, G.M. (1976) Pollen analysis and the origin of cave sediments in the Central Kentucky Karst. *Bulletin of the National Speleological Society* **38**(3), 53–8.

Peterson, J.A. (1982) Limestone pedestals and denudation estimates from Mt. Jaya, Irian Jaya. *Australian Geographer*, **15**, 170–3.

Petit, J.R., Jouzel, J., Raynaud, D., *et al.* 1999, Climate and atmospheric history of the past 420,000 years from the Vostok ice core, Antarctica. *Nature*, **399**, 429–36.

Pfeffer, K.-H. (1993) Zur Genese tropischer Karstgebiete auf den Westindischen Inseln. *Zeitschrift für Geomorphologie, N.F. Supplement-Band*, **93**, 137–58.

Pfeiffer, D. (1963) Die geschichtliche Entwicklung der Anschauungen uber das karstgrundwasser. *Beihefte zum Geologischen Jahrbuch* (Hannover), **57**, 111 pp.

Pham, K. (1985) The development of karst landscapes in Vietnam. *Acta Geologica Polonica* **35**(3/4), 305–19.

Phillips, W.M. (2004) Cosmogenic dating, in *Encyclopedia of Geomorphology* (ed. A.S. Goudie), Routledge, London, pp. 192–4.

Pielou, E.C. (1969) *An Introduction to Mathematical Ecology*, J. Wiley & Sons, New York.

Piccini, L., Romeo, A. and Badino, G. (1999) Moulins and marginal contact caves in the Gornergletscher, Switzerland. *Nimbus*, **23/24**, 94–9.

Piccini, L., Drysdale, R. and Heijnis, H. (2003) Karst morphology and cave sediments as indictors of the uplift history in the Alpi Apuane (Tuscany, Italy). *Quaternary International*, **101/102**, 219–27.

Picknett, R.G., Bray, L.G. and Stenner, R.D. (1976) The chemistry of cave waters, in *The Science of Speleology* (eds T.D. Ford and C.H.D. Cullingford), Academic Press, London, pp. 212–66.

Pinault, J-L., Plagnes, V., Aquilina, L. and Bakalowicz, M. (2001) Inverse modeling of the hydrological and the hydrochemical behaviour of hydrosystems: characterization of karst system functioning. *Water Resources Research*, **37**(8), 2191–204.

Pinneker, E.V. and Shenkman, B.M. (1992) Karst hydrology in the zone of sporadic permafrost, taking the southern part of the Siberian Platform as the example, *Proceedings, 2nd International Symposium of Glacier Caves and Karst in Polar Regions*, University of Silesia, pp. 105–18.

Pitty, A.F. (1968a) The scale and significance of solutional loss from the limestone tract of the southern Pennines, *Proceedings of the Geologist's Association*, **79**(2), 153–77.

Pitty, A.F. (1968b) Calcium carbonate content of water in relation to flow-through time. *Nature*, **217**, 939–40.

Plagnes, V. and Bakalowicz, M. (2002) The protection of a karst water resource from the example of the Larzac karst plateau (south of France): a matter of regulation or a matter of process knowledge? *Engineering Geology*, **65**, 107–16.

Plan L. (2005) Factors controlling carbonate dissolution rates quantified in a field test in the Austrian alps. *Geomorphology*, **68**, 201–212.

Playford, P. (2002) Palaeokarst, pseudokarst sequence stratigraphy in Devonian reef complexes of the Canning Basin, Western Australia, in *The Sedimentary Basins of Western Australia 3* (eds M. Keep and S.J. Moss), *Proceedings* of the Petroleum Exploration Society of Australia Symposium, Perth, Western Australia, pp. 763–93.

Pluhar, A. and Ford, D.C. (1970) Dolomite karren of the Niagara Escarpment, Ontario, Canada. *Zeitschrift für Geomorphologie*, **14**(4), 392–410.

Plummer, L.N. (1975) Mixing of seawater with calcium carbonate ground water: quantitative studies in the geological sciences. *Geological Society of America Memoir*, **142**, 219–36.

Plummer, L.N. and E. Busenberg (1982) The solubilities of calcite, aragonite and vaterite in solutions between $0°$ and $90°C$ an evaluation of the aqueous model for the system $CaCO_3$–CO_2–H_2O. *Geochimica et Cosmochimica Acta*, **46**, 1011–40.

Plummer, L.N. and Wigley, T.M.L. (1976) The dissolution of calcite in CO_2 saturated solutions at $25°C$ and 1 atmosphere total pressure. *Geochimica et Cosmochimica Acta*, **40**, 191–202.

Plummer, L.N., Wigley, T.M.L. and Parkhurst, D.L. (1978) The kinetics of calcite dissolution in CO_2 – water systems at $5°$ to $60°C$ and 0.0 to 1.0 atm CO_2. *American Journal of Science*, **278**, 179–216.

Plummer, L.N., Parkhurst, D.C. and Wigley, T.M.L. (1979) Critical review of the kinetics of calcite dissolution and precipitation, in *Chemical Modelling in Aqueous Systems* (ed. E. A. Jenne), American Chemical Society, Washington, DC, pp. 537–73.

Poiseuille, J.M.L. (1846) Recherches experimentales sur le mouvement des liquides dans les tubes de tres petits diametres. *Académie des Sciences, Paris Memoir Sav. Etrang.*, **9**, 433–545.

Pollard, W., Omelon, C., Andersen, D. and McKay, C. (1999. Perennial spring occurrence in the Expedition Fiord area of western Axel Heiberg Island, Canadian High Arctic. *Canadian Journal of Earth Sciences*, **36**, 105–120.

Polyak, V.J., McIntosh, W.C., Güven, N. and Provencio, P. (1998) Age and origin of Carlsbad Cavern and related caves from $^{40}Ar/^{39}Ar$ of Alunite. *Science*, **279**, 1919–1922.

Pouyllan, M. and M. Seurin (1985) Pseudo-karst dans les roches gres quartzitiques de la formation Roraima. *Karstologia*, **5**(1), 45–52.

Prestor, J. and Veselic, M. (1993). Effects of long term precipitation variability on water balance assessment of a karst basin, in *Proceedings of the International Symposium on Water Resources in Karst with Special Emphasis on Arid and Semi Arid Zones*, (ed. A. Afrasiabian), Shiraz, Iran, pp. 887–99.

Prestwich, J. (1854) Swallow holes on the chalk hills near Canterbury. *Quarterly Journal of the Geological Society* (London), **C**, 222–4.

Price, M., Downing, R.A. Edmunds, W.M. (1993) The Chalk as an aquifer, in *Hydrogeology of the Chalk of North-West Europe* (eds R.A. Downing, M. Price and G.P. Jones), Clarendon Press, Oxford, pp. 35–8.

Priesnitz, K. (1974) Losungsraten und irhe geomorphologische Relevanz. *Abhandlungen Akademie der Wissenschaften in Göttingen, Mathematisch-Physikalische Klasse 3*, **Folge 29**, 68084.

Pringle J.M. (1973) *Morphometric analysis of surface depressions in the Mangapu karst*. MSc thesis University of Auckland, New Zealand.

Proctor, C.J., Baker, A. and Barnes, W.L. (2002) A three thousand year record of North Atlantic climate. *Climate Dynamics*, **19**, 449–54.

Pulido-Bosch, A. (1986) Le karst dans les gypses de Sorbas (Almeria): aspects morphologiques et hydrogeologiques. *Karstologia Memoires*, **1**, 27–35.

Pulido-Bosch, A., Molina, L., Navarrete, F. and Martinez Vidal, J.L. (1993) Nitrate content in the groundwater of the Campo de Dalias (Almeria), in *Some Spanish Karstic Aquifers* (ed. A. Pulido-Bosch), University of Granada Press, Granada, pp. 183–94.

Pulina, M. (1971) Observations on the chemical denudation of some karst areas of Europe and Asia. *Studia Geomorphologica Carpatho-Balcanica*, **5**(B752), 79–92.

Pulina, M. (1992) Glacio-karsts gypseux de la zone polaire et périglaciaire: Exemple du Spitzberg et de Sibérie orientale, in *Karsts et Evolutions Climatiques* (eds J-N.

Salomon and R. Maire), Presses Universitaire de Bordeaux, pp. 266–83.

Pulina, M. (2005) Le karst et les phenomenes karstiques similaires des regions froides, in Salomon, J.-N et Pulina, M. Les karsts des régions climatiques extrêmes. *Karstologia Mémoires*, **14**, 11–100.

Pulinowa, M.Z. and Pulina, M. (1972) Phnomenes cryogenes dans les grottes et gouffres des Tatras. *Biuletyn Peryglacjalny*, **21**, 201–35.

Purdy, E.G. (1974) Reef configurations: cause and effect, in *Reefs in Time and Space* (ed. L.F. Laporte), Special Publication 18, Society of Economic Palaeontologists and Mineralogists, Tulsa, OK, pp. 9–76.

Purdy, E.G. and Winter, E.L. (2001) Origin of atoll lagoons. *Geological Society of America Bulletin*, **113**(7), 837–54.

Purser, B., Tucker, M. and Zeuger, D. (eds) (1994) *Dolomites: a Volume in Honour of Dolomieu*, Oxford, Blackwell, 451 pp.

Qian, W. and Zhu, Y. (2002) Little Ice Age climate near Beijing, China, inferred from historial and stalagmite records. *Quaternary Research*, **57**, 109–19.

Quinif, Y. (1996) Enregistrement et datation des effets sismotectoniques par l'étude des spéléothemes. *Annales de la Société Géologique de Belgique*, **119**(1), 1–13.

Quinlan, J.F. (1976) New Fluorescent direct dye suitable for tracing groundwater and detection with cotton, *Proceedings of the 3rd International Symposium on Underground Water Tracing*, Vol. 1, Institute of Karst Research, Postojna, Ljubljana, pp. 257–62.

Quinlan, J.F. (1978) *Types of karst, with emphasis on cover beds in their classification and development*. Univ. of Texas at Austin PhD thesis, 323 pp.

Quinlan, J.F. (1983) Groundwater pollution by sewage, creamery waste heavy metals in the Horse Cave area, Kentucky, in *Environmental Karst* (ed. P.H. Dougherty), Geology and Speleology Publications, Cincinnati, p. 52.

Quinlan, J.F. (1986) Legal aspects of sinkhole development and flooding in karst terranes: 1. review and synthesis. *Environmental Geology and Water Science*, **8**(1/2), 41–61.

Quinlan, J.F. and Ewers, R.O. (1981) Hydrogeology of the Mammoth Cave Region, Kentucky, in *Geological Society of America Cincinnati 1981 Field Trip Guidebooks, Vol. 3*, (ed.T.G. Roberts), pp. 457–506.

Quinlan, J.F. and Ewers, R.O. (1985) Ground water flow in limestone terranes: strategy, rationale and procedure for reliable, efficient monitoring of ground water quality in karst areas, *National Symposium and Exposition on Aquifer Restoration and Ground Water Monitoring, Proceedings*, National Water Well Association, Worthington, OH, pp. 197–234.

Quinlan, J.F. and Ewers, R.O. (1986) Reliable monitoring in karst terranes: it can be done, but not by an EPA-approved method. *Ground Water Monitoring Review*, **6**(1), 4–6.

Quinlan, J.F. and Ray, J.A. (1991) Ground-water remediation may be achievable in some karst aquifers that are contaminated, but it ranges from unlikely to impossible in most: I. Implications of long-term tracer tests for universal failure by scientists, consultants, and regulators, in *Proceedings of the 3rd Conference on Hydrology, Ecology, Monitoring, and Management of Ground Water in Karst Terranes (Nashville, Tennessee)* (eds J.F. Quinlan and A. Stanley), National Ground Water Association, Dublin, Ohio, pp. 553–8.

Quinlan, J.F., Smith, A.R. and Johnson, K.S. (1986) Gypsum karst and salt karst of the United States of America. *Le Grotte d'Italia*, **4**(13), 73–92.

Quinlan, J.F., Smart, P.L., Schindel, G.M., Alexander, E.C., Edwards, A.J. and Smith, A.R. (1991) Recommended administrative/regulatory definition of karst aquifer, principles for classification of carbonate aquifers, practical evaluation of vulnerability of karst aquifers, and determination of optimum sampling frequency at springs, in *Proceedings of the 3rd Conference on Hydrology, Ecology, Monitoring, and Management of Ground Water in Karst Terranes (Nashville, Tennessee)* (eds J.F. Quinlan and A. Stanley), National Ground Water Association, Dublin, Ohio, pp. 573–635.

Rachlewicz, G. and Szczuciński, W. (2004) Seasonal, annual and decadal ice mass balance changes in Jaskinia Lodowaw Ciemniaku, the Tatra Mountains, Poland. *Theoretical and Applied Karstology*, **17**, 16–21.

Racoviţă, G. and Onac, B.P. (2000) *Scărişoara Glacier Cave*, Editura Carpatica, Cluj-Napoca, 139 pp.

Racoviţă, G., Moldovan, O. and Onac, B. (eds) (2002) *Monografia carstului din Munţii Pădurea Craiului*, Institut de Speleologie 'Emil Racoviţă', Cluj-Napoca, 264 pp.

Racovitza, Gh. (1972) Sur la correlation entre l'evolution du climat et la dynamique des dpots souterrains de glace de la grotte Scarisoara. *Travaux de l'Institut de Speologie 'Emile Racovitza*, **XI**, 373–92.

Ragozin, A.L., Yolkin, V.A. and Chumachenko, S.A. (2005) Experience of regional karst hazard and risk assessment in Russia, in *Sinkholes and the Engineering and Environmental Impacts of Karst* (ed. B.E. Beck), Geotechnical Special Publication No. 144, American Society of Civil Engineers, pp. 72–81.

Railsback, L.B. (1999) Patterns in the compositions, properties and geochemistry of carbonate minerals. *Carbonates and Evaporites*, **14**, 1–20.

Railsback, L.B. (2000) *An Atlas of Speleothem Microfabrics*, University of Georgia, Athens, GA. http:// www.gly.uga.edu/ railsback/speleoatlas/.html

Railsback, L.B., Brook, G.A., Chen, J., *et al.* (1994) Environmental controls on the petrology of a late Holocene speleothem from Botswana with annual layers of aragonite and calcite. *Journal of Sedimentary Research*, **64**, 147–55.

Rauch, H.W. and White, W.B. (1970) Lithologic controls on the development of solution porosity in carbonate aquifers. *Water Resources Research*, **6**, 1175–92.

Ray, J.A. and Currens, J.C. (1998a) *Mapped Karst Ground-water Basins in the Beaver Dam 30 × 60 Minute Quadrangle*, Map and Chart Series 19, Kentucky Geological Survey.

Ray, J.A. and Currens, J.C. (1998b) *Mapped Karst Ground-water Basins in the Campbellsville 30 × 60 Minute Quadrangle*, Map and Chart Series 17, Kentucky Geological Survey.

Ray, J.A., Webb, J.S. and O'Dell, P.W. (1994) *Groundwater Sensitivity Regions of Kentucky 1:500 000*, Groundwater

Branch, Division of Water, Kentucky Department for Environmental Protection.

Reams, M.W. (1968) *Cave sediments and the geomorphic history of the Ozarks*, Washington Univ. PhD thesis, Missouri.

Reardon, E.J. (1992) Problems and approaches to the prediction of the chemical composition in cement/water systems. *Waste Management*, **12**, 221–39.

Reilly, T.E. and Goodman, A.S. (1985) Quantitative analysis of saltwater-freshwater relationships in groundwater systems – a historical perspective. *Journal of Hydrology*, **80**, 125–60.

Renault, P. (1968) Contribution a l'etude des actions mechaniques et sedimentologiques dans la speleogenese. *Annales de Speleologie*, **22**, 5–21, 209–67; **23**, 259–307; 529–96; **24**, 313–37.

Renault, P. (1970 *La formation des caverns*. Presses universitaire de France, Paris, 127 pp.

Renault, P. (1982) CO_2 atmospherique karstique et speleomorphologie. *Revue belge Geographique*, **106**, 121–30.

Renaut, R.W. and Jones, B. (2003) Sedimentology of hot spring systems. *Canadian Journal of Earth Sciences*, **40**, 1439–42.

Rhoades, R. and Sinacori, N.M. (1941) Patterns of groundwater flow and solution. *Journal of Geology*, **49**, 785–94.

Rhodes, D., Lantos, E.A., Lantos, J.A., *et al.* (1984) Pine Point orebodies and their relationship to structure, dolomitization and karstification of the Middle Devonian barrier complex. *Economic Geology*, **70**, 91–1055.

Rice, J.E. and Hartowicz, E. (2003) Biota as water quality indicators in springs at Fort Campbell, Kentucky, in *Sinkholes and the Engineering and Environmental Impacts of Karst* (ed. B.E. Beck), *Proceedings* of 9th Multidisciplinary Conference, Huntsville, Alabama, Geotechnical Special Publication 122, American Society of Civil Engineers, pp. 339–48.

Richards, D.A. and Dorale, J.A. (2003) Uranium-series chronology and environmental applications of speleothems, in *Uranium-series Geochemistry* (eds B. Bourdon, G.M. Henderson, C.C. Lundstrom and S.P. Turner). *Reviews in Mineralogy and Geochemistry*, **52**, 407–61.

Richards, D.A., Smart, P.L. and Edwards, R.L. (1994) Maximum sea levels for the last glacial period from U-series ages of submerged speleothems. *Nature*, **367**, 357–60.

Richards, D.A., Bottrell, S.H., Cliff, R.A. and Strohle, K.D. (1996) U–Pb dating of Quaternary age speleothems, in *Climate Change: The Karst Record* (ed. S.-E. Lauritzen), Special Publication 2, Karst Waters Institute, Charles Town, WV, pp. 136–7.

Richardson, J.J. (2001) Legal impediments to utilizing groundwater as a municipal water supply source in karst terrain in the United States, in *Geotechnical and Environmental Applications of Karst Geology and Hydrology* (eds B.F. Beck and J.G. Herring), Lisse, Balkema, pp. 253–58.

Richter, E. (1907) Beitrage zur Landeskunde Bosniens und der Herzegowina. *Wissenschaftliche Mitteilungen Bosnien Herzegowina*, **10**, 383–545.

Rickard, D.T. and Sjoberg, E.L. (1983) Mixed kinetic control of calcite dissolution rates. *American Journal of Science*, **238**, 815–30.

Riding, R. (2003) Structure and composition of organic reefs and carbonate mud mounds: concepts and categories. *Earth Science Reviews*, **58**(1–2), 233–41.

Riggs, A.C., Carr, W.J., Kolesar, P.T. and Hoffman, R.J. (1994) Tectonic speleogenesis of Devil's Hole, Nevada implications for hydrogeology and the development of long, continuous paleoenvironmental records. *Quaternary Research*, **42**, 241–54.

Ristic, D.M. (1976) Water regime of flooded poljes, in *Karst Hydrology and Water Resources: Vol. 1 Karst Hydrology* (ed. V. Yevjevich), Water Resources Publications, Colorado, pp. 301–18.

Roberge, J. (1979 *Geomorphologie du karst de la Haute-Saumons, Ile d'Anticosti, Quebec.* McMaster Univ. MSc thesis, 217 pp.

Roberge, J. and Caron, D. (1983) The occurrence of an unusual type of pisolite: the cubic cave pearls of Castleguard Cave, Columbia Icefields, Alberta, Canada. *Arctic and Alpine Research*, **15**(4), 517–22.

Roberts, M.S., Smart, P.L. and Baker, A. (1998) Annual trace element variations in a Holocene speleothem. *Earth and Planetary Science Letters*, **154**, 237–46.

Robinson, D.A. and Williams, R.G.B. (1994) Sandstone weathering and landforms in Britain and Europe, in *Rock Weathering and Landform Evolution* (eds D.A. Robinson and R.G.B. Williams), J. Wiley & Sons, Chichester, pp. 371–91.

Robinson, V.D. and Oliver, D. (1981) Geophysical logging of water wells, in *Case Studies in Groundwater Resources Evaluation* (ed. J. W. Lloyd), Clarendon, London, pp. 45–64.

Rodet, J. (1996) Une nouvelle organisation géométrique du drainage karstique des craies: le labyrinthe d'altération, l'example de la grotte de la Mansonnière (Bellou-sur-Huisene, Orne, France). *Comptes Rendus, Académie des Sciences (Paris)*, **322**(série IIa), 1039–45.

Rodriguez, R. (1995). Mapping karst solution features by the integrated geophysical method, in *Karst Geohazards* (ed. B. F. Beck), Balkema, Rotterdam, pp. 443–9.

Roehl, P.O. and Choquette, P.W. (1985) *Carbonate Petroleum Reservoirs*, Springer-Verlag, New York, 622 pp.

Roglić, J. (1960) Das Verhältnis der Flusserosion zum Karstprozess. *Zeitschrift für Geomorphologie*, **4**(2), 116–238.

Roglić, J. (1972) Historical review of morphological concepts, in *Karst: Important Karst Regions of the Northern Hemisphere* (eds M. Herak and V.T. Stringfield), Elsevier, Amsterdam, pp. 1–18.

Roglić, J. (1974) Les caracteres specifiques du karst Dinarique. *Centre National de la Recherche Scientifique, Memoirs et Documents*, **15**, 269–78.

Romani, L., Vasseur, F. and Viala, C. (1999) Le systeme des Fontanilles. *Spelunca*, **75**, 31–8.

Romero, J.C. (1970) The movement of bacteria and virus through porous media. *Ground Water*, **8**(2), 37–48.

Roques, H. (1962) Considerations theoriques sur la chimie des carbonates. *Annales de Speleologie*, **17**, 1–41, 241–84, 463–7.

Roques, H. (1964) Contribution a l'etude statique et cintique des systemes gaz carbonique-eau-carbonate. *Annales de Speleologie*, **19**, 255–484.

Roques, H. (1969) Problemes de transfert de masse poss par l'evolution des eaux souterraines. *Annales de Speleologie*, **24**, 455–94.

Rose, G. (1837) Uber die Buildung des Kalkspaths und Aragonite. *Annales de Chimie et de Physique*, **2**(42), 353–67.

Ross, J.-H., Serefiddin, F., Hauns, M. and Smart.C.C. (2001) 24 h tracer tests on diurnal parameter variability in a subglacial karst conduit: Small River valley, Canada. *Theoretical and Applied Karstology*, **13–14**, 93–9.

Rossi, G. (1986) Karst and structure in tropical areas, in *New Directions in Karst* (eds K. Paterson and M.M. Sweeting), Geo Books, Norwich, pp. 189–212.

Rossi, C., Munoz, A. and Cortel, A. (1997) Cave development along the water table in the Cobre System (Sierra de Penalabra, Cantabrian Mountains, Spain, *Proceedings of the 12th International Congress of Speleology*, Vol. 1, pp. 179–85.

Rothlisberger, H. (1998. The physics of englacial and subglacial meltwater drainage, *Proceedings, 4th International Symposium on Glacier Caves and Cryokarst in Polar and High Mountain Regions, Salzburg*, pp. 13–23.

Rozanski, K. and Florkowski, T. (1979) Krypton-85 dating of groundwater, in *Isotope Hydrology 1978*, Vol. 2, International Atomic Energy Agency, Vienna, 949 pp.

Rozanski, K., Araguas-Araguas, L. and Gonfiantini, R. (1993) Isotopic patterns in modern global precipitation, in *Climate Change in Continental Isotopic Records*, Geophysical Monograph 78, American Geophysical Union, Washington, DC, pp. 1–36.

Salomon, J.-N. (2000) *Précis de karstologie*, Presses Universitaires, Bordeaux, 250 pp.

Salomon, J.-N. (2003a) Cenotes et trous bleus, sites remarquables menacés par l'écotourisme. *Cahiers d'Outres-Mer*, **56**(223), 327–52.

Salomon, J.-N. (2003b) Karst system response in volcanically and tectonically active regions. *Zeitschrift für Geomorphologie, N.F. Supplement-Band*, **131**, 89–112.

Salomon, J.-N. (2005) Les karsts des zones arides et semi-arides, in *Les karsts des régions climatiques extrêmes* (eds J.-N Salomon and M. Pulina). *Karstologia Mémoires*, **14**, 159–91.

Salomon, J.-N. and Bustos, R. (1992) Le karst du gypse des Andes de Mendoza-Neuquen. *Karstologia*, **20**, 11–22.

Salomon, J.-N. and Maire, R. (eds) (1992) *Karst et évolutions climatiques*, Presses Universitaires, Bordeaux, 520 pp.

Salomon, J.-N. and Pulina, M. (eds) (2005) Les karsts des régions climatiques extrêmes. *Karstologia Mémoires*, **14**, 220 pp.

Salomon, J.-N., Pomel, S. and Nicod, J. (1995) L'évolution des cryptokarsts: comparaison entre le Périgord-Quercy (France) et le Franken Alb (Allemagne). *Zeitschrift für Geomorphologie*, **39**(4), 381–409.

Salomons, W. and Mook, W.G. (1980) Isotope geochemistry of carbonates in the weathering zone, in *Handbook of Environmental Isotope Geochemistry*, Vol. 2 (eds P. Fritz and J. Fontes), Elsevier, Amsterdam, pp. 239–69.

Salvamoser, J. (1984) Krypton–85 for groundwater dating, in *Isotope Hydrology 1983*, International Atomic Energy Agency, Vienna, pp. 831–2.

Salvigsen O. and Elgersma, A. (1985) Large-scale karst features and open taliks at Valdeborgsletta, outer Isfjorden, Svalbard. *Polar Research*, **3**(2), 145–53.

Sanchez, J.A., Perez, A., Coloma, P. and Martinez-Gil, J. (1998) Combined effects of groundwater and aeolian processes in the formation of the northernmost closed saline depressions of Europe: north-east Spain. *Hydrological Processes*, **12**, 813–20.

Sandvik, P.O. and Erdosh, G. (1977) *Geology of the Cargill Phosphate Deposit in Northern Ontario*, Bulletin, Canadian Institute of Mining, pp. 90–6.

Sangster, D.F. (1988) Breccia-hosted lead–zinc deposits in carbonate rocks, in *Paleokarst* (eds N.P. James and P.W. Choquette), Springer-Verlag, New York, pp. 102–16.

Sarg, J.F. (2001) The sequence stratigraphy, sedimentology economic importance of evaporite-carbonate transitions: a review. *Sedimentary Geology*, **140**, 9–42.

Sasowsky, I.D. and Mylroie, J. (eds) (2004) *Studies of Cave Sediments: Physical and Chemical Records of Paleoclimate*, Kluwer Academic, New York, 329 pp.

Sasowsky, I. D. and Wicks, C.M. (eds) (2000) *Groundwater Flow and Contaminant Transport in Carbonate Aquifers*, Balkema, Rotterdam, 193 pp.

Sasowsky, I.D., White, W.B. and Schmidt, V.A. (1995) Determination of stream-incision rate in the Appalachian plateaus by using cave sediment magnetostratigraphy. *Geology*, **23**(5), 415–8.

Saunderson, H.C. (1977) The sliding bed facies in sands and gravels: a criterion for full-pipe (tunnel) flow. *Sedimentology*, **24**, 623–38.

Sauro, U. (1996) Geomorphological aspects of gypsum karst areas with special emphasis on exposed areas. *International Journal of Speleology*, **25**(3–4), pp. 105–114.

Sauro, U., Martello, V. and Frigo, G. (1991) Karst environment and human impact in the Sette Communi Plateau (Venetian Pre-Alps), in *Proceedings of the International Conference on Environmental Changes in Karst Areas* (eds U. Sauro, A. Bondesan and M. Meneghel), I.C.E.C.K.A., Universita di Padova, pp. 269–78.

Sauter, M. (1992) Quantification and forecasting of regional groundwater flow and transport in a karst aquifer. *Tübinger Geowissenschaftlichen Abhandlungen, Reihe C*, **13**, 150 pp.

Sauter, M. (1997) Differentiation of flow components in a karst aquifer using the $\delta^{18}O$ signature, in *Tracer Hydrology 97* (ed. A. Kranjc), Rotterdam, Balkema, pp. 435–41.

Sawicki, L.R. von (1909) Ein Beitrag zum geographischen Zyklus im Karst. *Zeitschrift für Geographie*, **15**, 185–204, 259–81.

Sawkins, J., *et al.* (1869) *Reports on the Geology of Jamaica*, Memoir of the Geological Survey, Longman, Green, London (cited in Sweeting, 1972).

Scanlon, B.R., Mace, R.E., Barrett, M.E. and Smith, B. (2003) Can we simulate regional groundwater flow in a karst system using equivalent porous media models? Case study, Barton Springs Edwards aquifer, USA. *Journal of Hydrology*, **276**, 137–158.

Schellmann, G., Radtke, U., Potter, E.-K., *et al.* (2004) Comparison of ESR and TIMS U/Th dating of marine isotope stage

(MIS) 5e, 5c, 5a coral from Barbados – implications for palaeo sea-level changes in the Caribbean. *Quaternary International*, **120**, 41–50.

Schillat, B. (1977) Conservation of tectonic waves in the axes of stalagmites over long periods, *Proceedings of the 7th International Congress of Speleology, Sheffield*, pp. 377–9.

Schmidl, A. (1854) *Die Grotten und Hohlen von Adelsberg, Lueg, Planina und Laas*, Wien, 316 pp.

Schmidt, K.-H. (1979) Karstmorphodynamik und ihre hydrologische Steuerung. *Erdkunde*, **33**(3), 169–78.

Schmidt, V.A. (1982) Magnetostratigraphy of clastic sediments from caves within the Mammoth Cave National Park, Kentucky. *Science*, **217**, 827.

Schmotzer, J.K., Jester, W.A. and Parizek, R.R. (1973) Groundwater tracing with post sampling activation analysis. *Journal of Hydrology*, **20**(B823), 217–36.

Schoeller, H. (1962) *Les Eaux Souterraines*, Masson et Cie, Paris, 642 pp.

Scholle, P.A. and James, N.P. (eds) (1995, 1996) *Photo CD-Series 1, 2, 7, 8*, Society of Economic Paleontologists and Mineralogists, Denver.

Scholle P.A., Bebout, D.G. and Moore, C.H. (eds) (1983) *Carbonate Depositional Environments*, Memoir 33, American Association of Petroleum Geologists, Tulsa, OK, 761 pp.

Schott, J., Brantley, S., Drear, D., *et al.* (1989) Dissolution kinetics of strained calcite. *Geochimica et Cosmochimica Acta*, **53**, 373–82.

Schroeder, J. (1979) *Le developpement des grottes dans la region du Premier Canyon de la Riviere Nahanni, Sud, T.N.O.*, Univ. Ottawa PhD thesis, 265 pp.

Schroeder, J. (1999) Le drainage latèral d'un glacier subpolaire. *Nimbus*, **23/24**, 100–3.

Schroeder, J. and Ford, D.C. (1983) Clastic sediments in Castleguard Cave, Columbia Icefields, Alberta, Canada. *Arctic and Alpine Research*, **15**(4), 451–61.

Schroeder, J., Beaupré, M. and Cloutier, M. (1990) Substrat glaciotectonisé et till syngénétique à Pont-Rouge, Québec. *Géographie physique et Quaternaire*, **44**(1), 33–42.

Schultz, G.A. and Engman, E.T. (eds) (2000). *Remote Sensing in Hydrology and Water Management*, Springer-Verlag, Berlin.

Schulz, M. and Mudelsee, M. (2002) REDFIT: estimating red-noise spectra directly from unevenly spaced palaeoclimatic time series. *Computers and Geosciences*, **28**, 421–6.

Schulz, M. and Statteger, K. (1997) SPECTRUM: spectral analysis of unevenly spaced paleoclimate time series. *Computers and Geosciences*, **23**(9), 929–45.

Schwabe, S. and Herbert, R.A. (2004) Black Holes of the Bahamas: what they are and why they are black. *Quaternary International*, **121**, 3–11.

Schwarcz, H.P. (1980) Absolute age determination of archaeological sites by uranium series dating of travertine. *Archaeometry*, **22**, 3–25.

Schwarcz, H.P. (1982) Absolute dating of travertine from archaeological sites, in *Nuclear and Chemical Dating Techniques* (ed. L.A. Currie), Symposium Series 176, American Chemical Society, pp. 475–90.

Schwarcz, H.P. (1993) Uranium series dating and the origin of modern man, in The Origin of Modern Humans and the Impact of Chronometric Dating (eds M. Aitken, P.A. Mellard, C.B. Stringer and Paul Mellars). Princeton University Press, Princeton, NJ, pp. 12–26.

Schwarcz, H.P. (2000) Dating bones and teeth: the beautiful and the dangerous, in *Humanity from African Naissance to Coming Millennia* (eds P. V. Tobia, M.A. Raath, J. Moggi-Cecci, and G.A. Doyle), Firenze University Press, Florence, pp. 249–56.

Schwarcz, H.P. (in press) Stable isotopes in speleothems, in *Encyclopedia of Quaternary Science* (ed. S. Elias), Elsevier, New York.

Schwarcz, H.P. and Lee, Hee-Kwon (2000) Electron spin resonance dating of fault rocks, in *Quaternary Geochronology: Applications in Quaternary Geology and Paleoseismology* (eds J. Sowers, J. Noller and W.J. Lettis), Monograph, American Geophysical Union, Washington, DC, pp. 177–86.

Schwarcz, H.P., Harmon, R.S., Thompson, P. and Ford, D.C. (1976) Stable isotope studies of fluid inclusions in speleothems and their paleoclimatic significance. *Geochimica et Cosmochimica Acta*, **40**, 657–665.

Schwartz, J.H., Vu, T.L., Nguyen, L.C., *et al.* (1995) *A Review of the Pleistocene Hominoid Fauna of the Socialist Republic of Vietnam (excluding Hylobatidae)*. Anthropological Papers 76, American Museum of Natural History, 24 pp.

Šebela, S. (2003) The use of structural geological terms and their importance for karst caves. *Acta Carsologica*, **32**(2), 53–64.

Selby, M.J. (1980) A rock mass strength classification for geomorphic purposes: with tests from Antarctica and New Zealand. *Zeitschrift für Geomorphologie*, **24**, 31–51.

Selby, M.J. and Hodder, A.P.W. (1993) *Hillslope Materials and Processes*, 2nd edn, Oxford University Press, Oxford, 451 pp.

Self, C.A. and Hill, C.A. (2003) How speleothems grow: a guide to the ontogeny of cave minerals. *Journal of Cave and Karst Studies*, **65**(2), 130–51.

Senior, K. (2004) Di Feng Dong, China, in *Encyclopedia of Caves and Karst Science* (ed. J. Gunn), Fitzroy Dearborn, New York, pp. 285–7.

Serefiddin, F., Schwarcz, H.P., Ford, D.C. and Baldwin, S. (2004) Late Pleistocene paleoclimate in the Black Hills of South Dakota from isotopic records in speleothems. *Palaeogeography, Palaeoclimatology, Palaeoecology*, **203**, 1–17.

Serefiddin, F., Schwarcz, H.P. and Ford, D.C. (2005) Use of hydrogen isotope variations in speleothem fluid inclusions as an independent measure of paleoclimate. *Geological Society of America, Special Paper*, **395**, 43–53.

Shackleton, N.J. (2000) The 100,000-year Ice-Age cycle identified and found to lag temperature, carbon dioxide orbital eccentricity. *Science*, **289**, 1897–902.

Shaw, J. (1988) Subglacial erosional marks, Wilton Creek, Ontario. *Canadian Journal of Earth Sciences*, **25**, 1256–67.

Shaw, T.R. (1992) *History of Cave Science: the Exploration and Study of Limestone Caves to 1900*, Sydney Speleological Society, Sydney, 338 pp.

Shen, G., Gao, X., Zhao, J-X Collerson, K.D. (2004) U-series dating of Locality 15 at Zhoukoudian, China implications for hominid evolution. *Quaternary Research*, **62**, 208–13.

Shevenell, L. (1996) Analysis of well hydrographs in a karst aquifer: estimates of specific yields and continuum transmissivities. *Journal of Hydrology*, **174**, 331–55.

Shopov, Y.Y. (1987) Laser luminescent micro-zonal analysis: a new method for investigation of the alterations of the climate and solar activity during the Quaternary, in *Problems of Karst Studies of Mountainous Countries* (ed. T. Kiknadze), Metsniereba, Tbilisi, Georgia, pp 104–8.

Shopov, Y.Y. (1997) Luminescence of cave minerals, in *Cave Minerals of the World*, 2nd edn (eds C.A. Hill and P. Forti), National Speleological Society of America, Huntsville, AL, pp. 244–8.

Shopov, Y.Y., Ford, D.C. and Schwarcz, H.P. (1994) Luminescent microbanding in speleothems: High resolution chronology and paleoclimate. *Geology*, **22**(5), 407–10.

Shopov, Y.Y., Tsankov, L.T., Yonge, C.J., Krouse, H.P.R. Jull, A.J.T. (1997) Influence of the bedrock CO_2 on stable isotope records in cave calcites, *Proceedings of the 12th International Congress of Speleology, Switzerland*, Vol. 1, pp. 65–8.

Short, M.B., Baygents, J.C. and Goldstein, R.E. 2005. Stalactite growth as a free-boundary problem. *Physics of Fluids*, **17**, 083101-12.

Siegenthaler, U., U. Schotterer and I. Muller (1984) Isotopic and chemical investigations of springs from different karst zones in the Swiss Jura, in *Isotope Hydrology 1983*, International Atmoc Energy Agency, Vienna, pp. 153–72.

Siemers, J. and Dreybrodt, W. (1998) Early development of karst aquifers on percolation networks of fractures in limestone. *Water Resources Research*, **34**, 409–19.

Simms, M.J. (2003) The origin of enigmatic tubular lake-shore karren: A mechanism for rapid dissolution of limestone in carbonate-saturated waters. *Physical Geography*, **23**, 1–20.

Simms, M.J. (2004) Tortoises and hares: dissolution, erosion and isostasy in landscape evolution. *Earth Surface Processes and Landforms*, **29**, 477–94.

Simsek, S. (1999) Pamukkale travertine area, in *Karst Hydrogeology and Human Activities* (eds D. Drew and H. Hötzl), Balkema, Rotterdam, pp. 172–74.

Simsek, S., Günay, G., Elhatip, H. and Ekmekçi, M. (2000) Environmental protection of geothermal waters and travertines at Pamukkale, Turkey. *Geothermics*, **29**, 557–72.

Sjoberg, E.L. and Rickard, D.T. (1983) The influence of experimental design on the rate of calcite dissolution. *Geochimica et Cosmochimica Acta*, **47**, 2281–6.

Slabe, T. (1995) *Cave Rocky Relief and its Speleological Significance*, Znanstvenoraziskovalni Center SAZU, Lubljana, 128 pp.

Sletov, V.A. (1985) On ontogeny of crystalictite and helictite aggregates of calcite and aragonite from the caves of southern Fergana. *Novye Dannye o Mineralogii*, **32**, 119–127. [In Russian.]

Sloss, L.L. (1963) Sequences in the cratonic interior of North America. *Geological Society of America, Bulletin*, **74**, 93–114.

Smart, C.C. (1983a) *Hydrology of a Glacierised Alpine Karst*, McMaster Univ. PhD thesis, 343 pp.

Smart, C.C. (1983b) The hydrology of the Castleguard Karst, Columbia Icefields, Alberta, Canada. *Arctic and Alpine Research*, **15**(4), 471–86.

Smart, C.C. (1984) The hydrology of the Inland Blue Holes, Andros Island. *Cave Science*, **11**(1), 23–9.

Smart, C.C. (1997) Hydrogeology of glacial and subglacial karst aquifers: Small River, British Columbia, Canada, *Proceedings of the 6th Conference on Limestone Hydrology and Fissured Media*, pp. 315–8.

Smart, C.C. and Brown, M.C. (1981) Some results and limitations in the application of hydraulic geometry to vadose stream passages, *Proceedings of the 8th International Congress of Speleology, Bowling Green, Kentucky*, pp. 724–5.

Smart, C. and Simpson, B. (2001) An evaluation of the performance of activated charcoal in detection of fluorescent compounds in the environment, in *Geotechnical and Environmental Applications of Karst Geology and Hydrology* (eds B.F. Beck, and J.G. Herring), Balkema, Lisse, pp. 265–70.

Smart, C. and Worthington, S.R.H. (2004) Springs, in *Encyclopedia of Caves and Karst Science* (ed. J. Gunn), Fitzroy Dearborn, New York, pp. 699–703.

Smart, P.L. (1986) Origin and development of glacio-karst closed depressions in the Picos de Europa, Spain. *Zeitschrift für Geomorphologie, NF*, **30**(4), 423–43.

Smart, P.L. and Christopher, N.S.J. (1989) Ogof Ffynon Ddu, in *Limestones and Caves of Wales* (ed. T.D. Ford), Cambridge University Press, pp. 177–89.

Smart, P.L. and Friederich, H. (1987) Water movement and storage in the unsaturated zone of a maturely karstified carbonate aquifer, Mendip Hills, England, *Proceedings of Conference on Environmental Problems in Karst Terranes and their Solutions*. National Water Well Association, Dublin, Ohio, 59–87.

Smart, P.L. and Hodge, P. (1980) Determination of the character of the Longwood sinks to Cheddar resurgence conduit using an artificial pulse wave. *Transactions of the British Cave Research Association*, **7**(4), 208–211.

Smart, P.L. and Laidlaw, I.M.S. (1977) An evaluation of some fluorescent dyes for water tracing. *Water Resources Research*, **13**, 15–23.

Smart, P.L. and Brown, M.C. (1973) The use of activated carbon for the detection of the tracer dye Rhodamine WT, *Proceedings of the 6th International Congress of Speleology, Olomouc, CSSR*, Vol. 4, pp. 285–92.

Smart, P.L. and Richards, D.A. (1992) Age estimates for the Late Quaternary high sea-stands. *Quaternary Science Reviews*, **11**, 687–96.

Smart, P.L. and Smith, D.I. (1976) Water tracing in tropical regions, the use of fluorometric techniques in Jamaica. *Journal of Hydrology*, **30**, 179–95.

Smart, P.L, Waltham, T., Yang, M. and Zhang, Y. (1986) Karst geomorphology of western Guizhou, China. *Transactions of the British Cave Research Association*, **13**(3), 89–103.

Smart, P.L., Smith, B.W., Chandra, H., *et al.* (1988) An inter-comparison of ESR and uranium series ages for Quaternary speleothem deposits. *Quaternary Science Reviews*, **7**, 411–6.

Šmida, B., Audy, M. and Vlček, C. (2003) *Roraima 2003 Cueva Ojos de Cristal*, Expedition Report, Czech Speleological Society, Slovak Speleological Society, 28 pp.

Šmida, B., Brewer-Carias, C. and Audy, M. (2005) *Cueva Charles Brewer*, Vol. 3, Spravodaj, 178 pp.

Smith, D.I. (1972) The solution of limestone in an Arctic environment, in *Polar Geomorphology* (ed. D.E. Sugden), Special Publication, **4**, Institute of British Geographers, pp. 187–200.

Smith, D.I. and Newson, M.D. (1974) The dynamics of solutional and mechanical erosion in limestone catchments on the Mendip Hills, Somerset, in *Fluvial Processes in Instrumented Watersheds* (eds K.J. Gregory and D.E. Walling), Special Publication 6, Institute of British Geographers, pp. 155–67.

Smith, D.I. and Atkinson, T.C. (1976) Process, landforms and climate in limestone regions, in *Geomorphology and Climate* (ed. E. Derbyshire), John Wiley & Sons, Chichester, pp. 369–409.

Smith, D.I., Atkinson, T.C. and Drew, D.P. (1976) The hydrology of limestone terrains, in *The Science of Speleology* (eds T.D. Ford and C.H.D. Cullingford), Academic Press, London, pp. 179–212.

Smith, M.W. and Riseborough, D.W. (2002) Climate and the limits of permafrost: a zonal analysis. *Permafrost and Periglacial Processes*, **13**(1), 1–15.

Sneed, E.D. and Folk, R.L. (1958) Pebbles in the Lower Colorado River, Texas. A study of particle morphogenesis. *Journal of Geology*, **66**, 114–50.

Snow, D.T. (1968) Rock fracture spacings, openings porosities. *Journal of the Soil Mechanics and Foundation Division, American Society of Civil Engineers*, **94**, 73–91.

Soderberg, A.D. (1979) Expect the unexpected: foundations for dams in karst. *Bulletin, Association of Engineering Geologists*, **16**(3), 409–25.

Solomon, D.K., Cook, P.G. and Sanford, W.E. (1998) Dissolved gases in subsurface hydrology, in *Isotope Tracers in Catchment Hydrology* (eds C. Kendall, and J.J. McDonnell), Elsevier, Amsterdam, pp. 291–318.

Sondag, F., van Ruymbeke, M., Soubiès, F., *et al.* (2003) Monitoring present day climatic conditions in tropical caves using an Environmental Data Acquisition System (EDAS). *Journal of Hydrology*, **273**(1–4), 103–18.

Song, L. (1981) Some characteristics of karst hydrology in Guizhou plateau, China, *Proceedings of the 8th International Congress of Speleology, Bowling Green, Kentucky*, Vol. 1, pp. 139–42.

Song, L. (1986) Karst geomorphology and subterranean drainage in south Dushan, Guizhou Province, China. *Transactions of the British Cave Research Association*, **13**(2), 49–63.

Song, L. (1987) Pumping subsidence of surface in some karst areas of China, *Proceedings of the International Symposium on Karst and Man, Lubljana*, pp. 49–64.

Song, L. (1999) Sustainable development of agriculture in karst areas, south China. *International Journal of Speleology*, **28B**(1/4), 139–48.

Song, L. and Lin, J. (2004) Hongshui River fengcong karst, in *Encyclopedia of Caves and Karst Science* (ed. J. Gunn), Fitzroy Dearborn, New York, pp. 422–3.

Song, L., Deng, Z., Mangin, A. *et al.* (1993) *Structure, Functioning and Evolution of Karst Aquifers and Landforms in Conical Karst, Guizhou, China*, Sino-French Karst Hydrogeology Collaboration, Moulis, France, 213 pp.

Song, L., Waltham, T., Cao, N. and Wang, F (eds.) (1997) *Stone Forest: a Treasure of Natural Heritage*, China Environmental Science Press, 136 pp.

Sorriaux, P. (1982) *Contribution a l'tude de la sedimentation en milieu karstique: Le systeme de Niaux-Lombrives-Sabart, Pyrenees Arigeoises.* Univ. Paul Sabatier Thesis, 3rd cycle, Toulouse, 255 pp.

Soudet, H.J., Sorriaux, P. and Rolando, J.P. (1994) Relationship between Fractures and Karstification – the Oil- Bearing Paleokarst of Rospo Mare (Italy). *Bulletin des Centres de Recherches Exploration-Production Elf Aquitaine*, **18**(1), 257–97.

Sowers, G.F. (1984) Correction and protection in limestone terrace. in *Sinkholes: their Geology, Engineering and Environmental Impact* (ed. B.F. Beck), Balkema, Boston, pp. 373–8.

Sowers, G.F. (1996) *Building on Sinkholes*. ASCE Press, American Society of Civil Engineers, New York, 202 pp.

Spate, A.P., Jennings, J.N., Smith, D.I. and Greenaway, M.A. (1985) The micro-erosion meter: use and limitations. *Earth Surface Processes and Landforms*, **10**, 427–40.

Spencer, T. (1985) Weathering rates on a Caribbean reef limestone: results and implications. *Marine Geology*, **69**, 195–201.

Spencer, T. and Viles, H. (2002) Bioconstruction, bioerosion and disturbance on tropical coasts: coral reefs and rocky limestone shores. *Geomorphology*, **48**, 23–50.

Spörli, K.B., Craddock, C., Rutford, R.H. and Craddock, J.P. (1992) Breccia bodies in deformed Cambrian limestones, Heritage Range, Ellsworth Mountains, West Antarctica, in *Geology and Paleontology of the Ellsworth Mountains, West Antarctica* (eds G.F. Webers, C. Craddock and J.F. Splettstoesser), Memoir 170, Geological Society of America, Boulder, CO, pp. 365–74.

Spötl, C. and Mangini, A. (2002) Stalagmite from the Austrian Alps reveals Dansgaard-Oeschger events during isotope stage 3: implications for the absolute chronology of Greenland ice cores. *Earth and Plantary Science Letters*, **203**, 507–18.

Spötl, C., Fairchild, I.J. and Tooth, A.F. (2005) Cave air control on dripwater geochemistry, Obir Caves (Austria): Implications for speleothem deposition in dynamically ventilated caves. *Geochimica et Cosmochim Acta*, 2451–2468.

Spring, U. and Hutter, K. (1981a) Conduit flow of a fluid through its solid phase and its application to intraglacial channel flow. *International Journal of Engineering Science*, **20**(2), 327–63.

Spring, U. and Hutter, K. (1981b) Numerical studies of Jokulhaups. *Cold Regions Science and Technology*, **4**, 227–44.

Spring, W. and Prost, E. (1883) Etude sur les eaux de la Meuse. *Annales de la Societe Géologique de Belgique*, **XI**, 123–220.

Springer, G.S. and Kite, J.S. (1997) River-derived slackwater sediments in caves along Cheat River, West Virginia. *Geomorphology*, **18**, 91–100.

Stanton, R.J. (1966) The solution brecciation process. *Geological Society of America, Bulletin*, **77**, 843–48.

Stanton, W.I. and P.L. Smart (1981) Repeated dye traces of underground streams in the Mendip Hills, Somerset, *Proceedings of the University of Bristol Speleological Society*, **16**(1), 47–58.

Stauffer, B., Blunier, T., Dallenbach, A., *et al.* (1998) Atmospheric CO_2 concentration and millennial-scale climate change during the last glacial period. *Nature*, **392**, 59–62.

Stenson, R.E. (1990) *The morphometry and spatial distribution of surface depressions in gypsum, with examples from Nova Scotia, Newfoundland and Manitoba.* McMaster Univ. MSc thesis, 134 pp.

Stern, L., Engel, A.S., Bennet, P.C. and Porter, M.L. (2002) Subaqueous and subaerial speleogenesis in a sulfidic cave, in *Hydrogeology and Biology of Post-Paleozoic Carbonate Aquifers* (eds J.B. Martin, C.M. Wicks and I.D. Sasowsky), Special Publication 7, Karst Waters Institute, Charles Town, WV, pp. 89–91.

Stevanović, Z. and Dragašić, V. (1995) Some cases of Accidental Karst Water Pollution in the Serbian Carpathians. *Theoretical and Applied Karstology*, **8**, 137–44.

Stevanović, Z. and Milanović, P. (eds) (2005) *Water Resources and Environmental Problems in Karst*, National Committee of the International Association of Hydrogeologists of Serbia and Montenegro, Belgrade, 888 pp.

Stevanović, Z. and Mijatović, B. (eds) (2005) *Cvijić and Karst: Cvijić et Karst*, Serbian Academy of Science and Arts, Belgrade, 405 pp.

Stewart, G.R., Turnbull, M.H., Schmidt, S. and Reskine, P.D. (1995) ^{13}C natural abundance in plant communities along a rainfall gradient: a biological indicator of water availability. *Australian Journal of Plant Physiology*, **22**, 51–55.

Stewart, M. and Williams, P.W. (1981) Environmental isotopes in New Zealand hydrology 3: isotope hydrology of the Waikoropupu Springs and Takaka River, Northwest Nelson. *New Zealand Journal of* Science, **24**, 323–37.

Stewart, M.K. and Downes, C.J. (1982) Isotope hydrology of Waikoropupu Springs, New Zealand, in *Isotope Studies of Hydrologic Processes* (eds E.C. Perry and C.W. Montgomery), Northern Illinois University Press, DeKalb, pp. 15–23.

Stichler, W., Trimborn, P., Maloszewski, P., *et al.* (1997) Isotopic investigations, in *Karst Hydrogeological Investigations in South-Western Slovenia* (ed. A. Kranjc). *Acta Carsologica*, **XXVI**(1), 213–59.

Stierman, D.J. (2004) Geophysical detection of caves and karstic voids, in *Encyclopedia of Caves and Karst Science* (ed. J. Gunn), Fitzroy Dearborn, New York, pp. 377–80.

Stirling, C.H., Esat, T.M., Lambeck, K. and McCulloch, M.T. (1995) High-precision U-series dating of corals from Western Australia and implications for the timing and duration of the Last Interglacial. *Earth and Planetary Science Letters*, **135**, 115–30.

Stirling, C.H., Esat, T.M., Lambeck, K. and McCulloch, M.T. (1998) Timing and duration of the Last Interglacial: evidence for a restricted interval of widespread coral reef growth. *Earth and Planetary Science Letters*, **160**, 745–62.

Stoddart, D.R., Spencer, T. and Scoffin, T.P. (1985) Reef growth and karst erosion on Mangaia, Cook Islands: a reinterpretation. *Zeitschrift für Geomorphologie, Supplement-Band*, **57**, 121–40.

Stoddart, D. R., Woodroffe, C.D. and Spencer, T. (1990) Mauke, Mitiaro and Atiu: geomorphology of makatea islands in the southern Cooks. *Atoll Research Bulletin*, **341**, 1–65.

Stone, J., Allan, G.L., Fifield, L.K., *et al.* (1994) Limestone erosion measurements with cosmogenic chlorine-36 in calcite – preliminary results from Australia. *Nuclear Instruments and Methods in Physics Research, Series B*, **92**, 311–16.

Stone, J., Evans, J.M., Fifield, L.K., *et al.* (1998) Cosmogenic chlorine-36 production in calcite by muons. *Geochimica et Cosmochimica Acta*, **62**(3), 433–54.

St-Onge, D.A. and McMartin, I (1995) *Quaternary Geology of the Inman River Area, Northwest Territories*, Bulletin 446, Geological Survey of Canada, 59 pp.

Strecker, M.R., Bloom, A.L. Gilpin, L.M. and Taylor, F.W. (1986) Karst morphology of uplifted Quaternary coral limestone terraces: Santo island, Vanuatu. *Zeitschrift für Geomorphologie, NF*, **30**(4), 387–405.

Stringfield, V.T. and LeGrand, H.E. (1971) Effects of karst features on circulation of water in carbonate rocks in coastal areas. *Journal of Hydrology*, **14**, 139–57.

Stumm, W. and Morgan, J.J. (1996) *Aquatic Chemistry*, 3rd edn, John Wiley & Sons, New York, 980 pp.

Sunartadirdja, M.A. and Lehmann, H. (1960) Der Tropische Karst von Maros und Nord-Bone im SW-Celebes (Sulawesi). *Zeitschrift für Geomorphologie, Supplement-Band*, **2**, 49–65.

Sundborg, A. (1956) The River Klaralven, a study in fluvial processes. *Geografiska Annaler*, **38**, 125–316.

Sundquist, E.T. (1993) The global carbon dioxide budget. *Science*, **259**, 934–41.

Šušteršic, F. (1979) Some principles of cave profile simulation. *Actes de la Symposium International sur l'erosion karst, Aix-en-Provence*, pp. 125–31.

Šušteršic, F. (1994) Classic dolines of classical site. *Acta Carsologica*, **XXIII**(10), 123–54.

Šušteršic, F. (2000) Speleogenesis in the Lubljanica River drainage basin, Slovenia, in *Speleogenesis; Evolution of Karst Aquifers* (eds A.V. Klimchouk, D.C. Ford, A.N. Palmer and W. Dreybrodt), National Speleological Society of America, Huntsville, AL, pp. 397–406.

Šušteršic, F. (2006) A power function model for the basic geometry of solution dolines: considerations from the classical karst of south-central Slovenia. *Earth Surface Processes and Landforms*, **31**, 293–302.

Šušteršic. F. and Šušteršic, S. (2003) Formation of the Cerkniščica and flooding of Cerkniško Polje. *Acta Carsologica*, **32**(2); 121–36.

Svensson, U. and Dreybrodt, W. (1992) Dissolution kinetics of natural calcite minerals in CO_2-water systems approaching calcite equilibrium. *Chemical Geology*, **100**, 129–45.

Swarzenski P.W., Reich C.D., Spechler R.M., *et al.* (2001) Using multiple geochemical tracers to characterize the hydrogeology of the submarine spring off Crescent Beach, Florida. *Chemical Geology*, **179**(1–4), 187–202.

Sweeting, M.M. (1966) The weathering of limestones, with particular reference to the Carboniferous Limestones of northern England, in *Essays in Geomorphology* (ed. G.H. Dury), Heinemann, London, pp. 177–210.

Sweeting, M.M. (1972) *Karst Landforms*, Macmillan, London, 362 pp.

Sweeting, M.M. (1995) *Karst in China: its Geomorphology and Environment*, Springer-Verlag, Berlin.

Sweeting, M.M. (ed.) (1981) *Karst Geomorphology*, Benchmark Papers in Geology 59, Hutchinson-Ross. Stroudsburg, PA.

Sweeting, M.M. and Sweeting, G.S. (1969) Some aspects of the Carboniferous limestone in relation to its landforms with particular reference to N.W. Yorkshire and County Clare. *Recherche Mediterranee*, **7**, 201–8.

Swinnerton, A.C. (1932) Origin of limestone caverns. *Bulletin, Geological Society of America*, **43**, 662–693.

Szunyogh, G. (1989) Theoretical investigation of the development of spheroidal niches of thermal water origin, *Proceedings of the 10th International Congress of Speleology*, Vol. III, pp. 766–8.

Szunyogh, G. (2000) The theoretical-physical study of the process of karren development. *Karsztfejlödés*, **IV**, 125–50.

Taborosi, D., Jenson, J.W. and Mylroie, J.E. (2004) Karren features in island karst: Guam, Mariana Islands. *Zeitschrift fur Geomorphologie*, N.F., **48**(3), 369–89.

Tam, V.T., De Smedt, F., Batelaan, O. Dassargues, A. (2004) Study on the relationship between lineaments and borehole specific capacity in a fractured and karstified limestone area in Vietnam. *Hydrogeology Journal*, **12**, 662–73.

Tan, M. (1992) Mathematical modelling of catchment morphology in the karst of Guizhou, China. *Zeitschrift fur Geomorphologie*, N.F., **36**(1), 37–51.

Tan, M., Tungsheng, L., Xiaoguang, Q. Xianfeng, W. (1998) Signification chrono-climatique de spéléothèmes laminés de Chine du Nord. *Karstologia*, **32**(2), 1–6.

Tan, M., Liu, T., Hou, J., *et al.* (2003) Cyclic rapid warming on centennial-scale revealed by a 2650-year stalagmite record of warm season temperature. *Geophysical Research Letters*, **30**(12), 1617.

Tan, M., Hou, J. and Liu, T. (2004) Sun-coupled climate connection between eastern Asia and northern Atlantic. *Geophysical Research Letters*, **31**, L07207.

Tanahara, A., Taira, H., Yamakawa, K. and Tsuha, A. (1998) Application of excess ^{210}Pb dating method to stalactites. *Geochemical Journal*, **32**, 183–7.

Tang, T. (2002) Surface sediment characteristics and tower karst dissolution, Guilin, southern China. *Geomorphology*, **49**, 231–54.

Tang, T. and Day, M.J. (2000) Field survey and analysis of hillslopes on tower karst in Guilin, southern China. *Earth Surface Processes and Landforms*, **25**, 1221–35.

Tarhule-Lips, R. and Ford, D.C. (1998a) Condensation corrosion in caves on Cayman Brac and Isla de Mona, P.R. *Journal of Caves and Karst Studies*, **60**(2), 84–95.

Tarhule-Lips, R. and Ford, D.C. (1998b) Morphometric studies of bellhole development on Cayman Brac. *Cave and Karst Science*, **25**(3), 119–30.

Tate, T. (1879) The source of the R. Aire, *Proceedings of the Yorkshire Geological Society*, **VII**, 177–87.

Teal, L. and Jackson, M. (1997) Geologic overview of the Carlin Trend gold deposits and descriptions of recent discoveries. *Society of Economic Geologists Newsletter*, **31**, 13–25.

Telbisz, T. (2001. Töbrös felszínfejlödes számítógépes modellezése. *Karsztfeljlödés* (Szombathely), **VI**, 27–43.

Terjesen, S.C., Erga, O. and Ve, A. (1961) Phase boundary processes as rate determining steps in reactions between solids and liquids. *Chemical Engineering Science*, **74**, 277–88.

Terry, J.P. and Nunn, P.D. (2003) Interpreting features of carbonate geomorphology on Niue Island, a raised coral atoll. *Zeitschrift für Geomorphologie, Supplment-Band*, **131**, 43–57.

Teutsch, G. (1993) An extended double-porosity concept as a practical modelling approach for a karstified terrain, in *Hydrogeological Processes in Karst Terranes* (eds G. Günay, A.I. Johnson and W. Back), Publication 207, International Association of Hydrological Sciences, Wallingford, pp. 281–92.

Teutsch, G. and Sauter, M. (1998). Distributed parameter modelling approaches in karst-hydrological investigations. *Bulletin d'Hydrogéologie* (Neuchâtel), **16**, 99–110.

Tharp, T.M. (1995) Design against collapse of karst caverns, in *Karst Geohazards* (ed. B. F. Beck), Balkema, Rotterdam, pp. 397–406.

Tharp, T.M. (2001) Cover-collapse sinkhole formation and piezometric surface drawdown. in *Geotechnical and Environmental Applications of Karst Geology and Hydrology* (eds B.F. Beck and J.G. Herring). Balkema, Lisse, pp. 53–8.

Theis, C.V. (1935) The relation between the lowering of the piezometric surface and the rate and duration of discharge of a well using ground water storage. *Transactions, American Geophysical Union*, **2**, 519–24.

Therond, R. (1972) *Recherche sur l'etancheite des lacs de barrage en pays karstique*, Eyrolles, Paris, 443 pp.

Thomas, T.M. (1974) The South Wales interstratal karst. *Transactions of the British Cave Research Association*, **1**, 131–52.

Thorp, J. (1934) The asymmetry of the pepino hills of Puerto Rico in relation to the Trade Winds. *Journal of Geology*, **42**, 537–45.

Thrailkill, J. (1968) Chemical and hydrological factors in the excavation of limestone caves. *Bulletin, Geological Society of America*, **79**(B916), 19–46.

Thrailkill, J. (1985) Flow in a limestone aquifer as determined from water tracing and water levels in wells. *Journal of Hydrology*, **78**, 123–36.

Tinkler, K.J. and Stenson, R.E. (1992) Sculpted bedrock forms along the Niagara Escarpment, Niagara Peninsula, Ontario. *Géographie physique et Quaternaire*, **46**(2), 195–207.

Tintilozov, Z.K. (1983) *Akhali Atoni Cave System*, Metsniereba, Tbilisi, USSR, 150 pp.

Todd, D.K. (1980) *Groundwater Hydrology*, John Wiley & Sons, New York.

Tolmachev, V.V., Troitzky, G.M. and Khomenko, V.P. (1986) *Engineering/Building Development on Karst Terrains*, Moscow, Stroyizdat. Moscow, 177 pp. [In Russian.]

Tolmachev, V., Maximova, O. and Mamonova, T. (2005) Some new approaches to assessment of collapse risks in covered karsts, in *Sinkholes and the Engineering and Environmental Impacts of Karst* (ed. B.E. Beck), Geotechnical Special Publication No. 144, American Society of Civil Engineers, pp. 66–71.

Tooth, A.F. and Fairchild, I.J. (2003) Soil and karst aquifer hydrologic controls on the geochemical evolution of speleothem-forming drip waters, Crag Cave, southwest Ireland. *Journal of Hydrology*, **273**, 51–68.

Torbarov, K. (1976) Estimation of permeability and effective porosity in karst on the basis of recession curve analysis, in *Karst Hydrology and Water Resources: Vol. 1 Karst Hydrology* (ed. V. Yevjevich), Water Resources Publications, Colorado, pp. 121–36.

Tranter, J., Gunn, J., Hunter, C. and Perkins, J. (1997) Bacteria in the Castleton karst, Derbyshire, England. *Quarterly Journal of Engineering Geology*, **63**, 171–8.

Treble, P., Shelley, J.M.J. and Chappell, J. (2003) Comparison of high resolution sub-annual records of trace elements in a modern (1911–1992) speleothem with instrumental climate data from southwest Australia. *Earth and Planetary Science Letters*, **216**, 141–153.

Treble, P.C., Chappell, J., Gagan, M.K., *et al.* (2005) *In situ* measurement of seasonal $\delta^{18}O$ variations and analysis of isotopic trends in a modern speleothem from southwest Australia. *Earth and Planetary Science Letters*, **233**, 17–32.

Tricart, J. and Cailleux, A. (1972) *Introduction to Climatic Geomorphology*, Longman, London.

Tripathi, J.K and Rajamani, V. (2003) Weathering control over geomorphology of supermature Proterozoic Delhi Quartzites of India. *Earth Surface Processes and Landforms*, **28**, 1379–87.

Trišič, N. Bat, M., Polajnar, J. and Pristov, J. (1997). Water balance investigations in the Bohinj region, in Kranjc, A. (ed.), *Tracer Hydrology*. Balkema, Rotterdam, 295–298.

Trišič, N.(1997). Hydrology and Investigations of the Water Balance, in *Karst Hydrogeological Investigations in South-Western Slovenia* (ed. A. Kranjc). *Acta Carsologica*, **XXVI**(1), 19–30 and 123–41.

Troester, J.W., White, E.L. and White, W.B. (1984) A comparison of sinkhole depth frequency distributions in temperate and tropical karst regions, in *Sinkholes: their Geology, Engineering and Environmental Impact* (ed. B.F. Beck), Balkema, Rotterdam, pp. 65–73.

Trombe, F. (1952) *Traité de Spéléologie*, Payot, Paris, 376 pp.

Trudgill, S. (1985) *Limestone Geomorphology*, Longman, London.

Trudgill, S., High, C.J. and Hanna, F.K. (1981) Improvements to the micro-erosion meter. *British Geomorphological Research Group Technical Bulletin*, **29**, 3–17.

Trudgill, S.T. (1976) The marine erosion of limestones on Aldabra Atoll, Indian Ocean. *Zeitschrift für Geomorphologie, Supplement-Band*, **26**, 164–200.

Trudgill, S.T. and Inkpen, R. (1993) Impact of acid rain on karst environments, in *Karst Terrains; Environmental Changes and Human Impacts* (ed. P.W. Williams). *Catena Supplement*, **25**, 199–218.

Tsui, P.C. and Cruden, D.M. (1984) Deformation associated with gypsum karst in the Salt River Escarpment, northeastern Alberta. *Canadian Journal of Earth Science*, **21**, 949–59.

Tsykin, R.A. (1990) *Karst Sibiri*. Krasnoyarsk University Publishing House, Krasnoyarsk (cited by Filippov 2004).

Tucker, M.E. and Wright, V.P. (1990) *Carbonate Sedimentology*, Blackwell Science, Oxford, 482 pp.

Tudhope, A.W. and Risk, M.J. (1985) Rate of dissolution of carbonate sediments by microboring organisms, Davies Reef, Australia. *Journal of Sedimentary Petrology*, **55**, 440–447.

Tulipano, L. and Fidelibus, M.D. (1999) Groundwater salinization in the Apulia region, southern Italy, in *Karst hydrogeology and human activities: impacts, consequences and implications* (eds D. Drew and H. Hötzl), Balkema, Rotterdam, pp. 251–5.

Turkmen, S., Özgüler, E., Taga, H. and Karaogullarindan, T. (2002) Seepage problems in the karstic limestone foundation of the Kalecik Dam (south Turkey). *Engineering Geology*, **63**, 247–57.

TVA (1949) *Geology and Foundation Treatment*, Tennessee Valley Authority Projects, Technical Report No. 22, 550 pp.

Twidale, C.R. (1976) *Analysis of Landforms*, John Wiley & Sons, Sydney, 572 pp.

Twidale, C.R. (1984) The enigma of the Tindal Plain, Northern Territory. *Transactions of the Royal Society of South Australia* **108**(2), 95–103.

Twidale, C.R. and Bourne, J.A. (2000) Dolines of the Pleistocene dune calcarenite terrain of western Eyre Peninsula, South Australia: a reflection of underprinting? *Geomorphology*, **33**, 89–105.

UNESCO (1970) *International Legend for Hydrogeological Maps*, UNESCO, Paris.

UNESCO/IAHS (1983) *Methods and Instrumentation for the Investigation of Groundwater Systems, International Symposium Proceedings, Noordwijkerhout, Netherlands*.

Urai, J.L., Spiers, C.J., Zwart, H.J. and Lister, G.S. (1986) Weakening of rock salt by water during long-term creep. *Nature*, **324**, 554–7.

Urbani, F. (1994) Cavidades estudias en la expedicion al Macizo de Chimanta. *Bolletino Sociedad Venezolana Espeleologia*, **28**, 33–50.

Urbani, F. (2005) Quartzite caves: the Venezuelan perspective. *National Speleology Society News*, **63**(7), 20–1.

Urbani, F. and Szcerban, E. (1974) Venezuelan caves in non-carbonate rocks: a new field in karst research. *National Speleological Society News*, **32**, 233–5.

Urich, P.B. (1989) Tropical karst management and agricultural development: example from Bohol, Philippines. *Geografiska Annaler*, **71**B(2), 95–108.

Urich, P.B. (1993) Stress on tropical karst cultivated with wet rice: Bohol, Philippines. *Environmental Geology*, **21**, 129–36.

Urich, P.B. and Reeder, P. (1996) Environmental degradation in the Loboc River watershed, Bohol Province, Philippines. *Asia Pacific Viewpoint*, **37**(3), 283–93.

Urushibara-Yoshino, K. (1991) Land use and soils in karst areas of Java, Indonesia, *Proceedings, International Conference on Environmental Changes in Karst Areas*, International Geographical Union/International Speleological Union, pp. 61–7.

Urushibara-Yoshino, K. (2003) Karst terrain of raised coral islands, Minamidaito and Kikai in the Nansei Islands of Japan. *Zeitschrift für Geomorphologie, Supplement-Band*, **131**, 17–31.

Urushibara-Yoshino, K., Miotke, F.-D., Kashima, N., *et al.* (1999) Solution rate of limestone in Japan. *Physics and Chemistry of the Earth, Series A,* **24**(10), 899–903.

US Bureau of Land Management (1987) *Cave Resources Management,* US Department of the Interior, Washington, DC, 12 pp.

US Department of the Interior (1981) *Ground Water Manual,* John Wiley & Sons, New York.

US EPA (1993) *A Review of Methods of Assessing Aquifer Sensitivity and Groundwater Vulnerability to Pesticide Contamination,* Environmental Protection Agency, Washington, DC, 147 pp.

US EPA (1997) *Guidelines for Wellhead and Springhead Protection Area Delineation in Carbonate Rocks,* EPA 904-B-97-003, Environmental Protection Agency, Washington, DC, 120 pp.

US EPA (2000) National Primary Drinking Water Regulations: Ground Water Rule. *Federal Register,* **65**(91), 82 pp.

US EPA(2002) *The QTRACER2 Program for Tracer-Breakthrough Curve Analysis for Tracer Tests in Karst Aquifers and Other Hydrologic Systems,* EPA/600/R-02/001, US Environmental Protection Agency, 179 pp. plus diskette.

US Forest Service (1986) *Forest Service Manual: Directive 2356, Cave Management,* US Forest Service, Washington, DC, 6 pp.

Vacher, H.L. (1988) Dupuit–Ghyben–Herzberg analysis of strip island lenses. *Geological Society of America Bulletin,* **100**, 580–91.

Vacher, H.L. and Mylroie, J.E. (2002) Eogenetic karst from the perspective of an equivalent porous medium. *Carbonates and Evaporites,* **17**(2), 182–96.

Vacher, H.L. and Quinn, T.M. (eds) (1997) *Geology and Hydrogeology of Carbonate Islands,* Developments in Sedimentology 54. Elsevier, Amsterdam, 948 pp.

Vajoczki, S. and Ford, D.C. (2000) Underwater dissolutional pitting on dolostones, Lake Huron-Georgian Bay, Ontario. *Physical Geography,* **21**(5), 418–32.

Valen, V., Lauritzen, S-E. and Løvlie, R. (1997) Sedimentation in a high-latitude karst cave: Sirijordgrotta, Nordland, Norway. *Norsk Geologisk Tidsskrift,* **77**, 233–50.

Valvasor, J.W. (1687) *Die Ehre des Hertzogthums Crain,* 4 Vols, Endter, Lubljana.

Van Beynen, P.E., Ford, D.C. and Schwarcz, H.P. (2000) Seasonal variability in organic substances in surface and cave waters at Marengo Cave, Indiana. *Hydrological Processes,* **14**, 1177–97.

Van Beynen, P.E., Bourbonniere, R., Ford, D.C. and Schwarcz, H.P. (2001) Causes of colour and fluorescence in speleothems. *Chemical Geology,* **175**(3–4), 319–41.

Van Everdingen, R.O. (1981) *Morphology, hydrology and hydrochemistry of karst in permafrost near Great Bear Lake, Northwest Territories,* Paper 11, National Hydrological Research Institute of Canada.

Van Gassen, W. and Cruden, D.M. (1989) Momentum transfer and friction in the debris of rock avalanches. *Canadian Geotechnical Journal,* **26**, 623–8.

Vanara, N. (2000) Le fonctionnement actuel du rèseau Nèbèlè. *Spelunca,* **77**, 35–8.

Vandycke, S and Quinif, Y. (1998) Live faults in Belgian Ardenne revealed in Rochefort karstic network (Belgium). *Terra Nova,* **8**, 16–19.

Veni, G. and DuChene, H. (2001) *Living with Karst: a Fragile Foundation,* Environmental Awareness Series 4, American Geological Institute, 64 pp.

Veress, M. (2000a) The main types of karren development of limestone surfaces without soil covering. Karsztfejlödés, **IV**, 7–30.

Veress, M. (2000b) The history of the development of a karren trough based on its terraces. *Karsztfejlödés,* **IV**, 31–40.

Veress, M. (2000c) The morphogenetics of the karren meander and its main types. *Karsztfejlödés,* **IV**, 41–76.

Veress, M. and Toth, G. (2004) Types of meandering karren. *Zeitschrift fur Geomorphologie NF,* **48**(1), 53–77.

Veress, M. and Zentai, Z (2004) Karren development on Triglav. *Karsztfejlödés,* **IX**, 177–96. [In Hungarian.]

Veress, M., Pèntek, K. and Horvàth, T. (1992) Evolution of Corrosion Caverns: Ördög-lik, Bakony, Hungary. *Cave Science,* **19**(2), 41–50.

Veress, M., Toth, G., Zentai, Z. and Czöpek, I. (2003) Vitesse de recul d'un escarpement lapiazé (Ile Diego de Almagro, Patagonie, Chili. *Karstologia,* **41**(1); 23–6.

Verstappen, H. (1964) Karst morphology of the Star Mountains (central New Guinea) and its relation to lithology and climate. *Zeitschrift für Geomorphologie,* **8**, 40–9.

Vesper, D.J., Loop, C.M. and White, W.B. (2000) Contaminant transport in karst aquifers. *Theoretical and Applied Karstology,* **13**, 101–11.

Viles, H. (1984) Biokarst: review and prospect. *Progress in Physical Geography,* **8**(4), 523–42.

Viles, H.A. (1987) Blue-green algae and terrestrial limestone weathering on Aldabra Atoll: an S.E.M. and light microscope study. *Earth Surface Processes and Landforms,* **12**, 319–30.

Viles, H.A. (ed.) (2000) Recent advances in field and laboratory studies of rock weathering. *Zeitschrift für Geomorphologie, Supplement-Band,* 120, 193 pp.

Viles, H.A. (2003) Biokarst, in *Encyclopedia of Geomorphology* (ed. A.S. Goudie), Routledge, London, pp. 86–7.

Viles, H.A. and Pentecost, A. (1999). Geomorphological controls on tufa deposition at Nash Brook, South Wales, United Kingdom. *Cave and Karst Science,* **26**, 61–8.

Viles, H.A. and T. Spencer (1986) 'Phytokarst', blue-green algae and limestone weathering, in *New Directions in Karst* (eds K. Paterson and M.M. Sweeting), Geo Books, Norwich, pp. 115–40.

Villar, E., Bonet A., Diaz-Caneja, B., *et al.* (1984) Ambient temperature variations in the hall of the paintings of Altamira cave due to the presence of visitors. *Cave Science,* **1120**, 99–104.

Villar, E., Fernandez, P.L., Gutierrez, I., *et al.* (1986) Influence of visitors on carbon concentrations in Altamira Cave. *Cave Science,* **13**(1), 21–3.

Vincent, P.J. (1987) Spatial dispersion of polygonal karst sinks. *Zeitschrift fur Geomorphologie, N.F.,* **31**, 65–72.

Vincent, P. (2004) Polygenetic origin of limestone pavements in northern England. *Zeitschrift fur Geomorphologie*, N.F., **48**(4), 481–90.

Wagener, F.V.M. and Day, P.W. (1986) Construction on dolomite in South Africa. *Environmental Geology and Water Science*, **8**(1/2), 83–9.

Waltham, A.C. (1970) Cave development in the limestone of the Ingleborough district. *Geographical Journal*, **136**, 574–84.

Waltham, A.C. (1996) Ground subsidence over underground cavities. *Journal of the Geological Society of China*, **39**(4), 605–26.

Waltham, A.C. and. Brook, D.B. (1980) Geomorphological observations in the limestone caves of Gunung Mulu National Park, Sarawak. *Transactions of the British Cave Research Association*, **7**(3), 123–40.

Waltham, A.C. and Fookes, P.G. (2003) Engineering classification of karst ground conditions. *Quaterly Journal of Engineering Geology and Hydrogeology*, **36**, 101–18.

Waltham, A.C. and Hamilton-Smith, E. (2004) Ha Long Bay, Vietnam, in *Encyclopedia of Caves and Karst Science* (ed. J. Gunn), Fitzroy Dearborn, New York, pp. 413–13.

Waltham, A.C., Vandeven, G. and Ek, C.M. (1986) Site investigations on cavernous limestone for the Remouchamps Viaduct, Belgium. *Ground Engineering*, **19**(8), 16–18.

Waltham, A.C., Bell, F. and Culshaw, M. (2005) *Sinkholes and Subsidence: Karst and Cavernous Rocks in Engineering and Construction*. Praxis Publishing, Chichester, 382 pp.

Wang, F., Li, H., Zhu, R. and Qin, F. (2004) Late Quaternary downcutting rates of the Qianyou River from U/Th speleothem dates, Qinling mountains, China. *Quaternary Research*, **62**, 194–200.

Wang, Y.J., Chen, H., Edwards, R.L., *et al.* (2001) A high-resolution absolute-dated late Pleistocene monsoon record from Hulu Cave, China. *Science*, **294**, 2345–8.

Ward, R.C. (1975) *Principles of Hydrology*, 2nd edn, McGraw Hill, London, 403 pp.

Ward, R.C. and Robinson, M. (2000) *Principles of Hydrology*, 4th edn, McGraw-Hill, London, 448 pp.

Ward, R.S, Harrison, I., Leader, R.U. and Williams, A.T. (1997a) Fluorescent polystyrene microspheres as tracers of colloidal and particulate materials: examples of their use and developments in analytical technique, in *Tracer Hydrology 97* (ed. A. Kranjc), Rotterdam, Balkema, pp. 99–103.

Ward, R.S, Williams, A.T and Chadha, D.S. (1997b) The use of groundwater tracers for assessment of protection zones around water supply boreholes – a case study, in *Tracer Hydrology 97* (ed. A. Kranjc), Rotterdam, Balkema, pp. 369–76.

Warren, J.K. (1989) *Evaporite Sedimentology*, Prentice Hall. New Jersey, 285 pp.

Warren, J.K. (2000) Dolomite: occurrence, evolution and economically important associations. *Earth Science Reviews*, **52**, 1–81.

Waterhouse, J.D. (1984) Investigation of pollution of the karstic aquifer of the Mount Gambier area in South Australia, in *Hydrogeology of Karstic Terrains*, Vol. 1(1) (eds A. Burger and L. Dubertret), International Union of Geological Sciences, Heise, Hannover, pp. 202–5.

Waters, A. and Banks, D. (1997) The Chalk as a karstified aquifer: closed circuit television images of macrobiota. *Quarterly Journal of Engineering Geology*, **30**, 143–6.

Watson, J., Hamilton-Smith, E., Gillieson, D. and Kiernan, K. (eds) (1997) *Guidelines for Cave and Karst Protection*, International Union for the Conservation of Nature and Natural Resources, Gland, 63 pp.

Webb, G.E., Jell, J.S. and Baker, J.C. (1999) Cryptic intertidal microbialites in beach rock, Heron Island, Great Barrier Reef: implications for the origin of microcrystalline beach rock sediment. *Sedimentary Geology*, **126**, 317–34.

Webb, J.A., Fabel, D., Finlayson, B.L., *et al.* (1992) Denudation chronology from cave and river terrace levels: the case of the Buchan Karst, southeastern Australia. *Geological Magazine*, **129**(3), 307–17.

Webb, J.A., Grimes, K. and Osborne, A. (2003) Black holes: caves in the Australian landscape, in *Beneath the Surface: a Natural History of Australian Caves* (eds B. Finlayson and E. Hamilton-Smith), UNSW Press, Sydney, pp. 1–52.

Werner, A., Hotzl, H., Maloszewski, P. and Kass, W. (1998) Interpretation of tracer tests in karst systems with unsteady flow conditions, in *Karst Hydrology* (eds C. Leibundgut, J. Gunn and A. Dassargues), Publication 247, International Association of Hydrological Sciences, Wallingford, pp. 15–26.

Weyl, P.K. (1958) Solution kinetics of calcite. *Journal of Geology*, **66**, 163–76.

Wheeler, C. and Aharon, P. (1997) Geology and hydrogeology of Niue, in *Geology and Hydrogeology of Carbonate Islands* (eds H.L. Vacher and T. Quinn), Developments in Sedimentology 54, Elsevier, pp. 537–64.

Whitaker, F.F. and Smart, P.L. (1994) Bacterially mediation of organic matter: a major control on groundwater geochemistry and porosity generation in oceanic carbonate terrains, in *Breakthroughs in Karst Geomicrobiology and Redox Chemistry* (eds I.D. Sasowsky and M.V. Palmer), Karst Waters Institute of America, Special Publication 1, 72–4.

White, E.L. and White, W.B. (1969) Processes of cavern breakdown. *Bulletin of the National Speleological Society*, **31**, 83–96.

White, E.L., Aron, G. and White, W.B. (1984) The influence of urbanization on sinkhole development in central Pennsylvania, in *Sinkholes: their Geology, Engineering and Environmental Impact* (ed. B.F. Beck), Balkema, Rotterdam, pp. 275–81. [Also published in *Environmental Geology and Water Science*, **8**(1/2), 91–7 (1986).]

White, S. (1994) Speleogenesis in aeolian calcarenites: A case study in western Victoria. *Environmental Geology*, **23**, 248–55.

White, W.B. (1969) Conceptual models for carbonate aquifers. *Ground Water*, **7**(3), B97515–21.

White, W.B. (1976) Cave minerals and speleothems, in *The Science of Speleology* (eds T. D. Ford and C.H.D. Cullingford), Academic Press, London, pp. 267–327.

White, W.B. (1977a) The role of solution kinetics in the development of karst aquifers, in *Karst Hydrogeology* (eds J.S. Tolson and F.L. Doyle), Memoir 12, International Association of Hydrogeologists, pp. 503–17.

White, W.B. (1977b) Conceptual models for carbonate aquifers: revisited, in *Hydrologic Problems in Karst Regions* (eds R.R. Dilamarter and S.C. Csallany), Western Kentucky University, Bowling Green, pp. 176–87.

White, W.B. (1984) Rate processes: chemical kinetics and karst landform development, in *Groundwater as a Geomorphic Agent* (ed. R. G. LaFleur), Allen & Unwin, London, pp. 227–48.

White, W.B. (1997a) Thermodynamic equilibrium, kinetics, activation barriers reaction mechanisms for chemical reactions in Karst Terrains. *Environmental Geology*, **30**(1/2), 46–58.

White, W.B. (1997b) Color of speleothems, in *Cave Minerals of the World*, 2nd edn (eds C.A. Hill and P. Forti), National Speleological Society of America, Huntsville, AL, pp. 239–4.

White, W.B. (2000) Dissolution of limestone from field observations, in *Speleogenesis; Evolution of Karst Aquifers* (eds A.V. Klimchouk, D.C. Ford, A.N. Palmer and W. Dreybrodt), National Speleological Society of America, Huntsville, AL, pp. 149–55.

White, W.B. (2002) Karst hydrology: recent developments and open questions. *Engineering Geology*, **65**, 85–105.

White, W.B. (2004) Paleoclimate records from speleothems in limestone caves, in *Studies of Cave Sediments: Physical and Chemical Records of Paleoclimate* (eds I.D. Sasowsky and J. Mylroie), Kluwer Academic, New York, pp. 135–76.

White, W.B. and White, E.L. (eds) (1989) *Karst Hydrology: Concepts from the Mammoth Cave Area*. Van Nostrand Reinhold, New York, 346 pp.

White, W.B. and White, E.L. (1995) Correlation of contemporary karst landforms with paleokarst landforms: the problem of scale. *Carbonates and Evaporites*, **10**(2), 131–7.

White, W.B. and White, E.L. (2003) Gypsum wedging and cavern breakdown: studies in the Mammoth Cave System, Kentucky. *Journal of Cave and Karst Studies*, **65**(1), 43–52.

Wicks, C., Kelley, C. and Peterson, E. (2004) Estrogen in a karstic aquifer. *Groundwater*, **42**(3), 384–9.

Wigley, T.M.L. (1971) Ion pairing and water quality measurements. *Canadian Journal of Earth Science*, **8**(4), 468–76.

Wigley, T.M.L. and Brown, M.C. (1976) The physics of caves, in *The Science of Speleology* (eds T.D. Ford and C.H.D. Cullingford), Academic Press, London, pp. 329–58.

Wilcock, J.D. (1997) Simulation of cave hydrology using a conventional computer spreadsheet, in *Tracer Hydrology 97* (ed. A. Kranjc), Rotterdam, Balkema, pp. 443–8.

Wilford, G.E. (1966) 'Bell holes' in Sarawak caves. *Bulletin of the National Speleological Society*, **28**(4), 179–82.

Wilk, Z. (1989) Hydrogeological problems of the Cracow–Silesia Zn–Pb ore, in *Paleokarst – a Systematic and Regional Review* (eds P. Bosák, D.C. Ford, J. Głazek and I. Horáček), Academia Praha/Elsevier, Prague/Amsterdam, pp. 513–31.

Williams, K.M. and Smith, G.G. (1977) A critical evaluation of the application of amino acid racemization to geochronology and geothermometry. *Origins of Life*, **8**, 1–44.

Williams, P.W. (1963) An initial estimate of the speed of limestone solution in County Clare. *Irish Geography*, **4**, 432–41.

Williams, P.W. (1966) Limestone pavements: with special reference to western Ireland. *Transactions of the Institute of British Geographers*, **40**, 155–72.

Williams, P.W. (1968) An evaluation of the rate and distrubution of limestone solution and deposition in the River Fergus basin, western Ireland, in *Contributions to the Study of Karst* (eds P.W. Williams & J.N. Jennings), Publication G5, Research School for Pacific Studies, Australian National University, pp. 1–40.

Williams, P.W. (1969) The geomorphic effects of groundwater, in *Water, Earth and Man* (ed. R. J. Chorley), Methuen, London, pp. 269–84.

Williams, P.W. (1970) Limestone morphology in Ireland, in *Irish Geographical Studies* (eds N. Stephens and R. E. Glasscock), Queens University, Belfast, pp. 105–24.

Williams, P.W. (1971) Illustrating morphometric analysis of karst with examples from New Guinea. *Zeitschrift für Geomorphologie*, **15**, 40–61.

Williams, P.W. (1972a) Morphometric analysis of polygonal karst in New Guinea. *Geological Society of America Bulletin*, **83**, 761–96.

Williams, P.W. (1972b) The analysis of spatial characteristics of karst terrains, in *Spatial Analysis in Geomorphology* (ed. R. J. Chorley), Methuen, London, pp. 136–63.

Williams, P.W. (1977) Hydrology of the Waikoropupu Springs: a major tidal karst resurgence in northwest Nelson (New Zealand). *Journal of Hydrology*, **35**, 73–92.

Williams, P.W. (1978) Interpretations of Australasian karsts, in *Landform Evolution in Australasia* (eds J.L. Davies and M.A.J. Williams), ANU Press, Canberra, pp. 259–86.

Williams, P.W. (1982a) Karst landforms in New Zealand, in *Landforms of New Zealand* (eds J. Soons and M.J. Selby), Longman Paul, Auckland, pp. 105–25.

Williams, P.W. (1982b) Speleothem dates, Quaternary terraces and uplift rates in New Zealand. *Nature*, **298**, 257–60.

Williams, P.W. (1983) The role of the subcutaneous zone in karst hydrology. *Journal of Hydrology*, **61**, 45–67.

Williams, P.W. (1985) Subcutaneous hydrology and the development of doline and cockpit karst. *Zeitschrift für Geomorphologie*, **29**(4), 463–82.

Williams, P.W. (1987) Geomorphic inheritance and the development of tower karst. *Earth Surface Processes and Landforms*, **12**(5), 453–65.

Williams, P.W. (1988a) Hydrological control and the development of cockpit and tower karst, *Proceedings of the 21 Congress of the International Association of Hydrogeologists, Guilin, China*, Vol. XXI(Part 1), pp. 281–90.

Williams, P.W. (1988b) Karst water resources, their allocation the determination of ecologically acceptable minimum flows: the case of the Waikoropupu Springs, New Zealand, *Proceedings of the 21 Congress of the International Association of Hydrogeologists, Guilin, China*, Vol. XXI(Part 2), pp. 719–23.

Williams, P.W. (1992) Karst hydrology, in *Waters of New Zealand* (ed. M.P. Mosley), New Zealand Hydrological Society, Wellington, pp. 187–206.

Williams, P.W. (ed.) (1993) *Karst Terrains; Environmental Changes and Human Impacts. Catena Supplement*, **25**, 268 pp.

Williams, P.W. (1996) A 230 ka record of glacial and interglacial events from Aurora Cave, Fiordland, New Zealand. *New Zealand Journal of Geology and Geophysics*, **39**, 225–41.

Williams, P.W. (2004a) Polygonal karst and palaeokarst of the King Country, North Island, New Zealand. *Zeitschrift für Geomorphologie*, Suppl.-Vol 136, 45–67.

Williams, P.W. (2004b) Karst systems, in *Freshwaters of New Zealand* (eds J. Harding, P. Mosley, C. Pearson and B. Sorrell), New Zealand Hydrological Society and New Zealand Limnological Society, Christchurch, pp. 31.1–31.20.

Williams, P.W. (2004c) Dolines, in *Encyclopedia of Caves and Karst Science* (ed. J. Gunn), Fitzroy Dearborn, New York, pp. 304–10.

Williams, P.W. and Dowling, R.K. (1979) Solution of marble in the karst of the Pikikiruna Range, northwest Nelson, New Zealand. *Earth Surface Processes*, 4(B1010), 15–36.

Williams, P.W. and Fowler, A. (2002) Relationship between oxygen isotopes in rainfall, cave percolation waters and speleothem calcite at Waitomo, New Zealand. *New Zealand Journal of Hydrology*, 41(1), 53–70.

Williams, P.W., Lyons, R.G., Wang, X., *et al.* (1986) Interpretation of the paleomagnetism of cave sediments from a karst tower at Guilin. *Carsologica Sinica*, 5(2), 113–26.

Williams, P.W., King, D.N.T., Zhao, J.-X. and Collerson, K.D. (2004) Speleothem master chronologies: combined Holocene $\delta^{18}O$ and $\delta^{13}C$ records from the North Island of New Zealand and their palaeo-environmental interpretation. *The Holocene*, 14(2), 194–208.

Williams, P.W., King, D.N.T., Zhao, J.-X. and Collerson, K.D. (2005) Late Pleistocene to Holocene composite speleothem chronologies from South Island, New Zealand – did a global Younger Dryas really exist? *Earth and Planetary Science Letters*, 230(3–4), 301–17.

Williams, R.G.B. and Robinson, D.A. (1994) Weathering flutes on siliceous rocks in Britain and Europe, in (eds *Rock Weathering and Landform Evolution* D.A. Robinson and R.G.B. Williams), J. Wiley & Sons, Chichester, pp. 413–32.

Wilson, A.M., Sanford, W.E., Whitaker, F.F. Smart, P.L. (2001) Spatial patterns of diagenesis during geothermal circulation in carbonate platforms: *American Journal of Science*, 301, 727–52.

Wilson, J.F. (1968) Fluorometric procedures for dye tracing, in *Techniques of Water Resources Investigations of the United States Geological Survey*, Book 3, Chapter A12.

Wilson, J.L. (1974) Characteristics of carbonate platforms margins. *American Association of Petrologists, Bulletin*, 58, 810–24.

Wilson, W.L. and Beck, B.F. (1988) *Evaluating Sinkhole Hazards in Mantled Karst Terrane*, American Society of Civil Engineers, Nashville, 23 pp.

Wilson, W.L. and Morris, T.L. (1994) Cenote Verde: a meromictic karst pond, Quintana Roo, Mexico. in *Breakthroughs in Karst Geomicrobiology and Redox Geochemistry* (eds I.D. Sasowsky and M.V. Palmer), Special Publication 1, Karst Waters Institute, pp. 77–79.

Winograd, I.J., Coplen, T.B., Landwehr, J.M., *et al.* (1992) Continuous 500,000-year climate record from vein calcite in Devils Hole, Nevada. *Science*, 258, 255–60.

Witherspoon, P.A., Amick, C.H., Gale, J.E. and Iwai, K. (1979) Observations of a potential size effect in experimental deter-

mination of the hydraulic properties of fractures. *Water Resources Research*, 15, 1142–6.

Wolfe, T.E. (1973) *Sedimentation in karst drainage basins along the Allegheny Escarpment in southeastern West Virginia, U.S.A.* McMaster Univ. PhD thesis, 455 pp.

Wolman, M.G. and Miller, J.P. (1960) Magnitude and frequency of forces in geomorphic processes. *Journal of Geology*, 68, 54–74.

Wong, T., Hamilton-Smith, E., Chape, S. and Friederich, H. (eds) (2001) *Proceedings of the Asia-Pacific Forum on Karst Ecosystems and World Heritage*, UNESCO World Heritage Centre, Sarawak, Malaysia.

Woo, M.-K. and Marsh, P. (1977) Effect of vegetation on limestone solution in a small high Arctic basin. *Canadian Journal of Earth Science*, 14(4), 571–81.

Woodhead, J., Hellstrom, J., Maas, R., *et al.* (2006) U–Pb geochronology of speleothems by MC–ICPMS, in *Archives of Climate Change in Karst* (eds B.P. Onac, T. Tudor, S. Constantin and A. Persoiu) Special Publication 10, Karst Waters Institute, Charles Town, WV, pp. 69–71.

Worthington, S.R.H. (1984) *The paleodrainage of an Appalachian Fluviokarst: Friars Hole, West Virginia*. McMaster Univ. MSc. thesis, 218 pp.

Worthington, S.R.H. (1994) Flow velocities in unconfined carbonate aquifers. *Cave and Karst Science*, 21(1), 21–2.

Worthington, S.R.H. (1999) A comprehensive strategy for understanding flow in carbonate aquifers, in *Karst Modelling* (eds A.N. Palmer, M.V. Palmer and I.D. Sasowsky), Special Publication 5, Karst Waters Institute, Charles Town, WV, pp. 30–7.

Worthington, S.R.H. (2001) Depth of conduit flow in unconfined carbonate aquifers. *Geology*, 29(4), 335–338.

Worthington, S.R.H. (2002) Test methods for characterizing contaminant transport in a glaciated carbonate aquifer. *Environmental Geology*, 42, 546–51.

Worthington, S.R.H. and Smart, C.C. (2003) Empirical determination of tracer mass for sink to spring tests in karst. In: Sinkholes and the Engineering and Environmental Impacts of Karst; Proceedings of the ninth multidisciplinary conference, Huntsville, Alabama, Ed. B.F. Beck, *Americal Society of Civil Engineers, Geotechnical Special*, Publication No. 122, 287–295.

Worthington, S.R.H. (2004) Hydraulic and geologic factors influencing conduit flow depth. *Caves and Karst Science*, 31(3), 123–34.

Worthington, S.R.H. and Ford, D.C. (2001) *Chemical Hydrogeology of the Carbonate Bedrock at Smithville*, Smithville Phase IV Bedrock Remediation Program, Ministry of the Environment, Ontario.

Worthington, S.R.H., Davies, G.J. and Quinlan, J.F. (1992) Geochemistry of springs in temperate carbonate aquifers: recharge type explains most of the variation. Colloque d'Hydrologie en Pays Calcaire et en Milieu Fissuré (5th, Neuchâtel, Switzerland), *Proceedings. Annales Scientifiques de l'Université de Besancon, Géologie – Mémoires Hors Série*, 11, 341–7.

Worthingtom, S.R.H., Smart, C.C. and Ruland, W.W. (2003) Assessment of Groundwater Velocities to the Municipal Wells at Walkerton. in Stolle, D., Piggott, A.R. and Crowder, J.J.

(eds.) Ground to Water: *Theory and Practice; Proceedings of the 55th Canadian Geotechnical Conference,*1081–6. See also www.worthingtongroundwater.com/Walkerton.htm

Worthington, S.R.H., Ford, D.C. and Beddows, P.A. (2000) Porosity and permeability enhancement in unconfined carbonate aquifers as a result of solution, in *Speleogenesis; Evolution of Karst Aquifers* (eds A.V. Klimchouk, D.C. Ford, A.N. Palmer and W. Dreybrodt), National Speleological Society of America, Huntsville, AL, pp. 220–3.

Worthington, S.R.H., Schindel, G.M. and Alexander, E.C. (2002) Techniques for investigating the extent of karstification in the Edwards Aquifer, Texas, in *Hydrogeology and Biology of Post-Paleozoic Carbonate Aquifers* (eds J.B. Martin, C.M. Wicks and I.D. Sasowsky), Special Publication 7, Karst Waters Institute, Charles Town, WV, pp. 173–5.

Worthy, T.H. and Holdaway, R. (2002) *The Lost World of the Moa: Prehistoric Life of New Zealand*, Indiana University Press, Bloomington, Indianapolis, 718 pp.

Wray, R.A.L. (1997) A global review of solutional weathering forms on quartz sandstones. *Earth-Science Reviews* **42**, 137–160.

Wright, V.P. and Tucker, M.E. (eds) (1991) *Calcretes*, Blackwell Scientific, Oxford, 352 pp.

Wright, V.P., Esteban, M. and Smart, P. L. (1991) *Paleokarsts and Paleokarstic Reservoirs*, PRIS Contribution 152, University of Reading, 157 pp.

Xia, Q.K., Zhao, J.-X. and Collerson, K.D. (2001) Early-mid Holocene climatic variations in Tasmania, Australia: multi-proxy records in a stalagmite from Lynds Cave. *Earth Planetary Science Letters*, **194**: 177–187.

Xiong, K. (1992) Morphometry and evolution of fenglin karst in the Shuicheng area, western Guizhou, China. *Zeitschrift fur Geomorphologie*, N.F., **36**(2), 227–48.

Yanes, C.E and Briceno, H.O. (1993) Chemical weathering and the formation of pseudo-karst topography in the Roraima Group, Gran Sabana, Venezuela. *Chemical Geology*, **107**, 341–343.

Yeap, E.B. (1987) Engineering geological site investigation of former mining areas for urban development in Peninsular Malaysia in *The Role of Geology in Urban Development. Geological Society of Hong Kong Bulletin*, **3**, 319–34.

Yonge, C.J. (1982) *Stable isotope studies of water extracted from speleothems*. McMaster Univ. PhD thesis, 298 pp.

Yonge, C.J., Ford, D.C., Gray, J. and Schwarcz, H.P. (1985) Stable isotope studies of cave seepage water. *Chemical Geology*, **58**, 97–105.

Yoshikawa, M. and Shokohifard, G. (1993) Underground dam: a new technology for groundwater resource development, *Proceedings of the International Symposium on Water Resources in Karst with Special Emphasis on Arid and Semiarid Zones, Shiraz, Iran*, pp. 205–27.

Yoshimura, K., Liu, Z., Cao, J., Yuan, D., Inokura, Y. and Noto, M. (2004) Deep source CO_2 in natural waters and its role in extensive tufa deposition in the Huanglong Ravines, Sichuan, China. *Chemical Geology*, **205**(1–2), 141–53.

Young, R.W. and Young, A. (1992) *Sandstone Landforms*, Springer-Verlag, Berlin, 163 pp.

Younger, P.L., Teutsch, G., Custodio, E., *et al.* (2002) Assessments of the sensitivity to climate change of flow and natural water quality in four major carbonate aquifers of Europe, in *Sustainable Groundwater Development* (eds K.M. Hiscock, M.O. Rivett and R.M. Davison), Special Publication 193, Geological Society Publishing House, Bath, pp. 303–23.

Yuan Daoxian (1981) *A Brief Introduction to China's Research in Karst*, Institute of Karst Geology, Guilin, Guangxi, China.

Yuan Daoxian (1983) *Problems of Environmental Protection of Karst Areas*, Institute Karst Geology, Guilin, Guangxi, China, 15 pp.

Yuan, D. (1986) New observations on tower karst, in International Geomorphology, Part 2, (ed. V. Gardiner), J. Wiley & Sons, pp. 1109–23.

Yuan, D. (1996) Rock desertification in the subtropical karst of South China. *Zeitschrift für Geomorphologie, Supplement-Band*, **108**, 81–90.

Yuan, D. (ed.) (2001) *Guidebook for Ecosystems of Semiarid Karst in North China and Subtropical Karst in Southwest China*, IGCP 448, Karst Dynamics Laboaratory, Guilin, 94 pp.

Yuan, D. (2004) Yangshuo karst, China, in *Encyclopedia of Caves and Karst Science* (ed. J. Gunn), Fitzroy Dearborn, New York, pp. 781–3.

Yuan, D. and Liu, Z. (eds) (1998) *Global Karst Correlation (IGCP 299)*, Science Press, Beijing, 308 pp.

Yuan, D. and nine others (1991) *Karst of China*, Geological Publishing House, Beijing, 224 pp.

Yuan, D., Cheng, H., Edwards, R.L., *et al.* (2004) Timing, duration transitions of the last interglacial Asian monsoon. *Science*, **304**, 575–8.

Yurtsever, Y. (1983) Models for tracer data analysis, in *Guidebook on Nuclear Techniques in Hydrology*, Technical Report Series No. 91, International Atomic Energy Agency, Vienna, pp. 381–402.

Zámbó, L. (2004) Hyrological and geochemical characteristics of the epikarst based on field monitoring, in *Epikarst* (eds W.K. Jones, D.C. Culver and J.S. Herman), Special Publication 9, Karst Waters Institute, Charles Town, WV, pp. 135–9.

Zámbó, L. and Ford, D.C. (1997) Limestone dissolution processes in Beke doline, Aggtelek National Park, Hungary. *Earth Surface Processes and Landforms*, **22**, 531–43.

Zenis, P. and Gaal, L. (1986) Magnesite karst in the Slovenske Rudohorie Mts, Czechoslovakia, *Comunicacions, 9⁰ Congreso Internacional de Espeleologia*, Vol. 2, pp. 36–9.

Zhang, D. (1997) Contemporary karst solution processes on the Tibetan Plateau. *Mountain Research and Development*, **17**(2), 135–44.

Zhang, Z. (1980) Karst types in China. *Geological Journal*, **4**(6), 541–70.

Zhang, Z. (1996) Impacts of rice field irrigation on water budget in South China Karst. *Quaternary Research*, **2**, 10–17. [In Chinese, with English abstract.]

Zhao, J.-X., Xia, Q.K. and Collerson, K.D. (2000) Timing and duration of the Last Interglacial inferred from high resolution U-series chronology of stalagmite growth in Southern Hemisphere. *Earth Planetary Science Letters*, **184**, 633–44.

Zhao, J.-X., Wang, Y-J., Collerson, K.D. and Gagan, M.K. (2003) Speleothem U-series dating of semi-synchronous climate oscillations during the last deglaciation. *Earth Planetary Science Letters*, **216**, 155–61.

Zhong, S. and Mucci, A. (1993) Calcite precipitation in seawater using a constant addition technique: A new overall reaction kinetic expression. *Geochimica et Cosmochimica Acta*, **57**, 647–59.

Zhu Dehau (1982) Evolution of peak cluster depressions in the Guilin area and morphometric measurement. *Carsologica Sinica*, **10**(2), 127–34.

Zhu, R., An, Z., Potts, R. Hoffman, K.A. (2003) Magnetostratigraphic dating of early humans in China. *Earth-Science Reviews*, **61**, 341–59.

Zhu Xuewen (1988) *Guilin Karst*, Shanghai Scientific and Technical Publishers, 188 pp.

Zhu, X. and Chen, W. (2005) Tiankengs in the karst of China. *Cave and Karst Science*, **32**(2), 55–66.

Zibret, Z. and Z. Simunic (1976) A rapid method for determining water budget of enclosed and flooded karst plains, in *Karst Hydrology and Water Resources: Vol. 1 Karst Hydrology* (ed. V. Yevjevich), Water Resources Publications, Colorado, pp. 319–339.

Zisman, E.D. (2001) Application of a standard method of sinkhole detection in the tampa, Florida, area, in *Geotechnical and Environmental Applications of Karst Geology and Hydrology* (eds B.F. Beck, and J.G. Herring), Balkema, Lisse, pp. 187–192.

Zötl, J. (1974) *Karsthydrogeologie*, Springer-Verlag, Vienna.

Zupan Hajna, N. (2003) *Incomplete Solution: Weathering of Cave Walls and the Production, Transport and Deposition of Carbonate Fines*, Carsologica Series, Zalozba ZRC, 167 pp.

Index

Printed and bound in the UK by
CPI Antony Rowe, Eastbourne

Printed and bound by CPI Group (UK) Ltd, Croydon, CR0 4YY

27/10/2024

14580310-0005